Advances in Pharmaceutical Sciences: In Honor of Dr. Jean Jacques Vanden Eynde (JJ)

Advances in Pharmaceutical Sciences: In Honor of Dr. Jean Jacques Vanden Eynde (JJ)

Editors

Amélia Pilar Rauter
Klaus Kopka
Thomas Efferth
Joachim Jose
Maria Emília De Sousa
Guangshun Wang

MDPI

Basel • Beijing • Wuhan • Barcelona • Belgrade • Novi Sad • Cluj • Manchester

Editors

Amélia Pilar Rauter
Universidade de Lisboa
Lisboa
Portugal

Klaus Kopka
Institute of
Radiopharmaceutical Cancer
Research
Dresden
Germany

Thomas Efferth
Johannes Gutenberg
University
Mainz
Germany

Joachim Jose
Universität Münster
Münster
Germany

Maria Emília De Sousa
University of Porto
Porto
Portugal

Guangshun Wang
University of Nebraska
Medical Center
Omaha
USA

Editorial Office
MDPI AG
Grosspeteranlage 5
4052 Basel, Switzerland

This is a reprint of articles from the Special Issue published online in the open access journal *Pharmaceuticals* (ISSN 1424-8247) (available at: https://www.mdpi.com/journal/pharmaceuticals/special_issues/VN5JPSXZW9).

For citation purposes, cite each article independently as indicated on the article page online and as indicated below:

Lastname, A.A.; Lastname, B.B. Article Title. *Journal Name* **Year**, *Volume Number*, Page Range.

ISBN 978-3-7258-2461-8 (Hbk)
ISBN 978-3-7258-2462-5 (PDF)
doi.org/10.3390/books978-3-7258-2462-5

Contents

Preface

On the occasion of the 20th anniversary of Pharmaceuticals, we wish to honor Dr. Jean Jacques Vanden Eynde (JJ) for his seminal contributions to the success of our journal.

Editor-in-Chief from 2015 until July 2022, JJ introduced a wonderful atmosphere among editors and authors, bringing the journal to a prestigious level of scientific communication.

It is our pleasure to dedicate this Special Issue, which covers the most recent advances in pharmaceutical sciences, to JJ. The first paper authored by Kadoglou and collaborators addresses the cardiovascular dysfunction caused by chemotherapy in women with breast cancer. The next paper comprises a study leading to the preparation of stable pharmaceutical formulations for cardiovascular diseases, namely of atenolol, a beta blocker for the treatment of hypertension and arrhythmia, and of enalapril maleate, used to treat heart failure and chronic hypertension.

Meegan and coworkers report the synthesis of novel ethanoanthracenes with pro-apoptotic and antiproliferative effects in chronic lymphocytic leukemia, the most common leukemia in the western world. The next contribution to this Issue focuses on the natural product solidagenone with anti-inflammatory activity, a mechanism of action and a druggability profile which make it a promising agent for future investigation.

The development of novel styrylpyridinium compounds is also described, comprising their synthesis and properties as near-infrared fluorescent probes with potential for biomedical applications.

Lucarini and co-workers synthesized 6-*O*-sucrose monoesters and tested them, demonstrating their antifungal effects. The next contribution by de Sousa and collaborators covers photopharmacology and highlights the importance of porphyrins with antibacterial activity and the synergistic effect of a photodynamic treatment with antibiotic therapy against bacterial resistance.

The following paper by Rauter and co-workers describes the first nucleosides, cholinesterase inhibitors and copper chelators, with the perspective of generating water-soluble nucleoside drug candidates against Alzheimer's disease.

Future strategies for cancer treatment, the study of the impact of intranasal deposition on intranasal drug delivery, and a novel flow cytometry-based assay to identify ligands of the cyclic nucleotide-binding domain within hyperpolarization-activated cyclic nucleotide-gated channel 4 (HCN4) are the following topics discussed in this Issue. The last paper is a review on the development of compounds, which can specifically bind the BRD9 protein to modulate gene expression and to control cellular processes. Inhibition of BRD9 is particularly relevant in cancer research as BRD9 is expressed in a variety of cancer forms.

We hope to excite the readership with this Issue, especially all generations of scientists, including students, postdocs and senior researchers. With the latest findings described here, we would like to inspire them to make their own contributions in the discovery of the secrets of nature still unknown regarding novel treatments for uncurable diseases that our community still faces globally.

A final acknowledgment goes, of course, to JJ for his immense dedication and commitment to the journal Pharmaceuticals!

On the 29th November 2024

Amélia Pilar Rauter, Klaus Kopka, Thomas Efferth, Joachim Jose, Maria Emília De Sousa, and Guangshun Wang

Editors

pharmaceuticals

Article

The Impact of Chemotherapy on Arterial Stiffness and Ventricular–Arterial Coupling in Women with Breast Cancer

Nikolaos P. E. Kadoglou [1,*], Alexandriani Dimopoulou [1], Irene Tsappa [2], Pampina Pilavaki [2] and Anastasia Constantinidou [1,2]

1 Medical School, University of Cyprus, Nicosia 2029, Cyprus
2 Department of Medical Oncology, Bank of Cyprus Oncology Centre, Nicosia 2029, Cyprus
* Correspondence: nikoskad@yahoo.com; Tel.: +357-22895260

Abstract: Background: The cardiac toxicity of chemotherapy for breast cancer is not uncommon and has been associated with elevated morbidity and mortality. In the present study, we assessed the impact of chemotherapy on cardiovascular function by assessing the cardio–ankle vascular index (CAVI), global longitudinal strain (GLS) and ventricular–arterial coupling (VAC: CAVI/GLS ratio) in chemotherapy-treated women. Methods: This prospective study enrolled 78 women with breast cancer who were receiving anthracycline-based chemotherapy +/− anti-HER2 therapy (trastuzumab +/− pertuzumab). Forty-one age-matched healthy women served as controls. We comparatively evaluated left ventricular ejection fraction (LVEF), CAVI, GLS and VAC, between the chemotherapy and control groups. We also assessed their changes over time (baseline, 3-month and 6-month time point) and their independent association with the incidence of cancer therapy-related cardiovascular dysfunction (CTRCD) in the chemotherapy group. Results: In comparison to healthy controls, women receiving chemotherapy presented with significantly higher GLS (from −21.02 ± 2.09% to −19.01 ± 2.81%, $p < 0.001$) and VAC (−0.36 ± 0.06 to −0.41 ± 0.11, $p < 0.001$). The presence of CTRCD was associated with a further increase in GLS and CAVI and a significant decline in LVEF and VAC compared to CTRCD-free women ($p < 0.001$). Baseline, CAVI, GLS and VAC were independently associated with CTRCD development during follow-up. Conclusion: Women with breast cancer undergoing chemotherapy displayed abnormal levels of CAVI, VAC and GLS, compared to healthy individuals. Those effects on VAC and CAVI were more exaggerated among women with CTRCD, implicating their potential use to refine screening and therapeutic strategies for this specific population.

Keywords: chemotherapy; arterial stiffness; cardio–ankle vascular index; global longitudinal strain; breast cancer; ventricular–arterial coupling

Citation: Kadoglou, N.P.E.; Dimopoulou, A.; Tsappa, I.; Pilavaki, P.; Constantinidou, A. The Impact of Chemotherapy on Arterial Stiffness and Ventricular–Arterial Coupling in Women with Breast Cancer. *Pharmaceuticals* **2024**, *17*, 1115. https://doi.org/10.3390/ph17091115

Academic Editor: Hsin-Sheng Yang

Received: 28 June 2024
Revised: 1 August 2024
Accepted: 16 August 2024
Published: 23 August 2024

1. Introduction

Anthracycline-based chemotherapy is still the cornerstone of breast cancer patients, unless clinically contraindicated, leading to unambiguous improvement in survival rates [1]. Roughly 20% of breast cancers are human epidermal growth factor receptor 2 (HER2)-positive [2]. In those patients, trastuzumab, a monoclonal antibody, has been incorporated into breast-cancer treatment, and it is used sequentially following anthracyclines. Even in trials examining the efficacy of trastuzumab, almost 80% of participants received anthracycline-based chemotherapy [3]. The adverse impact of chemotherapy on myocardial function has been extensively described in the recent European Society of Cardiology Guidelines and related analysis [4]. Long-term results in breast-cancer survivors have reported the development of cardiovascular diseases in up to 16% of chemotherapy receivers in a 10-year follow-up period [5]. The early detection of the adverse impact of anthracyclines and/or trastuzumab administration in patients with breast cancer on myocardial dysfunction is of paramount prognostic value [6,7]. Speckle tracking in echocardiography and its index, global longitudinal strain (GLS), has been proposed as a more sensitive index of myocardial dysfunction at early stages in breast-cancer patients receiving chemotherapy [8,9]. Therefore, GLS monitoring has been established as an effective approach in those patients.

In parallel, there is growing evidence of a negative impact of chemotherapy on arterial stiffness in patients with malignancies [10]. In the case of breast cancer, anthracyclines and trastuzumab seem to increase arterial stiffness as well [11,12]. Regarding its independent association with future cardiovascular diseases [13], arterial stiffness could serve as a prognostic index of cardiovascular system impairment in breast-cancer patients under chemotherapy. However, the underlying mechanisms of arterial stiffening after chemotherapy remain mostly elusive. Pulse-wave velocity (PWV) constitutes a non-invasive, simple measurement of arterial stiffness, and its measurement is easy and repeatable. Notably, its main disadvantage is the high dependence on blood pressure [14]. To overcome this drawback, another formula of PWV calculation, termed the cardio–ankle vascular index (CAVI), has been proposed [15]. We and other investigators have documented the power of CAVI in patients with [16,17] and without cardiovascular diseases [18]. Despite previous evidence in other populations, there are no data about the relationship of subclinical vascular damage assessed by CAVI with the chemotherapy in women with breast cancer.

The role of ventricular–arterial coupling (VAC) in the pathophysiology of CVDs has long been described [19]. Increasing evidence supports its prognostic power in CVD [20]. Very limited data have indicated VAC impairment due to anti-cancer chemotherapy [21]. Regarding the sensitivity of VAC to promptly detect cardiovascular dysfunction, it is noteworthy to examine the clinical application of it in cancer patients under chemotherapy. Current data are scarce [22], and several formulas have been proposed for VAC calculation. Most of them are not easily applicable in daily clinical practice, and cut-off values are absent. Therefore, the interpretation of VAC results is challenging. Most recently, the PWV/GLS ratio has emerged as a promising, easily repeatable measure of VAC [22]. It has been tested as a prognostic marker of left ventricular dysfunction [23] or organ damage [24] in patients with myocardial infarction and hypertension, respectively. From the theoretical point of view, the arterial stiffening and the concomitant cardiac dysfunction induced by chemotherapy will lead to VAC impairment. Thereby, VAC could become ideally a sensitive and objective marker of cardiovascular dysfunction developed in chemotherapy-treated patients. However, there is little evidence supporting the negative impact of chemotherapy on VAC in cancer patients [25]. More and larger studies are required to validate whether chemotherapy impairs VAC in cancer patients [26]. The validation process has two significant limitations: previous studies included a great variance in cancer types, and methods of VAC calculation showed high diversity, since there is still no single widely accepted methos [27]. Our study focused on breast-cancer patients and used a modern formula of VAC calculation with less drawbacks compared to previous studies.

The aim of this study was to assess the alterations in arterial stiffness by calculating CAVI, after therapy with anthracyclines, either with or without anti-HER2 therapy (trastuzumab +/− pertuzumab) in women with breast cancer. In addition, we assessed GLS at echocardiogram and calculated VAC changes during the follow-up.

2. Results

2.1. General Characteristics

We initially enrolled 84 women with newly diagnosed or metastatic breast cancer treated with anthracycline based-chemotherapy with or without anti-HER2 therapy (trastuzumab +/− pertuzumab). In particular, 70 patients had early breast cancer: 18 were HER2+ and received trastuzumab treatment, 6 were triple-negative and all received anthracycline-based chemotherapy. The remaining were ER+ HER2−. There were 8 metastatic patients, of which one was HER2+, one was triple negative and three were ER+ HER2−.

During 6-month follow-up, three patients died due to cancer, two were lost from follow-up and one denied the second visit for personal reasons. At the end, 78 women completed all measurements and were analyzed, while 41 healthy women served as controls (2:1 ratio). At baseline, there were no significant differences between groups across all demographic and hemodynamic parameters (Table 1). A small percentage of women in the chemotherapy group had hypertension (8.5%) and/or dyslipidemia (6.1%) and only one woman was treated for diabetes. Notably, among hypertensive women, only half of them (four patients) were already on anti-hypertensive medications. In parallel, half of participants in both groups at baseline were already in the menopause period. We did not expect any impact of menopause on their cardiovascular system, since the vast majority of women had recently entered the menopause period. No significant differences were seen regarding the CAVI, GLS and VAC between groups at baseline ($p > 0.05$; Table 1).

Table 1. Comparison between groups at baseline.

Variables	Chemotherapy Group (N = 78)	Control Group (N = 41)	*p*-Value
Age (years)	53 ± 11	51 ± 6	0.121
Hypertension (n)	7 (8.9%)	0	-
Dyslipidemia (n)	5 (6.4%)	0	-
Diabetes (n)	1 (1.3%)	0	-
Smokers (n)	13 (16.7%)	6 (15%)	0.777
Menopause (n)	41 (52.6%)	19 (46.3%)	0.809
CAVI	7.10 ± 1.25	6.99 ± 1.05	0.758
GLS (%)	−21.02 ± 2.09	−21.16 ± 1.55	0.633
VAC1	−0.36 ± 0.06	−0.33 ± 0.05	0.290
BMI (kg/m^2)	27.51 ± 6.68	26.52 ± 4.65	0.156
SBP (mmHg)	131 ± 26	130 ± 13	0.767
DBP (mmHg)	83 ± 10	81 ± 8	0.204
LVEF (%)	63 ± 9	66 ± 7	0.105
E/A ratio	0.98 ± 0.22	1.00 ± 0.38	0.939
E/E′ ratio	7.17 ± 2.75	6.90 ± 2.85	0.806
LAVI (ml/m^2)	29.55 ± 7.68	29.35 ± 5.50	0.944
TAPSE (cm)	2.32 ± 0.32	2.59 ± 0.42	0.062
RVS′ (m/s)	0.14 ± 0.04	0.15 ± 0.02	0.361
TRVmax (m/s)	2.23 ± 0.38	2.85 ± 3.16	0.232
Pharmaceutical regimen			
ACEI (n)	3 (3.9%)	0	
CCB (n)	1 (1.3%)	0	
Statins (n)	2 (2.6%)	0	

ACEI, angiotensin-converting enzyme inhibitors; BMI, body-mass index; CAVI, cardio–ankle vascular index; CCBs, calcium channel blockers; DBP, diastolic blood pressure; E/A ratio, E transmitral flow velocity/A transmitral flow velocity; E/E′, E transmitral flow velocity/E′ tissue; GLS, global longitudinal strain; LAVI, left atrial volume index; LVEF, left ventricular ejection fraction; n, number; RVS′, right ventricular systolic excursion velocity by tissue Doppler; SBP, systolic blood pressure; TAPSE, tricuspid annular plane systolic excursion; TRVmax, maximal tricuspid regurgitation velocity; VAC, ventricular–arterial coupling. The terms of co-morbidities (hypertension, dyslipidemia and diabetes) are defined in Section 4.2.

2.2. Follow-Up Results

Significant reductions in BMI (from 27.51 ± 6.68 kg/m^2 to 26.02 ± 6.42 kg/m^2) and VAC (−0.36 ± 0.06 to −0.41 ± 0.11, $p < 0.001$) were documented between baseline and the 6-month time point, while GLS considerably increased (from −21.02 ± 2.09% to

−19.01 ± 2.81%, $p < 0.001$) in the chemotherapy group. Notably, there was a marginal non-significant reduction in LVEF (from 63 ± 9% to 59 ± 8%, $p = 0.098$) and a trend toward increased CAVI (from 7.10 ± 1.25 to 7.61 ± 1.45, $p = 0.157$).

During the follow-up, 20 patients achieved the CTRCD criteria. No acute cardiovascular events or cardiac-related hospitalization were reported during the follow-up period. At baseline, patients developing CRTCD did not differ from the CRTCD-free patients. However, at the end of the study, the CTRCD subgroup appeared to have a significant decline in LVEF and VAC and a remarkable increase in GLS and CAVI in comparison to the CTRCD-free patients and healthy controls ($p < 0.001$). No significant differences were detected between those groups in the rest of examined parameters, like changes in blood pressure or BMI. Notably, CTRCD-free patients showed a slight deterioration in CAVI, GLS and VAC, but those changes did not achieve a significant level.

The vast majority of smokers complied with oral instructions for smoking cessation, so the smoking rate dramatically decreased in both groups at follow-up. At the end of follow-up, one hypertensive woman stopped anti-hypertensive medications due to low levels of blood pressure, and another one developed hypertension and ultimately required therapy. Overall, the number of hypertensive women remained unaltered, and the same was true for their pharmaceutical regimen. Table 2 summarizes the measurements of CRTCD vs. CRTCD-free patients at baseline and the end of the study.

Table 2. Changes in variables during follow-up within subgroups of CTRCD vs. CTRCD-free patients.

Variables	CTRCD Subgroup (N = 20)		CTRCD-Free Subgroup (N = 58)		*p*-Value
	Baseline	End	Baseline	End	
Age (years)	53 ± 11		51 ± 6		-
Hypertension (n)	2 (10%)	1 (5%)	5 (8.6%)	6 (10.3%)	-
Dyslipidemia (n)	1 (5%)	1 (5%)	4 (6.9%)	5 (8.6%)	-
Diabetes (n)	0	1 (5%)	0	1 (1.8%)	-
Smokers (n)	3 (15%)	1 (5%)	10 (17.2%)	2 (3.2%)	-
Menopause (n)	10 (50%)	-	31 (53.4%)	-	-
CAVI	7.22 ± 1.34	9.51 ± 1.24 *	7.02 ± 0.91	7.51 ± 1.05	<0.001
GLS (%)	−21.01 ± 2.06	−16.43 ± 1.67 *	−21.09 ± 2.08	−20.22 ± 1.73	<0.001
VAC	−0.44 ± 0.09	−0.58 ± 0.09 *	−0.35 ± 0.06	−0.37 ± 0.06	<0.001
BMI (kg/m^2)	27.79 ± 5.56	26.18 ± 6.89 *	27.34 ± 7.05	26.02 ± 6.42 *	0.887
SBP (mmHg)	137 ± 15	137 ± 15	129 ± 27	129 ± 27	0.995
DBP (mmHg)	86 ± 9	86 ± 9	82 ± 10	82 ± 10	0.996
LVEF (%)	61 ± 4	52 ± 11 *	63 ± 7	61 ± 7	0.076
E/A ratio	0.98 ± 0.22	0.71 ± 0.31	0.99 ± 0.21	0.95 ± 0.20	0.101
E/E′ ratio	7.33 ± 3.16	8.52 ± 3.59	7.09 ± 3.01	7.99 ± 4.16	0.101
LAVI (mL/m^2)	30.29 ± 9.57	33.13 ± 11.82 *	27.15 ± 7.59	29.76 ± 8.10	0.554
TAPSE (cm)	2.21 ± 0.32	2.41 ± 0.24	2.40 ± 0.25	2.31 ± 0.39	0.675

Table 2. *Cont.*

Variables	CTRCD Subgroup (N = 20)		CTRCD-Free Subgroup (N = 58)		*p*-Value
	Baseline	**End**	**Baseline**	**End**	
RVS′ (m/s)	0.13 ± 0.03	0.16 ± 0.37	0.14 ± 0.05	0.14 ± 0.02	0.981
TRVmax (m/s)	2.47 ± 0.52	2.71 ± 0.46	2.10 ± 0.51	2.28 ± 0.30	0.429
Pharmaceutical regimen					
ACEI (n)	1	1	2	2	-
CCB (n)	0	0	1	1	-
Statins (n)	1	1	1	1	-

ACEI, angiotensin-converting enzyme inhibitors; BMI, body-mass index; CAVI, cardio–ankle vascular index; CCBs, calcium channel blockers; DBP, diastolic blood pressure; E/A ratio, E transmitral flow velocity/A transmitral flow velocity; E/E′, E transmitral flow velocity/E′ tissue; GLS, global longitudinal strain; LAVI, left atrial volume index; LVEF, left ventricular ejection fraction; n, number; RVS′, right ventricular systolic excursion velocity by tissue Doppler; SBP, systolic blood pressure; TAPSE, tricuspid annular plane systolic excursion; TRVmax, maximal tricuspid regurgitation velocity; VAC, ventricular–arterial coupling. The terms of co-morbidities (hypertension, dyslipidemia and diabetes) are defined in Section 4.2. * *Indicates a* $p < 0.05$ *change in variables within groups.*

2.3. Correlations

We examined all possible correlations with CTRCD development with all variables (anthropometric, clinical, echocardiographic and pharmaceutical) already listed in the tables at the end of the study. The values of CAVI, GLS and VAC significantly correlated with CTRCD development. LVEF was not included since it was almost the single major criteria for CRTCD identification. Medications were examined as covariates. Those parameters entered the linear regression analysis, and all of them were independently associated with CRTCD occurrence (Table 3).

Table 3. Associations between CRTCD and other variables within the chemotherapy group.

Variables	Univariate Analysis		Multivariate Analysis	
	β (SE)	***p***	**β (SE)**	***p***
CAVI	0.663 (0.099)	<0.001	0.268 (0.099)	0.008
GLS	−0.401 (0.051)	<0.001	0.273 (0.043)	0.041
VAC	−0.826 (0.077)	<0.001	−0.701 (0.145)	<0.001

CAVI, cardio–ankle vascular index; GLS, global longitudinal strain; VAC, ventricular–arterial coupling. Most important, VAC, at the end of the study, correlated with CTRCD ($r = -0.793$, $p < 0.001$) and LVEF ($r = 0.496$, $p < 0.001$), and those parameters remained independent determinants of VAC.

3. Discussion

The main findings of this prospective, observational study in female patients with breast cancer but without CV risk factors undergoing chemotherapy were (1) the significant decline in GLS; (2) the considerable elevation in CAVI, an index of aortic stiffness; and (3) the significant reduction in VAC, as expressed by the CAVI/GLS ratio after 6 months. Those effects were observed in the whole cohort, but they were predominantly accompanied by CTRCD development.

In agreement with the results of previous studies, a 6-month chemotherapy treatment with anthracyclines and/or anti-HER2 therapy considerably reduced GLS in the whole cohort of women with breast cancer who were undergoing chemotherapy, while an almost 20% of the cohort experienced CTRCD. The latter subgroup showed the most resounding decline in GLS, associated with significant impairment in LVEF. However, the rest of women—who were without clinically or echocardiographically evidenced CTRCD—showed a small but statistically significant reduction in GLS as well, compared to the baseline and controls, implicating a subtle systolic dysfunction. GLS is capable of unmasking even subtle reductions in cardiac output [28]. The accumulated data support the high incidence of cardiac dysfunction induced by chemotherapy, even when it is not clinically apparent or the LVEF reduction is slight [29,30]. The clinical significance of the GLS decline is paramount since it has been linked with an adverse prognosis [31]. Therefore, early detection and prompt therapy may prevent the deterioration of cardiac systolic function and thereby improve the prognosis of cancer patients [6].

Up to now, there have been scarce data on the impact of anti-cancer chemotherapeutic agents on the vascular system, showing a potential "vascular toxicity". Previous studies have demonstrated increased PWV after chemotherapy in breast-cancer patients receiving anthracyclines [11], but not trastuzumab [12], in the long-term. Therefore, anthracyclines may impair endothelial function and precipitate atherosclerosis in large arteries, predisposing patients to adverse cardiovascular events. This is the first study reporting the impact of chemotherapy on vascular function using CAVI, avoiding the interference of blood pressure.Vascular toxicity was previously implicated in a similar group of patients [32]. Contrary to studies on hypertensive patients, our study provides more robust evidence of the adverse effects of chemotherapy on the elasticity of arteries, since the blood pressure-lowering pharmaceutical agents remained unaltered in chemotherapy recipients. Therefore, the early diagnosis of chemotherapy-induced vascular toxicity should be the first priority, especially when classic cardiovascular risk factors remain "well-controlled" in women who have breast cancer and are under chemotherapy. The development of vascular cardiac injury, either with or without CTRCD occurrence, may not only affect chemotherapy continuation but also the short-term prognosis. The incorporation of CAVI measurement as a routine clinical practice of monitoring patients under chemotherapy could add a sensitive, non-invasive and objective tool to promptly identify patients at an early stage of CTRCD and tailor their therapy. Future studies will clarify the prognostic value of our findings.

The mechanical coupling of LV with the aorta demonstrates how the aortic-stiffness affects the LV function leading to both systolic and diastolic dysfunction, and eventually to the adverse remodeling of LV which follows the development of heart failure [33,34]. Regarding VAC as the crosstalk between the left ventricular function and arterial system, a plethora of studies have proposed it as an index of cardiovascular performance in several conditions without established CVD but at high risk for cardiovascular morbidity and mortality, like hypertension, diabetes mellitus, etc. [35,36]. Notably, the prognostic value of VAC has also been confirmed in patients with established CVD, like those after recent myocardial infarction [37]. There is a growing body of evidence supporting the usage of VAC as a therapeutic target that can help ameliorate the prognosis. In the context of acute heart failure, diuretics administration may improve arterial elasticity, rather than cardiac contractility, and it may yield significant benefits in the post-discharge period [19]. The pharmaceutical modification of VAC and its link with cardiovascular outcomes remain under investigation. Despite the pathophysiological link of VAC with cardiovascular alterations in several conditions, its major limitation remains the establishment of a gold-standard technique for its measurement [38]. The ideal method should be easy to perform and feasible, with high reproducibility and the ability to be validated.

Several indices have been proposed for VAC assessment. Some investigators have used the PWV/GLS ratio for VAC calculation, trying to limit potential confounders as much as possible. A reduced ratio has been linked with clinical outcomes in a small spectrum of cardiovascular disease [39], but, most importantly, it has been associated with greater severity of HF and a worse functional capacity [40]. In our study, we used CAVI as an index of arterial stiffness, as it demonstrates greater advantages than PWV. It remains to be proved whether this novel ratio represents the VAC. From the pathophysiology perspective, an altered CAVI/GLS ratio may reflect the net result of a cluster of disorders, including endothelial dysfunction, renin–angiotensin-system activation, abnormal collagen turnover, increased metalloproteinase activity and fibrosis, contributing to both cardiac and vascular dysfunction. Our study clearly demonstrated aortic stiffening and reduced VAC in patients with CTRCD, indicating an interplay between cardiac and vascular function. After extrapolating data from other clinical conditions, we noted that an impaired

VAC adversely affects the LV afterload, with apparent consequences like higher LV work and oxygen-supply demands [41]. We hypothesized that the cardiovascular toxicity of chemotherapeutic agents may not only explain their lower exercise capacity, but it can potentially predict their higher cardiovascular risk. Hence, CTRCD cannot entirely describe the cardiovascular complications of chemotherapy. Several cardiac alterations have been described in the past after anthracyclines or adjuvant therapy with trastuzumab, leading to a worse prognosis [42,43]. Therefore, the VAC measurement may unmask another phenotype of cardiovascular toxicity before the cardiac dysfunction becomes apparent. This may also explain the high cardiovascular risk of chemotherapy recipients in the long term in relation to myocardial injury and/or coronary artery disease.

There are several drawbacks to the present study. The most important of them is the small sample size. This was a small single-center study, which included a homogeneous cohort of female patients with breast cancer, and our preliminary findings cannot be generalized. The limited number of participants may also explain the existence of a few correlations between variables when the studied parameters were statistically different between the two groups. Therefore, a larger sample size is necessary to obtain robust evidence and detect further associations. Moreover, the cross-sectional design of our study prevented us from drawing firm conclusions about causality. Regarding the fact that some cardiotoxic effects may first become apparent after 6 months of follow-up, longer protocols should be carried out to better define and investigate the predictive value of arterial stiffness and VAC in women with breast cancer, as well as the impact of treatment on them.

In conclusion chemotherapy deteriorated arterial elasticity and reduced VAC in women with breast cancer during a 6-month follow-up. Therefore, this group has the potential for accelerated atherosclerosis and higher cardiovascular risk.

4. Materials and Methods

4.1. Research Design

This is a prospective case–control study comparing women with breast cancer who are eligible for chemotherapy (chemotherapy group) and healthy women (control group). At baseline, all participants underwent a medical history recording, clinical examination, arterial stiffness assessment and echocardiography to obtain measurements. All of those exams are described in detail in the following subsections and were performed before chemotherapy initiation. After the initial assessment, only women with breast cancer were followed up at 3 and 6 months. At those time-points, arterial stiffness and echocardiography measurements were obtained. Moreover, the chemotherapy group was further subdivided into subgroups of patients developing cancer therapy-related cardiovascular dysfunction (CTRCD) during the follow-up and CTRCD-free subgroup.

Chemotherapy group: We consecutively recruited 84 women with breast cancer who were eligible for chemotherapy. All women with breast cancer received anthracycline based-chemotherapy (FEC, AC or EC with or without Paclitaxel/Taxotere) either plus or not with anti-HER2 therapy (trastuzumab +/− pertuzumab).

Control group: A group of 41 age-matched healthy women without cancer, cardiovascular diseases or other chronic diseases which may affect the studied parameters served as the control group at baseline. The latter group was enrolled after an open invitation was sent to family members or friends of patients with breast cancer to participate in our study. Such chronic diseases included hypertension, diabetes, kidney (creatinine ≥2 mg/dl) or liver failure (AST ≥ 3 times the upper normal limit), systemic inflammatory diseases or autoimmune diseases, atrial fibrillation, acute infection or surgery.

Subgroups based on CTRCD development: According to current guidelines, we considered cancer therapy-related cardiovascular dysfunction (CTRCD) in asymptomatic patients with (1) new LVEF reduction <40%; (2) new LVEF reduction to 40–49%, when the amount of LVEF exceeds ≥10%; (3) new LVEF reduction to 40–49%, when the amount of LVEF is less than 10% and concomitant new relative decline in GLS by 15% from baseline; and (4) LVEF ≥ 50% and new relative decline in GLS by >15% from baseline. Biomarkers were not available to most patients, and they were not included in the analysis. We considered HF symptoms as an additional index, only when echocardiography findings were compatible to CRTCD, because fatigue and dyspnea were common complaints of patients under chemotherapy and could not distinguish those with CTRCD. When the latter diagnosis was set, the patients were referred to the cardio-oncology outpatient clinic for monitoring and HF-related therapy commitment. So, they were not further included in our study.

Written informed consent was obtained from all participants before enrollment. All procedures were performed according to the principles of Helsinki Declaration and were approved by the National Bioethics Committee of Cyprus (reference number of approval: EEBK/EEP/2021/34).

4.2. Participants' Clinical Examination

At each visit, we measured participants' weight to calculate the BMI and the blood pressure after sitting for 5 min. Based on a structured questionnaire, we recorded medications and co-morbidities, defined as follows: hypertension, BP $\geq$ 140/90 mmHg measured on repeated occasions or receipt of antihypertensive drugs; hyperlipidemia, fasting serum low-density-lipoprotein cholesterol (LDL-C) $\geq$130 mg/dL or statin therapy; active smokers, current or within the previous 6 months; diabetes mellitus, fasting plasma glucose (FPG) $\geq$ 126 mg/dL, or HbA1c $\geq$ 6%, or antidiabetic drugs; and coronary artery disease, history of stable or unstable angina, myocardial infarction, percutaneous or surgical myocardial revascularization.

4.3. Arterial-Stiffness Assessment

The arterial stiffness of the participants was calculated as the cardio–ankle vascular index (CAVI) (Vasera VS-1500, Fukuda Denshi, Tokyo, Japan). The calculation of the CAVI was based on the following formula: CAVI = a $\times$ PWV2/SPB$-$DPB + b (a and b are constants; PWV, pulse-wave velocity from the heart to the ankle; SBP, systolic blood pressure; and DBP, diastolic blood pressure). CAVI is a non-invasive and reproducible technique independent of blood pressure proportionally related to a higher risk of cardiovascular diseases. According to the recommended procedure, the participants were in a supine position for relaxation for 10 min. Four electrodes were attached to the participants' limbs for electrocardiogram (ECG) recording, and a microphone placed on participants' chest recorded phonocardiogram (PCG). Blood pressure measurements were obtained from cuffs wrapped around the four limbs. Basically, CAVI was calculated by recording the distance from the level of the aortic valve (i.e., brachial level) to each ankle and the time delay from the aortic valve closing to the detected notch in arterial pressure wave at unilateral ankle [44]. The right and left CAVI were measured and then averaged. Besides this, the device provided the ankle–brachial index (ABI) for both sides.

4.4. Global Longitudinal Stain (GLS) and Ventricular–Arterial Coupling (VAC)

Using a standard imaging system (GE E95 ultrasound scanner, Horten, Norway), two independent experienced cardiologists performed the echocardiographic studies at baseline and during follow-up. The offline imaging analysis was performed on a single PC workstation by both those cardiologists blinded to patients' data (GE Echopac 205 version software). Standard echocardiographic measurements included LVEF, LV end-diastolic volume (LVEDV) and LVESV. The LV myocardial deformation was measured using the LV global longitudinal stain (GLS) formula. In brief, analyzing the 3 apical views of echocardiography, the longitudinal strain was measured, each wall was subsequently divided into 3 segments (basal, mid and apical), and a total of 17 segmental strain curves were obtained. GLS was calculated as the average value of all strain peak values. The inter-observer variability was based on the intra-class correlation coefficient (ICC), which showed a high reproducibility [0.94 (0.84–0.97)]. The intra- and inter-observer reliability of strain analysis by our group was previously reported, and it is very low (<2.5%). Besides this, we calculated the echocardiography-derived myocardial work at rest by entering the blood pressure measurements to GLS analysis. Using vendor' software (GE echopac 205 version), we calculated the parameters of myocardial work. We then calculated the CAVI/GLS ratio as an index of VAC.

4.5. Statistical Analysis

Normally distributed continuous variables were presented as the mean ± SD. Normality of distribution was assessed by the Kolmogorov–Smirnov test. Comparisons of continuous and categorical variables were analyzed with Student's *t*-test and chi-square test, respectively. Changes in continuous variables within and between groups were assessed using paired samples and Student's *t*-tests, respectively. To test the univariate and multivariate associations of CAVI with any of the study population characteristics, we performed a Pearson correlation and multiple linear regression analysis for normally distributed variables. Medications were examined as covariates. A two-tailed $p < 0.05$ was considered to be significant. The computer software package SPSS (version 25.0; SPSS Inc., Chicago, IL, USA) was used for statistical analysis.

5. Conclusions

In conclusion, the combined therapy of anthracyclines plus trastuzumab in women with breast cancer reduced both GLS and VAC values and increased the arterial stiffness assessed by CAVI. Those effects were more pronounced among women developing CRTCD within 6 months from chemotherapy initiation. The clinical impact of cardiac and vascular toxicity of these chemotherapeutic agents and their potential consequences remains to be proved in the long term.

Author Contributions: Conceptualization, N.P.E.K.; methodology, N.P.E.K. and P.P.; validation, N.P.E.K.; formal analysis, A.D.; investigation, A.D. and P.P.; resources, I.T.; data curation, A.D. and I.T.; writing—original draft preparation, N.P.E.K. and I.T.; writing—review and editing, N.P.E.K. and A.C.; visualization, P.P.; supervision, N.P.E.K. and A.C.; project administration, N.P.E.K.; funding acquisition, N.P.E.K. All authors have read and agreed to the published version of the manuscript.

Funding: This research was funded by the University of Cyprus, as an internal Research project, grant number 128 IAT-2020, and the APC was funded by 128 IAT-2020.

Institutional Review Board Statement: The study was conducted in accordance with the Declaration of Helsinki and approved by Ethics Committee of Cyprus National (reference number of approval: EEBK/EEP/2021/34).

Informed Consent Statement: Informed consent was obtained from all subjects involved in the study.

Data Availability Statement: The data is unavailable due to ethical restrictions.

Acknowledgments: We would like to thank Lawko Ahmed for his contributions to the participants and data.

Conflicts of Interest: The authors declare no conflicts of interest.

References

1. Bansal, N.; Adams, M.J.; Ganatra, S.; Colan, S.D.; Aggarwal, S.; Steiner, R.; Amdani, S.; Lipshultz, E.R.; Lipshultz, S.E. Strategies to prevent anthracycline-induced cardiotoxicity in cancer survivors. *Cardiooncology* **2019**, *5*, 18. [CrossRef] [PubMed]
2. Serrano, J.M.; Mata, R.; González, I.; Del Castillo, S.; Muñiz, J.; Morales, L.J.; Espinosa, M.J.; Moreno, F.; Jiménez, R.; Cristobal, C.; et al. Early and late onset cardiotoxicity following anthracycline-based chemotherapy in breast cancer patients: Incidence and predictors. *Int. J. Cardiol.* **2023**, *382*, 52–59. [CrossRef]
3. Von Minckwitz, G.; Huang, C.S.; Mano, M.S.; Loibl, S.; Mamounas, E.P.; Untch, M.; Wolmark, N.; Rastogi, P.; Schneeweiss, A.; Redondo, A.; et al. KATHERINE Investigators. Trastuzumab Emtansine for Residual Invasive HER2-Positive Breast Cancer. *N. Engl. J. Med.* **2019**, *380*, 617–628. [CrossRef] [PubMed]
4. Leszek, P.; Klotzka, A.; Bartus, S.; Burchardt, P.; Czarnecka, A.M.; Dlugosz-Danecka, M.; Gierlotka, M.; Kosela-Paterczyk, H.; Krawczyk-Ozog, A.; Kubiatowski, T.; et al. A practical approach to the 2022 ESC cardio-oncology guidelines: Comments by a team of experts—Cardiologists and oncologists. *Kardiol. Pol.* **2023**, *81*, 1047–1063. [CrossRef] [PubMed]

5. Mauro, C.; Capone, V.; Cocchia, R.; Cademartiri, F.; Riccardi, F.; Arcopinto, M.; Alshahid, M.; Anwar, K.; Carafa, M.; Carbone, A.; et al. Cardiovascular Side Effects of Anthracyclines and HER2 Inhibitors among Patients with Breast Cancer: A Multidisciplinary Stepwise Approach for Prevention, Early Detection, and Treatment. *J. Clin. Med.* **2023**, *12*, 2121. [CrossRef] [PubMed]
6. Goulas, K.; Farmakis, D.; Constantinidou, A.; Kadoglou, N.P.E. Cardioprotective Agents for the Primary Prevention of Trastuzumab-Associated Cardiotoxicity: A Systematic Review and Meta-Analysis. *Pharmaceuticals* **2023**, *16*, 983. [CrossRef]
7. Alexandraki, A.; Papageorgiou, E.; Zacharia, M.; Keramida, K.; Papakonstantinou, A.; Cipolla, C.M.; Tsekoura, D.; Naka, K.; Mazzocco, K.; Mauri, D.; et al. New Insights in the Era of Clinical Biomarkers as Potential Predictors of Systemic Therapy-Induced Cardiotoxicity in Women with Breast Cancer: A Systematic Review. *Cancers* **2023**, *15*, 3290. [CrossRef]
8. De la Brassinne Bonardeaux, O.; Born, B.; Moonen, M.; Lancellotti, P. Mild Cardiotoxicity and Continued Trastuzumab Treatment in the Context of HER2-Positive Breast Cancer. *J. Clin. Med.* **2023**, *12*, 6708. [CrossRef]
9. Cronin, M.; Seher, M.; Arsang-Jang, S.; Lowery, A.; Kerin, M.; Wijns, W.; Soliman, O. Multimodal Imaging of Cancer Therapy-Related Cardiac Dysfunction in Breast Cancer—A State-of-the-Art Review. *J. Clin. Med.* **2023**, *12*, 6295. [CrossRef]
10. Mozos, I.; Borzak, G.; Caraba, A.; Mihaescu, R. Arterial stiffness in hematologic malignancies. *OncoTargets Ther.* **2017**, *10*, 1381–1388. [CrossRef]
11. Novo, G.; Di Lisi, D.; Manganaro, R.; Manno, G.; Lazzara, S.; Immordino, F.A.; Madaudo, C.; Carerj, S.; Russo, A.; Incorvaia, L.; et al. Arterial Stiffness: Effects of Anticancer Drugs Used for Breast Cancer Women. *Front. Physiol.* **2021**, *12*, 661464. [CrossRef] [PubMed]
12. Anastasiou, M.; Oikonomou, E.; Theofilis, P.; Papamikroulis, G.A.; Gazouli, M.; Kalogeras, K.; Lygkoni, S.; Pesiridis, T.; Goliopoulou, A.; Papatheodoridi, A.; et al. Prolonged Impact of Anti-Cancer Therapy on Endothelial Function and Arterial Stiffness in Breast Cancer Patients. *Vascul. Pharmacol.* **2023**, *152*, 107195. [CrossRef] [PubMed]
13. Triantafyllias, K.; Thiele, L.E.; Cavagna, L.; Baraliakos, X.; Bertsias, G.; Schwarting, A. Arterial Stiffness as a Surrogate Marker of Cardiovascular Disease and Atherosclerosis in Patients with Arthritides and Connective Tissue Diseases: A Literature Review. *Diagnostics* **2023**, *13*, 1870. [CrossRef] [PubMed]
14. Maga, M.; Wachsmann-Maga, A.; Batko, K.; Włodarczyk, A.; Kłapacz, P.; Krężel, J.; Szopa, N.; Sliwka, A. Impact of Blood-Flow-Restricted Training on Arterial Functions and Angiogenesis-A Systematic Review with Meta-Analysis. *Biomedicines* **2023**, *11*, 1601. [CrossRef] [PubMed]
15. Abdullah, N.; Blin, J.A.; Kamalul Arifin, A.S.; Abd Jalal, N.; Ismail, N.; Mohd Yusof, N.A.; Abdullah, M.S.; Husin, N.F.; Dauni, A.; Kamaruddin, M.A.; et al. Cardiovascular risk prediction with cardio-ankle vascular index in the malaysian cohort study. *Curr. Probl. Cardiol.* **2024**, *49*, 102192. [CrossRef]
16. Mylonas, S.N.; Moulakakis, K.G.; Kadoglou, N.; Antonopoulos, C.; Kotsis, T.E.; Kakisis, J.; Katsenis, K.; Liapis, C. Arterial Stiffness Assessed by Cardio-Ankle Vascular Index in Patients With Abdominal Aortic Aneurysm and Its Alterations after Treatment. *Vasc. Endovasc. Surg.* **2021**, *55*, 804–810. [CrossRef]
17. Kadoglou, N.; Moulakakis, K.G.; Mantas, G.; Spathis, A.; Gkougkoudi, E.; Mylonas, S.N.; Kakisis, J.; Liapis, C. Novel Biomarkers and Imaging Indices for the "Vulnerable Patient" with Carotid Stenosis: A Single-Center Study. *Biomolecules* **2023**, *13*, 1427. [CrossRef]
18. Tabata, T.; Sato, S.; Ohno, R.; Iwakawa, M.; Kiyokawa, H.; Morinaga, Y.; Tanji, N.; Kinoshita, T.; Shimizu, K. Cardio-Vascular Interaction Evaluated by Speckle-Tracking Echocardiography and Cardio-Ankle Vascular Index in Hypertensive Patients. *Int. J. Mol. Sci.* **2022**, *23*, 14469. [CrossRef]
19. Ikonomidis, I.; Aboyans, V.; Blacher, J.; Brodmann, M.; Brutsaert, D.L.; Chirinos, J.A.; De Carlo, M.; Delgado, V.; Lancellotti, P.; Lekakis, J.; et al. The role of ventricular-arterial coupling in cardiac disease and heart failure: Assessment, clinical implications and therapeutic interventions. A consensus document of the European Society of Cardiology Working Group on Aorta & Peripheral Vascular Diseases, European Association of Cardiovascular Imaging, and Heart Failure Association. *Eur. J. Heart Fail.* **2019**, *21*, 402–424. [CrossRef]
20. Scarlatescu, A.I.; Micheu, M.M.; Petre, I.G.; Oprescu, N.; Mihail, A.M.; Cojocaru, I.D.; Vatasescu, R.G. Left Ventricular-Arterial Coupling as an Independent Predictor of Adverse Events in Young Patients with ST Elevation Myocardial Infarction-A 3D Echocardiographic Study. *Biomedicines* **2024**, *12*, 105. [CrossRef]
21. Koelwyn, G.J.; Lewis, N.C.; Ellard, S.L.; Jones, L.W.; Gelinas, J.C.; Rolf, J.D.; Melzer, B.; Thomas, S.M.; Douglas, P.S.; Khouri, M.G.; et al. Ventricular-Arterial Coupling in Breast Cancer Patients After Treatment With Anthracycline-Containing Adjuvant Chemotherapy. *Oncologist* **2016**, *21*, 141–149. [CrossRef]
22. Holm, H.; Kruger, R.; Jujic, A.; Lamiral, Z.; Uys, A.S.; Girerd, N.; Magnusson, M. Ventricular-arterial coupling and cardiovascular risk among young adults: The African-PREDICT study. *Am. J. Physiol. Heart Circ. Physiol.* **2023**, *325*, H362–H371. [CrossRef] [PubMed]
23. Pavlidis, G.; Tsilivarakis, D.; Katogiannis, K.; Vlastos, D.; Katsanos, S.; Katsanaki, E.; Thymis, J.; Parissis, J.; Lambadiari, V.; Ikonomidis, I. Association of aortic stiffness early post myocardial infarction with left ventricular remodelling. *Eur. J. Clin. Investig.* **2024**, *54*, e14090. [CrossRef] [PubMed]
24. Camilli, M.; Birritella, L.; Delogu, A.B.; Lamendola, P.; De Vita, A.; Melita, V.; Romano, A.; Ruggiero, A.; Attinà, G.; Lanza, G.A.; et al. Left Ventricular-Arterial Coupling and Vascular Function in Childhood Cancer Survivors Exposed to Anthracycline Chemotherapy. *Rev. Cardiovasc. Med.* **2023**, *24*, 124. [CrossRef] [PubMed]

25. Climie, R.E.; Dillon, H.T.; Horne-Okano, Y.; Wallace, I.; Avery, S.; Kingwell, B.A.; La Gerche, A.; Howden, E.J. Vascular Aging Is Accelerated in Hematological Cancer Survivors Who Undergo Allogeneic Stem Cell Transplant. *Hypertension* **2023**, *80*, 1881–1889. [CrossRef]
26. Stoichescu-Hogea, G.; Buleu, F.N.; Nicusor Pop, G.; Duda-Seiman, D.; Ember, A.; Tudor, A.; Baneu, P.; Kundnani, N.R.; Christodorescu, R.; Dragan, S. Ventricular-arterial coupling assessed by PWV/GLS ratio in hypertensive patients. *Eur. Rev. Med. Pharmacol. Sci.* **2022**, *26*, 7024–7035. [CrossRef]
27. Dell'Angela, L.; Nicolosi, G.L. From ejection fraction, to myocardial strain, and myocardial work in echocardiography: Clinical impact and controversies. *Echocardiography* **2024**, *41*, e15758. [CrossRef]
28. Sulaiman, L.; Hesham, D.; Abdel Hamid, M.; Youssef, G. The combined role of NT-proBNP and LV-GLS in the detection of early subtle chemotherapy-induced cardiotoxicity in breast cancer female patients. *Egypt. Heart J.* **2021**, *73*, 20. [CrossRef]
29. Mele, D.; Malagutti, P.; Indelli, M.; Ferrari, L.; Casadei, F.; Da Ros, L.; Pollina, A.; Fiorencis, A.; Frassoldati, A.; Ferrari, R. Reversibility of Left Ventricle Longitudinal Strain Alterations Induced by Adjuvant Therapy in Early Breast Cancer Patients. *Ultrasound Med. Biol.* **2016**, *42*, 125–132. [CrossRef]
30. Muckiene, G.; Vaitiekus, D.; Zaliaduonyte, D.; Zabiela, V.; Verseckaite-Costa, R.; Vaiciuliene, D.; Juozaityte, E.; Jurkevicius, R. Prognostic Impact of Global Longitudinal Strain and NT-proBNP on Early Development of Cardiotoxicity in Breast Cancer Patients Treated with Anthracycline-Based Chemotherapy. *Medicina* **2023**, *59*, 953. [CrossRef]
31. Kerkhof, P.L.; Heyndrickx, G.R.; Li, J.K. Hemodynamic determinants and ventriculo-arterial coupling are sex-associated in heart failure patients. In Proceedings of the 38th Annual International Conference of the IEEE Engineering in Medicine and Biology Society (EMBC), Orlando, FL, USA, 18 October 2016; Volume 2016, pp. 3286–3289. [CrossRef]
32. Gasowski, J.; Piotrowicz, K.; Messerli, F.H. Arterial Hypertension after Age 65: From Epidemiology and Pathophysiology to Therapy Do We Know Where We Stand? *Kardiol. Pol.* **2018**, *76*, 723–730. [CrossRef] [PubMed]
33. Soulat, G.; Gencer, U.; Kachenoura, N.; Villemain, O.; Messas, E.; Boutouyrie, P.; Laurent, S.; Mousseaux, E. Changes in segmental pulse wave velocity of the thoracic aorta with age and left ventricular remodelling. An MRI 4D flow study. *J. Hypertens.* **2020**, *38*, 118–126. [CrossRef] [PubMed]
34. Saba, P.S.; Cameli, M.; Casalnuovo, G.; Ciccone, M.M.; Ganau, A.; Maiello, M.; Modesti, P.A.; Muiesan, M.L.; Novo, S.; Palmiero, P.; et al. Ventricular-vascular coupling in hypertension: Methodological considerations and clinical implications. *J. Cardiovasc. Med.* **2014**, *15*, 773–787. [CrossRef]
35. Gallo, G.; Savoia, C. Hypertension and Heart Failure: From Pathophysiology to Treatment. *Int. J. Mol. Sci.* **2024**, *25*, 6661. [CrossRef]
36. Aghezzaf, S.; Coisne, A.; Bauters, C.; Favata, F.; Delsart, P.; Coppin, A.; Seunes, C.; Schurtz, G.; Verdier, B.; Lamblin, N.; et al. Feasibility and prognostic significance of ventricular-arterial coupling after myocardial infarction: The RIGID-MI cohort. *Eur. Heart J. Cardiovasc. Imaging* **2024**, *25*, 668–677. [CrossRef] [PubMed]
37. Berthelot, E.; Bihry, N.; Brault-Melin, O.; Assayag, P.; Cohen-Solal, A.; Chemla, D.; Logeart, D. Changes in ventricular-arterial coupling during decongestive therapy in acute heart failure. *Eur. J. Clin. Investig.* **2014**, *44*, 982–988. [CrossRef]
38. Kadoglou, N.P.E.; Dimopoulou, A.; Gkougkoudi, E.; Parperis, K. Altered Arterial Stiffness, Ventricular-Arterial Coupling and Troponin Levels in Patients with Systemic Lupus Erythematosus. *Medicina* **2024**, *60*, 821. [CrossRef] [PubMed]
39. Saeed, S.; Holm, H.; Nilsson, P.M. Ventricular-arterial coupling: Definition, pathophysiology and therapeutic targets in cardiovascular disease. *Expert. Rev. Cardiovasc. Ther.* **2021**, *19*, 753–761. [CrossRef]
40. Pugliese, N.R.; Balletti, A.; Armenia, S.; De Biase, N.; Faita, F.; Mengozzi, A.; Paneni, F.; Ruschitzka, F.; Virdis, A.; Ghiadoni, L.; et al. Ventricular-Arterial Coupling Derived From Proximal Aortic Stiffness and Aerobic Capacity Across the Heart Failure Spectrum. *JACC Cardiovasc. Imaging* **2022**, *15*, 1545–1559. [CrossRef]
41. Ikonomidis, I.; Katsanos, S.; Triantafyllidi, H.; Parissis, J.; Tzortzis, S.; Pavlidis, G.; Trivilou, P.; Makavos, G.; Varoudi, M.; Frogoudaki, A.; et al. Pulse wave velocity to global longitudinal strain ratio in hypertension. *Eur. J. Clin. Investig.* **2019**, *49*, e13049. [CrossRef]
42. Grela-Wojewoda, A.; Niemiec, J.; Sas-Korczynska, B.; Zemelka, T.; Puskulluoglu, M.; Wysocki, W.M.; Wojewoda, T.; Pacholczak-Madej, R.; Adamczyk, A.; Mucha-Malecka, A.; et al. Adjuvant combined therapy with trastuzumab in patients with HER2-positive breast cancer and cardiac alterations: Implications for optimal cardio-oncology care. *Pol. Arch. Intern. Med.* **2022**, *132*, 16204. [CrossRef] [PubMed]
43. Yersal, O.; Eryilmaz, U.; Akdam, H.; Meydan, N.; Barutca, S. Arterial Stiffness in Breast Cancer Patients Treated with Anthracycline and Trastuzumab-Based Regimens. *Cardiol. Res. Pract.* **2018**, *2018*, 5352914. [CrossRef] [PubMed]
44. Kadoglou, N.P.E.; Moulakakis, K.G.; Mantas, G.; Kakisis, J.D.; Mylonas, S.N.; Valsami, G.; Liapis, C.D. The Association of Arterial Stiffness With Significant Carotid Atherosclerosis and Carotid Plaque Vulnerability. *Angiology* **2022**, *73*, 668–674. [CrossRef] [PubMed]

pharmaceuticals

MDPI

Article

Characterization of Liquid Dosage Forms of Atenolol and Enalapril Maleate for Oral and Enteral Feeding Administration

Sandra Mota [1,2], Ana Torres [1,2], Clara Quintas [1,3], António M. Peres [4], Nuno Ferreiro [4], Rebeca Cruz [5], Helena Ferreira [1,6,*], Isabel F. Almeida [1,2,*] and Susana Casal [5]

1 Associate Laboratory i4HB, Institute for Health and Bioeconomy, Faculty of Pharmacy, University of Porto, 4050-313 Porto, Portugal; up201608486@up.pt (S.M.); up201706122@up.pt (A.T.); claraquintas@ff.up.pt (C.Q.)
2 UCIBIO—Applied Molecular Biosciences Unit, Laboratory of Technology, Department of Drug Sciences, Faculty of Pharmacy, University of Porto, 4050-313 Porto, Portugal
3 UCIBIO—Applied Molecular Biosciences Unit, Laboratory of Pharmacology, Department of Drug Sciences, Faculty of Pharmacy, University of Porto, 4050-313 Porto, Portugal
4 CIMO, LA SusTEC, Instituto Politécnico de Bragança, 5300-253 Bragança, Portugal; peres@ipb.pt (A.M.P.); nuno.ferreiro@ipb.pt (N.F.)
5 Associated Laboratory for Green Chemistry (LAQV) of the Network of Chemistry and Technology (REQUIMTE), Laboratory of Bromatology and Hydrology, Department of Chemical Sciences, Faculty of Pharmacy, University of Porto, 4050-313 Porto, Portugal; rcruz@ff.up.pt (R.C.); sucasal@ff.up.pt (S.C.)
6 UCIBIO–Applied Molecular Biosciences, Laboratory of Microbiology, Department of Biological Sciences, Faculty of Pharmacy, University of Porto, 4050-313 Porto, Portugal
* Correspondence: hferr@ff.up.pt (H.F.); ifalmeida@ff.up.pt (I.F.A.)

Abstract: The limited availability of pharmaceutical formulations tailored for cardiovascular diseases in both pediatric and geriatric populations generates the need for compounded dosage forms to guarantee precise dosing and medication adherence. This study aimed to analyze the physicochemical properties and stability of formulations of atenolol and enalapril maleate prepared with a proprietary oral vehicle, SuspendIt®. To this end, palatability, injectability, pH, rheological behavior, and physical, microbiological, and chemical stability over a 180-day storage period at 25 °C and 5 °C were evaluated. Injectability tests confirmed the suitable use of both formulations for administration through enteral feeding tubes. By using a potentiometric electronic tongue, it was confirmed that the SuspendIt® vehicle effectively served as a bitter-blocking strategy for atenolol and enalapril maleate. Adequate stability throughout the storage period was confirmed in terms of the mechanical properties, pH, and effectiveness of the preservative system. The atenolol concentration remained above 90% of the initial amount, while the concentration of enalapril maleate decreased to 88% after 90 days of storage at 25 °C. In summary, the atenolol formulation maintained suitable chemical, physical, and microbiological stability after 180 days at both storage temperatures, while the enalapril maleate formulation remained stable up to 60 days at 25 °C and for 180 days at 5 °C.

Keywords: stability; compounding; oral formulations; atenolol; enalapril maleate; enteral feeding; liquid dosage forms

Citation: Mota, S.; Torres, A.; Quintas, C.; Peres, A.M.; Ferreiro, N.; Cruz, R.; Ferreira, H.; Almeida, I.F.; Casal, S. Characterization of Liquid Dosage Forms of Atenolol and Enalapril Maleate for Oral and Enteral Feeding Administration. *Pharmaceuticals* **2024**, *17*, 1052. https://doi.org/10.3390/ph17081052

Academic Editor: Guendalina Zuccari

Received: 5 June 2024
Revised: 26 July 2024
Accepted: 6 August 2024
Published: 9 August 2024

1. Introduction

Compounding medicines are important therapeutic options in all areas of medicine and are essential to the provision of healthcare [1]. Their use can overcome pitfalls related to limited dosage forms, limited dosages and strengths, limited orphan medicines, the need for alternative raw materials, and the need for specific combinations of drugs. In addition, they may serve as an alternative in cases of drug shortages or when pharmaceutical industries discontinue certain drugs due to waning interest. Extemporaneous preparations of targeted cardiovascular medications, such as enalapril maleate or atenolol, are typically necessary for particular demographic groups or niches, such as the elderly and children.

Enalapril is an ethyl ester prodrug converted to its active metabolite enalaprilat, which inhibits angiotensin-converting enzyme (ACE), preventing the formation of angiotensin II and inducing vasodilation. It reduces arterial pressure, afterload, blood volume, and preload, thereby inhibiting cardiac and vascular remodeling. Enalapril is used to treat hypertension, chronic heart failure [2], and chronic kidney disease with proteinuria [3]. Although not fully understood, atenolol appears to exert its therapeutic effect by preventing activation of β1 adrenoceptors in the heart, leading to negative inotropic and chronotropic effects, reducing cardiac output and arterial pressure. The reduction in cardiac workload makes it effective in ischemic coronary disease, while its negative chronotropic and dromotropic effects are the rational basis for the treatment of arrhythmias [4–6].

Both enalapril maleate and atenolol are commercially available in tablets at fixed dosages (enalapril, 2.5–20 mg; atenolol, 25–100 mg), which poses some challenges to ensuring accurate dose and medication adherence in specific populations, such as pediatrics or the elderly. The pediatric population is heterogeneous, ranging from preterm newborns to adolescents, necessitating dose adjustments based on their weight and age [2,7–9]. Additionally, regardless of the patient's age, the presence of renal insufficiency also demands dose reductions for both enalapril maleate and atenolol [7,8,10]. Therefore, providing alternative options to oral solid dosage forms, such as the development of liquid dosage forms to deliver these drugs, is necessary to ensure accurate dose adjustments based on a patient's weight, age, and renal function, as well as flexibility as the child grows, thereby reducing the risk of underdosing or overdosing. Furthermore, many children have limited or inconsistent swallowing abilities, as do many elderly individuals, due to age- or disease-related dysphagia or odynophagia [11]. Liquid pharmaceutical forms for oral administration may enhance medication adherence due to the facility of administration, which reduces the risk of choking [12]. In addition to oral administration, patients with neurological conditions that affect swallowing, severe dysphagia, or who are critically ill, such as those who are comatose, may require the administration of medications via enteral feeding tubes [13,14]. Flexibility in dose adjustment is crucial for these patients, who may need smaller doses. Liquid forms also facilitate drug absorption in these patients, often with compromised digestive processes, and reduce the risk of clogging the feeding tube, ensuring that they can receive their necessary medications safely and effectively, decreasing the complications associated with the cross-contamination from sharing of tablet crushing devices and the risks of occupational exposure to drug powders through inappropriate handling [15]. Despite the benefits, liquid oral formulations can also present challenges that must be addressed. These can include stability issues due to their aqueous nature, bioavailability concerns, palatability, and the convenience of transport and administration [16,17]. Overcoming these challenges requires the meticulous selection of excipients and innovative formulation strategies.

In liquid dosage form preparations, the vehicle consists of one or more excipients that carry the active substances [18]. Several oral vehicles are available, differing in their physicochemical and organoleptic properties [19,20]. The selection of an optimal oral vehicle is critical to safeguarding the stability, palatability, and therapeutic effectiveness of drugs [21,22]. Additionally, ongoing research and development of new oral vehicles are needed to address the specific requirements of different patient populations. For instance, there is currently a lack of commercially available liquid drug products specifically formulated for tube-feeding purposes, and compounding medicines often present short beyond-use dates [23,24]. As mentioned above, both enalapril and atenolol are often prescribed for chronic conditions and therefore require prolonged treatment. Therefore, the stability of extemporaneous formulations becomes a critical factor, and the development of products with extended shelf life is an advantage. Furthermore, it is important to ensure that these active substances are correctly solubilized in liquid formulations and remain so under physiological conditions. Atenolol, with a pKa value of 9.6, is sparingly to slightly soluble in water [25]. Its solubility increases in acidic media at pH 1.2 compared to slightly basic media at pH 7.4 [26]. Enalapril maleate is also sparingly soluble in water, exhibiting greater solubility in more basic media (pH 7.2) compared to acidic media (pH 4) [25,27].

SuspendIt® is an aqueous oral vehicle compatible with all feeding tubes that employs a synergistic polymer complex exhibiting a thixotropic behavior [28]. This characteristic facilitates the redispersion of drugs upon agitation, thereby minimizing sedimentation [28]. Comprising a natural sweetener derived from monk fruit, SuspendIt® is particularly well-suited for patients needing formulations devoid of sugar, casein, gluten, and parabens and other potentially harmful excipients for the pediatric population [28,29]. Several stability studies of oral formulations using SuspendIt® as a vehicle have already been conducted. Notable examples include oral suspensions combining trimethoprim and sulfadiazine, as well as oral suspensions of spironolactone, metronidazole, ursodiol, amlodipine besylate, and allopurinol [30–35]. Additionally, oral solutions containing clindamycin and naltrexone hydrochloride have also been investigated [36,37].

The aim of this study was to determine the chemical, physical, and microbiological stability of atenolol and enalapril maleate formulations prepared with SuspendIt® when stored at 25 °C and 5 °C for an extended period (180 days). Suitability for oral (palatability) and enteral feeding administration (injectability) was also evaluated.

2. Results

2.1. Preliminary Characterization of the SuspendIt® Vehicle and the Cardiovascular Oral Formulations

2.1.1. pH

The pH values were similar for the vehicle SuspendIt® (5.00 ± 0.03) and enalapril maleate formulation (4.95 ± 0.01), while a slightly lower pH was found for atenolol liquid preparation (4.65 ± 0.00). The cardiovascular formulations were prepared with the mentioned vehicle.

2.1.2. Rheological Properties

Flow Behavior

The apparent viscosity decreased with increased shear rate for SuspendIt® and for both atenolol and enalapril maleate formulations, which is typical of shear-thinning liquids (Figures 1 and 2).

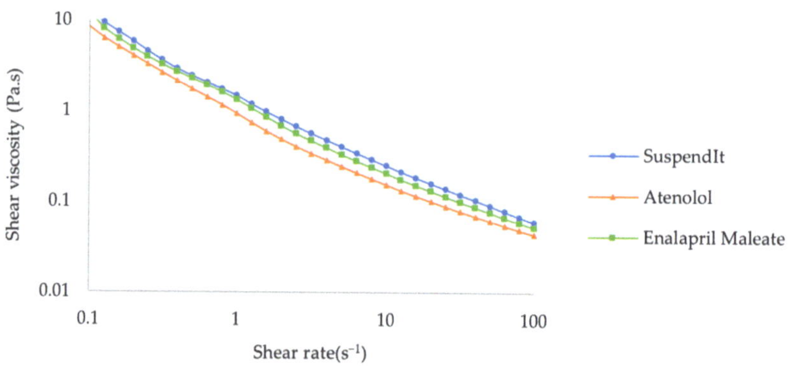

Figure 1. Flow curve of SuspendIt®, atenolol, and enalapril maleate formulations.

The Hershel-Bulkley model provided a good fit for flow curves. The consistency coefficient was lower for the cardiovascular formulations compared with the vehicle (Table 1). The flow indexes were lower than 1, thus confirming the shear-thinning behavior. Both formulations and the vehicle presented an apparent yield stress. The yield stress represents the stress threshold necessary to induce flow, playing an important role in the processing, physical stability, and practical application of liquid products [38].

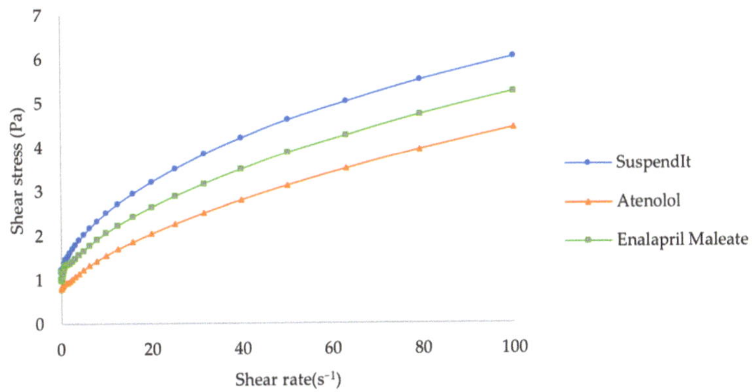

Figure 2. Rheogram of SuspendIt®, atenolol, and enalapril maleate formulations.

Table 1. Hershel-Bulkley model parameters for atenolol and enalapril maleate formulations and SuspendIt® vehicle (mean ± standard deviation, $n = 3$).

Sample	K (Pa.s^n)	n	τ_0 (Pa)	R^2
Atenolol formulation	0.238 ± 0.061	0.633 ± 0.030	0.756 ± 0.042	0.999 ± 0.000
Enalapril Maleate formulation	0.273 ± 0.028	0.594 ± 0.027	0.980 ± 0.068	0.998 ± 0.000
SuspendIt®	0.483 ± 0.007	0.516 ± 0.003	0.954 ± 0.003	0.999 ± 0.001

Thixotropy

SuspendIt® and both atenolol and enalapril maleate formulations are thixotropic fluids since their initial viscosity decreases over time and after applying a shear [39]. After the external force is removed, the initial viscosity is recovered in a time-dependent way [39,40]. Thixotropic behavior is invariably associated with shear-thinning flow behavior [39,40]. The recovery period for both formulations was slightly higher than for SuspendIt® alone, while the recovery rate was slightly lower (Figure 3, Table 2).

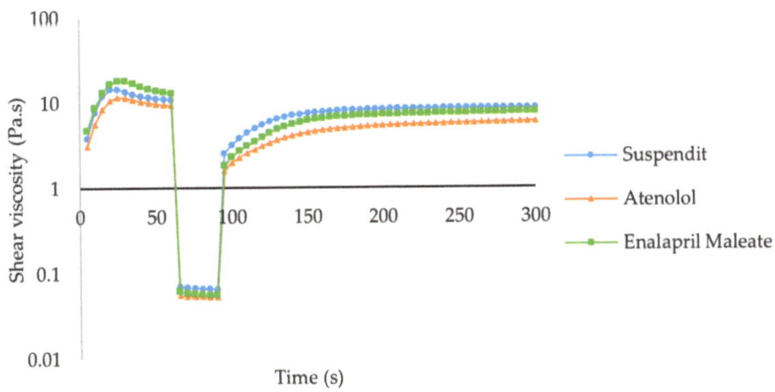

Figure 3. Thixotropic behavior of SuspendIt®, atenolol, and enalapril maleate formulations.

Table 2. Recovery period (50%) and recovery rate of SuspendIt®, atenolol, and enalapril maleate formulations (mean ± standard deviation, *n* = 3).

Sample	Recovery Period (s)	Recovery Rate (%)
Atenolol formulation	140.00 ± 13.64	69.12 ± 2.34
Enalapril Maleate formulation	150.56 ± 17.10	64.10 ± 6.99
SuspendIt®	120.0 ± 13.5	78.4 ± 3.1

Amplitude Sweep—Linear Viscoelastic Region (LVER)

The amplitude sweep test allows the determination of the linear viscoelastic region (LVER). Beyond a critical strain level, the material's behavior is non-linear, and the storage modulus (G′) declines. Increasing the strain above the critical strain disrupts the network structure. For both atenolol and enalapril maleate formulations, the LVER was similar to the vehicle—SuspendIt® (Figure 4). From 10% strain, the material becomes progressively more fluid-like, G′ declines, and thus further tests to study the viscoelasticity of these formulations should be performed at lower deformations within LVER. Below 10% strain, the structure is intact, the material behaves solid-like, and G′ > G″, indicating that the material is highly structured. The strength of the colloidal forces is reflected by tan δ = (G″/G′). For the atenolol formulation, the tan δ was about 0.58, while for SuspendIt® and for the enalapril formulation, the tan δ was about 0.47. A tan δ value of less than 1, as observed in the formulations under study, indicates strong particle association driven by colloidal forces [38]. For a system to maintain stability, an intermediate value of tan δ is optimal [38].

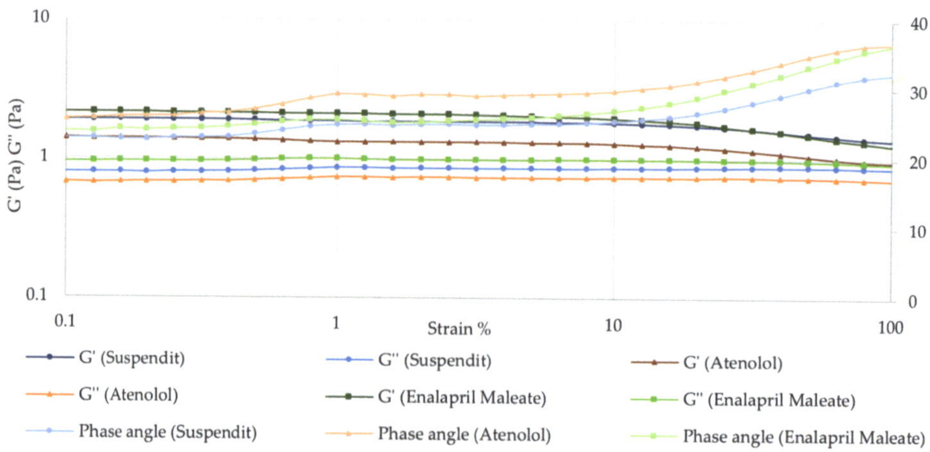

Figure 4. Amplitude sweep for SuspendIt®, atenolol, and enalapril maleate formulations.

Frequency Sweep (Mechanical Spectrum)

The atenolol and enalapril maleate formulations presented a solid-like behavior in the fragment of the spectrum observed, where the storage modulus (G′) was higher than the loss modulus (G″) over the whole frequency range, and the phase angle was lower than 45° [41]. This is the typical behavior of viscoelastic polymeric materials in the plateau zone. No crossover point was observed across the tested frequencies, indicating that the formulations exhibit entangled polymeric networks [41,42]. The observed frequency-dependence of G′ reflects a moderately structured material [41]. The results also demonstrate that the compounding formulation has similar viscoelastic properties as the SuspendIt® vehicle, meaning that their inner structure is similar (Figure 5).

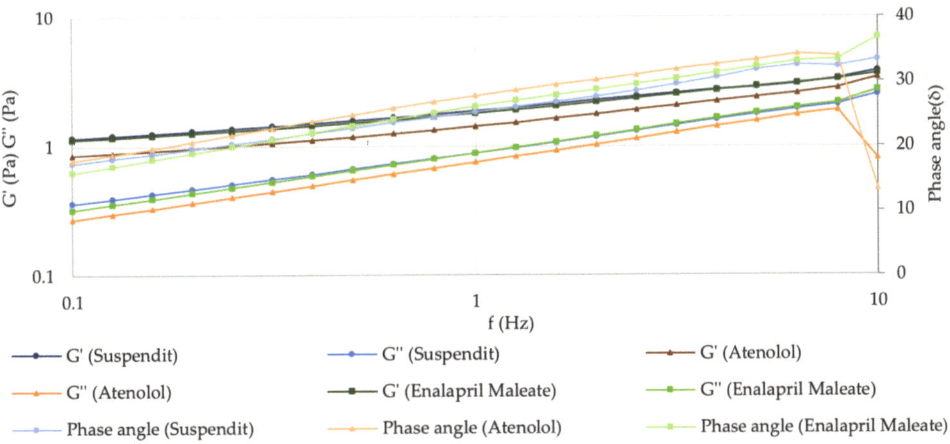

Figure 5. Mechanical spectrum of SuspendIt®, atenolol, and enalapril maleate formulations.

2.1.3. Injectability

The injectability test consisted of measuring the force required to promote the flow of both vehicles and suspensions using a syringe connected to a feeding tube (NeoMed® Enteral Feeding). The selected feeding tube was specifically designed for administration to neonatal and pediatric patients, facilitating nutritional delivery via nasal or oral gastric routes. Simple syrup was chosen as a comparator due to its wide use as a vehicle for dissolving active pharmaceutical ingredients in the preparation of compounded oral medications, which is relevant in hospital settings [43]. The force needed to inject simple syrup was about seven times higher than for the formulations in which the vehicle is SuspendIt®. The shear-thinning properties of the SuspendIt® formulations facilitated injection through the feeding tube, in contrast to simple syrup, which is a Newtonian fluid. The force needed to inject SuspendIt® was slightly higher than for the oral formulations of atenolol and enalapril maleate (Figure 6 and Table 3). The variation in injectability among the different SuspendIt® formulations can be attributed to their distinct consistency indexes, with the SuspendIt® vehicle exhibiting a higher consistency index. The tested cardiovascular formulations are thus easy to administer through feeding tubes, even those with a low diameter (outer diameter: 2.2 mm) adapted to neonates and infants.

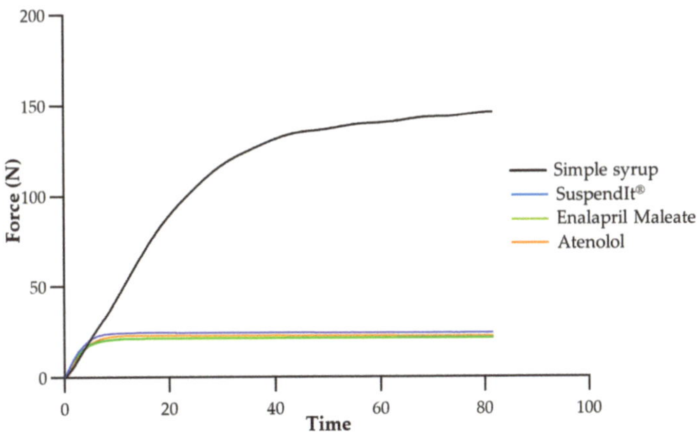

Figure 6. Injectability texturogram obtained for the studied formulations.

Table 3. Mean force and standard deviation of SuspendIt®, simple syrup, USP vehicle, atenolol, and enalapril maleate formulations (n = 3).

Sample	Mean Force ± Standard Deviation (N)
Atenolol formulation	22.5 ± 0.2
Enalapril Maleate formulation	21.6 ± 1.5
Simple syrup	146.8 ± 2.0
SuspendIt®	22.8 ± 1.4

2.1.4. Evaluation of the SuspendIt® Oral Suspending Vehicle as a Bitter Bloc for Atenolol and Enalapril Maleate

The effectiveness of SuspendIt® and the atenolol vehicle in masking the bitterness of atenolol and enalapril maleate was assessed qualitatively using a laboratory-developed potentiometric electronic tongue (E-tongue), which was previously built by the research team [17]. The potentiometric signal profiles generated by the 40 lipid sensor membranes of the E-tongue during the analysis of atenolol or enalapril maleate aqueous or vehicle solutions were used to construct supervised multivariate classification models. Signal intensity varied across sensors (40 sensors ranging from S1:1 to S1:20 and S2:1 to S2:20) and solutions under study: 0.1–336 mV for atenolol aqueous solution, 2–337 mV for atenolol in atenolol vehicles, and 0.7–381 mV for atenolol vehicles; 1–410 mV for enalapril maleate aqueous solution, 0.7–388 mV for enalapril maleate, and 0.5–386 mV for SuspendIt® vehicles. The varying signal profiles and intensities may be ascribed to the diverse electrostatic and/or hydrophobic interactions potentially established between the E-tongue sensor membranes, characterized by negative/positive polarities and polar/non-polar regions, and the hydroxyl, carboxyl, and/or amine functional groups present in atenolol and enalapril maleate, as well as in the constituents of the SuspendIt® vehicles.

Different linear discriminant (LDA) models were established based on the E-tongue fingerprints, seeking to assess the bitter block effect of the atenolol vehicle or SuspendIt® vehicle on atenolol or enalapril maleate formulations. This aimed to compare them with the bitter taste of their respective aqueous solutions (used as positive controls) and the pure vehicles (used as negative controls, i.e., sweet solutions). In both cases, LDA models comprising two discriminant functions explained 100% of the total data variability. For atenolol, the model used potentiometric signals from nine sensors (S1:6, S1:13, S1:17, S1:18, S2:1, S2:5, S2:8, S2:12, and S2:15) selected by the simulated annealing (SA) variable selection algorithm. The LDA-SA-E-tongue model exhibited a sensitivity of 100% for both training (Figure 7A) and internal validation (leave-one-out cross-validation, LOO-CV) procedures. For enalapril maleate, the supervised classification model employed signals from eight sensors (S1:3, S1:11, S1:13, S1:19, S2:9, S2:11, S2:12, and S2:15) selected by the SA algorithm. The LDA-SA-E-tongue model correctly classified all train set solutions (100% sensitivity, Figure 7B) and 95% of solutions in the LOO-CV procedure, with only one enalapril maleate solution in the SuspendIt® vehicle being misclassified as if it were a SuspendIt® vehicle without the drug.

From the data depicted in Figure 7, it is evident that both atenolol and enalapril maleate, represented by their aqueous solutions (which are typically bitter), are consistently positioned in the negative region of the respective first discriminant function (DF1). Conversely, solutions containing these drugs in the atenolol vehicle or SuspendIt® vehicle, as well as the pure vehicle solutions (known to be sweeter), are consistently placed in the positive region of DF1. This observation confirms the efficacy of the used vehicles in masking or blocking the bitter taste associated with atenolol or enalapril maleate. Additionally, by analyzing the coordinates of each drug's group centroid (aqueous versus vehicle solution), it became feasible to compute the distance between them. For atenolol, the distance between the groups was measured at 1197 arbitrary units, whereas for enalapril maleate, it was significantly lower at 356 arbitrary units, indicating a more than threefold difference. This difference suggests that the atenolol vehicle was more successful in suppressing the bitterness of atenolol compared to the SuspendIt® vehicle for enalapril maleate.

Furthermore, the placement of the groups (negative versus positive region of DF1) and the positive/negative contribution of each selected sensor included in DF1 enable us to deduce that certain sensors (S1:3, S1:19, S2:1, S2:8, S2:9, and S2:15) are more prone to detecting bitterness, whereas others (S1:6, S1:11, S1:13, S1:17, S1:18, S2:5, and S2:12) demonstrate a higher affinity for detecting sweetness. It is worth noting that sensors S1:11 and S2:11, although having the same composition, were included in the same LDA model, possibly due to differences in their physical properties, such as porosity and transparency, which could arise due to the drop-by-drop technique employed during the device fabrication, as previously discussed [17].

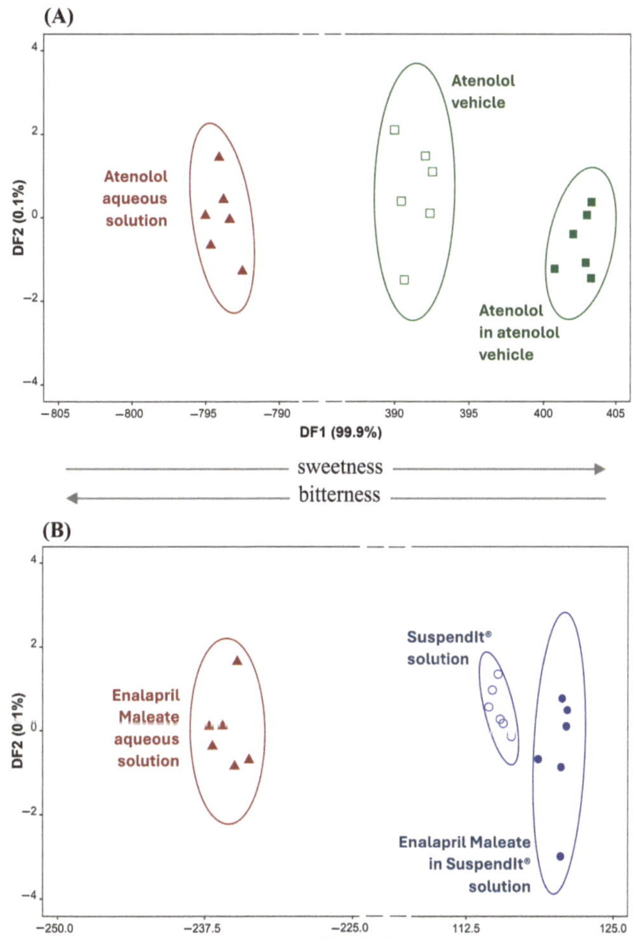

Figure 7. (**A**) Atenolol: 2D LDA plot based on the signals' profiles of nine sensors of the E-tongue (S1:6, S1:13, S1:17, S1:18, S2:1, S2:5, S2:8, S2:12, and S2:15). (**B**) Enalapril Maleate: 2D LDA plot based on the signals' profiles of eight sensors of the E-tongue (S1:3, S1:11, S1:13, S1:19, S2:9, S2:11, S2:12, and S2:15).

Lastly, it was explored the capability of the developed taste sensor device to distinguish between aqueous solutions of two drugs, atenolol and enalapril maleate, as well as their solutions in SuspendIt® vehicles simultaneously. To achieve this, a novel LDA-SA-E-tongue classification model was developed. This model, employing potentiometric signals from three E-tongue sensors (S1:10, S2:3, and S2:5) selected via the SA algorithm, comprised three DFs that explained 81.8%, 18.1%, and 0.1% of the data variability. The model allowed accurate classification of all samples during both training (Figure 8) and LOO-CV procedures. These findings demonstrate the effectiveness of the E-tongue in distinguishing the varying bitterness/sweetness levels among the four oral formulations investigated: atenolol and enalapril maleate in water (likely differing in bitterness intensity) and the oral suspensions of both drugs in SuspendIt® vehicles (with differing sweetness intensities). Analysis of sensor contributions to the DFs suggests that S2:3 (negatively contributing to DF1) primarily detected bitterness, while S1:10 and S2:5 (positively contributing to DF1) exhibited higher sensitivity to sweetness. Particularly, the substantial difference between the centroids of atenolol aqueous solution and atenolol in the atenolol vehicle (distance = 49 arbitrary units) compared to that of enalapril maleate aqueous solution and enalapril maleate in the SuspendIt® vehicle (distance = 16 arbitrary units) suggests that while both vehicles effectively block bitterness, the atenolol vehicle demonstrates superior effectiveness in masking bitterness compared to the SuspendIt® vehicle, as evidenced by the larger centroid distance. In addition to the natural sweetener derived from monk fruit found in SuspendIt®, the atenolol vehicle contains two additional sweeteners: acesulfame potassium and steviol glycosides (95%). These added sweeteners likely contribute to its enhanced capability to mask the bitter taste of the active ingredients.

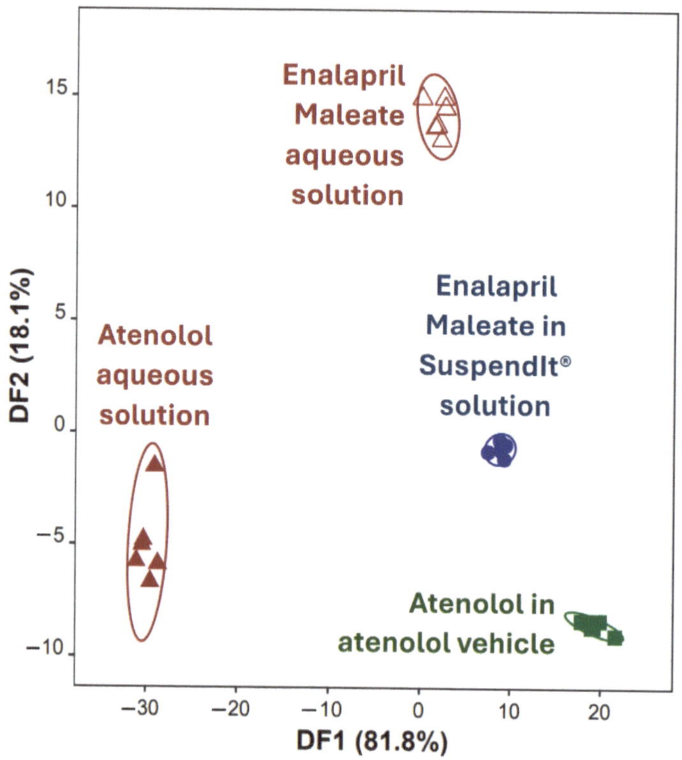

Figure 8. Discrimination of atenolol and enalapril maleate solutions using the E-tongue: 2D LDA plot based on the signals' profiles of three sensors of the E-tongue (S1:10, S2:3, and S2:5).

2.2. *Stability Study of the Oral Formulations*

2.2.1. Organoleptic Characteristics

The appearance and odor of the formulations remained unchanged throughout the period of analysis except for a slight yellowing observed at 25 °C after 180 days of storage for both atenolol and enalapril maleate formulations (Figures S6 and S7) compared to their appearance and odor at time zero.

2.2.2. pH

After 180 days of storage, the pH of both formulations remained stable with less than 5% variation (Table 4). Also, no relevant differences were observed between the two storage temperatures tested.

Table 4. pH of atenolol and enalapril maleate formulations (mean ± standard deviation, *n* = 3).

	Atenolol Formulation		Enalapril Maleate Formulation	
	5 °C	25 °C	5 °C	25 °C
Day 0	4.65 ± 0.00		4.95 ± 0.01	
Day 14	4.67 ± 0.01	4.66 ± 0.03	4.97 ± 0.01	4.95 ± 0.01
Day 30	4.73 ± 0.01	4.71 ± 0.01	4.95 ± 0.01	4.96 ± 0.01
Day 60	4.73 ± 0.01	4.74 ± 0.00	4.93 ± 0.01	4.98 ± 0.02
Day 90	4.69 ± 0.02	4.67 ± 0.02	4.90 ± 0.02	4.94 ± 0.06
Day 120	4.76 ± 0.02	4.76 ± 0.02	4.96 ± 0.01	4.98 ± 0.02
Day 180	4.72 ± 0.00	4.74 ± 0.02	4.94 ± 0.02	4.93 ± 0.03

2.2.3. Rheological Properties

Flow Behavior

The shear-thinning behavior of the oral formulations remained unchanged throughout the period of analysis, with slight variations in apparent viscosity (Figures 9 and S8–S10). After 180 days of storage at both temperatures, the variation in all Hershel-Bulkley parameters was roughly lower than 15% (Tables 5–8).

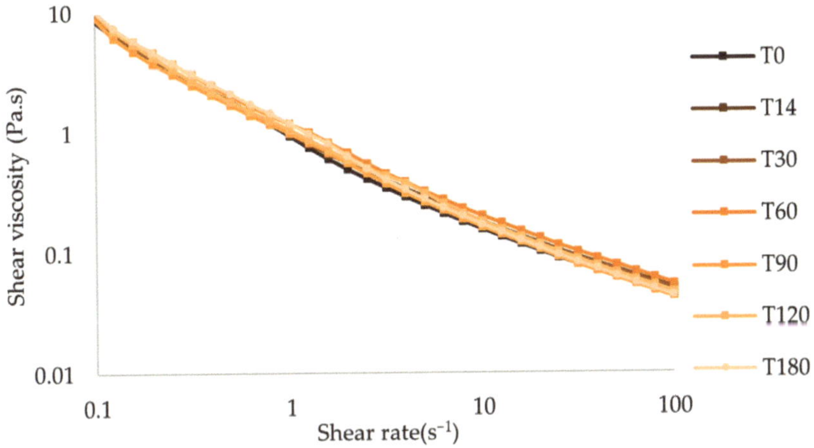

Figure 9. Flow curve of atenolol oral formulation after storage at 25 °C.

Table 5. Hershel-Bulkley model parameters for atenolol formulation after storage at 25 °C (mean ± standard deviation, *n* = 3).

Day	0	14	30	60	90	120	180
K ($Pa.s^n$)	0.238 ± 0.061	0.245 ± 0.031	0.257 ± 0.017	0.255 ± 0.055	0.221 ± 0.019	0.221 ± 0.053	0.239 ± 0.052
n	0.633 ± 0.030	0.629 ± 0.023	0.620 ± 0.004	0.618 ± 0.032	0.609 ± 0.023	0.608 ± 0.035	0.584 ± 0.048
τ_0 (Pa)	0.756 ± 0.042	0.811 ± 0.088	0.763 ± 0.009	0.835 ± 0.050	0.786 ± 0.085	0.822 ± 0.098	0.827 ± 0.100
R^2	0.999 ± 0.000	0.999 ± 0.001	1.000 ± 0.000	0.999 ± 0.001	0.999 ± 0.001	0.997 ± 0.002	0.995 ± 0.005

Table 6. Hershel-Bulkley model parameters for atenolol formulation after storage at 5 °C (mean ± standard deviation, *n* = 3).

Day	14	30	60	90	120	180
K (Pa.s^n)	0.241 ± 0.037	0.239 ± 0.032	0.225 ± 0.038	0.247 ± 0.007	0.229 ± 0.024	0.221 ± 0.011
N	0.635 ± 0.034	0.639 ± 0.022	0.645 ± 0.025	0.611 ± 0.008	0.628 ± 0.025	0.613 ± 0.012
τ_0 (Pa)	0.817 ± 0.112	0.821 ± 0.068	0.857 ± 0.050	0.767 ± 0.036	0.826 ± 0.083	0.738 ± 0.050
R^2	0.999 ± 0.000	0.999 ± 0.000	0.999 ± 0.000	0.999 ± 0.000	0.999 ± 0.000	0.999 ± 0.000

Table 7. Hershel-Bulkley model parameters for enalapril maleate formulation after storage at 25 °C (mean ± standard deviation, *n* = 3).

Day	0	14	30	60	90	120	180
K (Pa.s^n)	0.273 ± 0.028	0.298 ± 0.030	0.290 ± 0.075	0.265 ± 0.019	0.210 ± 0.040	0.216 ± 0.040	0.229 ± 0.002
n	0.594 ± 0.027	0.586 ± 0.018	0.584 ± 0.048	0.606 ± 0.014	0.614 ± 0.027	0.635 ± 0.036	0.586 ± 0.007
τ_0 (Pa)	0.980 ± 0.068	1.086 ± 0.015	1.076 ± 0.052	1.116 ± 0.030	1.077 ± 0.038	1.209 ± 0.068	1.128 ± 0.089
R^2	0.998 ± 0.000	0.995 ± 0.002	0.997 ± 0.001	0.996 ± 0.004	0.993 ± 0.000	0.988 ± 0.005	0.985 ± 0.009

Table 8. Hershel-Bulkley model parameters for enalapril maleate formulation after storage at 5 °C (mean ± standard deviation, *n* = 3).

Day	14	30	60	90	120	180
K (Pa.s^n)	0.226 ± 0.035	0.250 ± 0.118	0.203 ± 0.017	0.226 ± 0.029	0.262 ± 0.022	0.192 ± 0.026
N	0.641 ± 0.029	0.592 ± 0.064	0.655 ± 0.016	0.621 ± 0.022	0.593 ± 0.009	0.641 ± 0.024
τ_0 (Pa)	1.115 ± 0.069	1.012 ± 0.092	1.143 ± 0.062	1.063 ± 0.026	1.003 ± 0.044	1.113 ± 0.107
R^2	0.998 ± 0.001	0.999 ± 0.001	0.997 ± 0.001	0.997 ± 0.002	0.998 ± 0.001	0.993 ± 0.006

Thixotropy

Analyzing the graphs of the thixotropic behavior of both atenolol and enalapril maleate formulations at both temperatures, the recovery behavior after a shear period was similar throughout the period of analysis, with slight variations in the recovery rate and period (Figures 10 and S11–S13, Tables S3–S6).

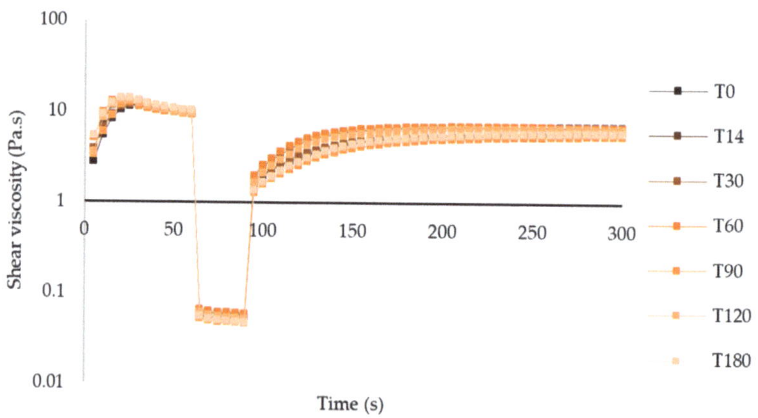

Figure 10. Thixotropic behavior of atenolol oral formulation after storage at 25 °C.

Amplitude Sweep—Linear Viscoelastic Region (LVER)

After 180 days, there were no relevant changes in the LVER of atenolol and enalapril maleate formulations for both temperatures, with critical strains around 10% (Figures 11 and S14–S16).

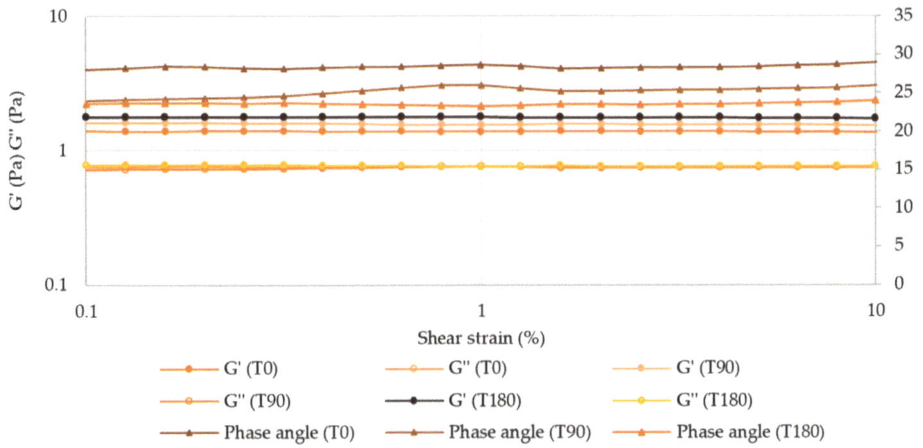

Figure 11. Amplitude sweep for atenolol oral formulation after 180 days of storage at 25 °C.

Frequency Sweep (Mechanical Spectrum)

After 180 days of storage at both temperatures, the mechanical spectrum of atenolol and enalapril maleate formulations did not suffer any relevant changes, which indicates that the inner structure of these structured fluids is preserved (Figures 12 and S17–S19).

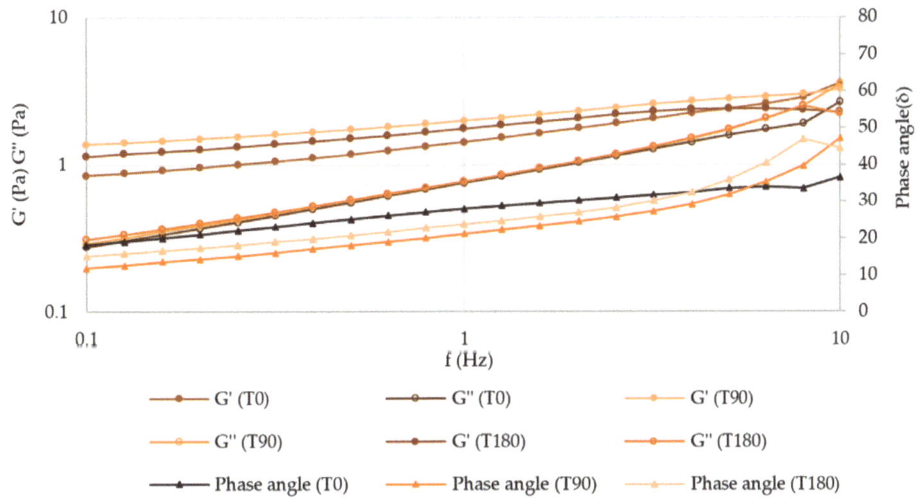

Figure 12. Mechanical spectrum of atenolol oral formulation after 180 days of storage at 25 °C.

2.2.4. Preservative Effectiveness

After 180 days of storage, the reduction in microbial count in all the microorganisms tested (*Escherichia coli* ATCC 8739, *Pseudomonas aeruginosa* ATCC 9027, *Staphylococcus aureus* ATCC 6538, *Candida albicans* ATCC 10231, and *Aspergillus brasiliensis* ATCC 16404) confirmed that the preservative system in atenolol and enalapril maleate formulations successfully prevented the growth of the challenge organisms, in accordance with USP Chapter <51> requirements, thus conferring the desired microbiological stability when stored either at 5 °C or 25 °C (Tables 9 and S7) [44].

Table 9. United States Pharmacopeia Chapter <51> Antimicrobial Effectiveness Test of atenolol oral formulation.

Challenge Microorganism	Day	5 °C	25 °C
Escherichia coli	0	Pass *	
Pseudomonas aeruginosa		Pass **	
Staphylococcus aureus		Pass *	
Candida albicans		Pass **	
Aspergillus brasiliensis		Pass **	
Escherichia coli	30	Pass *	Pass *
Pseudomonas aeruginosa		Pass *	Pass *
Staphylococcus aureus		Pass *	Pass *
Candida albicans		Pass *	Pass *
Aspergillus brasiliensis		Pass *	Pass *
Escherichia coli	90	Pass *	Pass *
Pseudomonas aeruginosa		Pass *	Pass *
Staphylococcus aureus		Pass *	Pass *
Candida albicans		Pass *	Pass *
Aspergillus brasiliensis		Pass *	Pass *
Escherichia coli	180	Pass *	Pass *
Pseudomonas aeruginosa		Pass *	Pass *
Staphylococcus aureus		Pass *	Pass *
Candida albicans		Pass *	Pass *
Aspergillus brasiliensis		Pass *	Pass *

* >1 log reduction after 14 days; ** 1 log reduction after 14 days.

2.2.5. Active Substance Assay

The HPLC-UV method was developed and validated for each drug, according to international guidelines to support the stability indicative assay. The chromatographic conditions were previously adjusted to clearly separate degradation products, while the use of two detectors (UV and fluorescence) further enhanced the method's capacity to accurately quantity the unaltered drugs.

The concentration of atenolol was above 90% during the entire assay, with similar concentrations under both storage conditions (Figure 13, Tables 10 and S8). According to the US Pharmacopoeia, compounded preparations are considered to be stable if the drug concentration remains within 90–110% of the initial value (day 0).

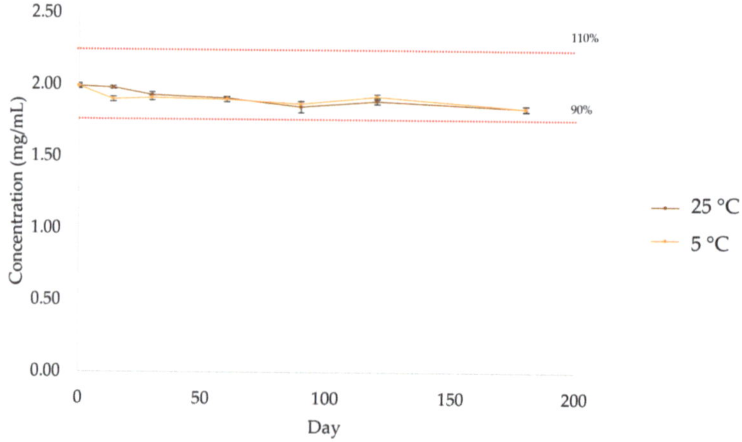

Figure 13. Quantification by HPLC-UV of atenolol in the formulation after storage at 5 °C and 25 °C.

Table 10. Quantification of atenolol by HPLC-UV after storage at room temperature (n = 3, 25 °C).

Day	0	14	30	60	90	120	180
Mean (mg/mL)	1.99	1.98	1.93	1.91	1.85	1.89	1.84
SD	0.01	0.01	0.02	0.01	0.04	0.02	0.02
%	99	99	97	95	93	95	92

There was a progressive decrease in the amount of enalapril maleate (Figure 14, Tables 11 and S9) in the formulation kept at room temperature (25 °C), while the amount of enalapril maleate in the formulation kept at refrigerated temperature (5 °C) remained within the 90–110% range. For the formulation stored at room temperature, the percentage of active substance decreased below 90% from day 90 onwards.

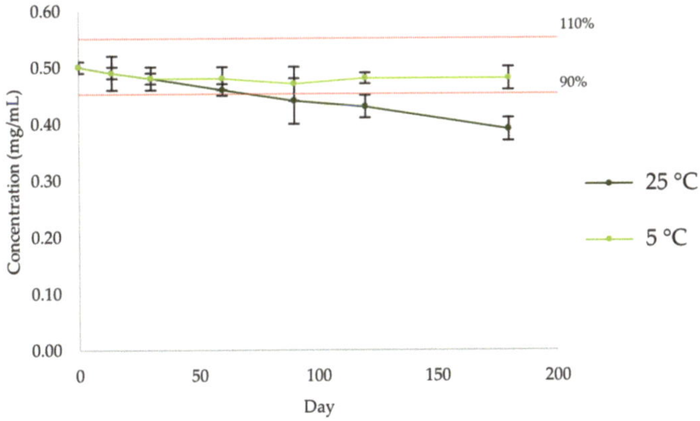

Figure 14. Quantification by HPLC-UV of enalapril maleate in the formulation after storage at 5 °C and 25 °C.

Table 11. Quantification of enalapril by HPLC-UV after storage at room temperature (n = 3, 25 °C).

Day	0	14	30	60	90	120	180
Mean (mg/mL)	0.50	0.49	0.48	0.46	0.44	0.43	0.39
SD	0.01	0.00	0.01	0.00	0.00	0.00	0.01
%	101	97	96	91	88	85	78

3. Discussion

Compounded medicines must meet specific quality standards to ensure safety and therapeutic effectiveness. Despite not requiring regulatory approval, pharmaceutical compounding is governed by national or state boards of pharmacy and must comply with USP compounding requirements or the European Pharmacopeia (EP) monograph regarding pharmaceutical preparations [1]. When developing compounded medicines, it is crucial to consider the harmful potential of excipients for the target population as well as their organoleptic properties and suitability for the intended type of administration, thereby potentiating patients' acceptability and medication adherence.

The palatability of a pharmaceutical formulation refers to its level of pleasantness concerning several organoleptic properties, in particular taste [17,45]. Since unpalatable oral dosage forms can impact patients' acceptability of the compounded medicine and medication adherence, palatability is a key parameter [46]. Therefore, the bitterness-blocking potential of the atenolol vehicle or SuspendIt® vehicle has been verified through E-tongue assays. These assays showed that the bitterness of atenolol or enalapril maleate aqueous solutions could be effectively suppressed by dissolving these drugs in the oral vehicles under study. Previously, the research team successfully used a similar chemometric-sensor approach to evaluate the ability of various vehicles to mask the bitterness of oral solutions of ranitidine [17].

The mechanical properties of liquid dosage forms should allow easy administration, accurate measurement, and proper mixing and ensure adequate physical stability. Whenever desired, viscosity modifiers should be smartly selected to provide the desired rheological behavior. Using hydrophilic polymers as thickening agents is a common strategy to increase the viscosity of oral pharmaceuticals. In the case of SuspendIt®, amorphophallus konjac root powder and xanthan gum are synergistically combined to provide a three-dimensional structure that imparts a viscosity increase. This increase should be controlled to be within the desired range to facilitate all operations during product handling and administration and maintain physical stability [21]. Furthermore, while many proprietary oral vehicles are commonly classified as suspensions due to their high viscosity, the tested formulations in this study are solutions from a physical standpoint since atenolol and enalapril maleate are water-soluble at the concentrations used. The formulations are clear, without any opacity or residue. The atenolol and enalapril maleate formulations exhibited shear-thinning behavior, which means that their apparent viscosity decreases when subjected to stress. Polymers are characterized by an entanglement with a random orientation of the inner structure when at rest [21,47,48]. In shear-thinning behavior, when sufficient shear is applied, the polymer chains begin to disentangle [48]. Individual molecules exhibit less flow resistance than entangled structures, resulting in shear-thinning flow behavior with decreasing viscosity at higher shear rates [48]. Upon removal of the shear stress, the internal structure of the material gradually reassembles, resulting in the recovery of viscosity to its original level over time [48]. This time-dependent behavior is called thixotropy. The shear-thinning behavior shown by the atenolol and enalapril maleate formulations facilitates their administration orally and through enteral feeding tubes. The active substances are dissolved in the vehicle SuspendIt®, which minimizes any impact on the mechanical properties. Consequently, no differences in the type of flow behavior between the vehicle and the formulations were observed. Shear-thinning behavior is typical of polymeric dispersions used in pharmaceutical formulations such as konjac and xanthan gum (the thickeners used in SuspendIt®), carbomer, and poloxamer dispersions [49–52]. The formulations become more fluid upon shaking, thus allowing a convenient administration. Also, when injected through a feeding tube, progressive fluidification ensures an effortless administration. Yield stress, defined as the minimum shear stress required to initiate flow, is a helpful parameter to evaluate the product's performance and processability and predict the product's long-term stability and shelf life [45]. Yield stress reduces the ability to flow under shipping vibrations and gravity [53]. Since the existence of apparent yield stress indicates a more structured fluid, an influence on the swallowing process and mouthfeel is also expected [54]. Both liquid dosage forms presented an apparent yield stress, as described by the Heschel-Bulkley rheological model, thus reinforcing their suitability for oral administration and predicting good physical stability. The thixotropic behavior exhibited by the cardiovascular oral liquids implies their capacity to undergo thinning upon agitation and time-dependent recovery of the viscosity upon settling [55]. The recovery plays a pivotal role in ensuring enhanced physical stability of the formulations [55]. Also, as the recovery of the viscosity is not immediate, the formulations can easily be injected through the feeding tube. The tested formulations also presented viscoelastic behavior typical of polymeric dispersions [48]. Polymers are viscoelastic materials due to their ability to provide mechanical responses across different length scales, from short-range to long-range, when subjected to an applied force [56]. The amplitude sweep test stands as a relevant tool for understanding the dynamics of material behavior as it progresses from solid-like to liquid-like under increasing deformation [57]. This test offers insights into the flexibility and structural integrity of materials, thereby facilitating a comprehensive characterization of their viscoelastic properties [57]. The linear viscoelastic region (LVER) is obtained as the region wherein stress varies proportionally to deformation until a critical strain is attained [57]. The larger this region, the greater the material's resistance to deformation. The critical strain was reached at around 10% for both formulations, which is similar in comparison with other oral liquids such as xanthan gum, guar gum, locust bean gum, and konjac dispersions [47,58,59], foreseeing a good resistance to deformation for the atenolol and enalapril maleate formulations. Below the critical strain, the storage modulus of the formulations was higher than the loss modulus. This behavior is consistent with that observed in other polymeric dispersions, such as xanthan gum [47]. The frequency sweep test is used to obtain the mechanical spectrum of a material that represents its "fingerprint" since it is unique to each material [60]. This test is performed in the LVER and is employed to investigate the viscoelastic properties of a material [60]. For both cardiovascular formulations, the storage modulus was higher than the loss modulus, implying a strong association between the hydrophilic polymers and a solid-like behavior in the measured frequency span. No marked differences were observed in the oscillatory measurements (amplitude and frequency sweep) between the formulations and the vehicle SuspendIt®. This is likely due to the molecular dissolution of the active substances in the vehicle, which does not affect the inner structure afforded by the thickeners. Taken together, the results of the mechanical properties of the cardiovascular formulations support their use for oral and enteral feeding tube administration, predicting good physical stability. The avoidance of potentially harmful excipients is recommended wherever possible in pediatric formulations, especially for neonates [61]. The proposed cardiovascular formulations are free of excipients such as parabens, sorbitol, propylene glycol, and benzyl alcohol. Vehicles used to compound oral liquid medications of many drugs reported in the literature usually contain parabens, since aqueous liquid formulations present a risk of microbiological instability [62–64].

Stability tests ensure that compounded medicines remain effective and safe for use throughout their intended shelf life, thereby providing accurate and reliable beyond-use dates. HPLC-UV methods were developed and validated for each drug, according to international guidelines. Both chromatographic systems were previously tested for their capacity to separate compounds resulting from forced degradation from the main peaks using drug standards submitted to acidic (HCl 1 M), basic (NaOH 0.1 M), heat (100 °C, 1 h), and light exposure (500 W, 1 h) conditions. Their stability-indicating capability was confirmed by verifying that degradation products (resulting from subjection to heat, acid and basic conditions, and oxidation) did not coelute with the drugs. These stability-indicating methods are important to properly investigate chemical stability over time, especially for pediatric patients, because they have immature organs, slower metabolism, and elimination capabilities and are more susceptible to the deleterious consequences that may arise from the presence of harmful degradation products [65]. Regarding the organoleptic characteristics, a yellowish darkening occurred over time in both formulations when stored at 25 °C. The same happened with SuspendIt®, which may suggest that this color change is inherent to the vehicle itself and not due to the degradation of the active substances. On the contrary, refrigeration preserved the original light color for all preparations. Based on the appearance of the formulations, no apparent precipitation of the drugs was observed over the storage period. It would also be important to test the formulations in biorelevant media, such as FaSSIF and FeSSIF, to determine if any drug precipitation occurs, as this is essential to ensure oral bioavailability.

The stability-indicating assay suggested the feasibility of using liquid dosage forms of atenolol (stored at both room and refrigerated temperatures) and enalapril maleate (stored at refrigerated temperature) with SuspendIt® within a 180-day period after preparation. The only exception was the enalapril maleate preparation stored at room temperature, which showed a decrease in drug content below 90% from 90 days onwards. Therefore, compounding pharmacists can recommend patients and healthcare professionals store enalapril maleate formulation in a refrigerator to ensure optimal preservation. Significant variations in pH notably impact the solubility and stability of drugs in formulations, as they can reflect the chemical alteration of the compounded medicines, underscoring the importance of monitoring this parameter over time [66]. During the 180-day storage period, the pH of the atenolol and enalapril maleate formulations showed negligible variation, suggesting that no marked chemical alterations occurred. The shear-thinning behavior of the formulation was maintained over the 180 days of analysis for both room and refrigerated storage temperatures. Thixotropic behavior was also maintained over time, with slight variations in recovery rate and recovery time for both oral formulations at both temperatures. The LVER of the oral formulations did not change over the 180-day storage period, and the critical strain remained around 10%. The characteristic mechanical spectrum of both formulations at different storage temperatures also did not change over the analysis time. The observed mechanical stability confirmed the prediction of good stability resulting from the rheological analysis previously performed.

The evaluation of the preservative effectiveness, following USP Chapter <51> requirements, confirmed that the preservative system in atenolol and enalapril maleate remained effective throughout the storage period, either at 5 °C or 25 °C. The preservative systems successfully inhibited the growth of challenge microorganisms, both bacteria and fungus, which is particularly relevant in multidose aqueous liquid formulations that present a higher risk of microbiological contamination. This result reflects that the preservative system is adequate for these preparations, ensuring microbiological quality, safety, and extended shelf life over analysis time.

The poor stability of atenolol and enalapril maleate in aqueous solutions presents a notable obstacle to the creation of liquid dosage forms [67–69]. Degradation was observed as soon as 30 min for enalapril maleate [70] and may be influenced by pH changes as well as by the water sorption activity of excipients in contact with this active substance [71]. The degradation of atenolol in an aqueous solution is also affected by pH variations, with increased stability observed at lower (acidic) pH levels [72]. Consequently, there is a need to develop oral vehicles capable of preserving the stability of these active substances while masking the bitterness characteristic of atenolol and enalapril maleate [73,74]. Previous studies on the chemical stability of atenolol and enalapril maleate in other oral vehicles reported beyond-use dates between 30–90 days for enalapril and 9 to 40 days for atenolol [67,69,75–77]. In this work, the use of SuspendIt® as a vehicle for atenolol and enalapril maleate, stored under refrigerated conditions, led to an extended shelf life of the products to 180 days, which is a result much greater than the aforementioned stability studies.

4. Materials and Methods

4.1. Chemicals and Reagents

SuspendIt®, atenolol USP, enalapril maleate USP, citric acid USP, sodium citrate USP, acesulfame, and steviol glycosides (95%) were provided by the Professional Compounding Centers of America (PCCA) (Houston, TX, USA). Sucrose was obtained from Acofarma (Barcelona, Spain). Lyophilized microorganisms, *Escherichia coli* ATCC 8739, *Pseudomonas aeruginosa* ATCC 9027, *Staphylococcus aureus* ATCC 6538, *Candida albicans* ATCC 10231, and *Aspergillus brasiliensis* ATCC 16404, were obtained from Liofilchem (ATCC licensed derivatives) by Frilabo (Maia, Portugal). Tryptic Soy Agar (TSA) and Sabouraud Dextrose Agar culture mediums were also obtained from Frilabo (Maia, Portugal).

The simple syrup is a 66.7% aqueous sucrose solution prepared according to the Portuguese Galenic Form using 66.7% (w/w) sucrose and 33.3% (w/w) purified water, and it was used in the injectability test [43]. The sucrose was weighed and added to the purified water while being stirred manually. The mixture was heated in a water bath at a temperature of 70–80 °C to help dissolve the sucrose and stirred throughout the process. Finally, sufficient purified water, previously heated to 50 °C, was added to complete the volume, and then the solution was filtered through a Chardin paper filter.

The oral formulations of atenolol (Table 12) and enalapril maleate (Table 13) in SuspendIt® were prepared according to formulations developed by the supplier of SuspendIt® (PCCA). The atenolol vehicle corresponds to SuspendIt® with additional excipients. For the atenolol formulation, atenolol, citric acid monhydrate, sodium citrate dihydrate, acesulfame potassium, and steviol glycosides (95%) were triturated to reduce particle size. SuspendIt® was then added until it formed a smooth paste. More SuspendIt® was added in portions with magnetic stirring for 20 min. Finally, the volume of the atenolol formulation was topped up with SuspendIt®. For the enalapril maleate formulation, SuspendIt® and enalapril maleate were mixed with magnetic stirring for 20 min. Finally, the volume of the enalapril maleate formulation was topped up with SuspendIt®.

Table 12. Atenolol 2 mg/mL oral formulation composition.

	Atenolol USP	Active Substance	0.200 g
Atenolol vehicle	Citric acid USP monohydrate	pH buffer/chelating agent	0.935 g
	Sodium citrate dihydrate USP	pH buffer/chelating agent	1.632 g
	Acesulfame potassium FCC	Sweetener	0.100 g
	Steviol glycosides 95%	Sweetener	0.100 g
	SuspendIt®	Suspending agent	q.s. 100 mL

Table 13. Enalapril Maleate 0.5 mg/mL oral formulation composition.

Enalapril Maleate USP	Active Substance	0.05 g
SuspendIt®	Suspending agent	q.s. 100 mL

4.2. *Preliminary Characterization of SuspendIt® Vehicle and the Oral Suspensions*

A comparison study was carried out between the vehicle used—SuspendIt®—and the atenolol and enalapril maleate oral formulations to be studied. To this purpose, pH, rheological properties, injectability, and palatability were analyzed as described below.

4.2.1. pH

The pH was measured in triplicate on a HI 2211 pH/ORP/°C meter (Hanna Instruments, Woonsocket, RI, USA) at 25 °C.

4.2.2. Rheological Properties

The rheological properties (flow curve, thixotropy, linear viscoelastic region, and mechanical spectrum) were determined on a Kinexus Prime lab+ Rheometer (Malvern Panalytical, Malvern, UK) at 25 °C. The cone and plate geometry (cone angle: 4°; diameter: 40 mm; working gap: 0.150 mm) was used in all tests. For the determination of the flow curve, measurements were performed in the shear rate range of 0.1 s^{-1} to 100 s^{-1}. A rheological model was fitted to the results, namely the Herschel-Bulkley model. This model is employed to characterize the rheological behavior of non-Newtonian fluids, where the stress-strain relationship is nonlinear [78]. This model incorporates a yield stress parameter and a power law index, effectively capturing the shear-thinning properties of these fluids [78]. The shear rate induced by a specific shear stress applied to the fluid can be quantified using the following equation:

$$\tau = \tau_0 + K\left(\frac{du}{dy}\right)^n$$

where du/dy is the shear rate, τ is the shear stress, τ_0 is the yield stress, K is the consistency coefficient, and n is the flow index.

For the determination of the linear viscoelastic region, an amplitude sweep was carried out in the shear strain range of 0.1% to 100% at a frequency of 1.0 Hz. For the mechanical spectrum, a frequency sweep was carried out from 10 to 0.1 Hz. at a shear strain of 0.4%. To assess the thixotropic behavior, a three-step test was performed under the following conditions: shear rate, 0.1 s^{-1}; time in the first phase, 1 min; shear rate, 100 s^{-1}; time in the second phase, 30 s; shear rate: 0.1 s^{-1}; time in the third phase, 10 min. All measurements were obtained in triplicate.

4.2.3. Injectability

The injectability test was performed using a texture analyzer TA.XT.plus (Stable Micro Systems, Godalming, UK) at 25 °C (Figure 15). The injectability test aims to measure the force required to promote the flow of the oral suspensions through a 50 mL syringe coupled to NeoMed®® Radiopaque, Polyurethane (PUR)—Enteral Feeding Tube with Connector (outer diameter: 6.5FR; length: 90 cm). The syringe was filled using with the sample (30 mL), and the whole device was mounted on the texture analyzer working in compression mode. During extrusion tests, samples were assayed in triplicate, applying the force vertically and using a plunger displacement rate of 0.2 mm/s for a distance of 25 mm. The simple syrup was also tested for comparison.

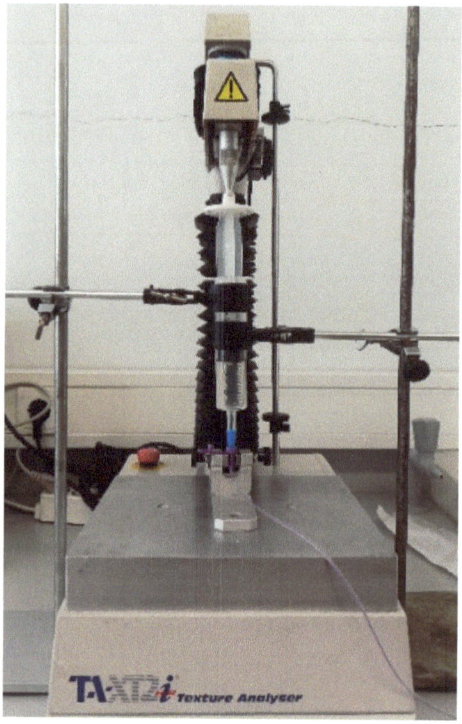

Figure 15. Injectability test set up in the texture analyzer TA.XT.plus (Stable Micro Systems, Godalming, UK).

4.2.4. E-tongue Apparatus and Potentiometric Analysis of Atenolol and Enalapril Maleate Solutions

A custom-designed potentiometric E-tongue with 40 lipid polymeric sensor membranes was employed as a taste-sensing tool to assess the ability of the atenolol vehicle and the SuspendIt® vehicle to mask bitterness in atenolol and enalapril maleate formulations. The specifications of the E-tongue device, number of sensor arrays, sensor membrane types, and data acquisition apparatus and software were previously outlined in a study by the research team [17]. For the experimental procedures, six independent samples (100 mL) of each oral formulation were prepared: atenolol in aqueous solution (2 mg/mL), atenolol in atenolol vehicle (2 mg/mL), enalapril maleate in aqueous solution (0.5 mg/mL), and enalapril maleate in SuspendIt® vehicle (0.5 mg/mL). Additionally, six independent 100 mL samples of each pure vehicle (without any drug) were also prepared. Due to the challenge of conducting potentiometric assays in non-conductive liquids with high viscosity, the initial solutions were diluted ten times with deionized water to facilitate electrochemical analysis and prevent damage to the lipid sensor membranes of the E-tongue. The solutions were analyzed at ambient temperature (≈20 °C) using the E-tongue. Each analysis lasted 5 min, allowing for a pseudo-equilibrium to be reached between the E-tongue's non-specific lipid polymeric membranes and the constituents of each solution. Following each assay, the E-tongue was rinsed with deionized water, and after four to five assays, it was submerged in a 0.01 mol/L HCl aqueous solution to assess signal repeatability and ensure effective cleaning of the sensor arrays. At the conclusion of each day, the E-tongue was stored at room temperature and immersed in a 0.01 mol/L HCl cleaning solution to maintain the integrity and functionality of the lipid polymeric membranes over extended periods.

Statistical Analysis

The discrimination capability of the E-tongue was evaluated using chemometric techniques. Specifically, Linear Discriminant Analysis (LDA) combined with the simulated annealing (SA) variable selection algorithm, as detailed previously [17], was employed. This algorithm enabled the identification of the most pertinent subset of non-redundant E-tongue sensors. Assessment of LDA performance was conducted on both the original grouped data (training) and through leave-one-out cross-validation (LOO-CV). The visualization of the former involved 2D plots of the two most influential discriminant functions (DFs), while the latter was evaluated based on sensitivity values, representing the percentage of accurately classified samples within predetermined groups. LOO-CV, commonly used for small datasets lacking an external validation dataset, was employed in this study. All statistical analyses were performed using the packages of the open-source statistical software *R* (RStudio 2024.04.0 Build 735), at a 5% significance level.

4.3. Stability Study

Afterwards, a six-month stability study was conducted for atenolol and enalapril maleate oral formulations prepared with SuspendIt® by evaluating the chemical, physical, and microbiological stability. Three independent batches (1.7 L each) of the oral suspensions of atenolol and enalapril maleate were prepared. Samples were stored in amber plastic prescription bottles at two temperature conditions (5 °C and 25 °C). Samples were assayed initially and at pre-determined time intervals over a six-month period. Together with the drug assays, physicochemical properties such as pH and rheological behavior were measured. These measurements were performed initially and after 14, 30, 60, 90, 120, and 180 days of storage. Additionally, formulations were tested for microbiological stability, namely the preservative effectiveness test (initially and after 30, 90, and 180 days of storage), according to USP Chapter <51> antimicrobial effectiveness testing [44]. The appearance and odor of the formulations were also recorded over time. The odor analysis was conducted using olfactory sensory evaluation by a human assessor. Although small changes can go undetected, marked odor modification is likely to be perceived [79].

4.3.1. Preservative Effectiveness

SuspendIt® contains potassium sorbate and sodium benzoate as antimicrobial preservatives. The standard challenge organisms described in the USP (USP Chapter <51> antimicrobial effectiveness testing includes *Escherichia coli* ATCC 8739, *Pseudomonas aeruginosa* ATCC 9027, *Staphylococcus aureus* ATCC 6538, *Candida albicans* ATCC 10231, and *Aspergillus brasiliensis* ATCC 16404 [44]. This test provides the following criteria to determine antimicrobial effectiveness:

- Bacteria: not less than 1 log reduction from the initial count at 14 days, and no increase from the 14 days' count at 28 days;
- Yeasts and molds: no increase from the initial calculated count at 14 days and 28 days.

A suitability test of the counting method in the presence of the products was performed for each product as described in USP Chapter <51>. The suitability test is intended to verify the capacity of microorganisms to recover 50% growth compared with saline control. The products with no dilution, with one dilution (1 mL of product without dilution in 9 mL of sterile saline), and with two dilutions (1 mL of product with one dilution in 9 mL of sterile saline) were tested. Since the inhibitory capacity of the preservatives was very pronounced, preventing counting, it was necessary to dilute the products. For all the microorganisms, except for *P. aeruginosa*, one dilution (1 mL of product in 9 mL of sterile saline) of the products was needed.

4.3.2. Active Substance Assay

Quantification of atenolol and enalapril maleate was performed using high-performance liquid chromatography (HPLC). The method was developed based on both EU and US Pharmacopeias, with minor adjustments, particularly with the inclusion of a precipitation step during sample preparation to avoid clogging the chromatographic column, and gradient adjustments to grant complete separation from matrix components. On each sampling day, the flasks for analysis were delivered to the HPLC lab. The vials were left at room temperature, and all samples and standard solutions were processed as a batch. The sample preparation procedure was also applied to all standard solutions.

Briefly, 750 μL of homogenized preparations were transferred to Eppendorf tubes, followed by the addition of 500 μL of acetonitrile. After a brief vortexing with visual confirmation of precipitation, the tubes were centrifuged at 13,000 rpm for 5 min at room temperature. A 500-μL portion was transferred to the injection vials containing 1000 μL of buffer (monobasic sodium phosphate buffer prepared with 2.8 g per liter of water and pH adjusted to 2.5 with phosphoric acid). After homogenization, the vials were ready for injection. Under this procedure, the samples and standards were injected with a 5-fold dilution in comparison to their original concentration, corresponding to 0.1 mg/mL for enalapril and 0.4 mg/mL for atenolol preparations. This final dilution is equivalent to the one described in monographs.

Quantification was based on external calibration curves, with standards prepared daily in deionized water at 4.0 mg/mL for atenolol and at 1.0 mg/mL for enalapril, with dilutions in the working range (25 to 150% of the expected concentrations). Both water-based and matrix-based (SuspendIt®) working standards calibration curves were previously tested, with no differences between the two approaches. In addition to UV detection (228 nm for atenolol and 215 nm for enalapril), a second channel with fluorescence detection was included that, due to its specificity, allowed us to confirm the results obtained.

Table S1 resumes the HPLC method characteristics for both quantifications, including column, gradient, injection, etc. The methods were optimized to grant no coelution with other sample components, with the shortest analysis time, providing stable elution conditions, and with the lowest consumption of solvents due to our green-chemistry policy.

Both methods were validated to determine their adequacy and stability to support a 6-month study (Table S2). The tests included a definition of an adequate working range, repeatability over consecutive injections and consecutive extractions, as well as its accuracy testing of three spiking levels within the working range (75%, 100%, and 125%) and presenting its average.

5. Conclusions

This study underscored the comprehensive characterization needed when selecting a suitable vehicle for compounded liquid dosage forms suitable for oral and enteral feeding tube administration. For this purpose, palatability, injectability, rheological properties, pH, and physical, chemical, and microbiological stability were studied. Both cardiovascular formulations were found suitable for administration through enteral feeding tubes. Injectability tests showed both formulations required less force for injection through a syringe coupled to an enteral feeding tube compared to simple syrup. Moreover, the bitterness of drugs could be effectively masked by using the atenolol vehicle or SuspendIt® vehicle, as confirmed by the taste sensor device, thus supporting adequate palatability and predicting good medication adherence. These results should, however, confirmed with the target patients. While both atenolol and enalapril maleate preparations showed overall stability in terms of mechanical properties and pH during the 180 days of storage at both temperatures, chemical stability presented a different scenario. A decrease in drug content below 90% was observed for enalapril maleate from 90 days onwards when stored at 25 °C. The effectiveness of the preservative system was maintained throughout the storage period, as per USP Chapter <51> requirements. In summary, the atenolol oral formulation maintained suitable chemical, physical, and microbiological stability for 180 days of storage at 5 °C and 25 °C, while enalapril maleate remained stable up to 60 days at 25 °C and for 180 days at 5 °C. The use of the proposed oral compounding formulations provides a customized, reliable, and less costly solution for the long-term treatment of cardiovascular patients.

Supplementary Materials: The following supporting information can be downloaded at: https://www.mdpi.com/article/10.3390/ph17081052/s1, Figure S1: Overlaid chromatograms from standard solutions prepared in HPLC water from a calibration curve prepared for atenolol (A) and enalapril (B) analysis; Figure S2: Chromatogram from an enalapril standard solution detected under UV and fluorescence detection.; Figure S3: Chromatograms from six atenolol samples at T90: atenolol eluting at 4.4 min, completely separated from the other matrix components under UV and FLD detections.; Figure S4: Chromatograms from a triplicate set of enalapril samples analyzed at day 30, with the UV reading above and the FLD below.; Figure S5: Overlay chromatograms from the degradation studies, with the UV reading above and the FLD below; Figure S6: Appearance of SuspendIt® (left) and atenolol (middle) and enalapril maleate (right) formulations after 180 days storage at 25 °C; Figure S7: Appearance of SuspendIt® (left) and atenolol (middle) and enalapril maleate (right) formulations after 180 days storage at 5 °C; Figure S8: Flow curve of atenolol oral formulation after storage at 5 °C; Figure S9: Flow curve of enalapril maleate oral formulation after storage at 25 °C; Figure S10: Flow curve of enalapril maleate oral formulation after storage at 5 °C; Figure S11: Thixotropic behavior of atenolol oral formulation after storage at 5 °C; Figure S12: Thixotropic behavior of enalapril maleate oral formulation after storage at 25 °C; Figure S13: Thixotropic behavior of enalapril maleate oral formulation after storage at 5 °C; Figure S14: Amplitude sweep for atenolol oral formulation after 180 days storage at 5 °C; Figure S15: Amplitude sweep for enalapril maleate oral formulation after 180 days storage at 25 °C; Figure S16: Amplitude sweep for enalapril maleate oral formulation after 180 days storage at 5 °C; Figure S17: Mechanical spectrum of atenolol oral formulation after 180 days storage at 5 °C; Figure S18: Mechanical spectrum of enalapril maleate oral formulation after 180 days storage at 25 °C; Figure S19: Mechanical spectrum of enalapril maleate oral formulation after 180 days storage at 5 °C. Table S1: HPLC method characteristics.; Table S2: Overall performance of the analytical methods; Table S3: Recovery period (s) (50%) of atenolol formulation (mean ± standard deviation).; Table S4: Recovery rate (%) of atenolol formulation (mean ± standard deviation).; Table S5: Recovery period (s) (50%) of enalapril maleate formulation (mean ± standard deviation).; Table S6: Recovery rate (%) of enalapril maleate oral formulation (mean ± standard deviation).; Table S7: United States Pharmacopeia Chapter <51> Antimicrobial Effectiveness Test of enalapril maleate oral formulation; Table S8: Quantification of atenolol by HPLC-UV after storage at refrigerated temperature (n = 3, 5 °C).; Table S9: Quantification of enalapril by HPLC-UV after at refrigerated temperature (n = 3, 5 °C).

Author Contributions: Conceptualization, I.F.A., S.C. and H.F.; methodology, I.F.A., S.C. and H.F.; software, A.M.P.; formal analysis, A.M.P., N.F. and R.C.; investigation, S.M., A.T., N.F. and R.C.; writing—original draft preparation, S.M., A.T. and A.M.P.; writing—review and editing, C.Q., I.F.A., S.C. and H.F.; supervision, I.F.A. All authors have read and agreed to the published version of the manuscript.

Funding: This study was financed by the Professional Compounding Centers of America (PCCA).

Institutional Review Board Statement: Not applicable.

Informed Consent Statement: Not applicable.

Data Availability Statement: Data is contained within the article and Supplementary Materials.

Acknowledgments: The authors received support and help from the Foundation for Science and Technology (FCT, Portugal) by national funds FCT/MCTES (PIDDAC) to Associate Laboratory Institute for Health and Bioeconomy—i4HB (DOI 10.54499/LA/P/0140/2020), to Research Unit on Applied Molecular Biosciences—UCIBIO (DOI 10.54499/UIDB/04378/2020 and DOI 10.54499/UIDP/04378/2020), to CIMO (UIDB/00690/2020, https://doi.org/10.54499/UIDB/00690/2020; and UIDP/00690/2020, https://doi.org/10.54499/UIDP/00690/2020), to LA SusTEC (LA/P/0007/2020) and to LAQV-REQUIMTE

(DOI 10.54499/LA/P/0008/2020, DOI 10.54499/UIDP/50006/2020 and DOI 10.54499/UIDB/50006/2020). Nuno Ferreiro acknowledges the Ph.D. research grant (2022.10072.BD) provided by FCT. Rebeca Cruz is gratefully supported by FCT (2022.00965.CEECIND). S. Mota acknowledges her Ph.D. research grant (2022.12487.BD) and A. Torres also acknowledges her Ph.D. research grant (2023.03685.BD), both fully supported by FCT.

Conflicts of Interest: The authors declare no conflicts of interest.

References

1. Carvalho, M.; Almeida, I.F. The Role of Pharmaceutical Compounding in Promoting Medication Adherence. *Pharmaceuticals* **2022**, *15*, 1091. [CrossRef]
2. Smeets, N.J.L.; Schreuder, M.F.; Dalinghaus, M.; Male, C.; Lagler, F.B.; Walsh, J.; Laer, S.; de Wildt, S.N. Pharmacology of enalapril in children: A review. *Drug Discov. Today* **2020**, *25*, 1957–1970. [CrossRef] [PubMed]
3. Schoolwerth, A.C.; Sica, D.A.; Ballermann, B.J.; Wilcox, C.S. Renal considerations in angiotensin converting enzyme inhibitor therapy: A statement for healthcare professionals from the Council on the Kidney in Cardiovascular Disease and the Council for High Blood Pressure Research of the American Heart Association. *Circulation* **2001**, *104*, 1985–1991. [CrossRef]
4. Heel, R.C.; Brogden, R.N.; Speight, T.M.; Avery, G.S. Atenolol: A Review of its Pharmacological Properties and Therapeutic Efficacy in Angina Pectoris and Hypertension. *Drugs* **1979**, *17*, 425–460. [CrossRef] [PubMed]
5. Khan, M.G. Beta-Blockers. In *Encyclopedia of Heart Diseases*; Khan, M.G., Ed.; Academic Press: Burlington, VT, USA, 2006; pp. 159–167.
6. Gorre, F.; Vandekerckhove, H. Beta-blockers: Focus on mechanism of action Which beta-blocker, when and why? *Acta Cardiol.* **2010**, *65*, 565–570. [CrossRef] [PubMed]
7. INFARMED, I.P. Resumo das características do medicamento: Enalapril. Infomed. Available online: https://extranet.infarmed.pt/INFOMED-fo/ (accessed on 5 August 2024).
8. Medicamento, C.T.d. *Formulário Galénico Português*; Associação Nacional das Farmácias: Lisboa, Portugal, 2001.
9. Siddiqi, N.; Shatat, I.F. Antihypertensive agents: A long way to safe drug prescribing in children. *Pediatr. Nephrol.* **2020**, *35*, 2049–2065. [CrossRef] [PubMed]
10. INFARMED, I.P. Resumo das características do medicamento: Atenolol. Infomed. Available online: https://extranet.infarmed.pt/INFOMED-fo/ (accessed on 5 August 2024).
11. Breitkreutz, J.; Boos, J. Paediatric and geriatric drug delivery. *Expert Opin. Drug Deliv.* **2007**, *4*, 37–45. [CrossRef] [PubMed]
12. Lam, M.S. Extemporaneous compounding of oral liquid dosage formulations and alternative drug delivery methods for anticancer drugs. *Pharmacotherapy* **2011**, *31*, 164–192. [CrossRef] [PubMed]
13. Stavroulakis, T.; McDermott, C.J. Enteral feeding in neurological disorders. *Pract. Neurol.* **2016**, *16*, 352–361. [CrossRef]
14. Ferreira-Neto, C.J.B.; de Andrade, R.A.; Tonin, F.S.; Wiens, A. Solid Oral Dosage Forms Use in Adults with Neurological Disorders and Swallowing Difficulties: A Scoping Review. *Dysphagia* **2022**, *37*, 909–922. [CrossRef]
15. Kurien, M.; Penny, H.; Sanders, D.S. Impact of direct drug delivery via gastric access devices. *Expert Opin. Drug Deliv.* **2015**, *12*, 455–463. [CrossRef]
16. Malkawi, W.A.; AlRafayah, E.; AlHazabreh, M.; AbuLaila, S.; Al-Ghananeem, A.M. Formulation Challenges and Strategies to Develop Pediatric Dosage Forms. *Children* **2022**, *9*, 488. [CrossRef]
17. Ogbonna, J.D.; Cunha, E.; Attama, A.A.; Ofokansi, K.C.; Ferreira, H.; Pinto, S.; Gomes, J.; Marx, Í.M.; Peres, A.M.; Lobo, J.M.; et al. Overcoming Challenges in Pediatric Formulation with a Patient-Centric Design Approach: A Proof-of-Concept Study on the Design of an Oral Solution of a Bitter Drug. *Pharmaceuticals* **2022**, *15*, 1331. [CrossRef]
18. INFARMED–Instituto Nacional da Farmácia e do Medicamento. *Farmacopeia Portuguesa 9: Edição Oficial*; INFARMED, I.P.: Lisboa, Portugal, 2009.
19. Leon Lachman, H.A.L.; Joseph, L.K. *The Theory and Practice of Industrial Pharmacy*, 3rd ed.; Lea & Febiger Philadelphia: Philadelphia, PA, USA, 1970.
20. Loyd, V.; Allen, N.G.P.; Howard, C.A. *Ansel's Pharmaceutical Dosage Forms and Drug Delivery Systems*, 9th ed.; Wolters Kluwer Health/Lippincott Williams & Wilkins: Philadelphia, PA, USA, 2011.
21. Cutaia, K.; Chablani, L.; Zhao, F. Basics of Compounding: Vehicles for Compounded Oral Liquid Medications: A Review. *Int. J. Pharm. Compd.* **2018**, *22*, 480–489. [PubMed]
22. Freerks, L.; Sucher, W.; Tarnow, M.-J.; Eckert, C.; Klein, S. Vehicles for Drug Administration to Children: Results and Learnings from an In-Depth Screening of FDA-Recommended Liquids and Soft Foods for Product Quality Assessment. *Pharm. Res.* **2022**, *39*, 497–509. [CrossRef] [PubMed]
23. Blaszczyk, A.; Brandt, N.; Ashley, J.; Tuders, N.; Doles, H.; Stefanacci, R.G. Crushed Tablet Administration for Patients with Dysphagia and Enteral Feeding: Challenges and Considerations. *Drugs Aging* **2023**, *40*, 895–907. [CrossRef] [PubMed]
24. Phillips, M.S. Handbook of Drug Administration via Enteral Feeding Tubes on Behalf of the British Pharmaceutical Nutrition Group. *Am. J. Pharm. Educ.* **2007**, *71*, 99.
25. European Directorate for the Quality of Medicines & HealthCare. *European Pharmacopoeia*, 10th ed.; Council of Europe: Strasbourg, France, 2019.

26. Vogelpoel, H.; Welink, J.; Amidon, G.L.; Junginger, H.E.; Midha, K.K.; Möller, H.; Olling, M.; Shah, V.P.; Barends, D.M. Biowaiver monographs for immediate release solid oral dosage forms based on biopharmaceutics classification system (BCS) literature data: Verapamil hydrochloride, propranolol hydrochloride, and atenolol. *J. Pharm. Sci.* **2004**, *93*, 1945–1956. [CrossRef]
27. Bontha, L.; Sriramaneni, R.N. New Dissolution Method for the Evaluation of Enalapril Maleate Tablets using pH 7.2 Phosphate Buffer in vitro and Determination of its Content by Validated UV Spectrophotometric Method. *JASA* **2007**, *2*, 34–37.
28. Professional Compounding Centers of America (PCCA). Similar Names. Different Products—SuspendIt® & SuspendIt® Anhydrous. Available online: https://www.pccarx.com/Blog/similar-names-different-products-suspendit-suspendit-anhydrous (accessed on 15 November 2023).
29. Ip, K.; Shan, A.; Carvalho, M.; Baker, S.; Banov, D. Physicochemical Stability of Extemporaneously Prepared Oral Suspension of Fluconazole 50 mg/mL in SuspendIt™. *Pharm. Technol. Hosp. Pharm.* **2018**, *3*, 101–112. [CrossRef]
30. Graves, R.; Phan, K.V.; Bostanian, L.A.; Mandal, T.K.; Pramar, Y.V. Stability of Spironolactone Oral Suspension in PCCA Base, SuspendIt. *Int. J. Pharm. Compd.* **2017**, *21*, 334–338. [PubMed]
31. Graves, R.A.; Phan, K.V.; Bostanian, L.A.; Mandal, T.K.; Pramar, Y.V. Physicochemical Stability of an Oral Suspension of Trimethoprim 20 mg/mL in Combination with Sulfadiazine 200 mg/mL in PCCA Base SuspendIt. *Int. J. Pharm. Compd.* **2017**, *21*, 430–435. [PubMed]
32. Pramar, Y.V.; Mandal, T.K.; Bostanian, L.A.; Le, G.; Morris, T.C.; Graves, R.A. Physicochemical and Microbiological Stability of Compounded Metronidazole Suspensions in PCCA SuspendIt. *Int. J. Pharm. Compd.* **2021**, *25*, 169–175.
33. Pramar, Y.V.; Mandal, T.K.; Bostanian, L.A.; Nguyen, A.T.; Miller, V.; Morris, T.C.; Graves, R.A. Stability of Compounded Ursodiol Suspensions in PCCA Base, SuspendIt. *Int. J. Pharm. Compd.* **2019**, *23*, 70–76.
34. Graves, R.A.; Mandal, T.K.; Bostanian, L.A.; Nguyen, A.T.; Swopes, D.; Morris, T.C.; Pramar, Y.V. Physicochemical Stability of Compounded Amlodipine Besylate Suspensions in PCCA Base, SuspendIt. *Int. J. Pharm. Compd.* **2019**, *23*, 519–527. [PubMed]
35. Pramar, Y.V.; Mandal, T.K.; Bostanian, L.A.; Le, G.; Morris, T.C.; Graves, R.A. Physicochemical Stability of Compounded Allopurinol Suspensions in PCCA Base, SuspendIt. *Int. J. Pharm. Compd.* **2020**, *24*, 413–419. [PubMed]
36. Pramar, Y.V.; Graves, R.A.; Ledet, G.A.; Phan, K.V.; Bostanian, L.A.; Mandal, T.K. Stability of Clindamycin Hydrochloride in PCCA Base SuspendIt. *Int. J. Pharm. Compd.* **2016**, *20*, 421–425. [PubMed]
37. Pramar, Y.V.; Mandal, T.K.; Bostanian, L.A.; Nguyen, A.T.; Morris, T.C.; Graves, R.A. Physicochemical Stability of Compounded Naltrexone Hydrochloride Solutions in PCCA Base SuspendIt. *Int. J. Pharm. Compd.* **2019**, *23*, 157–162. [PubMed]
38. TA Instruments. Understanding Rheology of Structured Fluids. Available online: https://www.tainstruments.com/pdf/literature/AAN016_V1_U_StructFluids.pdf (accessed on 27 May 2024).
39. Peter Herh, J.T.; Wu, S.; Bernzen, M.; Rudolph, B. The rheology of pharmaceutical and cosmetic semisolids. *Am. Lab.* **1998**, *30*, 12–14.
40. Paar, A. Basics of Thixotropy. Available online: https://wiki.anton-paar.com/en/basics-of-thixotropy/ (accessed on 27 May 2024).
41. NETZSCH-Gerätebau GmbH. *Rotational Rheology: Interpretation of Data by Application*, 2nd ed.; NETZSCH-Gerätebau GmbH: Selb, Germany, 2022.
42. Stojkov, G.; Niyazov, Z.; Picchioni, F.; Bose, R. Relationship between Structure and Rheology of Hydrogels for Various Applications. *Gels* **2021**, *7*, 255. [CrossRef]
43. Medicamento, C.T.d. Xarope Simples, BP2000 (FGP B.7.). In *Formulário Galénico Português*; Associação Nacional de Farmácias: Lisboa, Portugal, 2007.
44. Pharmacopeia, U.S. Antimicrobial Effectiveness Testing <51>. In *USP–NF*; United States Pharmacopeia: Rockville, MD, USA, 2020.
45. Chiclana-Rodríguez, B.; Garcia-Montoya, E.; Romero-Obon, M.; Rouaz-El-Hajoui, K.; Nardi-Ricart, A.; Suñé-Pou, M.; Suñé-Negre, J.M.; Pérez-Lozano, P. Palatability and Stability Studies to Optimize a Carvedilol Oral Liquid Formulation for Pediatric Use. *Pharmaceutics* **2024**, *16*, 30. [CrossRef] [PubMed]
46. Elgammal, A.; Ryan, J.; Bradley, C.; Crean, A.; Bermingham, M. The impact of drug palatability on prescribing and dispensing of antibiotic formulations for paediatric patients: A cross-sectional survey of general practitioners and pharmacists. *Fam. Pract.* **2023**. [CrossRef]
47. Mrokowska, M.M.; Krztoń-Maziopa, A. Viscoelastic and shear-thinning effects of aqueous exopolymer solution on disk and sphere settling. *Sci. Rep.* **2019**, *9*, 7897. [CrossRef] [PubMed]
48. Ramli, H.; Zainal, N.F.A.; Hess, M.; Chan, C.H. Basic principle and good practices of rheology for polymers for teachers and beginners. *Chem. Teach. Int.* **2022**, *4*, 307–326. [CrossRef]
49. Wang, C.; Xu, M.; Lv, W.-P.; Qiu, P.; Gong, Y.-Y.; Li, D.-S. Study on Rheological Behavior of Konjac Glucomannan. *Phys. Procedia* **2012**, *33*, 25–30. [CrossRef]
50. Zhong, L.; Oostrom, M.; Truex, M.J.; Vermeul, V.R.; Szecsody, J.E. Rheological behavior of xanthan gum solution related to shear thinning fluid delivery for subsurface remediation. *J. Hazard. Mater.* **2013**, *244–245*, 160–170. [CrossRef]
51. Al-Malah, K. Rheological properties of carbomer dispersions. *Annu. Trans. Nord. Rheol. Soc.* **2006**, *14*, 123.
52. Miller, S.C.; Drabik, B.R. Rheological properties of poloxamer vehicles. *Int. J. Pharm.* **1984**, *18*, 269–276. [CrossRef]
53. Ross, A.I.V.; Tyler, P.; Borgognone, M.G.; Eriksen, B.M. Relationships between shear rheology and sensory attributes of hydrocolloid-thickened fluids designed to compensate for impairments in oral manipulation and swallowing. *J. Food Eng.* **2019**, *263*, 123–131. [CrossRef]
54. Chen, T.T. *Rheological Techniques for Yield Stress Analysis*; TA Instruments: New Castle, DE, USA, 2000.

55. Banov, D.; Liu, Y.; Ip, K.; Shan, A.; Vu, C.; Zdoryk, O.; Bassani, A.S.; Carvalho, M. Analysis of the Physical Characteristics of an Anhydrous Vehicle for Compounded Pediatric Oral Liquids. *Pharmaceutics* **2023**, *15*, 2642. [CrossRef]
56. Sangroniz, L.; Fernández, M.; Santamaria, A. Polymers and rheology: A tale of give and take. *Polymer* **2023**, *271*, 125811. [CrossRef]
57. Öhrlund, Å. Evaluation of Rheometry Amplitude Sweep Cross-Over Point as an Index of Flexibility for HA Fillers. *J. Cosmet. Dermatol. Sci. Appl.* **2018**, *8*, 47–54. [CrossRef]
58. Carmona, J.A.; Lucas, A.; Ramírez, P.; Calero, N.; Muñoz, J. Nonlinear and linear viscoelastic properties of a novel type of xanthan gum with industrial applications. *Rheol. Acta* **2015**, *54*, 993–1001. [CrossRef]
59. Ghebremedhin, M.; Schreiber, C.; Zielbauer, B.; Dietz, N.; Vilgis, T. Interaction of xanthan gums with galacto- and glucomannans. Part II: Heat induced synergistic gelation mechanism and their interaction with salt. *J. Phys. Mater.* **2020**, *3*, 034014. [CrossRef]
60. Paar, A. Frequency Sweeps. Available online: https://wiki.anton-paar.com/en/frequency-sweeps/ (accessed on 11 April 2024).
61. European Medicines Agency. Use of Methyl- and Propylparaben as Excipients in Human Medicinal Products for Oral use—Scientific Guideline. Available online: https://www.ema.europa.eu/en/use-methyl-propylparaben-excipients-human-medicinal-products-oral-use-scientific-guideline (accessed on 28 May 2024).
62. Saito, J.; Nadatani, N.; Setoguchi, M.; Nakao, M.; Kimura, H.; Sameshima, M.; Kobayashi, K.; Matsumoto, H.; Yoshikawa, N.; Yokoyama, T.; et al. Potentially harmful excipients in neonatal medications: A multicenter nationwide observational study in Japan. *J. Pharm. Health Care Sci.* **2021**, *7*, 23. [CrossRef]
63. Álvaro-Alonso, E.A.; Lorenzo, M.P.; Gonzalez-Prieto, A.; Izquierdo-García, E.; Escobar-Rodríguez, I.; Aguilar-Ros, A. Physicochemical and Microbiological Stability of Two Oral Solutions of Methadone Hydrochloride 10 mg/mL. *Molecules* **2022**, *27*, 2812. [CrossRef] [PubMed]
64. Binson, G.; Beuzit, K.; Migeot, V.; Marco, L.; Troussier, B.; Venisse, N.; Dupuis, A. Preparation and Physicochemical Stability of Liquid Oral Dosage Forms Free of Potentially Harmful Excipient Designed for Pediatric Patients. *Pharmaceutics* **2019**, *11*, 190. [CrossRef] [PubMed]
65. Belayneh, A.; Tadese, E.; Molla, F. Safety and Biopharmaceutical Challenges of Excipients in Off-Label Pediatric Formulations. *Int. J. Gen. Med.* **2020**, *13*, 1051–1066. [CrossRef] [PubMed]
66. Vázquez-Blanco, S.; González-Freire, L.; Dávila-Pousa, M.C.; Crespo-Diz, C. pH determination as a quality standard for the elaboration of oral liquid compounding formula. *Farm. Hosp.* **2018**, *42*, 221–227. [CrossRef]
67. Foppa, T.; Murakami, F.S.; Silva, M.A. Development, validation and stability study of pediatric atenolol syrup. *Pharmazie* **2007**, *62*, 519–521.
68. Garner, S.S.; Wiest, D.B.; Reynolds, E.R., Jr. Stability of atenolol in an extemporaneously compounded oral liquid. *Am. J. Hosp. Pharm.* **1994**, *51*, 508–511. [CrossRef] [PubMed]
69. Nahata, M.C.; Morosco, R.S.; Hipple, T.F. Stability of enalapril maleate in three extemporaneously prepared oral liquids. *Am. J. Health Syst. Pharm.* **1998**, *55*, 1155–1157. [CrossRef] [PubMed]
70. Diego, M.; Godoy, G.; Mennickent, S.; Godoy, R. Chemical stability of enalapril maleate drug substance and tablets by a stability-indicating liquid chromatographic method. *Química Nova* **2010**, *34*, 450–454. [CrossRef]
71. Bout, M.R.; Vromans, H. Influence of commonly used excipients on the chemical degradation of enalapril maleate in its solid state: The role of condensed water. *Eur. J. Pharm. Sci.* **2022**, *171*, 106121. [CrossRef] [PubMed]
72. Hatem, O.; Suhail, F.; Juda, A. Determination of chemical potential for carvedilol, atenolol and Propranolol diffusion through SDS micelle solution. *Der Pharma Chem.* **2016**, *8*, 276–287.
73. Rafik, K. Prodrugs for Masking the Bitter Taste of Drugs. In *Application of Nanotechnology in Drug Delivery*; Ali Demir, S., Ed.; IntechOpen: Rijeka, Croatia, 2014; p. 12.
74. Georgieva, Y.; Kassarova, M.; Kokova, V.; Apostolova, E.; Pilicheva, B. Taste masking of enalapril maleate by microencapsulation in Eudragit EPO(®) microparticles. *Pharmazie* **2020**, *75*, 61–69. [PubMed]
75. Patel, D.; Doshi, D.H.; Desai, A. Short-term stability of atenolol in oral liquid formulations. *Int. J. Pharm. Compd.* **1997**, *1*, 437–439.
76. Allen, L.V., Jr.; Erickson, M.A., III. Stability of alprazolam, chloroquine phosphate, cisapride, enalapril maleate, and hydralazine hydrochloride in extemporaneously compounded oral liquids. *Am. J. Health-Syst. Pharm.* **1998**, *55*, 1915–1920. [CrossRef]
77. Sosnowska, K.; Winnicka, K.; Czajkowska-Kośnik, A. Stability of extemporaneous enalapril maleate suspensions for pediatric use prepared from commercially available tablets. *Acta Pol. Pharm.* **2009**, *66*, 321–326.
78. Vajravelu, K.; Sreenadh, S.; Devaki, P.; Prasad, K. Mathematical model for a Herschel-Bulkley fluid flow in an elastic tube. *Open Phys.* **2011**, *9*, 1357–1365. [CrossRef]
79. Sharma, A.; Kumar, R.; Aier, I.; Semwal, R.; Tyagi, P.; Varadwaj, P. Sense of Smell: Structural, Functional, Mechanistic Advancements and Challenges in Human Olfactory Research. *Curr. Neuropharmacol.* **2019**, *17*, 891–911. [CrossRef]

pharmaceuticals

MDPI

Article

Synthesis and Biochemical Evaluation of Ethanoanthracenes and Related Compounds: Antiproliferative and Pro-Apoptotic Effects in Chronic Lymphocytic Leukemia (CLL)

James P. McKeown [1,2], Andrew J. Byrne [1,2], Sandra A. Bright [3], Clara E. Charleton [1,2], Shubhangi Kandwal [4,5,6], Ivan Čmelo [5,6], Brendan Twamley [7], Anthony M. McElligott [8], Darren Fayne [5,6], Niamh M. O'Boyle [1,2], D. Clive Williams [3] and Mary J. Meegan [1,2,*]

1 School of Pharmacy and Pharmaceutical Sciences, Panoz Institute, Trinity College, The University of Dublin, East End 4/5, Dublin 2, D02 PN40 Dublin, Ireland
2 School of Pharmacy and Pharmaceutical Sciences, Trinity Biomedical Sciences Institute, Trinity College Dublin, 152-160 Pearse St, Dublin 2, D02 R590 Dublin, Ireland
3 School of Biochemistry and Immunology, Trinity Biomedical Sciences Institute, Trinity College Dublin, 152-160 Pearse St, Dublin 2, D02 R590 Dublin, Ireland; clive.williams@tcd.ie (D.C.W.)
4 Molecular Design Group, School of Biochemistry and Immunology, Trinity Biomedical Sciences Institute, Trinity College Dublin, 152-160 Pearse St, Dublin 2, D02 R590 Dublin, Ireland
5 Molecular Design Group, School of Chemical Sciences, Dublin City University, Glasnevin, D09 V209 Dublin, Ireland
6 DCU Life Sciences Institute, Dublin City University, Glasnevin, D09 V209 Dublin, Ireland
7 School of Chemistry, Trinity College Dublin, Dublin 2, D02 P3X2 Dublin, Ireland
8 Discipline of Haematology, School of Medicine, Trinity Translational Medicine Institute, St. James's Hospital and Trinity College, Dublin 8, D08 W9RT Dublin, Ireland; tony.mcelligott@tcd.ie
* Correspondence: mmeegan@tcd.ie

Citation: McKeown, J.P.; Byrne, A.J.; Bright, S.A.; Charleton, C.E.; Kandwal, S.; Čmelo, I.; Twamley, B.; McElligott, A.M.; Fayne, D.; O'Boyle, N.M.; et al. Synthesis and Biochemical Evaluation of Ethanoanthracenes and Related Compounds: Antiproliferative and Pro-Apoptotic Effects in Chronic Lymphocytic Leukemia (CLL). *Pharmaceuticals* **2024**, *17*, 1034. https://doi.org/10.3390/ph17081034

Academic Editor: Michael Danilenko

Received: 1 July 2024
Revised: 25 July 2024
Accepted: 30 July 2024
Published: 5 August 2024

Correction Statement: This article has been republished with a minor change. The change does not affect the scientific content of the article and further details are available within the backmatter of the website version of this article.

Abstract: Chronic lymphocytic leukemia (CLL) is a malignancy of mature B cells, and it is the most frequent form of leukemia diagnosed in Western countries. It is characterized by the proliferation and accumulation of neoplastic B lymphocytes in the blood, lymph nodes, bone marrow and spleen. We report the synthesis and antiproliferative effects of a series of novel ethanoanthracene compounds in CLL cell lines. Structural modifications were achieved via the Diels–Alder reaction of 9-(2-nitrovinyl)anthracene and 3-(anthracen-9-yl)-1-arylprop-2-en-1-ones (anthracene chalcones) with dienophiles, including maleic anhydride and *N*-substituted maleimides, to afford a series of 9-(*E*)-(2-nitrovinyl)-9,10-dihydro-9,10-[3,4]epipyrroloanthracene-12,14-diones, 9-(*E*)-3-oxo-3-phenylprop-1-en-1-yl)-9,10-dihydro-9,10-[3,4]epipyrroloanthracene-12,14-diones and related compounds. Single-crystal X-ray analysis confirmed the structures of the novel ethanoanthracenes **23f**, **23h**, **24a**, **24g**, **25f** and **27**. The products were evaluated in HG-3 and PGA-1 CLL cell lines (representative of poor and good patient prognosis, respectively). The most potent compounds were identified as **20a**, **20f**, **23a** and **25n** with IC_{50} values in the ranges of 0.17–2.69 μM (HG-3) and 0.35–1.97 μM (PGA-1). The pro-apoptotic effects of the potent compounds **20a**, **20f**, **23a** and **25n** were demonstrated in CLL cell lines HG-3 (82–95%) and PGA-1 (87–97%) at 10 μM, with low toxicity (12–16%) observed in healthy-donor peripheral blood mononuclear cells (PBMCs) at concentrations representative of the compounds IC_{50} values for both the HG-3 and PGA-1 CLL cell lines. The antiproliferative effect of the selected compounds, **20a**, **20f**, **23a** and **25n**, was mediated through ROS flux with a marked increase in cell viability upon pretreatment with the antioxidant NAC. **25n** also demonstrated sub-micromolar activity in the NCI 60 cancer cell line panel, with a mean GI_{50} value of 0.245 μM. This ethanoanthracene series of compounds offers potential for the further development of lead structures as novel chemotherapeutics to target CLL.

Keywords: ethanoanthracene; chalcone; B cell; chronic lymphocytic leukemia (CLL); antiproliferative; pro-apoptotic

1. Introduction

Chronic lymphocytic leukemia (CLL) is a malignancy of mature B cells, and it is the most frequent form of leukemia diagnosed in Western countries. The incidence of CLL varies by race and geographical location [1,2], with Ireland having one of the highest incidence rates [3]. It is typically a slow-growing cancer, and it is characterized by the proliferation and accumulation of neoplastic B lymphocytes in the blood, lymph nodes, bone marrow and spleen. Traditional chemoimmunotherapy approaches have been the mainstay of CLL treatment. However, with an evolving understanding of the biology of CLL and the development of targeted therapies, the treatment of patients diagnosed with CLL has changed dramatically over the past 5–10 years [4–7].

The most commonly used chemotherapy drugs approved by the Food and Drug Administration (FDA) for the treatment of CLL include the alkylating agents bendamustine **1** [8] and cyclophosphamide, together with the nucleoside fludarabine phosphate **2,** while the adenosine deaminase (ADA) inhibitor pentostatin **3** is also used in CLL patients who have relapsed (Figure 1). The development of small molecules targeting the B-cell receptor (BCR) signaling pathway has revolutionized the treatment of CLL [9]. Bruton's tyrosine kinase (BTK) [10] and the phosphoinositide 3-kinase δ (PI3Kδ) isoform [11] are essential for BCR signaling [12]. The targeted orally active inhibitor ibrutinib **4** was the first kinase inhibitor to be approved for CLL treatment, and it is now used to treat patients with CLL regardless of their treatment history [13]. Ibrutinib **4** [14], acalabrutinib **5** [15], zanubrutinib **6** [16] and tirabrutinib **7** [17,18] interact with BTK to form a covalent bond with the cysteine residue Cys481, resulting in the inhibition of BTK activity. While these inhibitors have had a significant impact on the treatment of B-cell malignancies, acquired resistance frequently emerges in patients, leading to clinical relapse and disease progression [19]. Non-covalent inhibitors of BTK, such as pirtobrutinib **8** [20] and fenebrutinib **9** [21], are also effective in the treatment of CLL [22], and they have been evaluated in relapsed or refractory B-cell non-Hodgkin lymphoma (NHL) and CLL [23]. Many promising small-molecule, irreversible BTK inhibitors are under clinical evaluation in targeted cancer therapy development [24].

The PI3Kδ inhibitor idelalisib **10** is an inhibitor of B-cell receptor signaling, and it is useful in patients with 17p deletion or TP53 mutation, who usually have a poor outcome, and in relapsed CLL [25,26]. Duvelisib **11** is an inhibitor of PI3Kδ and PI3Kγ, and it has been used in the treatment of high-risk CLL patients [27]. Venetoclax **12** is an orally bioavailable selective inhibitor of the anti-apoptotic B-cell lymphoma 2 (Bcl-2) protein, and it provides targeted therapy for the treatment of CLL patients, with a high overall response (80%) [28,29] (Figure 2); however, the development of clinical resistance is associated with its use [30]. Additionally, immunotherapies are now successfully used in the clinical treatment of CLL, e.g., the CD20 targeting monoclonal antibodies including rituximab, obinutuzumab and ofatumumab [27,31]. The programmed cell death protein 1 (PD-1) blocking antibodies pembrolizumab and nivolumab [27] can provide selective efficacy for CLL patients with Richter's transformation (RT), an aggressive complication of CLL [28]. Many of the oral agents are more tolerable than the traditional combination of chemotherapy and immunotherapy programs, and they are suitable for the treatment of older and more frail patients [32].

Small molecules discovered in preclinical development studies for CLL include the glutaminase inhibitor CB-839 (telaglenastat **13**) alone and in combinations with the widely used CLL drugs venetoclax, ibrutinib or MCL-1 inhibitor AZD-5991 [33]. The novel compound NX-2127 **14,** currently in Phase 1b clinical trials, combines the activity of a targeted BTK degrader with a second compound that degrades the Ikaros zinc finger transcription factors [34] (Figure 2). SGR-1505 **15** is an investigational mucosa-associated lymphoid tissue–lymphoma translocation protein 1 (MALT1) inhibitor, and it is currently in Phase 1 clinical development for refractory B-cell neoplasms [35]. The pre-clinical modeling of novel therapeutics in CLL has resulted in the discovery of new agents with potential for targeted therapies [36]. Recently reported molecules that target the BCR-associated kinases

BTK, PI3K and spleen tyrosine kinase (Syk), together with inhibitors of Bcl-2, are currently either approved or in additional clinical trials [22,37].

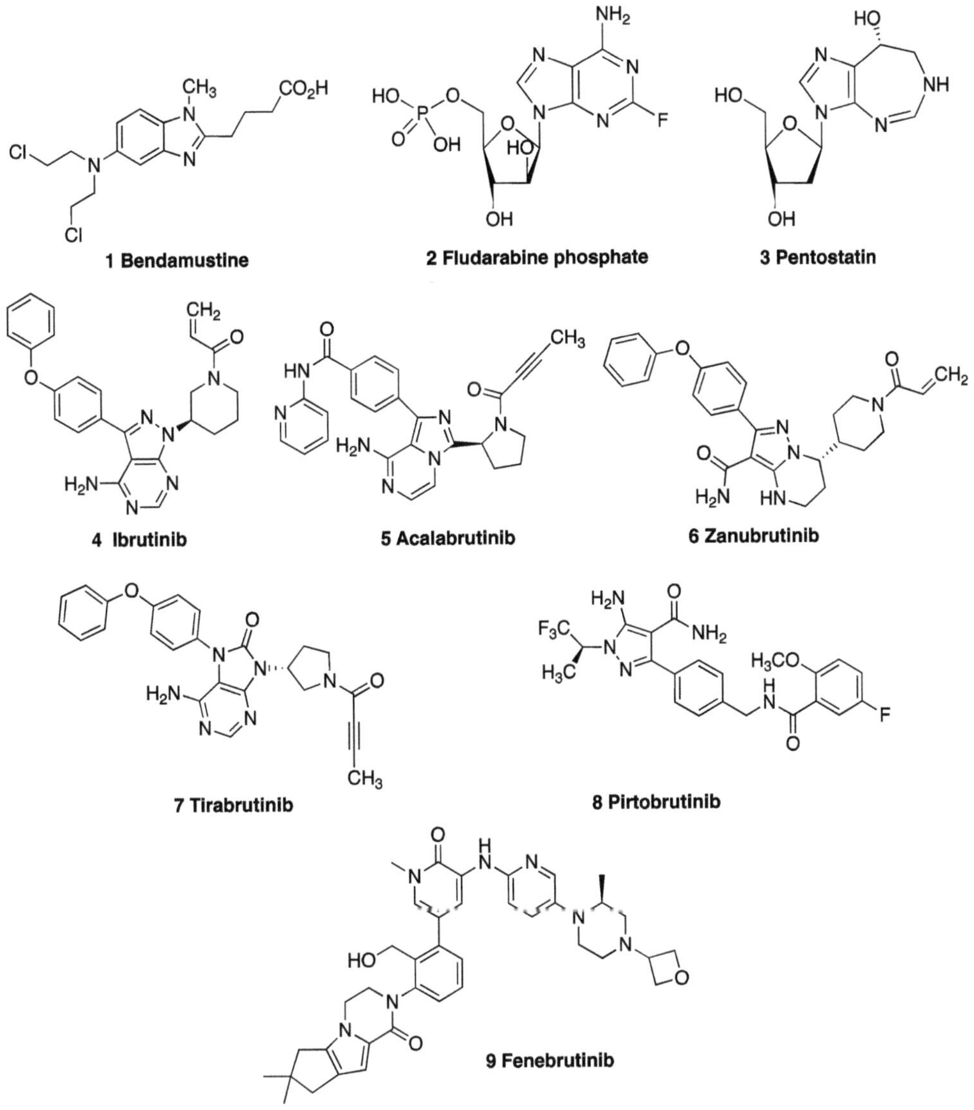

Figure 1. Drugs used in the treatment of CLL: alkylating agents **1** bendamustine, **2** fludarabine phosphate and **3** pentostatin; covalent BTK inhibitors **4** ibrutinib, **5** acalabrutinib, **6** zanubrutinib and **7** tirabrutinib; and non-covalent BTK inhibitors **8** pirtobrutinib and **9** fenebrutinib.

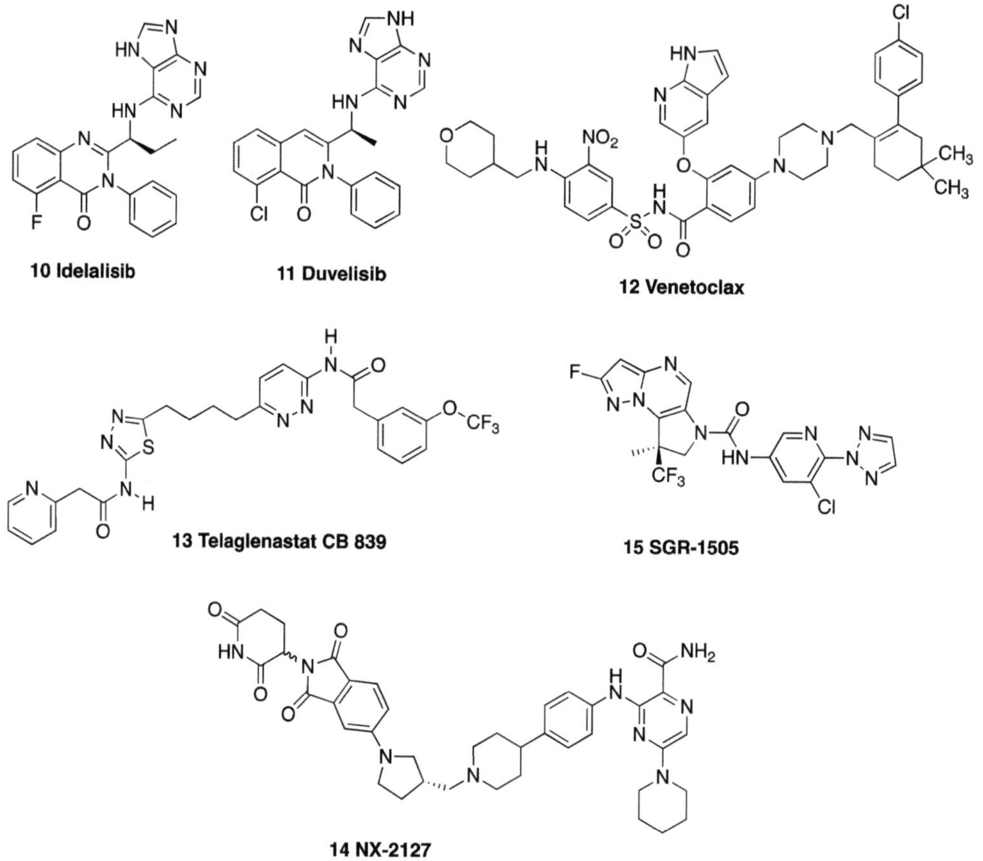

Figure 2. Drugs targeting CLL: PI3Kδ inhibitor idelalasilib **10**; PI3Kδ and PI3Kγ inhibitor duvelisib **11**; Bcl-2 inhibitor venetoclax **12**; glutaminase inhibitor telaglenastat CB 839 **13**; and dual BTK degrader NX-2127 **14** and MALT-1 inhibitor SGR-1505 **15**.

Previously, we identified a series of (*E*)-nitrostyrenes, e.g., **17a**, **17b** and (*E*)-9-(2-nitrovinyl)anthracenes **18a–e** (Figure 3), which potently reduced cell viability in the Burkitt's lymphoma (BL) cell lines MUTU-I and DG-75 [38]. In CLL cells associated with a poor patient prognosis, they demonstrated IC_{50} values of <10 μM [38–40] and were significantly more potent than fludarabine phosphate [39]. Anti-cancer and apoptotic effects were reported for nitrostyrenes and nitrovinyl compounds in oral and colon cancers, together with the modulation of tumorigenesis in colon and breast cancers via reactive oxygen species (ROS) effects [41,42]. We also demonstrated the potent antiproliferative and pro-apoptotic effects of (*E*)-9-(2-nitrovinyl)-9,10-dihydro-9,10-ethanoanthracenes in BL cells [43], indicating that the nitrovinylanthracene pharmacophore may have potential for design of CLL targeting compounds. The tetracyclic ethanoanthracene scaffold is similar to the antidepressant maprotiline **16** (Figure 3), which demonstrated antiproliferative effects in B-cell malignancies [44]. Maleimide-based ethanoanthracenes have demonstrated selective cannabinoid receptor CB2 [45], antimicrobial [46] and anti-inflammatory activities [47], as well as the inhibition of S100P/RAGE interaction in pancreatic cancer cells [48]. Glucocorticoid receptor modulation [49] and neuronal calcium channel and glycogen synthase kinase-3 beta (GSK-3 beta) modulation targeting Alzheimer's disease [50] have also been reported for ethanoanthracene derivatives.

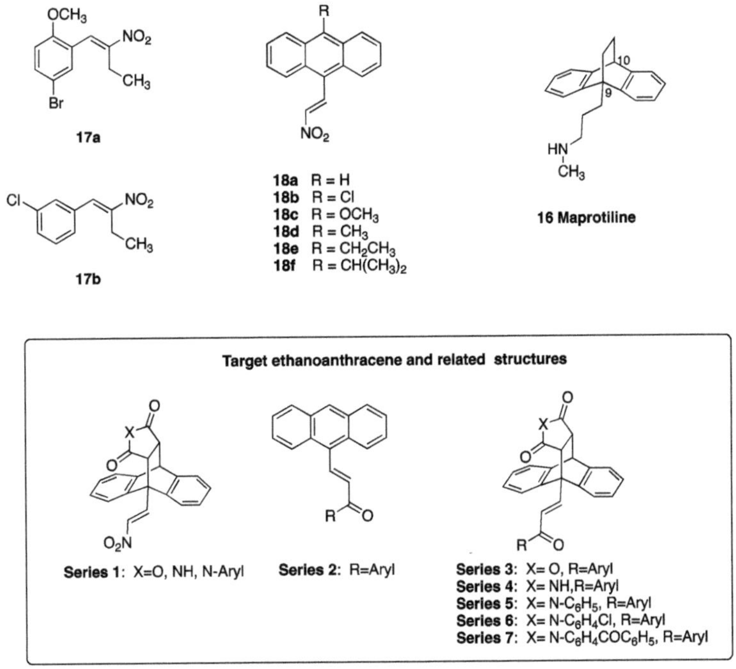

Figure 3. Nitrostyrenes **17a**, **17b**, nitrovinylanthracenes **18a–e** and maprotiline **16**; target ethanoanthracene structures, Series 1–7.

In the present work, we report the design and synthesis of 110 novel compounds that are based on the 9,10-dihydro-9,10-ethanoanthracene and related scaffolds and that are grouped into seven series. Our objective was the identification of potent and selective compounds to target CLL with pro-apoptotic effects. The cycloaddition reaction of anthracene dienes with diverse dienophiles affords structurally varied ethanoanthracene products [43,45–52], while stereoselective Diels–Alder reactions of chiral C-9-substituted anthracenes with various maleimides have been reported [53]. Anthracenes suitably substituted with a nitrovinyl or α,β-unsaturated ketone group at C-9 were reacted with the dienophiles maleic anhydride, maleimide, *N*-phenylmaleimides, acrylonitrile and acetylenedicarboxylate to afford products containing a modified ethanoanthracene-bridgehead structure (Figure 3, target structures, Series 1–7). The ethanoanthracene compounds were biologically evaluated in the CLL cell lines HG-3 (containing unmutated immunoglobulin heavy-chain variable-region gene (IGHV), representing poor-prognosis CLL) and PGA-1 (containing mutated IGVH, representing good-prognosis disease).

2. Results and Discussion

2.1. Chemistry

2.1.1. Series 1: (E)-9-(2-Nitrovinyl)-9,10-dihydro-9,10-[3,4]epipyrroloanthracene-12, 14-diones (**20a–20h**)

Based on our previous research with (*E*)-9-(2-nitrovinyl)-9,10-dihydro-9,10-[3,4]epipyrroloanthracene-12,14-diones in BL [43], a focused series of the most potent compounds was identified for further evaluation and testing in related B-cell malignancy CLL, (Scheme 1). (*E*)-9-(2-Nitrovinyl)anthracenes **18a** and **18b** were prepared via a piperidine-catalyzed Henry–Knoevenagel condensation reaction of 9-anthraldehyde with nitromethane [43]. The dienophiles acrylonitrile, maleimide, maleic anhydride, *N*-phenylmaleimides **19a–c**, dimaleimide **19d** and dimethyl acetylenedicarboxylate were chosen for Diels–Alder cycloaddition. The maleimide dienophiles **19b**, **19c** and **19d** were prepared from 2,5-furandione via amic acid intermediates [43]. The cycloaddition of dienophiles **19a–c** with (*E*)-9-(2-nitrovinyl)anthracene **18a** afforded the adducts **20a–20c**, whereas that with **18b** provided compound **20g** (30–51%), Scheme 1. Adducts **20d**, **20e** and **20h** were obtained upon the reaction of maleimide, acrylonitrile and maleic anhydride, respectively, with diene **18a**, while the dimer **20f** was obtained upon the cycloaddition of anthracene **18a** with the dimaleimide **19d**, as previously reported [43] (Scheme 1). These lead compounds **20a–20h** from BL studies bear a structural similarity to the tetracyclic antidepressant maprotiline **16**, containing the three main characteristics of an electrophilic functional group chain at position 9 of the anthracene-based core, the polyaromatic core itself and the two carbon-aliphatic bridge structure-linking positions 9 and 10 of the center ring.

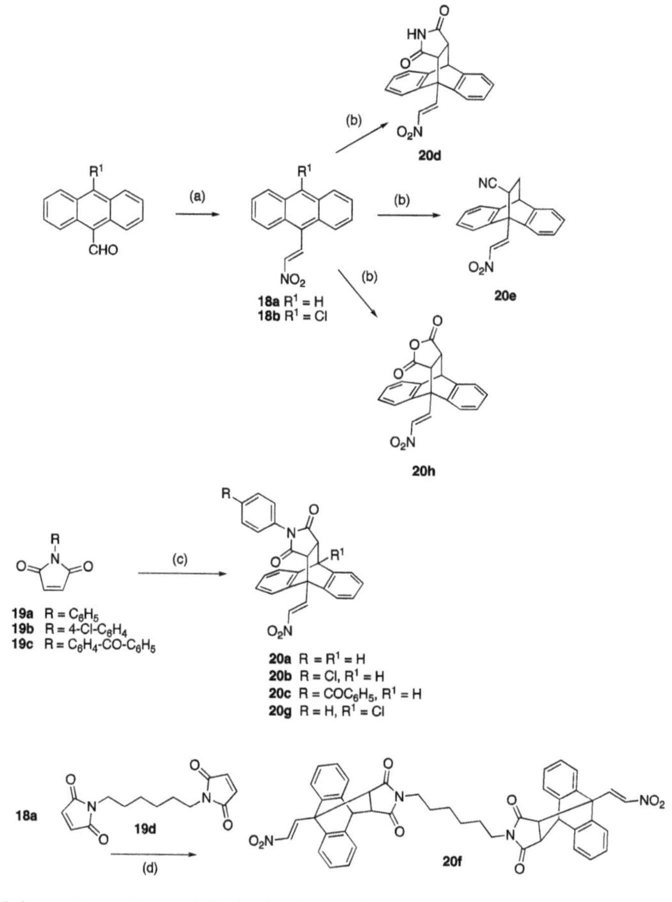

Scheme 1. Synthesis of Series 1 ethanoanthracenes **20a–g** reagents and conditions: (**a**) piperidine acetate, excess nitromethane (CH_3NO_2), 90 °C, N_2, 1.5 h (71–99%); (**b**) dienophile (maleic anhydride) for **20h**, maleimide for **20d**, NCHC=CH_2 for **20e**, toluene, 90 °C, 48 h (30–80%); (**c**) dienophile **19a** for *N*-arylmaleimides **20a** and **20g**, **19b** for **20g**, **19c** for **20c**, toluene, 90 °C, 48 h, (15–51%); and (**d**) toluene, 90 °C, 48 h (10%).

2.1.2. Series 2: (E)-3-(Anthracen-9-yl)-1-phenylprop-2-en-1-ones (**21a–q**)

The α,β-unsaturated carbonyl bioisosteric alternatives to the nitrovinyl group (located at C9 of the anthracene core) of our initial lead compounds (Series 1) were next investigated, principally due to potential nitrovinyl group genotoxicity and mutagenicity arising from biological metabolites of the nitro group [54]. Carcinogenicity, hepatotoxicity, mutagenicity and bone marrow suppression have been the major adverse toxicity issues associated with nitro-containing drugs [55]. The α,β-unsaturated ketone (chalcone) offers an alternative to the nitrovinyl group as an electrophilic system if required as a potential target for the enzymatic action of the compounds [56]. Covalent interactions of chalcones with biological targets can occur via the Michael-acceptor activity of the α,β-unsaturated carbonyl system, or they can also be mediated through radical-scavenging or reduction. [57] It is of interest that electrophilic compounds such as α,β-unsaturated ketones with CLL-selective cytotoxicity can interact with redox-sensitive proteins in primary CLL cells, resulting in the induction of nuclear factor erythroid 2–related factor 2 (Nrf2) signaling [58].

In the present work, a series of (*E*)-3-(anthracen-9-yl)-1-phenylprop-2-en-1-ones (Series 2) **21a–21q** was prepared using a Claisen–Schmidt condensation reaction between 9-anthraldehyde and a panel of substituted aromatic and heterocyclic acetophenone derivatives in the presence of sodium hydroxide (Scheme 2, Table 1, Step a, yields 38–88%). The chalcones contained the diverse substitution of the aryl ring in both positions and atom types, including halogens, alkyl, methoxy groups and an unsubstituted benzoyl ring. The 2-naphthyl, 2 and 4-pyridyl rings, together with 2-furyl and 2-thienyl heterocycles, were also investigated (Scheme 2, Table 1). In the ^{1}H nmR spectrum of **21a**, H15 is identified as a doublet at δ 8.80 (*J* = 15.8 Hz), and it confirms the *E* alkene configuration [59]. The ^{13}C nmR spectrum of compound **21a** identified the alkene C-15 and C-16 signals at 142.5 and 130.4 ppm, respectively (see Supplementary Information). The *E* configuration of alkene **21k** was confirmed via X-ray crystallography (Table 2) [60,61].

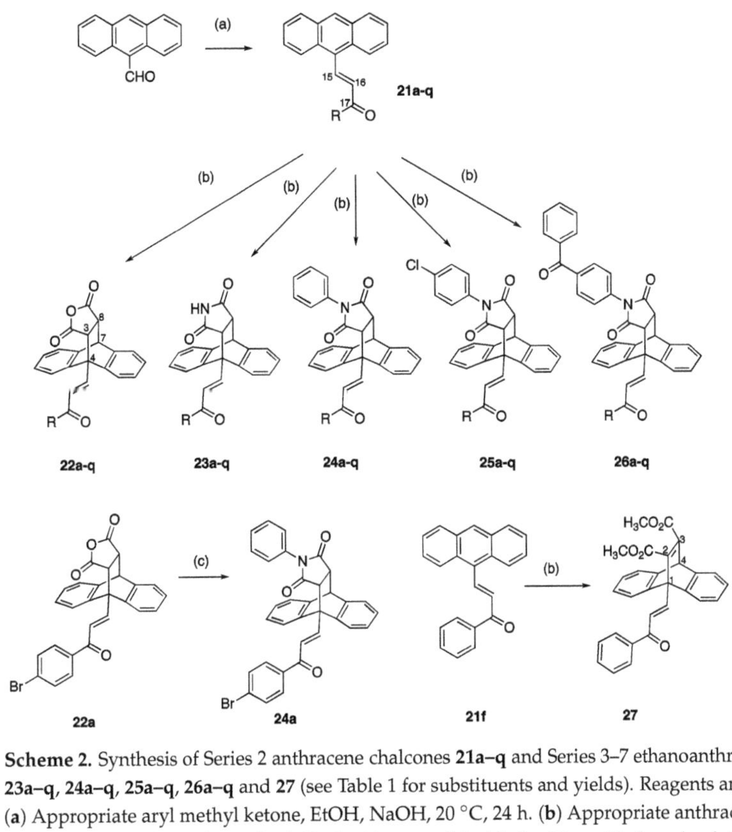

Scheme 2. Synthesis of Series 2 anthracene chalcones **21a–q** and Series 3–7 ethanoanthracenes **22a–q**, **23a–q**, **24a–q**, **25a–q**, **26a–q** and **27** (see Table 1 for substituents and yields). Reagents and conditions: (**a**) Appropriate aryl methyl ketone, EtOH, NaOH, 20 °C, 24 h. (**b**) Appropriate anthracene chalcone **21a–q**, dieneophile (maleic anhydride for **22a–q**, maleimide for **23a–q**, *N*-phenylmaleimide for **24a–q**, *N*-(4-chlorophenyl)maleimide for **25a–q**, *N*-(4-benzoylphenyl)maleimide for **26a–q**, dimethyl acetylenedicarboxylate for **27**), toluene, 90 °C, 48 h. (**c**) Aniline, acetic acid 120 °C, 2–3 h (72%).

Table 1. Anthracene chalcones **21a–q** and ethanoanthracenes **22a–q**, **23a–q**, **24a–q**, **25a–q** and **26a–q**.

R	Series 2		Series 3		Series 4		Series 5		Series 6		Series 7	
	Cpd No.	Yield %	Cpd No.	Yield %	Cpd No.	Yield %	Cpd No.	Yield %	Cpd No.	Yield %	Cpd No.	Yield %
4-Br-C_6H_4	**21a**	65	**22a**	86	**23a**	91	**24a**	84	**25a**	95	**26a**	71
4-NO_2-C_6H_4	**21b**	78	**22b**	77	**23b**	87	**24b**	86	**25b**	82	**26b**	75
4-C_2H_5-C_6H_4	**21c**	61	**22c**	30	**23c**	82	**24c**	90	**25c**	3	**26c**	76
2,4-$(Cl)_2$-C_6H_3	**21d**	75	**22d**	76	**23d**	82	**24d**	75	**25d**	87	**26d**	63
3,4-$(OCH_3)_2$-C_6H_3	**21e**	85	**22e**	83	**23e**	83	**24e**	93	**25e**	4	**26e**	72
C_6H_5	**21f**	44	**22f**	91	**23f**	61	**24f**	42	**25f**	48	**26f**	49
4-I- C_6H_4	**21g**	70	**22g**	81	**23g**	93	**24g**	20	**25g**	92	**26g**	85
4-F-C_6H_4	**21h**	88	**22h**	79	**23h**	100	**24h**	88	**25h**	88	**26h**	77
4-OCH_3-C_6H_4	**21i**	58	**22i**	30	**23i**	70	**24i**	56	**25i**	63	**26i**	50
4-CH_3-C_6H_4	**21j**	61	**22j**	73	**23j**	45	**24j**	51	**25j**	60	**26j**	59
3,4,5-$(OCH_3)_3$-C_6H_2	**21k**	86	**22k**	78	**23k**	53	**24k**	62	**25j**	35	**26k**	44
4-Cl-C_6H_4	**21l**	65	**22l**	89	**23l**	65	**24l**	20	**25l**	39	**26l**	30
4-(4-Pyridyl)	**21m**	48	**22m**	99	**23m**	100	**24m**	82	**25m**	92	**26m**	62
4-(2-Pyridyl)	**21n**	58	**22n**	81	**23n**	72	**24n**	91	**25n**	57	**26n**	75
2-Furyl	**21o**	67	**22o**	81	**23o**	81	**24o**	87	**25o**	46	**26o**	60
2-Naphthyl	**21p**	38	**22p**	86	**23p**	86	**24p**	65	**25p**	88	**26p**	55
2-Thienyl	**21q**	86	**22q**	31	**23q**	80	**24q**	81	**25q**	70	**26q**	51

Table 2. X-ray crystal structure of compounds **21k**, **23f**, **23h**, **24a**, **24g**, **25f** and **27** [a,b].

Compound	X-Ray Structure	Compound	X-Ray Structure
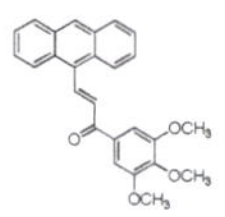 **21k**	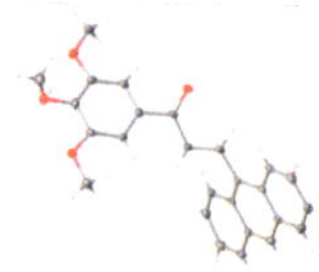	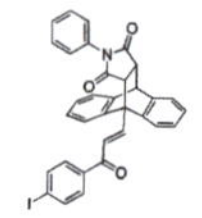 **24g**	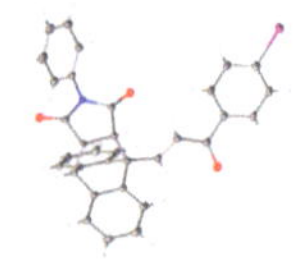
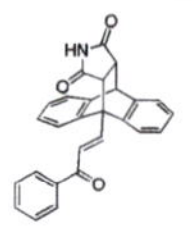 **23f**	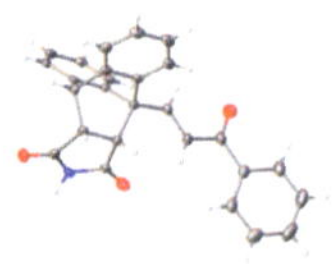	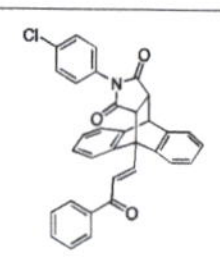 **25f**	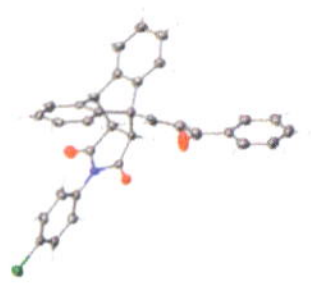
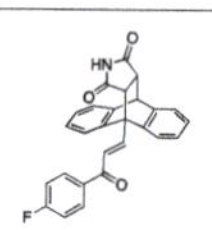 **23h**	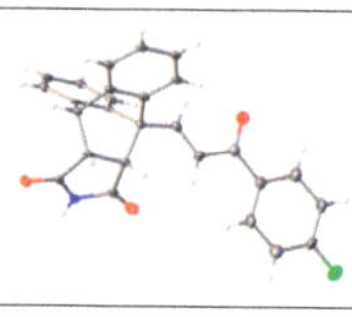	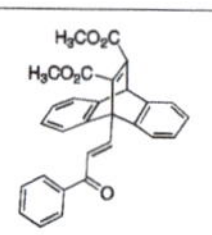 **27**	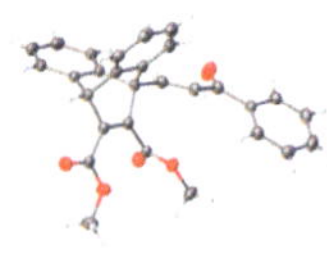
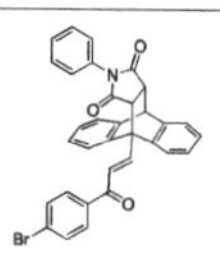 **24a**	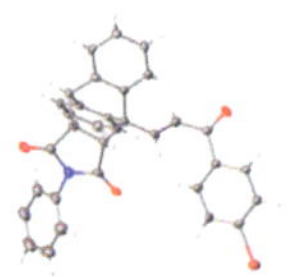		

[a] X-ray crystal structure of compounds **21k**, **23f**, **23h**, **24a**, **24g**, **25f** and **27** with the thermal ellipsoids set to 50% probability. [b] Crystallographic data deposited with the Cambridge Crystallographic Data Centre (CCDC) **21k** (Deposition Number 2341799), **23f** (Deposition Number 2341800), **23h** (Deposition Number 2341801), **24a** (Deposition Number 2341802), **24g** (Deposition Number 2341803), **25f** (Deposition Number 2341804) and **27** (Deposition Number 2341805).

2.1.3. Series 3: Maleic Anhydride-Substituted Ethanoanthracenes **22a–22q**

The preparation of compounds **22a–22q** was achieved using a Diels–Alder reaction, with the panel of substituted anthracene-chalcones **21a–21q** acting as a diene system and maleic anhydride acting as the dienophile [51] in yields of 30–99% (Scheme 2, step (b), Table 1). The products contained three distinct structural features: a dihydrofuran-2,5-dione bridgehead, an anthracene-derived central core and an α,β-unsaturated ketone with a substituted aryl ring. The IR spectrum of **22e** confirmed absorbances at 1669 cm^{-1} (α,β-unsaturated ketone) and 1774 cm^{-1} (maleic anhydride carbonyl). The ^{1}H nmR spectrum of **22e** confirmed the anthracene core protons at δ 3.63 (dd, *J* = 9.5, 3.1 Hz), δ 3.81 (d, *J* = 9.5 Hz) and δ 4.85 (d, *J* = 3.3 Hz) assigned to H8, H3 and H7, respectively. The alkene H22 is identified at δ 7.87 (d, *J* = 16.2 Hz). In the ^{13}C nmR spectrum of **22e**, C8, C3, C7 and quaternary C4 were identified at 54.1, 54.3, 49.4 and 56.33 ppm, respectively. The maleic anhydride C2, C9 and ketone C23 carbonyls were assigned at 175.4, 176.2 and 192.8 ppm. The alkenes C21 (136.8 ppm) and C22 (144.5 ppm) and ethanoanthracene bridges C3 (54.3, ppm), C7 (49.4 ppm) and C8 (54.1 ppm) were identified from the DEPT 90, HSQC and HMBC spectra (see Supplementary Information).

2.1.4. Series 4: Maleimide-Substituted Ethanoanthracenes (**23a–23q**)

The ethanoanthracene compound panel **23a–23q** was prepared using the Diels–Alder reaction of the substituted chalcones **21a–21q** with maleimide (Scheme 2, Table 1). In the ^{1}H nmR spectrum of **23a**, the signals at δ 3.32 (m), 3.82 (d, *J* = 8.7 Hz) and 4.81 (*J* = 3.3 Hz) were assigned to the aliphatic H8, H3 and H7, respectively. The maleimide NH was identified at δ 10.87. The ^{13}C nmR spectrum of **23a** shows C8 of the ethanoanthracene bridge at 49.3 ppm, while the C3 was observed at 49.1 ppm. The signals at 45.0 and 51.7 ppm represent C7 and C4 of the anthracene core. The DEPT 90 spectrum was used to confirm the assignment of the alkene carbons C21 (143.1 ppm), C22 (131.5 ppm) and ethanoanthracene aliphatic bridge carbons C3 (49.1 ppm), C8 (49.3 ppm) and C7 (45 ppm) (Supplementary Information). The *E* configuration of the ethanoanthracene chalcone alkene of products **23f** and **23h** was confirmed via X-ray crystallography (Table 2) [61–63].

2.1.5. Series 5: *N*-Phenylmaleimide-Substituted Ethanoanthracenes (**24a–24q**)

Compounds **24a–24q** were obtained using the Diels–Alder reaction of the anthracene chalcones **21a–21q** with *N*-phenylmaleimide (Table 1, Scheme 2). The *N*-phenylmaleimide adduct **24a** was alternatively prepared by reacting the corresponding maleic anhydride chalcone adduct **22f** with aniline in acid-catalyzed dehydration and the subsequent cyclization of the amic acid intermediate (Scheme 2, step c). This method resulted in a lower yield (72%) compared to the direct Diels–Alder cycloaddition of the chalcone **21f** and *N*-phenylmaleimide (84%). In the ^{1}H nmR spectrum of **24i**, the characteristic aliphatic protons H8, H3 and H7 were identified at δ 3.55 (dd, *J* = 8.3, 3.3 Hz), 3.99 (d, *J* = 8.3 Hz) and 4.94 (d, *J* = 3.7 Hz), respectively, while the alkene protons H28 and H27 were observed at δ 7.81 and 7.89 (*J* = 16.3 Hz), confirming the *E* configuration. In the ^{13}C nmR spectrum of **24i**, the signals at 45.4, 48.2 and 48.3 ppm were assigned to C7, C3 and C8, respectively, of the aliphatic ethanoanthracene bridge using the HMBC and HSQC spectra, while C4 was identified at 52.04 ppm.

The stereo- and regioselectivity of Diels–Alder reaction and retention of the alkene *E* configuration was obtained from the X-ray crystallography of **24a** and **24g** (Table 2) [61–63]. Interestingly, the reaction of *N*-cyclohexylmaleimide and 9-hydroxymethylanthracene in the confined space inside carbon nanotubes resulted in the 1,4-exo Diels–Alder cycloaddition product instead of the 9,10-endo adduct obtained in conventional conditions [64].

2.1.6. Series 6: *N*-(4-Chlorophenyl)maleimide-Substituted Ethanoanthracenes (**25a–25q**)

The ethanoanthracenes **25a–25q** were prepared using the Diels–Alder reaction as above, with the substituted chalcones **21a–21q** acting as the diene and *N*-(4-chlorophenyl) maleimide **19b** as the dienophile (yields 3–95%) (Scheme 2, Table 1). The ^{1}H nmR spectrum of **25k** shows H-8, H-3 and H-7 at δ 3.53 (dd, J = 8.3, 3.3 Hz), δ 4.00 (d, J = 8.3 Hz) and δ 4.94 (J = 3.3 Hz), while the alkene H-26 and H-27 were identified as doublets at δ 7.80 and δ 7.88 (J = 16.2 Hz), respectively. In the ^{13}C nmR spectrum of **25k**, C-8, C-3, C-7 and C-4 of the ethanoanthracene bridge were identified at 45.4, 48.3, 48.6 and 52.11 ppm, respectively. (See Supplementary Information for HSQC, DEPT 90 and HMBC spectra.) The X-ray structure of **25f** is displayed in Table 2.

2.1.7. Series 7: *N*-(4-Benzoylphenyl)maleimide-Substituted Ethanoanthracenes (**26a–26q**) and Adduct 27

The *N*-(4-benzoylphenyl)maleimide adducts **26a–26q** were obtained using the Diels–Alder cycloaddition reaction of the anthracene chalcones **21a–21q** with the *N*-(4-benzoylphenyl) maleimide **19c** (Scheme 2, Step b, Table 1). In the ^{1}H nmR spectrum of **26l**, the ethanoanthracene carbon bridge protons H8, H3 and H7 were identified at δ 3.59 (J = 3.3, 8.3 Hz), 4.04 (d, J = 8.3 Hz) and δ 4.97 (d, J = 2.9 Hz). In the ^{13}C nmR spectrum of **26l**, the ethanoanthracene bridge carbons C3, C8 and C7 and alkenes C27 and C28 were identified at 48.5, 45.4, 52.2, 142.8 and 131.8 ppm, respectively (see Supplementary Information).

The Diels–Alder reaction of the chalcone **21f** with dimethyl acetylenedicarboxylate afforded the adduct **27**, containing an alkene on the central carbon bridge (Scheme 2, Step b) [65]. H-4 (center ring of the polycyclic core) was identified at δ 5.80 ppm, while H-10 and H-13 of the polycyclic core, together with alkene H-18, were identified at δ 7.55 as a multiplet; the doublet δ 7.73, J = 16.1 Hz, corresponded to alkene H-17. In the ^{13}C nmR spectrum, the signals at 49.9, 52.9, 53.2 and 57.8 ppm were assigned to C1, C4, C2 and C3 (bridgehead alkene). The structure of the adduct **27** was confirmed via X-ray crystallography (Table 2) [61,62].

2.2. *X-ray Crystallography Study of Ethanoanthracenes*

The X-ray-crystallographic analysis of a series of the novel ethanoanthracene adducts was carried out, and it confirmed the regioisomer obtained from the Diels–Alder reactions (Table 2). Initially, the *E* configuration of the alkene **21k** was observed via X-ray crystallography (Table 2). The C15–C16 bond length was consistent with an alkene (1.336 Å), and the C17–O18 bond length was consistent with a carbonyl group (1.228 Å) [60,61]. The trigonal–planar nature of the alkene bond was also evident in bond angles of 120.1° and 124.6° for C15-C16-C17 and C1-C15-C16, respectively (Supplementary Information, Table S1 and Figure S3 for crystal data and structure refinement for **21k**).

The retention of the *E* configuration for the alkene double bond of the maleimide-substituted ethanoanthracenes products **23f** and **23h** was confirmed via X-ray crystallography with a C6–C7 alkene bond length 1.321 Å and a O3–C8 carbonyl bond length of 1.219 Å [61,62] (Table 2). Bond angles of 128.06° and 120.82° for C5–C6–C7 and C6–C7–C8, respectively, indicated the trigonal–planar configuration of the alkene. The core dihydroethanoanthracene moiety was rigid, with the flat succinimide ring fused to the ethano-bridge; the presence of the imide was supported by the carbonyl bond lengths of 1.205 Å for O1–C1 and O2–C2. In addition, the N1–C1 and N1–C2 bond lengths of 1.375 Å and 1.380 Å, respectively, confirm the C-N imide bond [63]. The X-ray crystal showed the presence of a racemic mixture, with the two stereocenters of the ethanoanthracene carbon bridge in the (*R*,*R*) and (*S*,*S*) stereochemical configuration (Supplementary Information, Table S1, and Figures S4, S7 and S8).

Additional evidence to confirm the stereo- and regioselectivity of the Diels–Alder reaction and retention of the alkene *E* configuration was obtained from the X-ray crystallography of the *N*-phenylmaleimide-substituted ethanoanthracenes **24a** and **24g** (Table 2). For compound **24a**, the C10–C11 (1.335 Å) and O9–C8 (1.227 Å) bond lengths were consistent with alkene and carbonyl functional groups. Bond angles of 123.9° and 125.6° for C10–C11–C12 and C8–C10–C11, respectively supported the trigonal–planar configuration [61,62]. Bond lengths of 1.18 Å and 1.19 Å for O32–C31 and O29–C28 are typical of the maleimide carbonyl groups. Furthermore, the N30–C33 bond (1.36 Å) was within the expected range for conjugated C–N bonds [63]. (See Figure S5 and Table S1, Supplementary Information, for crystal data and structure refinement for **24a** and **24g.**)

The X-ray structure of the *N*-(4-chlorophenyl)maleimide-substituted ethanoanthracene **25f** is displayed in Table 2, which confirms the structural assignment, showing the core dihydroethanoanthracene moiety to be rigid, with similar conformations observed in the various X-ray structures discussed above. (See Table 2 and Table S1, Supplementary Information, for crystal data and structure refinement for compound **25f.**)

The structure of the adduct **27** obtained from the chalcone diene **21f** with dienophile dimethyl acetylenedicarboxylate was also confirmed via X-ray crystallography (Table 2). The bond length of C15–C16 was consistent with an alkene (1.3269 Å), and the bond length of 1.2263 Å was consistent with a carbonyl (O28–C27) [61,62]. The bond angles for C14–C25–C26 and C27–C26–C25 were observed as 126.47° and 120.01°, respectively, indicating the trigonal–planar structure. The ester carbonyl bond lengths of 1.2073 Å and 1.2021 Å were observed for O18–C17 and O22–C21, while the O23–C24 and O19–C20 bond lengths of 1.45 Å and 1.4464 Å are typical of the ester C–O bond. The distinctive ethenoanthracene alkene bridgehead was observed with a bond length of 1.34 Å (see Figure S7 and Table S1, Supplementary Information, for crystal data and structure refinement for compound **27**).

2.3. *Stability Study for Chalcones* **21a**, **21i**, *Anhydride* **22h**, *and Maleimides* **23a**, **23g**, **23n**, **24a**, **24h**, **26a** *and* **26n**

A preliminary stability study was carried out on selected chalcones, maleic anhydride and maleimide adducts at acidic, neutral and basic conditions (pH 4, 7.4 and 9) using HPLC to determine suitability for further preclinical investigations. Chalcones **21a** and **21i**, anhydride **22h**, and maleimides **23a**, **23g** and **24a** demonstrated poor stability over the pH range studied, with <50% recovery at 24 h (Figure 4 and Supplementary Information Table S2, Figure S1). At pH 4.0, the rate of degradation of 2-pyridyl-substituted ethanoanthracene **23n** was minimal up to 8 h, with a half-life ($t_{1/2}$) of 19 h, while at pH 7.4, degradation was slower; $t_{1/2}$ >24 h. The compound remained stable at pH 9.0, with an 11% decrease in compound concentration over 24 h and $t_{1/2}$> 24 h. Compound **23n** was most stable at pH 9.0 (89% drug remaining at 24 h) in comparison with 70% and 39% remaining at pH 7.4 and 4.0 after 24 h. A possible degradation route for the maleimide **23n** could be through the hydrolysis of the cyclic imide of the maleimide ring, leading to the resultant amic acid product [66,67]. Compound **23n** was the most stable of the compounds evaluated at pH 9, and it would be suitable for further preclinical investigation. The *N*-phenylmaleimide adduct **24h** demonstrated notable stability at pH 7.4 and pH 9, with 80% and 81%, respectively, remaining at 24 h, and $t_{1/2}$ >24 h. At pH 4, it was less stable with 25% remaining ($t_{1/2}$ = 14.8 h). The *N*-benzoylmaleimide **26a** demonstrated good stability with 64%, 61% and 66% remaining after 24 h in pH conditions 4, 7.4 and 9, respectively, and $t_{1/2}$> 24 h, while **26n** was less stable with 12% remaining at pH 4 (24 h).

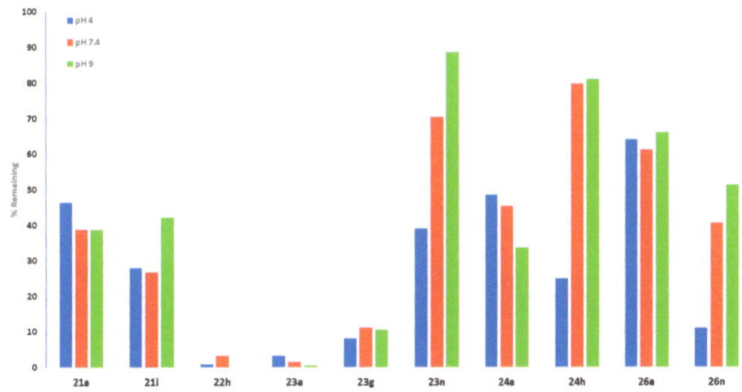

Figure 4. Stability study for compounds **21a**, **21i**, **22h**, **23a**, **23g**, **23n**, **24a**, **24h**, **26a** and **26n** at pH 4.0, pH 7.5 and pH 9.0 over 24 h.

All samples were analyzed using acetonitrile–water (80:20%, 70:30%, 60:40% isocratic) as the mobile phase over 10 min and a flow rate of 1 mL/min. Stock solutions were prepared by dissolving 5 mg of compounds in a mobile phase ([(acetonitrile (80%): water (20%)], 10 mL). Anhydride **22h** was not detected at its retention time of 5.2 min at pH 9.0. The data presented represent the results of single experiments.

2.4. Biological Results and Discussion

2.4.1. Evaluation of In Vitro Antiproliferative Activity of Chalcone and Nitrovinyl-Substituted Ethanoanthracenes in CLL Cell Lines HG-3 and PGA-1

The phenotypic cellular responses of the synthesized ethanoanthracenes were then investigated in the CLL cell lines HG-3 and PGA-1. The HG-3 cell line is representative of a poor patient prognosis; the cell line was established via EBV (Epstein–Barr virus) infection from an IGVH1–2 unmutated B1 lymphocyte-origin CLL patient clone [68]. The PGA-1 cell line, which is representative of a good patient prognosis, was established from leukemic B cells of a CLL patient with mutated IGVH1-2 [69]. The compounds were evaluated at 10-μM and 1-μM concentrations using an alamarBlue viability assay, and fludarabine was used as a clinically relevant comparative positive control [39]. Vehicle-treated cells [(DMSO (1% *v*/*v*)] were considered 100% viable, and the viabilities of each compound were calculated accordingly. The structure–activity relationships for the series were determined, and the most potent ethanoanthracene compounds were identified for further investigation. The present study was arranged according to structural type (Series 1–7).

2.4.2. Series 1: Nitrovinyl-Substituted Ethanoanthracenes (**20a–20h**)

The antiproliferative activities of the nitrovinyl-substituted ethanoanthracenes **20a–20h** in HG-3 and PGA-1 cell lines are shown in Figure 5, with reference compounds maprotiline and fludarabine. These were among the most potent compounds initially identified in our previous work in the related B-cell malignancy BL cell lines (e.g., compound **20a** IC_{50} = 0.21 μM in a chemosensitive MUTU-1 cell line and IC_{50} = 0.71 μM in a chemoresistant DG-75 cell line [43]). In HG-3 cells, at 10 μM treatment concentration, the most potent compounds were identified as the *N*-phenylmaleimide adducts **20g** (2% cell viability), **20b** (8%), and **20c** (11%); the dimer **20f**, *N*-phenylmaleimide **20a**, maleimide **20d** and acrylonitrile adduct **20e** show reduced cell viability (14–43%). In PGA-1 cells, the most effective compounds were identified as **20g** (1% viable cells), **20h** (2%), **20b** (3%), **20c** (4%) and **20e** (3%) at 10 μM concentration, while compounds **20a**, **20d** and **20f** were less potent (11%, 17% and 17%, respectively). The compounds were more effective in PGA-1 compared to HG-3 cells with the *p*-chloro- and benzophenone-based *N*-phenylmaleimides **20b** and **20c,** as well as the acrylonitrile adduct **20e**, initially identified as promising antiproliferative agents.

(A)

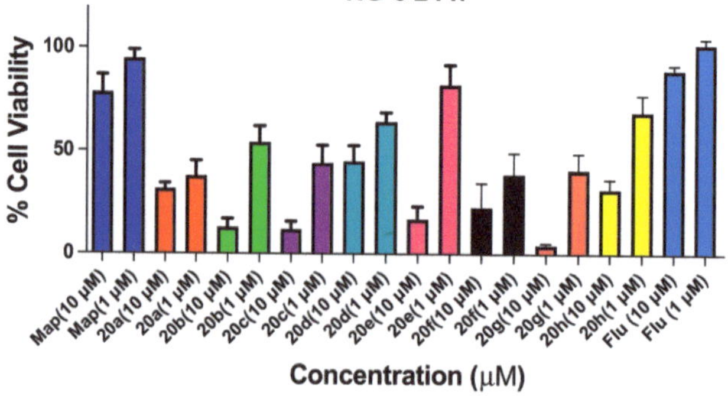

Figure 5. *Cont.*

(**B**)

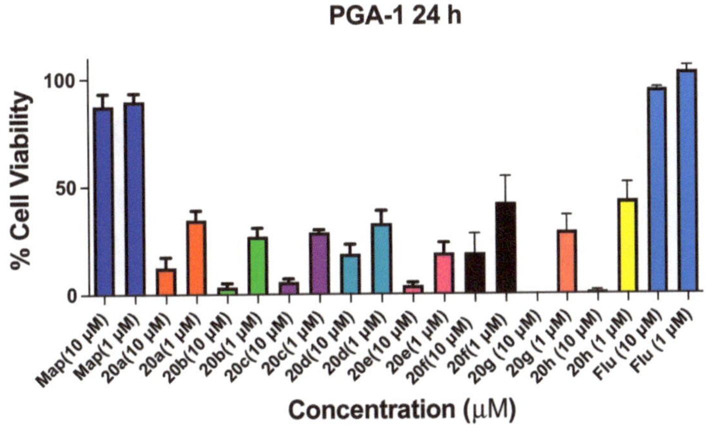

Figure 5. Cell viability data for ethanoanthracenes **20a–e**, **20g**, **20h** and **20f**. Cell viability data for (*E*)-9-(2-Nitrovinyl)-9,10,11,15-tetrahydro-9,10-[3,4]epipyrroloanthracene-12,14-diones **20a–e**, **20g** and **20h** and the related dimer **20f** in CLL: (**A**) HG-3 cells (1 and 10 μM) and (**B**) PGA-1 cells (1 and 10 μM). The cell proliferation of HG-3 and PGA-1 cells was determined with an alamarBlue assay. Compound concentrations of either 1 μM or 10 μM for 24 h were used to treat the cells (in triplicate) with control wells containing vehicle DMSO (1% *v*/*v*). Map = maprotiline; Flu = fludarabine. The mean value for three experiments is shown.

2.4.3. Series 2: (E)-3-(Anthracen-9-yl)-1-phenylprop-2-en-1-ones (21a-21q)

The biochemical activity of the anthracene-based chalcones **21a–21q** in the HG-3 and PGA-1 cell lines are shown in Figure 6A,B, with fludarabine and nitrovinylanthracene **18a** acting as internal standards. These structures contain diverse functional group substitutions on the α,β-unsaturated ketone, ranging from ring-activating groups (4-nitro aryl, **21b**) to ring-deactivating groups (3,4-dimethoxy aryl, **21e**) and polyaromatic systems (2-naphthyl, **21p**). From the initial HG-3 cell line screen, the most potent compounds observed were **21e** (3,4-dimethoxy), **21n** (2-pyridyl), **21k** (3,4,5-trimethoxy), **21d** (2,4-dichloro) and **21m** (4-pyridyl) with % viability in the range of 37–54% at a 10 μM concentration and a 50–61% viability range at 1 μM (Figure 6A). In the PGA-1 cell line, the most potent compounds at a 10-μM treatment concentration were **21m** (4-pyridyl, 43% cell viability), **21k** (3,4,5-trimethoxy, 43%), **21n** (2-pyridyl, 47%), **21i** (4-methoxy, 49%), **21d** (2,4-dichloro, 50%) and **21p** (2-naphthyl, 50%). The halogenated compounds were weakly active at 10 μM with **21h** (4-fluoro) and **21l** (4-chloro) (70% viability) (Figure 6B). The anthracene chalcones from Series 2 were not evaluated further due to their poor activity, although the antiproliferative activity of chalcones was previously reported in various leukemia cell lines [70].

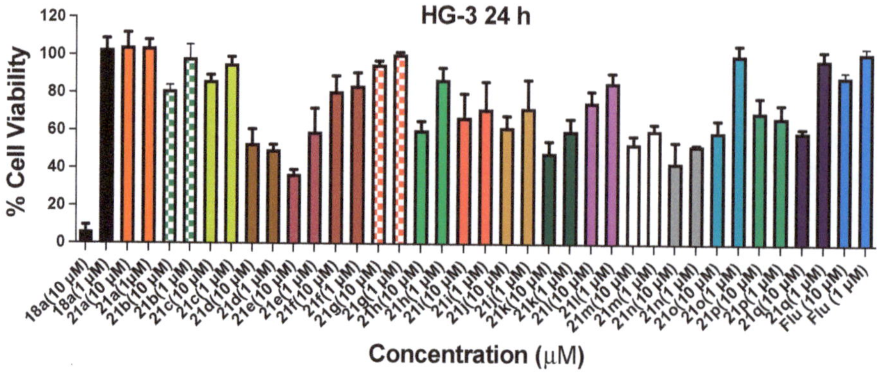

(A)

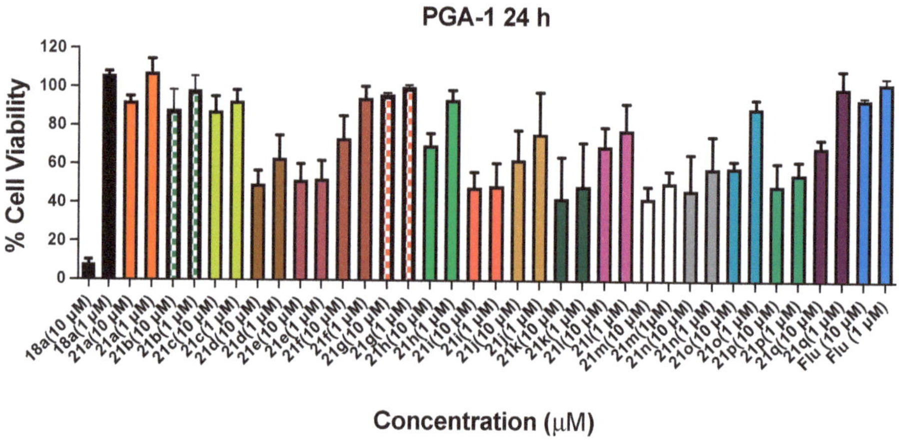

(B)

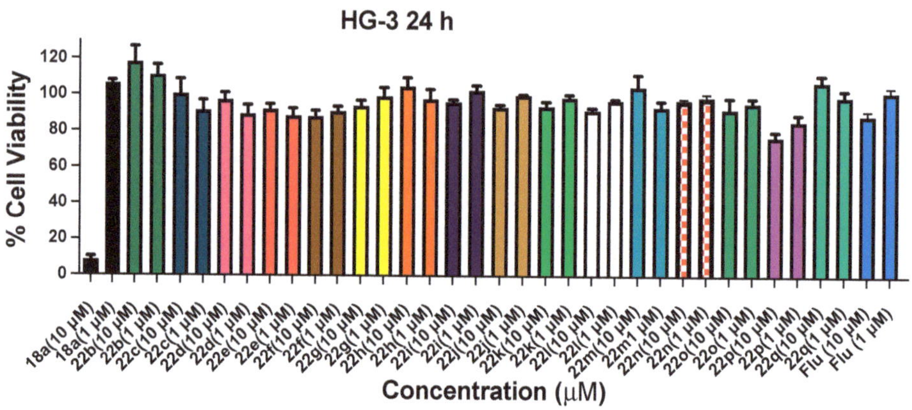

(C)

Figure 6. *Cont.*

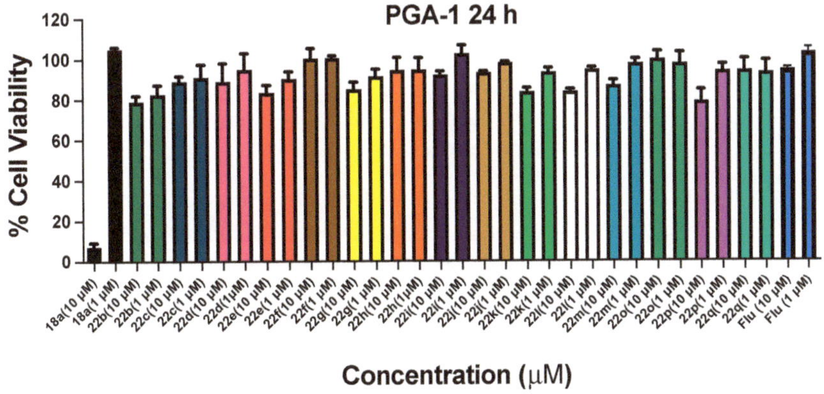

(D)

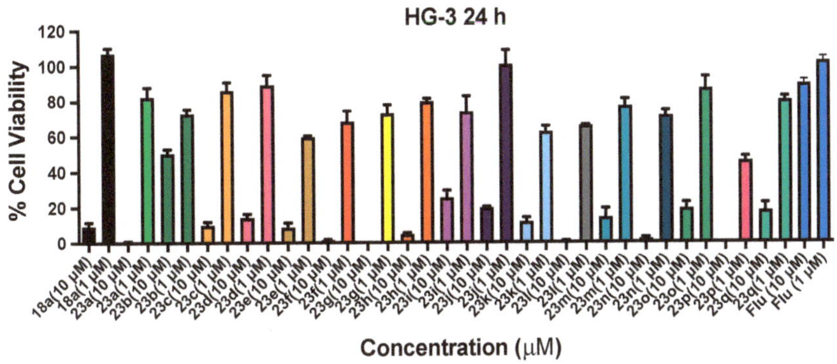

(E)

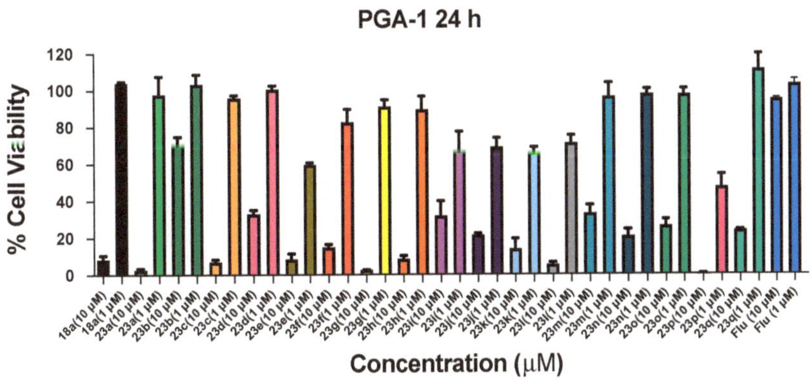

(F)

Figure 6. Cell viability data for chalcones **21a–q** and ethanoanthracenes **22a–22q** and **23a–23q** in CLL cell lines HG-3 and PGA-1. Cell viability data for chalcones **21a–q** (**A**,**B**), maleic anhydride ethanoanthracene adducts **22a–22q** (**C**,**D**) and maleimide ethanoanthracene adducts **23a–23q** (**E**,**F**) in CLL cell lines: the cell proliferation of HG-3 and PGA-1 cells was determined with an alamarBlue assay. Compound concentrations of either 1 µM or 10 µM for 24 h were used to treat the cells (in triplicate) with control wells containing vehicle DMSO (1% v/v). Flu = fludarabine. The mean value for three independent experiments is shown.

2.4.4. Series 3: Maleic Anhydride-Substituted Ethanoanthracenes (**22a–22q**)

The cell viability of compounds **22a–22q** in the HG-3 and PGA-1 cell lines is shown in Figure 6C,D. The majority of this compound series was weakly active or inactive across both treatment concentrations with the exception of **22p** (2-naphthyl) with cell viability of 77% in HG-3 and 79% in PGA-1 cells. The poor activity displayed for these compounds in series 3 did not merit further investigation.

2.4.5. Series 4: Maleimide-Substituted Ethanoanthracenes (**23a–23q**)

The antiproliferative activities of the maleimide-substituted ethanoanthracenes compounds **23a–23q** in the HG-3 cell line are shown in Figure 6E. In contrast to the maleic anhydride Series 3 compounds, the maleimide adducts demonstrated promising activity with the most potent of the panel identified as **23p** (2-naphthyl, 0% viable cells), **23g** (4-iodo 0%), **23a** (4-bromo, 0.5%), **23l** (4-chloro, 0.7%), **23f** (phenyl, 2%) and **23n** (2-pyridyl, 2%) when screened at a 10-μM concentration. Similar trends were observed at the 1-μM concentration, with the most potent compound identified as **23p** (2-naphthyl derivative, 46%). The most potent of the panel screened in the PGA-1 cell line at 10 μM were identified as **23p** (2-naphthyl, 0% viable cells), **23g** (4-iodo, 2%), **23a** (4-bromo, 3%), **23l** (4-chloro, 5%), **23c** (4-ethyl, 7%) and **23h** (4-fluoro, 9%); see Figure 6F. With an increasing size of the halogen substituent and increasing electronegativity on the *para* position of the acetophenone-derived aryl ring, an increase in compound potency was observed (I > Br > Cl > F), as also noted for the HG-3 cells. Compared to maleic anhydride (Series 3), the maleimide-derived adducts elicited greater biological activity at both concentrations and in both cell lines tested. These diverse sets of aryl substituents (alkyl, halogens and nitro) for heterocycles eliciting a similar degree of antiproliferative activity in the CLL cells suggests a moderate tolerance for substitution on the acetophenone-derived portion of the molecules in relation to their anticancer effect. The Series 4 compounds are generally more effective in the aggressive HG-3 cell type than in PGA-1 cells; e.g., compounds **23n** (2-pyridyl) and **23f** (phenyl) were notably less effective at 10 μM in PGA-1 cells (19% higher viability for **23n** and 13% higher viability for **23f** compared to HG-3 cells).

2.4.6. Series 5: *N*-Phenylmaleimide Substituted Ethanoanthracenes (**24a–24q**)

The cell viability results for the *N*-phenylmaleimide compounds **24a–24q** in the HG-3 cell line are shown in Figure 7A. All compounds were effective at a 10-μM concentration, with the most potent compounds identified in the screen as follows: **24o** (2-furyl, 3% cell viability), **24n** (2-pyridyl, 4%), **24b** (4-nitro, 10%), **24a** (4-bromo, 14%,), **24l** (4-chloro, 24%,) and **24q** (2-thienyl, 24%). The remaining compounds were also moderately effective, with cell viability in the range of 29–65%. At the lower treatment concentration (1 μM), **24n** (2-pyridyl) showed the greatest activity (48% viable cells remaining). In the PGA-1 cell line (Figure 7B), a similar trend was observed with the most potent compounds identified as **24o** (2-furyl, 9% viable cells), **24l** (4-chloro, 17%), **24b** (4-nitro, 22%), **24a** (4-bromo, 25%) and **24f** (phenyl, 26%).

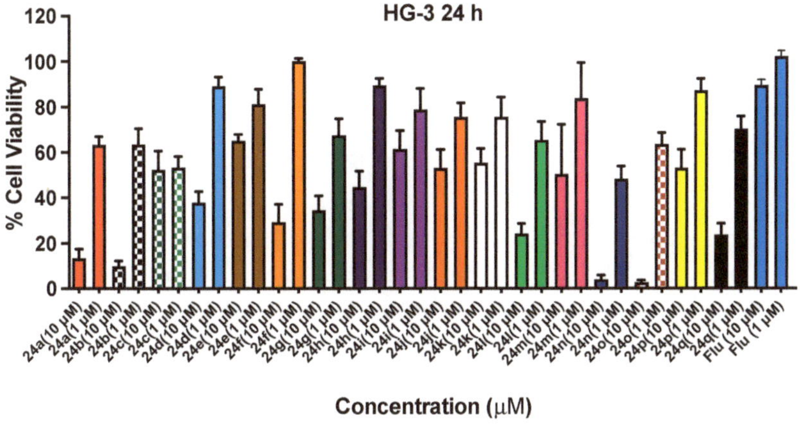

(A)

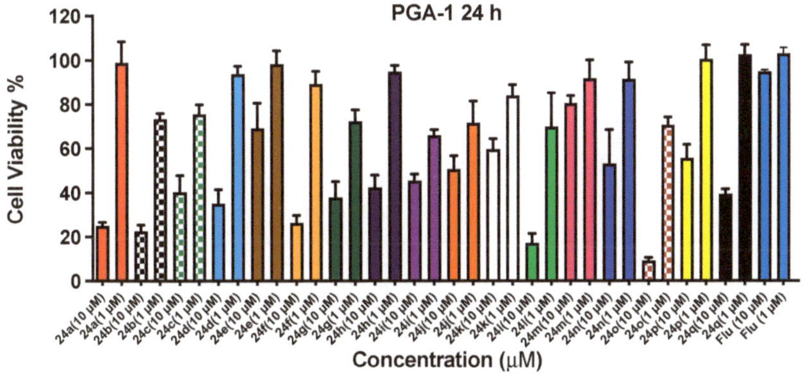

(B)

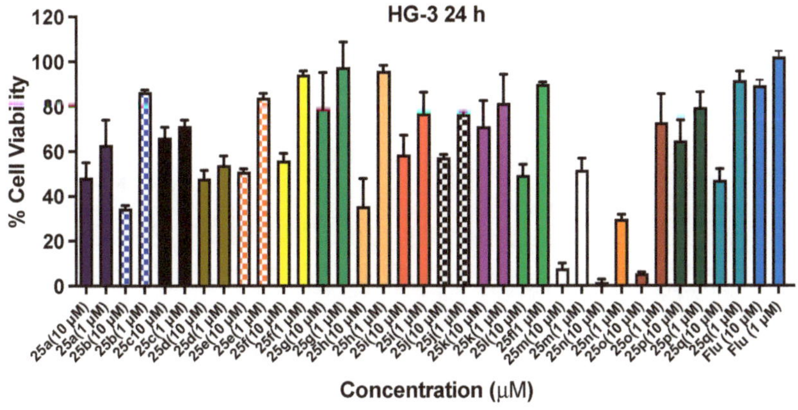

(C)

Figure 7. *Cont.*

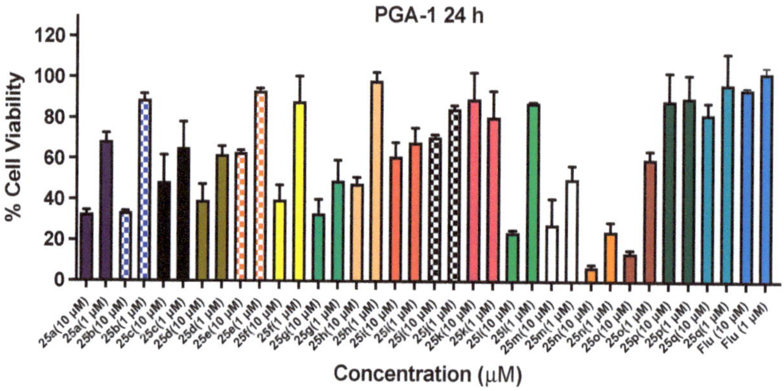

(D)

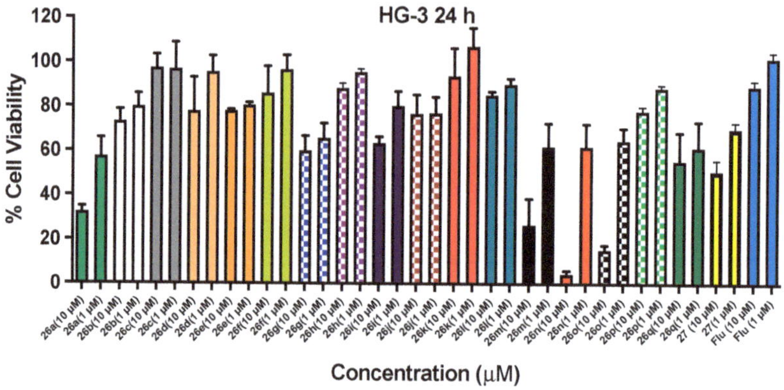

(E)

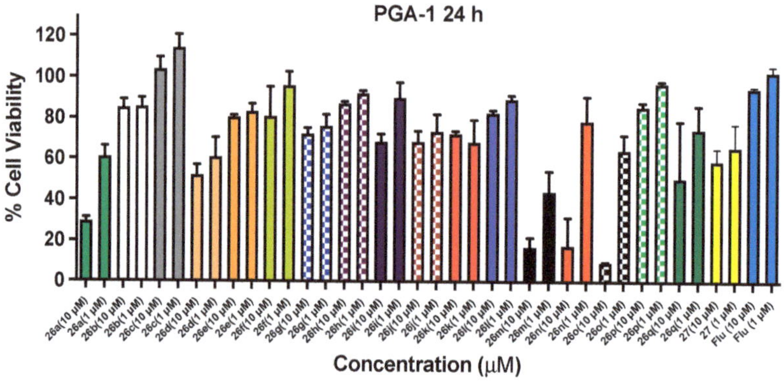

(F)

Figure 7. Cell viability data for ethanoanthracenes **24a–q**, **25a–q**, **26a–q** and **27** in CLL cell lines HG-3 and PGA-1. Cell viability data for *N*-phenylmaleimide-substituted ethanoanthracenes (**24a–q**, Panels **A**,**B**), *N*-(4-chlorophenyl)maleimide-substituted ethanoanthracenes (**25a–q**, Panels **C** and **D**) and *N*-(4-benzoylphenyl)maleimide-substituted ethanoanthracenes (**26a–q**, Panels **E**,**F**) and **27** (**E**,**F**) were determined in CLL cells HG-3 cells (1 and 10 μM) and PGA-1 cells (1 and 10 μM). The cell proliferation of HG-3 and PGA-1 cells was determined with an alamarBlue assay. Compound concentrations of either 1 μM or 10 μM for 24 h were used to treat the cells (in triplicate) with control wells containing vehicle DMSO (1% *v*/*v*). Flu = fludarabine. The mean value for three independent experiments is shown.

2.4.7. Series 6: *N*-(4-Chlorophenyl)maleimide-Substituted Ethanoanthracenes (**25a–25q**)

The antiproliferative activity of the *N*-(4-chlorophenyl)maleimide series **25a–25q** in the HG-3 cell line is shown in Figure 7C. The most potent compounds of the series screened were identified as **25n** (2-pyridyl, 2% viable cells), **25o** (2-furyl, 5%,), **25m** (4-pyridyl, 8%), **25h** (4-fluoro, 35%,), **25b** (4-nitro, 35%) and **25q** (2-thienyl, 47%). **25n** (2-pyridyl) was also identified as the most potent at 1 μM with 30% cell viability. The *N*-(4-chlorophenyl)maleimides identified with promising activity from the panel screened in the PGA-1 cell at a 10-μM concentration were **25n** (2-pyridyl, 7% viable cells), **25o** (2-furyl, 14%), **25l** (24%, 4-chloro), **25m** (4-pyridyl, 28%), **25a** (4-bromo, 33%), **25g** (4-iodo, 33%) and **25b** (4-nitro, 34%). At the lower treatment concentration (1 μM), **25n** was also observed to have the greatest activity (25% cell viability), (Figure 7D).

2.4.8. Series 7: *N*-(4-Benzoylphenyl)maleimide-Substituted Ethanoanthracenes (**26a–26q**)

The preliminary biochemical evaluation of the *N*-(4-benzoylphenyl)maleimide compound series **26a–26q** in the HG-3 cell line is shown in Figure 7E, with the most potent compounds of the panel screened at 10 μM identified as **26n** (2-pyridyl, 4% viable cells), **26o** (2-furyl, 15%), **26m** (4-pyridyl, 27%), **26q** (2-thienyl, 56%) and **26a** (4-bromo, 31%). At a 1-μM treatment concentration, **26a** (4-bromo) was the most effective (55% cell viability). In the PGA-1 cell line (Figure 7F), **26o** (2-furyl, 9% viable cells remaining), **26m** (4-pyridyl, 17%), **26n** (2-pyridyl, 18%) and **26a** (4-bromo, 29%) were also identified as potent lead compounds. At a lower treatment concentration of 1 μM, **26m** (4-pyridyl) demonstrated the greatest activity with 43% cell viability.

Following this preliminary evaluation of the panel of ethanoanthracene compounds described above, a number of key structural features were identified as critical for antiproliferative activity in CLL cells. The nitrovinyl and 1-phenyl-2-propen-1-one pharmacophores were demonstrated to be critical for cytotoxic effect. The maleimide-derived adducts (Series 4–7) were also found to produce a more potent effect than the corresponding furan-based products (Series 3). However, the nitrovinyl-based maleic anhydride adduct **20h** (Series 1) demonstrated significantly greater efficacy in both cell lines (31% viability in HG-3; 2% in PGA-1 at 10 μM) when compared with the most potent maleic anhydride chalcone adduct **22p** (>77% viability in both cell lines at 10 μM), thus confirming the requirement for the nitrovinyl pharmacophore for potency in the maleic anhydride-based series. For the maleimide-based compounds (Series 4–7), a comparison of the potent 2-pyridylchalcones compounds **23n** (Series 4), **24n** (Series 5), **25n** (Series 6) and **26n** (Series 7) confirmed the *N*-(4-benzoylphenyl)maleimide substitution as the most effective of the structural series evaluated.

In a more detailed analysis of the structure–activity relationships of the *N*-(4-benzoylphenyl) maleimide compounds (Series 7), it was observed that the replacement of the unsubstituted benzoyl aryl ring of **26f** with the heterocycles in **26m** (4-pyridyl), **26n** (2-pyridyl) and **26q** (2-thienyl) on the molecular scaffold resulted in an increase in potency for the compounds across both cell lines. The 4-pyridyl derivative **26m** showed a notable decrease in cell viability of 64% (to 17% viable cells) at a 1-μM treatment concentration in comparison to the unsubstituted **26f** in the PGA-1 cell line. In HG-3 cells, a 59% reduction in cell viability was observed at 10 μM for **26m** and a 64% decrease at the 1-μM concentration compared to **26f**. With **26n** (2-pyridyl derivative), a significant reduction in cell viability was noted in PGA-1 cells, showing decreases of 63% and 20% at 10 μM and 1 μM, respectively. Similarly, in HG-3 cells, **26n** showed substantial decreases in cell viability compared to **26f** (82% greater decrease in cell viability at 10 μM and 34% greater decrease at 1 μM). In addition, the presence of bromine at C-4 of the benzoyl aryl ring of **26a** resulted in a significant increase in biological activity when compared to **26f**, leading to 20% and 70% reduction in cell viability at 10 μM and 1 μM in PGA-1 cells. In HG-3 cells, a similar decrease in cell viability of 29% and 64% at 10- and 1-μM treatment concentrations, respectively, was observed. The acetylene adduct **27** showed moderate activity (50% viability in HG-3 and 58% in PGA-1 cells) at 10 μM. This result indicated that the maleimide structure present in the maleimide series 4 adducts resulted in superior efficacy for corresponding compound **23f** (0.5%, viability in HG-3, 13.9% viability in PGA-1 cells) when compared with the alkene-type bridge structure in compound **27**.

The more potent compounds identified from Series 1 and Series 4–7 were then progressed to further biological investigations.

2.4.9. In vitro IC_{50} Determination of Selected Nitrovinyl-Based Ethanoanthracene and Chalcone-Based Ethanoanthracene Derivatives in CLL Cells

The IC_{50} values of the most promising anti-proliferative nitrovinyl-based ethanoanthracene and chalcone-based ethanoanthracene compounds identified from preliminary biological screening at 10-μM and 1-μM concentrations were obtained for both HG-3 and PGA-1 cell lines. Fludarabine was used as a clinically relevant positive control; IC_{50} = 28.1 μM (HG-3) and 32.0 μM (PGA-1).

2.4.10. (E)-9-(2-Nitrovinyl)-Substituted Ethanoanthracenes (Series 1)

Compounds **20a–20f** were selected and tested across a concentration range of 5–0.01 μM and assessed for viability at 24 h using the alamarBlue assay. The compounds exerted a more potent effect than the fludarabine control (18–59-fold greater in HG-3 cells, 29–53-fold greater in PGA-1 cells) across both HG-3 and PGA-1 cell lines with IC_{50} ranges of 0.48–1.6 μM and 0.61–1.1 μM, respectively. In HG-3 cells, the most potent compounds identified were **20a** (IC_{50} 0.48 μM, *N*-phenylmaleimide derivative) and **20c** (IC_{50} 0.71 μM, *N*-(4-benzoylphenyl)maleimide; while in PGA-1 cells, the most potent compounds were **20a** (IC_{50} 0.61 μM) and **20b** (IC_{50} 0.66 μM, *N*-(4-chlorophenylmaleimide). Overall, the two most potent compounds across both cell lines were **20a** (mean IC_{50}: 0.55 μM) and **20b** (mean IC_{50}: 0.76 μM). In both cell lines, **20e** (cyano-derivative) had the highest IC_{50} values (Table 3) with the remaining compounds having sub-micromolar activity. This suggests that the general maleimide structure plays an important role in the biological activity observed, as we previously reported in BL [71]. In addition, the presence of an unsubstituted aromatic ring on the maleimide-derived functional group **20a** leads to better activity over the unsubstituted maleimide **20d**, *N*-(4-chloromaleimide **20b** and the *N*-(4-benzoylphenyl)maleimide **20c** adducts. The novel dimeric compound **20f,** which we identified with potent anti-proliferative activity in BL [43] [71], was also demonstrated to exert a more potent effect than the fludarabine control (165-fold greater in HG-3 cells, 91-fold greater in PGA-1 cells) across both HG-3 and PGA-1 cell lines with IC_{50} values of 0.17 μM and 0.35 μM, respectively, and a mean IC_{50} value of 0.53 μM.

Table 3. IC_{50} values of selected (*E*)-9-(2-nitrovinyl)ethanoanthracenes **20a–20f** in HG-3 and PGA-1 cell lines.

Compound Number	HG-3 IC_{50} (μM) [a]	PGA-1 IC_{50} (μM) [a]	Mean IC_{50} (μM) [a]	logP [b]
20a	0.48	0.61	0.55	3.29
20b	0.85	0.66	0.76	3.81
20c	0.71	0.88	0.8	3.54
20d	0.78	0.76	0.77	1.86
20e	1.6	1.1	1.4	2.98
20f	0.17	0.35	0.26	4.30
Fludarabine	28.1	32	30.1	−2.5 [c]

[a] The cell proliferation of HG-3 and PGA-1 cells was determined with an alamarBlue assay (seeding density: 1–5 × 10^4 cells/mL per well for 96-well plates). IC_{50} values are half-maximal inhibitory concentrations required to block the growth stimulation of HG3 and PGA-1 cells. The values represent the mean ± SEM (error values × 10^{-6}) for at least three experiments performed in triplicate at 24 h. Treatment at seven different concentrations (0.01–50 μM) was used for the determination of the IC_{50} values for each compound. [b] Predicted logP (consensus) property using Swiss ADME; [c] predicted property using Chemaxon.

2.4.11. In Vitro IC_{50} Determination of the Most Potent Chalcone-Based Ethanoanthracene Derivatives in HG-3 and PGA-1 Cells (Series 4–7)

The most potent compounds from the preliminary screening of Series **2–7** (compounds **23a**, **23c**, **23f–23i**, **23k**, **23l**, **23n**, **23p**, **24f**, **24l**, **25n** and **26n**) were chosen for IC_{50} determination in the PGA-1 and HG-3 cell lines. The compounds were tested over a concentration range of 50–0.01 μM and assessed at 24 h via an alamarBlue cell viability assay. The results are presented in Table 4. All compounds exerted a more potent effect than the fludarabine control (3–22-fold greater in HG-3 cells, 3–37-fold greater in PGA-1 cells) across both HG-3 and PGA-1 cell lines with IC_{50} ranges of 1.31–10.28 μM and 0.31–12.6 μM, respectively. In HG-3 cells, the most potent compounds identified were **25n** (IC_{50} 1.31 μM, 2-pyridyl, *N*-(4-chlorophenyl)maleimide derivative) and **23p** (IC_{50} 1.41 μM, 2-naphthyl, maleimide derivative), while in PGA-1 cells, the most potent compounds were **24l** (IC_{50}, 0.31 μM, 4-chloro, maleimide derivative) and **26n** (IC_{50} 0.39 μM, 2-pyridyl, N-(4-benzoylphenyl)maleimide derivative). Overall, the two most potent compounds across both cell lines were **25n** (mean IC_{50} 1.09 μM) and **26n** (mean IC_{50} 1.12 μM).

Table 4. IC_{50} values of selected ethanoanthracene derivatives in HG-3 and PGA-1 CLL cell lines.

Compound Number	HG-3 IC_{50} (μM) [a]	PGA-1 IC_{50} (μM) [a]	Mean IC_{50} (μM) [a]	logP [b]
23a	2.69	1.97	2.33	4.35
23c	4.61	2.72	3.67	4.36
23f	3.55	11.3	7.43	3.73
23g	4.47	2.77	3.62	4.38
23h	4.65	2.88	3.77	4.05
23i	10.28	12.6	11.4	3.72
23k	4.9	5.38	5.19	3.66
23l	2.23	3.21	2.72	4.27
23n	2.39	1.48	1.94	3.03
23p	1.41	1.30	1.36	4.62
24f	7.35	7.13	7.24	3.69
24l	3.45	0.31	1.88	5.66
25n	1.31	0.87	1.09	4.98
26n	1.85	0.39	1.12	5.51
Fludarabine	28.1	32.0	30.1	−2.5 [c]

[a] The cell proliferation of HG-3 and PGA-1 cells was determined with an alamarBlue assay (seeding density: 1–5 × 10^4 cells/mL per well for 96-well plates). IC_{50} values are half-maximal inhibitory concentrations required to block the growth stimulation of HG-3 and PGA-1 cells. Values represent the mean ± SEM (error values × 10−6) for at least three experiments performed in triplicate at 24 h. Treatment at seven different concentrations (0.01–50 μM) was used for the determination of the IC_{50} values for each compound. [b] Predicted logP (consensus) property using Swiss ADME; [c] predicted logP property using Chemaxon.

2.4.12. National Cancer Institute (NCI) 60-Cell-Line Panel Screening

A number of our most promising compounds identified through preliminary cell viability testing (**23c**, **23d**, **23f**, **23h**, **23l**, **23m**, **23n**, **23o**, **24f**, **24l**, **25n** and **26n**) were selected for NCI drug screening in the 60-cell-line panel. The panel includes nine subcategories based on cancer tissue types, which include leukemia, ovarian, central nervous system, breast, lung, prostate, skin, renal and colon cancer. The compounds were initially assessed in a one-dose assay at a fixed treatment concentration of 10 μM. The compounds selected for further investigation were progressed to a five-dose assay across the concentration ranges of 0.01, 0.1, 1, 10 and 100 μM [72]. The endpoints assessed in the screening protocol were GI_{50}, TGI and LC_{50} [72].

As the main focus of this work is the antiproliferative and anti-cancer effects of ethanoanthracenes in CLL, attention was focused primarily on leukemia cell line results within the wider 60-cell-line panel. The leukemia cell lines used for the assessment were as follows: CCRF-CEM (acute lymphoblastic leukemia), HL-60 (TB) (acute promyelocytic leukemia), K-562 (chronic myeloid leukemia), MOLT-4 (acute lymphoblastic leukemia), RPMI-8226 (plasmacytoma/myeloma) and SR (large cell immune blastic lymphoma). The resultant GI_{50} values obtained are presented in Table 5. All compounds demonstrated GI_{50} values in the low and sub-micromolar range across all leukemia cell lines, with compound **25n** identified as the most potent with a mean GI_{50} of 0.29 μM, indicating the potential for the compound for further investigation. The mean GI_{50} values across the leukemia cell panel were determined in the range of 0.29–2.08 μM. Compounds **25n** (0.29 μM), **24l** (0.98 μM), **23h** (1.00 μM) and **23n** (1.32 μM) were the most promising overall anti-leukemic agents screened. These results are in good alignment with those determined in the HG-3 and PGA-1 CLL cell lines (Table 4), with compound **25n** identified as the most potent compound in both CLL cell lines (HG-3, IC_{50} = 1.31 μM; PGA-1, IC_{50} = 0.87 μM and mean IC_{50} = 1.09 μM). They warrant further study as lead compounds for the development of more selective and potent anti-cancer agents.

Table 5. Summary of NCI 60-cell-line screening results for selected ethanoanthracene compounds in the leukemia cell panel [a].

Cell Line	**GI_{50} μM** [b]											
Compound Number	**23c** [c]	**23d** [d]	**23f** [e]	**23h** [f]	**23l** [g]	**23m** [h]	**23n** [i]	**23o** [j]	**24f** [k]	**24l** [l]	**25n** [m]	**26n** [n]
CCRF-CEM	2.66	1.24	2.16	2.45	1.66	2.39	1.44	1.89	1.45	0.57	0.32	2.59
HL-60 (TB)	0.07	1.58	1.5	0.02	1.47	1.76	1.02	0.32	1.02	0.8	0.29	1.9
K-562	2.42	2.19	1.94	1.42	1.99	1.88	1.37	2.57	1.39	1.55	0.29	1.77
MOLT-4	2.23	2.16	2.17	0.46	1.7	2.57	2.11	1.77	1.23	0.76	0.32	2.29
RPMI-8226	1.87	1.1	1.52	1.62	1.66	2.11	1.22	1.68	1.55	1.16	0.25	2.49
SR	1.37	1.25	1.57	0.29	2.15	1.76	0.78	1.59	2.03	1.07	0.28	1.37
Mean	1.77	1.59	1.81	1	1.77	2.08	1.32	1.64	1.45	0.98	0.29	2.07

[a] Data obtained from NCI in vitro, human-tumor-cell-screen, five-dose assay [72]. [b] GI50 is the molar concentration of the compound causing a 50% inhibition to the growth of the tumor cells; [c] NSC 803357; [d] NSC 803361; [e] NSC 815555; [f] NSC 803358; [g] 815556; [h] 803363; [i] 803355; [j] 803360; [k] 815557; [l] 815558; [m] 815561; and [n] 815562.

In addition to the leukemia results above, promising anticancer activity was observed across the NCI 60-cell-line panel with particularly interesting activity in melanoma and breast cancer cell lines and notable GI_{50} value ranges of **25n** (melanoma: 0.194–1.08 μM; breast: 0.295–0.627 μM), **23h** (melanoma: 1.60–2.48 μM; breast: 0.885–1.88 μM), **23n** (melanoma: 1.12–2.14 μM; breast: 0.295–0.627 μM) and **23l** (melanoma: 1.40–1.81 μM; breast: 1.34–1.94 μM) (Tables 6 and 7). These values suggest the potential exploration of these compounds in breast and skin cancers in the future.

Table 6. Antitumor evaluation of compounds **23h, 23l** and **25n** in NCI 60-cell-line screen [a].

Cell Line		**Compound 23h**	**Compound 23l**	**Compound 25n**
		GI_{50} (μM) [b,c]	**GI_{50} (μM)** [b,d]	**GI_{50} (μM)** [b,e]
NSC Lung Cancer	A549/ATCC	2.37	2.40	0.390
	EKVX	1.85	1.43	0.491
	HOP-62	2.45	1.69	0.587
	HOP-92	1.33	1.32	1.19
	NCI-H226	2.04	1.92	0.293
	NCI-H23	2.61	1.78	0.446
	NCI-H332M	2.64	1.42	0.821
	NCI-H460	1.64	1.80	0.241
	NCI-H552	1.74	1.46	0.331
Colon Cancer	COLO205	1.79	1.91	1.16
	HCT-2998	1.70	1.69	0.335
	HCT-116	0.537	1.23	0.212
	HCT-15	1.19	1.47	0.213
	HT29	1.86	2.27	0.335
	KM12	2.18	2.14	0.344
	SW-620	2.27	1.89	0.299

Table 6. *Cont.*

Cell Line		Compound 23h	Compound 23l	Compound 25n
		GI_{50} (μM) [b,c]	GI_{50} (μM) [b,d]	GI_{50} (μM) [b,e]
CNS Cancer	SF-268	4.56	1.99	0.699
	SF295	nd [f]	nd[f]	0.379
	SF539	1.69	1.51	0.316
	SNB-19	2.68	1.58	0.425
	SNB-75	1.37	1.10	nd[f]
	U251	1.84	1.52	0.358
Prostate Cancer	PC-3	1.42	1.39	0.436
	DU-145	1.77	1.67	0.410
Melanoma	LOX IMV1	1.65	1.60	0.194
	MALME-3M	1.60	1.51	0.316
	M14	2.14	1.65	0.344
	MDA-MB-435	1.93	1.68	0.298
	SK-MEL-2	2.48	1.81	0.670
	SK-MEL-28	1.98	1.80	0.455
	SK-MEL-5	1.95	1.70	1.08
	UACC-257	2.13	1.56	0.496
	UACC-62	1.66	1.40	0.400
Ovarian Cancer	IGROV1	2.10	1.67	0.236
	OVCAR-3	1.44	1.63	0.222
	OVCAR-4	1.96	1.42	0.764
	OVCAR-5	1.60	1.58	0.568
	OVCAR-8	1.43	1.13	0.409
	NCI/ADR-RES	2.16	1.77	0.356
	SK-OV-3	5.28	1.85	3.30
Renal Cancer	786-0	2.44	1.73	0.216
	A498	1.16	1.28	0.539
	ACHN	1.39	1.65	0.299
	CAKI-1	1.02	1.24	1.07
	RXF 393	1.27	1.20	0.222
	SN12C	2.34	1.56	0.415
	TK-10	3.18	1.92	1.46
	UO-31	1.12	1.28	0.208
Breast Cancer	MCF-7	0.885	1.34	0.326
	MDA-MB-231/ATCC	1.64	1.88	0.463
	HS 578T	1.82	1.94	0.458
	BT-549	1.37	1.34	0.295
	T-47D	nd[f]	nd[f]	0.312
	MDA-MB-468	1.88	1.70	0.627
Leukemia	CCRF-CEM	0.566	1.66	0.321
	HL-60(TB)	0.795	1.47	0.292
	K-562	1.55	1.99	0.293
	MOLT-4	0.759	1.70	0.366
	RPMI-8226	1.16	1.66	0.252
	SR	1.07	2.15	0.284
	MID GI_{50} [g]	1.698	1.621	0.245
	MID TGI [h]	4.265	4.265	1.905
	MID LC_{50} [i]	14.45	13.183	18.62

[a] Data obtained from NCI in vitro human tumor cell screen 5 dose assay [72]. [b] GI50 is the molar concentration of the compound causing a 50% inhibition of the growth of the tumor cells; [c] NSC 815558; [d] NSC 815556; [e] NSC 815561; [f] Nd: Not determined; [g] MID GI_{50}: MG-MID full-panel mean-graph mid-point; [h] TGI (total growth inhibition) is the mean concentration required to completely inhibit the growth of all cells over the total NCI 60 cell line panel; [i] LC_{50} is the mean concentration required to kill 50% of the cells over the total NCI 60-cell-line panel in the assay.

Table 7. Summary of NCI 60-cell lines' mean screening results (GI_{50}, TGI and LC_{50}) for selected compounds.

NCI Reference Number	Compound Number	Mean GI_{50} (μM) [a]	Mean TGI (μM) [b]	Mean LC_{50} (μM) [a]
803357	**23c**	1.905	4.670	16.98
803361	**23d**	2.511	7.580	29.51
815555	**23f**	1.819	4.570	14.79
803358	**23h**	1.698	4.265	14.45
815556	**23l**	1.621	4.265	13.18
803363	**23m**	2.511	7.760	28.18
803355	**23n**	1.659	4.570	14.79
803360	**23o**	1.778	3.890	10.97
815557	**24f**	2.089	19.95	72.40
815558	**24l**	1.698	10.72	43.70
815561	**25n**	0.245	1.905	18.62
815562	**26n**	7.585	100.0	100.0

[a] GI_{50} and LC_{50} are the mean concentrations required to inhibit growth and kill 50% of the cells in the assay (all cell lines), respectively [73]. [b] TGI is the mean concentration required to completely inhibit the growth of all cells [73].

The *N*-(4-chlorophenyl)maleimide ethanoanthracene adduct **25n** was identified as the most potent compound evaluated in the NCI panel, with a mean GI_{50} value of 0.245 nM determined across the NCI-60 panel of cell lines screened. (See Table 6: NCI-60 cell line screen, Figure 8: Heatmap for the activity of compound **25n** in the cell lines of the NCI-60 screen). Significantly, the GI_{50} values for **25n** were in the sub-micromolar range for all but six of the cell lines investigated. **25n** displayed significant potency in all breast cancer cell lines evaluated in the panel, with GI_{50} values in the range of 312–0.627 μM. GI_{50} values below 300 nm were obtained for compound **25n** in 17 of the panel cell lines tested with activity against non-small-cell lung, colon, CNS, ovarian and prostate cell lines tested. Figure 8 displays a heatmap of the activity of compound **25n** across the cell lines in the NCI-60 screen. In addition, compound **25n** was particularly effective in the chemoresistant HT-29 human colorectal adenocarcinoma cell line with an IC_{50} value of 0.335 μM. The mean GI_{50} values for the panel of 60 cell lines for the most potent compounds evaluated were determined to be in the range of 0.245–2.511 μM (apart from compound **26n** GI_{50} = 7.585 μM), and they are shown in Table 7.

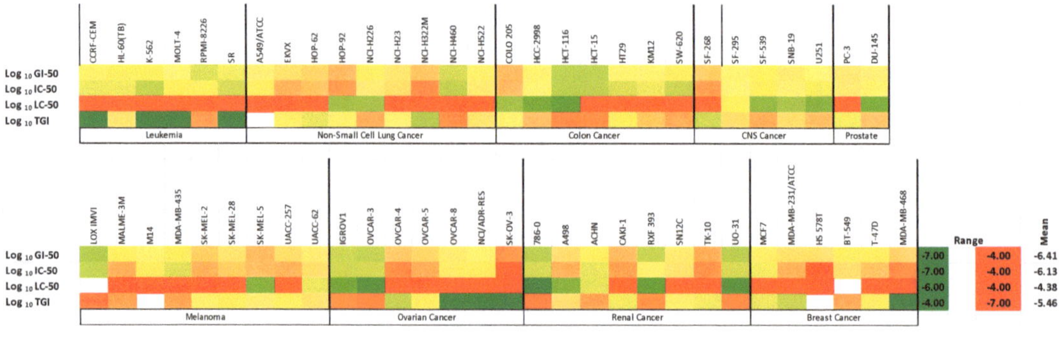

Figure 8. Heatmap for compound **25n** across cell lines in the NCI-60 screen.

Heatmap for the antiproliferative activity of compound **25n** across cell lines in the NCI-60 screen using three different values (growth-inhibitory effect, GI_{50}; drug concentration at which the response is reduced by half, IC_{50}; cytostatic effect, TGI; cytotoxic effect, LC_{50}; concentration in molar).

2.4.13. COMPARE Analysis of Compounds **23h, 23n, 24l** and **25n**

The COMPARE algorithm [74], allows for the comparison of compound activity profiles across the NCI 60-cell-line panel. Compounds that display similar activity profiles often result in cell growth inhibition through related mechanisms of action and display robust correlations in mean profiles across the screening assays. Representative potent compounds **23h, 23n, 24l and 25n** were used as COMPARE seeds, with the NCI Standard Agents database selected for comparison. The COMPARE program was then used to probe the mechanism of action of the series of ethanoanthracene maleimide adducts [75]. The antiproliferative profiles (GI_{50}, TGI and LC_{50} values) of potent compounds **23h, 23n, 24l** and **25n** were compared with compounds with a known mechanism of antiproliferative action in the NCI Standard Agents Database. The top five ranked compounds for each seed based on Pearson's correlation coefficient were identified. Based on GI_{50} values, the chemotherapeutic agents that had fair to strong positive Pearson's correlation coefficients (0.34–0.92) were identified (Supplementary Information, Tables S3–S6). Compound **23h** demonstrated a good correlation with the antitumor purine analog pentostatin, effective in lymphoid neoplasms and a potent inhibitor of ADA (r = 0.918). The correlation of **23h** with the anticancer ribonucleotide reductase inhibitor caracemide, (r = 0.663) was also observed. **24l** correlates with the DNA-alkylating drugs fluorodopan (r = 0.631) and melphalan (r = 0.538), while **23n** correlated with dichloroallyl lawsone (r = 0.505), an antineoplastic dihydroorotate dehydrogenase (DHOD) inhibitor (see Supplementary Information, Tables S3–S6). Many of the positively correlated chemotherapeutic agents identified in the COMPARE correlation analysis mediate cytotoxicity through DNA synthetic pathways and resultant DNA functionality, which may be relevant in determining the mechanism of action of the ethanoanthracene compounds screened.

2.4.14. Cheminformatics Analysis of Lead Ethanoanthracene Compounds

Estimation of ADME (absorption, distribution, metabolism and excretion) parameters early in the preclinical discovery phase can potentially reduce the number of pharmacokinetics-related drug failures in the clinical phases of drug development [76]. Based on the preliminary antiproliferative studies, a panel of novel compounds (**20a–20e, 23a, 23c, 23d, 23f–23i, 23k–23p, 24f, 24l, 25n** and **26n**) was selected for further investigation of their ADME properties using the Swiss ADME cheminformatics webtool [77] (Supplementary Information Figure S2 and Tables S7–S9). The majority of the compounds satisfy the criteria for effective oral drug development, demonstrating logP values in the range of 1.86–4.98, HBD of 0–1, HBA of 3–5, rotatable bonds of 2–6 and TPSA of 54–91 Å^2. The maleimide and *N*-phenylmaleimide series of compounds are not ionized at a physiological pH, (e.g., **20d**, theoretical pKa value of 10.4, calculated from Chemicalise). High blood–brain barrier (BBB) absorption levels and high GI absorption were predicted for the compounds, indicating the potential of these compounds as lead compounds for further development. **20a** was identified as the most potent analog evaluated in the nitrovinylanthracene series with an IC_{50} value of 0.48 μM (HG-3) and 0.61 μM (PGA-1) cell lines (MW = 422, HBD = 0, HBA = 4, Rot bonds = 3, cLogP 3.29, TPSA = 83.20 Å^2). The lead compounds tested were not predicted to inhibit the metabolic activity of CYP2D6 or CYP3A4.

The aqueous solubility predicted for the compounds was in the range of 0.011–61.4 μg/mL, with the compounds **20d** and **23m** predicted to have the highest solubility in the series (61.4 μg/mL and 29.8 μg/mL, respectively). Potential correlations were assessed between the estimated physicochemical properties and biological activity. A negative correlation was determined for molar refractivity, log P and the skin permeation coefficient, while a positive correlation was observed for water solubility and may contribute to the identification of suitable lead compounds.

Pan Assay Interference Compounds (PAINS) contain functional groups or fragments that contribute to high reactivity. Lead compounds (**20a–20e, 23a, 23c, 23d, 23f–23i, 23k–23p, 24f, 24l, 25n** and **26n**) were screened according to PAINS filters to reduce the possibility of such compounds being selected for progression and optimization. Despite the presence of the α,β-unsaturated ketone and nitrovinyl group (Michael acceptors) in the lead molecules considered, no PAINS alerts were flagged (Supplementary Information Table S9) [77]. The Brenk filter (used to identify fragments known to induce toxicity directly or post-metabolic activation for highly reactive and metabolically labile functional groups) identified two structural alerts in the series: the maleimide bridgehead structure and the α,β-unsaturated ketone (Michael acceptor) [77].

2.4.15. In vitro LDH Cytotoxicity Investigation of Selected Lead Compounds **20a**, **20f**, **23a** and **25n**

Further biochemical studies were then investigated for selected lead compounds **20a**, **20f**, **23a** and **25n** based on their potency in the antiproliferative assays in CLL cells. The lactate dehydrogenase (LDH) assay is used to assess the membrane integrity of cells as a function of the cytoplasmic LDH released. The loss of cell membrane integrity (due to cytotoxic insult) is detectable through LDH release into the cell growth medium. The assay is initiated via the reduction in cellular co-factor NAD^+ to NADH catalyzed via the LDH enzyme. Newly formed NADH catalyzes the acidic reduction in iodonitrotetrazolium chloride (INT) to the highly colored formazan (red) (mediated via the diaphorase enzyme) [78,79]. The results for the LDH cytotoxicity assay of the HG-3 cell line for the selected compounds **20a**, **20f**, **23a** and **25n** (which were representative of the most potent ethanoanthracene compounds identified in preliminary biochemical screening) are shown in Figure 9A. At the 10-μM treatment concentration, mild to moderate cytotoxicity (19–47%) was observed. The nitrovinyl dimer **20f** displayed the lowest LDH release at 19%, with **23a** (chalcone maleimide, 25%), **20a** (nitrovinyl *N*-phenylmaleimide 33%) and **25n** (chalcone, *N*-(4-chlorophenyl)maleimide) having the highest cytotoxicity (47%). At the 1-μM treatment concentration, cytotoxicity remained mild to moderate with an observed range of 12–37% with **20f** displaying the lowest % LDH release (12%). In the PGA-1 cell line (Figure 9B), **20a** displayed the lowest LDH release (9%) at 10 μM, followed by **25n** (15%), **20f** (19%) and **23a** (23%). Cytotoxicity remained low to mild at 1 μM (4–18%) with **23a** displaying the lowest LDH release (4%). Interestingly, the assessed compounds were observed to cause a greater degree of LDH release from HG-3 cells compared to PGA-1, suggesting greater cytotoxicity against more aggressive variants of CLL. In addition, the mean TGI (total growth inhibition) value for the potent compound **25n** (mean GI_{50}: 0.245 μM) over the NCI 60-cancer-cell-line panel was 18.62 μM, indicating a wide therapeutic window for the compound (Table 7).

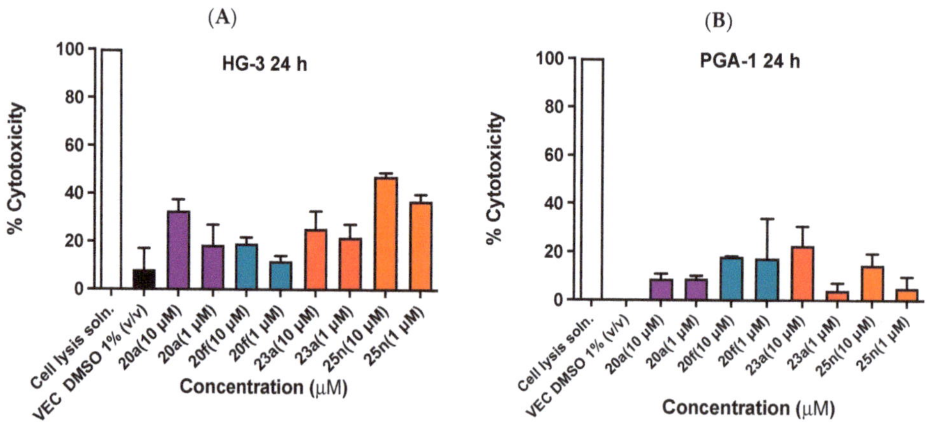

Figure 9. LDH assay for cytotoxicity of compounds **20a**, **20f**, **23a** and **25n** in the HG-3 (Panel **A**) and PGA-1 (Panel **B**) cell lines.

Following the treatment of the HG-3 and PGA-1 cells at 10 μM for 24 hr, the amount of LDH released was determined using an LDH assay kit from Promega (G1780). Cell lysis solution = 100% cell lysis and LDH release). Data are presented as % cell death at a concentration of 10 μM. Cells were treated with ethanoanthracene compounds **20a**, **20f**, **23a** and **25n** (1 μM and 10 μM) for 24 h. Values are shown as the mean of three independent experiments.

2.4.16. Pro-Apoptotic Effects of Selected Compounds **20a**, **20f**, **23a** and **25n** in HG-3 Cells and PGA-1 Cells

The pro-apoptotic effects of selected nitrovinylethanoanthracene compounds **20a** and **20f** and anthracene chalcones **23a** and **25n** in HG-3 cells and PGA-1 cells was then investigated using FITC (fluorescein isothiocyanate), Annexin V/PI (propidium iodide) and FACS (fluorescence-activated cell sorting) analysis to characterize the mode of cellular death induced. Apoptosis was assessed as % total apoptosis (total of early (Q3) and late (Q2) apoptosis).

The results from the Annexin V/PI studies of representative potent lead ethanoanthracene compounds **20a, 20f, 23a** and **25n** are shown in Figure 10, together with our previously reported lead nitrovinylanthracene compound **18a** [40]. In HG-3 cells, all compounds produced significant apoptosis that was largely concentration-dependent (Figure 10A,C). At the 10-μM concentration, the most active compounds with 95% total apoptosis were **23a** (4-bromophenylchalcone, maleimide) and **20a** (nitrostyrene, *N*-phenylmaleimide), followed by **25n** (2-pyridylchalcone, *N*-(4-chlorophenyl)maleimide) and **20f** (nitrostyrene dimer) with 93% and 82% apoptosis, respectively. At a 1-μM treatment concentration, the most potent pro-apoptotic effects were identified for the nitrostyrene derivatives **20a** and **20f** (both inducing 82% total apoptosis), while the chalcone ethanoanthracenes **25n** (37%) and **23a** (33% apoptosis) were less effective. The pro-apoptotic effect of nitrostyrene dimer **20f** remained constant across all three concentrations, possibly associated with the potent antiproliferative effect in HG-3 cells (IC_{50} value of 0.17 μM). In PGA-1 cells, all compounds produced significant concentration-dependent apoptosis (Figure 10B,D). At the 10-μM treatment concentration, the nitrostyrene **20a** induced the greatest apoptosis (97% total apoptosis), followed by **23a** (89%), **20f** (88%) and **25n** (87%). At the 1-μM treatment concentration, the nitrovinyl compounds **20a** and **20f** produced the most potent effect with 83% and 76% apoptosis, respectively. These results from compounds **20a, 23a, 25n** and **20f** compare favorably with the effect of the clinically used drug fludarabine phosphate, which was found to induce a poorer increase in apoptosis (24.6%) for cancer cells isolated from CLL patients [39]. In summary, the results across both cell lines indicate that compounds **20a, 20f, 23a** and **25n** induce significant apoptosis in a largely concentration-dependent manner and support a pro-apoptotic mechanism of action for this class of ethanoanthracene compounds.

(A) (B)

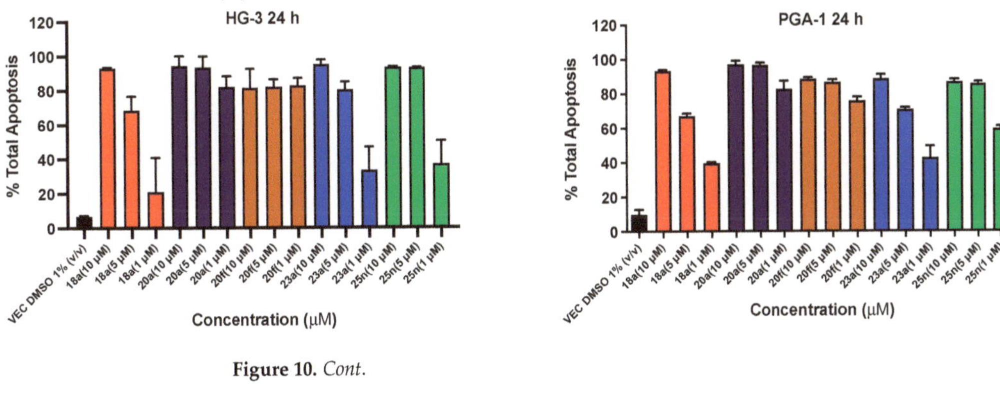

Figure 10. *Cont.*

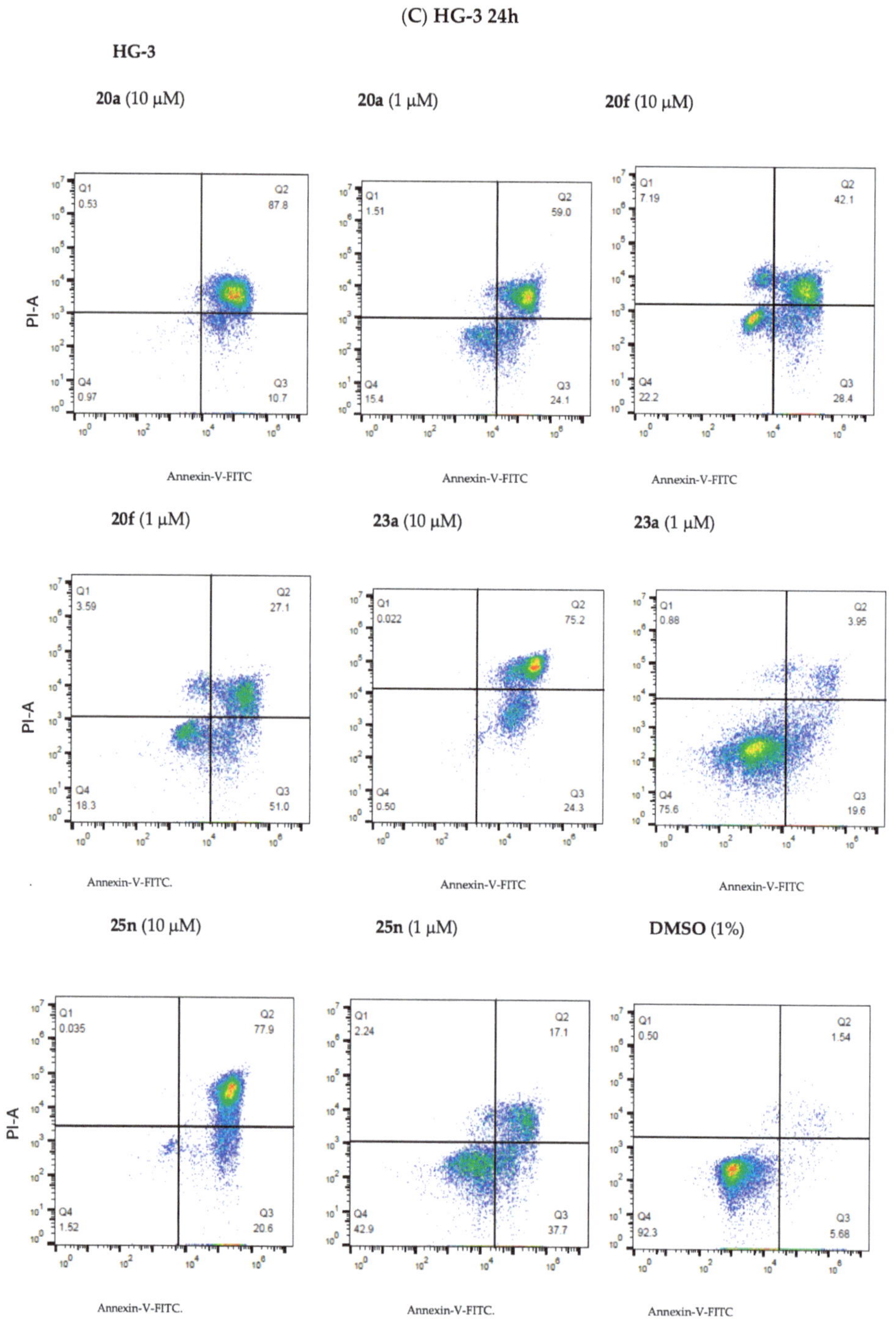

Figure 10. *Cont.*

(D) PGA-1 24 h

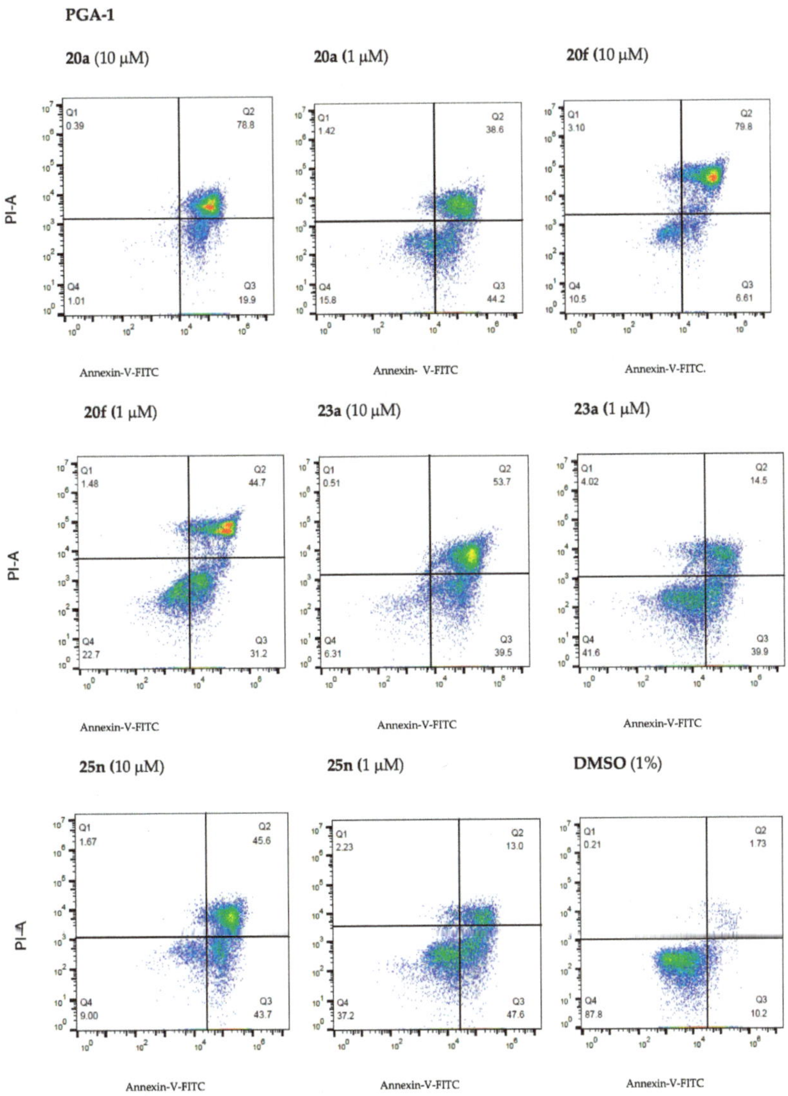

Figure 10. Ethanoanthracene nitrostyrene compounds **18a**, **20a**, **20f**, **23a** and **25n** induce apoptosis in HG-3 and PGA-1 CLL cells. Compounds **18a**, **20a**, **20f**, **23a** and **25n** potently induce apoptosis in HG-3 and PGA-1 cell lines (Annexin V/PI FACS). HG-3 and PGA-1 leukemia cells were treated with **18a**, **20a**, **20f**, **23a** and **25n** (10 μM, 5 μM and 1 μM) and a control vehicle [(1% DMSO (*v*/*v*))] at 48 h for panels **A** and **B**, respectively. The % of apoptotic cells was determined by staining with Annexin V-FITC and PI (Panels **C** and **D** show compounds **20a**, **20f**, **23a** and **25n** at 1- and 10-μM concentrations). The lower left quadrant cells are negative for both Annexin V-FITC and PI, and the upper left shows PI cells that are necrotic. The lower right quadrant shows Annexin-positive cells in the early apoptotic stage, and the upper right shows both Annexin- and PI-positive cells in the late apoptosis stage. The experiment was replicated on three independent days.

2.4.17. Ex Vivo Assessment of Selected Compounds **20a**, **23a** and **25n** in Donors' Peripheral Blood Mononuclear Cells (PBMCs)

Representative examples of the most potent ethanoanthracene compounds **20a**, **23a** and **25n** were assessed in donor peripheral blood mononuclear cells (PBMCs). Whole blood samples from healthy donors (×5) were isolated, prepared and treated following the protocol previously described [40]. The percentage of apoptosis observed in the treatment of isolated donor PBMCs with nitrovinylethanoanthracene **20a** is concentration-dependent (Figure 11A). **20a** exhibited moderate toxicity (40%) towards the healthy-donor lymphocytes at the highest concentration tested of 1.25 μM. When compared to the IC_{50} values in both HG-3 and PGA-1 CLL cell lines of 0.48 μM and 0.61 μM, respectively, relatively low to moderate amounts of apoptotic death in healthy cells was noted (10–35% at concentrations of 0.34 and 0.68 μM). For ethanoanthracene **23a**, the percentage of total apoptosis observed is illustrated in Figure 11B. **23a** exhibited the lowest toxicity of the three compounds evaluated in healthy-donor PBMCs at its highest treatment concentration of 5 μM with observed total apoptosis of 28%. When these results were compared to the IC_{50} values of **23a** in both HG-3 and PGA-1 CLL cell lines of 2.69 μM and 1.97 μM, respectively, very low amounts of lymphocyte cell death were detected (5–14% at 1.25–2.5 μM). The percentage of total apoptosis observed in the treatment of isolated donor PBMCs with the anthracene–chalcone **25n** (normalized against vehicle 0.5% *v*/*v* DMSO) is illustrated in Figure 11C, and the dose-response behavior was also concentration-dependent, with compound **25n** moderately toxic to healthy-donor PBMCs when treated at a 5.00 μM concentration (39%). Relatively low toxicity (12–16%) was observed at concentrations representative of the IC_{50} values in both HG-3 and PGA-1 CLL cell lines, which were 1.31 μM and 0.87 μM, respectively.

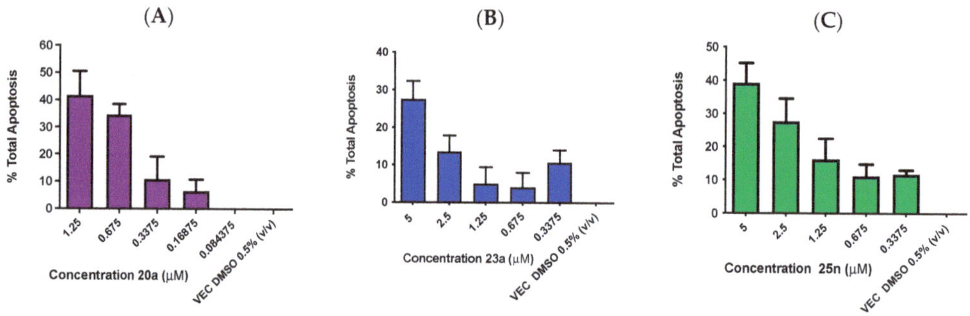

Figure 11. Percentage of total apoptosis observed upon treatment of isolated donor PBMCs with compounds **20a**, **23a** and **25n**. Ethanoanthracene compounds **20a**, **23a** and **25n** induced apoptosis upon the treatment of isolated human-donor peripheral blood mononuclear cells (PBMCs). Isolated donor PBMCs were treated with compound **20a** (Panel **A**) at 1.25 μM, 0.68 μM, 0.34 μM, 0.17 μM and 0.08 μM concentrations, compounds **23a** (Panel **B**) and **25n** (Panel **C**) at 5 μM, 2.5 μM, 1.25 μM, 0.68 μM and 0.34 μM concentrations, and all were normalized against a control vehicle (0.5% DMSO (*v*/*v*)) at 48 h. The % of apoptotic cells was determined via staining with Annexin V-FITC and PI. The experiment was performed individually and replicated on three independent days.

2.4.18. Effect of Pre-Treatment with Antioxidants on Cellular Viability of HG-3 and PGA-1 CLL Cells

Due to the pivotal role that reactive oxygen species (ROS) play in malignant cell transformation and progression, a wide variety of chemotherapeutic agents that rely on ROS modulation (either directly or indirectly) have been identified e.g., the quinone-based anticancer agents doxorubicin, mitomycin C and bleomycin used clinically for breast, small-cell lung and hematologic cancers [80–82].

Compounds **20a** and **20f** (potent nitrovinyl-based ethanoanthracenes), and **23a** and **25n** (potent chalcone-based ethanoanthracenes), were assessed at treatment concentrations of 10 μM and 1 μM in the absence and presence of the ROS inhibitor N-acetylcysteine (NAC) to assess the potential for the synthesized compounds to mediate their antiproliferative effect through ROS flux. In HG-3 cells, NAC addition impacts the ability of all compounds to reduce cell viability to varying degrees (Figure 12A). With the nitrovinyl **20a** and chalcone compound **23a,** a dramatic increase of 103% and 78%, respectively, in cell viability was noted at a 10-μM treatment concentration, following pre-treatment with NAC. Smaller relative increases in cell viability were observed for chalcone **25n** and nitrovinyl dimer **20f** (17% and 14%, respectively). A similar trend in viability was observed at the 1-μM treatment concentration. In PGA-1 cells, NAC addition also impacts on the ability of all compounds to reduce cell viability (Figure 12B). For **20a,** viability was increased to 120% upon pre-treatment with NAC, while a moderate increase of 48% in cell viability was observed with chalcone **23a**; **25n** and **20f** showed little change in cell viability. At the lower 1-μM treatment concentration, **20a** was the most effective with a 92% increase in cell viability. These results strongly suggest a potential ROS-dependent mechanism for chalcone-based compound **23a** and nitrovinyl-based compound **20a**, and they support our previous reports of the ROS-dependent action of nitrovinyl-based ethanoantracenes in BL cell lines [43]; meanwhile, the antiproliferative action of chalcone **25n** and nitrovinyl dimer **20f** showed less dependence on ROS.

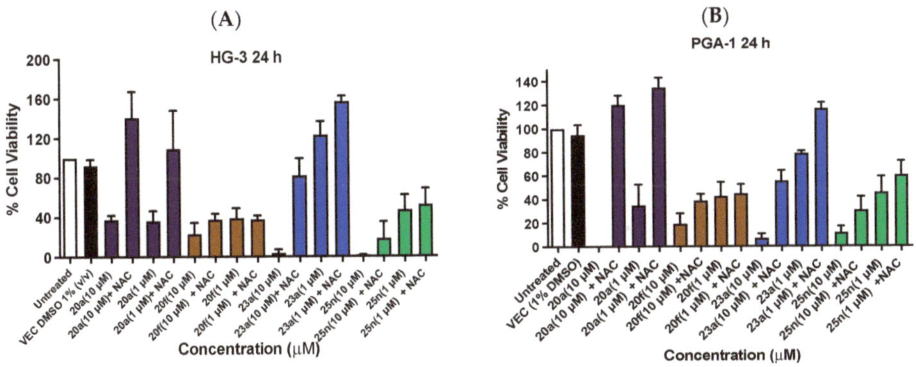

Figure 12. Effect of antioxidant pre-treatment (N-acetylcysteine, NAC) on the viability of HG-3 and PGA-1 CLL cells treated with compounds **20a**, **20f**, **23a** and **25n.** The cell viability of HG-3 and PGA-1 cells was determined with an alamarBlue assay (seeding density: 2×10^5 cells/mL per well for 96-well plates). Compound concentrations of either 1 μM or 10 μM for 24 h were used to treat the HG-3 and PGA-1 CLL cells (in triplicate) with control wells containing vehicle DMSO (1% v/v). The cells were pre-treated with NAC (2 μL, 5 mM) for 1 h, (Panel **A**,**B**) and protected from light before then being treated with the compound. The mean value for three independent experiments is shown.

2.4.19. Caspase Inhibition Assay

Caspases (cysteine-aspartic proteases) are a family of aspartic acid residue-specific regulatory cysteine proteases that play an essential role in key cellular homeostatic processes such as inflammation and programmed cell death through the induction of apoptosis [83,84]. To assess whether the biological activity of both major ethanoanthracene classes (nitrostyrene and chalcone-based) is caspase-dependent, compounds **20a** and **23a** (representing Series 1 and Series 4, respectively) (Figure 13) were selected for evaluation at treatment concentrations of 10 μM and 1 μM across both CLL cell lines using a caspase assay protocol. In HG-3 cells, pre-treatment with pan-caspase inhibitor (CI) Z-VAD-FMK [85] (20 μM) resulted in an increase in the viable cells remaining of 24% for compound **23a** (10 μM) and 21% at 1 μM (Figure 13A). In contrast to this, with compound **20a**, a decrease of 10% in cell viability was noted at 10 μM and 2% at 1 μM. Little difference was found between cell viability responses (with or without the caspase inhibitor) at both 10-μM and 1-μM concentrations for compound **20a**. In the PGA-1 cell lines, pre-treatment with

Z-VAD-FMK (20 μM) resulted in a 68% increase in viable cells for compound **23a** at 10 μM and an increase of 79% at 1 μM (Figure 13B). For compound **20a**, an increase of 17% at 1 μM was observed, suggesting a correlation between antiproliferative activity and the induction of apoptotic cell death for nitrostyrene compound **20a**. The involvement of caspases in the ability of compound **23a** to reduces cell viability in HG-3 and PGA-1 cells was demonstrated, together with the pro-apoptotic effects of **20a** in PGA-1 cells, suggesting the selective roles of the nitrovinyl and α,β-unsaturated ketone pharmacophores in the induction of apoptosis via caspases.

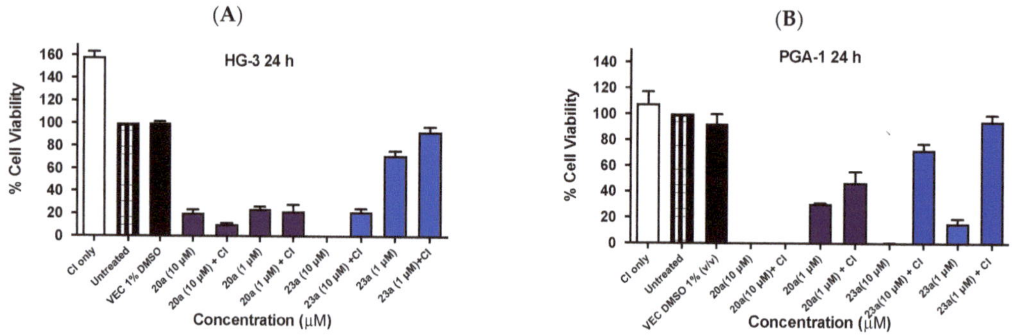

Figure 13. Effect of pre-treatment with caspase inhibitor Z-VAD-FMK on HG-3 and PGA-1 cell viability for compounds **20a** and **23a**. Cell viability analysis (24 h) for inhibitor studies of compounds **20a** and **23a** in HG-3 (Panel **A**) and PGA-1 (Panel **B**) CLL cell lines: the HG-3 and PGA-1 CLL cells (2×10^5 cells/mL) were pre-treated at 37 °C with 40 μM of caspase inhibitor (CI) (Z-VAD-FMK) for 4 h prior to compound treatment at 1 μM and 10 μM for 24 h. The cell proliferation of HG-3 and PGA-1 cells was determined with an alamarBlue assay (CI = caspase inhibitor, n = 2).

2.5. Computational Study

2.5.1. Molecular Modeling Study

The synthesized ethanoanthracene compounds (series 1–7) are structurally related to the antidepressant maprotiline **16,** which displays antiproliferative effects in B-cell malignancies and may have similar cellular antiproliferative effects. To examine the structural similarities, selected compounds (**20a–20f, 23a, 23c, 23f–23i, 23k, 23l, 23n, 23p, 24f, 24l, 25n** and **26n**) were overlaid on maprotiline using MOE flexible alignment, as described in our previous work; this approach is based on similarity terms such as hydrogen bond donor/acceptor, aromaticity and partial charge [40]. Similar functional groups in each molecule are flexibly aligned and superimposed using a stochastic search procedure while the full conformational flexibility for each structure is samples. All databases and reference structures are provided in the Supplementary Information as .sdf or .mdb files.

An overlay of the most potent ethanoanthracenes **20a, 20b, 20d, 20f, 23a** and **25n** (displayed as green in their respective overlays with maprotiline pink) and reference nitrovinylanthracene compound **18b**, together with the IC_{50} values in the HG-3 and PGA-1 CLL cell lines, is shown in Table 8. The overlay scores of the enthanoanthracene compounds were lower (better) than the lead nitrostyrene compound **18b**. Common molecular features were clearly identified, e.g., the ethanoanthracene structure that overlays with the 9,10-dihydroanthracene core structure of maprotiline, whereas the nitrostyrene and α,β-unsaturated ketone mapped closely to the cyclic core of maprotiline. The MOE flexible alignment demonstrated a lack of correlation between the lowest-(best-) scored compounds and those with the most promising cellular data for the CLL cell lines (Table 8). The overlay results obtained for all the ethanoanthracene compounds in this study are inconclusive (Table S10, Supplementary Information), and they suggest that these compounds may have a different mechanism of action when compared with maprotiline.

Table 8. Overlay of ethanoanthracenes (**20a**, **20b**, **20d**, **20f**, **23a** and **25n**) and nitrostyrene **18b** on maprotiline with their overlay scores [a].

Compound No.	Overlay of Compounds (Green) with Maprotiline (Pink)	Overlay (F) Score	Overlay (S) Score	Mean IC_{50} (μM)
20a	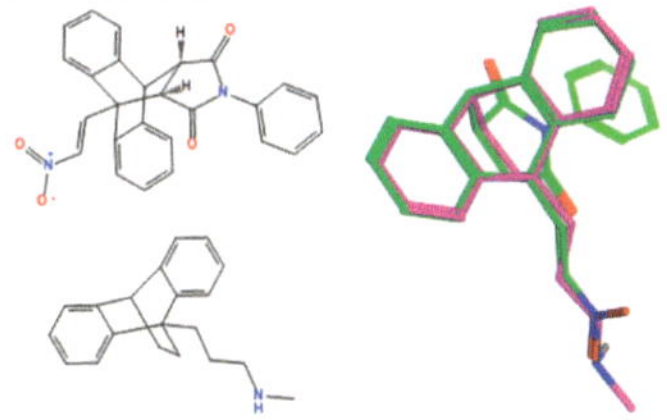	−157.66	−66.46	0.55
20b	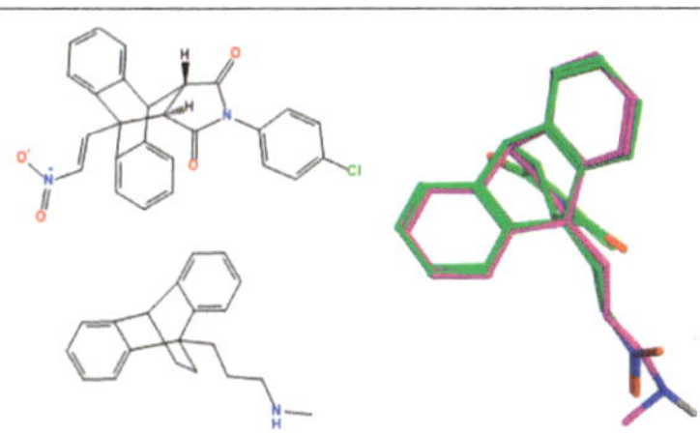	−158.37	−48.91	0.76
20d	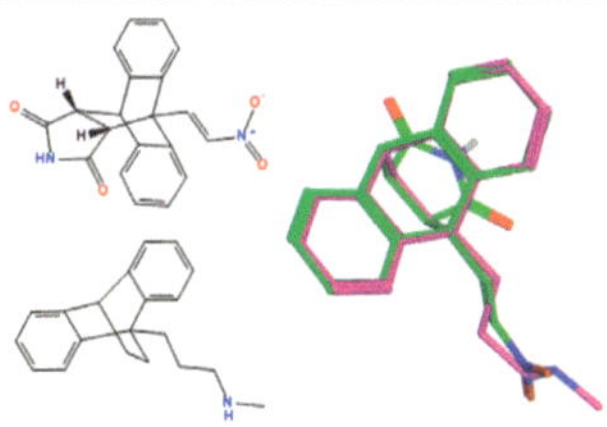	−156.30	−86.35	0.77
20f	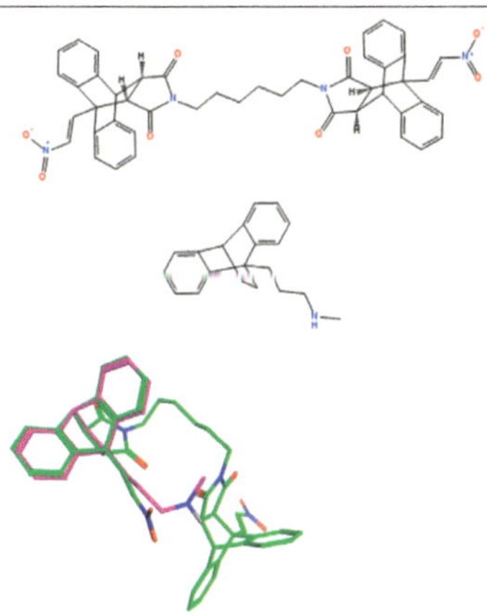	−163.50	−65.11	0.53
23a	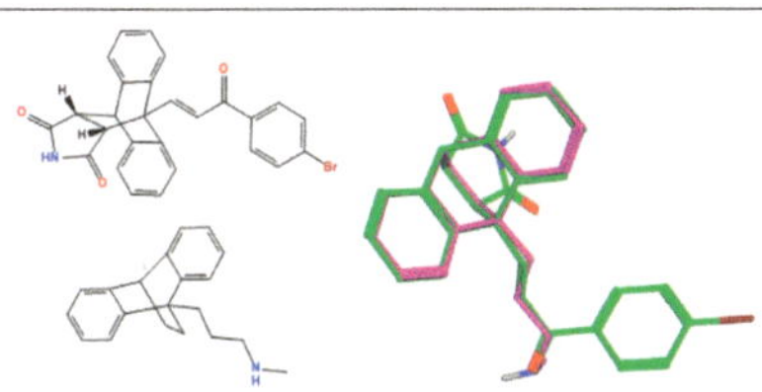	−165.19	−79.80	2.33

Table 8. *Cont.*

Compound No.	Overlay of Compounds (Green) with Maprotiline (Pink)	Overlay (F) Score	Overlay (S) Score	Mean IC_{50} (μM)
25n	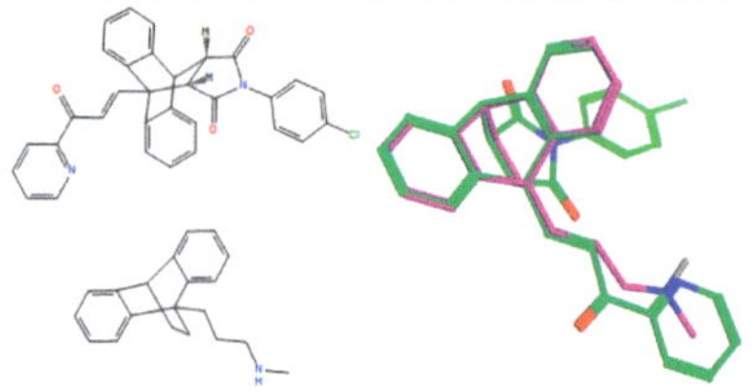	−169.14	−58.04	1.09
Anthracene 18b	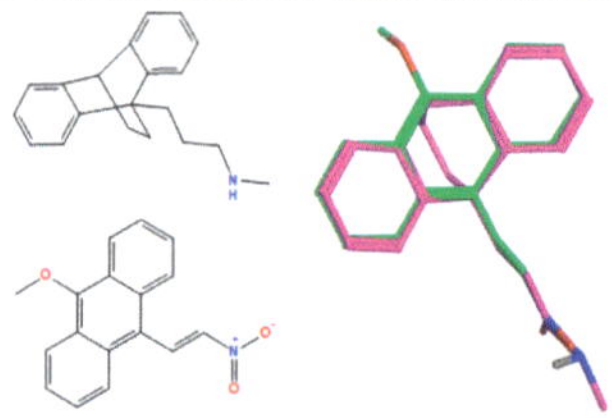	−132.50	−36.05	0.73

[a] Overlay of ethanoanthracene compounds series and nitrostyrene compound **18b** on maprotiline with their overlay scores (F and S) in MOE 2022, 2D images and mean IC_{50} values. The F column contains the similarity score (the lower, the better) of the alignment. The S column has the sum of the U (the average internal energy of the ligand) and F values. The compounds shown represent those with the best mean IC_{50} values on the HG-3 and PGA-1 cell lines.

2.5.2. In Silico Target Prediction

Computational modeling was investigated to gain insight into the possible mechanisms of action of the ethanoanthracenes. Selected novel compounds (**20a**, **20b**, **20d–20f**, **23a**, **23c**, **23f–23i**, **23k**, **23l**, **23n**, **23p**, **24f**, **24l**, **25n** and **26n**), together with lead anthracene **18b** and maprotiline, were submitted to the SwissTargetPrediction service (STP) [86] for evaluation. The STP methodology is conceptually rooted in the SAR principle, which states that similar compounds tend to have similar biological activity. Thus, for a given query structure, STP finds structurally similar compounds with known activities and uses them to compile a list of potential targets for the given query structure.

STP generated a list of potential targets for each of the submitted compounds. Overall, all predictions were assigned relatively low internal confidence scores via STP, mostly at around 0.05–0.15. However, this was consistent with most predictions for maprotiline itself, except for a full 1.00 prediction confidence score for its several already experimentally confirmed targets. For dimeric compound **20f**, STP was not able to provide predictions with any degree of confidence. The tested compounds share similar in vitro cytotoxic effects, and it could be assumed that these observed effects are caused by the compounds hitting the same unknown targets(s). Thus, targets predicted for the majority of the tested compounds became the focus of the STP-based analysis. To identify the most prevalent predicted targets among the 21 tested compounds, the STP target lists were aggregated by summing the individual prediction confidence scores for each compound. The resulting target list thus contains the cumulative prediction confidences for each given target, within the 21 tested compounds (Figure 14).

Since the score contributions of individual compounds are in the 0.1-score range, these targets were found to be a hit by STP for most of the 21 tested compounds. For a more general view of the target types predicted for the tested compounds, the cumulative probability scores for individual targets were further aggregated by summing based on their general protein family groups. To obtain a direct comparison between target groups predicted for the tested compounds and target groups predicted for maprotiline, both results were individually converted to standard z-scores (Figure 15).

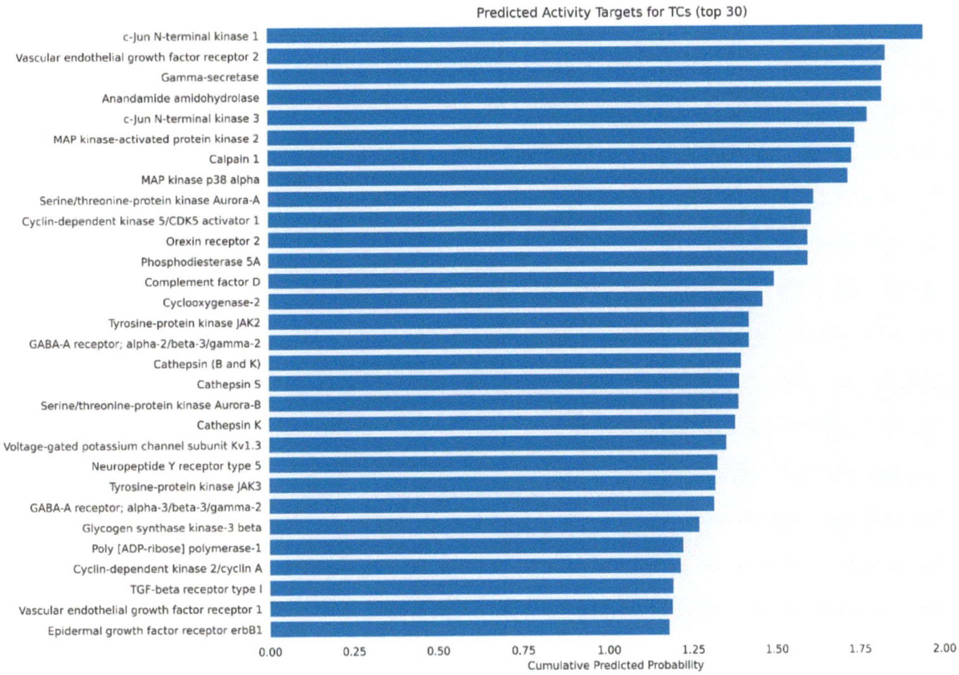

Figure 14. Cumulative probability scores for compounds **20a**, **20b**, **20d–20f**, **23a**, **23c**, **23f–23i**, **23k**, **23l**, **23n**, **23p**, **24f**, **24l**, **25n** and **26n** for the 30 targets most strongly indicated via STP.

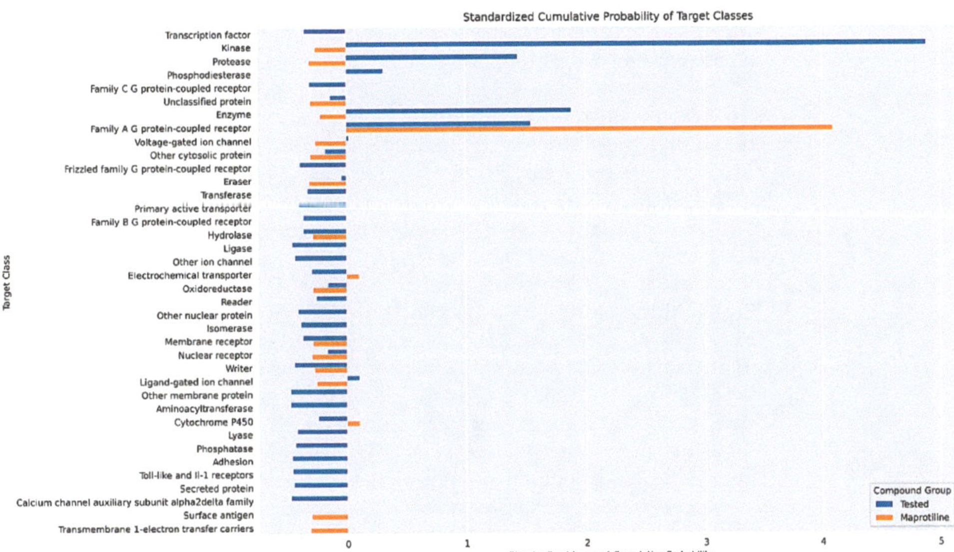

Figure 15. Standardized cumulative probability scores of target groups indicated via STP for the tested compounds **20a**, **20b**, **20d–20f**, **23a**, **23c**, **23f–23i**, **23k**, **23l**, **23n**, **23p**, **24f**, **24l**, **25n** and **26n** compared to Maprotiline. Scores were standardized using z-scores (i.e., differences in standard deviations from their mean).

The aggregated STP outputs indicate that the tested compounds might, indeed, share some family A GPCR activity with maprotiline, which is a known antagonist of muscarinic acetylcholine receptors, having antiproliferative activity [44]. However, in contrast with maprotiline, there are stronger ligand-similarity hits against a variety of kinases, especially the c-Jun N-terminal (JNK), MAP and VEGFR kinases. These predicted kinase targets are reasonable, as each has a role in promoting cellular proliferation, so blocking them would promote the observed functional phenotype. It is interesting that the COMPARE analysis indicated that the compounds mediate cytotoxicity through DNA synthetic pathways. The in silico target evaluation remains inconclusive, but it offers insight into future investigations for these compounds, e.g., molecular docking with select family-A GPCRs, as well as JNK, MAP and VEGFR kinases.

3. Materials and Methods

3.1. Chemistry

All reagents were commercially available and were used without any further purification unless otherwise indicated. Melting points were measured on a Stuart Melting Point Apparatus, SMP20, and they are uncorrected. Infra-red (IR) spectra were recorded on a Perkin Elmer FT-IR Paragon 1000 spectrometer. ^{1}H and ^{13}C nuclear magnetic resonance spectra (NMR) were recorded at 25 °C on a Bruker DPX 400 spectrometer (400.13 MHz, ^{1}H; 100.61 MHz, ^{13}C), Bruker Avance III 400 or Avance II 600 (400.13 MHz/600.13 MHz, ^{1}H; 100.61 MHz/150.61 MHz, ^{13}C) in either $CDCl_3$ (internal standard tetramethylsilane (TMS)) or DMSO-d_6. For $CDCl_3$, ^{1}H-NMR spectra were assigned relative to the TMS peak at 0.00 ppm, and ^{13}C-NMR spectra were assigned relative to the middle $CDCl_3$ peak at 77.0 ppm. For CD_3OD, ^{1}H and ^{13}C-NMR spectra were assigned relative to the center peaks of the CD_3OD multiplets at 3.30 ppm and 49.00 ppm, respectively. Coupling constants are reported in Hertz. High-resolution mass spectrometry (HRMS) was carried out by Dr. Gary Hessman of the School of Chemistry, Trinity College Dublin, using atmospheric pressure chemical ionization (APCI) with the occasional use of electrospray ionization (ESI-MS) in both positive and negative modes. ESI mass spectra were acquired using a Bruker micrOTOF-Q III spectrometer interfaced to a Dionex UltiMate 3000 LC in positive and negative modes, as required. Masses were recorded over the range of 100–1400 m/z. APCI experiments were carried out on a Bruker micrOTOF-Q III spectrometer interfaced with a Dionex UltiMate 3000 LC or direct insertion probe. The instrument was operated in positive or negative mode, as required. Masses were recorded over a range of 100–1600 m/z. Mass measurement accuracies of $<\pm5$ ppm were obtained. TLC was carried out on silica gel on aluminum foils with fluorescent indicator F-254 nm. Flash-column chromatography was carried out on Merck Kieselgel 60 (particle size: 0.040–0.063 mm). Analytical high-performance liquid chromatography (HPLC) was performed using a Waters 2487 Dual Wavelength Absorbance detector, a Waters 1525 binary HPLC pump and a Waters 717plus Autosampler with a Thermo Scientific Hypersil GOLD C18 reverse-phase 5 μM 150 × 4.6 mm chromatography column. Samples were detected using wavelengths of 232 nm compounds. All samples were analyzed using acetonitrile (75%)–TFA (0.1%) in water (25%) over 15 min and a flow rate of 1 mL/min. Microwave experiments were carried out using the Discover CEM microwave synthesizer on standard power setting (300 watts) unless otherwise stated. Maleimides **19a–c** and (*E*)-9-(2-nitrovinyl)-9,10,11,15-tetrahydro-9,10-[3,4]epipyrroloanthracene-12,14-diones **20a–h** were prepared as we previously reported [43] (see Supplementary Information). (*E*)-3-(Anthracen-9-yl)-1- phenylprop-2-en-1-ones **21a–d**, **21f–j** and **21l–q** were prepared as previously reported [87–94] (see Supplementary Information).

3.2. General Procedure for the Preparation of (E)-3-(anthracen-9-yl)-1-phenylprop-2-en-1-ones **21a–q**

To a solution of NaOH (6 mmol, 0.24 g) in 50% aqueous EtOH (20 mL) was added the appropriate acetophenone (5.02 mmol). After the dissolution of the acetophenone, 9-

anthracenecarboxaldehyde (5.02 mmol, 1.035 g) was added, resulting in a colored solution. This solution was stirred at room temperature for 24 h to achieve a colored suspension that was filtered, washed with minimal cold EtOH and dried. The crude product was then recrystallized from the minimal amount of EtOH or MeOH to afford a pure product.

(*E*)-3-(Anthracen-9-yl)-1-(3,4-dimethoxyphenyl)prop-2-en-1-one (**21e**) was synthesized from 3′,4′-dimethoxyacetophenone (5.02 mmol, 0.904 g) and 9-anthracenecarboxaldehyde (5.02 mmol, 1.035 g), according to the general procedure above; yellow powder (85%) Mp. 138–139 °C. ^{1}H nmR (400 MHz, $CDCl_3$) δ 3.97 (d, *J* = 11.61 Hz, 6 H, 2xCH_3), 6.91 (d, *J* = 8.71 Hz, 1 H), 7.40–7.53 (m, 4 H, ArH), 7.54–7.58 (d, *J* = 15.76 Hz, 1 H,CH=C), 7.65–7.72 (m, 2 H, ArH), 8.01–8.05 (m, 2H, ArH), 8.29–8.32 (m, 2H, ArH), 8.47 (br.s, 1H, ArH), 8.77 (d, *J* = 15.76 Hz, 1H, ArH). ^{13}C nmR (101 MHz, $CDCl_3$) ppm 53.07, 53.11, 110.07, 110.85, 123.34, 125.39, 126.31, 128.20, 128.87, 129.61, 130.80, 131.03, 131.31, 141.00, 187.81 (C=O). IR ν max (KBr): 3081.44, 3049.16, 2968.80 (Ar C-H), 1656.98 (C=O), 1622.51 (*trans* C=C), 1591.24, 1580.66, 1515.04, 1442.23, 1416.45 (Ar C=C), 1266.73, 1014.80 (C-O) cm^{-1}. HRMS (ESI) calculated for $C_{25}H_{20}NaO_3$ [M^++Na] 391.1305: found 391.1307.

(*E*)-3-(Anthracen-9-yl)-1-(3,4,5-trimethoxyphenyl)prop-2-en-1-one (**21k**) was synthesized from 1-(3,4,5-trimethoxyphenyl)ethan-1-one (5.02 mmol, 0.65 mL) and 9-anthracenecarboxaldehyde (5.02 mmol, 1.035 g) according to the general procedure above; yellow crystals (86%), Mp. 160–164 °C. ^{1}H nmR (400 MHz, $CDCl_3$) δ 3.90–3.99 (m, 9H), 7.35 (s, 2 H), 7.49–7.56 (m, 5H), 8.03–8.08 (m, 2H), 8.28–8.35 (m, H), 8.50 (s, 1 H), 8.79 (d, *J* = 15.76 Hz, 1H). ^{13}C nmR (101 MHz, $CDCl_3$) ppm. 56.35, 61.00, 61.45, 62.52, 106.22, 125.38, 126.40, 128.37, 128.92, 129.62, 130.27, 130.70–131.49, 133.10, 141.70, 153.24, 188.33 (C=O). IR ν max (KBr): 3045.78 (Ar C-H), 1657.91 (C=O), 1592.99 (Ar C=C), 1268.62 (C-O) cm^{-1}. HRMS (APCI) calculated for $C_{26}H_{22}NaO_4$ [M^++Na]: 421.1421; found: 421.1410.

3.3. General Preparation for (E)-9-(3-oxo-3-phenylprop-1-en-1-yl)-9,10-dihydro-9,10-[3,4]furanoanthracene-12,14-diones **22a-q**

To a solution of the appropriate chalcone anthracene analog **21a-q** (1.0 mmol) in toluene (2 mL) was added dienophile maleic anhydride (1.3 mmol). The mixture was stirred and heated at 90 °C for 48 h. The reaction mixture was then cooled to room temperature, and the resulting solid was isolated via filtration. The solid product was washed with toluene (2 mL) and diethyl ether (2 mL) and then recrystallized from toluene.

9-(*E*)-3-(4-Bromophenyl)-3-oxoprop-1-en-1-yl)-9,10-dihydro-9,10-[3,4]furanoanthracene-12,14-dione (**22a**) was prepared from (*E*)-3-(anthracen-9-yl)-1-(4-bromophenyl)prop-2-en-1-one (1.0 mmol, 0.3873 g) and maleic anhydride (1.3 mmol, 0.13 g) according to the general procedure; light yellow powder (86%), Mp. 254–256 °C [51]. ^{1}H nmR (400 MHz, $CDCl_3$) δ 3.65 (dd, *J* = 9.12, 3.32 Hz, 1 H, CH), 3.81 (d, *J* = 9.12 Hz, 1 H, CH), 4.87 (d, *J* = 3.32 Hz, 1 H, CH), 7.16–7.36 (m, 6 H,5 x ArH, 1xC=CH), 7.39–7.44 (m, 2 H), 7.66–7.76 (m, 3 H, ArH), 7.95 (d, *J* = 16.17 Hz, 1 H, C=CH), 8.03 (d, *J* = 8.71 Hz, 2 H, ArH)^{13}C nmR (101 MHz, $CDCl_3$) ppm 45.46, 48.96, 49.28, 51.93, 123.52, 123.87, 124.18, 127.12, 127.63, 130.39, 130.92, 132.16, 136.24, 137.36, 138.85, 140.22, 141.51, 168.82, 169.74, 188.80 (C=O). IR ν max (KBr): 3328.89, 3074.11 (Ar C-H), 1773.04, 1665.51 (C=O), 1617.10, *trans* (C=C), 1581.33, 1458.65, 1397.18, 1076.35 (C-O) cm^{-1}. LRMS (APCI) 485.04 [M^++H].

9-(*E*)-3-(4-Nitrophenyl)-3-oxoprop-1-en-1-yl)-9,10-dihydro-9,10-[3,4]furanoanthracene-12,14-dione (**22b**) was prepared from (*E*)-3-(anthracen-9-yl)-1-(4-nitrophenyl)prop-2-en-1-one (1.0 mmol, 0.3534 g) and maleic anhydride (1.3 mmol, 0.13 g) according to the general procedure above; cream powder (77%), Mp. >200 °C. ^{1}H nmR (400 MHz, DMSO-d_6) δ 3.80 (dd, *J* = 9.12, 3.32 Hz, 1 H), 4.25 (d, *J* = 9.12 Hz, 1 H), 4.95 (d, *J* = 3.32 Hz, 1 H), 7.16–7.28 (m, 3 H), 7.31–7.36 (m, 1 H), 7.41 (dd, *J* = 5.60, 3.11 Hz, 1 H), 7.53–7.57 (m, 1 H), 7.55 (d, *J* = 7.05 Hz, 1 H), 7.86 (s, 1 H), 8.34–8.44 (m, 4 H). ^{13}C nmR (101 MHz, DMSO-d_6) 49.73, 49.78, 51.71, 56.89, 109.99, 123.47, 124.01, 124.46, 124.51, 125.00, 125.74, 125.77, 127.03, 127.53, 127.94, 130.55, 132.01, 138.92, 139.80, 141.19, 142.06, 142.92, 150.52, 170.66, 171.37, 189.02 (chalcone C=O) ppm. IR ν max (KBr): 2972.12, (C-H), 1770.72 (C=O), 1667.96, (Ar C=C),

1596.43, (NO_2), 1516.78, 1346.74 (C-O) cm^{-1}. HRMS (APCI) calculated for $C_{27}H_{16}NO_6$ [M^+−H]: 450.0983; found: 450.0982.

(*E*)-3-(4-Ethylphenyl)-3-oxoprop-1-en-1-yl)-9,10-dihydro-9,10-[3,4]furanoanthracene-12,14-dione (**22c**) was prepared from (*E*)-3-(anthracen-9-yl)-1-(4-ethylphenyl)prop-2-en-1-one (1.0 mmol, 0.336 g) and maleic anhydride (1.3 mmol, 0.13 g) according to the general procedure above; cream powder (70%), Mp. 195–200 °C. ^{1}H nmR (400 MHz, DMSO-d_6) δ 1.16–1.23 (m, 3H), 2.70 (q, *J* = 7.46 Hz, 7H), 3.79 (dd, *J* = 9.12, 3.32 Hz, 4H), 4.25 (d, *J* = 9.12 Hz, 4H), 4.94 (d, *J* = 3.32 Hz, 4H), 7.16–7.20 (m, 8H), 7.20–7.23 (m, H), 7.23–7.27 (m, 9H), 7.28–7.33 (m, 1H), 7.37–7.42 (m, 4H), 7.43 (d, *J* = 8.29 Hz, 2H), 7.54 (d, *J* = 6.63 Hz, 4H), 7.74–7.80 (m, 4H), 7.82–7.88 (m, 4H), 8.09 (d, *J* = 8.29 Hz, 7H). ^{13}C nmR (101 MHz, DMSO-d_6) ppm: 15.65, 28.70, 44.80, 49.37, 49.60, 51.58, 123.37, 123.91, 125.77, 127.03, 127.47, 127.79, 129.40, 132.06, 135.37, 138.95, 140.03, 140.65, 141.25, 142.31, 150.55, 170.57, 171.42, 189.00 (chalcone C=O) IR ν max (ATR): 3071.36, 2882.33 (Ar C-H), 1778.67, 1670.45 (C=O), 1619.17 (C=C), 1604.57, 1580.44 (Ar C=C), 1465.80 (CH), 1374.93 (CH), 1224.51 (C-O) cm^{-1}. HRMS (APCI): calculated for $C_{29}H_{23}O_4$: 435.1591 [M^++H]; found: 435.1595.

9-(*E*)-3-(2,4-Dichlorophenyl)-3-oxoprop-1-en-1-yl)-9,10-dihydro-9,10-[3,4]furanoanthracene-12,14-dione (**22d**) was prepared from (*E*)-3-(anthracen-9-yl)-1-(2,4-dichlorophenyl)prop-2-en-1-one (1.0 mmol, 0.377 g) and maleic anhydride (1.3 mmol, 0.13 g) according to the general procedure above; beige powder (76%), Mp. 232–236 °C. ^{1}H nmR (400 MHz, $CDCl_3$) δ 3.60 (dd, *J* = 9.12, 3.32 Hz, 1H), 3.72 (d, *J* = 9.54 Hz, 1H), 4.84 (d, *J* = 2.90 Hz, H), 7.20–7.28 (m, 5H, ArH), 7.28–7.34 (m, 2 H, ArH, 1, CH=C), 7.35–7.39 (m, 1H, ArH), 7.39–7.44 (m, 2H, ArH), 7.52 (d, *J* = 1.66 Hz, 1H, ArH), 7.68 (d, *J* = 8.29 Hz, 1H, ArH), 7.75 (d, *J* = 16.59 Hz, 1H, CH=C). ^{13}C nmR (101 MHz, $CDCl_3$) ppm 45.49, 49.12, 51.74, 123.44, 123.73, 124.30, 125.64, 127.11, 127.63, 127.92, 128.16, 130.46, 130.95, 134.60, 140.21, 141.14, 143.00, 168.53, 169.72, 191.11 (chalcone C=O) ppm. IR ν max (KBr): 3071.19 (Ar C-H), 1776.62, 1662.20 (C=O), 1619.95 (C=C), 1580.28, 1456.96 (Ar C=C), 1212.92 (C-O) cm^{-1}. HRMS (APCI) calculated for $C_{27}H_{17}Cl_2O_4$: 475.0498 [M^++H]; found 475.0491.

9-(*E*)-3-(3,4-Dimethoxyphenyl)-3-oxoprop-1-en-1-yl)-9,10-dihydro-9,10-[3,4]furanoanthracene-12,14-dione (**22e**) was prepared from (*E*)-3-(anthracen-9-yl)-1-(3,4-dimethoxyphenyl)prop-2-en-1-one (1.0 mmol, 0.3684 g) and maleic anhydride (1.3 mmol, 0.13 g) according to the general procedure above; cream/yellow powder (83%), Mp. 243–246 °C. ^{1}H nmR (400 MHz, $CDCl_3$) δ 3.63 (dd, *J* = 9.33, 3.11 Hz, 1H, CH), 3.81 (d, *J* = 9.54 Hz, 1H, CH), 3.87 (s, 3H, OCH_3), 3.89 (s, 3H, OCH_3), 4.85 (d, *J* = 3.32 Hz, 1H, CH), 6.96 (d, *J* = 8.29 Hz, 1H, ArH), 7.19–7.27 (m, 4H, ArH), 7.31–7.37 (m, 2H, ArH), 7.37–7.43 (m, 2H, ArH), 7.73–7.76 (m, 2H, ArH, CH=C), 7.78 (s, 1 H), 7.82 (dd, *J* = 8.29, 2.07 Hz, 1 H, ArH), 7.87 (d, *J* = 16.2 Hz, 1H, CH=C) ppm. ^{13}C nmR (101 MHz, $CDCl_3$) ppm 49.30, 54.1, 54.38, 56.31, 60.74, 61.03, 110.27, 111.02, 123.59, 123.70, 124.07, 125.53, 127.08, 127.54, 127.89, 128.04, 130.68, 131.47, 137.43, 139.09, 139.75, 140.26, 141.75, 149.30, 153.67, 175.36, 176.17, 192.77 (chalcone C=O). IR ν max (ATR): 3071.19 (Ar C-H), 1777.39, 1668.50 (C=O), 1619.46 (C=C), 1580.35, 1456.83 (Ar C=C), 1213.57, 1018.24 (C-O) cm^{-1}. HRMS (ESI): calculated for $C_{29}H_{22}NaO_6$; 489.1309 [M^+−H] found: 489.1305.

(*E*)-9-(3-Oxo-3-phenylprop-1-en-1-yl)-9,10-dihydro-9,10-[3,4]furanoanthracene-12,14-dione (**22f**) was synthesized from (*E*)-3-(anthracene-9-yl)-1-phenylprop-2-en-1-one (1.0 mmol, 0.308 g) and maleic anhydride, (1.3 mmol, 0.13 g) according to the general procedure above; white powder (91%), Mp 214–218 °C. ^{1}H nmR (400 MHz, DMSO-d_6) δ 3.81 (dd, *J* = 9.12, 3.32 Hz, 1H), 4.27 (d, *J* = 9.12 Hz, 1H), 4.96 (d, *J* = 3.32 Hz, 1H), 7.19–7.29 (m, 5H), 7.30–7.35 (m, 1H), 7.39–7.45 (m, 1H), 7.56 (d, *J* = 6.63 Hz, 1H), 7.58–7.65 (m, 2H), 7.68–7.74 (m, 1H), 7.77–7.92 (m, 2H), 8.14–8.21 (m, 2H). ^{13}C nmR (101 MHz, DMSO-d_6) 38.57–41.35, 44.84, 48.69–50.08, 51.64, 122.68–124.27,124.99, 125.48–125.98, 127.06, 127.40–128.13, 128.64, 128.91–129.60, 132.07, 134.01, 138.96,141.21, 142.29, 170.61, 171.43, 189.75 (chalcone C=O) ppm. IR ν max (ATR): 1774.31 (C=O), 1634.26 (C=C), 1221.80 (C-O) cm^{-1}. HRMS (APCI): calculated for $C_{27}H_{18}O_4$ [M^++Na]:429.1097; found 429.1101.

9-(*E*)-3-(4-Iodophenyl)-3-oxoprop-1-en-1-yl)-9,10-dihydro-9,10-[3,4]furanoanthracene-12,14-dione (**22g**) was prepared from (*E*)-3-(anthracen-9-yl)-1-(4-iodophenyl)prop-2-en-1-

one (1.0 mmol, 0.434g) and maleic anhydride (1.3 mmol, 0.13 g) according to the general procedure above, light yellow powder (81%), Mp. 228–231 °C. ^{1}H nmR (400 MHz, $CDCl_3$) δ 3.63 (dd, *J* = 9.12, 2.90 Hz, 1H, CH), 3.79 (d, *J* = 9.12 Hz, 1H, CH), 4.86 (d, *J* = 3.32 Hz, 1H, CH), 7.15–7.34 (m, 6H, ArH), 7.36–7.44 (m, 2 H, ArH), 7.70 (d, *J* = 16.17 Hz, 1H, CH=C), 7.83–7.97 (m, 5 H, ArH, CH=C). ^{13}C nmR (101 MHz, $CDCl_3$) ppm 45.46, 48.97, 49.27, 51.93, 123.53, 123.88, 124.20, 125.59, 127.12, 127.63, 127.95, 128.13, 130.24, 130.88, 136.77, 137.38, 138.17, 138.86, 140.22, 141.53, 168.83, 189.12 (chalcone C=O). IR ν max (ATR): (C-H) 3016.49, 2876.80, (*trans* C=C) 1861.84, (C=O) 1771.38, (Ar C=C) 1619.73, 1579.29, 1479.43, 1456.94, (C-O) 1292.32, (C-I) 619.66 cm^{-1}. LRMS (APCI): 531.20 [$M^{+}-H$]. HRMS (APCI): calculated for $C_{27}H_{16}IO_4$ [$M^{+}-H$]: 531.0099; found 531.0102.

9-(*E*)-3-(4-Fluorophenyl)-3-oxoprop-1-en-1-yl)-9,10-dihydro-9,10-[3,4]furanoanthracene -12,14-dione (**22h**) was prepared from (*E*)-3-(anthracen-9-yl)-1-(4-fluorophenyl)prop-2-en-1-one (1.0 mmol, 0.3264 g) and maleic anhydride (1.3 mmol, 0.13 g) according to the general procedure above; cream/yellow powder (79%), Mp. 242–247 °C. (HPLC: 97.8%, RT 6.17 min). ^{1}H nmR (400 MHz, $CDCl_3$) δ 3.64 (dd, *J* = 9.54, 3.32 Hz, 1H, CH), 3.81 (d, *J* = 9.12 Hz, 1H, CH), 4.87 (d, *J* = 3.32 Hz, 1H, CH), 7.18–7.28 (m, 6H, ArH), 7.29–7.36 (m, 2 H, ArH), 7.36–7.45 (m, 2H, ArH), 7.75 (d, *J* = 16.59 Hz, 1H, CH=C), 7.93 (d, *J* = 16.59 Hz, CH=C), 8.16–8.24 (m, 2H, ArH). ^{13}C nmR (101 MHz, $CDCl_3$) ppm 45.47, 48.96, 49.30, 51.91, 115.73–116.31, 123.54, 123.91, 124.17, 125.58, 127.12, 127.61, 128.03, 131.06, 131.59, 137.38, 140.23, 141.09, 141.58,168.86, 169.78, 188.26 (chalcone C=O). IR ν max (ATR): 3064.79, 2968.92 (Ar C-H), 1773.01 (C=O),1667.75 (C=C), 1625.39, 1585.13, 1481.57, 1458.78 (Ar C=C), 1292.88 (C-F), 1069.82 (C-O) cm^{-1}. LRMS (APCI): 423.23 [$M^{+}-H$]. HRMS (APCI): calculated for $C_{27}H_{16}FO_4$ [$M^{+}-H$]: 423.103811; found 423.103417.

(*E*)-9-(3-(4-Methoxyphenyl)-3-oxoprop-1-en-1-yl)-9,10-dihydro-9,10-[3,4]furanoanthracene -12,14-dione (**22i**) was synthesized from (*E*)-3-(anthracen-9-yl)-1-(4-methoxyphenyl)prop-2-en-1-one (1.0 mmol, 0.338 g) and maleic anhydride (1.3 mmol, 0.128 g) according to the general procedure above; (30%), Mp. 200–205 °C. ^{1}H nmR (400 MHz, $CDCl_3$) δ 3.81 (dd, *J* = 8.92, 3.11 Hz, 1H) 3.87 (s, 3H) 4.27 (d, *J* = 9.12 Hz, 1H) 4.96 (d, *J* = 3.32 Hz, 1H) 7.10–7.16 (m, 2H) 7.20 (d, *J* = 3.73 Hz, 2H) 7.21 -7.29 (m, 4H) 7.29–7.33 (m, 1H) 7.42 (dd, *J* = 4.98, 3.73 Hz, 1H) 7.56 (d, *J* = 6.63 Hz, 1H) 7.73–7.91 (m, 1H) 8.15–8.20 (m, 2H)^{13}C nmR (101 MHz, $CDCl_3$) ppm 40.15–40.69, 44.50, 49.19–50.26, 51.26, 51.57, 55.08, 57.04–58.17, 114.72, 122.69, 127.03, 130.49, 131.59, 138.97, 140.11, 141.27, 142.39, 170.58, 171.44, 175.90, 185.10, 187.87 (chalcone C=O). IR ν max (ATR): 1670.35 (C=O), 1600.48 (Ar C=C), 1229.19 (C-O) cm^{-1}. HRMS (APCI): calculated for $C_{28}H_{19}O_5$ [$M^{+}-H$]: 435.1238; found: 435.1241.

(*E*)-9-(3-Oxo-3-(*p*-tolyl)prop-1-en-1-yl)-9,10-dihydro-9,10-[3,4]furanoanthracene-12,14-dione (**22j**) was synthesized from (*E*)-3-(anthracen-9-yl)-1-(*p*-tolyl)prop-2-en-1-one (1.0 mmol, 0.3384 g) and maleic anhydride (1.3 mmol, 0.13 g) according to the general procedure above; white powder (73%), Mp. 223–227 °C [51]. ^{1}H nmR (400 MHz, $CDCl_3$) δ 2.41 (s, 3H) 3.81 (dd, *J* = 8.92, 3.11 Hz, 1H) 4.27 (d, *J* = 9.12 Hz, 1H) 4.96 (d, *J* = 2.90 Hz, 1H) 7.20 (d, *J* = 3.73 Hz, 2H) 7.21–7.29 (m, 3H) 7.29–7.34 (m, H) 7.42 (d, *J* = 7.88 Hz, 3H) 7.56 (d, *J* = 6.63 Hz, 1H) 7.82 (q, *J* = 16.17 Hz, 2H) 8.08 (d, *J* = 7.88 Hz, 2H). ^{13}C nmR (101 MHz, $CDCl_3$) ppm 21.70, 44.81, 49.00–50.00, 123.38, 124.98, 127.04, 127.49, 127.85, 129.30, 129.99, 132.05, 135.15, 138.96, 140.06, 140.69, 142.33, 144.52, 170.59, 171.43, 189.16 (chalcone C=O). IR ν max (ATR): 1769.23 (C=O), 1603.35 (Ar C=C), 1458.64 (C-CH_3), 1228.91 (C-O) cm^{-1}. LRMS (APCI): 419.30 [$M^{+}+H$]. HRMS (APCI): calculated for $C_{29}H_{24}NaO_5$ [$M^{+}+Na$]: 475.1516; found 475.1512.

(*E*)-9-(3-Oxo-3-(3,4,5-trimethoxyphenyl)-prop-1-en-1-yl)-9,10-dihydro-9,10-[3,4]furanoanthracene-12,14-dione (**22k**) was synthesized from (*E*)-3-(anthracene-9-yl)-1-(4-chorophenyl) prop-2-en-1-one (1.0 mmol, 0.398 g) and maleic anhydride (1.3 mmol, 0.13 g) according to the general procedure above; white powder (78%), Mp. 262–266 °C. ^{1}H nmR (400 MHz, $CDCl_3$) δ 3.77 (s, 3H), 3.80 (dd, *J* = 9.12, 3.32 Hz, 1H), 3.87 (s, 6H) 4.27 (d, *J* = 9.12 Hz, 1H), 4.96 (d, *J* = 3.32 Hz, 1H) 7.20–7.23 (m, 2H) 7.23–7.28 (m, 3H) 7.32–7.35 (m, 1H) 7.40–7.43 (m, 1H) 7.48 (s, 2H) 7.56 (d, *J* = 6.22 Hz, 1H) 7.72–7.91 (m, 2 H). ^{13}C nmR (101 MHz, $CDCl_3$) ppm 43.55, 44.81, 49.23–49.97, 56.55, 60.70, 106.84, 123.39, 123.96, 125.00, 125.78, 127.50,

132.58, 132.95, 138.99, 139.83–140.53, 153.45, 171.40, 188.92 (chalcone C=O). IR *ν* max (ATR): 1662.19 (C=O), 1571.62 (Ar C=C), 1231.78 (C-O) cm^{-1}. HRMS (APCI): calculated $C_{30}H_{25}O_7$ [M$^+$+H]: 495.1449; found: 495.1449.

(*E*)-9-(3-(4-Chlorophenyl)-3-oxoprop-1-en-1-yl)-9,10-dihydro-9,10[3,4]furanoanthracene-12,14-dione (**22l**) was synthesized from (*E*)-3-(anthracen-9-yl)-1-(4-chlorophenyl)prop-2-en-1-one (1.0 mmol, 0.3428 g) and maleic anhydride (1.3 mmol, 0.13 g) according to the general procedure above; white powder (89%), Mp. 248–252 °C. ^{1}H nmR (400 MHz, $CDCl_3$) δ 3.81 (dd, *J* = 8.71, 3.32 Hz, 1H), 4.27 (d, *J* = 9.12 Hz, 1H), 4.96 (d, *J* = 3.32 Hz, H), 7.18–7.29 (m, 5H), 7.30–7.35 (m, 1H), 7.42 (dd, *J* = 5.60, 3.11 Hz, 1H), 7.56 (d, *J* = 7.05 Hz, 1H), 7.67–7.72 (m, 2H), 7.79–7.90 (m, 2H), 8.16–8.21 (m, 2H). ^{13}C nmR (101 MHz, $CDCl_3$) ppm 39.13–40.72, 44.90, 125.51, 128.81, 129.41–129.54, 130.18, 131.84, 134.30, 135.66, 142.34, 143.69, 148.97, 149.27, 150.80, 171.10, 191.34 (chalcone C=O). IR *ν* max (ATR): 1773.83 (C=O), 1666.32 (C=C), 1618.07 (Ar C=C), 1221.94 (C-O), 712.12 (C-Cl) cm^{-1} HRMS (APCI): calculated for $C_{27}H_{17}ClO_4$ [M$^+$+H]: 439.0743; found: 439.0739.

9-(*E*)-3-Oxo-3-(pyridin-4-yl)prop-1-en-1-yl)-9,10-dihydro-9,10-[3,4]furanoanthracene-12,14-dione (**22m**) was prepared from (*E*)-3-(anthracen-9-yl)-1-(pyridin-4-yl)prop-2-en-1-one (1.0 mmol, 0.309 g) and maleic anhydride (1.3 mmol, 0.13 g) according to the general procedure above; light brown solid (99%), Mp. >200 °C. ^{1}H nmR (400 MHz, $CDCl_3$) δ 3.65 (dd, *J* = 9.12, 2.90 Hz, 1H, CH), 3.80 (d, *J* = 9.12 Hz, 1H, CH), 4.87 (d, *J* = 2.49 Hz, 1H, CH), 7.10–7.23 (m, 3H, ArH), 7.25–7.34 (m, 3H, ArH), 7.36–7.46 (m, 2H, ArH), 7.70 (d, *J* = 16.17 Hz, 1H, CH=C), 7.94 (d, *J* = 4.98 Hz, 2 H, ArH), 8.01 (d, *J* = 16.17 Hz, CH=C). ^{13}C nmR (101 MHz, $CDCl_3$) ppm 45.44, 48.99, 49.24, 51.99, 123.42, 123.71, 124.33, 125.70, 127.18, 127.77, 128.02, 128.28, 130.41, 140.20, 143.49, 150.40, 168.91, 169.60, 188.91 (chalcone C=O). IR *ν* max (KBr): 3064.92, 2969.14 (Ar C-H), 1773.15 (C=O), 1638.35, 1677.99 (C=C), 1625.35, 1585.28, 1458.89 (Ar C=C), 1069.92 (C-N) cm^{-1}. HRMS (APCI) calculated for $C_{26}H_{18}NO_4$ [M$^+$+H]: 408.1230 found: 408.1231.

9-(*E*)-3-Oxo-3-(pyridin-2-yl)prop-1-en-1-yl)-9,10-dihydro-9,10-[3,4]furanoanthracene-12,14-dione (**22n**) prepared from (*E*)-3-(anthracen-9-yl)-1-(pyridin-2-yl)prop-2-en-1-one (1.0 mmol, 0.3094 g) and maleic anhydride (1.3 mmol, 0.13 g) according to the general procedure above; white powder (100%), Mp. 225–227 °C. ^{1}H nmR (400 MHz, $CDCl_3$) δ 3.62 (dd, *J* = 9.12, 3.32 Hz, 1H, CH), 3.93 (d, *J* = 9.54 Hz, 1H, CH), 4.85 (d, *J* = 3.32 Hz, 1H, CH), 7.13–7.24 (m, 3H, ArH), 7.25–7.31 (m, 3 H,ArH), 7.36–7.45 (m, 3 H ArH), 7.51 (ddd, *J* = 7.57, 4.67, 1.04 Hz, 1 ArH), 7.93 (td, *J* = 7.67, 1.66 Hz, 1H, ArH), 8.13 (d, *J* = 16.59 Hz, 1 H, CH=C), 8.29 (dt, *J* = 7.88, 1.04 Hz, 1H, ArH), 8.42 (d, *J* = 16.59 Hz, 1H, CH=C), 8.70–8.73 (m, 1H, ArH). IR *ν* max (ATR): 3087.05, 3064.46, 2968.93 (Ar C-H), 1770.84 (C=O), 1676.99 (C=C), 1625.37, 1584.71, 1479.76, 1456.99 (Ar C=C), 1325.60 (C-N), 1084.30 (C-O) cm^{-1}. HRMS (APCI): calculated for $C_{26}H_{18}NO_4$: 408.1230 [M$^+$+H]; found: 408.1237.

9-(*E*)-3-(Furan-2-yl)-3-oxoprop-1-en-1-yl)-9,10-dihydro-9,10-[3,4]furanoanthracene-12,14-dione (**22o**) was prepared from (*E*)-3-(anthracen-9-yl)-1-(furan-2-yl)prop-2-en-1-one (1.0 mmol, 0.298 g) and maleic anhydride (1.3 mmol, 0.13 g) according to the general procedure above; cream powder (81%), Mp. 222–225 °C. ^{1}H nmR (400 MHz, $CDCl_3$) δ 3.63 (dd, *J* = 9.54, 3.32 Hz, 1H, CH), 3.82 (d, *J* = 9.12 Hz, 1H, CH), 4.85 (d, *J* = 3.32 Hz, 1H, CH), 6.63 (dd, *J* = 3.73, 1.66 Hz, 1H, CH=C), 7.15–7.29, 5H, ArH, CH=C), 7.33–7.40 (m, 2H, ArH), 7.40–7.46 (m, 2H, ArH), 7.63 (d, *J* = 16.59 Hz, 1H, CH=C), 7.69 (d, *J* = 1.24 Hz, 1H, CH=C), 8.02 (d, *J* = 16.17 Hz, 1H, CH=C). ^{13}C nmR (101 MHz, $CDCl_3$) ppm 45.50, 49.08, 49.26, 51.82, 112.78, 118.91, 123.63, 123.95, 124.15, 125.53, 127.05, 127.56, 127.91, 128.04, 130.36, 137.38, 138.96, 140.29, 140.61, 141.55, 147.26, 153.19, 168.65, 169.93, 177.01 (chalcone C=O). IR *ν* max (ATR): 3071.48, 2882.79 (Ar C-H), 1776.90, 1668.01 (C=O), 1580.26, 1405.03 (Ar C=C), 1226.47, 1071.29 (C-O) cm^{-1}. HRMS (APCI): calculated for $C_{25}H_{17}O_5$: 397.1071 [M$^+$+H]; found: 397.1074.

9-(*E*)-3-(Naphthalen-2-yl)-3-oxoprop-1-en-1-yl)-9,10-dihydro-9,10-[3,4]furanoanthracene-12,14-dione (**22p**) was prepared from (*E*)-3-(anthracen-9-yl)-1-(naphthalen-2-yl)prop-2-en-1-one (1.0 mmol, 0.3584 g) and maleic anhydride (1.3 mmol, 0.13 g) according to the general procedure above; green crystals (86%), Mp. 226–231 °C. ^{1}H nmR (400 MHz, $CDCl_3$) δ 3.66

(dd, *J* = 9.12, 3.32 Hz, 1H), 3.86 (d, *J* = 9.54 Hz, 1 H), 4.88 (d, *J* = 3.32 Hz, 1H), 7.17–7.31 (m, 4H), 7.34–7.46 (m, 4H), 7.54–7.65 (m, 2H), 7.88–7.94 (m, 2H), 7.96–8.03 (m, 3H), 8.22 (dd, *J* = 8.71, 1.66 Hz, 1H), 8.71 (s, 1 H). ^{13}C nmR (101 MHz, $CDCl_3$) ppm 55.53–56.38, 60.88, 69.77–71.12, 106.07, 106.66, 107.75, 110.78–111.13, 114.44, 114.76, 123.43, 126.97, 127.76, 128.45. IR ν max (ATR): 3064.36, 2968.90 (Ar C-H), 1771.08 (C=O), 1676.62, 1625.39 (C=C), 1584.75, 1458.10 (Ar C=C),1 325.74 (C-O) cm^{-1}. HRMS (APCI): calculated for $C_{31}H_{21}O_4$ [M$^+$+H]: 457.1434; found: 457.1438.

9-(*E*)-3-Oxo-3-(thiophen-2-yl)prop-1-en-1-yl)-9,10-dihydro-9,10-[3,4]furanoanthracene-12,14-dione (**22q**) was prepared from (*E*)-3-(anthracen-9-yl)-1-(thiophen-2-yl)prop-2-en-1-one (1.0 mmol, 0.3144 g) and maleic anhydride (1.3 mmol, 0.13 g) according to the general procedure above; cream powder (76%), Mp. 214–218 °C. ^{1}H nmR (400 MHz, $CDCl_3$) δ 3.64 (dd, *J* = 9.12, 2.90 Hz, 1H, CH), 3.81 (d, *J* = 9.12 Hz, 1H, CH), 4.86 (d, *J* = 2.90 Hz, 1H, CH), 7.11–7.28 (m, 5 H, ArH, CH=C), 7.28–7.46 (m, 4H, ArH), 7.66 (d, *J* = 16.17 Hz, 1H, CH=C), 7.75 (d, *J* = 4.98 Hz, 1H, CH=C), 7.91–8.01 (m, 2H, CH=C). ^{13}C nmR (101 MHz, $CDCl_3$) ppm 45.40, 48.99, 49.63, 52.01, 56.07, 110.20, 111.07, 123.48, 123.70, 123.93, 125.41, 126.70, 127.24, 127.41, 130.88, 131.48, 139.42, 140.66–140.97, 142.44, 149.24, 153.52, 174.93, 175.64, 188.72 (chalcone C=O). IR ν max (ATR): 3089.66, 2882.60 (Ar C-H), 1777.72, 1664.07 (C=O) 1616.88 (C=C), 1579.42, 1512.85, 1465.10, 1456.62, 1418.52 (Ar C=C), 1213.16 (C-O) cm-^{1}. LRMS (APCI): 411.22 [M$^+$−H]. HRMS (APCI): calculated for $C_{25}H_{16}NaO_4S$ [M$^+$+Na]: 435.0662; found: 435.0658.

3.4. General Preparation for (E)-9-(3-oxo-3-phenylprop-1-en-yl)-9,10-dihydro-9,10-[3,4]epipyrroloanthracene-12,14-diones **23a-q**

To a solution of the appropriate chalcone anthracene analog **21a–21q** (1.0 mmol) in toluene (2 mL) was added dienophile maleimide (1.3 mmol). The mixture was heated and stirred at 90 °C for 48 h. The reaction mixture was then cooled to room temperature, and the resulting product was isolated via filtration and washed with toluene (2 mL) and diethyl ether (2 mL). The product was then recrystallized from toluene.

9-(*E*)-3-(4-Bromophenyl)-3-oxoprop-1-en-1-yl)-9,10-dihydro-9,10-[3,4]epipyrroloanthracene-12,14-dione (**23a**) was prepared from (*E*)-3-(anthracen-9-yl)-1-(4-bromophenyl)prop-2-en-1-one (1.0 mmol, 0.387 g) and maleimide (1.3 mmol, 0.13g) according to the general procedure above; light yellow powder (91%), Mp. 295–297 °C. (HPLC: 87.8%, RT 5.02 min). ^{1}H nmR (400 MHz, DMSO-d_6) δ 3.32 (m, 1H, CH), 3.82 (d, *J* = 8.82 Hz, 1H, CH), 4.81 (d, *J* = 3.32 Hz, 1H, CH), 7.11–7.22 (m, 5 H, ArH, CH=C), 7.23–7.29 (m, 1H, ArH), 7.30–7.35 (m, 1H, ArH), 7.51 (d, *J* = 7.05 Hz, 1 H, ArH), 7.78–7.85 (m, 4H), 8.06–8.12 (m, 2H), 10.87 (s, 1H). ^{13}C nmR (101 MHz, $CDCl_3$) ppm 44.96, 49.09, 49.30, 51.70, 123.24, 123.66, 124.61, 126.65, 127.16, 127.28, 131.14, 131.47, 132.47, 136.76, 139.29, 140.24, 142.15, 143.10, 143.11, 177.48, 177.74, 189.08 (chalcone C=O). IR ν max (ATR): 3341 (N-H), 3096.65 (C-H), 1720 (C=O), 1629.57 (Ar C=C), 1603.21, 1522.86, 1458.09, (C-N) 1311, 1165, (C-Br) 1003.06 cm^{-1}. HRMS (EI) calculated for $C_{27}H_{17}BrNO_3$ [M$^+$−H]: 482.0397; found 482.0391.

9-(*E*)-3-(4-Nitrophenyl)-3-oxoprop-1-en-1-yl)-9,10-dihydro-9,10-[3,4]epipyrroloanthracene-12,14-dione (**23b**) was prepared from (*E*)-3-(anthracen-9-yl)-1-(4-nitrophenyl)prop-2-en-1-one (1.0 mmol, 0.353 g) and maleimide (1.3 mmol, 0.13 g) according to the general procedure above; beige powder (87%), Mp. >300 °C (dec). ^{1}H nmR (400 MHz, DMSO-*d6*) δ 3.35 (d, *J* = 3.32 Hz, 1H, CH), 3.82 (d, *J* = 8.71 Hz, 1H, CH), 4.82 (d, *J* = 2.90 Hz, 1H, CH), 7.15–7.23 (m, 4H, ArH), 7.28–7.33 (m, 1 H, ArH), 7.34–7.39 (m, 1H, ArH), 7.52–7.57 (m, 1H, ArH), 7.82–7.89 (d, *J* = 15.8 Hz, 1 H, CH=C), 7.89–8.01 (d, *J* = 16.2 Hz, 1H, CH=C,), 8.37–8.47 (m, 4H, ArH), 10.88 (s, 1H). ^{13}C nmR (101 MHz, DMSO-d_6) 20.05, 24.86, 44.97, 49.24, 51.78, 63.44, 73.92, 78.91, 88.39, 109.61, 111.59, 118.64, 123.30, 123.71, 124.64, 125.65, 126.66, 127.25, 130.51, 131.69, 139.28, 140.12, 142.13, 142.52, 142.96, 144.40, 150.46,177.52, 177.70, 189.16 (chalcone C=O) ppm. IR ν max (ATR): 3343 (N-H), 3058.45 (C-H), 1719 (C=O), 1677.02, 1638.82 (C=C), 1599.84, 1458.22 (Ar C=C), 1517, 1340 (NO_2), 1317, 1164 (C-N) cm-^{1}. HRMS (EI): calculated for $C_{27}H_{17}N_2O_3$ [M$^+$−H]: 449.1143; found: 449.1146.

9-((*E*)-3-(4-Ethylphenyl)-3-oxoprop-1-en-1-yl)-9,10-dihydro-9,10-[3,4]epipyrroloanthracene-12,14-dione (**23c**) was prepared from (*E*)-3-(anthracen-9-yl)-1-(4-ethylphenyl)prop-2-en-1-one (1.0 mmol, 0.336 g) and maleimide (1.3 mmol, 0.13 g) according to the general procedure above; white powder (82%), Mp. 265–269 °C. ^{1}H nmR (400 MHz, DMSO-d_6) δ 1.24 (t, *J* = 7.58 Hz, 3H), 2.73 (q, *J* = 7.83 Hz, 2H), 3.32 (d, *J* = 3.42 Hz, 1H), 3.82 (d, *J* = 8.31 Hz, 1H), 4.82 (d, *J* = 3.42 Hz, 1H), 7.11–7.26 (m, 5H), 7.27–7.32 (m, 6H), 7.34 (d, *J* = 6.85 Hz, 1H), 7.80–7.84 (m, 1H), 7.80–7.84 (m, 1H), 7.87 (d, *J* = 16.14 Hz, 1H), 8.13 (d, *J* = 8.31 Hz, 2H), 10.87 (br. s., 1H). IR ν max (ATR): (N-H) 3343 (C-H) 3088.25, (*trans* C=C), 1776.11, 1720 (C=O), 1626.04, 1603.36 (Ar C=C), 1522.96, 1457.88, (C-N) 1313, 1165 cm^{-1}. HRMS (EI) calculated for $C_{29}H_{23}NNaO_3$ [M$^+$+Na]: 456.1570; found: 456.1562.

9-(*E*)-3-(2,4-Dichlorophenyl)-3-oxoprop-1-en-1-yl)-9,10-dihydro-9,10-[3,4]epipyrroloanthracene-12,14-dione (**23d**) was prepared from (*E*)-3-(anthracen-9-yl)-1-(2,4-dichlorophenyl) prop-2-en-1-one (1.0 mmol, 0.377 g) and maleimide (1.3 mmol, 0.13 g) according to the general procedure above; beige powder (82%), Mp. 265–268 °C. ^{1}H nmR (400 MHz, DMSO-d_6) δ 3.24 (dd, *J* = 8.6, 3.2 Hz, 1H), 3.61 (d, *J* = 8.6 Hz, 1H), 4.76 (d, *J* = 3.2 Hz, 1H), 7.23–7.10 (m, 7H), 7.31 (dq, *J* = 4.7, 2.6 Hz, 1H), 7.49 (dd, *J* = 6.6, 1.8 Hz, 1H), 7.67–7.53 (m, 2H), 7.86–7.79 (m, 2H), 10.81 (s, 1H) ^{13}C nmR (101 MHz, $CDCl_3$) ppm 44.92, 49.24, 51.71, 123.17, 123.47, 124.73, 125.68, 126.63, 127.12, 127.26, 127.38, 128.23, 130.31, 131.41, 131.87, 135.33, 136.45, 137.33, 139.23, 139.82, 142.09, 142.59, 146.63, 177.28, 177.68, 192.42 (chalcone C=O) IR ν max (ATR): 3336 (N-H), 3062.21 (C-H), 1778.31, 1720 (C=O), 1601.84 (Ar C=C), 1578.57, 1521.96, 1457.78, (C-N) 1315, 1171 cm^{-1}. HRMS (ESI): calculated for $C_{27}H_{16}Cl_2NO_3$ [M$^+$−H]: 472.0513; found: 472.0519.

9-(E)-3-(3,4-Dimethoxyphenyl)-3-oxoprop-1-en-1-yl)-9,10-dihydro-9,10-[3,4]epipyrroloanthracene-12,14-dione (**23e**) was prepared from (*E*)-3-(anthracen-9-yl)-1-(3,4-dimethoxyphenyl)prop-2-en-1-one (1.0 mmol, 0.368 g) and maleimide (1.3 mmol, 0.13 g) according to the general procedure above, cream/yellow powder (83%), Mp. 269–271 °C. ^{1}H nmR (400 MHz, $CDCl_3$) δ 3.37 (dd, *J* = 8.5, 3.2 Hz, 1H), 3.56 (d, *J* = 8.6 Hz, 1H), 3.98 (d, *J* = 9.3 Hz, 6H), 4.81 (d, *J* = 3.2 Hz, 1H), 6.97 (d, *J* = 8.4 Hz, 1H), 7.14–7.24 (m, 4H), 7.35 (ddd, *J* = 10.9, 6.8, 2.5 Hz, 3H), 7.38–7.45 (m, 1H), 7.47 (s, 1H), 7.75–7.84 (m, 2H), 7.85 (dd, *J* = 8.4, 2.0 Hz, 1H), 7.91 (d, *J* = 16.2 Hz, 1H). ^{13}C nmR (101 MHz, $CDCl_3$) ppm 45.41, 48.99, 49.63, 52.00, 56.05, 56.09, 110.20, 111.07, 123.48, 123.70, 123.90, 123.95, 125.41, 126.70, 127.20, 127.27, 127.41, 130.88, 131.48, 137.96, 139.42, 140.87, 142.44, 149.24, 153.52, 174.95, 175.67, 188.72 (chalcone C=O). IR ν max (ATR): 3342 (N-H), 3061.45 (C-H), 1774.72, 1719 (C=O), 1626.19 (Ar C=C), 1598.13, 1517.42, 1458.15, (C-N) 1317, 1169, (C-O) 1268.92, 1017.26 cm^{-1}. HRMS (ESI): calculated for $C_{29}H_{22}NO_5$ [M$^+$−H] 464.1504; found: 464.1504.

(*E*)-9-(3-oxo-3-phenylprop-1-en-yl)-9,10-dihydro-9,10-[3,4]epipyrroloanthracene-12,14-dione (**23f**) was synthesized from (*E*)-3-(anthracen-9-yl)-1-phenylprop-2-en-1-one (1.0 mmol, 0.308 g) and maleimide (1.3 mmol, 0.13 g) according to the general procedure above; (61%), Mp. 218–222 °C. ^{1}H nmR (400 MHz, DMSO-d_6) δ 3.80 (d, *J* = 8.71 Hz, 1H) 4.79 (d, *J* = 3.32 Hz, 1H) 7.10–7.24 (m, 6H) 7.24–7.29 (m, 1H) 7.32–7.37 (m, 1H) 7.52 (d, *J* = 6.63 Hz, 1H) 7.57–7.65 (m, 2H) 7.67 -7.74 (m, 1H) 7.84 (dd, *J* = 16.60 Hz, 2H) 8.12–8.21 (m, 2H) 10.84 (s, 1H). ^{13}C nmR (101 MHz, DMSO-d_6) 123.23, 123.64, 124.62, 125.66, 125.76, 126.67 127.24, 128.64, 129.15, 129.30–129.46 131.65–131.81, 133.89, 137.78, 139.32, 140.34, 142.55, 143.20, 177.31–177.97, 189.92 (chalcone C=O) ppm. IR ν max (ATR): 3347.06 (N-H), 1720.18 (C=O), 1674.84, 1628.01 (C=C), 1000.09 (CN) cm^{-1}. HRMS (APCI): calculated for $C_{27}H_{20}NO_3$ [M$^+$+H]: 406.1438; found: 406.1450.

9-(*E*)-3-(4-Iodophenyl)-3-oxoprop-1-en-1-yl)-9,10-dihydro-9,10-[3,4]epipyrroloanthracene-12,14-dione (**23g**) was prepared from (*E*)-3-(anthracene-9-yl)-1-(4-iodophenyl)prop-2-en-1-one (1.0 mmol, 0.449 g) and maleimide (1.3 mmol, 0.13 g) according to the general procedure above; yellow powder (93%), Mp. 288–293 °C. (HPLC: 98.22, RT 6.72 min). ^{1}H nmR (400 MHz, DMSO-d_6) δ 3.79 (d, J = 8.6 Hz, 1H), 3.31–3.29 (m,1H). 4.80 (d, J = 3.2 Hz, 1H), 7.29–7.24 (m,1H), 7.23–7.13 (m, 5H), 7.37–7.31 (m, 1H), 7.52 (dt, J = 7.0, 1.1 Hz, 1H), 7.82 (d, J = 6.7 Hz, 2H), 7.95–7.91 (m, 2H), 8.03–7.98 (m, 2H), 10.85 (s, 1H). ^{13}C nmR (101 MHz, DMSO-d_6) 44.98, 49.13, 49.32, 51.72, 102.63, 123.26, 123.67, 124.62, 125.65, 126.66, 127.17,

127.29, 130.84, 131.45, 137.06, 138.35, 139.30, 140.27, 142.17, 143.05, 143.12, 177.48, 177.74, 189.40. ppm IR ν max (ATR): 3339 (N-H), 2987.83 (C-H),1688.48 (C=O) 1633, 1621.45 (Ar C=C), 1557.57, 1485.56, 687.18 cm^{-1}. HRMS (APCI): calculated $C_{27}H_{19}INO_3$ [M^{+}+H]: 532.0404, found 532.0402.

9-(*E*)-3-(4-Fluorophenyl)-3-oxoprop-1-en-1-yl)-9,10-dihydro-9,10-[3,4]epipyrroloanthracene-12,14-dione (**23h**) was prepared from (*E*)-3-(anthracen-9-yl)-1-(4-fluorophenyl)prop-2-en-1-one (1.0 mmol, 0.326 g) and maleimide (1.3 mmol, 0.13 g) according to the general procedure above; cream powder (100%), Mp. 260–268 °C. ^{1}H nmR (400 MHz, $CDCl_3$) δ 3.36 (dd, *J* = 8.5, 3.2 Hz, 1H), 3.54 (d, *J* = 8.5 Hz, 1H), 4.80 (d, *J* = 3.2 Hz, 1H), 7.12–7.26 (m, 6H), 7.32 (m, 3H), 7.37–7.44 (m, 1H), 7.49 (s, 1H), 7.76 (d, *J* = 16.2 Hz, 1H), 7.94 (d, *J* = 16.2 Hz, 1H), 8.16–8.26 (m, 2H). IR ν max (ATR): (N-H) 3340 (C-H) 2984.37, 1776.16, (C=O) 1718, 1630.99 (Ar C=C), 1595.20, 1457.80, (C-N) 1311, 1157, (C-F) 1224.84 cm^{-1}. HRMS (APCI): calculated for $C_{27}H_{19}FNO_3$ [M^{+}+H]: 412.1002; found: 412.1004.

(*E*)-9-(3-(4-Methoxyphenyl)-3-oxoprop-1-en-1-yl)-9,10-dihydro-9,10-[3,4]epipyrrolo-anthracene-12,14-dione (**23i**) was synthesized from (*E*)-3-(anthracen-9-yl)-1-(4-methoxyphenyl) prop-2-en-1-one (1.0 mmol, 0.338g) and maleimide (1.3 mmol, 0.13 g) according to the general procedure above; white solid (70%), Mp. 281–284 °C. ^{1}H nmR (400 MHz, $CDCl_3$) δ 3.80 (d, *J* = 8.29 Hz, 1H), 3.87 (s, 3H), 4.79 (d, *J* = 2.90 Hz, 1H), 7.10–7.19 (m, 5H), 7.19–7.23 (m, 3H), 7.25 (app t, *J* = 5.39 Hz, 2H), 7.32–7.36 (m, 1H), 7.52 (d, *J* = 6.63 Hz, 1H), 7.74–7.88 (m, 2H), 8.17 (d, *J* = 9.12 Hz, 2H). ^{13}C nmR (101 MHz, $CDCl_3$) ppm 39.15–40.72, 44.97, 49.37, 56.07, 123.21, 124.60, 125.65, 127.01–127.38, 130.62, 131.37–132.01, 140.43, 141.47, 143.30, 177.79, 188.12 (C=O chalcone). IR ν max (ATR): 3343.73 (N-H), 1670.36 (C=O), 1601.24 (Ar C=C), 1230.33 (C-O), 1019.60 (C-N) cm^{-1}. LRMS (APCI): 434.30 [$M^{+}-$H]. HRMS (APCI): calculated $C_{28}H_{20}NO_4$ [$M^{+}-$H]: 434.1398; found: 434.1392.

(*E*)-9-(3-Oxo-3-(*p*-tolyl)prop-1-en-1-yl)-9,10-dihydro-9,10-[3,4]epipyrroloanthracene-12,14-dione (**23j**) was synthesized from (*E*)-3-(anthracen-9-yl)-1-(*p*-tolyl)prop-2-en-1-one (1.0 mmol, 0.338 g) and maleimide (1.3 mmol, 0.13 g) according to the general procedure above; white powder (46%), Mp. 129–134 °C. ^{1}H nmR (400 MHz, $CDCl_3$) δ 2.41 (s, 3H) 3.80 (d, *J* = 8.71 Hz, 1H) 4.79 (br. s., 1H) 7.10–7.30 (m, 7H) 7.30–7.45 (m, 3H) 7.52 (d, *J* = 6.63 Hz, 1H) 7.75–7.89 (m, 2H) 8.03–8.04 (m, 1H) 8.04–8.13 (m, 2H)^{13}C nmR (101 MHz, $CDCl_3$) ppm 21.69, 39.07–40.73, 44.99, 48.86–49.53, 55.35, 123.23, 123.64, 126.93–127.51, 129.04–129.58, 129.96, 135.30, 140.39, 141.79–142.33, 144.37, 177.47, 189.36 (chalcone C=O)IR ν max (ATR): 1710.75 (C=O), 1625.28 (Ar C=C), 1458.68 (C-CH_3), 1231.30 (C-O), 1072.28 (C-N) cm^{-1}. HRMS (APCI): calculated $C_{28}H_{22}NO_3$ [$M^{+}-$H]: 420.1594; found: 420.1606.

(*E*)-9-(3-Oxo-3-(3,4,5-trimethoxyphenyl)prop-1-en-1-yl)-9,10-dihydro-9,10-[3,4] epipyrrolo-anthracene-12,14-dione (**23k**) was synthesized from (*E*)-3-(anthracen-9-yl)-1-(3,4,5-trimethoxyphenyl)prop-2-en-1-one (1.0 mmol, 0.398 g) and maleimide (1.3 mmol, 0.13 g) according to the general procedure above; white powder (53%), Mp. 257–260 °C. ^{1}H nmR (400 MHz, $CDCl_3$) δ 3.30 (d, *J* = 2.90 Hz, 1H), 3.76–3.79 (m, 3H), 3.81 (s, 1H), 3.87 (s, 6H), 4.79 (d, *J* = 3.32 Hz, 1H), 7.16–7.25 (m, 5H), 7.28 (d, *J* = 4.98 Hz, 1H), 7.32 -7.36 (m, 1H), 7.47–7.54 (m, 3H), 7.72–7.87 (m, 2H), 10.85 (s, 1H). ^{13}C nmR (101 MHz, $CDCl_3$) ppm 41.66, 43.00, 46.27, 49.08–49.42, 51.67, 56.51, 65.87, 106.82, 123.23, 123.67, 124.66, 125.66, 126.63,127.39, 132.39, 133.09, 139.33, 141.76, 142.56, 143.27,177.96, 189.32 (C=O chalcone). IR ν max (ATR): 1670.40 (C=O), 1600.25 (Ar C=C), 1229.98 (C-O), 1017.32 (C-N) cm^{-1}. HRMS (APCI): calculated for $C_{30}H_{24}NO_6$ [$M^{+}-$H]: 494.1609; found: 494.1605.

(*E*)-9-(3-(4-Chlorophenyl)-3-oxoprop-1-en-1-yl)-9,10-dihydro-9,10-[3,4]epipyrroloan-thracene-12,14-dione (**23l**) was synthesized from (*E*)-3-(anthracen-9-yl)-1-(4-chlorophenyl) prop-2-en-1-one (1.0 mmol, 0.343 g) and maleimide (1.3 mmol, 0.13 g) according to the general procedure above; white solid (65%), Mp. 280–285 °C. (HPLC: 97.72%, RT 7.23 min). ^{1}H nmR (400 MHz, $CDCl_3$) δ 3.80 (d, *J* = 8.71 Hz, 1H) 4.80 (d, *J* = 2.90 Hz, 1H) 7.09–7.39 (m, 8H) 7.52 (d, *J* = 7.05 Hz, 1H) 7.68 (d, *J* = 8.29 Hz, H) 7.85 (s, 2 H) 8.19 (d, *J* = 8.29Hz, 2H) 10.85 (s, 1H). ^{13}C nmR (101 MHz, $CDCl_3$) ppm 44.98, 49.11, 49.32, 51.71, 123.26, 123.67,124.13, 124.62, 125.65, 125.75, 126.66, 127.18, 127.29, 128.64, 129.34, 129.54, 136.45, 140.28, 142.17, 143.11, 177.12–178.10, 198.62 (chalcone C=O). IR ν max (ATR): 1717.64 (C=O), 754.55 (C-Cl)

cm^{-1}. LRMS (APCI): 438.2284 [M^+-H]. HRMS (APCI): calculated for $C_{27}H_{18}ClNNaO_3$ [M^++Na]: 462.0867; found 462.0864.

9-(*E*)-3-Oxo-3-(pyridin-4-yl)prop-1-en-1-yl)-9,10-dihydro-9,10-[3,4]epipyrroloanthracene-12,14-dione (**23m**) was prepared from (*E*)-3-(anthracen-9-yl)-1-(pyridin-4-yl)prop-2-en-1-one (1.0 mmol, 0.3094 g) and maleimide (1.3 mmol, 0.13g) according to the general procedure above; powder (100%), Mp. 256–258 °C. ^{1}H nmR (400 MHz, $CDCl_3$) δ 3.36 (dd, *J* = 8.29, 3.32 Hz, 1H, CH), 3.54 (d, *J* = 8.29 Hz, 1H, CH), 4.81 (d, *J* = 3.32 Hz, 1H, CH), 7.12–7.25 (m, 6H, ArH), 7.26–7.31 (m, 1 H, ArH), 7.33–7.38 (m, 1H, ArH), 7.39–7.43 (m, 1H, ArH), 7.73 (d, *J* = 16.17 Hz, 1 H, CH=C), 7.95 (d, *J* = 5.39 Hz, 2H, ArH), 8.01 (d, *J* = 16.17 Hz, 1H, CH=C), 8.86 (br.s., 1 H, NH). ^{13}C nmR (101 MHz, $CDCl_3$) ppm 45.37, 48.96, 49.54, 52.07, 121.87, 123.29, 123.57, 124.14, 125.55, 126.76, 127.35, 127.60, 130.63, 1 (N-H), 3010.28 (C-H), 1776.21, 1721 (C=O), 1634.91 (Ar C=C), 1457.77, 1407.25, (C-N) 1317, 1166 cm^{-1}. HRMS (ESI): calculated $C_{26}H_{17}N_2O_3$ [M^+-H] 405.1245; found: 405.1247.

9-(*E*)-3-Oxo-3-(pyridin-2-yl)prop-1-en-1-yl)-9,10-dihydro-9,10-[3,4]epipyrroloanthracene-12,14-dione (**23n**) was prepared from (*E*)-3-(anthracen-9-yl)-1-(pyridin-2-yl)prop-2-en-1-one (1.0 mmol, 0.309 g) and maleimide (1.3 mmol, 0.13 g) according to the general procedure above; brown powder (72%), Mp. > 200 °C. (HPLC: 92.20%, RT 3.80 min). ^{1}H nmR (400 MHz, DMSO-d_6) δ 3.33 (dd, *J* = 8.50, 3.11 Hz, 1H, CH), 3.64 (d, *J* = 8.29 Hz, 1H, CH), 4.78 (d, *J* = 3.32 Hz, 1H, CH), 7.10–7.22 (m, 4H, ArH), 7.29 (d, *J* = 7.46 Hz, 1H, ArH), 7.31–7.35 (m, 1H, ArH), 7.37–7.43 (m, 2H, ArH), 7.49 (ddd, *J* = 7.57, 4.66, 1.04 Hz, 2H, ArH), 7.90 (td, *J* = 7.67, 1.66 Hz, 1H, ArH), 8.11–8.19 (m, 1 H, ArH), 8.15 (d, *J* = 16.59 Hz, 1H, CH=C), 8.28 (d, *J* = 7.88 Hz, 1H, ArH), 8.40 (d, *J* = 16.59 Hz, 1H, CH=C), 8.67–8.72 (m, 1H, ArH). ^{13}C nmR (101 MHz, DMSO-d_6) 44.99, 49.37, 49.74–50.02, 51.79, 123.23, 123.29, 123.61, 124.67, 126.62, 127.12, 127.17, 127.30, 130.11, 135.68, 139.39, 140.2 (C-H), 1776.26, 1719 (C=O), 1623.69, 1603.44 (Ar C=C), 1523.46, 1458.07, (C-N) 1323, 1170 cm^{-1}. HRMS (ESI): calculated for $C_{26}H_{19}N_2O_3$ [M^+-H]: 407.1390; found: 407.1396.

9-(*E*)-3-(Furan-2-yl)-3-oxoprop-1-en-1-yl)-9,10-dihydro-9,10-[3,4]epipyrroloanthracene-12,14-dione (**23o**) was prepared from (*E*)-3-(anthracen-9-yl)-1-(furan-2-yl)prop-2-en-1-one (1.0 mmol, 0.293g) and maleimide (1.3 mmol, 0.13 g) according to the general procedure above; light yellow powder (81%), Mp. 268–270 °C. ^{1}H nmR (400 MHz, $CDCl_3$) δ 3.34 (dd, *J* = 8.6, 3.2 Hz, 1H), 3.54 (d, *J* = 8.6 Hz, 1H), 4.79 (d, *J* = 3.2 Hz, 1H), 6.61 (dd, *J* = 3.6, 1.7 Hz, 1H), 7.10–7.24 (m, 4H), 7.27 (dd, *J* = 6.9, 1.2 Hz, 1H), 7.34 (dd, *J* = 5.5, 3.2 Hz, 2H), 7.36–7.47 (m, 3H), 7.59–7.71 (m, 2H), 8.04 (d, *J* = 16.3 Hz, 1H)· ^{13}C nmR (101 MHz, $CDCl_3$) ppm 45.42, 49.06, 49.55, 51.92, 67.18, 112.64, 118.71, 123.52, 123.82, 123.99, 125.41, 126.67, 127.22, 127.29, 127.43, 130.42, 137.89, 139.25, 140.95, 141.73, 142.21, 147.10, 153.32, 174.77, 175.62, 177.38. IR ν max (ATR): 3343 (N-H), 3033.42 (C-H), 1776.09, 1719 (C=O), 1626.86, 1602.84 (Ar C=C), 1522.49, 1458.42 (C-N), 1330, 1168, (C-O), 1287.87, 1010.65 cm^{-1} HRMS (ESI): calculated for $C_{25}H_{18}NO_4$ [M^++H]: 407.1390; found: 407.1396.

9-(*E*)-3-(Naphthalen-2-yl)-3-oxoprop-1-en-1-yl)-9,10-dihydro-9,10-[3,4]epipyrrolo-anthracene-12,14-dione (**23p**) was prepared from (*E*)-3-(anthracen-9-yl)-1-(naphthalen-2-yl)prop-2-en-1-one (1.0 mmol, 0.358 g) and maleimide (1.3 mmmol, 0.13 g) according to the general procedure above; cream/yellow powder (71%), Mp. 209–215 °C. ^{1}H nmR (400 MHz, $CDCl_3$) δ 3.42 (dd, *J* = 8.56, 3.18 Hz, 1H), 3.64 (d, *J* = 8.56 Hz, 1H), 4.86 (d, *J* = 3.18 Hz, 1H), 7.23–7.25 (m, 1H), 7.25–7.28 (m, 2H), 7.37–7.42 (m, 2H), 7.43–7.47 (m, 2H), 7.57–7.62 (m, 1H), 7.63–7.68 (m, 1H), 7.91–7.98 (m, 3H), 8.00–8.08 (m, 3H), 8.26 (dd, *J* = 8.56, 1.71 Hz, 3H), 8.76 (s, 3H). ^{13}C nmR (101 MHz, $CDCl_3$) ppm 45.5, 49.1, 49.6, 123.6, 123.9, 124.1, 124.6, 125.5, 126.8, 126.9, 127.3, 127.4, 127.5, 127.9, 128.6, 128.8, 129.8, 131.0, 131.6, 135.6, 142.1, 193.0. IR ν max (ATR): 3343 (N-H), 3062 (C-H), 1776.30 (C=O), 1719, (Ar C=C),1625.58, 1602.54 (C=C), 1522.97, 1457.95, (C-N) 1312, 1164 cm^{-1}. HRMS (APCI): calculated for $C_{31}H_{22}NO_3$ [M^++H]: 456.1594; found: 456.1599.

9-(*E*)-3-Oxo-3-(thiophen-2-yl)prop-1-en-1-yl)-9,10-dihydro-9,10-[3,4]epipyrrolo-anthracene-12,14-dione (**23q**) was prepared from (*E*)-3-(anthracen-9-yl)-1-(thiophen-2-yl)prop-2-en-1-one (1.0 mmol, 0.3144 g) and maleimide (1.3 mmol, 0.13 g) according to the general procedure above; beige/yellow powder (80%), Mp. 263–268 °C. ^{1}H nmR (400 MHz, $CDCl_3$)

δ 3.35 (ddd, J = 8.5, 3.2, 0.6 Hz, 1H), 3.53 (d, J = 8.5 Hz, 1H), 4.80 (d, J = 3.1 Hz, 1H), 7.12–7.23 (m, 5H), 7.27–7.31 (m, 1H), 7.34 (ddd, J = 7.9, 4.1, 1.9 Hz, 2H), 7.38–7.44 (m, 1H), 7.63–7.78 (m, 2H), 7.94–8.03 (m, 2H). ^{13}C nmR (101 MHz, $CDCl_3$) ppm 45.40, 49.05, 49.60, 51.95, 123.51, 123.82, 124.00, 125.42, 126.71, 127.25, 127.31, 127.45, 128.42, 131.06, 132.95, 134.55, 137.91, 139.26, 140.93, 141.51, 142.25, 145.09, 174.85, 175.61, 181.86 (chalcone C=O). IR ν max (ATR): 3343 (N-H), 2963.67 (C-H), 1776.30, 1718, (C=O), 1623.56 (Ar C=C), 1582.14, 1524.12, 1457.91, 1324 (C-N), 1169, 720.73 cm^{-1}. HRMS (APCI): calculated for $C_{25}H_{18}NO_3S$ [M$^+$+H]: 412.1002; found: 412.1004.

3.5. General Preparation for (E)-9-(3-oxo-3-phenylprop-1-en-yl)-13-phenyl-9,10-dihydro-9,10-[3,4]epipyrroloanthracene-12,14-diones **24a–q**

To a solution of the appropriate chalcone anthracene **21a–q** (1.0 mmol) in toluene (2 mL) was added dienophile *N*-phenylmaleimide (1.3 mmol). The mixture was heated and stirred at 90 °C for 48 h. The reaction was then cooled to room temperature, and the crude solid product was isolated via filtration, washed with toluene (2 mL) and diethyl ether (2 mL), and then recrystallized from toluene.

(*E*)-3-(4-Bromophenyl)-3-oxoprop-1-en-1-yl)-13-phenyl-9,10-dihydro-9,10-[3,4]epipyrroloanthracene-12,14-dione (**24a**) was obtained from (*E*)-3-(anthracen-9-yl)-1-(4-bromophenyl)prop-2-en-1-one (1.0 mmol, 0.387 g) and *N*-phenylmaleimide (1.3 mmol, 0.23 g) according to the general procedure above; white powder (84%), Mp. 218 °C. (HPLC: 94.98%: RT 1.90 min). ^{1}H nmR (400 MHz, $CDCl_3$) δ 3.47 (dd, J = 8.29, 3.32 Hz, 2H), 3.64 (d, J = 8.29 Hz, 2H), 4.92 (d, J = 2.90 Hz, 2H), 6.46–6.53 (m, 3H), 7.19–7.23 (m, 2H), 7.25–7.30 (m, 8H), 7.30–7.35 (m, 3H), 7.35–7.43 (m, 4H), 7.45 (dd, J = 7.05, 1.24 Hz, 2H), 7.61–7.65 (m, 3H), 7.80 (d, J = 16.17 Hz, 1H), 7.97–8.05 (m, 3H). ^{13}C nmR (101 MHz, $CDCl_3$) ppm 45.94, 47.95, 48.48, 52.49, 123.51, 123.92, 124.17, 125.55, 126.05, 126.44, 126.82, 127.38, 127.55, 127.94, 128.36, 128.91, 129.11, 129.13, 130.49, 131.14, 131.25, 132.06, 134.17, 136.45, 138.17, 139.39, 140.80, 142.17, 142.8 (C=C), 1490, 1455 (Ar C=C), 1380.41 (C-N), 1180 (C-O), 691.05 cm^{-1}. LRMS (APCI) 560.08 [M$^+$+H]. HRMS (ESI) calculated for $C_{33}H_{22}BrNNaO_3$: 582.0675 [M$^+$+Na]; found: 582.0668. *Alternative preparation of* **24a**: 9-(*E*)-3-(4-Bromophenyl)-3-oxoprop-1-en-1-yl)-9,10-dihydro-9,10-[3,4]furanoanthracene-12,14-dione (**22a**) (0.575 mmol) was treated with aniline (0.07 g, 0.748 mmol) in acetic acid (5 mL) at 120 °C for 2–3 h. The reaction mixture was cooled in an ice bath, and then deionized water was added. The colorless, solid product was filtered and dried (72%). The product was identical when compared with the sample obtained above.

9-(*E*)-3-(4-Nitrophenyl)-3-oxoprop-1-en-1-yl)-13-phenyl-9,10-dihydro-9,10-[3,4]epipyrroloanthracene-12,14-dione (**24b**) was prepared from (*E*)-3-(anthracen-9-yl)-1-(4-nitrophenyl)prop-2-en-1-one (1.0 mmol, 0.353 g) and *N*-phenylmaleimide (1.3 mmol, 0.23 g) according to the general procedure above; cream powder (86%), Mp. 242–246 °C. ^{1}H nmR (400 MHz, $CDCl_3$) δ 3.50 (dd, J = 8.4, 3.2 Hz, 1H), 3.67 (d, J = 8.5 Hz, 1H), 4.94 (d, J = 3.2 Hz, 1H), 6.54–6.45 (m, 2H), 7.40–7.17 (m, 9H), 7.46–7.37 (m, 1H), 7.51–7.43 (m, 1H), 7.85 (d, J = 16.2 Hz, 1H), 8.06 (d, J = 16.2 Hz, 1H), 8.38–8.26 (m, 4H). ^{13}C nmR (101 MHz, $CDCl_3$) ppm 45.92, 47.95, 48.43, 52.55, 123.39, 123.77, 123.95, 124.28, 125.65, 126.42, 126.87, 127.42, 127.51, 127.67, 128.99, 129.16, 129.96, 131.04, 131.17, 138.16, 139.18, 140.77, 141.95, 144.32, 174.72, 175.20 (chalcone C=O). IR ν max (ATR): 3335.77 (Ar C-H), 1771.68, 1700.63, 1671.17 (C=O), 1519, 1383 (NO_2), 1179 (C-N) cm^{-1}. LRMS (APCI): [M$^+$+H] 527.14. HRMS (ESI) calculated for $C_{33}H_{22}N_2NaO_5$, 549.1421 [M$^+$+Na]; found: 549.1424.

(*E*)-9-(3-(4-Ethylphenyl)-3-oxoprop-1-en-1-yl)-13-phenyl-9,10-dihydro-9,10-[3,4]epipyrroloanthracene-12,14-dione (**24c**) was synthesized from (*E*)-3-(anthracen-9-yl)-1-(4-ethylphenyl) 2-en-1-one (1.0 mmol, 0.336 g) and *N*-phenylmaleimide (1.3 mmol, 0.23 g) according to the general procedure above; light green powder (90%), Mp. 207 °C. IR ν max (ATR): 2969.28, 2922.38, 1708.43, 1380.94, 1178.76, 715.69, 753.23, 689.68 cm^{-1}. ^{1}H nmR (400 MHz, DMSO-d_6) δ 3.57 (dd, J = 8.31, 3.42 Hz, 3H), 4.01 (d, J = 8.31 Hz, H), 4.97 (d, J = 2.93 Hz, 3H), 6.38–6.46 (m, 6H), 7.26–7.36 (m,2H), 7.36–7.39 (m, 3H), 7.40–7.43 (m, 1H), 7.59–7.65 (m, 1 H), 7.83 (d, J = 17.12 Hz, 1 H), 7.96 (d, J = 16.63 Hz, 1H), 8.01 (d, J = 5.87 Hz, 2 H),

8.89 (d, J = 5.87 Hz, 2H). ^{13}C nmR (101 MHz, DMSO-d_6) 15.6, 25.9, 28.7, 45.4, 48.3, 52.1, 123.4, 123.9, 124.8, 125.7, 126.9, 127.0, 127.2, 127.3, 127.4, 127.5, 128.2, 128.6, 128.8, 129.0, 129.3, 129.4, 132.0, 132.2, 135.1, 135.5, 139.2, 140.2, 141.7, 141.9, 142.9, 150.4, 170.4, 175.3, 175.7, 189.4 (chalcone C=O) ppm. LRMS (APCI) 510.20 [M+H]. HRMS (ESI) calculated for $C_{35}H_{27}KNO_3$, 548.1623 [M$^+$+K]; found: 548.1627.

(*E*)-9-(3-(2,4-Dichlorophenyl)-3-oxoprop-1-en-1-yl)-13-phenyl-9,10-dihydro-9,10-[3,4] epipyrroloanthracene-12,14-dione (**24d**) was synthesized from (*E*)-3-(anthracen-9-yl)-1-(2,4-dichlorophenyl)prop-2-en-1-one (1.0 mmol, 0.377 g) and *N*-phenylmaleimide (1.3 mmol, 0.23 g) according to the general procedure above; white powder (75%), Mp. 208–210 °C. ^{1}H nmR (400 MHz, $CDCl_3$) δ 3.44 (dd, J = 8.29, 3.32 Hz, 1H), 3.55 (d, J = 8.29 Hz, 1H), 4.91 (d, J = 3.32 Hz, 1H), 6.46 (m, 2H), 7.24 (m, 7H), 7.35 (m, 5H), 7.46 (m, 1H), 7.50 (d, J = 1.66 Hz, H), 7.66 (d, J = 8.29 Hz, 1 H), 7.83 (d, J = 17.00 Hz, 1H), 7.79 (d, J = 17.00 Hz, 1H). ^{13}C nmR (101 MHz, $CDCl_3$) ppm 45.97, 48.23, 48.34, 52.30, 123.43, 123.76, 124.28, 125.57, 126.38, 126.81, 127.33, 127.43, 127.45, 127.57, 128.88, 129.08, 130.87, 131.16, 132.66, 134.88, 136.80, 138.15, 139.07, 140.78, 141.69, 144.77, 174.43, 175.25, 191.55 (chalcone C=O). IR ν max (ATR): 3067.32 (C-H), 1776.64, 1710.54, 1630.86 (C=O), 1381.66 (C-N), 1183.14 (C-O) cm^{-1}. HRMS (ESI) calculated for $C_{33}H_{20}Cl_2NO_3$ [M$^+$−H]: 548.0826; found: 548.0825.

9-(*E*)-3-(3,4-Dimethoxyphenyl)-3-oxoprop-1-en-1-yl)-13-phenyl-9,10-dihydro-9,10-[3,4] epipyrroloanthracene-12,14-dione (**24e**) was prepared from (*E*)-3-(anthracen-9-yl)-1-(3,4-dimethoxyphenyl)prop-2-en-1-one (1.0 mmol, 0.368 g) and *N*-phenylmaleimide (1.3 mmol, 0.23 g) according to the general procedure above; white powder (93%), Mp. 242–244 °C. ^{1}H nmR (400 MHz, $CDCl_3$) δ 3.47 (dd, J = 8.4, 3.3 Hz, 1H), 3.65 (d, J = 8.4 Hz, 1H), 3.94 (d, J = 9.4 Hz, 6H), 4.92 (d, J = 3.2 Hz, 1H), 6.45–6.54 (m, 2H), 6.91 (d, J = 8.4 Hz, 1H), 7.20 (td, J = 7.5, 1.6 Hz, 1H), 7.20–7.28 (m, 3H), 7.24–7.32 (m, 3H), 7.34–7.44 (m, 3H), 7.42–7.48 (m, 1H), 7.75 (d, J = 2.0 Hz, 1H), 7.80–7.89 (m, 2H), 7.95 (d, J = 16.1 Hz, 1H). ^{13}C nmR (101 MHz, $CDCl_3$) ppm 45.96, 47.99, 48.51, 52.48, 56.02, 56.08, 110.18, 110.99, 123.60, 123.84, 124.08, 125.49, 126.43, 126.79, 127.30, 127.33, 127.47, 128.82, 129.05, 130.89, 131.54, 138.21, 139.62, 140.83, 141.01, 142.42, 153.49, 188.72 (chalcone C=O). IR ν max (ATR): 3011.75 (Ar C-H), 1775.53, 1708.96, 1676.49 (C=O), 1379.80 (C-N), 1181.33 (C-O) cm^{-1}. LRMS (APCI): 542.28 [M$^+$+H]. HRMS (ESI) calculated for $C_{35}H_{27}NNaO_5$ [M$^+$+Na]: 564.1781; found 564.1783.

(*E*)-9-(3-Oxo-3-phenylprop-1-en-yl)-13-phenyl-9,10-dihydro-9,10-[3,4]epipyrroloanthracene-12,14-dione (**24f**) was synthesized from (*E*)-3-(anthracen-9-yl)-1-phenylprop-2-en-1-one (1.0 mmol, 0.308 g) and *N*-phenylmaleimide (1.3 mmol, 0.23 g) according to the general procedure above; white powder (42%). Mp. 103–105 °C. ^{1}H nmR (400 MHz, DMSO-d_6) δ 3.54 (dd, J = 8.29, 3.32 Hz, 1H) 4.00 (d, J = 8.29 Hz, 1H) 4.94 (d, J = 3.32 Hz, H) 6.39–6.43 (m, 2H) 7.19–7.27 (m, 3H) 7.27–7.34 (m, 6H) 7.36–7.41 (m, 1H) 7.57–7.62 (m, 3H) 7.66–7.71 (m, 1H) 7.88 (d, J = 1.66 Hz, 2H) 8.13–8.17 (m, 2H). ^{13}C nmR (101 MHz, DMSO-d_6) 45.41, 48.33, 52.15, 122.71, 123.42, 123.90, 124.86, 125.72, 126.88, 127.06, 127.44, 128.95, 129.20, 129.36, 132.01, 132.18, 133.88, 137.74, 139.24, 140.16, 141.70, 142.35, 142.82, 175.35, 175.76, 189.96 (chalcone C=O) ppm. IR ν max (ATR): 1707.50 (C=O), 1673.57 (C=C), 1636.62 (Ar C=C),1332.23 (C-N) cm^{-1}. LRMS (APCI) 482.20 [M$^+$+H].

9-(*E*)-3-(4-Iodophenyl)-3-oxoprop-1-en-1-yl)-13-phenyl-9,10-dihydro-9,10-[3,4]epipyrroloanthracene-12,14-dione (**24g**) was prepared from (*E*)-3-(anthracen-9-yl)-1-(4-iodophenyl) prop-2-en-1-one (1.0 mmol, 0.434 g) and *N*-phenylmaleimide (1.3 mmol, 0.23 g) according to the general procedure above; white crystals (20%), Mp. 214 °C. ^{1}H nmR (400 MHz, DMSO-d_6) δ ppm 3.51 (dd, J = 8.3, 3.2 Hz, 1H), 3.97 (d, J = 8.4 Hz, 1H), 4.92 (d, J = 3.2 Hz, 1H), 6.34–6.43 (m, 2H), 7.09–7.16 (m, 1H), 7.18–7.23 (m, 3H), 7.25–7.33 (m, 7H), 7.33–7.39 (m, 1H), 7.41–7.50 (m, 1H), 7.57 (dt, J = 7.1, 1.2 Hz, 1H), 7.78–7.85 (m, 2H), 7.86–7.91 (m, 2H), 7.94–8.01 (m, 2H). ^{13}C nmR (101 MHz, DMSO-d_6) 44.98, 49.13, 49.32, 51.72, 102.63, 123.26, 123.67, 124.62,125.65, 126.66, 127.17, 127.29, 130.84,131.45, 137.06, 138.35, 139.30, 140.27, 142.17, 143.05, 143.12, 177.48, 177.74, 189.40 (chalcone C=O) ppm. IR ν max (ATR): 3088.53 (C-H), 1772.98, 1705 (C=O), 1633.80 (Ar C=C), 1587.05, 1499.05, (C-N) 1467.03, 1381, 1180, 691.17 cm^{-1}. LRMS (APCI) [M$^+$+H]: found: 608.03. HRMS (ESI) calculated for $C_{33}H_{21}INO_3$, 606.0572 [M$^+$−H]; found: 606.0567.

9-(*E*)-3-(4-Fluorophenyl)-3-oxoprop-1-en-1-yl)-13-phenyl-9,10-dihydro-9,10-[3,4]epi-pyrroloanthracene-12,14-dione (**24h**) was prepared from (*E*)-3-(anthracen-9-yl)-1-(4-fluorophenyl) prop-2-en-1-one (1.0 mmol, 0.434 g) and *N*-phenylmaleimide (1.3 mmol, 0.23 g) according to the general procedure above; green solid (88%), Mp. 79 °C. (HPLC: 96.41%, RT 10.04 min). ^{1}H nmR (400 MHz, $CDCl_3$) δ 3.48 (dd, *J* = 8.4, 3.2 Hz, 1H), 3.66 (d, *J* = 8.4 Hz, 1H), 4.93 (d, *J* = 3.2 Hz, 1H), 6.44–6.54 (m, 2H), 7.13–7.21 (m, 2H), 7.22 (dd, *J* = 3.5, 1.6 Hz, 1H), 7.25–7.31 (m, 5H), 7.31–7.42 (m, 4H), 7.43–7.49 (m, 1H), 7.83 (d, *J* = 16.1 Hz, 1H), 8.00 (d, *J* = 16.1 Hz, 1H), 8.15–8.24 (m, 2H). ^{13}C nmR (101 MHz, $CDCl_3$) ppm 45.39, 48.93, 49.61, 52.01, 115.81, 116.03, 123.41, 123.79, 124.03, 125.47, 126.74, 127.28, 127.32, 127.49, 131.19, 131.58, 131.67, 134.09, 134.12, 137.94, 139.26, 140.92, 142.18, 142.28, 175.17, 175.73, 188.70 (chalcone C=O). LRMS (APCI) 500.20 [M$^+$+H]. HRMS (ESI) calculated for $C_{33}H_{22}FNNaO_3$, 522.1476 [M$^+$+H]; found 522.1473.

(*E*)-9-(3-(4-Methoxyphenyl)-3-oxoprop-1-en-1-yl)-13-phenyl-9,10-dihydro-9,10-[3,4] epipyrroloanthracene-12,14-dione (**24i**) was synthesized from (*E*)-3-(anthracen-9-yl)-1-(4-methoxyphenyl)prop-2-en-1-one (1.0 mmol, 0.338 g) and *N*-phenylmaleimide (1.3 mmol, 0.23 g) according to the general procedure above; white powder (56%), Mp. 207–212 °C. ^{1}H nmR (400 MHz, $CDCl_3$) δ 3.55 (dd, *J* = 8.29, 3.32 Hz, 1H), 3.85 (s, 3H), 3.99 (d, *J* = 8.29 Hz, 1H), 4.94 (d, 1H, *J* = 3.7 Hz, 1H), 7.09–7.18 (m, 5H), 7.20–7.34 (m, 11H), 7.81 (d, J = 16.3 Hz, 1H), 7.89 (d, *J* = 16.3 Hz, 1H), 8.16 (d, *J* = 9.12Hz, 2H). ^{13}C nmR (101 MHz, $CDCl_3$) ppm 45.42, 48.27, 48.34, 52.04, 56.03, 114.67, 122.22, 123.41, 123.90, 124.83, 125.71, 126.85, 127.07, 127.26–127.55, 128.64, 129.00, 129.32, 130.59, 131.53, 131.92, 132.19, 139.24, 140.28, 141.20, 141.72, 142.95, 163.88, 175.34, 175.76, 177.93, 188.14 (chalcone C=O). IR ν max (ATR): 3343.74 (N-H), 1670.84 (C=O), 1600.33 (Ar C=C), 1380.68, 1230.69 (C-O), 1024.50 (C-N) cm^{-1}. LRMS (APCI) found: 512.19, [M$^+$+H]. HRMS (ESI) calculated for $C_{34}H_{24}NO_4$ [M$^+$−H]: 510.1705; found: 510.1708.

(*E*)-9-(3-Oxo-3-(*p*-tolyl)prop-1-en-1-yl)-13-phenyl-9,10-dihydro-9,10-[3,4]epipyrrolo-anthracene-12,14-dione (**24j**) was synthesized from (*E*)-3-(anthracen-9-yl)-1-(*p*-tolyl)prop-2-en-1-one (1.0 mmol, 0.338 g) and *N*-phenylmaleimide (1.3 mmol, 0.23 g) according to the general procedure above; white powder (51%), Mp. 185–190 °C. ^{1}H nmR (400 MHz, $CDCl_3$) δ 2.39 (s, 3H), 3.53 (dd, *J* = 8.50, 3.11 Hz, 1H), 3.99 (d, *J* = 8.71 Hz, 1H), 4.94 (d, *J* = 2.90 Hz, 1H), 6.38–6.44 (m, 2H), 7.19–7.26 (m, 3H), 7.26–7.34 (m, 6H), 7.36–7.42 (m, 3H), 7.59 (d, *J* = 6.63 Hz, 1H), 7.86 (d, *J* = 3.32 Hz, 2H), 8.06 (d, *J* = 8.29 Hz, 2H). ^{13}C nmR (101 MHz, $CDCl_3$) ppm 21.69, 39.22–40.74, 48.34, 52.13, 123.89, 125.38 -125.98, 126.76–127.19, 127.26–127.61, 128.92–129.49, 129.95, 131.97, 135.25, 140.21, 141.91, 144.37, 175.34, 189.38 (chalcone C=O). IR ν max (ATR): 2917.28 (Ar C-H), 1772.13 (C=O), 1624.76 (Ar C=C), 1458.34 (C-CH_3), 1230.50 (C-O), 1072.21 (C-N) cm^{-1} LRMS (APCI) 496.19 [M$^+$+H]. HRMS (ESI) calculated for $C_{34}H_{25}KNO_3$ [M$^+$+K] 534.1466; found 534.1460.

(*E*)-9-(3-Oxo-3-(3,4,5-trimethoxyphenyl)prop-1-en-1-yl)-13-phenyl-9,10-dihydro-9,10-[3,4] epipyrroloanthracene-12,14-dione (**24k**) was synthesized from (*E*)-3-(anthracen-9-yl)-1-(3,4,5-trimethoxyphenyl)prop-2-en-1-one (1.0 mmol, 0.398 g) and *N*-phenylmaleimide (1.3 mmol, 0.23 g) according to the general procedure above; white powder (62%), Mp. 242–246 °C. ^{1}H nmR (400 MHz, $CDCl_3$) δ 3.53 (dd, *J* = 8.50, 3.11 Hz, 1H), 3.76 (s, 3H), 3.68 (s, 6H), 4.00 (d, *J* = 8.29 Hz, 1H), 4.94 (d, *J* = 3.32 Hz, 1H), 6.42 (dd, *J* = 7.46, 2.07 Hz, 2H), 7.23–7.26 (m, 2H), 7.26–7.33 (m, 6H), 7.34 (s, 1H), 7.37–7.40 (m, H), 7.47 (s, 2H), 7.58–7.61 (m, 1H), 7.78–7.91 (m, 2H). ^{13}C nmR (101 MHz, $CDCl_3$) ppm 45.43, 48.1–48.62, 56.54, 60.67, 106.88, 123.42, 123.95, 124.88, 125.71, 126.75 -126.96, 127.29–127.48, 128.94, 129.26, 132.13, 132.50, 133.05, 140.18, 141.65, 153.41, 175.39, 175.73, 189.11 (chalcone C=O) IR ν max (ATR): 1702.71 (C=.O), 1123.37 (C-O) cm^{-1}. LRMS (APCI) 572.19 [M$^+$+H]. HRMS (ESI) calculated for $C_{36}H_{28}NO_6$ [M$^+$−H]: 570.1922; found: 570.1920

(*E*)-9-(3-(4-Chlorophenyl)-3-oxoprop-1-en-1-yl)-13-phenyl-9,10-dihydro-9,10-[3,4]epi-pyrroloanthracene-12,14-dione (**24l**) was synthesized from (*E*)-3-(anthracen-9-yl)-1-(4-chlorophenyl) prop-2-en-1-one (1.0 mmol, 0.343 g) and *N*-phenylmaleimide (1.3 mmol, 0.23 g) according to the general procedure above; white powder (20%), Mp. 207–211 °C. ^{1}H nmR (400 MHz, DMSO-d_6) δ 3.53 (dd, *J* = 8.29, 3.32 Hz, 1H), 3.99 (d, *J* = 8.29Hz, 1H), 4.94

(d, *J* = 3.32 Hz, 1H), 6.37–6.44 (m, 2H), 7.23 (d, *J* = 0.83 Hz, 3H), 7.27–7.35 (m, 6H), 7.35–7.40 (m, 1H), 7.59 (d, *J* = 6.63 Hz, 1H), 7.67 (d, *J* = 8.29 Hz, 2H), 7.87 (s, 2H), 8.16 (d, *J* = 8.29 Hz, 2H). ^{13}C nmR (101 MHz, $CDCl_3$) ppm: 39.12–40.77, 45.42, 52.17, 123.44, 123.92, 124.86, 125.72, 126.72 -127.66, 129.24–129.66, 131.03, 131.75, 132.16, 135.09, 138.84, 139.21, 141.68, 142.74, 142.92, 175.37, 175.76, 188.95 (chalcone C=O). IR *ν* max (ATR): 3058.71 (Ar C-H), 1708.71(C=O), 1674.04 (C=C), 1591.39 (Ar C=C), 1455.69 (C-C), 1015.46 (C-N), 743.22 (C-Cl) cm^{-1}. LRMS (APCI) 516.13 [M^++H]. HRMS (ESI) calculated for $C_{33}H_{23}ClNO_3$ [M^++H]: 516.1361; found: 516.1361.

9-(*E*)-3-Oxo-3-(pyridin-4-yl)prop-1-en-1-yl)-13-phenyl-9,10-dihydro-9,10-[3,4]epipyrroloanthracene-12,14-dione (**24m**) was prepared from (*E*)-3-(anthracen-9-yl)-1-(pyridin-4-yl)prop-2-en-1-one (1.0 mmol, 0.309 g) and *N*-phenylmaleimide (1.3 mmol, 0.23 g) according to the general procedure above; cream crystals (82%), Mp. 203–206 °C. ^{1}H nmR (400 MHz, DMSO-d_6) δ 3.57 (dd, *J* = 8.4, 3.2 Hz, 1H, H-8), 4.01 (d, *J* = 8.4 Hz, 1H, H-3), 4.97 (d, *J* = 3.2 Hz, 1H, H7), 6.42 (d, *J* = 15.1 Hz, 1H), 7.16–7.62 (m, 14H), 8.01 (m, 2H), 8.88 (m, 2H). ^{13}C nmR (101 MHz, DMSO-d_6) ppm 44.93 (C-7), 47.78 (C-8), 48.04 (C-3), 51.76 (C-4), 121.73, 123.03, 123.50, 124.43, 126.41, 126.58, 126.79, 126.93, 127.06, 128.55, 128.86, 131.30, 131.67, 134.67, 138.74, 139.46, 141.18, 142.09, 143.29, 143.96, 150.89, 174.93, 175.28, 189.58 (chalcone C=O). ^{15}N (400 MHz, DMSO-d_6) 194.8 maleimide), 329.7 (pyridine) ppm. HRMS (ESI) calculated for $C_{32}H_{22}N_2NaO_3$ [M^++Na]: 505.1523; found 505.1519.

9-(*E*)-3-Oxo-3-(pyridin-2-yl)prop-1-en-1-yl)-13-phenyl-9,10-dihydro-9,10-[3,4]epipyrroloanthracene-12,14-dione (**24n**) was prepared from (*E*)-3-(anthracen-9-yl)-1-(pyridin-2-yl)prop-2-en-1-one (1.0 mmol, 0.309 g) and *N*-phenylmaleimide (1.3 mmol, 0.23 g) according to the general procedure above; gray powder (91%), Mp. >256 °C (dec). ^{1}H nmR (400 MHz, $CDCl_3$) δ 3.47 (dd, *J* = 8.4, 3.2 Hz, 1H), 3.75 (d, *J* = 8.4 Hz, 1H), 4.92 (d, *J* = 3.2 Hz, 1H), 6.44–6.53 (m, 2H), 7.14–7.30 (m, 7H), 7.31–7.42 (m, 2H), 7.46 (ddt, *J* = 7.2, 6.1, 1.1 Hz, 3H), 7.88 (td, *J* = 7.7, 1.8 Hz, 1H), 8.21 (d, *J* = 16.6 Hz, 1H), 8.26 (dt, *J* = 7.9, 1.1 Hz, 1H), 8.49 (d, *J* = 16.5 Hz, 1H), 8.69 (ddd, *J* = 4.8, 1.8, 1.0 Hz, 1H). ^{13}C nmR (101 MHz, $CDCl_3$) ppm 46.06, 48.18, 48.39, 52.51, 123.15, 123.74, 124.07, 124.15, 125.42, 126.41, 126.70, 127.00, 127.21, 127.31, 127.38, 128.69, 128.94, 130.07, 131.31, 137.00, 138.22, 139.70, 140.93, 142.30, 142.60, 149.08, 174.40, 175.50, 188.99 (chalcone C=O). IR *ν* max (ATR): 3054.46 (Ar C-H), 1779.59, 1713.79, 1674.55 (C=O), 1201.95 (C-N), 1181.33 (C-O) cm^{-1}. LRMS (APCI) 483.23 [M^++H].

9-(*E*)-3-(Furan-2-yl)-3-oxoprop-1-en-1-yl)-13-phenyl-9,10-dihydro-9,10-[3,4]epipyrroloanthracene-12,14-dione (**24o**) was prepared from (*E*)-3-(anthracen-9-yl)-1-(furan-2-yl)prop-2-en-1-one (1.0 mmol, 0.298 g) and *N*-phenylmaleimide according to the general procedure above; cream powder (87%), Mp. >200 °C. ^{1}H nmR (400 MHz, $CDCl_3$) δ 3.47 (dd, *J* = 8.4, 3.2 Hz, 1H), 3.65 (d, *J* = 8.4 Hz, 1H), 4.92 (d, *J* = 3.2 Hz, 1H), 6.43–6.53 (m, 2H), 6.58 (dd, *J* = 3.6, 1.7 Hz, 1H), 7.13–7.29 (m, 5H), 7.25–7.31 (m, 2H), 7.33 (dd, *J* = 7.4, 1.5 Hz, 1H), 7.36–7.44 (m, 2H), 7.41–7.49 (m, 2H), 7.66 (dd, *J* = 1.7, 0.8 Hz, 1H), 7.71 (d, *J* = 16.3 Hz, 1H), 8.09 (d, *J* = 16.3 Hz, 1H). ^{13}C nmR (101 MHz, $CDCl_3$) ppm 45.96, 48.07, 48.46, 52.41, 112.61, 118.83, 123.63, 123.99, 124.13, 125.49, 126.40, 126.78, 127.34, 127.50, 128.81, 129.04, 130.61, 131.25, 138.15, 139.45, 140.83, 141.76, 142.19, 147.13, 175.35. IR *ν* max (ATR): 1738.78, 1695.65, 1665.48, 1646.52, 1616.16, 1562.62 cm^{-1}. LRMS (APCI) 472.19 [M^++H]. HRMS (ESI) calculated for $C_{31}H_{20}NO_4$ [M^+−H]: 470.1398; found: 470.1394

9-(*E*)-3-(Naphthalen-2-yl)-3-oxoprop-1-en-1-yl)-13-phenyl-9,10-dihydro-9,10-[3,4]epipyrroloanthracene-12,14-dione (**24p**) was prepared from (*E*)-3-(anthracen-9-yl)-1-(naphthalen-2-yl)prop-2-en-1-one (1.0 mmol, 0.358 g) and *N*-phenylmaleimide (1.3 mmol, 0.23g) according to the general procedure above; light green powder (65%), Mp. 202–206 °C. ^{1}H nmR (400 MHz, $CDCl_3$) δ 3.52 (dd, *J* = 8.4, 3.2 Hz, 1H), 3.69 (d, *J* = 8.5 Hz, 1H), 4.94 (d, *J* = 3.2 Hz, 1H), 6.64–6.71 (m, 2H), 7.10–7.22 (m, 2H), 7.19–7.25 (m, 1H), 7.22–7.31 (m, 3H), 7.33–7.42 (m, 2H), 7.38–7.48 (m, 3H), 7.44–7.64 (m, 2H), 7.67–7.75 (m, 4H), 7.77 (s, 1H), 7.97 (dd, *J* = 3.8, 1.1 Hz, 1H), 8.05 (d, *J* = 16.1 Hz, 1H). IR *ν* max (ATR): 3068.21 (Ar C-H), 1773.41, 1710.67, 1675.13 (C=O), 1380.71 (C-N), 1195.89 (C-O) cm^{-1}. LRMS (APCI) 530.31 [M^+−H]. HRMS (ESI) calculated for $C_{37}H_{24}NO_3$ [M^+−H]: 530.1762; found: 530.1751.

9-(*E*)-3-Oxo-3-(thiophen-2-yl)prop-1-en-1-yl)-13-phenyl-9,10-dihydro-9,10-[3,4]epipyrrolo-anthracene-12,14-dione (**24q**) was prepared from (*E*)-3-(anthracen-9-yl)-1-(thiophen-2-yl) prop-2-en-1-one (1.0 mmol, 0.314 g) and *N*-phenylmaleimide (1.3 mmol, 0.23 g) according to the general procedure above; cream powder (81%), Mp. 233–236 °C. ^{1}H nmR (400 MHz, $CDCl_3$) δ 3.48 (dd, *J* = 8.4, 3.2 Hz, 1H), 3.65 (d, *J* = 8.4 Hz, 1H), 4.93 (d, *J* = 3.2 Hz, 1H), 6.45–6.54 (m, 2H), 7.14–7.19 (m, 1H), 7.19–7.24 (m, 1H), 7.24–7.31 (m, 6H), 7.38 (ddd, *J* = 13.1, 6.4, 2.5 Hz, 3H), 7.44–7.49 (m, 1H), 7.70 (dd, *J* = 4.9, 1.1 Hz, 1H), 7.77 (d, *J* = 16.1 Hz, 1H), 7.98 (dd, *J* = 3.8, 1.1 Hz, 1H), 8.04 (d, *J* = 16.1 Hz, 1H). ^{13}C nmR (101 MHz, $CDCl_3$) ppm 45.96, 48.06, 48.51, 52.44, 123.63, 124.01, 124.14, 125.51, 126.43, 126.82, 127.36, 127.52, 128.43, 128.85, 129.08, 131.20, 133.10, 134.54, 138.17, 139.46, 140.82, 141.60, 142.24, 174.57. IR ν max (ATR): 1729.12, 1660.73, 1393.28, 1327.06, 1204.75 cm^{-1}. LRMS (APCI) [M$^+$+Na]: 510.30. HRMS (ESI) calculated for $C_{31}H_{21}NNaO_3S$ [M$^+$+Na]: 510.1134; found: 510.1132.

3.6. General Preparation for (E)-13-(4-Chlorophenyl)-9-(3-oxo-3-phenylprop-1-en-1-yl)-9,10-dihydro-9,10-[3,4]epipyrroloanthracene-12,14-diones **25a–q**

To a solution of the appropriate chalcone anthracene **21a–q** (1.0 mmol) in toluene (2 mL) was added dienophile 1-(4-chlorophenyl)-1*H*-pyrrole-2,5-dione (1.3 mmol). The mixture was stirred and heated at 90 °C for 48 h. The reaction was then cooled to room temperature, and the resulting solid was isolated via filtration. The crude product was washed with toluene (2 mL) and diethyl ether (2 mL) and then recrystallized from toluene.

(*E*)-9-(3-(4-Bromophenyl)-3-oxoprop-1-en-1-yl)-13-(4-chlorophenyl)-9,10-dihydro-9,10-[3,4]epipyrroloanthracene-12,14-dione (**25a**) was prepared from (*E*)-3-(anthracen-9-yl)-1-(4-bromophenyl)prop-2-en-1-one (1.0 mmol, 0.387 g) and 1-(4-chlorophenyl)-1*H*-pyrrole-2,5-dione (1.3 mmol, 0.27 g) according to the general procedure above; white powder (95%) Mp. 253–254 °C. ^{1}H nmR (400 MHz, $CDCl_3$) δ 3.52 (dd, J = 8.32, 3.15 Hz, 1H), 4.01 (d, *J* = 8.33 Hz, 1H), 4.94 (d, *J* = 3.22 Hz, 1H), 6.47 (d, *J* = 8.28 Hz, 2H), 7.14–7.32 (m, 9H), 7.45–7.56 (m, 1H), 7.60 (d, *J* = 6.3 Hz, 1H), 7.82 (m, 2H), 7.85 (d, J = 16.12 Hz, 1 H), 7.97 (d, *J* = 8.4 Hz, 2H). ^{13}C nmR (101 MHz, $CDCl_3$) 45.96, 48.00, 48.47, 52.52, 123.53, 123.98, 124.22, 125.56, 126.92, 127.43, 127.48, 127.62, 127.71, 128.46, 129.38, 129.65, 130.50, 131.21, 132.12, 134.86, 136.44, 138.13,139.37, 140.66, 142.05, 142.68, 174.42, 175.10, 189.26 ppm. IR ν max (ATR): 1708.82, 1673.68, 1624.31, 1585.32, 1492.81, 1386.87, 1315.79, 1189.52, 1067.58, 1005.54, 850.69, 830.15 cm^{-1}. HRMS (ESI) calculated for $C_{33}H_{21}BrClNNaO_3$ [M$^+$+Na]: 616.0286; found: 616.0287.

(*E*)-13-(4-Chlorophenyl)-9-(3-(4-nitrophenyl)-3-oxoprop-1-en-1-yl)-9,10-dihydro-9,10-[3,4]epipyrroloanthracene-12,14-dione (**25b**) was synthesized from (*E*)-3-(anthracen-9-yl)-1-(4-nitrophenyl)prop-2-en-1-one (1.0 mmol, 0.353 g) and 1-(4-chlorophenyl)-1*H*-pyrrole-2,5-dione (1.3 mmol, 0.27 g) according to the general procedure above; white powder (82%), ^{1}H nmR (400 MHz, $CDCl_3$) δ 3.51 (dd, *J* = 8.50, 3.11 Hz, 4 H), 3.67 (d, *J* = 8.29 Hz, 1 H), 4.94 (d, *J* = 3.32 Hz, 1 H), 6.47 (d, *J* = 9.12 Hz, 2 H), 7.20–7.31 (m, 6 H), 7.31–7.38 (m, 2 H), 7.39–7.45 (m, 1 H), 7.46–7.51 (m, 1 H), 7.84 (d, *J* = 16.17 Hz, 1 H), 8.06 (d, *J* = 16.17 Hz, 1 H), 8.32 (d, *J* = 8.71 Hz, 2 H), 8.37 (d, *J* = 8.71 Hz, 2 H. IR ν max (ATR): 1739.76, 1665.28, 1646.57, 1562.09, 1494.06 cm^{-1}. ^{13}C nmR (101 MHz, $CDCl_3$) 121.93, 123.44, 123.80, 125.15, 125.30, 125.58, 125.62, 126.00, 126.39, 126.51, 127.27, 127.35, 127.52, 127.72, 127.77, 128.06, 128.56, 128.73, 128.80, 129.03, 130.79, 131.16, 133.12, 134.17, 136.03, 138.44. HRMS (ESI) calculated for $C_{33}H_{21}Cl_2N_2O_5$ [M$^+$+Cl]: 595.0833; found: 595.0820.

(*E*)-13-(4-Chlorophenyl)-9-(3-(4-ethylphenyl)-3-oxoprop-1-en-1-yl)-9,10-dihydro-9,10-[3,4]epipyrroloanthracene-12,14-dione (**25c**) was synthesized from (*E*)-3-(anthracen-9-yl)-1-(4-ethylphenyl)prop-2-en-1-one (1 mmol, 0.336 g) and 1-(4-chlorophenyl)-1*H*-pyrrole-2,5-dione (1.3 mmol, 0.27 g) according to the general procedure above; white powder (3%), Mp. >220 (dec). ^{1}H nmR (400 MHz, $CDCl_3$) δ 1.19–1.25 (m, 3H), 2.37 (m, 2H), 3.54 (dd, *J* = 8.27, 3.39 Hz, 1H), 3.98 (d, *J* = 8.27 Hz, 1 H), 4.97 (d, *J* = 2.91 Hz, 1 H), 6.39–6.47 (m, 2H), 7.23–7.32 (m, 5H), 7.34–7.36 (m, 1 H), 7.37–7.43 (m, 3H), 7.47–7.59 (m, 2 H), 7.78 (d, *J* = 16.2 Hz, 1 H), 7.83 (d, *J* = 5.87 Hz, 2 H), 7.98 (d, *J* = 5.87 Hz, 2 H). IR ν max (ATR): 1733.88, 1695.56, 1665.62, 1646.02, 1616.24, 1562.03, 1520.74 cm^{-1}. ^{13}C nmR (101 MHz, $CDCl_3$) 15.22, 29.04, 46.00, 48.05, 48.49, 52.51, 123.66, 124.12, 124.14, 125.49, 126.89, 127.39, 127.53, 127.72, 128.31,

129.22, 129.34, 129.36, 129.71, 131.71, 134.78, 135.42, 138.15, 139.56, 140.70, 141.53, 142.24, 150.28, 174.32, 175.15, 189.78 ppm. HRMS (ESI) calculated for $C_{35}H_{26}ClNNaO_3$ [M^++Na]: 566.1493; found: 556.1500.

(*E*)-13-(4-Chlorophenyl)-9-(3-(2,4-dichlorophenyl)-3-oxoprop-1-en-1-yl)-9,10-dihydro-9,10-[3,4]epipyrroloanthracene-12,14-dione (**25d**) was synthesized from (*E*)-3-(anthracen-9-yl)-1-(2,4-dichlorophenyl)prop-2-en-1-one (1.0 mmol, 0.309 g) and 1-(4-chlorophenyl)-1*H*-pyrrole-2,5-dione (1.3 mmol, 0.27 g) according to the general procedure above; white powder (87%), Mp. 251 (dec) °C. ^{1}H nmR (400 MHz, $CDCl_3$) δ 3.45 (dd, *J* = 8.50, 3.11 Hz, 3H), 3.56 (d, *J* = 8.29 Hz, 2H), 4.91 (d, *J* = 3.32 Hz, 2H), 6.43 (d, *J* = 9.12 Hz, 2H), 7.19–7.23 (m, 1H), 7.25 -7.30 (m, 4H), 7.30–7.37 (m, 3H), 7.41 (d, *J* = 2.07 Hz, 1H), 7.43 -7.48 (m, 1H), 7.52 (d, *J* = 1.66 Hz, 1H), 7.67 (d, *J* = 8.29 Hz, 1H), 7.81 (d, *J* = 16.59 Hz, 1H). IR *ν* max (ATR): 1738.28, 1695.68, 1648.38, 1616.26, 1562.04, 1520.75, 1451.55 cm^{-1}. ^{13}C nmR (101 MHz, $CDCl_3$) 46.00, 48.23, 48.37, 52.33, 123.46, 123.83, 124.34, 125.59, 126.92, 127.40, 127.53, 27.64, 127.67, 129.36, 129.59, 130.45, 130.91, 132.71, 134.84, 134.94, 136.81, 137.53, 138.13, 139.06, 140.66, 141.59, 144.60, 174.21, 175.04, 191.56.ppm. HRMS (ESI) calculated for $C_{32}H_{20}Cl_4NO_3$ [M^++Cl]: 618.0203; found: 618.0193.

(*E*)-13-(4-Chlorophenyl)-9-(3-(3,4-dimethoxyphenyl)-3-oxoprop-1-en-1-yl)-9,10-dihydro-9,10-[3,4]epipyrroloanthracene-12,14-dione (**25e**) was synthesized from (*E*)-3-(anthracen-9-yl)-1-(3,4-dimethoxyphenyl)prop-2-en-1-one (1.0 mmol, 0.368 g) and 1-(4-chlorophenyl)-1*H*-pyrrole-2,5-dione (1.3 mmol, 0.27 g) according to the general procedure above; white powder (4%) Mp. 167–169 °C. ^{1}H nmR (400 MHz, $CDCl_3$) δ 2.83 (d, *J* = 4.98 Hz, 1H), 2.88 (d, *J* = 4.98 Hz, 1H), 3.49 (dd, *J* = 8.09, 3.11 Hz, 1H), 3.71 (d, *J* = 9.12 Hz, 1H), 4.93 (d, *J* = 3.73 Hz, 1H), 6.48 (d, *J* = 9.12 Hz, 1H), 7.50–7.67 (m, 5H), 7.87–7.91 (m, 1H), 7.93–7.99 (m, 1H), 8.06 (d, *J* = 15.76 Hz, 1H), 8.21 (d, *J* = 9.54 Hz, 1H). IR *ν* max (ATR): 1713.42, 1604.96, 1542.00, 1493.58, 1397.57, 1247.61, 1184.72 1091.31 cm^{-1}. ^{13}C nmR (101 MHz, $CDCl_3$) 35.72, 67.62, 123.46, 123.83, 124.34, 125.59, 126.92, 127.40, 127.53, 127.64, 127.67, 129.36, 129.59, 130.45, 130.91, 132.71, 134.84, 134.94, 136.81, 137.53, 138.13, 139.06, 140.66, 141.60, 144.40, 169.94, 171.92, 172.24 ppm; LRMS (APCI) 514.20. HRMS 621.1041, found 621.1008.

(*E*)-13-(4-Chlorophenyl)-9-(3-oxo-3-phenylprop-1-en-1-yl)-9,10-dihydro-9,10-[3,4] epipyrroloanthracene-12,14-dione (**25f**) was synthesized from (*E*)-3-(anthracen-9-yl)-1-phenylprop-2-en-1-one (1.0 mmol, 0.308g) and 1-(4-chlorophenyl)-1*H*-pyrrole-2,5-dione (1.3 mmol, 0.27 g) according to the general procedure above; white powder (48%), Mp. 205–210 °C. ^{1}H nmR (400 MHz, DMSO-d_6) δ 3.53 (dd, *J* = 8.50, 3.11 Hz, 1H) 4.00 (d, *J* = 8.71 Hz, 1H) 4.94 (d, *J* = 3.32 Hz, 1H) 6.43–6.49 (m, 2H) 7.20–7.29 (m, 5H) 7.29–7.35 (m, 1H) 7.36–7.44 (m, 3H) 7.55–7.62 (m, 3H) 7.68 (d, *J* = 7.46 Hz, 1H) 7.86 (d, *J* = 1.24 Hz, 2H) 8.13–8.18 (m, 2H). ^{13}C nmR (101 MHz, DMSO-d_6) 39.13–40.81, 48.39, 52.14, 123.39, 123.92, 124.88, 125.69, 127.46, 129.11, 129.44, 130.91, 132.03, 133.89, 137.73, 139.18, 140.10, 141.60, 142.26, 142.73, 175.14, 175.56, 189.95 (chalcone C=O) ppm. IR *ν* max (ATR): 3059.03 (Ar C-H), 1707.71 (C=O), 1673.57 (C=C), 1637.61 (Ar C=C), 1327.46 (C-N), 767.80 (C-Cl) cm^{-1}. LRMS (APCI) 514.21 [M^+−H]. HRMS (ESI) calculated for $C_{33}H_{21}ClNO_3$ [M^+−H]: 514.1209; found: 514.1191.

(*E*)-13-(4-Chlorophenyl)-9-(3-(4-iodophenyl)-3-oxoprop-1-en-1-yl)-9,10-dihydro-9,10-[3,4] epipyrroloanthracene-12,14-dione (**25g**) was synthesized (*E*)-3-(anthracen-9-yl)-1-(4-iodophenyl) prop-2-en-1-one (1.0 mmol, 0.4493 g) and 1-(4-chlorophenyl)-1*H*-pyrrole-2,5-dione (1.3 mmol, 0.27 g) according to the general procedure above; white powder (92%), Mp. 225–228 °C. ^{1}H nmR (400 MHz, DMSO-d_6) δ 3.52 (dd, *J* = 8.29, 3.32 Hz, H), 3.98 (d, *J* = 8.29 Hz, 2H), 4.93 (d, *J* = 3.32 Hz, 2H), 6.45 (d, *J* = 8.29 Hz, 2H), 7.16–7.23 (m, 2H), 7.23–7.28 (m, 2H), 7.29–7.34 (m, 1H), 7.36 (d, *J* = 3.73 Hz, H), 7.37–7.41 (m, H), 7.54–7.59 (m, 1H), 7.81 (d, *J* = 16.59 Hz, 1H), 7.84–7.92 (m, H), 7.97 (d, *J* = 8.29 Hz, 2H). IR *ν* max (ATR): 1716.83, 1661.60, 1493.51, 1392.83, 1325.76 cm^{-1}. LRMS (APCI) [M^++H]: 642.04. HRMS (ESI) calculated for $C_{33}H_{21}ClINNaO_3$ [M^+−H]: 664.0147; found: 664.0133.

(*E*)-13-(4-Chlorophenyl)-9-(3-(4-fluorophenyl)-3-oxoprop-1-en-1-yl)-9,10-dihydro-9,10[3,4] epipyrroloanthracene-12,14-dione (**25h**) was synthesized from (*E*)-3-(anthracen-9-yl)-1-(4-fluorophenyl)prop-2-en-1-one (1.0 mmol, 0.326 g) and 1-(4-chlorophenyl)-1*H*-pyrrole-2,5-

dione (1.3 mmol, 0.27 g) according to the general procedure above; white powder (88%) Mp. >250 °C. ^{1}H nmR (400 MHz, DMSO-d_6) δ 3.49 (dd, J = 8.50, 3.11 Hz, 4 H), 3.67 (d, J = 8.29 Hz, 5 H), 4.93 (d, J = 3.32 Hz, 1 H), 6.47 (d, J = 8.71 Hz, 2 H), 7.16–7.23 (m, 3 H), 7.23–7.31 (m, 6 H), 7.34–7.43 (m, 4 H), 7.47 (d, J = 6.63 Hz, 1 H), 7.83 (d, J = 15.76 Hz, 1 H), 8.00 (d, J = 15.76 Hz, 1 H), 8.21 (dd, J = 8.71, 5.39 Hz, 5 H). ^{13}C nmR (101 MHz, DMSO-d_6) 45.4, 48.4, 48.4, 52.1, 116.4, 116.6, 123.4, 123.9, 124.9, 125.7, 126.9, 127.4, 127.4, 127.5, 128.7, 129.5, 130.9, 131.9, 132.1, 132.2, 133.6, 134.4, 134.5, 139.2, 140.1, 141.6, 142.4, 142.7, 175.2, 175.5, 188.5 ppm. IR ν max (ATR): 1709.12, 1681.34, 1638.55, 1596.79, 1494.89, 1480.52, 1390.28, 1324.82 cm^{-1}. HRMS (ESI) calculated for $C_{33}H_{21}ClFNNaO_3$ [M$^+$+Na]: 556.1086; found: 556.1081.

(*E*)-13-(4-Chlorophenyl)-9-(3-(4-methoxyphenyl)-3-oxoprop-1-en-1-yl)-9,10-dihydro-9,10-[3,4] epipyrroloanthracene-12,14-dione (**25i**) was synthesized from (*E*)-3-(anthracen-9-yl)-1-(4-methoxyphenyl)prop-2-en-1-one (1.0 mmol, 0.338 g) and 1-(4-chlorophenyl)-1*H*-pyrrole-2,5-dione (1.3 mmol, 0.27 g) according to the general procedure above; white powder (63%), Mp. 220–225 °C. ^{1}H nmR (400 MHz, $CDCl_3$) δ 3.53 (dd, J = 8.50, 3.11 Hz, 1H), 3.85 (s, 3H), 3.99 (d, J = 8.29 Hz, 1H), 4.94 (d, J = 3.32 Hz, 1H), 6.44–6.48 (m, 2H), 7.11 (d, J = 8.71 Hz, 2H), 7.19–7.34 (m, 6H), 7.35–7.40 (m, 1H), 7.39–7.43 (m, 2H), 7.59 (d, J = 7.05 Hz, 1H), 7.77–7.91 (m, 2H), 8.15 (d, J = 9.12 Hz, 2H). ^{13}C nmR (101 MHz, $CDCl_3$) ppm 39.21–40.74, 45.39, 48.40, 56.07, 123.39, 123.92, 124.86, 125.68, 126.88, 127.23–127.57, 128.72, 129.46, 131.53, 131.94, 140.21, 141.11, 142.86, 175.13, 175.56, 188.11 (chalcone C=O). IR ν max (ATR): 3344.59 (N-H), 1669.34 (C=O), 1599.27 (Ar C=C), 1230.74 (C-O), 1021.65 (C-N), 766.69 (C-Cl) cm^{-1}. LRMS (APCI) 544.24 [M$^+$−H]. HRMS (ESI) calculated for $C_{34}H_{24}ClNNaO_4$ [M$^+$+Na]: 568.1286; found 568.1283.

(*E*)-13-(4-Chlorophenyl)-9-(3-oxo-3-(*p*-tolyl)prop-1-en-1-yl)-9,10-dihydro-9,10-[3,4] epipyrroloanthracene-12,14-dione (**25j**) was synthesized from (*E*)-3-(anthracen-9-yl)-1-(*p*-tolyl)prop-2-en-1-one (1.0 mmol, 0.338 g) and 1-(4-chlorophenyl)-1*H*-pyrrole-2,5-dione (1.3 mmol, 0.27 g) according to the general procedure above; white powder (60%), Mp. 215–220 °C. ^{1}H nmR (400 MHz, $CDCl_3$) δ 2.39 (s, 3H), 3.53 (dd, J = 8.29, 3.32 Hz, 1H), 3.99 (d, J = 8.29 Hz, 1H), 4.94 (d, J = 2.90 Hz, 1H), 6.43–6.49 (m, 2H, 7.20–7.29 (m, 5H), 7.29–7.34 (m, 1H), 7.35–7.43 (m, 5H), 7.55–7.61 (m, 1H), 7.85 (d, J = 2.49 Hz, 2H), 8.06 (d, J = 7.88 Hz, 2H). ^{13}C nmR (101 MHz, $CDCl_3$) ppm 21.69, 39.15–40.74, 48.40, 52.13, 123.39, 123.92, 124.87, 125.72, 126.88, 127.18–127.58, 128.44–128.84, 129.05 -129.59, 129.95, 132.00, 135.24, 140.15, 141.53–141.91, 142.78, 175.13, 189.36 (chalcone C=O) IR ν max (ATR): 3340.19 (N-H), 2970.43 (C-H), 1709.33 (C=O), 1623.92 (Ar C=C), 1464.76 (C-C), 1224.81 (C-O), 1014.05 (C-N), 728.59 (C-Cl) cm^{-1}. LRMS (APCI) 528.33 [M$^+$−H]. HRMS (ESI) calculated for $C_{34}H_{24}ClNNaO_3$ [M$^+$+Na]: 552.1337; found 552.1337.

(*E*) 13 (4 Chlorophenyl)-9-(3-oxo-3-(3,4,5-trimethoxyphenyl)-prop-1-en-1-yl)-9,10-dihydro-9,10-[3,4]epipyrroloanthracene-12,14-dione (**25k**) was synthesized from (*E*)-3-(anthracen-9-yl)-1-(3,4,5-trimethoxyphenyl)prop-2-en-1-one (1.0 mmol, 0.398 g) and 1-(4-chlorophenyl)-1*H*-pyrrole-2,5-dione (1.3 mmol, 0.27 g) according to the general procedure above; white powder (35%), Mp. 160–163 °C. ^{1}H nmR (400 MHz, $CDCl_3$) δ 3.53 (dd, J = 8.29, 3.32 Hz, 1H), 3.76 (s, 3H), 3.85 (s, 6H), 4.00 (d, J = 8.29 Hz, 1H), 4.94 (d, J = 3.32 Hz, 1H), 6.45–6.49 (m, 2H), 7.20–7.31 (m, 5H), 7.31–7.43 (m, 4H), 7.47 (s, 2H), 7.57–7.61 (m, 1H), 7.80 (d, J = 16.0 Hz, 1H), 7.88 (d, J = 16.0 Hz, 1H). ^{13}C nmR (101 MHz, $CDCl_3$) ppm 45.41, 48.3, 48.6, 52.11, 56.54, 60.67, 106.87, 124.90, 125.68, 126.85, 127.24–127.60, 128.55, 129.30–129.66, 132.54, 141.50, 142.60, 142.78, 153.42, 175.19, 189.10 (chalcone C=O). IR ν max (ATR): 2940.25 (C-H), 1707.34 (C=O), 1579.08 (Ar C=C), 1234.05 (C-O), 1090.23 (C-N), 766.71 cm^{-1}. LRMS (APCI): 606.21 [M$^+$+H].

(*E*)-13-(4-Chlorophenyl)-9-(3-(4-chlorophenyl)-3-oxoprop-1-en-1-yl)-9,10-dihydro-9,10-[3,4] epipyrroloanthracene-12,14-dione (**25l**) was synthesized from (*E*)-3-(anthracen-9-yl)-1-(4-chlorophenyl)prop-2-en-1-one (1.0 mmol, 0.343 g) and 1-(4-chlorophenyl)-1*H*-pyrrole-2,5-dione (1.3 mmol, 0.27 g) according to the general procedure above; white powder (39%), Mp. 268–271 °C. ^{1}H nmR (400 MHz, $CDCl_3$) δ 3.53 (dd, J = 8.29, 3.32 Hz, 1H), 4.00 (d, J = 8.29 Hz, 1H), 4.94 (d, J = 3.32 Hz, 1H), 6.42–6.48 (m, 2H), 7.20–7.31 (m, 6H), 7.32 (s, H), 7.36–7.43 (m, 3H), 7.59 (d, J = 7.05 Hz, 1H), 7.65–7.70 (m, 2H), 7.87 (s, 2H), 8.14–8.19

(m, 2H). ^{13}C nmR (101 MHz, $CDCl_3$) ppm 39.07–40.76, 45.39, 48.39, 52.16, 123.42, 123.94, 124.89, 125.72, 127.46, 128.68, 129.32–129.73, 130.69–131.19, 131.78, 133.58, 138.84, 139.17, 140.04, 141.59, 142.74, 175.16, 175.55, 188.92 (chalcone C=O). IR ν max (ATR): 1707.42 (C=O), 1673.82 (C=C), 1589.01 (Ar C=C), 1015.71 (C-N), 746.57 cm^{-1}. LRMS (APCI) 548.09 [M^+−H]. HRMS Found 645.4544 [M^++H_3Cl_2Na}; $C_{33}H_{24}Cl_4NO_3Na$ requires 645.0408.

(*E*)-13-(4-Chlorophenyl)-9-(3-oxo-3-(pyridin-4-yl)prop-1-en-1-yl)-9,10-dihydro-9,10-[3,4] epipyrroloanthracene-12,14-dione (**25m**) was synthesized from (*E*)-3-(anthracen-9-yl)-1-(pyridin-2-yl)prop-2-en-1-one (1.0 mmol, 0.309 g) and 1-(4-chlorophenyl)-1*H*-pyrrole-2,5-dione (1.3 mmol, 0.27 g) according to the general procedure (4), (92%), powder, Mp. 197–198 °C. IR ν max (ATR): 1703.19, 1493.90, 777.18 cm^{-1}. ^{1}H nmR (400 MHz, DMSO-d_6) δ 3.52 (dd, *J* = 8.29, 3.32 Hz, 1H), 3.97 (d, *J* = 8.71 Hz, 1H), 4.93 (d, *J* = 2.90 Hz, 1H), 6.44 (d, *J* = 8.71 Hz, 2H), 7.24–7.29 (m, 4H), 7.30–7.35 (m, 1H), 7.35–7.42 (m, 3H), 7.54–7.60 (m, 2H), 7.78 (d, *J* = 16.17 Hz, 1H), 7.91 (d, *J* = 15.76 Hz, 1H), 7.97 (d, *J* = 6.63 Hz, 2H), 8.85 (d, *J* = 6.22 Hz, 2H). ^{13}C nmR (101 MHz, $CDCl_3$) 44.89. 47.79, 48.23, 51.82, 123.07, 123.53, 123.59, 124.47, 125.23, 126.40, 126.89, 126.97, 127.11, 128.22, 129.00, 130.39, 131.39, 133.12, 138.68, 139.29, 141.07, 141.88, 145.03, 147.47 ppm. HRMS (ESI) calculated for $C_{32}H_{22}Cl_2NO_3$ [M^++H]: 551.0935; found: 551.0924.

(*E*)-13-(4-Chlorophenyl)-9-(3-oxo-3-(pyridin-2-yl)prop-1-en-1-yl)-9,10-dihydro-9,10-[3,4] epipyrroloanthracene-12,14-dione (**25n**) was synthesized from (*E*)-3-(anthracen-9-yl)-1-(pyridin-2-yl)prop-2-en-1-one (1.0 mmol, 0.309 g) and 1-(4-chlorophenyl)-1*H*-pyrrole-2,5-dione (1.3 mmol, 0.27 g), according to the general procedure above; cream powder, 57%. IR ν max (ATR): 1710.53, 1677.50, 1624.02, 1490.52, 1388.35, 1324.42, 1457.44, 1190.22, 1085.27, 1015.50, 996.38 cm^{-1}. ^{1}H nmR (400 MHz, DMSO-d_6) δ 3.50 (dd, *J* = 8.29, 3.32 Hz, 3H), 3.84 (d, *J* = 8.71 Hz, H), 4.94 (d, *J* = 2.90 Hz, 3 H), 6.45 (d, *J* = 8.29 Hz, H), 7.11–7.22 (m, 1 H), 7.22–7.30 (m, 4H), 7.30–7.35 (m, 1H), 7.36–7.42 (m, 3H), 7.59 (d, *J* = 7.05 Hz, 1H), 7.70 (dd, *J* = 7.46, 4.98 Hz, 1H), 8.04 (d, *J* = 17.00 Hz, 1H), 8.07–8.12 (m, 1H), 8.21 (d, *J* = 7.88 Hz, 1H), 8.27 (d, *J* = 16.59 Hz, 1H), 8.76 (d, *J* = 4.98 Hz, 1H). ^{13}C nmR (101 MHz, DMSO-d_6) 45.41, 48.40, 48.96, 52.25, 127.37, 127.41, 127.54, 128.31, 128.64, 129.4, 130.28, 130.89, 133.51, 138.34, 139.29, 140.07, 141.66, 142.60, 143.09, 149.75, 153.68, 174.99, 175.58, 188.85 ppm. HRMS (ESI) calculated for $C_{23}H_{22}ClN_2O_3$ [M^++H]: 517.1313; found: 517.1032.

(*E*)-13-(4-Chlorophenyl)-9-(3-(furan-2-yl)-3-oxoprop-1-en-1-yl)-9,10-dihydro-9,10-[3,4] epipyrroloanthracene-12,14-dione (**25o**) was prepared from (*E*)-3-(anthracen-9-yl)-1-(furan-2-yl)prop-2-en-1-one (1.0 mmol, 0.298 g) and 1-(4-chlorophenyl)-1*H*-pyrrole-2,5-dione (1.3 mmol, 0.27 g) according to the general procedure above; white powder (46%) Mp. 256 °C (dec). ^{1}H nmR (400 MHz, $CDCl_3$) δ 3.54 (dd, *J* = 8.33, 3.18 Hz, 1H), 4.03 (d, *J* = 8.21 Hz, 1H), 4.94 (d, *J* = 3.3 Hz, 1H), 6.61 (m, 2H), 7.12–7.32 (m, 6H), 7.35–7.43 (m, 2H), 7.45–7.49 (m, 2H), 7.55–7.63 (m, 2H), 7.66 (m, 1H), 7.75 (d, *J* = 16.18 Hz, 1H), 8.13 (d, *J* = 16.18 Hz, 1H). ^{13}C nmR (101 MHz, $CDCl_3$) ppm 45.4, 48.3, 48.8, 52.0, 113.4, 120.6, 123.4, 123.9, 124.9, 125.7, 125.7, 126.9, 127.4, 127.5, 128.6, 129.3, 129.4, 130.9, 131.4, 133.5, 139.2, 140.0, 141.5, 141.6, 142.6, 149.2, 153.1, 167.3, 175.1, 175.5, 176.8. IR ν max (ATR): 1709.72, 1675.02, 1626.02, 1586.77, 1494.51, 1480.32, 1388.67 cm^{-1}. LRMS (APCI): 506.14 [M^++H].

(*E*)-13-(4-Chlorophenyl)-9-(3-(naphthalen-2-yl)-3-oxoprop-1-en-1-yl)-9,10-dihydro-9,10-[3,4] epipyrroloanthracene-12,14-dione (**25p**) was prepared from (*E*)-3-(anthracen-9-yl)-1-(naphthalen-2-yl)prop-2-en-1-one (1.0 mmol, 0.358 g) and 1-(4-chlorophenyl)-1*H*-pyrrole-2,5-dione (1.3 mmol, 0.27 g) according to the general procedure above; gray powder (88%), Mp. 243–244 °C. ^{1}H nmR (400 MHz, $CDCl_3$) δ 3.51 (dd, *J* = 8.50, 3.11 Hz, 3H), 3.72 (d, *J* = 8.29 Hz, 3H), 4.95 (d, *J* = 3.32 Hz, 1H), 6.51 (d, *J* = 8.71 Hz, 2H), 7.24–7.32 (m, 6H), 7.38–7.43 (m, 2H), 7.43–7.50 (m, 2H), 7.52–7.58 (m, 1H), 7.58–7.64 (m, 1H), 7.91 (d, *J* = 7.88 Hz, 1H), 7.94–8.02 (m, H), 8.05–8.11 (m, 1H), 8.23 (dd, *J* = 8.50, 1.45 Hz, 1H), 8.71 (s, 1H). ^{13}C nmR (101 MHz, $CDCl_3$) ppm 46.00, 48.1, 48.47, 52.5, 123.63, 124.08, 124.16, 124.57, 125.50, 126.78, 126.89, 127.40, 127.54, 127.70, 127.82, 128.57, 128.71, 129.31, 129.48, 129.71, 129.74, 130.81, 131.62, 132.59, 134.74, 135.05, 135.68, 138.16, 140.69, 142.05, 142.19, 174.34, 175.15, 189.99. IR ν max (ATR): 1709.81, 1668.28, 1645.28, 1584.47, 1459.45, 1420.04, 1389.07, 1493.53, 1442.07 cm^{-1}.

LRMS (APCI): [M^++Cl] 602.31. HRMS calculated for $C_{37}H_{24}Cl_2NO_3$ [M+Cl-2H] 598.0977; found 598.1371.

(*E*)-13-(4-Chlorophenyl)-9-(3-oxo-3-(thiophen-2-yl)prop-1-en-1-yl)-9,10-dihydro-9,10-[3,4] epipyrroloanthracene-12,14-dione (**25q**) was prepared from (*E*)-3-(anthracen-9-yl)-1-(thiophen-2-yl)prop-2-en-1-one (1.0 mmol, 0.314 g) and 1-(4-chlorophenyl)-1*H*-pyrrole-2,5-dione (1.3 mmol, 0.27 g) according to the general procedure above; white powder (70%), Mp. 180–181 °C. ^{1}H nmR (400 MHz, DMSO-d_6) δ 3.56 (dd, *J* = 8.29, 3.32 Hz, 1H), 3.99 (d, *J* = 8.29 Hz, 1H), 4.95 (d, *J* = 2.90 Hz, 1H), 6.48 (d, *J* = 8.71 Hz, 2H), 7.09–7.17 (m, 2H), 7.18–7.32 (m, H), 7.32–7.36 (m, 1 H), 7.36–7.43 (m, 3H), 7.59 (d, *J* = 6.63 Hz, 1H), 7.84 (d, *J* = 15.76 Hz, H), 7.93 (d, *J* = 16.59 Hz, 1H), 8.10 (d, *J* = 4.56 Hz, 1H), 8.21 (d, *J* = 3.32 Hz, 1H). ^{13}C nmR (101 MHz, DMSO-d_6) 45.45, 48.35, 48.68, 52.08, 123.42, 123.9, 124.88, 125.7, 125.74, 126.88, 127.3, 127.5, 129.33, 129.44, 129.4, 130.90, 131.70, 133.58, 134.6, 136.50, 137.79, 139.18, 140.05, 141.60, 142.67, 145.27 ppm. IR ν max (ATR): 1733.84, 1708.84, 1699.66, 1662.25, 1616.18, 1582.06, 1546.74, 1562.06 cm^{-1}. LRMS (APCI): 520.30 [M^+−H]. HRMS (APCI) calculated for $C_{31}H_{20}ClNNaO_3S$ [M^++Na]: 544.0745; found: 544.0735.

3.7. General Preparation for (E)-13-(4-Benzoylphenyl)-9-(3-oxo-3-phenylprop-1-en-1-yl)-9,10-dihydro-9,10-[3,4] Epipyrroloanthracene-12,14-diones 26a–q

To a solution of the appropriate chalcone anthracene **21a–q** (1.0 mmol) in toluene (2 mL) was added dienophile 1-(4-benzoylphenyl)-1*H*-pyrrole-2,5-dione (1.3 mmol). The mixture was heated with stirring at 90 °C for 48 h. The reaction mixture was then cooled to room temperature, and the resulting crude solid product was isolated via filtration. The product was washed with toluene (2 mL) and diethyl ether (2 mL) and then recrystallized from toluene.

(*E*)-13-(4-Benzoylphenyl)-9-(3-(4-bromophenyl)-3-oxoprop-1-en-1-yl)-9,10-dihydro-9,10-[3,4] epipyrroloanthracene-12,14-dione (**26a**) was synthesized from (*E*)-3-(anthracen-9-yl)-1-(4-bromophenyl)prop-2-en-1-one (1.0 mmol, 0.387 g) and 1-(4-benzoylphenyl)-1*H*-pyrrole-2,5-dione (1.3 mmol, 0.361 g) according to the general procedure above; white powder (71%), Mp. 209–210 °C. (HPLC: 93.25%, RT 2.57 min). ^{1}H nmR (400 MHz, DMSO-d_6) δ 3.57 (dd, *J* = 8.29, 2.90 Hz, 3H), 4.03 (d, *J* = 8.71 Hz, 3H), 4.96 (d, *J* = 2.90 Hz, 3H), 6.65 (d, *J* = 8.29 Hz, 6H), 7.20–7.30 (m, 15H), 7.32–7.36 (m, 3H), 7.36–7.41 (m, 3H), 7.48–7.54 (m, 6H), 7.56–7.61 (m, 3H), 7.61–7.69 (m, 15H), 7.80 (d, *J* = 8.71 Hz, 6H), 7.83–7.87 (m, 2H), 7.88–7.93 (m, 1H), 8.08 (d, *J* = 8.71 Hz, 2H). ^{13}C nmR (101 MHz, DMSO-d_6) 45.4, 48.4, 48.5, 52.2, 123.4, 124.0, 124.9, 125.7, 126.9, 127.0, 127.4, 127.5, 127.6, 128.0, 129.1, 130.1, 130.6, 131.1, 131.8, 132.5, 133.4, 135.5, 136.7, 137.0, 137.4, 139.2, 140.0, 141.6, 142.7, 142.8, 175.1, 175.5, 189.1, 195.3 ppm. LRMS (APCI) [M^++H]: 664.16.

(*E*)-13-(4-Benzoylphenyl)-9-(3-(4-nitrophenyl)-3-oxoprop-1-en-1-yl)-9,10-dihydro-9,10-[3,4]epipyrroloanthracene-12,14-dione (**26b**) was prepared from (*E*)-3-(anthracen-9-yl)-1-(4-nitrophenyl)prop-2-en-1-one (1.0 mmol, 0.353 g) and 1-(4-benzoylphenyl)-1*H*-pyrrole-2,5-dione (1.3 mmol, 0.361 g) according to the general procedure above; white powder (75%); Mp. 175–177 °C. ^{1}H nmR (400 MHz, DMSO-d_6) δ 3.59 (dd, *J* = 8.50, 3.11 Hz, 2H), 4.04 (d, *J* = 8.71 Hz, 2H), 4.98 (d, *J* = 3.32 Hz, 2H), 6.66 (d, *J* = 8.29 Hz, 3H), 7.09–7.17 (m, 4H), 7.19–7.26 (m, 5H), 7.26–7.32 (m, 6H), 7.34–7.42 (m, 4H), 7.50–7.55 (m, 4H), 7.58–7.64 (m, 2H), 7.65–7.70 (m, 8H), 7.92 (dd, *J* = 16.60 Hz, 2H), 8.34–8.43 (m, 4H). ^{13}C nmR (101 MHz, DMSO-d_6) 45.45, 48.43, 48.66, 52.30, 123.51, 124.02, 124.49, 124.94, 125.75, 126.91, 127.01, 127.48, 127.63, 128.6, 129.07, 129.34, 130.08, 130.50, 130.61, 132.03, 133.36, 135.54, 136.98, 137.40, 137.79, 139.17, 139.91, 141.60, 142.52, 142.54, 144.14, 150.47, 175.15, 175.51, 189.23 (chalcone C=O), 195.28 (benzophenone C=O) ppm. IR ν max (ATR): 1714.57, 1687.42, 1639.11, 1653.38, 1604.55, 1557.12, 1506.50, 1449.13, 1409.43, 1377.97, 1275.37, 1173.59 cm^{-1}. LRMS (APCI): 631.19 [M^++H]. HRMS (APCI) calculated for $C_{40}H_{26}KN_2O_6$ [M^++K]: 669.1423; found: 669.1418.

(*E*)-13-(4-Benzoylphenyl)-9-(3-(4-ethylphenyl)-3-oxoprop-1-en-1-yl)-9,10-dihydro-9,10-[3,4]epipyrroloanthracene-12,14-dione (**26c**) was prepared from (*E*)-3-(anthracen-9-yl)-1-(4-ethylphenyl)prop-2-en-1-one (1.0 mmol, 0.336 g) and 1-(4-benzoylphenyl)-1*H*-pyrrole-2,5-

dione (1.3 mmol, 0.361 g) according to the general procedure above; white powder (76%), Mp. 246–247 °C. ^{1}H nmR (400 MHz, DMSO-d_6) δ 3.58 (dd, *J* = 8.50, 3.11 Hz, 1H), 3.84 (s, 2H), 3.85 (s, 3H), 4.04 (d, *J* = 8.71 Hz, 1H), 4.97 (d, *J* = 3.32 Hz, 1H), 6.67 (d, *J* = 8.29 Hz, 2H), 7.13 (d, *J* = 8.29 Hz, H), 7.19–7.32 (m, 5), 7.32–7.36 (m, 1H), 7.37–7.42 (m, 1H), 7.49–7.56 (m, 2H), 7.60 (d, *J* = 6.63 Hz, 1H), 7.63–7.72 (m, 5H), 7.81 (d, *J* = 16.17 Hz, 1H), 7.78–7.84 (m, 1H), 7.87 (dd, *J* = 8.71, 2.07 Hz, 1H), 7.91 (d, *J* = 16.17 Hz, 1H). ^{13}C nmR (101 MHz, DMSO-d_6) 45.43,48.47, 48.55, 52.14, 56.01, 56.29, 111.31, 111.51,123.39, 123.93, 124.08, 124.89, 125.71, 126.88, 126.97, 127.37, 127.45, 127.55, 129.08, 130.08, 130.60, 132.11, 133.36, 135.56, 136.97, 137.35, 139.20, 140.21, 141.65, 149.34, 175.10, 175.52, 188.25 (chalcone C=O), 195.30 (benzophenone C=O) ppm. IR ν max (ATR): 1716.45, 1660.71, 1391.17, 1325.22 cm^{-1}. LRMS (APCI) 614.14 [M$^+$+H].

(*E*)-13-(4-Benzoylphenyl)-9-(3-(2,4-dichlorophenyl)-3-oxoprop-1-en-1-yl)-9,10-dihydro-9,10-[3,4]epipyrroloanthracene-12,14-dione (**26d**) was prepared from (*E*)-3-(anthracen-9-yl)-1-(2,4-dichlorophenyl)prop-2-en-1-one (1.0 mmol, 0.377 g) and 1-(4-benzoylphenyl)-1*H*-pyrrole-2,5-dione (1.3 mmol, 0.361 g) according to the general procedure above; white powder (63%). ^{1}H nmR (400 MHz, DMSO-d_6) δ 3.58 (dd, *J* = 8.29, 3.32 Hz, 1H), 4.04 (d, *J* = 8.29 Hz, 1H), 4.97 (d, *J* = 2.90 Hz, 1H), 6.67 (d, *J* = 8.29 Hz, 2H), 7.09–7.16 (m, 1H), 7.19–7.32 (m, 5H), 7.32–7.36 (m, 1H), 7.37–7.43 (m, 1H), 7.50–7.56 (m, 2H), 7.60 (d, *J* = 6.22 Hz, 1H), 7.63–7.71 (m, 6H), 7.81 (d, *J* = 16.17 Hz, 1H), 7.86 (d, *J* = 1.66 Hz, 1H), 7.91 (d, *J* = 16.17 Hz, 1H). ^{13}C nmR (101 MHz, DMSO-d_6) 48.5, 52.1, 56.0, 56.3, 111.3, 111.5, 123.9, 124.1, 124.9, 126.9, 127.0, 127.5, 129.1, 130.1, 130.6, 132.1, 133.4, 135.6, 137.0, 137.4, 139.2, 140.2, 140.9, 141.7, 142.9, 144.5, 145.7, 149.3, 153.9, 175.1, 175.5, 188.2 (chalcone C=O), 195.3 (benzophenone C=O) ppm. LRMS (APCI): 614.15 [M$^+$-C_3H_3].

(*E*)-13-(4-Benzoylphenyl)-9-(3-(3,4-dimethoxyphenyl)-3-oxoprop-1-en-1-yl)-9,10-dihydro-9,10-[3,4]epipyrroloanthracene-12,14-dione (**26e**) was prepared from (*E*)-3-(anthracen-9-yl)-1-(3,4-dimethoxyphenyl)prop-2-en-1-one (1.0 mmol, 0.368 g) and 1-(4-benzoylphenyl)-1*H*-pyrrole-2,5-dione (1.3 mmol, 0.361 g) according to the general procedure above; white powder (72%), Mp. 161–164 °C. IR ν max (ATR): 2969, 1714.39, 1651.24, 1603.13, 1507.90, 1458.19, 1378.67, 1275.61, 1177.36, 929.79, 925.10 cm^{-1}. ^{1}H nmR (400 MHz, DMSO-d_6) δ 1.20 (t, *J* = 7.46 Hz, 2H), 2.28 (s, 2H), 3.58 (dd, *J* = 8.50, 3.11 Hz, 1H), 4.03 (d, *J* = 8.29 Hz, 1H), 4.97 (d, *J* = 3.32 Hz, 1H), 6.67 (d, *J* = 8.29 Hz, 1H), 7.09–7.18 (m, 2H), 7.19–7.25 (m, 3H), 7.26–7.32 (m, 2H), 7.32–7.36 (m, H), 7.38–7.45 (m, 2H), 7.50–7.55 (m, 2H), 7.58–7.63 (m, 1H), 7.63–7.71 (m, 4H), 7.82–7.93 (m, 2H), 7.87 (dd, *J* = 16.20 Hz, 2H), 8.10 (d, *J* = 7.88 Hz, 1H). ^{13}C nmR (101 MHz, DMSO-d_6) 21.49, 28.70, 45.4, 48.47, 48.52, 52.18, 123.41, 123.93, 124.89, 126.90, 127.02, 127.38, 127.48, 127.57, 128.64, 128.80, 129.07, 129.34, 129.39, 130.08, 130.60, 132.09, 133.36, 135.50, 135.56, 136.97, 137.38, 137.78, 139.18, 140.14, 141.63, 141.76, 142.80, 150.41, 175.09, 175.51, 189.38 (chalcone C=O), 195.29 (benzophenone C=O) ppm. LRMS (APCI) 614.19 [M$^+$-OCH_3].

(*E*)-13-(4-Benzoylphenyl)-9-(3-oxo-3-phenylprop-1-en-1-yl)-9,10-dihydro-9,10-[3,4] epipyrroloanthracene-12,14-dione (**26f**) was synthesized from (*E*)-3-(anthracen-9-yl)-1-phenylprop-2-en-1-one (1.0 mmol, 0.308 g) and 1-(4-benzoylphenyl)-1*H*-pyrrole-2,5-dione (1.3 mmol, 0.361 g) according to the general procedure above; white powder (49%), Mp. 240–241 °C. ^{1}H nmR (400 MHz, DMSO-d_6) δ 3.58 (dd, *J* = 8.50, 3.11 Hz, 1H), 4.04 (d, *J* = 8.29 Hz, 1H), 4.97 (d, *J* = 2.90 Hz, 1H), 6.67 (d, *J* = 8.29 Hz, 2H), 7.19–7.32 (m, 6H), 7.34 (s, 1H), 7.37–7.42 (m, 1H), 7.49–7.56 (m, 2H), 7.57–7.64 (m, 4H), 7.64–7.72 (m, 6H), 7.89 (s, 2H), 8.16 (d, *J* = 7.05 Hz, 2H). ^{13}C nmR (101 MHz, DMSO-d_6) 39.21–40.63, 48.48, 52.20, 55.34, 125.75, 127.03, 127.48, 128.94–129.60, 130.08, 130.61, 131.99, 136.97, 137.39, 139.17, 142.75, 174.93–175.59, 176.01, 189.96 (chalcone C=O), 195.31 (benzophenone C=O) ppm. IR ν max (ATR): 1708.57 (C=O), 1670.41 (C=C), 1603.92 (Ar C=C), 1308.22 (C-N) cm^{-1}. HRMS (APCI) calculated for $C_{40}H_{27}NNaO_4$ [M$^+$+Na]: 608.1832; found: 608.1830.

(*E*)-13-(4-Benzoylphenyl)-9-(3-(4-iodophenyl)-3-oxoprop-1-en-1-yl)-9,10-dihydro-9,10-[3,4]epipyrroloanthracene-12,14-dione (**26g**) was prepared from (*E*)-3-(anthracen-9-yl)-1-(4-iodophenyl)prop-2-en-1-one (1.0 mmol, 0.338 g) and 1-(4-benzoylphenyl)-1*H*-pyrrole-2,5-dione (1.3 mmol, 0.361 g) according to the general procedure above; white powder

(85%), Mp. 186–188 °C. ^{1}H nmR (400 MHz, DMSO-d_6) δ 3.59 (dd, *J* = 8.29, 2.90 Hz, 1H), 4.04 (d, *J* = 8.71 Hz, 1H), 4.98 (d, *J* = 3.32 Hz, 1H), 6.67 (d, *J* = 8.29 Hz, 2H), 7.10–7.17 (m, 2H), 7.19–7.32 (m, 6H), 7.33–7.37 (m, 1H), 7.38–7.43 (m, 1H), 7.50–7.56 (m, 2H), 7.61 (d, *J* = 7.05 Hz, 1H), 7.63–7.71 (m, 5H), 7.86 (d, *J* = 16.59 Hz, 1H), 7.89–7.95 (m, 3H), 7.96–8.03 (m, 2H). ^{13}C nmR (101 MHz, DMSO-d_6) 45.45, 48.4, 48.55, 52.23, 102.63, 123.45, 123.97, 124.90, 125.72, 125.7, 126.91, 127.03, 127.40, 127.48, 127.59, 128.64, 129.07, 129.34, 130.09, 130.61, 130.82, 131.76, 133.35, 135.56, 136.98, 137.04, 137.39, 137.79, 138.36, 139.17, 140.04, 141.61, 142.69, 142.77, 175.11, 175.50, 189.44, 195.28 ppm. IR ν max (ATR): 1715.20, 1687.34, 1638.37, 1653.19, 1604.04, 1448.80, 1379.55, 1275.10, 1173.70, 1190.33 cm^{-1}. LRMS (APCI) 710.13 [M$^+$−H]. HRMS (ESI) calculated for $C_{40}H_{25}INO_4$ [M$^+$−H]: 710.0834; found: 710.0823.

(*E*)-13-(4-Benzoylphenyl)-9-(3-(4-fluorophenyl)-3-oxoprop-1-en-1-yl)-9,10-dihydro-9,10-[3,4] epipyrroloanthracene-12,14-dione (**26h**) was synthesized from (*E*)-3-(anthracen-9-yl)-1-(4-fluorophenyl)prop-2-en-1-one (1.0 mmol, 0.387 g) and 1-(4-benzoylphenyl)-1*H*-pyrrole-2,5-dione (1.3 mmol, 0.361 g) according to the general procedure above, white powder (77%). ^{1}H nmR (400 MHz, DMSO-d_6) δ 3.51 (dd, *J* = 8.22, 3.17 Hz, 1H), 4.02 (d, *J* = 8.4 Hz, 1H), 4.98 (d, *J* = 3.28 Hz, 1H), 6.49–6.54 (m, 2H), 7.15–7.32 (m, 5H), 7.35 (m, 1H), 7.36–7.40 (m, 4H), 7.41–7.46 (m, 4H), 7.47–7.49 (m, 1H), 7.83 (m, 3H), 8.02 (d, *J* = 16.18 Hz, 1H), 8.13–8.23 (m, 2H).

(*E*)-13-(4-Benzoylphenyl)-9-(3-(4-methoxyphenyl)-3-oxoprop-1-en-1-yl)-9,10-dihydro-9,10-[3,4]epipyrroloanthracene-12,14-dione (**26i**) was prepared from (*E*)-3-(anthracen-9-yl)-1-(4-methoxyphenyl)prop-2-en-1-one (1.0 mmol, 0.338 g) and 1-(4-benzoylphenyl)-1*H*-pyrrole-2,5-dione (1.3 mmol, 0.361 g) according to the general procedure above; white powder (50%), Mp. 224–228 °C. ^{1}H nmR (400 MHz, $CDCl_3$) δ 3.58 (dd, *J* = 8.29, 3.32 Hz, 1H), 3.86 (s, 3H), 4.04 (d, *J* = 8.29 Hz, 1H), 4.97 (d, *J* = 3.32 Hz, H), 6.66 (d, *J* = 8.29 Hz, 2H), 7.12 (d, *J* = 8.71 Hz, 2H), 7.20–7.31 (m, 5H), 7.31–7.35 (m, 1H), 7.38–7.41 (m, 1H), 7.51–7.56 (m, 2H), 7.60 (d, *J* = 6.63 Hz, 1H), 7.64–7.71 (m, 5H), 7.79–7.93 (m, 2H), 8.16 (d, *J* = 8.71 Hz, 2H). ^{13}C nmR (101 MHz, DMSO-d_6) 39.26, 40.72, 48.48, 52.15, 56.07, 114.67, 123.42, 123.93, 124.87, 125.71, 126.76, 127.69, 130.08, 131.53, 133.36, 136.97, 139.18, 141.10, 141.64, 142.88, 163.89, 175.09, 175.52, 188.12 (chalcone C=O), 195.29 (benzophenone C=O) ppm. IR ν max (ATR): 3344.36 (N-H), 1659.32 (C=O), 1599.79 (Ar C=C), 1231.15 (C-O), 1020.42 (C-N) cm^{-1}. HRMS (ESI) calculated for $C_{41}H_{29}NNaO_5$ [M$^+$+Na]: 638.1938; found: 638.1943.

(*E*)-13-(4-Benzoylphenyl)-9-(3-oxo-3-(*p*-tolyl)prop-1-en-1-yl)-9,10-dihydro-9,10-[3,4] epipyrroloanthracene-12,14-dione (**26j**) was prepared from (*E*)-3-(anthracen-9-yl)-1-(*p*-tolyl)prop-2-en-1-one (1.0 mmol, 0.3384 g) and 1-(4-benzoylphenyl)-1*H*-pyrrole-2,5-dione (1.3 mmol, 0.361g) according to the general procedure above; white powder (59%), Mp. 256–260 °C. ^{1}H nmR (400 MHz, $CDCl_3$) δ 2.40 (s, 3H), 3.57 (dd, *J* = 8.29, 2.90 Hz, 1H), 4.03 (d, *J* = 8.29 Hz, 1H), 4.96 (d, *J* = 2.90 Hz, 1H), 6.66 (d, *J* = 8.29 Hz, 2H), 7.20–7.31 (m, 5H), 7.31–7.36 (m, 1H), 7.40 (d, *J* = 7.88 Hz, 3H), 7.50–7.56 (m, 2H), 7.56–7.71 (m, 6H), 7.86 (d, *J* = 3.32 Hz, 2H), 8.06 (d, *J* = 7.88 Hz, 2H). ^{13}C nmR (101 MHz, $CDCl_3$) 21.69, 39.21–40.78, 48.49, 55.35, 123.41, 127.03, 128.83–129.51, 130.02, 136.97, 137.38, 139.18, 142.80, 144.38, 175.51 (chalcone C=O), 189.37 (benzophenone C=O) ppm. IR ν max (ATR): 2965 (O-H), 1712.89 (C=O), 1626.70 (Ar C=C), 1458.41 (C-C), 1231.54 (C-O), 1128.77 (C-N) cm^{-1}. HRMS (ESI) calculated for $C_{41}H_{29}NNaO_4$ [M$^+$+Na]: 622.1989; found: 622.1982.

(*E*)-13-(4-Benzoylphenyl)-9-(3-oxo-3-(3,4,5-trimethoxyphenyl)-prop-1-en-1-yl)-9,10-dihydro-9,10-[3,4]epipyrroloanthracene-12,14-dione (**26k**) was prepared from (*E*)-3-(anthracen-9-yl)-1-(3,4,5-trimethoxyphenyl)prop-2-en-1-one (1 mmol, 0.398 g) and 1-(4-benzoylphenyl)-1*H*-pyrrole-2,5-dione (1.3 mmol, 0.361 g) according to the general procedure above; white powder (44%), Mp. 210–216 °C. ^{1}H nmR (400 MHz, $CDCl_3$) δ 3.58 (dd, *J* = 8.50, 3.11 Hz, 1H), 3.76 (s, 3H), 3.85 (s, 6 H), 4.04 (d, *J* = 8.71 Hz, 1H), 4.97 (d, *J* = 3.32 Hz, 1H), 6.65–6.70 (m, 2H), 7.21–7.31 (m, 5H), 7.35–7.38 (m, 1H), 7.38–7.42 (m, 1H), 7.48 (s, 2H), 7.51–7.56 (m, 2H), 7.59–7.62 (m, 1H), 7.63–7.70 (m, 5H), 7.80–7.93 (m, 2H). ^{13}C nmR (101 MHz, $CDCl_3$) 39.24, 40.78, 48.40, 52.16, 56.56, 60.68, 106.89, 125.53, 125.91, 126.85, 127.40, 128.64, 129.07, 129.33, 130.07, 130.56, 132.54, 133.05, 133.34, 137.00, 140.12, 141.56, 153.43, 175.03, 175.61, 189.07 (chalcone C=O), 195.29 (benzophenone C=O) ppm. IR ν max (ATR): 2941.10 (O-H), 1708.03

(C=O), 1579.54 (Ar C=C), 1234.15 (C-O), 1091.54 (C-N) cm^{-1}. HRMS (ESI) calculated for $C_{43}H_{33}NNaO_7$ [M$^+$+Na]: 698.2149; found: 698.2141.

(*E*)-13-(4-Benzoylphenyl)-9-(3-(4-chlorophenyl)-3-oxoprop-1-en-1-yl)-9,10-dihydro-9,10-[3,4] epipyrroloanthracene-12,14-dione (**26l**) was synthesized from (*E*)-3-(anthracen-9-yl)-1-(4-chorophenyl)prop-2-en-1-one (1.0 mmol, 0.343 g) and 1-(4-benzoylphenyl)-1*H*-pyrrole-2,5-dione (1.3 mmol, 0.361 g) according to the general procedure above; white powder (30%), Mp. 217–220 °C. ^{1}H nmR (400 MHz, DMSO-d_6) δ 3.59 (dd, *J* = 3.32, 8.29 Hz, 1H), 4.04 (d, *J* = 8.29 Hz, 1H), 4.97 (d, *J* = 2.90 Hz, 1H), 6.66 (d, *J* = 8.71 Hz, 2H), 7.20–7.31 (m, 6H), 7.34 (m, 1H), 7.39 (m, 1H), 7.50–7.56 (m, 2H), 7.60 (d, *J* = 6.63 Hz, 1H), 7.66 (m, 6H), 7.89 (m, 2H), 8.18 (d, *J* = 8.0 Hz, 2H) ^{13}C nmR (101 MHz, DMSO-d_6) 42.31, 45.42, 48.48, 52.20, 123.44, 123.95, 124.90, 125.64, 125.78, 126.85, 127.14, 127.35, 127.66, 128.64, 129.08, 129.55, 130.08, 130.61, 131.03, 131.81, 136.41, 138.85, 139.16, 142.68, 142.81, 175.12, 175.51, 195.31 (chalcone C=O), 197.42 (benzophenone C=O) ppm. IR *ν* max (ATR): 3024.65 (Ar C-H), 1712.90 (C=O), 1672.35 (C=C), 1603.46 (Ar C=C), 1013.52 (C-N), 769.64 cm^{-1}. HRMS (ESI) calculated for $C_{40}H_{25}ClNO_4$ [M$^+$−H]: 618.1478; found: 618.1487.

(*E*)-13-(4-Benzoylphenyl)-9-(3-oxo-3-(pyridin-4-yl)prop-1-en-1-yl)-9,10-dihydro-9,10-[3,4] epipyrroloanthracene-12,14-dione (**26m**) was synthesized from (*E*)-3-(anthracen-9-yl)-1-(pyridin-4-yl)prop-2-en-1-one (1.0 mmol, 0.309 g) and 1-(4-benzoylphenyl)-1*H*-pyrrole-2,5-dione (1.3 mmol, 0.361 g) according to the general procedure above; white powder (62%), Mp. 230–231 °C. ^{1}H nmR (400 MHz, DMSO-d_6) δ 3.58 (dd, *J* = 8.29, 3.32 Hz, 2H), 4.02 (d, *J* = 8.29 Hz, 2H), 4.97 (d, *J* = 3.32 Hz, 2H), 6.66 (d, *J* = 8.71 Hz, 4H), 7.19–7.32 (m, 10H), 7.33–7.38 (m, 2 H), 7.38–7.42 (m, 2 H), 7.49–7.56 (m, 4H), 7.59–7.64 (m, 3H), 7.64–7.69 (m, 9H), 7.78–7.85 (m, 2H), 7.81 (d, *J* = 16.17 Hz, 1H), 7.91–7.97 (m, 2H), 7.94 (d, *J* = 16.17 Hz, 3H), 7.99 (d, *J* = 5.81 Hz, 4H). ^{13}C nmR (101 MHz, DMSO-d_6) 45.41, 48.40, 48.67, 52.27, 122.20, 123.48, 123.99, 124.94, 125.73, 126.93, 127.02, 127.48, 127.63, 129.08, 130.08, 130.61, 131.80, 133.38, 135.51, 136.96, 137.40, 139.14, 139.85, 141.57, 142.48, 143.77, 144.35, 151.35, 175.14, 175.51, 190.04 (chalcone C=O), 195.32 (benzophenone C=O) ppm. IR *ν* max (ATR): 1714.95, 1687.33, 1654.19, 1604.43, 1449.30, 1377.44, 1275.15, 1173.54, 1190.08 cm^{-1}. LRMS (APCI) 587.37 [M$^+$+H]. HRMS (ESI) calculated for $C_{39}H_{25}N_2O_4$ [M$^+$−H]: 585.1820; found: 585.1818.

(*E*)-13-(4-Benzoylphenyl)-9-(3-oxo-3-(pyridin-2-yl)prop-1-en-1-yl)-9,10-dihydro-9,10-[3,4] epipyrroloanthracene-12,14-dione (**26n**) was prepared from (*E*)-3-(anthracen-9-yl)-1-(pyridin-2-yl)prop-2-en-1-one (1.0 mmol, 0.309 g) and 1-(4-benzoylphenyl)-1*H*-pyrrole-2,5-dione (1.3 mmol, 0.361 g) according to the general procedure above; gray powder (75%), Mp. > 240 °C (dec). HPLC: 96.29%, RT 8.88 min). ^{1}H nmR (400 MHz, DMSO-d_6) δ 3.54 (dd, *J* = 4.98, 3.32 Hz, 1H), 3.86–3.91 (m, 1H), 4.96 (d, *J* = 3.32 Hz, 1H), 6.66 (d, *J* = 8.71 Hz, 2H), 7.19–7.31 (m, 6H), 7.32–7.38 (m, 1H), 7.38–7.42 (m, 1H), 7.52 (s, 2H), 7.58–7.64 (m, 2H), 7.64–7.73 (m, H), 8.08 (d, *J* = 16.59 Hz, 1H), 8.09–8.12 (m, H), 8.20–8.23 (m, 1H), 8.27 (d, *J* = 16.59 Hz, 1H). ^{13}C nmR (101 MHz, DMSO-d_6) 45.44, 48.50, 49.04, 52.29, 123.30, 123.42, 123.89, 124.96, 125.75, 126.89, 126.95, 127.44, 128.31, 128.64, 129.07, 129.33, 130.06, 130.32, 130.56, 133.36, 135.53, 136.97, 137.33, 137.79, 138.34, 139.23, 140.06, 141.67, 142.62, 143.08, 149.7, 153.68, 174.9, 175.54, 188.8 (chalcone C=O), 195.32 (benzophenone C=O) ppm. LRMS (APCI) 587.49 [M$^+$+H]. IR *ν* max (ATR): 1710.44, 1678.62, 1651.53, 1622.75, 1372.38, 1278.33, 1323.21, 1022.35, 999.00, 926.58 cm^{-1}. HRMS (ESI) calculated for $C_{39}H_{26}N_2NaO_4$ [M$^+$+Na]: 609.1785; found: 609.1786.

(*E*)-13-(4-Benzoylphenyl)-9-(3-(furan-2-yl)-3-oxoprop-1-en-1-yl)-9,10-dihydro-9,10-[3,4] epipyrroloanthracene-12,14-dione (**26o**) was prepared from (*E*)-3-(anthracen-9-yl)-1-(furan-2-yl)prop-2-en-1-one (1.0 mmol, 0.298 g) and 1-(4-benzoylphenyl)-1*H*-pyrrole-2,5-dione (1.3 mmol, 0.361 g) according to the general procedure above; white powder (60%), Mp. >230 °C (dec). IR *ν* max (ATR) 2961.18, 1709.80, 1667.15, 1661.00, 1652.47, 1620.31, 1604.15, 1577.76, 1461.92, 1373.31, 1276.89, 1156.72, 940.89, 724.18 cm^{-1}. ^{1}H nmR (400 MHz, DMSO-d_6) δ 3.60 (dd, *J* = 8.56, 3.18 Hz, 1H), 3.98 (d, *J* = 8.31 Hz, 1H), 4.99 (d, *J* = 3.42 Hz, 3H), 6.68 (d, *J* = 8.31 Hz, 2H), 7.21–7.28 (m, 2H), 7.28–7.33 (m, 2 H), 7.33–7.38 (m, 1H), 7.39–7.44 (m,

1H), 7.52–7.60 (m, 2H), 7.60–7.64 (m, 1H), 7.65–7.74 (m, 6H), 7.75–7.80 (m, 1H), 7.95 (d, *J* = 16.63 Hz, 1H), 8.14 (d, *J* = 1.47 Hz, 1H). LRMS (APCI) 576.11 [M^++H].

(*E*)-13-(4-Benzoylphenyl)-9-(3-(naphthalen-2-yl)-3-oxoprop-1-en-1-yl)-9,10-dihydro-9,10-[3,4]epipyrroloanthracene-12,14-dione (**26p**) was synthesized from (*E*)-3-(anthracen-9-yl)-1-(naphthalen-2-yl)prop-2-en-1-one (1.0 mmol, 0.358 g) and 1-(4-benzoylphenyl)-1*H*-pyrrole-2,5-dione (1.3 mmol, 0.361 g) according to the general procedure above; gray powder (88%). ^{1}H nmR (400 MHz, DMSO-d_6) δ 3.58 (dd, *J* = 8.31, 3.42 Hz, 1H), 3.91 (d, *J* = 8.80 Hz, 1H), 5.00 (d, *J* = 3.42 Hz, 1H), 6.69 (d, *J* = 8.31 Hz, 2H), 7.19–7.34 (m, 5H), 7.38 (d, *J* = 4.89 Hz, 1H), 7.41–7.45 (m, H), 7.51–7.58 (m, 2 H), 7.60–7.67 (m, 1H), 7.67–7.77 (m, 5H), 8.11 (d, *J* = 16.14 Hz, H), 8.24 (d, *J* = 7.83 Hz, 1H), 8.32 (d, *J* = 16.63 Hz, 1H), 8.79 (d, *J* = 4.40 Hz, 1H). LRMS (APCI): 636.24 [M^++H].

(*E*)-13-(4-Benzoylphenyl)-9-(3-oxo-3-(thiophen-2-yl)prop-1-en-1-yl)-9,10-dihydro-9,10-[3,4]epipyrroloanthracene-12,14-dione (**26q**) was synthesized from (*E*)-3-(anthracen-9-yl)-1-(thiophen-2-yl)prop-2-en-1-one (1.0 mmol, 0.3144 g) and 1-(4-benzoylphenyl)-1*H*-pyrrole-2,5-dione (1.3 mmol, 0.361 g) according to the general procedure above; white solid (51%). Mp. 255–257 °C. ^{1}H nmR (400 MHz, DMSO-d_6) δ 3.59 (dd, *J* = 5.39, 3.73 Hz, 1H), 4.02 (d, *J* = 8.71 Hz, 1H), 4.97 (d, *J* = 3.32 Hz, 1H), 6.66 (d, *J* = 8.29 Hz, 2H), 7.19–7.32 (m, 6H), 7.34 (s, 1H), 7.38–7.42 (m, 1H), 7.54 (d, *J* = 7.46 Hz, 2H), 7.60 (s, 1H), 7.65–7.71 (m, 5H), 7.85 (d, *J* = 15.76 Hz, 1H), 7.89 (d, *J* = 16.59 Hz, 1H), 8.12 (dd, *J* = 4.98, 0.83 Hz, 1H), 8.21 (dd, *J* = 3.73, 0.83 Hz, 1H). ^{13}C nmR (101 MHz, DMSO-d_6) 45.4, 48.4, 48.8, 52.1, 123.4, 124.0, 124.9, 125.7, 127.0, 127.4, 127.5, 127.6, 128.6, 129.1, 129.3, 129.5, 130.1, 130.6, 131.7, 133.3, 134.6, 135.5, 136.5, 137.0, 137.4, 139.2, 140.0, 141.5, 141.6, 142.7, 145.2, 175.1, 175.5, 182.0, 195.3 ppm. IR ν max (ATR): 1734.00, 1699.51, 1665.19, 1646.71, 1618.07, 1520.87, 1493.38, 1460.27 cm^{-1}. LRMS (APCI) [M^+−H]: 590.54. HRMS (ESI) calculated for $C_{38}H_{24}NO_4S$ [M^+−H] 590.1431; Found 590.1430.

Dimethyl (*E*)-9-(3-oxo-3-phenylprop-1-en-1-yl)-9,10-dihydro-9,10-ethenoanthracene-11,12-dicarboxylate (**27**) was synthesized from (*E*)-3-(anthracen-9-yl)-1-phenylprop-2-en-1-one (1 mmol, 0.3084 g) and dimethyl acetylenedicarboxylate (1.3 mmol, 0.16 mL) according to the general procedure above; orange/yellow crystalline solid (62%), [65], Mp. >250 °C. IR ν max (ATR): 1726.80, 1714.47, 1679.84, 1644.84, 1601.46, 1448.65, 1325.56, 1294.56, 1276.25, 947.76 cm^{-1}. ^{1}H nmR (400 MHz, DMSO-d_6) δ 3.64 (s, 3H), 3.72 (s, 3H), 5.80 (s, 1H), 7.07–7.15 (m, 4H), 7.45–7.50 (m, 2H), 7.52–7.58 (m, 3H), 7.61 (m, 2H), 7.73 (d, *J* = 16.12 Hz, 2H), 8.08 (d, *J* = 7.77 Hz, 2H). ^{13}C nmR (101 MHz, DMSO-d_6) 49.98, 52.95, 53.23, 57.83, 122.94, 124.84, 125.77, 126.31, 129.13, 129.56, 133.42, 134.17, 137.31, 137.80, 142.62, 143.97, 145.09, 154.91, 163.57, 167.01, 189.21 ppm. HRMS (ESI) calculated for $C_{29}H_{22}NaO_5$ [M^++Na]: 473.1360; found: 473.1351.

3.8. Stability Study of Compounds **21a**, **21i**, **22h**, **23a**, **23g**, **23n**, **24a**, *24h*, *26a* and *26n*

The HPLC stability studies for compounds **21a**, **21i**, **22h**, **23a**, **23g**, **23n**, **24a**, **24h**, **26a** and **26n** were performed using a Symmetry® column (C18, 5 μM, 4.6 × 150 mm), a Waters 2487 Dual Wavelength Absorbance detector, and an HPLC pump: Waters 1525 binary and Autosampler; Waters 717 plus (Waters Corporation, Milford, MA, USA), mobile phase [(acetonitrile (80%) and water (20%)] with a typical run time of 10 min and a flow rate of 1 mL/min with detection at 254 nm. The selected compounds **21a**, **21i**, **22h**, **23a**, **23g**, **23n**, **24a**, **24h**, **26a** and **26n** (5 mg) in the mobile phase (10 mL) provided the stock solution for the analysis. The phosphate buffers were prepared at pH values of 4, 7.4 and 9, as described in the *British Pharmacopoeia* monograph 2019. The stock solution (30 μL) was added to the appropriate buffer (1 mL) and retained at 37 °C. Samples were then analyzed over the following 24-h period at selected time intervals (*t* = 0 min, 5 min, 30 min, 60 min, 90 min, 120 min and 24 h).

3.9. X-ray Crystallography Analysis

Samples were mounted on a MiTeGen micromount with NVH immersion oil. Data for **21k**, **23f**, **23h**, **24a** and **24g** were collected from a shock-cooled single crystal at 100(2)

K on a Bruker D8 Quest ECO three-circle diffractometer with a standard sealed X-ray tube using a graphite monochromator and a CMOS area detector. The diffractometer was equipped with an Oxford Cryostream 800 low-temperature device, and it used MoK_α radiation (λ = 0.71073 Å). Data for **25f** and **27** were collected from a shock-cooled single crystal at 100(2) K on a Bruker APEX2 Kappa Duo Kappa diffractometer with a microfocus sealed X-ray tube using mirror optics as a monochromator and an APEX2 detector. The diffractometer was equipped with an Oxford Cobra low-temperature device, and it used Cu K_α radiation (λ = 1.54178 Å). All data were integrated with SAINT, and a multi-scan absorption correction using SADABS was applied [95,96]. Structures were solved with dual methods using XT and refined with full-matrix least-squares methods against F^2 by XL using Olex2 [97–99]. All non-hydrogen atoms were refined with anisotropic displacement parameters. All hydrogen atoms were refined in an isotropic manner on calculated positions using a riding model with their U_{iso} values constrained to 1.5 times the U_{eq} of their pivot atoms for terminal sp^3 carbon atoms and 1.2 times for all other carbon atoms. Disordered moieties were refined using bond lengths' restraints and displacement parameter restraints.

In **23f**, the N-H hydrogen was located on the difference map and refined with restraints (DFIX). The molecule has a chiral center at C3 and C4; however, it is a racemate, and the R configuration is shown (centrosymmetric space group). In **23h**, the pendant fluorophenyl group was modeled as disordered over two positions (C24–F1) with 50% occupancy. Restraints (SADI, DFIX and RIGU) and constraints (EADP) were used to model the disorder.

In **24a**, the majority of the molecule was disordered over two positions with approx. 61:39% occupancy. Restraints (SIMU and ISOR) and some constraints (EADP) were used in the model for L.S. convergence. **25f** was a poorly diffracting small crystal with weak high-angle data and incomplete Friedel pair collection. The model has chirality at C25 and C31, S (Polar Space Group).

The crystallographic data for the structures reported here were deposited with the Cambridge Crystallographic Data Centre [100] (CCDC) for **21k** (Deposition Number 2341799), **23f** (Deposition Number 2341800), **23h** (Deposition Number 2341801), **24a** (Deposition Number 2341802), **24g** (Deposition Number 2341803), **25f** (Deposition Number 2341804) and **27** (Deposition Number 2341805), and they contain the supplementary crystallographic data for this paper. These data can be obtained free of charge from the Cambridge Crystallographic Data Centre via www.ccdc.cam.ac.uk/structures.

3.10. *In Silico Computational Study*

Two-dimensional structures of the 21 novel anthracene compounds and maprotiline were exported from MOE post-standardization in the SMILES format [101]. The exported structures were then individually submitted as queries to the SwissTargetPrediction (STP) web service [86]. For each of the 22 submitted structures, STP automatically generated its structural fingerprints and estimated its 3D shape. Based on these characteristics, STP then searched its internal database of known active compounds (based mainly on ChEMBL data) [102] for ones that are similar in structure and/or shape [103]. All compounds discovered through the similarity search were then tracked via STP to their specific biological targets. All targets hit by these similar compounds were then compiled into a prioritized list. This potential target list was generated and downloaded in a tabular format for each of the 21 + 1 query structures. The resulting target lists were aggregated and further processed in a standard Python/Pandas data analysis environment [104].

3.11. *Molecular Modeling: Computational Overlay Study*

All the compounds were opened in a database viewer. The compounds were washed with default values, and explicit hydrogens were added. For each compound, MMFF94x partial charges were calculated, and each was minimized to a gradient of 0.001 kcal/mol/Å. The compounds were then overlaid individually on a 3D structure of maprotiline using flexible alignment on MOE with default values [105].

3.12. Biochemistry

3.12.1. Materials

The EBV-transformed CLL PGA-1 (M-IGVH, good prognosis) and HG-3 (UM-IGVH, poor prognosis) cell lines were provided by Professor Anders Rosén (Linköping University, Linköping, Sweden) [106]. alamarBlue was obtained from BioSource, Nivelles, Belgium, and fetal bovine serum (FBS) was sourced from Invitrogen, U.K. RPMI 1640 medium, HEPES, gentamycin (G418) and glutamine were sourced from Thermo Fisher Scientific, Dun Laoghaire, Ireland. Cell culture consumables were obtained from Greiner Bio-One Ltd., Stonehouse, U.K., while all other reagents used were obtained from Sigma-Aldrich (now Merck), Arklow, Ireland.

3.12.2. Cell Culture

The CLL PGA-1 and HG-3 leukemia cell lines were grown in RPMI-1640 (Glutamax; Thermo Fisher Scientific, Inc., Dun Laoghaire, Dublin, Ireland) medium that was supplemented with 10% (*v*/*v*) FBS, 50 µg/mL of streptomycin and 50 U/mL of penicillin. The cells were seeded at a density of 2×10^5 cells/mL. The cell cultures were maintained at 37 °C in a humidified atmosphere of 5% CO_2/95% O_2, and the cells were passaged at least twice weekly, according to their confluency.

3.12.3. AlamarBlue Cell Viability Assay

CLL cells (HG-3 and PGA-1) were seeded at a density of 2×10^5 cells/well (200 µL per well) in 96-well plates. The cells were treated with the appropriate concentrations of test compounds or control drugs (maprotiline and fludarabine), and the samples were incubated for 24 h. alamarBlue (20 µL) was then added to each well, and the samples were further incubated at 37 oC in the dark for 4 h. Wells containing only the reagent and cell culture medium (without cells) were used as blank controls, and the vehicle used was DMSO (1% *v/v*) in all experiments. The 96-well plates were analyzed on a fluorescence plate reader (SpectraMax Gemini, Molecular Devices) with emission and excitement wavelengths of 590 nm and 544 nm, respectively, and fluorescence was measured. The decrease in cell viability was obtained with reference to the vehicle (100% viability). An RPMI medium containing alamarBlue was used as a blank. The viabilities of the compounds were calculated with reference to the vehicle-treated cells (100% viable). A non-linear, sigmoidal dose–response curve was obtained from transformed data (Final Concentration = Log (Final Concentration); IC_{50} values (concentration of drug required for 50% reduction in cell survival) were obtained using Prism (Prism 10 for mac OS 10.0.2 GraphPad Software, Inc., La Jolla, CA, USA). The biochemical assays were performed in triplicate on at least three independent occasions, and the mean values were determined.

3.12.4. Annexin V/PI Apoptosis Assay

CLL cells PGA-1 and HG-3 (1×10^6 cells/mL) were treated at 37 °C with either the vehicle (1% (*v*/*v*) DMSO) or the ethanoanthracene compounds **20a**, **20f**, **23a** and **25n** (10 µM, 5 µM and 1 µM) for 48 h, following the protocol previously described [40]. The cells were then harvested via centrifugation at 400× *g* in a temperature-controlled Sorvall centrifuge. The cells were rinsed with Ca^{2+} Annexin-V-binding buffer (0.1 M HEPES, pH 7.4; 0.14 M NaCl; 25 mM $CaCl_2$, 0.5 mL). The samples were resuspended in FITC Annexin V (diluted 1:33 in Ca^{2+} Annexin V-binding buffer, 50 µL) and incubated for 10 min (on ice and protected from light). The samples were washed with Annexin-V-binding buffer and resuspended in a PI solution (0.5 µg/mL, 500 µL) and analyzed within 1 h using a BD Accuri C6 flow cytometer counting 10,000 cells using the FlowJo software Version 10 (FlowJo LLC, Ashland, OR, USA).

3.12.5. Inhibitor Studies: Reactive Oxygen Species (ROS)

The ROS cell viability assay included the pre-treatment of CLL cells PGA-1 and HG-3 with *N*-acetyl cysteine. *N*-acetyl cysteine (NAC) (Sigma, Berrinba, Queensland, Australia) was dissolved in sterile water and subsequently diluted to obtain a 5-mM stock solution concentration. Fresh solutions were prepared for each experiment. CLL cells PGA-1 and HG-3 were seeded at a density of 2×10^5 cells/mL in a 96-well plate. The cells were protected from light and pre-treated with NAC (2 μL) for 1 h before then being treated with compounds **20a, 20f, 23a** and **25n** at concentrations 1 μM or 10 μM for 24 h. The remainder of the assay was carried out as described previously for the alamarBlue cell viability assay.

3.12.6. Caspase Inhibitor Study

Compounds **20a** and **23a** (representing Series 1 and Series 4, respectively) were selected for evaluation at two treatment concentrations, 10 μM and 1 μM, across both HG-3 and PGA-1 CLL cells. The HG-3 and PGA-1 CLL cells were seeded at a density of 2×10^5 cells/mL and were subsequently pre-treated at 37 oC with 40 μM of pan-caspase inhibitor Z-VAD-FMK (G-Biosciences, Geno Technology Inc., St. Louis, MO, USA) per well for 4 h prior to compound treatment for 24 h. The caspase assay protocol was adapted from Bright et al. [85], and the remainder of the assay was carried out as described above for the alamarBlue cell viability assay and completed in triplicate on two independent days. The output was statistically analyzed using a one-way ANOVA with a Bonferroni multiple-comparison test. Statistical significance was defined as any comparison with a generated p-value of <0.05.

3.12.7. Cytotoxicity Assay

The cytotoxicity of the selected compounds **20a, 20f, 23a** and **25n** was determined using the CytoTox 96 non-radioactive cytotoxicity assay (Promega Corporation; 2800 Woods Hollow Road, Madison, WI, USA). CLL cells HG-3 and PGA-1 were incubated for 24 h and then treated with the selected compounds **20a, 20f, 23a** and **25n** at 10-μM and 1-μM concentrations, following the protocol outlined above in the cell viability assay. After 24 h, the lysis solution (10×) (20 μL) was added, and the cells were incubated for a further 1 h to ensure complete cell death. Supernatant (50 μL) was removed from each well and transferred to a clean 96-well plate. "Substrate mix" (50 μL) was added, and the plate was stored in the dark at 20 °C for 30 min. "Stop solution" (50 μL) was added to each well, and the absorbance at 490 nm was recorded using a Dynatech MR5000 plate reader; the % cell death was determined relative to the control cell lysis solution.

3.12.8. Evaluation of Selected Compounds **20a**, **23a** and **25n** in Donor Peripheral Blood Mononuclear Cells (PBMCs)

The human peripheral blood mononuclear cells (PBMCs) were generated as follows: peripheral blood was obtained from healthy donors (n = 2) after informed consent was received. The blood was then placed into a 50-mL falcon tube and diluted with an equal volume of phosphate-buffered saline (PBS). PBMCs were isolated using density-gradient centrifugation with LymphoPrep, as described previously [107]. The cells were resuspended in RPMI media containing 10% FBS and 1% penicillin/streptomycin (P/S). The cells were counted using a haemocytometer and seeded into 96 well plates at a density of 1×10^5 cells/well (200 μL), and they were incubated at 37 °C for 1 h prior to compound treatment. The Annexin V/PI assay was used as a preliminary method of assessing the selective toxicity of the above compounds when compared with a previous alamarBlue cell viability assay in CLL cell lines. The assay was carried out as described above for the Annexin V/PI assay previously described above, and it was completed in triplicate on two independent days. Approval for this study was obtained from the School of Pharmacy and Pharmaceutical Sciences Trinity College Dublin Research Ethics Committee (2020-06-01-MS).

4. Conclusions

Three new classes of targeted agents have been approved for the treatment of CLL in the last decade: BTK inhibitors (ibrutinib, acalabrutinib and zanubrutinib), the BCL2 inhibitor (venetoclax) and PI3K inhibitors (idelalisib and duvelisib). However, the development of resistance or intolerance to these multiple classes of drugs is now observed, and new therapeutic options for this growing population of patients is identified as an unmet medical need [7,30,34,36,108–111].

In our work, we report the synthesis of compounds that demonstrate a significant antiproliferative action in CLL cell lines and also promote apoptosis. However, we have not yet identified the molecular target(s) for these compounds. The further investigation of the mechanism of action of the compounds reported in our research (particularly against the targets identified in our in silico study) may offer an alternative approach to the development of effective drugs for the treatment of CLL patients who have development resistance to the targeted therapy drugs.

In the present work, we have identified a series of potent lead compounds that have nitrostyrene-ethanoanthracene and chalcone-ethanoanthracene scaffold structures. The structures of the novel ethanoanthracene compounds were confirmed via X-ray crystallography. According to preliminary biochemical screening in CLL cell lines HG-3 and PGA-1, the compounds displayed highly effective antiproliferative activity in the CLL cell lines. The lead compounds were identified as **20a**, **20f**, **23a** and **25n**, with IC_{50} values in the ranges of 0.17–2.69 μM (HG-3) and 0.35–1.97 μM (PGA-1, with a more potent effect than the fludarabine control drug). **25n** was the most potent compound in both CLL cell lines (HG-3, IC_{50} = 1.31 μM; PGA-1, IC_{50} = 0.87 μM). The lead compounds were subsequently found to induce cell death via a pro-apoptotic mechanism. The compounds also demonstrated potent antiproliferative activity in a panel of leukemia cell lines, with the mean GI_{50} values for compound **25n** across the leukemia cell panel determined as 0.29 μM. Evidence for ROS involvement in their antiproliferative activity suggested a structure-dependent factor linked to the potential degree of ROS involvement. Caspase dependence was indicated for the representative anthracene–chalcone lead compound **23a**. The compounds were demonstrated to be relatively non-toxic in LDH assays and also in PBMCs. While the biological macromolecular target(s) for the antiproliferative action of these compounds has yet to be identified, they may be effective through the disruption of signaling pathways or tumor-microenvironmental signaling. The antiproliferative nitroalkenes and chalcones identified in this study may act as electrophilic substrates for Michael-addition reactions of nucleophiles occurring in CLL cells and biological systems. The in silico target prediction remains inconclusive, as no singular target responsible for the activity has been identified. However, the target prediction indicated some potential targets for future investigations of molecular docking with selected family-A GPCRs, together with JNK, MAP and VEGFR kinases. This docking campaign, followed by molecular dynamics confirmation of the simulated protein–ligand interactions, could indicate which of these targets, if any, should be selected for in vitro experiments.

Overall, the identified lead compounds **20a**, **23a**, **25n** and **20f** represent promising potential scaffolds for CLL drug discovery, and they warrant further preclinical study as lead compounds for the development of more selective and potent antileukemic agents.

Supplementary Materials: The following supplementary information is available online at https://www.mdpi.com/article/10.3390/ph17081034/s1: Experimental details for the preparation of nitrovinylethanoanthracenes **20a–h** and (E)-3-(anthracen-9-yl)-1-phenylprop-2-en-1-ones **21a-q**; Figure S1: Stability analysis for **23n** at pH 4.0, pH 7.5 and pH 9.0; Figure S2: Bioavailability analysis for compound 23a; Figures S3-S8; X-ray analysis for **21k, 23f, 24a, 23h** and **27**; Figures S9–49: ^{1}H nmR, ^{13}C nmR, DEPT-90, HSQC and HMBC for ethanoanthracenes; Table S1: X-ray data for compounds **21k, 23f, 23h**, 24a, 24g, 25f and 27; Table S2: Stability data for ethanoanthracene 23n at pH 4.0, pH 7.5 and pH 9.0; Tables S3–S6; COMPARE analysis for compounds **23n, 23h, 25n** and **24l**; Tables S7–S9: Tier-1 profiling screen, physicochemical descriptors, Lipinski properties pharmacokinetic, ADMET and drug-likeness predictions for selected ethanoanthracenes; and Table S10: Overlay of ethanoanthracene compounds series with maprotiline with their overlay scores.

Author Contributions: Conceptualization, M.J.M., D.C.W., S.A.B. and A.M.M. formal analysis, A.J.B., J.P.M., B.T., D.F., S.K., I.Č., D.C.W., S.A.B. and M.J.M.; funding acquisition, M.J.M.; investigation, A.J.B., J.P.M., S.A.B., C.E.C., S.K., I.Č., D.F., B.T. and A.J.B.; methodology, A.J.B., S.A.B., J.P.M., D.F.,

I.Č. and S.K.; supervision, D.C.W., S.A.B., A.M.M., N.M.O., M.J.M. and D.F.; writing—original draft, M.J.M. and J.P.M.; writing—review and editing, M.J.M., J.P.M., A.J.B., N.M.O., S.A.B., B.T. and D.F. All authors have read and agreed to the published version of the manuscript.

Funding: Part of the research conducted in this publication was funded by the Irish Research Council under grant numbers GOIPG/2021/954 and GOIPD/2023/1294 and Trinity College Dublin Postgraduate research scholarships (AJB, JPMcK).

Institutional Review Board Statement: Approval for this study was obtained from the School of Pharmacy and Pharmaceutical Sciences Trinity College Dublin Research Ethics Committee (2020-06-01-MS).

Informed Consent Statement: Informed consent was obtained from all subjects involved in the study.

Data Availability Statement: The original contributions presented in the study are included in the article/Supplementary Materials, further inquiries can be directed to the corresponding author.

Acknowledgments: Postgraduate research scholarships from Trinity College Dublin (AJB, JPMcK) and Irish Research Council Postgraduate Fellowship (GOIPG/2021/954 and GOIPD/2023/1294; SK and IČ) are gratefully acknowledged. We thank John O'Brien and Manuel Ruether for nmR spectroscopy, Gary Hessman for high-resolution mass spectrometry, Brian Talbot for HPLC and Barry Moran for flow cytometry. We also thank Thomas Fraisse, Christopher Teodorescu and Zaid Alobaidi for HPLC analysis, Adam Bergin and Keith Ferris for biochemical analysis and Peadar Grant for manuscript preparation. The Trinity Biomedical Sciences Institute (TBSI) is supported by a capital infrastructure investment from Cycle 5 of the Irish Higher Education Authority's Programme for Research in Third Level Institutions (PRTLI). This study was also co-funded under the European Regional Development Fund. D.F., I.C. and S.K. thank the software vendors for their continuing support of academic research efforts, in particular the contributions of the Chemical Computing Group (CCG) and OpenEye, Cadence Molecular Sciences. The support and provisions of Dell Ireland, the Trinity Centre for High Performance Computing (TCHPC) and the Irish Centre for High-End Computing (ICHEC) are also gratefully acknowledged.

Conflicts of Interest: The authors declare no conflicts of interest.

Abbreviations

ADA	Adenosine deaminase
ADME	Absorption, distribution, metabolism and excretion
BBB	Blood–brain barrier
Bcl-2	B-cell lymphoma-2 protein
BCR	B-cell receptor
BTK	Bruton's tyrosine kinase
BL	Burkitt's lymphoma
CB2	Cannabinoid receptor
CLL	Chronic lymphocytic leukemia
CYP450	Cytochrome P450
CYP24A1	25-Hydroxyvitamin D-24-hydroxylase
DHOD	Dihydroorotate dehydrogenase
DEPT	Distortionless enhancement via polarization transfer
DMSO	Dimethylsulfoxide
EBV	Epstein–Barr virus
FACS	Fluorescence-activated cell sorting
FBS	Fetal bovine serum
FDA	The United States Food and Drug Administration
FITC	Fluorescein isothiocyanate
GSK-3 beta	Glycogen synthase kinase-3 beta
HBD	Hydrogen bond donors
HBA	Hydrogen bond acceptors
HEPES	N-2-Hydroxyethylpiperazine-N′-2-ethanesulfonic acid
HMBC	Heteronuclear multiple-bond correlation

HSQC	Heteronuclear single quantum coherence
HRMS	High-resolution mass spectrometry
IC_{50}	Half-maximal inhibitory concentration
IGVH	Immunoglobulin heavy-chain gene
INT	Iodonitrotetrazolium chloride
IR	Infrared
LDH	Lactate dehydrogenase
MALT1	Mucosa-associated lymphoid tissue–lymphoma translocation protein 1
MEM	Eagle's minimum essential medium
NAC	N-acetylcysteine
NCI	National Cancer Institute
Nrf2	Nuclear factor erythroid 2–related factor 2
NHL	Non-Hodgkin lymphoma
NMR	Nuclear magnetic resonance
PAINS	Pan-assay interference compounds
PBMCs	Peripheral blood mononuclear cells
PI3Kδ	Phosphoinositide 3-kinase δ isoform
PI	Propidium iodide
PD-1	Programmed cell death protein 1
ROS	Reactive oxygen species
RPMI 1640	Roswell Park Memorial Institute 1640 medium
RT	Richter's transformation
Syk	Spleen tyrosine kinase
TPSA	Topological polar surface area

References

1. Cancer Stat Facts: Leukemia—Chronic Lymphocytic Leukemia (CLL). National Cancer Institute. 2022. Available online: https://seer.cancer.gov/statfacts/html/clyl.html (accessed on 26 April 2024).
2. Cronin, K.A.; Scott, S.; Firth, A.U.; Sung, H.; Henley, S.J.; Sherman, R.L.; Siegel, R.L.; Anderson, R.N.; Kohler, B.A.; Benard, V.B.; et al. Annual report to the nation on the status of cancer, part 1: National cancer statistics. *Cancer* **2022**, *128*, 4251–4284. [CrossRef] [PubMed]
3. Waldron, C.; O'Brien, D.; Brophy, S.; Perera, K.; Crotty, G.M.; Dunlea, E.; Walsh, A.; Connolly, M.; Clifford, R.; O'Leary, H.; et al. Epidemiology of chronic lymphocytic leukaemia in an Irish subpopulation with total case ascertainment: An additional tool for health economic planning. *Br. J. Haematol.* **2022**, *196*, e47–e49. [CrossRef] [PubMed]
4. Hallek, M. Chronic lymphocytic leukemia: 2020 update on diagnosis, risk stratification and treatment. *Am. J. Hematol.* **2019**, *94*, 1266–1287. [CrossRef] [PubMed]
5. Shadman, M. Diagnosis and treatment of chronic lymphocytic leukemia: A review. *JAMA* **2023**, *329*, 918–932. [CrossRef] [PubMed]
6. Hampel, P.J.; Parikh, S.A. Chronic lymphocytic leukemia treatment algorithm 2022. *Blood Cancer J.* **2022**, *12*, 161. [CrossRef] [PubMed]
7. Bennett, R.; Seymour, J.F. Update on the management of relapsed/refractory chronic lymphocytic leukemia. *Blood Cancer J.* **2024**, *14*, 33. [CrossRef]
8. Lalic, H.; Aurer, I.; Batinic, D.; Visnjic, D.; Smoljo, T.; Babic, A. Bendamustine: A review of pharmacology, clinical use and immunological effects (review). *Oncol. Rep.* **2022**, *47*, 1–16. [CrossRef] [PubMed]
9. Lamanna, N.; O'Brien, S. Novel agents in chronic lymphocytic leukemia. *Hematol. Am. Soc. Hematol. Educ. Program.* **2016**, *2016*, 137–145. [CrossRef] [PubMed]
10. McDonald, C.; Xanthopoulos, C.; Kostareli, E. The role of Bruton's tyrosine kinase in the immune system and disease. *Immunology* **2021**, *164*, 722–736. [CrossRef] [PubMed]
11. Hus, I.; Pula, B.; Robak, T. PI3K inhibitors for the treatment of chronic lymphocytic leukemia: Current status and future perspectives. *Cancers* **2022**, *14*, 1571. [CrossRef]
12. Wiestner, A. BCR pathway inhibition as therapy for chronic lymphocytic leukemia and lymphoplasmacytic lymphoma. *Hematol. Am. Soc. Hematol. Educ. Program.* **2014**, *2014*, 125–134. [CrossRef] [PubMed]
13. Zhou, H.; Hu, P.; Yan, X.; Zhang, Y.; Shi, W. Ibrutinib in chronic lymphocytic leukemia: Clinical applications, drug resistance, and prospects. *Onco Targets Ther.* **2020**, *13*, 4877–4892. [CrossRef] [PubMed]
14. Davids, M.S.; Brown, J.R. Ibrutinib: A first in class covalent inhibitor of Bruton's tyrosine kinase. *Future Oncol.* **2014**, *10*, 957–967. [CrossRef] [PubMed]
15. Abbas, H.A.; Wierda, W.G. Acalabrutinib: A selective Bruton tyrosine kinase inhibitor for the treatment of B-cell malignancies. *Front. Oncol.* **2021**, *11*, 668162. [CrossRef] [PubMed]

16. Tam, C.S.; Munoz, J.L.; Seymour, J.F.; Opat, S. Zanubrutinib: Past, present, and future. *Blood Cancer J.* **2023**, *13*, 141. [CrossRef] [PubMed]
17. Munakata, W.; Ando, K.; Yokoyama, M.; Fukuhara, N.; Yamamoto, K.; Fukuhara, S.; Ohmachi, K.; Mishima, Y.; Ichikawa, S.; Ogiya, D.; et al. Long-term safety profile of tirabrutinib: Final results of a Japanese phase I study in patients with relapsed or refractory B-cell malignancies. *Int. J. Hematol.* **2023**, *117*, 553–562. [CrossRef] [PubMed]
18. Walter, H.S.; Rule, S.A.; Dyer, M.J.; Karlin, L.; Jones, C.; Cazin, B.; Quittet, P.; Shah, N.; Hutchinson, C.V.; Honda, H.; et al. A phase 1 clinical trial of the selective BTK inhibitor ONO/GS-4059 in relapsed and refractory mature B-cell malignancies. *Blood* **2016**, *127*, 411–419. [CrossRef] [PubMed]
19. Woyach, J.A.; Ruppert, A.S.; Guinn, D.; Lehman, A.; Blachly, J.S.; Lozanski, A.; Heerema, N.A.; Zhao, W.; Coleman, J.; Jones, D.; et al. BTK(C481S)-mediated resistance to ibrutinib in chronic lymphocytic leukemia. *J. Clin. Oncol.* **2017**, *35*, 1437–1443. [CrossRef]
20. Mato, A.R.; Shah, N.N.; Jurczak, W.; Cheah, C.Y.; Pagel, J.M.; Woyach, J.A.; Fakhri, B.; Eyre, T.A.; Lamanna, N.; Patel, M.R.; et al. Pirtobrutinib in relapsed or refractory B-cell malignancies (BRUIN): A phase 1/2 study. *Lancet* **2021**, *397*, 892–901. [CrossRef]
21. Crawford, J.J.; Johnson, A.R.; Misner, D.L.; Belmont, L.D.; Castanedo, G.; Choy, R.; Coraggio, M.; Dong, L.; Eigenbrot, C.; Erickson, R.; et al. Discovery of GDC-0853: A potent, selective, and noncovalent Bruton's tyrosine kinase inhibitor in early clinical development. *J. Med. Chem.* **2018**, *61*, 2227–2245. [CrossRef]
22. Zhang, Q.; Wen, C.; Zhao, L.; Wang, Y. A comprehensive review of small-molecule inhibitors targeting Bruton tyrosine kinase: Synthetic approaches and clinical applications. *Molecules* **2023**, *28*, 8037. [CrossRef] [PubMed]
23. Byrd, J.C.; Smith, S.; Wagner-Johnston, N.; Sharman, J.; Chen, A.I.; Advani, R.; Augustson, B.; Marlton, P.; Renee Commerford, S.; Okrah, K.; et al. First-in-human phase 1 study of the BTK inhibitor GDC-0853 in relapsed or refractory B-cell NHL and CLL. *Oncotarget* **2018**, *9*, 13023–13035. [CrossRef] [PubMed]
24. Zhong, L.; Li, Y.; Xiong, L.; Wang, W.; Wu, M.; Yuan, T.; Yang, W.; Tian, C.; Miao, Z.; Wang, T.; et al. Small molecules in targeted cancer therapy: Advances, challenges, and future perspectives. *Signal Transduct. Target. Ther.* **2021**, *6*, 201. [CrossRef] [PubMed]
25. von Tresckow, J.; Heyl, N.; Robrecht, S.; Giza, A.; Aldaoud, A.; Schlag, R.; Klausmann, M.; Linde, H.; Stein, W.; Schwarzer, A.; et al. Treatment with idelalisib in patients with chronic lymphocytic leukemia—Real world data from the registry of the German CLL study group. *Ann. Hematol.* **2023**, *102*, 3083–3090. [CrossRef] [PubMed]
26. Sharman, J.P.; Coutre, S.E.; Furman, R.R.; Cheson, B.D.; Pagel, J.M.; Hillmen, P.; Barrientos, J.C.; Zelenetz, A.D.; Kipps, T.J.; Flinn, I.W.; et al. Final results of a randomized, phase III study of rituximab with or without idelalisib followed by open-label idelalisib in patients with relapsed chronic lymphocytic leukemia. *J. Clin. Oncol.* **2019**, *37*, 1391–1402. [CrossRef] [PubMed]
27. Iyer, P.; Wang, L. Emerging therapies in CLL in the era of precision medicine. *Cancers* **2023**, *15*, 1583. [CrossRef] [PubMed]
28. Souers, A.J.; Leverson, J.D.; Boghaert, E.R.; Ackler, S.L.; Catron, N.D.; Chen, J.; Dayton, B.D.; Ding, H.; Enschede, S.H.; Fairbrother, W.J.; et al. ABT-199, a potent and selective Bcl-2 inhibitor, achieves antitumor activity while sparing platelets. *Nat. Med.* **2013**, *19*, 202–208. [CrossRef] [PubMed]
29. Birkinshaw, R.W.; Gong, J.N.; Luo, C.S.; Lio, D.; White, C.A.; Anderson, M.A.; Blombery, P.; Lessene, G.; Majewski, I.J.; Thijssen, R.; et al. Structures of Bcl-2 in complex with venetoclax reveal the molecular basis of resistance mutations. *Nat. Commun.* **2019**, *10*, 2385. [CrossRef] [PubMed]
30. Furstenau, M.; Eichhorst, B. Novel agents in chronic lymphocytic leukemia: New combination therapies and strategies to overcome resistance. *Cancers* **2021**, *13*, 1336. [CrossRef]
31. Hagenbeek, A.; Gadeberg, O.; Johnson, P.; Pedersen, L.M.; Walewski, J.; Hellmann, A.; Link, B.K.; Robak, T.; Wojtukiewicz, M.; Pfreundschuh, M.; et al. First clinical use of ofatumumab, a novel fully human anti-CD20 monoclonal antibody in relapsed or refractory follicular lymphoma: Results of a phase 1/2 trial. *Blood* **2008**, *111*, 5486–5495. [CrossRef]
32. O'Reilly, A.; Murphy, J.; Rawe, S.; Garvey, M. Chronic lymphocytic leukemia: A review of front-line treatment options, with a focus on elderly CLL patients. *Clin. Lymphoma Myeloma Leuk.* **2018**, *18*, 249–256. [CrossRef] [PubMed]
33. Timofeeva, N.; Ayres, M.L.; Baran, N.; Santiago-O'Farrill, J.M.; Bildik, G.; Lu, Z.; Konopleva, M.; Gandhi, V. Preclinical investigations of the efficacy of the glutaminase inhibitor CB-839 alone and in combinations in chronic lymphocytic leukemia. *Front. Oncol.* **2023**, *13*, 1161254. [CrossRef] [PubMed]
34. Robbins, D.W.; Noviski, M.A.; Tan, Y.S.; Konst, Z.A.; Kelly, A.; Auger, P.; Brathaban, N.; Cass, R.; Chan, M.L.; Cherala, G.; et al. Discovery and preclinical pharmacology of NX-2127, an orally bioavailable degrader of Bruton's tyrosine kinase with immunomodulatory activity for the treatment of patients with B cell malignancies. *J. Med. Chem.* **2024**, *67*, 2321–2336. [CrossRef] [PubMed]
35. Study of SGR-1505 in Mature B-Cell Neoplasms. Schrödinger, Inc. 2023. Available online: https://classic.Clinicaltrials.Gov/ct2/show/nct05544019 (accessed on 27 March 2024).
36. Herman, S.E.; Wiestner, A. Preclinical modeling of novel therapeutics in chronic lymphocytic leukemia: The tools of the trade. *Semin. Oncol.* **2016**, *43*, 222–232. [CrossRef] [PubMed]
37. Lamanna, N. New oral small molecules in the treatment of chronic lymphocytic leukemia. *Cancer* **2015**, *121*, 1917–1926. [CrossRef] [PubMed]
38. Byrne, A.J.; Bright, S.A.; Fayne, D.; McKeown, J.P.; McCabe, T.; Twamley, B.; Williams, C.; Meegan, M.J. Synthesis, antiproliferative and pro-apoptotic effects of nitrostyrenes and related compounds in Burkitt's lymphoma. *Med. Chem.* **2018**, *14*, 181–199. [CrossRef] [PubMed]

39. Bright, S.A.; Byrne, A.J.; Vandenberghe, E.; Browne, P.V.; McElligott, A.M.; Meegan, M.J.; Williams, D.C. Selected nitrostyrene compounds demonstrate potent activity in chronic lymphocytic leukaemia cells, including those with poor prognostic markers. *Oncol. Rep.* **2019**, *41*, 3127–3136. [CrossRef] [PubMed]
40. Byrne, A.J.; Bright, S.A.; McKeown, J.P.; Bergin, A.; Twamley, B.; McElligott, A.M.; Noorani, S.; Kandwal, S.; Fayne, D.; O'Boyle, N.M.; et al. Synthesis and pro-apoptotic effects of nitrovinylanthracenes and related compounds in chronic lymphocytic leukaemia (CLL) and Burkitt's lymphoma (BL). *Molecules* **2023**, *28*, 8095. [CrossRef] [PubMed]
41. Hung, A.C.; Tsai, C.H.; Hou, M.F.; Chang, W.L.; Wang, C.H.; Lee, Y.C.; Ko, A.; Hu, S.C.; Chang, F.R.; Hsieh, P.W.; et al. The synthetic beta-nitrostyrene derivative CYT-Rx20 induces breast cancer cell death and autophagy via ROS-mediated MEK/ERK pathway. *Cancer Lett.* **2016**, *371*, 251–261. [CrossRef]
42. Tsai, C.H.; Hung, A.C.; Chen, Y.Y.; Chiu, Y.W.; Hsieh, P.W.; Lee, Y.C.; Su, Y.H.; Chang, P.C.; Hu, S.C.; Yuan, S.F. 3′-hydroxy-4′-methoxy-beta-methyl-beta-nitrostyrene inhibits tumorigenesis in colorectal cancer cells through ROS-mediated DNA damage and mitochondrial dysfunction. *Oncotarget* **2017**, *8*, 18106–18117. [CrossRef]
43. Byrne, A.J.; Bright, S.A.; McKeown, J.P.; O'Brien, J.E.; Twamley, B.; Fayne, D.; Williams, D.C.; Meegan, M.J. Design, synthesis and biochemical evaluation of novel ethanoanthracenes and related compounds to target Burkitt's lymphoma. *Pharmaceuticals* **2020**, *13*, 16. [CrossRef]
44. Cloonan, S.M.; Williams, D.C. The antidepressants maprotiline and fluoxetine induce type II autophagic cell death in drug-resistant Burkitt's lymphoma. *Int. J. Cancer* **2011**, *128*, 1712–1723. [CrossRef]
45. Bisi, A.; Mokhtar Mahmoud, A.; Allara, M.; Naldi, M.; Belluti, F.; Gobbi, S.; Ligresti, A.; Rampa, A. Polycyclic maleimide-based scaffold as new privileged structure for navigating the cannabinoid system opportunities. *ACS Med. Chem. Lett.* **2019**, *10*, 596–600. [CrossRef] [PubMed]
46. Bonvicini, F.; Manet, I.; Belluti, F.; Gobbi, S.; Rampa, A.; Gentilomi, G.A.; Bisi, A. Targeting the bacterial membrane with a new polycyclic privileged structure: A powerful tool to face Staphylococcus aureus infections. *ACS Infect. Dis.* **2019**, *5*, 1524–1535. [CrossRef]
47. Arya, S.K.; Kumar, S.; Rani, R.; Kumar, N.; Roy, P.; Sondhi, S.M. Synthesis, anti-inflammatory, and cytotoxicity evaluation of 9,10-dihydroanthracene-9,10-α,β-succinimide and bis-succinimide derivatives. *Med. Chem. Res.* **2013**, *22*, 4278–4285. [CrossRef]
48. Camara, R.; Ogbeni, D.; Gerstmann, L.; Ostovar, M.; Hurer, E.; Scott, M.; Mahmoud, N.G.; Radon, T.; Crnogorac-Jurcevic, T.; Patel, P.; et al. Discovery of novel small molecule inhibitors of S100P with in vitro anti-metastatic effects on pancreatic cancer cells. *Eur. J. Med. Chem.* **2020**, *203*, 112621. [CrossRef] [PubMed]
49. Yang, B.V.; Vaccaro, W.; Doweyko, A.M.; Doweyko, L.M.; Huynh, T.; Tortolani, D.; Nadler, S.G.; McKay, L.; Somerville, J.; Holloway, D.A.; et al. Discovery of novel dihydro-9,10-ethano-anthracene carboxamides as glucocorticoid receptor modulators. *Bioorganic Med. Chem. Lett.* **2009**, *19*, 2139–2143. [CrossRef] [PubMed]
50. Bisi, A.; Arribas, R.L.; Micucci, M.; Budriesi, R.; Feoli, A.; Castellano, S.; Belluti, F.; Gobbi, S.; de Los Rios, C.; Rampa, A. Polycyclic maleimide-based derivatives as first dual modulators of neuronal calcium channels and GSK-3beta for Alzheimer's disease treatment. *Eur. J. Med. Chem.* **2019**, *163*, 394–402. [CrossRef]
51. Sadeq, H.; Thiemann, T.; Graham, J.P.; al Jasem, Y.; Bugenhagen, B.; Al-Rawashdeh, N.; al Sulaibi, M. Preparation of 3-(9-anthryl)acrylates and 9-aroylethenylanthracenes as Pi-extended anthracenes and their Diels–Alder type adducts with electron-poor dienophiles. *Polycycl. Aromat. Compd.* **2017**, *37*, 148–160. [CrossRef]
52. Zhang, Q.; Lv, J.; Luo, S. Enantioselective Diels-Alder reaction of anthracene by chiral tritylium catalysis. *Beilstein J. Org. Chem.* **2019**, *15*, 1304–1312. [CrossRef]
53. Sanyal, A.S.; Snyder, J.K. Stereoselective Diels—Alder reactions of chiral anthracenes. *Org. Lett.* **2000**, *2*, 2527–2530. [CrossRef] [PubMed]
54. Bandowe, B.A.M.; Meusel, H. Nitrated polycyclic aromatic hydrocarbons (nitro-pahs) in the environment—A review. *Sci. Total Environ.* **2017**, *581*, 237–257. [CrossRef] [PubMed]
55. Nepali, K.; Lee, H.Y.; Liou, J.P. Nitro-group-containing drugs. *J. Med. Chem.* **2019**, *62*, 2851–2893. [CrossRef] [PubMed]
56. Zhuang, C.; Zhang, W.; Sheng, C.; Zhang, W.; Xing, C.; Miao, Z. Chalcone: A privileged structure in medicinal chemistry. *Chem. Rev.* **2017**, *117*, 7762–7810. [CrossRef]
57. Kenari, F.; Molnar, S.; Perjesi, P. Reaction of chalcones with cellular thiols. The effect of the 4-substitution of chalcones and protonation state of the thiols on the addition process. Diastereoselective thiol addition. *Molecules* **2021**, *26*, 4332. [CrossRef] [PubMed]
58. Wu, R.P.; Hayashi, T.; Cottam, H.B.; Jin, G.; Yao, S.; Wu, C.C.; Rosenbach, M.D.; Corr, M.; Schwab, R.B.; Carson, D.A. Nrf2 responses and the therapeutic selectivity of electrophilic compounds in chronic lymphocytic leukemia. *Proc. Natl. Acad. Sci. USA* **2010**, *107*, 7479–7484. [CrossRef]
59. Wijayanti, L.W.; Swasono, R.T.; Lee, W.; Jumina, J. Synthesis and evaluation of chalcone derivatives as novel sunscreen agent. *Molecules* **2021**, *26*, 2698. [CrossRef] [PubMed]
60. Vázquez-Vuelvas, O.F.; Enriquez-Figueroa, R.A.; Garcia-Ortega, H.; Flores-Alamo, M.; Pineda-Contreras, A. Crystal structure of the chalcone (*E*)-3-(furan-2-yl)-1-phenylprop-2-en-1-one. *Acta Crystallogr. Sect. E Crystallogr. Commun.* **2015**, *71*, 161–164. [CrossRef] [PubMed]
61. Sun, H.; Khan, A.; Usman, R.; Wang, M. Understanding relationship between stacking modes and optical properties of organic charge transfer cocrystals involving anthracyl chalcones and TCNB. *J. Photochem. Photobiol. A Chem.* **2019**, *371*, 315–326. [CrossRef]

62. Feng, Q.; Wang, J.; Huan, W.; Shen, C.; Guo, F.; Lu, J.; Diao, G. Polymorphism and configurational isomerism in 3-(9-anthryl)-1-(3-hydroxyphenyl)prop-2-en-1-one. *J. Mol. Struct.* **2019**, *1195*, 355–363.
63. Kemnitz, C.R.; Loewen, M.J. "Amide resonance" correlates with a breadth of C−N rotation barriers. *J. Am. Chem. Soc.* **2007**, *129*, 2521–2528. [CrossRef] [PubMed]
64. Smith, N.M.; Iyer, K.S.; Corry, B. The confined space inside carbon nanotubes can dictate the stereo- and regioselectivity of Diels-Alder reactions. *Phys. Chem. Chem. Phys.* **2014**, *16*, 6986–6989. [CrossRef] [PubMed]
65. Aswathia, C.S.; Unnikrishnana, P.A.; Saumya, T.S. Synthesis, characterization and DFT studies of a few enone appended dibenzobarrelenes. *Chem. Data Collect.* **2020**, *25*, 100340. [CrossRef]
66. Sato, Y.; Kojima, T.; Goto, T.; Oomikawa, R.; Watanabe, H.; Wakabayashi, K. Hydrolysis and phytotoxic activity of cyclic imides. *Agric. Biol. Chem.* **1991**, *55*, 2677–2681.
67. Shriner, R.; Kurosawa, T. Chalcones. II. Decomposition by alkali. *J. Am. Chem. Soc.* **1930**, *52*, 2538–2540. [CrossRef]
68. Rosen, A.; Bergh, A.C.; Gogok, P.; Evaldsson, C.; Myhrinder, A.L.; Hellqvist, E.; Rasul, A.; Bjorkholm, M.; Jansson, M.; Mansouri, L.; et al. Lymphoblastoid cell line with B1 cell characteristics established from a chronic lymphocytic leukemia clone by in vitro EBV infection. *Oncoimmunology* **2012**, *1*, 18–27. [CrossRef]
69. Lewin, N.; Aman, P.; Mellstedt, H.; Zech, L.; Klein, G. Direct outgrowth of in vivo Epstein-Barr virus (EBV)-infected chronic lymphocytic leukemia (CLL) cells into permanent lines. *Int. J. Cancer* **1988**, *41*, 892–895. [CrossRef]
70. Singh, P.; Anand, A.; Kumar, V. Recent developments in biological activities of chalcones: A mini review. *Eur. J. Med. Chem.* **2014**, *85*, 758–777. [CrossRef]
71. McNamara, Y.M.; Bright, S.A.; Byrne, A.J.; Cloonan, S.M.; McCabe, T.; Williams, D.C.; Meegan, M.J. Synthesis and antiproliferative action of a novel series of maprotiline analogues. *Eur. J. Med. Chem.* **2014**, *71*, 333–353. [CrossRef]
72. National Cancer Institute (NCI), NCI-60 screening methodology. Available online: https://dtp.cancer.gov/discovery_development/nci-60/methodology.htm (accessed on 26 April 2024).
73. Biological Testing Branch, National Cancer Institute, Bethesda, MD. 2018. Available online: https://dtp.Nci.Nih.Gov/branches/btb/hfa.Html (accessed on 26 April 2024).
74. Cleaveland, E.S.; Monks, A.; Vaigro-Wolff, A.; Zaharevitz, D.W.; Paull, K.; Ardalan, K.; Cooney, D.A.; Ford, H., Jr. Site of action of two novel pyrimidine biosynthesis inhibitors accurately predicted by the compare program. *Biochem. Pharmacol.* **1995**, *49*, 947–954. [CrossRef]
75. Compare Analysis. Available online: https://dtp.Cancer.Gov/databases_tools/compare.Htm (accessed on 26 April 2024).
76. Dahlin, J.L.; Inglese, J.; Walters, M.A. Mitigating risk in academic preclinical drug discovery. *Nat. Rev. Drug Discov.* **2015**, *14*, 279–294. [CrossRef] [PubMed]
77. Daina, A.; Michielin, O.; Zoete, V. SwissADME: A free web tool to evaluate pharmacokinetics, drug-likeness and medicinal chemistry friendliness of small molecules. *Sci. Rep.* **2017**, *7*, 42717. [CrossRef] [PubMed]
78. Decker, T.; Lohmann-Matthes, M.L. A quick and simple method for the quantitation of lactate dehydrogenase release in measurements of cellular cytotoxicity and tumor necrosis factor (TNF) activity. *J. Immunol. Methods* **1988**, *115*, 61–69. [CrossRef] [PubMed]
79. Nachlas, M.M.; Margulies, S.I.; Goldberg, J.D.; Seligman, A.M. The determination of lactic dehydrogenase with a tetrazolium salt. *Anal. Biochem.* **1960**, *1*, 317–326. [CrossRef] [PubMed]
80. Gorrini, C.; Harris, I.S.; Mak, T.W. Modulation of oxidative stress as an anticancer strategy. *Nat. Rev. Drug Discov.* **2013**, *12*, 931–947. [CrossRef] [PubMed]
81. Rahman, M.A.; Ahmed, K.R.; Haque, F.; Park, M.N.; Kim, B. Recent advances in cellular signaling interplay between redox metabolism and autophagy modulation in cancer: An overview of molecular mechanisms and therapeutic interventions. *Antioxidants* **2023**, *12*, 428. [CrossRef] [PubMed]
82. Nakamura, H.; Takada, K. Reactive oxygen species in cancer: Current findings and future directions. *Cancer Sci.* **2021**, *112*, 3945–3952. [CrossRef]
83. Hounsell, C.; Fan, Y. The duality of caspases in cancer, as told through the fly. *Int. J. Mol. Sci.* **2021**, *22*, 8927. [CrossRef] [PubMed]
84. Boice, A.; Bouchier-Hayes, L. Targeting apoptotic caspases in cancer. *Biochim. Biophys. Acta Mol. Cell Res.* **2020**, *1867*, 118688. [CrossRef]
85. Li, X.; Yao, X.; Zhu, Y.; Zhang, H.; Wang, H.; Ma, Q.; Yan, F.; Yang, Y.; Zhang, J.; Shi, H.; et al. The caspase inhibitor Z-VAD-FMK alleviates endotoxic shock via inducing macrophages necroptosis and promoting MDSCS-mediated inhibition of macrophages activation. *Front. Immunol.* **2019**, *10*, 1824.
86. Daina, A.; Michielin, O.; Zoete, V. SwisstTargetPrediction: Updated data and new features for efficient prediction of protein targets of small molecules. *Nucleic Acids Res.* **2019**, *47*, W357–W364. [CrossRef] [PubMed]
87. James, J.P.; Ishwar, B.K.; More, U.A.; Joshi, S.D. Design, synthesis, molecular modeling, and ADMET studies of some pyrazoline derivatives as shikimate kinase inhibitors. *Med. Chem. Res.* **2018**, *27*, 546–559. [CrossRef]
88. Becker, H.D.; Beckers, H.C.; Sandros, K.; Andersson, K. Photochemistry of the anthracene chromophore: Novel isomerization of 1-(9-anthryl)-2-benzoylethylenes. *Tetrahedron Lett.* **1985**, *26*, 1589–1592. [CrossRef]
89. Velagapudi, N.; Sharma, N.; Bandla, R. Synthesis, screening and QSAR analysis of chalcone derivatives as potential anti bacterial agents. *Int. J. Pharm. Chem. Biol. Sci.* **2017**, *7*, 211–218.

90. Gonzalez, J.J.; Ortega, E.; Rothemund, M.; Gold, M.; Vicente, C.; de Haro, C.; Bautista, D.; Schobert, R.; Ruiz, J. Luminescent gold(I) complexes of 1-pyridyl-3-anthracenylchalcone inducing apoptosis in colon carcinoma cells and antivascular effects. *Inorg. Chem.* **2019**, *58*, 12954–12963. [CrossRef] [PubMed]
91. Mishra, N.; Sasmal, D. Development of selective and reversible pyrazoline based MAO-B inhibitors: Virtual screening, synthesis and biological evaluation. *Bioorganic Med. Chem. Lett.* **2011**, *21*, 1969–1973. [CrossRef] [PubMed]
92. Kharbanda, C.; Alam, M.S.; Hamid, H.; Javed, K.; Dhulap, A.; Bano, S.; Ali, Y. Antidiabetic effect of novel benzenesulfonylureas as PPAR-γ agonists and their anticancer effect. *Bioorganic Med. Chem. Lett.* **2015**, *25*, 4601–4605. [CrossRef] [PubMed]
93. Levai, A.; Jeko, J. Synthesis of 1-substituted 3,5-diaryl-2-pyrazolines by the reaction of α,β-unsaturated ketones with hydrazines. *J. Heterocycl. Chem.* **2006**, *43*, 111–115. [CrossRef]
94. Buu-Hoi, N.P.; Hoan, N. Some anthracene derivatives of potential biological interest. *J. Org. Chem.* **1951**, *16*, 874–881. [CrossRef]
95. Bruker AXS Inc. *SAINT, v8.40A*; Bruker AXS Inc.: Madison, WI, USA, 2019.
96. Krause, L.; Herbst-Irmer, R.; Sheldrick, G.M.; Stalke, D. Comparison of silver and molybdenum microfocus x-ray sources for single-crystal structure determination. *J. Appl. Crystallogr.* **2015**, *48*, 3–10. [CrossRef]
97. Sheldrick, G.M. Crystal structure refinement with SHELXL. *Acta Crystallogr. C Struct. Chem.* **2015**, *71*, 3–8. [CrossRef] [PubMed]
98. Sheldrick, G.M. SHELXT—Integrated space-group and crystal-structure determination. *Acta Crystallogr. A Found. Adv.* **2015**, *71*, 3–8. [CrossRef] [PubMed]
99. Dolomanov, O.V.; Bourhis, L.J.; Gildea, R.J.; Howard, J.A.K.; Puschmann, H. OLEX2: A complete structure solution, refinement and analysis program. *J. Appl. Cryst.* **2009**, *42*, 339–341. [CrossRef]
100. Groom, C.R.; Bruno, I.J.; Lightfoot, M.P.; Ward, S.C. The Cambridge Structural Database. *Acta Crystallogr. B Struct. Sci. Cryst. Eng. Mater.* **2016**, *72*, 171–179. [CrossRef]
101. Chemical Computing Group Inc. *Molecular Operating Environment (MOE) Version 2022.02*, Chemical Computing Group Inc.: Montreal, QC, Canada, 2022.
102. Zdrazil, B.; Felix, E.; Hunter, F.; Manners, E.J.; Blackshaw, J.; Corbett, S.; de Veij, M.; Ioannidis, H.; Lopez, D.M.; Mosquera, J.F.; et al. The ChEMBL database in 2023: A drug discovery platform spanning multiple bioactivity data types and time periods. *Nucleic Acids Res.* **2024**, *52*, D1180–D1192. [CrossRef] [PubMed]
103. Gfeller, D.; Michielin, O.; Zoete, V. Shaping the interaction landscape of bioactive molecules. *Bioinformatics* **2013**, *29*, 3073–3079. [CrossRef] [PubMed]
104. The Pandas Development Team. Pandas-dev/pandas: Pandas. V2.2.2. 2024. Available online: https://doi.Org/10.5281/zenodo.3509134 (accessed on 23 April 2024).
105. Omega 4.2.2.0. OpenEye, Cadence Molecular Sciences, Santa Fe, nm, USA. Available online: http://www.eyesopen.com (accessed on 10 April 2024).
106. Lanemo Myhrinder, A.; Hellqvist, E.; Sidorova, E.; Soderberg, A.; Baxendale, H.; Dahle, C.; Willander, K.; Tobin, G.; Backman, E.; Soderberg, O.; et al. A new perspective: Molecular motifs on oxidized LDL, apoptotic cells, and bacteria are targets for chronic lymphocytic leukemia antibodies. *Blood* **2008**, *111*, 3838–3848. [CrossRef] [PubMed]
107. Brophy, S.; Amet, R.; Foy-Stones, H.; Gardiner, N.; McElligott, A.M. Isolation and cryopreservation of mononuclear cells from peripheral blood and bone marrow of blood cancer patients. *Methods Mol. Biol.* **2023**, *2645*, 179–187.
108. Timofeeva, N.; Gandhi, V. Ibrutinib combinations in CLL therapy: Scientific rationale and clinical results. *Blood Cancer J.* **2021**, *11*, 79. [CrossRef]
109. Zygmunciak, P.; Robak, T.; Pula, B. Treatment of double-refractory chronic lymphocytic leukemia-an unmet clinical need. *Int. J. Mol. Sci.* **2024**, *25*, 1589. [CrossRef]
110. Liang, E.C.; Albittar, A.; Huang, J.J.; Hirayama, A.V.; Kimble, E.L.; Portuguese, A.J.; Chapuis, A.; Shadman, M.; Till, B.G.; Cassaday, R.D.; et al. Factors associated with long-term outcomes of CD19 CAR T-cell therapy for relapsed/refractory CLL. *Blood Adv.* **2023**, *7*, 6990–7005. [CrossRef] [PubMed]
111. Borogovac, A.; Siddiqi, T. Transforming CLL management with immunotherapy: Investigating the potential of CAR T-cells and bispecific antibodies. *Semin. Hematol.* **2024**, *61*, 119–130. [CrossRef] [PubMed]

pharmaceuticals

Article

Solidagenone from *Solidago chilensis* Meyen Protects against Acute Peritonitis and Lipopolysaccharide-Induced Shock by Regulating NF-κB Signaling Pathway

Ivanilson Pimenta Santos [1], Laís Peres Silva [1], Dahara Keyse Carvalho Silva [1], Bruna Padilha Zurita Claro dos Reis [1], Temistocles Barroso de Oliveira [2], Andressa Maia Kelly [2], Edivaldo dos Santos Rodrigues [3], Claudia Valeria Campos de Souza [3], José Fernando Oliveira-Costa [4], Simone Sacramento Valverde [2], Osvaldo Andrade Santos-Filho [3], Milena Botelho Pereira Soares [1,5,*] and Cássio Santana Meira [1,5,*]

1 Gonçalo Moniz Institute, Oswaldo Cruz Foundation, FIOCRUZ, Salvador 40296-710, BA, Brazil
2 Pharmaceutical Technology Institute—FarManguinhos, Natural Products Department, Oswaldo Cruz Foundation, FIOCRUZ, Rio de Janeiro 21041-250, RJ, Brazil; andressa.kelly@far.fiocruz.br (A.M.K.)
3 Laboratory of Molecular Modeling and Computational Structural Biology, Natural Products Research Institute, Health Sciences Center, Federal University of Rio de Janeiro, Rio de Janeiro 21941-599, RJ, Brazil; claudiavaleria@ippn.ufrj.br (C.V.C.d.S.); osvaldo@ippn.ufrj.br (O.A.S.-F.)
4 Center for Infusions and Specialized Medicines of Bahia, Bahia State Health Department, Salvador 41745-900, BA, Brazil; josefernandocosta@hotmail.com
5 Institute of Innovation in Advanced Health Systems (ISI SAS), University Center SENAI/CIMATEC, Salvador 41650-010, BA, Brazil
* Correspondence: milena@bahia.fiocruz.br (M.B.P.S.); cassio.meira@fieb.org.br (C.S.M.)

Citation: Santos, I.P.; Silva, L.P.; Silva, D.K.C.; dos Reis, B.P.Z.C.; de Oliveira, T.B.; Kelly, A.M.; dos Santos Rodrigues, E.; de Souza, C.V.C.; Oliveira-Costa, J.F.; Valverde, S.S.; et al. Solidagenone from *Solidago chilensis* Meyen Protects against Acute Peritonitis and Lipopolysaccharide-Induced Shock by Regulating NF-κB Signaling Pathway. *Pharmaceuticals* **2024**, *17*, 273. https://doi.org/10.3390/ph17030273

Academic Editor: Roberta Rocca

Received: 28 December 2023
Revised: 9 February 2024
Accepted: 16 February 2024
Published: 21 February 2024

Abstract: Anti-inflammatory agents are widely used for the treatment of inflammatory diseases. Nevertheless, the associated side effects of the available drugs make it necessary to search for new anti-inflammatory drugs. Here, we investigated the anti-inflammatory activity of solidagenone. Initially, we observed that a single dose of 30, 60, or 90 mg/kg of solidagenone did not result in mortality or elicit any discernible signs of toxicity in mice. At the same doses, solidagenone promoted a significant reduction in the migration of neutrophils in an acute peritonitis model and decreased mortality in a lipopolysaccharide-induced endotoxic shock model. Interestingly, treatment with solidagenone conferred a protective effect against leukopenia and thrombocytopenia, hematological disorders commonly observed in sepsis conditions. In addition, treatment with all the doses of solidagenone promoted a significant reduction in nitric oxide, TNF-α, and IL-1β levels relative to the LPS-stimulated vehicle-treated cultures. Furthermore, gene expression and in silico analyses also supported the modulation of the NF-κB pathway by solidagenone. Finally, in silico pharmacokinetics predictions indicated a favorable drugability profile for solidagenone. Taken together, the findings of the present investigation show that solidagenone exhibits significant anti-inflammatory properties in acute experimental models, potentially through the modulation of the NF-κB signaling pathway.

Keywords: solidagenone; inflammation; endotoxic shock; NF-κB

1. Introduction

Inflammation has a pivotal role in rectifying imbalances to the body's homeostasis and is crucial for the repair, remodeling, and renewal of various tissues under diverse and adverse conditions [1]. Functioning as the primary line of defense, inflammation safeguards the host from infections induced by a spectrum of pathogens, including bacteria, fungi, parasites, and viruses [2]. Additionally, other stimuli, such as cellular damage, chemical agents, physical injuries, burns, radiation, freezing, ischemia, and reperfusion, can incite inflammatory responses [3].

Classically, the treatment of inflammation entails the administration of non-steroidal anti-inflammatory drugs or glucocorticoids [4,5]. Both exhibit varying degrees of effectiveness in alleviating inflammation, and their use is accompanied by a spectrum of adverse effects such as atherosclerosis, cardiac alterations, gastrointestinal disorders, hypertension, renal toxicity, type 2 diabetes, visceral obesity, and others [5,6].

Therefore, the development of new substances exhibiting anti-inflammatory properties holds significant importance for clinical use, with the goal of securing alternative treatments characterized by efficacy and diminished adverse effects. An attractive strategy in this search for new drugs involves the exploration of molecules with anti-inflammatory potential derived from medicinal plants, given the vast and yet undiscovered diversity of bioactive compounds within the plant kingdom [7,8]. Natural products stand out as a noteworthy reservoir for therapeutic interventions against various diseases, with the anti-inflammatory effects of traditional herbal medicines, crude plant extracts, natural compounds, and their derivatives having already demonstrated significant promise [7,9].

In this context, *Solidago chilensis* Meyen (Asteraceae) emerges as a highly promising reservoir for the discovery of anti-inflammatory agents. This plant species, with a historical use spanning over 700 years in South American folk medicine, has been recognized for its diverse therapeutic attributes, including anticancer, antidepressant, diuretic, gastroprotective, burn treatment, and anti-inflammatory activities [10–13].

Phytochemical analyses of *S. chilensis* have revealed a spectrum of chemical compounds, including flavonoids such as quercitrin, quercetin, and rutin, as well as diterpenes such as deoxysolidagenone, solidagolactol, and solidagenone [14]. Notably, the relatively underexplored compound solidagenone, isolated from the rhizomes, leaves, and inflorescences of *S. chilensis*, exhibits promising biological activities [13,15–17]. Solidagenone has demonstrated gastroprotective activity in a murine model of gastric injury induced by hydrochloric acid [15], attenuated skin inflammation in experimental models [16], exhibited antidepressant-like effects in mice [13], and a protective effect in a model of airway inflammation induced by ovalbumin [17]. Consequently, solidagenone holds considerable potential for further investigations into its anti-inflammatory activity. The present study was designed to assess the anti-inflammatory properties of solidagenone in acute experimental models of inflammation and its mechanism of action.

2. Results and Discussion

Firstly, the toxicity effects of a single dose of solidagenone were investigated. Administration of 30, 60, or 90 mg/kg of solidagenone did not result in mortality or elicit any discernible signs of toxicity in the animals (Table 1). Furthermore, no significant variance in body weight was noted among the animals treated with different doses of solidagenone when compared to mice treated only with the vehicle solution (Table 2).

Table 1. Effect of solidagenone on behavioral and general appearance of male BALB/c mice.

Behavior or General Appearance	Observations			
	Vehicle	Solidagenone (30 mg/kg)	Solidagenone (60 mg/kg)	Solidagenone (90 mg/kg)
Changes in the eyes	No changes	No changes	No changes	No changes
Changes in the fur	No changes	No changes	No changes	No changes
Changes in the skin	No changes	No changes	No changes	No changes
Coma	Absent	Absent	Absent	Absent
Convulsions	Absent	Absent	Absent	Absent
Diarrhea	Absent	Absent	Absent	Absent
Lethargy	Absent	Absent	Absent	Absent
Salivation	Absent	Absent	Absent	Absent
Sleep	Usual	Usual	Usual	Usual
Tremors	Absent	Absent	Absent	Absent

The animals were subjected to daily observations for a duration of 14 days.

Table 2. Body weight of BALB/c mice treated with compound solidagenone.

Days	Vehicle	Solidagenone (30 mg/kg)	Solidagenone (60 mg/kg)	Solidagenone (90 mg/kg)
0	24.3 (±0.6)	24.1 (±0.3)	23.8 (±1.6)	22.8 (±0.9)
7	24.2 (±0.3)	24.7 (±0.5)	24.0 (±1.8)	22.8 (±1.2)
14	24.4 (±0.5)	24.8 (±0.6)	24.5 (±1.5)	23.0 (±0.7)

Values represent the mean ± standard deviation of six animals per group.

Toxicological investigations of bioactive compounds derived from medicinal plants are essential to ensure the safety of using herbs in the treatment of a series of diseases [18]. Therefore, it is relevant to analyze the acute toxicity generated by treatment with solidagenone in mice. Here, we verified that the compound did not present toxicity at any of the administered doses used (30, 60, and 90 mg/kg). These findings are consistent with the literature; Rodriguez and collaborators [19] used a similar methodology to the one used in this study and verified that solidagenone, intraperitoneally injected, did not show any observable symptoms of toxicity or mortality in Swiss mice treated with doses ranging from 100 to 600 mg/kg.

A crucial index for evaluating the physiological or pathological state of mice subjected to experimental tests is body weight. Alterations in body weight often correlate with significant physiological changes, warranting a thorough analysis of this parameter [20]. In our study, we found that treatment with solidagenone did not induce significant changes in the weight of the animals.

Next, we assessed the anti-inflammatory activity of solidagenone, first in a mouse model of acute peritonitis induced by carrageenan. As revealed in Figure 1, mice subjected to carrageenan stimulation and treated with a vehicle solution exhibited a mean of 274.8 neutrophils in 300 counted cells, a significant increase ($p < 0.05$) compared to the naïve group which presented a mean of 1.8 neutrophils in 300 counted cells. In comparison to the vehicle-treated group, pre-treatment with solidagenone at doses of 30, 60, and 90 mg/kg resulted in a statistically significant reduction ($p < 0.05$) in neutrophil migration by 58.9%, 47.6%, and 41.9%, respectively. Under the same conditions, the administration of dexamethasone, at a dose of 30 mg/kg, also promoted a significant reduction ($p < 0.05$) of 59.6% in neutrophil migration (Figure 1). In agreement with these data, Liz and colleagues [21] demonstrated that an aqueous rhizome extract of *S. chilensis* and its two derived fractions promoted the inhibition of leukocyte migration, particularly neutrophil migration, and exudation in a mouse model of air pouch using carrageenan. Moreover, these effects were accompanied by a reduction in IL-1β, TNFα, and nitric oxide production, as well as a decrease in myeloperoxidase and adenosine deaminase activity [21]. Using a model of pleurisy induced by carrageenan, similar results were found with extracts from leaves or inflorescences of *S. chilensis* [22].

Next, the anti-inflammatory effect of solidagenone was further investigated in a mouse model of endotoxic shock. As revealed in Figure 2, in comparison with the vehicle group, the animals treated with solidagenone at 60 and 90 mg/kg exhibited an increased survival rate of 45.5 and 63.3%, respectively, which was significant ($p < 0.05$). Under the same conditions, dexamethasone, administered at a dose of 30 mg/kg, also promoted a significant increase of 63.3% ($p < 0.05$) in the survival rate.

As expected, LPS administration induced leukopenia and thrombocytopenia in the animals (Table 3). Interestingly, treatment with solidagenone also conferred a significant ($p < 0.05$) protective effect against these hematological disorders, which are commonly observed in sepsis conditions. Dexamethasone, while promoting a more pronounced leukocyte recovery, did not afford protection against thrombocytopenia (Table 3).

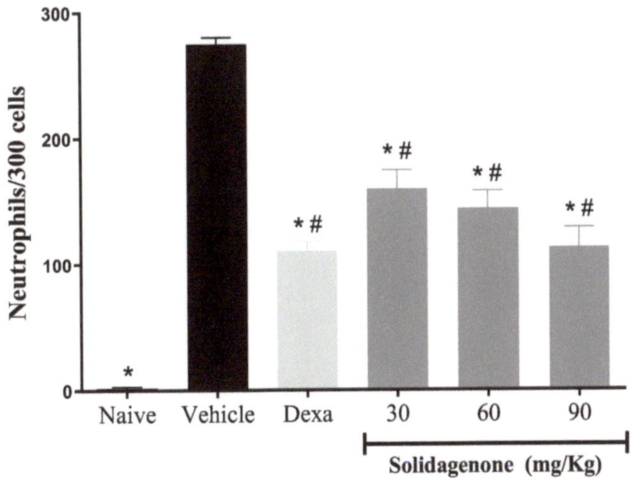

Figure 1. Effect of solidagenone in a model of acute peritonitis. BALB/c mice ($n = 6$/group) were submitted to a challenge with 1% carrageenan solution after treatment with solidagenone (30, 60, and 90 mg/kg) or dexamethasone (Dexa; 30 mg/kg) or vehicle (saline solution with 10% DMSO). Naïve group consisted of untreated and unchallenged animals. Values represent the means ± S.D. of six mice/group. * $p < 0.05$ compared to the vehicle group; # $p < 0.05$ compared to the naïve group.

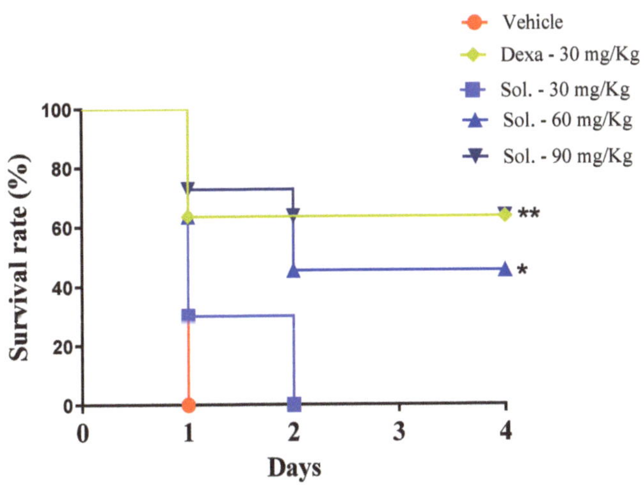

Figure 2. Survival curve of mice treated with solidagenone submitted to endotoxic shock. Mice were orally treated with solidagenone at doses of 30 mg/kg (■), 60 mg/Kg (▲), and 90 mg/kg (▼); dexamethasone at a dose of 30 mg/kg (♦); or vehicle (•). The results are from two experiments performed independently. * $p < 0.05$ compared to the vehicle group. ** $p < 0.01$ compared to the vehicle group. Statistical analyses were performed using the Logrank test (Mantel Cox).

For a better understanding of the anti-inflammatory effect of solidagenone, we measured the amount of nitric oxide and cytokines in the supernatant of resident macrophages from animals previously treated with solidagenone and stimulated in vitro with LPS plus IFNγ. As revealed in Figure 3, treatment with all the doses of solidagenone promoted a significant reduction in the levels of nitric oxide (Figure 3A), TNF-α (Figure 3B), and IL-1β (Figure 3C) relative to the LPS-stimulated vehicle-treated cultures. In a similar way, dexamethasone, administered at a dose of 30 mg/kg, also promoted a significant ($p <$

0.05) reduction in the levels of all the pro-inflammatory molecules evaluated (Figure 3). In agreement with these data, solidagenone also reduced NO, IL-1β, and TNFα production and gene expression of several inflammatory mediators (*NOS2, IL1β, TNFα,* and *Cox2*) in peritoneal macrophages stimulated in vitro with LPS plus IFNγ [17]. Moreover, the anti-inflammatory activity of solidagenone was also associated with nitric oxide, IL-6, and TNFα reductions in croton-oil-, arachidonic-acid-, and phenol-induced ear edema mouse models [16].

Table 3. Solidagenone attenuates leukopenia and thrombocytopenia in mice challenged with LPS.

Group	Dose (mg/kg)	Leukocytes (10^3 cells/μL)	Thrombocytes (10^3 mm^3)
Naive	-	3.8 ± 0.9 *	449.8 ± 49.2 *
Vehicle	-	1.7 ± 0.2	334.5 ± 36.7
Dexamethasone	30	3.8 ± 0.4 *	337.0 ± 26.8
Solidagenone	30	1.9 ± 0.3	365.8 ± 15.2
Solidagenone	60	2.2 ± 0.4	399.4 ± 3.4 *
Solidagenone	90	2.7 ± 0.5 *	403.5 ± 40.4 *

Values represent the mean ± standard deviation of six animals per group. * $p < 0.05$ compared to the vehicle group.

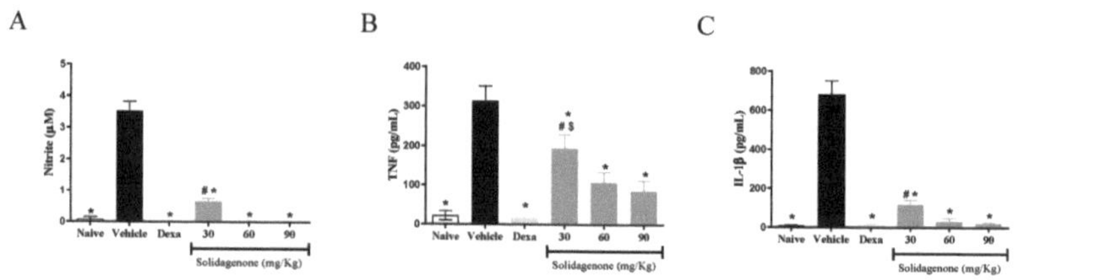

Figure 3. In vivo treatment with solidagenone decreases nitric oxide, TNF-α, and IL-1β production by LPS-stimulated macrophages. Concentrations of nitrite (**A**), TNF-α (**B**), and IL-1β (**C**). Values represent the means ± S.D. of six mice/group. * $p < 0.05$ compared to the vehicle group; # $p < 0.05$ compared to naïve group. $ $p < 0.05$ compared to dexamethasone group.

It is well established that LPS induces Toll-like receptor-4 (TLR4) activation, triggering intracellular signaling pathways, including NF-κB activation, a pivotal transcription factor that modulates diverse pro-inflammatory genes such as IL-1β, NOS2, and TNF-α [23,24]. Notably, as revealed in Figure 4, the treatment of activated macrophages in vitro with different concentrations of solidagenone results in a significant ($p < 0.05$) reduction in the gene expression of NF-κB. Similar results were obtained with dexamethasone treatment.

To understand the interaction between solidagenone and NF-κB, docking simulations were performed to investigate the interactions between solidagenone and key components involved in the NF-κB signaling pathway: the IκB kinase enzyme (IKK) and the p65 subunit of the NF-κB transcription factor. The NF-κB transcription factor is normally bound to the inhibitor of κB (IκB) in the cytosol, preventing its translocation to the nucleus and the subsequent expression of inflammatory mediators [25,26]. The activation of NF-κB is regulated by the IKK complex, which phosphorylates serine residues on IκB proteins, leading to their polyubiquitination and degradation [27]. This process results in the release of NF-κB, allowing it to translocate to the nucleus and initiate the transcription of genes associated with inflammation [28].

However, the regulation of NF-κB activity extends beyond its liberation from inhibitory subunits and nuclear translocation in the cytoplasm. The transactivation potential of NF-κB in the nucleus, which involves the recruitment of the transcriptional apparatus and the stimulation of target gene expression, is further influenced by post-translational modifications of the transcription factor and its chromatin environment [29]. In this scenario, the p65 subunit of NF-κB was identified as a target of several post-translational modifications such as acetylation, methylation, and phosphorylation [30–33]. For example, Wei and colleagues [34] demonstrated that the demethylation of arginine 30 (Arg30) in the DNA-binding domain of p65 by protein-arginine methyltransferase 5 (PRMT5) plays a crucial role in the ability of NF-κB to bind to κB elements and regulate gene expression.

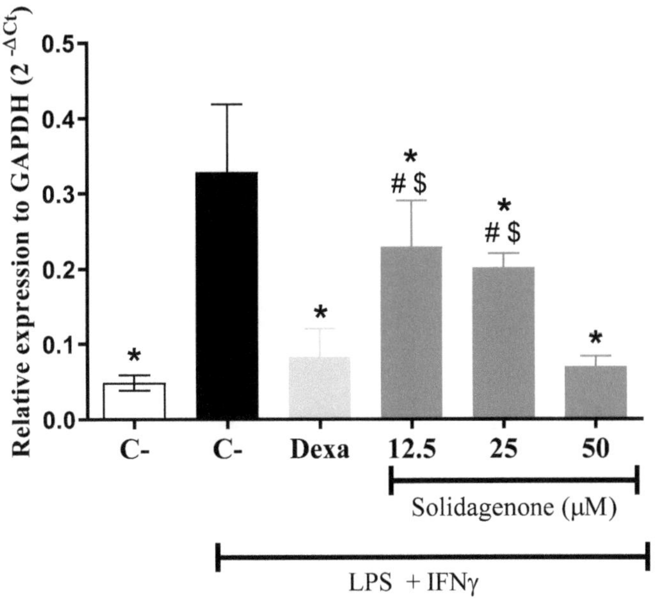

Figure 4. Gene expression of NF-κβ in untreated macrophages or macrophages treated with solidagenone. Values represent the means ± S.D. of four determinations obtained in one of two experiments performed. * $p < 0.05$ compared to stimulated and untreated cells; # $p < 0.05$ compared to unstimulated and untreated cells; $ $p < 0.05$ compared to dexamethasone-treated cells.

Certainly, the p65 subunit of NF-κB and IKK play critical roles in preventing the nuclear translocation and activation of NF-κB. They are involved in different regulatory mechanisms for this nuclear factor, contributing to the control of NF-κB activation [25,34]. In addition, both molecules are widely used as targets for the discovery of new anti-inflammatory drugs, and they are the targets of one of the most used inhibitors of the NF-κB pathway, BAY11-7082 [35,36]. Therefore, docking assays with these proteins are useful in validating new inhibitors of the NF-κB pathway.

In this context, we performed docking simulations of solidagenone in the binding site of the p65 subunit of NF-κB (Figure 5). As can be seen, solidagenone docks to the NF-κB system through seven van der Waals interactions (Lys-218, Arg-30, Phe-184, Asn-155, Pro189, Lys-79, and Thr-191), one alkyl interaction (Ala-192), one Pi-alkyl interaction (Ala-188), and one hydrogen bond interaction (Asn-190). The calculated docking energy was equal to −6.4 kcal/mol. Other images of the solidagenone-p65 subunit of NF-κB interaction are shown in Supplementary Material (Figure S1).

Regarding the IKK system, it is important to mention that it is organized in two chains (A and B), comprising four domains each: the C-terminal dimerization domain (SDD), the C-terminal kinase domain (KDC), the N-terminal kinase domain (KDN), and the ubiquitin-like domain (ULD) (Supplementary Material, Figure S2). Since, according to Liu and coworkers [37], chain B is an active conformation, this was the chain chosen as the protein target for docking solidagenone. Figure 6 shows the details of the simulated IKK–solidagenone complex. Solidagenone docks to the IKK system through six van der Waals interactions (Lys-44, Asp-166, Lys-147, Thr-23, Gly-22, and Asn-150), one Pi-Sigma interaction (Leu-2), four Pi-alkyl interactions (Ala-42, Va-l29, Ile-165, and Val-152), and two hydrogen bond interactions (Asn-28 and Glu-149). The calculated docking energy was equal to −7.8 kcal/mol. Consequently, the solidagenone–IKK interaction is more stable than the solidagenone–NF-κB interaction. Figure S3 shows some images of the solidagenone–IKK interaction. The docking results, as well as the observed reduction in NF-κB gene expression and decreased production of mediators related to NF-κB activation (such as Il-1β, nitric oxide, and TNFα), suggest that the anti-inflammatory effect of solidagenone is linked to the inhibition of the NF-κB pathway. This promising insight provides a foundation for future investigations to delve deeper into the specific mechanisms and interactions involved in the modulation of NF-κB by solidagenone, contributing to a more comprehensive understanding of its anti-inflammatory properties.

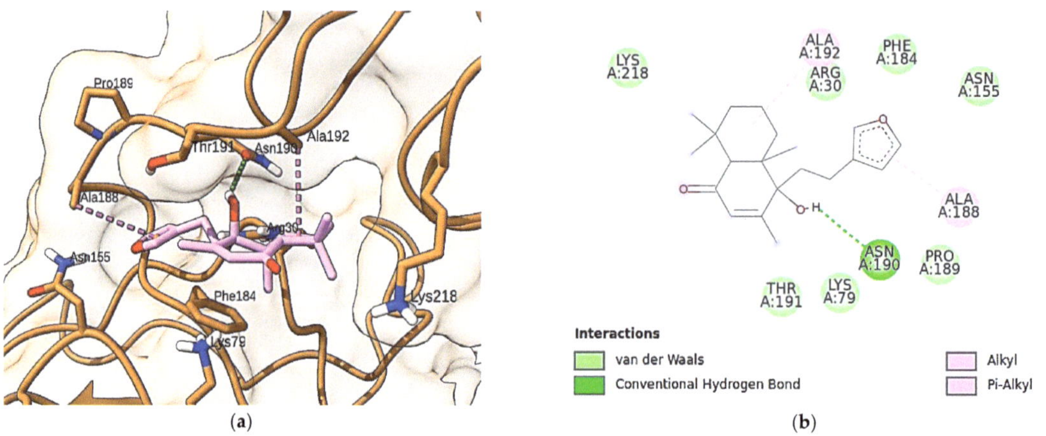

Figure 5. Docking analysis of solidagenone binding to the p65 subunit of NF-κB. (**a**) Residues of the selected protein (PDB ID: 1NFI) and their interactions with solidagenone. (**b**) Two-dimensional interactions of solidagenone with amino acid residues in the p65 subunit of NF-κB.

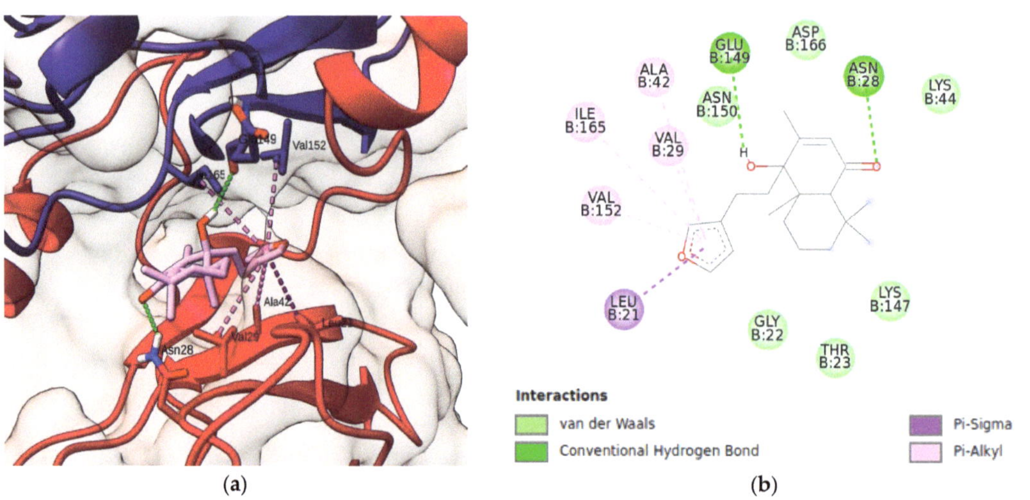

Figure 6. Docking analysis of solidagenone binding to IKK domain. (**a**) Three-dimensional structure of selected protein (PDB ID: 4KIK) and its interaction with solidagenone. (**b**) Two-dimensional interactions of solidagenone with amino acid residues in the IKK domain.

Lastly, the in silico ADME profile of solidagenone was generated and is summarized in Figure 7 and Table S1. The *Bioavailability Radar* (Figure 7) has six axes for the six relevant properties for oral bioavailability. Each property has a descriptor, and the range of optimal values is depicted as a pink area. For saturation, the ratio of sp^3 hybridized carbons over the total carbon count of the molecule (Fraction Csp3) should be at least 0.25. For size, the molecular weight should be between 150 and 500 g/mol. For polarity, the TPSA should be between 20 and 130 Å. For solubility, log *S* should not exceed six. For lipophilicity, XLOGP3 should be in the range from −0.7 to +6.0. For flexibility, the molecule should not have more than nine rotatable bonds. In this context, solidagenone can be estimated to be drug-like, since the red line is fully included in the pink area. Table S1 shows computed parameters for solidagenone, grouped into different sections (physico-chemical properties, lipophilicity, water solubility, pharmacokinetics, druglikeness, and medicinal chemistry).

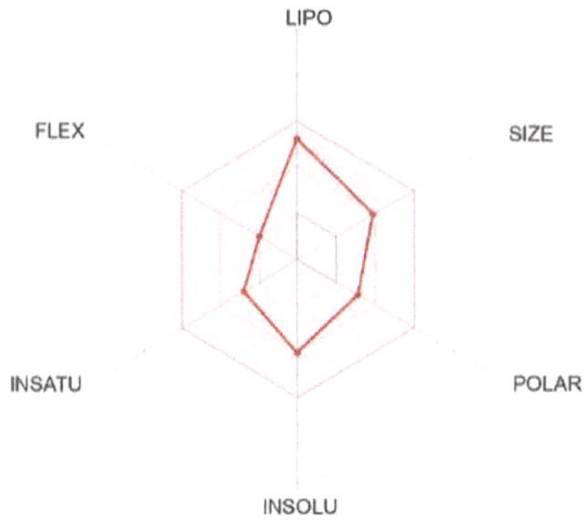

Figure 7. SwissADME plot of drug-likeness of solidagenone. The pink area represents the optimal range for each property. For saturation (INSATU), the ratio of sp^3 hybridized carbons over the total carbon count of the molecule (Fraction Csp3) should be at least 0.25. For size, the molecular weight should be between 150 and 500 g/mol. For polarity (POLAR), the TPSA should be between 20 and 130 Å. For solubility (INSOLU), log *S* should not exceed 6. For lipophilicity (LIPO), XLOGP3 should be in the range from −0.7 to +6.0. For flexibility (FLEX), the molecule should not have more than 9 rotatable bonds.

3. Materials and Methods

3.1. Drugs

Solidagenone was obtained from *Solidago chilensis* inflorescences from the Pharmaceutical Technology Institute (FarManguinhos, Oswaldo Cruz Foundation, Rio de Janeiro, Brazil), as previously described [16,38]. Dexamethasone (Sigma-Aldrich, St. Louis, MO, USA) was used as positive control in anti-inflammatory experiments. All compounds were solubilized in dimethyl sulfoxide (DMSO; PanReac, Barcelona, Spain) and diluted in culture medium for use in in vitro assays or saline for in vivo assays. The final concentration of DMSO did not exceed 0.1% in all in vitro assays or 10% in all in vivo analyses.

3.2. Animals

BALB/c mice (4 to 8 weeks old) were bred and housed at the Gonçalo Moniz Institute (Oswaldo Cruz Foundation, Salvador, Bahia, Brazil) in sterilized cages, under controlled environmental conditions, and provided with a balanced rodent diet and water ad libitum. All animal experiments and procedures were conducted in accordance with the institution's committee on the ethical handling of laboratory animals and were approved under the number L-IGM-29/2009.

3.3. Acute Toxicity in Mice

BALB/c mice (male; 6–8 weeks of age; n = 6/group) were divided into four experimental groups and treated orally with a single dose of solidagenone (30, 60, or 90 mg/kg) or vehicle (saline solution with 10% DMSO). Following the completion of treatment, the mice were monitored for general toxicity signs over a 15-day period. This involved the observation of morphological and behavioral changes. Additionally, the body mass of the animals was measured on days 0, 7, and 14.

3.4. Induction of Acute Peritonitis in Mice

BALB/c mice (males; 6–8 weeks old) were divided into six experimental groups (n = 6) and treated orally with solidagenone (30, 60, or 90 mg/kg), dexamethasone (30 mg/kg), or vehicle (saline solution with 10% DMSO) 24 and 1 h before the challenge. Subsequently, the animals were challenged through an intraperitoneal injection of 250 µL of carrageenan (Sigma-Aldrich, St. Louis, MO, USA; 1 mg/mL; 250 µL), as previously described [39]. After 4 h, the mice were euthanized and peritoneal exudates were harvested by peritoneal lavage using a saline solution. The cells were centrifuged at 400× g for 10 min at 4 °C. Cytospin preparations were stained with rapid panoptic and a differential count of 300 cells was performed by a blinded investigator.

3.5. LPS-Induced Endotoxin Shock

BALB/c mice (male; 4–5 weeks of age) were randomized into five experimental groups (n = 11/group) and treated orally with solidagenone (30, 60, or 90 mg/kg), dexamethasone (30 mg/kg), or vehicle (saline solution with 10% DMSO) 24 and 3 h before the challenge. Subsequently, the animals received a lethal dose of LPS (600 µg; from serotype 0111:B4 *Escherichia coli*, Sigma-Aldrich, St. Louis, MO, USA) in saline via the intraperitoneal route, as previously described [40]. The animals were monitored for 4 days to observe survival. In addition, a second set of experiments was performed, and heparinized blood samples were collected 6 h after the challenge with LPS (600 µg) to analyze the leukocytes and thrombocytes using the PE 7010VET Hematology Analyzer (Shenzhen Prokan Electronics Inc., Shenzhen, China).

3.6. Cytokine and Nitric Oxide Production by Resident Macrophages

To evaluate nitric oxide and cytokine production by resident macrophages, groups of male BALB/c mice were orally treated with solidagenone, dexamethasone, or vehicle in the doses described above. After 90 min, the mice were subjected to euthanasia for macrophage collection by means of peritoneal wash using cold Dulbecco's modified Eagle's medium (DMEM; Life Technologies, GIBCO-BRL, Gaithersburg, MD, USA). The cells were washed twice with DMEM, resuspended in DMEM supplemented with 10% fetal bovine serum (FBS; GIBCO) and 50 µg/mL of gentamycin (Life Technologies, Carlsbad, CA, USA), and plated into 96-well plates at a density of 2×10^5 cells/well. After 2 h of incubation at 37 °C, the plates were washed with a saline solution and new medium was added to remove non-adherent cells. The cells were then activated with LPS (500 ng/mL) and IFNγ (5 ng/mL) and further incubated at 37 °C and 5% CO_2. Cell-free supernatants were collected at two different time points after incubation for the quantification of TNFα (4 h) and for the quantification of nitric oxide and IL-1β (24 h) and kept at −80 °C until further use.

3.7. Real-Time Reverse Transcription–Polymerase Chain Reaction (qRT-PCR)

Peritoneal exudate macrophages were obtained as previously described [17]. Then, the cells were plated into a 24-well plate at 1×10^6 cells/well in a DMEM medium supplemented with FBS and gentamicin for 24 h at 37 °C and 5% CO_2. The cells were then pretreated with solidagenone (50, 25, and 12.5 µM) or dexamethasone (12.5 µM) for 1 h and then stimulated with LPS (500 ng/mL) and IFNγ (5 ng/mL) and incubated at 37 °C for 3 h. After treatment, the gene expression of NF-κB was measured as previously described [17]. The following primer sequences were used in real-time PCR assays: *NFkb*:5′-ATGGCAGACGATGATCCCTAC-3 and 3′-TGTTGACAGTGGTATTTCTGGTG-5.

3.8. Solidagenone Structure

For molecular modeling purposes, the solidagenone structure was obtained from PubChem, the public chemical database of the National Center for Biotechnology Information (NCBI) of the National Library of Medicine (NLM), an institute of the U.S. National Institutes of Health (NIH, Bethesda, MD, USA) [41]. This structure was energy-minimized using the quantum mechanical modeling method DFT B3LYP/def2-TZVP, which was implemented in the Orca 5.0.3 software [42].

3.9. Crystallographic Protein Structures

The crystallographic structures of human proteins p65 subunit of NF-κB, with a resolution of 2.70 Å (PDB ID: 1NFI) [43], and IκB kinase enzyme (IKK), with a resolution of 2.83 Å (PDB ID: 4KIK) [37], were both obtained from the Protein Data Bank [44] and used in the docking simulations with solidagenone. The protein preparation steps were carried out as follows: (1) non-essential water molecules were removed; (2) polar hydrogens were added to the protein; and (3) partial charges were calculated using both the Kollman [45] and Gasteiger's approaches [46].

3.10. Molecular Docking Simulations

Molecular docking simulations were performed with the AutoDock Vina 1.1.2 program [47]. Two molecular graphical programs, UCSF Chimera X [48] and BIOVIA Discovery Studio 2021 [49], were used for visualizing ligand–protein docking interactions in 3D and 2D representations, respectively.

3.11. In Silico ADME

The SwissADME platform [50] was used for computing the physicochemical and pharmacokinetics parameters related to the adsorption, distribution, metabolism, and excretion (ADME) properties of solidagenone, as well as its drug-likeness.

3.12. Statistical Analyses

Statistical analyses were performed using GraphPad Prism version 8.0 (GraphPad Software, San Diego, CA, USA). One-way analysis of variance (ANOVA) followed by Newman–Keuls multiple comparison tests were used for the comparison of groups. P-values less than 0.05 were considered to indicate statistical significance. The presented data are representative of at least two or three independent experiments.

4. Conclusions

Taken together, the findings of the present investigation show that solidagenone exhibits significant anti-inflammatory properties in acute experimental models, potentially through the modulation of the NF-κB signaling pathway. Furthermore, solidagenone shows no acute toxicity, and its in silico ADME profile suggests favorable drug-like characteristics. The presented data fortify the potential of solidagenone as a basis for the development of alternative treatments with reduced adverse effects for inflammatory conditions.

Supplementary Materials: The following supporting information can be downloaded at: https://www.mdpi.com/article/10.3390/ph17030273/s1, Figure S1: Solidagenone docked onto the p65 subunit of NF-κB; Figure S2: IKK system; Figure S3: IKK system and details of solidagenone–IKK interaction. Table S1: ADME parameters, drug-likeness, and medicinal chemistry of solidagenone.

Author Contributions: I.P.S.: Investigation, conceptualization, formal analysis, writing—original draft; L.P.S.: formal analysis, writing—original draft; D.K.C.S.: methodology, formal analysis; B.P.Z.C.d.R.: methodology, formal analysis; T.B.d.O.: methodology, formal analysis; A.M.K.: methodology, formal analysis; E.d.S.R.: methodology, formal analysis; C.V.C.d.S.: methodology, conceptualization, writing—review and editing; J.F.O.-C.: conceptualization, investigation, formal analysis; S.S.V.: conceptualization, writing—review and editing; O.A.S.-F.: conceptualization, resources, writing—review, supervision, and editing; M.B.P.S.: conceptualization, writing—review and editing, supervision, funding acquisition; C.S.M.: conceptualization, formal analysis, writing—review and editing, supervision. All authors have read and agreed to the published version of the manuscript.

Funding: This work was supported by grants from PRONEX (grant number 0002/2014). M.B.P.S is a recipient of the Conselho Nacional de Desenvolvimento Científico e Tecnológico (CNPq) fellowship. E.d.S.R. is grateful for the PhD scholarship from Coordenação de Aperfeiçoamento de Pessoal de Nível superior—Brasil (CAPES), Finance Code 001.

Institutional Review Board Statement: The study was approved by the Ethics Committee and Institutional Animal Care and Use Committee of FIOCRUZ (IGM 009/2009).

Informed Consent Statement: Not applicable.

Data Availability Statement: The data presented in this study are available on request from the corresponding author.

Conflicts of Interest: The authors declare no conflict of interest.

References

1. Rathinam, V.A.K.; Chan, F.K.-M. Inflammasome, Inflammation, and Tissue Homeostasis. *Trends Mol. Med.* **2018**, *24*, 304–318. [CrossRef]
2. Hawiger, J.; Zienkiewicz, J. Decoding inflammation, its causes, genomic responses, and emerging countermeasures. *Scand. J. Immunol.* **2019**, *90*, e12812. [CrossRef] [PubMed]
3. Roe, K. An inflammation classification system using cytokine parameters. *Scand. J. Immunol.* **2021**, *93*, e12970. [CrossRef]
4. Ronchetti, S.; Migliorati, G.; Bruscoli, S.; Riccardi, C. Defining the role of glucocorticoids in inflammation. *Clin. Sci.* **2018**, *132*, 1529–1543. [CrossRef]
5. Bindu, S.; Mazumder, S.; Bandyopadhyay, U. Non-steroidal anti-inflammatory drugs (NSAIDs) and organ damage: A current perspective. *Biochem. Pharmacol.* **2020**, *180*, 114147. [CrossRef] [PubMed]
6. Oray, M.; Abu Samra, K.; Ebrahimiadib, N.; Meese, H.; Foster, C.S. Long-term side effects of glucocorticoids. *Expert Opin. Drug Saf.* **2016**, *15*, 457–465. [CrossRef] [PubMed]
7. Newman, D.J.; Cragg, G.M. Natural Products as Sources of New Drugs over the nearly four decades from 01/1981 to 09/2019. *J. Nat. Prod.* **2020**, *83*, 770–803. [CrossRef] [PubMed]
8. Atanasov, A.G.; Zotchev, S.B.; Dirsch, V.M.; Supuran, C.T. Natural products in drug discovery: Advances and opportunities. *Nat. Rev. Drug Discov.* **2021**, *20*, 200–216. [CrossRef] [PubMed]
9. Azab, A.; Nassar, A.; Azab, A.N. Anti-Inflammatory Activity of Natural Products. *Molecules* **2016**, *21*, 1321. [CrossRef]
10. Mors, W.; Rizzini, C.; Pereira, N. *Medicinal Plants of Brazil*; Reference Publications, Inc.: Algonac, MI, USA, 2000.
11. Catarino, H.R.C.; de Godoy, N.P.; Scharlack, N.K.; Neves, L.M.G.; de Gaspi, F.O.; Esquisatto, M.A.M.; do Amaral, M.E.C.; Mendonça, F.A.S.; dos Santos, G.M.T. InGaP 670-nm laser therapy combined with a hydroalcoholic extract of *Solidago chilensis* Meyen in burn injuries. *Lasers Med. Sci.* **2015**, *30*, 1069–1079. [CrossRef]
12. De Barros, M.; Mota da Silva, L.; Boeing, T.; Somensi, L.B.; Cury, B.J.; de Moura Burci, L.; Santin, J.R.; de Andrade, S.F.; Monache, F.D.; Cechinel-Filho, V. Pharmacological reports about gastroprotective effects of methanolic extract from leaves of *Solidago chilensis* (Brazilian arnica) and its components quercitrin and afzelin in rodents. *Naunyn-Schmiedeberg's Arch. Pharmacol.* **2016**, *389*, 403–417. [CrossRef] [PubMed]
13. Locateli, G.; de Oliveira Alves, B.; Miorando, D.; Ernetti, J.; Alievi, K.; Zilli, G.A.L.; Serpa, P.Z.; Vecchia, C.A.D.; Mota da Silva, L.; Müller, L.G.; et al. Antidepressant-like effects of solidagenone on mice with bacterial lipopolysaccharide (LPS)-induced depression. *Behav. Brain Res.* **2020**, *395*, 112863. [CrossRef]
14. Valverde, S.S.; Oliveira, T.B.; Souza, S.P. *Solidago chilensis* Meyen (Asteraceae). *Rev. Fitos* **2012**, *7*, 131–136. [CrossRef]
15. Schmeda-Hirschmann, G.; Rodriguez, J.; Astudillo, L. Gastroprotective activity of the diterpene solidagenone and its derivatives on experimentally induced gastric lesions in mice. *J. Ethnopharmacol.* **2002**, *81*, 111–115. [CrossRef]
16. Valverde, S.S.; Santos, B.C.S.; de Oliveira, T.B.; Gonçalves, G.C.; de Sousa, O.V. Solidagenone from *Solidago chilensis* Meyen inhibits skin inflammation in experimental models. *Basic Clin. Pharmacol. Toxicol.* **2021**, *128*, 91–102. [CrossRef] [PubMed]
17. Vasconcelos, J.F.; Santos, I.P.; de Oliveira, T.B.; Kelly, A.M.; do Reis, B.P.Z.C.; Orge, I.D.; Meira, C.S.; Valverde, S.S.; Soares, M.B.P. The protective effect of solidagenone from *Solidago chilensis* Meyen in a mouse model of airway inflammation. *Basic Clin. Pharmacol. Toxicol.* **2022**, *130*, 44–55. [CrossRef]
18. Dashti, A.; Shokrzadeh, M.; Karami, M.; Habibi, E. Phytochemical identification, acute and subchronic oral toxicity assessments of hydroalcoholic extract of *Acroptilon repens* in BALB/c mice: A toxicological and mechanistic study. *Heliyon* **2022**, *8*, e08940. [CrossRef] [PubMed]

19. Rodríguez, J.A.; Bustamante, C.; Astudillo, L.; Schmeda-Hirschmann, G. Gastroprotective activity of solidagenone on experimentally-induced gastric lesions in rats. *J. Pharm. Pharmacol.* **2002**, *54*, 399–404. [CrossRef]
20. Ugwah-Oguejiofor, C.J.; Okoli, C.O.; Ugwah, M.O.; Umaru, M.L.; Ogbulie, C.S.; Mshelia, H.E.; Umar, M.; Njan, A.A. Acute and sub-acute toxicity of aqueous extract of aerial parts of *Caralluma dalzielii* N. E. Brown in mice and rats. *Heliyon* **2019**, *5*, e01179. [CrossRef]
21. Liz, R.; Vigil, S.V.G.; Goulart, S.; Izabel, M.; Moritz, G.; Schenkel, E.P.; Fröde, T.S. The anti-inflammatory modulatory role of *Solidago chilensis* Meyen in the murine model of the air pouch. *J. Pharm. Pharmacol.* **2008**, *60*, 515–521. [CrossRef]
22. Goulart, S.; Moritz, M.I.G.; Lang, K.L.; Liz, R.; Schenkel, E.P.; Fröde, T.S. Anti-inflammatory evaluation of *Solidago chilensis* Meyen in a murine model of pleurisy. *J. Ethnopharmacol.* **2007**, *113*, 346–353. [CrossRef]
23. Bartuzi, P.; Hofker, M.H.; van de Sluis, B. Tuning NF-κB activity: A touch of COMMD proteins. *Biochim. Biophys. Acta.* **2013**, *1832*, 2315–2321. [CrossRef] [PubMed]
24. Yu, H.; Lin, L.; Zhang, Z.; Zhang, H.; Hu, H. Targeting NF-κB pathway for the therapy of diseases: Mechanism and clinical study. *Signal Transduct. Target. Ther.* **2020**, *5*, 209. [CrossRef] [PubMed]
25. Baker, R.G.; Hayden, M.S.; Ghosh, S. NF-κB, inflammation, and metabolic disease. *Cell Metab.* **2011**, *13*, 11–22. [CrossRef] [PubMed]
26. Xu, G.; Lo, Y.-C.; Li, Q.; Napolitano, G.; Wu, X.; Jiang, X.; Dreano, M.; Karin, M.; Wu, H. Crystal structure of inhibitor of κB kinase β (IKKβ). *Nature* **2011**, *472*, 325–330. [CrossRef] [PubMed]
27. Scheidereit, C. IκB kinase complexes: Gateways to NF-κB activation and transcription. *Oncogene* **2006**, *25*, 6685–6705. [CrossRef] [PubMed]
28. Vallabhapurapu, S.; Karin, M. Regulation and function of NF-κB transcription factors in the immune system. *Annu. Rev. Immunol.* **2009**, *27*, 693–733. [CrossRef] [PubMed]
29. Vermeulen, L.; De Wilde, G.; Notebaert, S.; Vanden Berghe, W.; Haegeman, G. Regulation of the transcriptional activity of the nuclear factor-κB p65 subunit. *Biochem. Pharmacol.* **2002**, *64*, 963–970. [CrossRef] [PubMed]
30. Huang, B.; Yang, X.-D.; Lamb, A.; Chen, L.-F. Posttranslational modifications of NF-κB: Another layer of regulation for NF-κB signaling pathway. *Cell Signal.* **2010**, *22*, 1282–1290. [CrossRef]
31. Christian, F.; Smith, E.L.; Carmody, R.J. The Regulation of NF-κB Subunits by Phosphorylation. *Cells* **2016**, *5*, 12. [CrossRef]
32. Schmitz, M.L.; Bacher, S.; Kracht, M. IκB-independent control of NF-κB activity by modulatory phosphorylations. *Trends Biochem. Sci.* **2001**, *26*, 186–190. [CrossRef] [PubMed]
33. Wang, D.; Baldwin, A.S., Jr. Activation of nuclear factor-κB-dependent transcription by tumor necrosis factor-α is mediated through phosphorylation of RelA/p65 on serine 529. *J. Biol. Chem.* **1998**, *273*, 29411–29416. [CrossRef] [PubMed]
34. Wei, H.; Wang, B.; Miyagi, M.; She, Y.; Gopalan, B.; Huang, D.-B.; Ghosh, G.; Stark, G.R.; Lu, T. PRMT5 dimethylates R30 of the p65 subunit to activate NF-κB. *Proc. Natl. Acad. Sci. USA* **2013**, *110*, 13516–13521. [CrossRef]
35. Meng, X.; Martinez, M.A.; Raymond-Stintz, M.A.; Winter, S.S.; Wilson, B.S. IKK inhibitor bay 11-7082 induces necroptotic cell death in precursor-B acute lymphoblastic leukaemic blasts. *Br. J. Haematol.* **2010**, *148*, 487–490. [CrossRef]
36. Lee, J.; Rhee, M.H.; Kim, E.; Cho, J.Y. BAY 11-7082 is a broad-spectrum inhibitor with anti-inflammatory activity against multiple targets. *Mediat. Inflamm.* **2012**, *2012*, 416036. [CrossRef]
37. Liu, S.; Misquitta, Y.R.; Olland, A.; Johnson, M.A.; Kelleher, K.S.; Kriz, R.; Lin, L.L.; Stahl, M.; Mosyak, L. Crystal structure of a human IκB kinase β asymmetric dimer. *J. Biol. Chem.* **2013**, *288*, 22758–22767. [CrossRef] [PubMed]
38. Valverde, S.S.; Azevedo, S.S.; Tomassini, T.C.B. Utilização de CLAE, como paradigma na obtenção e controle do diterpeno solidagenona a partir de inflorescências de *Solidago chilensis* Meyen (arnica brasileira). *Rev. Bras. Farm.* **2009**, *90*, 196–199.
39. Daltro, S.R.T.; Santos, I.P.; Barros, P.L.; Moreira, D.R.M.; Tomassini, T.C.B.; Ribeiro, I.M.; Ribeiro-dos-Santos, R.; Meira, C.S.; Soares, M.B.P. In vitro and In Vivo Immunomodulatory Activity of *Physalis angulata* Concentrated Ethanolic Extract. *Planta Medica* **2021**, *87*, 160–168. [CrossRef] [PubMed]
40. Costa, J.F.O.; Barbosa-Filho, J.M.; Maia, G.L.d.A.; Guimarães, E.T.; Meira, C.S.; Ribeiro-Dos-Santos, R.; de Carvalho, L.C.P.; Soares, M.B.P. Potent anti-inflammatory activity of betulinic acid treatment in a model of lethal endotoxemia. *Int. Immunopharmacol.* **2014**, *23*, 469–474. [CrossRef]
41. Kim, S.; Chen, J.; Cheng, T.; Gindulyte, A.; He, J.; He, S.; Li, Q.; Shoemaker, B.A.; Thiessen, P.A.; Yu, B.; et al. PubChem 2023 update. *Nucleic Acids Res.* **2023**, *51*, D1373–D1380. [CrossRef]
42. Neese, F. The ORCA Program System. *Wiley Interdiscip. Rev. Comput. Mol. Sci.* **2012**, *2*, 73–78. [CrossRef]
43. Jacobs, M.D.; Harrison, S.C. Structure of an IκBα/NF-κB complex. *Cell* **1998**, *95*, 749–758. [CrossRef]
44. Berman, H.M.; Westbrook, J.; Feng, Z.; Gilliland, G.; Bhat, T.N.; Weissig, H.; Shindyalov, I.N.; Bourne, P.E. The Protein Data Bank. *Nucleic Acids Res.* **2000**, *28*, 235–242. [CrossRef] [PubMed]
45. Bayly, C.I.; Cieplak, P.; Cornell, W.; Kollman, P.A. A well-behaved electrostatic potential based method using charge restraints for deriving atomic charges: The RESP model. *J. Phys. Chem.* **1993**, *97*, 10269–10280. [CrossRef]
46. Gasteiger, J.; Marsili, M. Iterative partial equalization of orbital electronegativity—A rapid access to atomic charges. *Tetrahedron* **1980**, *36*, 3219–3228. [CrossRef]
47. Trott, O.; Olson, A.J. AutoDock Vina: Improving the speed and accuracy of docking with a new scoring function, efficient optimization, and multithreading. *J. Comput. Chem.* **2010**, *31*, 455–461. [CrossRef] [PubMed]

48. Pettersen, E.F.; Goddard, T.D.; Huang, C.C.; Couch, G.S.; Greenblatt, D.M.; Meng, E.C.; Ferrin, T.E. UCSF Chimera—A visualization system for exploratory research and analysis. *J. Comput. Chem.* **2004**, *25*, 1605–1612. [CrossRef] [PubMed]
49. BIOVIA; Dassault Systèmes. *Discovery Studio Visualizer, Version 21.1.0*; Dassault Systèmes: San Diego, CA, USA, 2021.
50. Daina, A.; Michielin, O.; Zoete, V. SwissADME: A free web tool to evaluate pharmacokinetics, drug-likeness and medicinal chemistry friendliness of small molecules. *Sci. Rep.* **2017**, *7*, 42717. [CrossRef]

pharmaceuticals

MDPI

Article

Styrylpyridinium Derivatives for Fluorescent Cell Imaging

Reinis Putralis [1,2,†], Ksenija Korotkaja [3,†], Martins Kaukulis [1,4], Zhanna Rudevica [3], Juris Jansons [3], Olga Nilova [3], Martins Rucins [1], Laura Krasnova [1], Ilona Domracheva [1], Mara Plotniece [2,4], Karlis Pajuste [1], Arkadij Sobolev [1], Felikss Rumnieks [3], Laura Bekere [1], Anna Zajakina [3], Aiva Plotniece [1,2,*] and Gunars Duburs [1,*]

1 Latvian Institute of Organic Synthesis, LV-1006 Riga, Latvia; reinis.putralis@osi.lv (R.P.); martins.kaukulis@osi.lv (M.K.); rucins@osi.lv (M.R.); chernova@osi.lv (L.K.); ilona@farm.osi.lv (I.D.); kpajuste@osi.lv (K.P.); arkady@osi.lv (A.S.); laura.bekere.8@gmail.com (L.B.)
2 Department of Pharmaceutical Chemistry, Faculty of Pharmacy, Riga Stradiņš University, LV-1007 Riga, Latvia; mara.plotniece@rsu.lv
3 Latvian Biomedical Research and Study Centre, LV-1067 Riga, Latvia; ksenija.korotkaja@biomed.lu.lv (K.K.); zanna.rudevica@biomed.lu.lv (Z.R.); jansons@biomed.lu.lv (J.J.); nilovaa.olga@gmail.com (O.N.); felikss.rumnieks@biomed.lu.lv (F.R.); anna@biomed.lu.lv (A.Z.)
4 Faculty of Materials Science and Applied Chemistry, Riga Technical University, LV-1048 Riga, Latvia
* Correspondence: aiva@osi.lv (A.P.); gduburs@osi.lv (G.D.); Tel.: +371-6701-4852 (A.P.)
† These authors contributed equally to this work.

Abstract: A set of styrylpyridinium (SP) compounds was synthesised in order to study their spectroscopic and cell labelling properties. The compounds comprised different electron donating parts (julolidine, *p*-dimethylaminophenyl, *p*-methoxyphenyl, 3,4,5-trimethoxyphenyl), conjugated linkers (vinyl, divinyl), and an electron-withdrawing *N*-alkylpyridinium part. Geminal or *bis*-compounds incorporating two styrylpyridinium (*bis*-SP) moieties at the 1,3-trimethylene unit were synthesised. Compounds comprising a divinyl linker and powerful electron-donating julolidine donor parts possessed intensive fluorescence in the near-infrared region (maximum at ~760 nm). The compounds had rather high cytotoxicity towards the cancerous cell lines HT-1080 and MH-22A; at the same time, basal cytotoxicity towards the NIH3T3 fibroblast cell line ranged from toxic to harmful. SP compound **6e** had IC_{50} values of 1.0 ± 0.03 μg/mL to the cell line HT-1080 and 0.4 μg/mL to MH-22A; however, the basal toxicity LD_{50} was 477 mg/kg (harmful). The compounds showed large Stokes' shifts, including 195 nm for **6a,b**, 240 nm for **6e**, and 325 and 352 nm for **6d** and **6c**, respectively. The highest photoluminescence quantum yield (PLQY) values were observed for **6a,b**, which were 15.1 and 12.2%, respectively. The PLQY values for the SP derivatives **6d,e** (those with a julolidinyl moiety) were 0.5 and 0.7%, respectively. Cell staining with compound **6e** revealed a strong fluorescent signal localised in the cell cytoplasm, whereas the cell nuclei were not stained. SP compound **6e** possessed self-assembling properties and formed liposomes with an average diameter of 118 nm. The obtained novel data on near-infrared fluorescent probes could be useful for the development of biocompatible dyes for biomedical applications.

Keywords: styrylpyridines; julolidine; self-assembly; near-infrared fluorescent dye; cell labelling; macrophages

Citation: Putralis, R.; Korotkaja, K.; Kaukulis, M.; Rudevica, Z.; Jansons, J.; Nilova, O.; Rucins, M.; Krasnova, L.; Domracheva, I.; Plotniece, M.; et al. Styrylpyridinium Derivatives for Fluorescent Cell Imaging. *Pharmaceuticals* **2023**, *16*, 1245. https://doi.org/10.3390/ph16091245

Academic Editor: Jean-Pierre Bazureau

Received: 15 June 2023
Revised: 25 August 2023
Accepted: 29 August 2023
Published: 4 September 2023

1. Introduction

The styryl fragment (Ph-CH=CH-) is a common moiety included in the structures of different pharmaceuticals, polymers, agrochemicals, fluorophores, dyes, and materials. It is one of the most common functional groups in biologically active organic compound structures [1,2]. Styryl dyes have high-tech applications as materials for light emission and photoelectric and photochemical activity [1,3]. Among various styryl compounds, two important classes of natural products are mentioned—stilbenes, including resveratrol derivatives [4], and chalcones, including curcumin derivatives [5]—as the main representatives. 4-*N*,*N*-Dimethylaminostyryl-1-methylpyridinium tosylate [6] is converted to an important optoelectronic material [7]. Styrylpyridinium tosylates (and also bromides or chlorides) are widely proposed as fluorescent probes for biochemical, biophysical, and molecular biology studies [6,8–11]. Large Stokes' shift styryl dyes were mentioned as bio-imaging agents and have been widely studied during the last decade. For example, cyanine–styryl derivatives with Stokes' shifts larger than 70 nm were proposed as DNA binding agents and also displayed anti-bacterial and anti-cancer properties [12]. A highly fluorescent styrylpyridinium–pyrene hybrid with a large Stokes' shift of around 130 nm showed remarkable selectivity to stain the nucleus in both live and fixed cells [13]. Other styrylpyridinium dyes with Stokes' shifts of more than 200 nm and improved fluorescence quantum yield have been proposed as useful reagents for the colorimetric detection of fluoride ions [14].

It was previously demonstrated that the fluorescence of a group of styrylpyridinium (SP) probes increased after an association with glucose oxidase (GOx), resulting in blue shifts up to 30 nm of the fluorescence emission maxima [15]. Pronounced changes in the fluorescence intensities of styrylpyridinium derivatives in the presence of GOx suggested that the enzyme readily binds to the probes via non-covalent interactions. Cationic styryl dyes show a wide range of F_{max} values with a change in the electron-withdrawing (EWG) properties of a cationic heteroaromatic ring and electron-donating force of another aromatic ring, making them good candidates for application as solid-state fluorescent materials [16]. It was confirmed that the strong electron acceptor–donor system that was provided by the electron-donating *N*,*N*-dialkyl group in the molecule increased the electron flow towards the electron acceptor moieties. The long π conjugation style, length of the conjugated bridge, and molecular planarity are essential preconditions for the design of functional chromophores [17–19]. The synthesis and evaluation of the physicochemical and self-assembling properties of styrylpyridinium moieties containing a 1,4-DHP amphiphile as a prospective delivery system were demonstrated by our research group [20]. It was demonstrated that the structure of SP or related derivatives affected their properties. Thus, a change in the nature of the donor and acceptor units and that of the π-linker led to an intense increase in hyperpolarisability after grooving of the molecular conjugation, whereas the hyperpolarisability was almost unaffected by an increase in the donor/acceptor strength [18].

Julolidine-2,3,6,7-tetrahydro-1H,5Hbenzo[1,2]quinolizine and its derivatives are a structural subunit widely used in synthesis of fluorescent dyes [21]. The structure of julolidine is formally proposed as dipropylaniline with two fixed moieties and is defined as an electron-donating group in fluorescent dyes. Julolidine derivatives have shown potential applications in various fields of science and technology due to their fluorescence properties [22,23].

In recent decades, styrylpyridinium has also been shown to have biological properties, including antimicrobial properties against bacteria [24–26] and antifungal activity for effective inhibition of cells of the pathogenic yeast *Candida albicans* [8,9]. The styrylpyridinium derivative 4-(4-cyanostyryl)-1-dodecylpyridin-1-ium (CSDP+) bromide has shown low toxicity on mammalian Hek-293 cells and reduces *Candida albicans* adhesion [27]. Additionally, various styryl dyes have been proposed as effective probes for DNA, RNA, β-amyloid peptides, chymotrypsin, heparin, and other bioanalytes [28,29]. *Bis*-piperazinyl-styrylpyridium-bromobutenyl salt (Br-PSP-PSP-Br) was synthesised and proposed as a sensitive and rapid diagnostic system for avian influenza virus recognition [30]. The cytotoxic assay of hydroxyl and nitro styryl-quinoline derivatives demonstrated that the hydroxy analogues exhibited better cytotoxicity than the nitro analogues [31]. It was concluded that the presence of an EWG at the styryl ring of the molecule is necessary for increasing the cytotoxicity levels [31]. Furthermore, julolidine derivatives were shown to inhibit amyloid B protein self-assembly, which is regarded as a prime strategy for Alzheimer's disease treatment [32].

The fluorescent properties of the julolidine-based compounds were studied to visualise the accumulation and distribution of these compounds within the cell, aiming to understand their potential anti-parasitic effect. Julolidine-based fluorescent compounds possessed trypanosome alternative oxidase (TAO) inhibition activity in the submicromolar range and accumulated specifically in the mitochondria of mammalian and trypanosome cells [33]. The fluorescent labelling of cells or drug formulations represents an advanced approach for monitoring the treatment results in many diseases, including cancer [34]. Cell tracking and cell migration analyses are important for cancer monitoring and tracking immune cell migration [35,36]. These analyses can provide valuable insights into the dynamic changes in the tumour microenvironment, metastasis, and the interaction between cancer cells and immune cells [37]. By understanding cell migration patterns, novel strategies for cancer treatment, immunotherapy, and targeted drug delivery can be developed.

In this work, we designed new styrylpyridinium (SP) compounds possessing electron-donating substituents at the phenyl moiety and variation in the length of the alkyl chain at the EWG *N*-alkylpyridinium fragment by introducing conjugated vinyl or divinyl linkers. Additionally, geminal or *bis*-styrylpyridinium (*bis*-SP) derivatives with two SP moieties conjugated through the 1,3-trimethylene unit were obtained, combining two geminal fluorescent moieties in the same molecule. In total, nine compounds were synthesised and their spectral, self-assembling, and biological properties were assessed. For evaluation of the biological properties, cytotoxicity assessment was performed. In addition, the cell labelling properties of one selected compound (**6e**) were evaluated using fluorescence microscopy and flow cytometry analysis.

2. Results and Discussion

2.1. Chemistry

2.1.1. Design

The design of the SP derivatives for this study was performed by changing the main structural parameters in the following ways (Figure 1):

1. Modification of the electron-donating part (*p*-methoxyphenyl, 3,4,5-trimethoxyphenyl, *p*-dimethylaminophenyl or julolidine moieties);
2. Modification of the conjugated linker (vinyl, divinyl);
3. Modification of the electron-withdrawing *N*-alkylpyridinium part, introducing dodecyl or hexadecyl chains;
4. Geminal or *bis*-SP derivatives with two styrylpyridinium moieties that are conjugated through the 1,3-trimethylene (propane) unit.

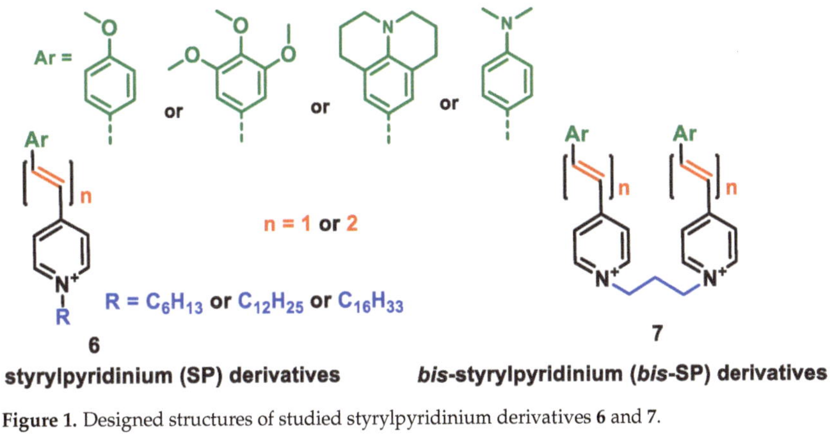

Figure 1. Designed structures of studied styrylpyridinium derivatives **6** and **7**.

In this way, we attempted to modify the structure of novel putative fluorescent probes that also possessed anticancer activity to obtain theranostic properties. We designed fluorescent probes for the near-infrared (NIR) region comprising strong electron-donating moieties possessing one or three methoxy groups with a dimethylamino group or fixed dipropylamino group included in the julolidine moiety. We evaluated the possibility of modifying the absorption and emission wavelengths by obtaining a modified conjugated chain (incorporating one or two vinyl groups).

2.1.2. Synthesis-Modified Conjugated Chain (Incorporating One or Two Vinyl Groups)

The designed styrylpyridinium derivatives **6** and **7** were obtained from corresponding aldehydes **4** or **5** and 4-picolinium bromides **2** or *bis*-4-methylpyridinium dibromide **3** in EtOH solution in the presence of a catalytic amount of piperidine. For compounds with a vinyl linker, the reaction mixture was refluxed for 24 h in analogue using the procedures described by Vaitkiene et al. [8]. For compounds with a divinyl linker, synthesis was performed in MeOH solution in the presence of 12.2 eq pyrrolidine; the reaction mixture was stirred at RT for 2 h. The schematic synthetic procedures are presented in Scheme 1. The substituents for the structures of the obtained SP **6** and *bis*-SP **7** derivatives are depicted in Table 1.

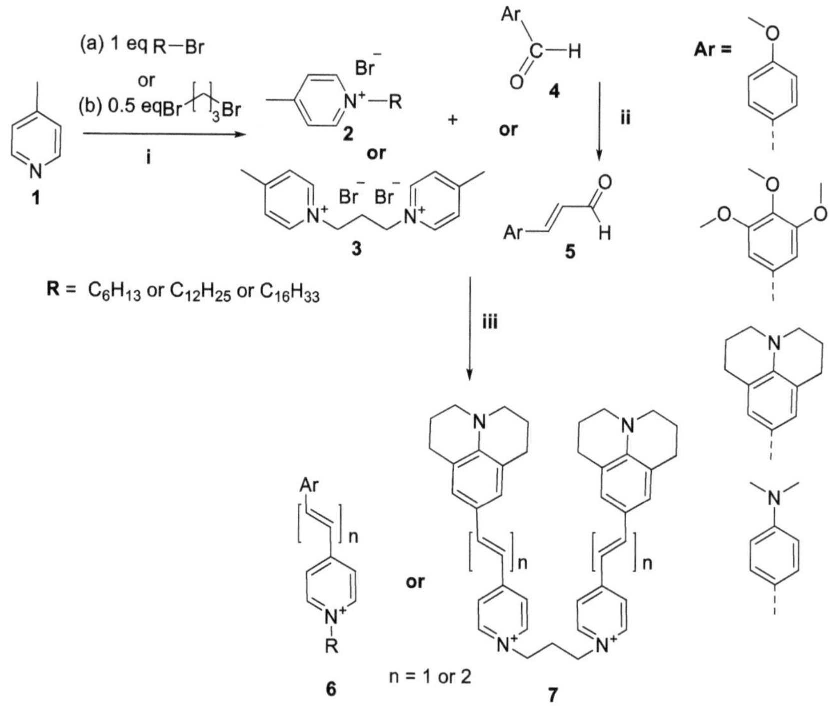

Scheme 1. Schematic synthetic procedure of the proposed styryl derivatives: **i**: 2-butanone, 80 °C, 24 h; **ii**: in pressure vessel, 5 eq triphenyl (1,3-dioxolan-2-ylmethyl)phosphonium bromide, 4 eq NaH, and 0.12 eq 18-crown-6, dry THF, under Ar, RT, 2 h; **iii**: 2 eq aldehyde, 1 eq picolinium **2** or **3**, EtOH, three drops of piperidine, 90 °C, 24 h for compounds **6a**, **b**, **d**, **f**, **g**, and **7a** with vinyl linkers and in MeOH solution in the presence of 12.2 eq pyrrolidine at RT for 2 h for compounds **6c**, **e**, and **7b** with divinyl linkers (for details see Supplementary Materials).

Table 1. Substituents of compounds **6a–6g** and **7a**,**b**.

Entry	Compound	Ar	R	n	Ref
1	**6a**	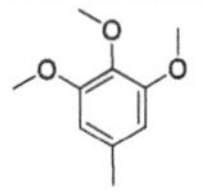	$C_{12}H_{25}$	1	
2	**6b**		$C_{16}H_{33}$	1	[38]
3	**6c**	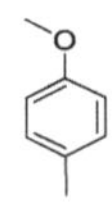	$C_{16}H_{33}$	2	
4	**6d**	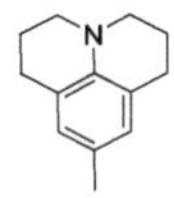	$C_{12}H_{25}$	1	
5	**6e**		$C_{16}H_{33}$	2	
6	**6f (DSP-6)** *	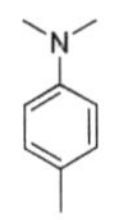	C_6H_{13}	1	[39]
7	**6g (DSP-12)** *		$C_{12}H_{25}$	1	[39]
8	**7a**	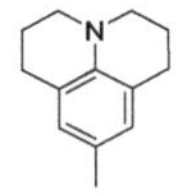	-	1	[40]
9	**7b** *		-	2	

* Counter ion is tosylate.

The full detailed description of the synthesis and characterisation of the new styrylpyridines **6a**,**c–e** and their structural analogue *bis*-SP **7b** is given in the supplementary data (see Supplementary Materials). The structures of original SP **6a**,**c–e** and *bis*-SP **7b** were established and confirmed based on the ^{1}H NMR, ^{13}CNMR, and HRMS data (Figures S1–S15, Supplementary Materials). The data for SP derivatives **6b** [38], **6f** [39], **6g** [39], and *bis*-SP derivative **7a** [40] were in good agreement with the literature data. According to high-performance liquid chromatography (HPLC) data, the purity of the studied compounds was at least 97%.

2.2. Cytotoxicity

The evaluation of the in vitro cytotoxicity (IC_{50}, half maximal inhibitory concentration) of the abovementioned compounds and their structural derivatives was assessed using the standard colorimetric 3-(4,5-dimethylthiazol-2-yl)-2,5-diphenyltetrazolium bromide (MTT) assay on two monolayer tumour cell lines, namely HT-1080 (human connective tissue fibrosarcoma) and MH-22A (mouse hepatocarcinoma). Additionally, the impact of the compound on non-cancerous "normal" cells—mouse fibroblasts (NIH3T3)—was estimated in order to measure the basal toxicity (LD_{50}, median lethal dose). The obtained results were used for investigation of the structure–activity relationships and exploration of the effect of substituents on cellular response and toxicity. The results are presented in Table 2. The estimated LD_{50} gives information regarding the compound properties and could potentially be used to select promising compounds by providing a starting dose for in vivo experiments.

Table 2. Cytotoxic effect of the compounds (IC_{50}, μg/mL) after 72 h treatment of HT-1080 (human connective tissue fibrosarcoma) and MH-22A (mouse hepatocarcinoma) cell lines and basal cytotoxicity (LD_{50}) towards the NIH3T3 (mouse fibroblast) cell line.

Entry	Compound	Ar	R	n	IC_{50} (μg/mL)		Basal Cytotoxicity (NIH3T3) LD_{50} mg/kg
					HT-1080	MH-22A	
1	**6a**	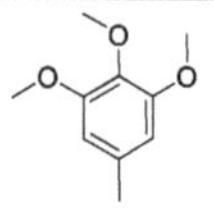	$C_{12}H_{25}$	1	0.22 ± 0.06	0.45 ± 0.08	249 ± 3
2	**6b**		$C_{16}H_{33}$	1	0.18 ± 0.02	1.08 ± 0.06	91 ± 6
3	**6c**	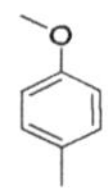	$C_{16}H_{33}$	2	0.20 ± 0.04	0.13 ± 0.03	85 ± 7
4	**6d**	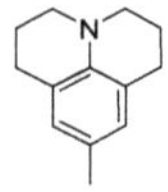	$C_{12}H_{25}$	1	1.4 ± 0.3	0.050 ± 0.003	169 ± 18
5	**6e**		$C_{16}H_{33}$	2	1.0 ± 0.03	0.40 ± 0.06	477 ± 11
6	**6f (DSP-6)** *	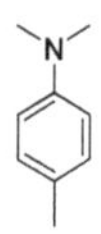	C_6H_{13}	1	0.62 ± 0.06	0.062 ± 0.008	413 ± 25
7	**6g (DSP-12)** *		$C_{12}H_{25}$	1	0.40 ± 0.1	0.50 ± 0.1	203 ± 32
8	**7a**	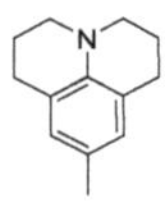	-	1	2.0 ± 0.3	4.0 ± 0.3	1102 ± 20
9	**7b** *		-	2	6.0 ± 0.2	5.0 ± 0.6	857 ± 13

* The counter ion is tosylate.

We determined the cytotoxicity of the styrylpyridinium compounds SP **6** and *bis*-SP **7** (Table 2). The characteristics of structure–property relationships were investigated by analysing the cytotoxicity towards cancerous cell lines (HT-1080, MH-22A) and basal toxicity (LD_{50}) for the non-cancerous NIH3T3 cell line.

2.2.1. The Effect of Alkyl Chain Length

The obtained data reveal that the compounds with long alkyl chains ($C_{12}H_{25}$ or $C_{16}H_{33}$) at the pyridinium nitrogen atom, in general, have high basal toxicity: from 249 mg/kg in the case of 4-(3,4,5-trimethoxystyryl)-*N*-dodecylpyridinium bromide (**6a**) (Table 2, **entry 1**), up to 85 mg/kg in the case of 4-methoxystyrylvinyl-*N*-hexadecylpyridinium bromide (**6c**) (Table 2, **entry 3**). As an exception, the LD_{50} = 477 mg/kg for 4-(julolidinyl-buta-1,3-dienyl-1)-*N*-hexadecyl pyridinium bromide (**6e**) (Table 2, **entry 5**) represented a medium LD_{50} value.

In the case of identical compounds with different lengths of alkyl chains (**6a** and **6b**), a longer alkyl chain increased the basal cytotoxicity from 249 to 91 mg/kg. Similar findings were observed in the case of **6f** and **6g**: the basal cytotoxicity increased from 413 to 203 mg/kg. Only in the case of **6d** and **6e** did a longer alkyl chain (in combination with divinyl linker) significantly decrease the basal cytotoxicity (from 169 to 477 mg/kg).

On the other hand, compounds possessing two styrylpyridinium or styrylvinylpyridinium (phenyldivinylpyridinium) moieties attached to a propane linker in positions 1 and 3 had significantly lower basal toxicity, 1102 mg/kg and 857 mg/kg, respectively (see compounds **7a** and **7b** (Table 2, **entries 8 and 9**)). These compounds can be regarded as geminal derivatives of **6d** and **6e**. Interestingly, the geminal compounds **7a** and **7b** were less toxic than **6d** and **6e** for all of the tested cell lines—both non-cancerous and cancerous ones.

2.2.2. The Effect of the Electron-Donating (Ar) Group

For selected compounds, **6a–6f**, we performed a determination of the cytotoxicity in a wider variety of cancer and non-cancerous cell lines (see Table S2, Supplementary Materials) in order to learn the characteristic features of selectivity or the similarity of biological properties.

Compounds **6e** and **6c** are structurally identical except for the presence of different electron-donating groups. Replacing the julolidinyl group in compound **6e** with 4-metoxyphenyl moiety (compound **6c**) led to increased cytotoxicity and basal toxicity (Table 2, **entry 3**). The IC_{50} values decreased from 1 μg/mL to 0.2 μg/mL for the HT-1080 cell line and from 0.4 μg/mL to 0.13 μg/mL for the MH-22A cell line. The LD_{50} value decreased from 477 mg/kg to 85 mg/kg. Moreover, the impact was consistently observed across all tested cell lines, as outlined in the supplementary materials (Table S2, Supplementary Materials).

The insertion of a 3,4,5-trimethoxyphenyl moiety in compound **6a** instead of the julolidinyl derivative in compound **6d** moderately decreased the cytotoxicity. Comparing identical compounds (**6a** and **6d**) with different electron-donating groups, the cytotoxicity was lower for compound **6a**: 12.5 times lower for cell line BHK-21, 21.9 times lower for cell line NIH3T3, and 8.5 times lower for cell line MH-22A. Furthermore, the basal toxicity decreased by 1.5 times (see Table S2, Supplementary Materials).

Nevertheless, among all the tested cell lines, compound **6e** (with a julolidinyl electron-donating group) demonstrated the least toxicity when compared with the others.

2.2.3. The Effect on Different Cell Lines

Interestingly, no significant similarities were found between the observed toxicities of the compounds on different cell lines. For example, the cytotoxicity of compound **6d** in the MH-22A cell line was very high (0.05 μg/mL) compared with the cell line HT-1080 (1.4 μg/mL). Additionally, compounds **6a** and **6b** had opposite comparative cytotoxicities in MH-22A (**6a** was 2.4 times more toxic) and basal toxicities (compound **6b** was 2.8 times more toxic). These findings suggest that the cytotoxic effects of the compounds are influenced by factors specific to each cell line rather than because of a common biological pattern. Thus, the cytotoxicity of these compounds cannot be easily predicted or generalised. Instead of cytotoxic promiscuity, there seems to be a level of specificity, indicating that the compounds may interact with distinct cellular targets or pathways in a cell-line-dependent manner. Similar observations have been previously demonstrated in the literature [41–43].

Certain compounds exhibited enhanced cytotoxicity against cancer cells. For instance, compound **6d** had 6.5 times higher basal cytotoxicity than compound **7a** and 80 times higher cytotoxicity towards the cell line MH-22A than compound **7a** (Table 2, **entry 4**). Therefore, the SP compound **6d** could represent a potential anticancer drug as it possesses cancer-cell-specific toxicity. Similarly, it was found that compound **6e** has high and comparatively selective cytotoxicity towards the MH-22A (mouse hepatocarcinoma) cell line: IC_{50} = 0.4 μg/mL. This is 3.75 times more toxic than against the non-cancerous Hek-293 (human embryonic kidney cells) cell line and eight times more cytotoxic than against the non-cancerous NIH3T3 cell line. If we compare identical compounds with varying lengths of the alkyl chain, i.e., **6d** and **6e**, the basal cytotoxicity was ~2.8 times lower in the case of **6e**, whereas the cytotoxicity towards the HT-1080 cell line was 1.4 times higher.

The obtained data support the selection of specific compounds for further in-depth studies. Practically, compounds **6e**, **7a**, and **7b** possesses the lowest cytotoxicity towards non-cancerous cell lines, making them attractive candidates for further investigation or development. Remarkably, the SP compound **6e** has moderate cytotoxicity towards non-cancerous cell lines and more expressive cytotoxicity towards several cancer cell lines (e.g., MH-22A).

2.3. Fluorescence

The spectroscopic properties of compounds **6a–e** and **7a**,**b** were studied, and the results are presented in Figure 2 and Table 3. The resonance structures of compounds **6d**,**e** (Figure 2a) show a possibly intramolecular charge transfer between the julolidine moiety and the pyridinium unit. The normalised fluorescence spectra of the tested compounds **6a–e** and **7a**,**b** in DMSO solution (Figure 2c) and the fluorescence spectra of compound **6e** in various media (Figure 2d) were measured using a spectrophotometer multiplate reader (Infinite M1000). The UV/Vis spectra in DMSO solution of compounds **6a–g** (Figure 2b) were reordered with a UV/Vis spectrophotometer, and the fluorescence spectra of compound **6e** in EtOH–glycerol mixtures (Figure 2e) were collected with a Spectrofluorometer FS5.

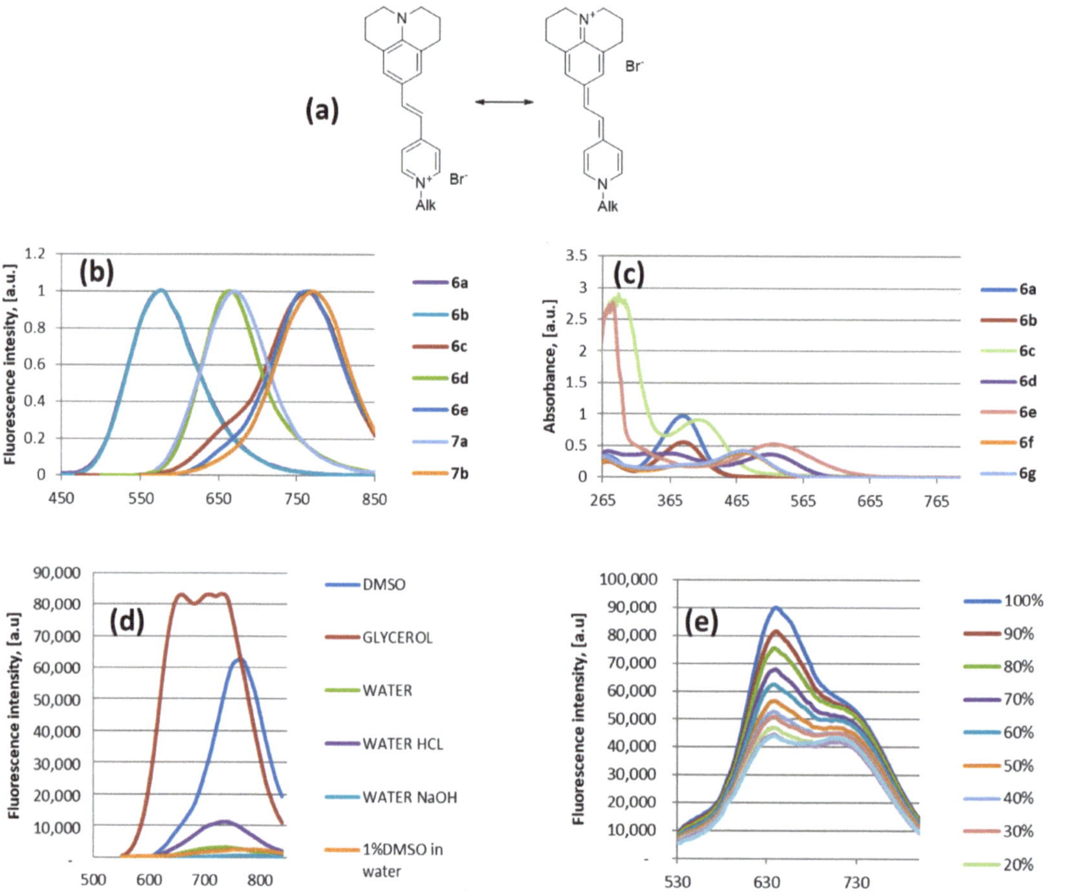

Figure 2. The spectroscopic properties of compounds **6a–e** and **7a**,**b**: (**a**) resonance structures of compounds **6d**,**e**; (**b**) normalised fluorescence spectra of compounds **6a–e** and **7a**,**b** in DMSO solution; (**c**) UV/Vis spectra of compounds **6a–g** in DMSO, the concentration for compounds **6a**,**b**,**d**,**f**,**g** is 0.33 mM and for compounds **6c**,**e** is 5 mM; (**d**) fluorescence spectra of compound **6e** in various media: DMSO (0.5 mM); glycerol (0.01 mM); water (0.5 mM); 0.1 M HCl (0.5 mM); 0.1 M NaOH (0.5 mM); and 1% DMSO/99% water (5 mM); (**e**) fluorescence spectra of compound **6e** (5 mM) in EtOH–glycerol mixtures (0%—pure EtOH:100%—pure glycerol).

Table 3. Spectral properties of compounds (fluorescence maximum (nm) in DMSO solution and excitation maximum (nm)) and UV data (extinction coefficient (log ε) and absorption maximum (nm)). The concentrations of compounds **6a**, **b**, **d**, **f**, and **g** were 0.33 mM and of compounds **6c** and **e** were 5 mM for UV measurements. Values of Stokes' shift (nm) and photoluminescence quantum yields (PLQY) (%).

Entry	Compound	6a	6b	6c	6d	6e	6f	6g	7a	7b
1	fluor. max. λ, nm (excitation. λ, nm)	580 (380)	580 (380)	760 (530)	660 (530)	760 (530)	n	n	660 (530)	770 (530)
2	log ε (abs. max., λ, nm)	5.46 (384)	5.23 (385)	4.26 (408)	5.06 (365) 5.05 (517)	4.02 (520)	5.09 (476)	5.10 (476)	n	n
3	Stokes' shift, nm	196	195	352	325	240	n	n	n	n
4	PLQY, %	15.1	12.2	0.1	0.5	0.7	1.6	1.4	n	n

n—not measured.

It was found that the maximum fluorescence intensity was observed when the samples were excited at 380 nm for compounds **6a**,**b** and at 530 nm for compounds **6c**–**e** and **7a**,**b**. Compounds **6a**,**b**, with 3,4,5-trimetoxyphenyl moieties and vinyl linkers, had a fluorescence maximum at 580 nm, whereas SP **6c**, with a 4-metoxyphenyl moiety and a divinyl linker had a fluorescence maximum at 760 nm. Compounds with julolidinyl groups and vinyl linkers (SP **6d** and *bis*-SP **7a**) showed a fluorescence maximum at 660 nm, whereas compounds with a divinyl linker (SP **6e** and *bis*-SP **7b**) showed fluorescence maximums at 760 nm and 770 nm, respectively (Figure 2b, Table 3, **entry 1**). The theoretical calculations using the TDDFT method (the details are given in Supplementary Materials) revealed that the excitation and emission mechanism for SP compound **6e** follows the mechanism proposed by Jablonski [44]. According to the calculations, the excitation occurs at 542 nm and the emission at 765 nm. The obtained theoretical calculation data are in good agreement with the experimental data, where the excitation occurs at 530 nm and the emission at 760 nm (Table 3).

We have also tested whether the characteristics of compound **6e** depended upon the viscosity of its surrounding environment. The fluorescence intensity dramatically increased in viscous media such as glycerol, but it was lower in water media and basic or acidic media, as well as in 1% DMSO solution in water as a mimic of in vivo conditions (Figure 2d). It was underlined that water media reduced the fluorescence quantum yields and the lifetime of organic fluorophores by up to threefold [45]. We also demonstrated fluorescence intensity changes for SP compound **6e** with different ethanol/glycerol ratios to show the viscosity-dependent results (Figure 2e). In pure glycerol, the maximum value of fluorescence was detected at 645 nm. As the amount of glycerol decreased, the fluorescence intensity of the sample decreased as well. The second fluorescence maximum at 720 nm was observed when the amount of ethanol had reached approximately 50% in the composition.

The spectral properties (absorption and fluorescence maximum in DMSO solution) of compounds **6a**–**e** and **7a**,**b**, UV spectral data (extinction coefficient (log ε) and absorption maximum), values of Stokes' shift, and fluorescence quantum yield are presented in Table 3.

The studied compounds have large Stokes' shift values (Table 3, **entry 3**)—around 195 nm for compounds **6a** and **b**; 240 nm for compound **6e**; 325 and 352 nm for compounds **6d** and **6c**, respectively. The revealed properties of the reported fluorescent compounds are in line with the results of recent articles on NIR styryl dyes [46], where it is indicated that very large Stokes' shifts are crucial in bioimaging applications. Synthesis and studies of styryl dyes are continuing with the purpose of searching for efficient plasma membrane probes [47], highly selective fluorescent probes for lysosome visualisation in live cells [48], probes to visualise the nucleus in live cells [49], and probes for mitochondria imaging in live cells [50].

Photoluminescence quantum yield (PLQY) values for the DMSO solution of compounds **6a–g** were measured (Table 3, **entry 4**). The highest PLQY values were observed for the SP compounds **6a,b** (each with a 3,4,5-trimetoxyphenyl moiety), which were 15.1% and 12.2%, respectively. The PLQY values for the SP derivatives **6f,g** (each with an *N*-dimethylaminophenyl moiety) were 1.6% and 1.4%, respectively. The PLQY value for the SP derivative **6c** (with a 4-dimethylaminophenyl moiety) was only 0.1%. The PLQY values for the SP derivatives **6d,e** (each with a julolidinyl moiety) were 0.5% and 0.7%, respectively. Previously, we designed naphthalene-2-yl)vinyl)pyridinium groups containing 1,4-dihydropyridine amphiphile, for which the PLQY value for the EtOH solution was 1% [20].

Thus, after analysis of the data regarding absorption and fluorescence maximum for the compounds, we proposed that the SP compounds **6c,e** and *bis*-SP **7b** have a useful pleiotropic combination of putative diagnostic and chemotherapeutic, i.e., theranostic, properties.

2.4. Self-Assembling Properties

Compounds with a cationic moiety and long alkyl chains in the molecule could possess self-assembling properties. The estimation of self-assembling properties for the selected compounds was performed using dynamic light scattering (DLS) measurements. The values for the polydispersity index (PDI) and the hydrodynamic average diameter (D_{av}) of nanoparticles formed by compounds **6a,c–e,g** were determined (with a final compound concentration of 0.1 mM in the aqueous medium). The nanoparticles were prepared by the ethanol injection method via the dissolution of the compound in ethanol and after that by the dispersion of the solution into water. The approach is rapid, simple, and can be easily scaled up [51]. This method was previously used successfully for the preparation and characterisation of lipid-like pyridinium amphiphiles [20,52,53]. The obtained results are summarised in Table 4. The DLS measurements were performed for freshly prepared samples at room temperature.

Table 4. Values for polydispersity index (PDI) and average diameter (D_{av}) of formed nanoparticles; the final concentration of samples was 0.1 mM.

Compound	6a	6c	6d	6e	6g (DSP-12)
D_{av}, nm	597 ± 115	212 ± 4	349 ± 4	118 ± 4	207 ± 6
PDI	0.331 ± 0.261	0.218 ± 0.026	0.286 ± 0.030	0.260 ± 0.011	0.254 ± 0.008

The obtained data demonstrate that all of the samples were sufficiently homogeneous, with PDI values in the range 0.218 to 0.331. Compound **6a** formed larger particles with an average diameter of around 600 nm, whereas for compound **6d** the particle diameter was around 350 nm. The other tested compounds, **6c**, **e**, and **g**, formed smaller nanoparticles with average diameters of 212, 118, and 207 nm, respectively. In the case of *bis*-SP **7**, we obtained unanalysable samples, probably due to the insufficient lipophilic characteristics of the compound molecule.

We analysed the obtained data regarding compound cytotoxicity, fluorescence, and self-assembly properties and selected compound **6e** for the detailed cell labelling studies. Compound **6e** possessed the lowest cytotoxicity towards non-cancerous cell lines and more expressive cytotoxicity towards several cancer cell lines, demonstrated a large Stokes' shift, and had self-assembling properties. By combining therapeutic and diagnostic functionalities, this dual-purpose compound demonstrates its theranostic capabilities. Theranostic approaches allow for precise and targeted therapy based on the specific characteristics of a patient's disease, while also enabling the ability to monitor the response to treatment and make adjustments if needed [54–58].

*2.5. SP Compound **6e** for Cell Staining*

SP compound **6e** was selected for fluorescent cell imaging analysis. First, the compound was used to label the BHK-21 cells that are widely used in molecular biology studies [59–61]. Commercially available CellTracker™ Deep Red dye (Invitrogen) was used as a control because it has a similar emission range (ex/em: 630/660). The BHK-21 cells were incubated with fluorescent dyes to stain the cells, then seeded on a 24-well plate and cultured for 24 h. SP compound **6e** was used at 25 µM concentration, providing a non-toxic effect on BHK-21 cells (Table S2, **entry 5**, Supplementary Materials). The Deep Red dye was used according to the manufacturer's instructions.

The day after staining, the cells were collected and washed several times with PBS buffer in order to evaluate the stability of the staining. The fluorescent signal was analysed by flow cytometry (Figure 3). Both compound **6e** and the Deep Red dye exhibited relatively high stability, as the fluorescent signals of both dyes remained unchanged after cell washing. However, only (61 ± 3)% of Deep Red stained cells were positive, whereas SP compound **6e** ensured 99.9% positive cells. Moreover, the signal intensity (fluorescence shift from unstained cells) of cells stained with compound **6e** was remarkably higher than for the Deep Red dye. The calculated stain index for SP compound **6e** was (92 ± 19) before cell washing and (77 ± 15) after cell washing. However, in the case of Deep Red, the calculated stain index was (2.27 ± 0.07) before cell washing and (2.21 ± 0.07) after cell washing. High intensity usually facilitates flow cytometry data acquisition by enhancing the signal-to-background ratio and enabling optimal compensation in complex multicolour fluorescence analysis.

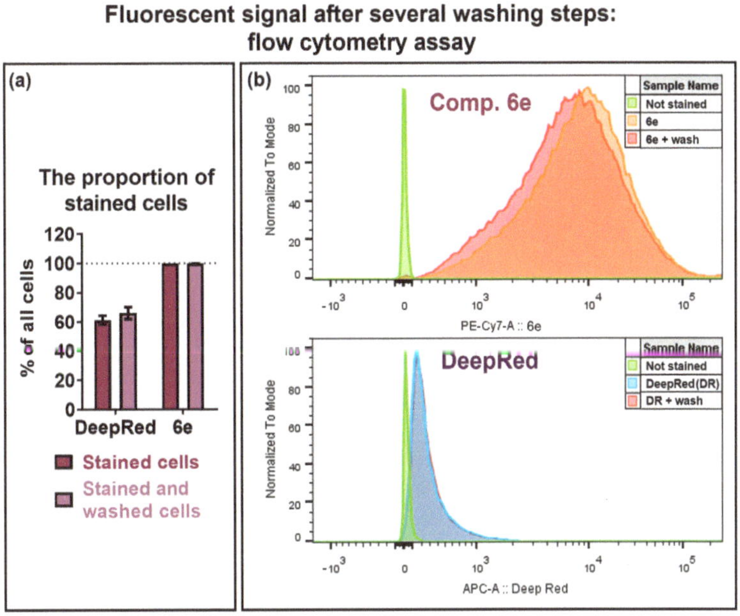

Figure 3. Fluorescence and stability of the fluorescent signal in BHK-21 cells stained with SP compound **6e** in comparison with the commercially available dye analogue CellTracker™ Deep Red (Cat. No. C34565; Invitrogen, Waltham, MA, USA). BHK-21 cells were treated with 1 µM Deep Red or 25 µM compound **6e** for 15 min. The cells were washed and cultured for 24 h in a fresh medium, after which the cells were collected and washed again 3 times with PBS (+wash) or analysed without washing. The number of fluorescent cells was quantified by flow cytometry. Unstained BHK-21 cells were used as a control. (**a**) The proportion of stained cells. Data are presented as mean ± SEM (n = 3). (**b**) Fluorescence intensity is presented as histograms of stained cells in respective channels.

In order to assess the stability of staining after prolonged cell culture, the stained BHK-21 cells (Day 1) were passaged twice (Days 2 and 4) and cultured for six days in total (Day 6, Figure 4). On day 6, the cells were collected and analysed by flow cytometry. The number of positive fluorescent cells decreased significantly: no positive cells were detected in the Deep Red sample, whereas (15.0 ± 0.7)% were detected in the SP-compound-**6e**-stained sample. The cells stained with SP compound **6e** were visualised by fluorescent microscopy, which showed the decrease in the red fluorescence signal on day 6 compared with day 1, demonstrating a low but detectable level for the positive cells (Figure 4). It can be concluded that, in contrast to the Deep Red cell tracker, SP compound **6e** effectively stains cells and can be used for up to 6 days of cell culture.

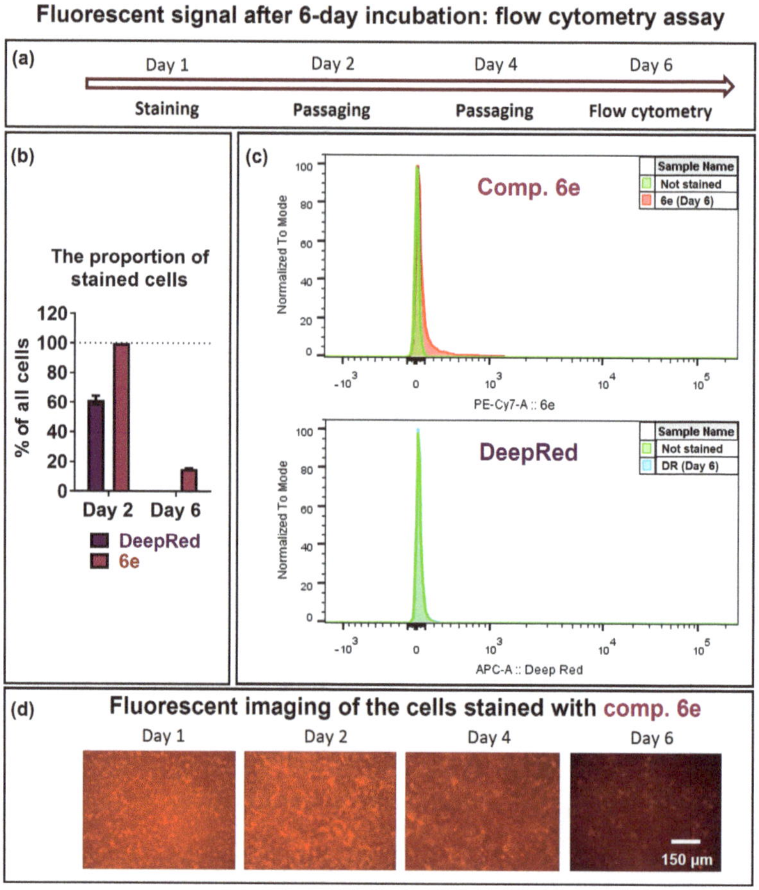

Figure 4. Fluorescence and stability of the fluorescent signal in BHK-21 cells stained with compound **6e** in comparison with the commercially available dye analogue CellTracker™ Deep Red (Cat. No. C34565; Invitrogen). BHK-21 cells were treated with 1 μM Deep Red or 25 μM compound **6e** for 15 min and washed with PBS. Then, the cells were cultured in a fresh media for six days (passaged twice), collected on day 6, and the % of fluorescent cells was measured by flow cytometry. (**a**) The timeline of the experimental design. (**b**) The proportion of stained cells. Data are presented as mean ± SEM (n = 3). (**c**) Fluorescence intensity is presented as histograms of stained cells in respective channels. (**d**) The cell fluorescence microscopy at indicated days after staining is shown for compound **6e**. The fluorescent imaging was performed using a Leica DM-IL fluorescent microscope with an N2.1 filter (excitation: 515–560 nm, emission: 580 nm).

2.6. Staining of Primary Cells

Primary cells, unlike many cancerous cell lines, undergo senescence processes and have limited potential for self-renewal. Primary immune cells are very sensitive to culture conditions and their staining is challenging due to the toxicity of dyes. In this study, we used murine bone-marrow-derived macrophages (BMDMs) to evaluate the staining potential of the 6e compound in primary immune cells. BMDMs represent immune cells with high plasticity that are used in many human disease-related studies, including those involving autoimmune diseases and cancer [62–64]. The first step was to examine whether SP compound **6e** has an impact on cell viability. Bone-marrow-derived macrophages are progenitors that are able to differentiate into pro-inflammatory M1 and anti-inflammatory M2 phenotypes, which ideally fit to demonstrate the functional activity of labelled cells in routine biological studies [65–69]. To assess toxicity, the BMDMs were stained with various concentrations (0.1–100 μM) of compound **6e**, and after 24 h, cell viability was measured using the CCK-8 assay (Figure 5). The calculated IC_{50} value was found to be (9.2 ± 1.5) μM; however, according to the fluorescence microscopy images, the staining was sufficient at a 5 μM concentration, corresponding to the rather high proportion (79.1 ± 1.2)% of viable cells.

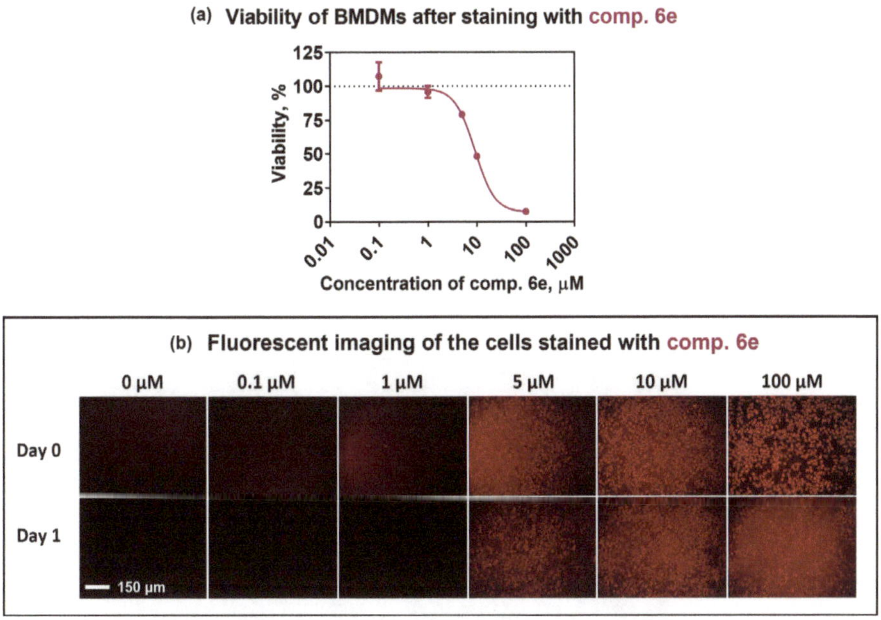

Figure 5. CCK-8 viability assay and fluorescence microscopy of BMDMs stained with SP compound **6e** at different concentrations. BMDMs were incubated in a medium supplemented with compound **6e** for 15 min at 37 °C, after which they washed with PBS and incubated overnight in a fresh media. (**a**) Viability of the BMDMs stained with compound **6e** at different concentrations. Data are presented as mean ± SEM (n = 3). (**b**) Fluorescent microscopy pictures were captured using a Leica DM-IL fluorescent microscope with an N2.1 filter (excitation: 515–560 nm, emission: 580 nm) immediately after staining (Day 0) and after the overnight incubation (Day 1).

The stained BMDMs were studied using confocal microscopy to verify the membrane, cytoplasmic, or nuclear localisation of the staining (Figure 6). Wheat germ agglutinin (WGA) Alexa Fluor 488 (Invitrogen, W11261) was chosen for selective plasma membrane labelling and DAPI (Invitrogen, D1306) was selected for nucleus labelling. It was found that the dye **6e** was localised in the cell cytoplasm and does not stain the cell nucleus (Figure 6a).

Furthermore, the SP compound **6e** was compared with commercially available CM-DiI Dye (Invitrogen, ex/em: 553/570, Figure 6b).

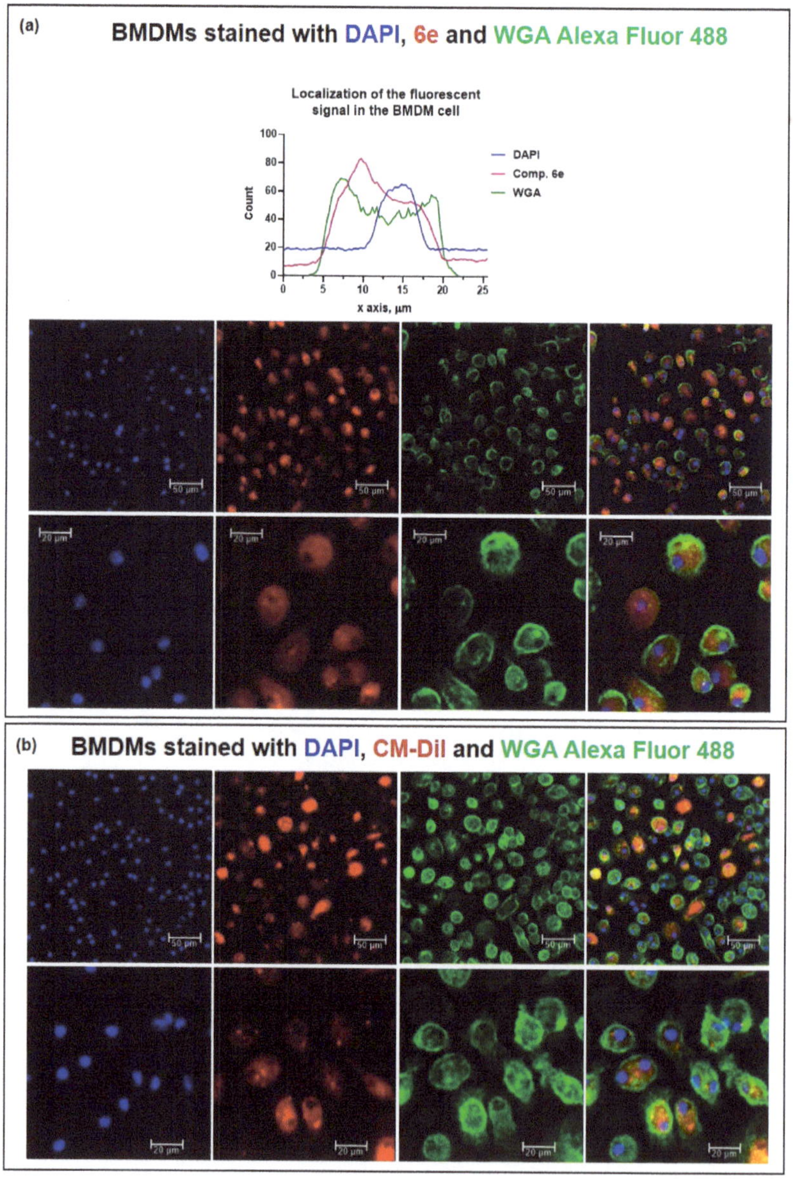

Figure 6. The stained BMDMs were studied using confocal laser scanning microscopy with a Leica TCS SP8. A DPSS (561 nm) laser was used for the excitation of compound **6e** and CM-Dil, an argon (488 nm) laser for WGA, and a diode (405 nm) laser for DAPI. For fluorescence detection, HyD was used for compound **6e** and CM-Dil (578 nm–636 nm) and PMT detectors were used for WGA (493 nm–739 nm) and DAPI (417 nm–498 nm), pinhole at 95 µm. (**a**) Confocal microscopy and localisation of the fluorescent signal of BMDMs stained with DAPI, 5 µM compound **6e**, and 1 µg/mL WGA Alexa Fluor 488. (**b**) Confocal microscopy of BMDMs stained with DAPI, 4 µM CM-Dil, and 1 µg/mL WGA Alexa Fluor 488.

The staining with compound **6e** appeared to be more evenly distributed and the level of noise was lower. These observations suggest that compound **6e** offers advantages in terms of achieving a clearer signal-to-noise ratio and better overall imaging quality. The homogeneous cytoplasmic distribution and lower noise level obtained with compound **6e** staining contribute to improved visualisation and accurate analysis of the sample, enhancing the reliability of data.

The impact of SP compound **6e** on the BMDMs' functional characteristics was also assessed by performing cell differentiation into the M1 phenotype in the presence of compound **6e**. The BMDMs were treated with 5–50 μM compound **6e** and the cell medium was replaced with a macrophage differentiation medium containing recombinant mouse interferon-gamma (IFNγ) and the TLR2/1 ligand PAM3SCK4 to ensure M1 polarisation [37,65]. Activated cells express proinflammatory molecules (cytokines and chemokines), as well as nitric oxide (NO) as the main features of their proinflammatory M1-like macrophage phenotype. To verify the NO level in cell culture media, nitrites, as nitric oxide oxidation products, were measured by the standard Griess test. The nitric oxide assay demonstrated that BMDMs stained with compound **6e** displayed strong NO production after treatment with IFNγ and PAM3SCK4 (Figure 7a). Although, the maximal NO production was observed for unstained cells (30 ± 2 μM), the stained BMDMs also revealed relatively high NO production (up to 21.5 μM). However, as the amount of compound 6e increased, the NO signal gradually decreased. These results strongly suggest a correlation with a reduction in cell viability, as depicted in Figure 5a. The standard fluorescence microscopy analysis demonstrated a specific and bright red fluorescence signal in stained BMDM cells at less toxic concentrations of compound **6e** (5 μM). We concluded that stained M1 macrophages produce NO; therefore, the pre-stained BMDMs can be used for functional programming, as well as for cell tracking and cell migration assays.

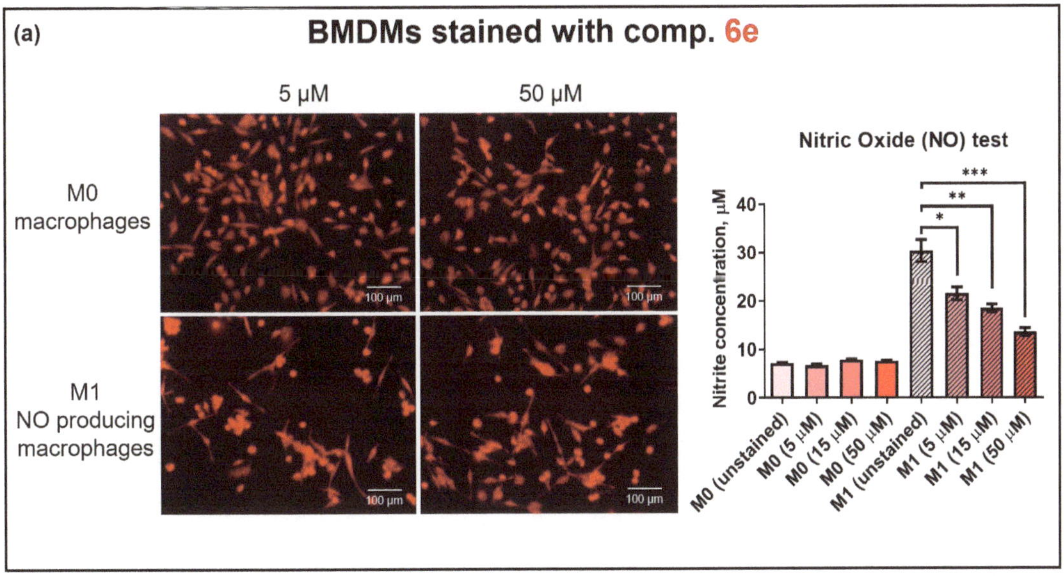

Figure 7. *Cont.*

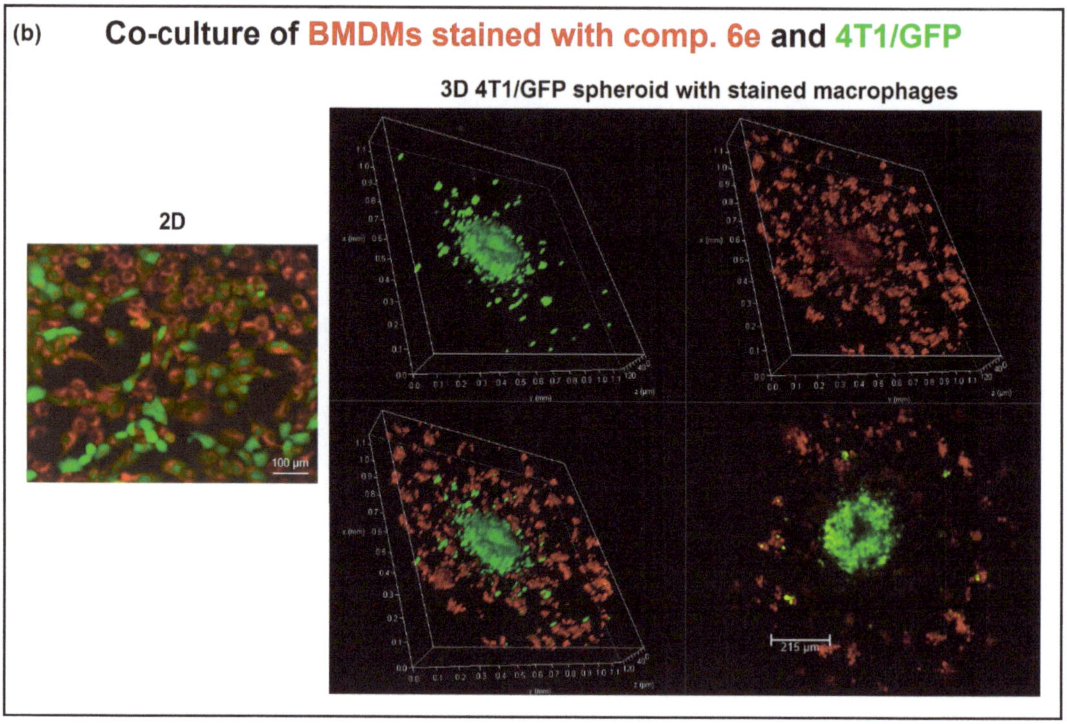

Figure 7. Activation of pre-stained macrophages to the M1 phenotype and co-cultivation with cancer cell spheroids. (**a**) Bone-marrow-derived macrophages (M0) were stained with compound **6e** at various concentrations (5, 15, and 50 μM) and used for fluorescent microscopy imaging. A DPSS (561 nm) laser was used for the excitation of compound **6e** and HyD was used for fluorescence detection. The level of nitric oxide (NO) was determined in cell culture supernatants after 2 days of BMDM activation with IFNγ/Pam3SCK4 ligand. The bars represent the mean values ± SEM. * p = 0.0292; ** p = 0.0083; *** p = 0.0023; n = 3. (**b**) Confocal microscopy of BMDMs stained with 5 μM compound **6e** co-cultivated with 4T1 mammary cancer cells expressing GFP (green): (left) monolayer; (right) 3D culture. A DPSS (561 nm) laser was used for the excitation of compound **6e** and an argon (488 nm) laser for GFP. For fluorescence detection, HyD was used for compound **6e** and a PMT detector was used for GFP (493–560 nm), pinhole at 95 μm.

Since the macrophages have a significant role in cancer progression, the co-cultivation of **6e**-stained BMDMs with murine mammary cancer cells expressing green fluorescent protein (4T1/GFP) was tested. In summary, the experiment involved staining BMDMs with compound **6e** and introducing them to 4T1/GFP cells cultured in a monolayer. The staining facilitated cell identification in co-culture (Figure 7b). Next, a cancer cell spheroid was formed by incubating 4T1/GFP cells in an ultra-low attachment plate. The pre-stained BMDMs were then added to the cancer cell spheroid, and the sample was analysed using confocal microscopy. The confocal microscopy revealed interactions between the cancer cell spheroid and macrophages. Remarkably, macrophage-induced cancer cell migration out of the spheroid could be detected by the co-localisation of BMDMs and cancer cells.

By employing pre-stained BMDMs in conjunction with cancer cell spheroids, researchers can gain valuable insights into the interactions between cancer cells and macrophages, shedding light on the processes of cell migration and phagocytosis in the context of cancer progression. The confocal imaging data clearly show the high potential of the **6e** probe for use in the imaging of co-cultures (BMDMs + 4T1/GFP) and the investigation of cell–cell interactions in both 2D and 3D cancer cell models [37,70] (Figure 7b).

3. Materials and Methods

3.1. Chemistry

More detailed descriptions of the synthetic procedures and characterisation of the original unpublished intermediates and compounds are described in the Supplementary Materials.

3.2. Cell Culture and Measurement of Cell Viability

The HT-1080 (human connective tissue fibrosarcoma, ATCC® CCL-121™), MH-22A (mouse hepatocarcinoma, ECACC Cat. No. 96121721), A-549 (human lung carcinoma, ATCC® CCL-185™), NIH3T3 (Swiss Albino mouse embryo fibroblast, ECACC Cat. No. 93061524), Hek293 (human embryonic kidney, ATCC® CRL-1573™), BHK-21 (Syrian hamster kidney, ATCC® CCL-10™), A-431 (human epidermoid carcinoma, ATCC® CRL-1555™), Saos-2 (human osteosarcoma, ATCC® HTB-85™), HeLa (human epitheloid cervix carcinoma, ATCC® CCL-2™), U937 (human myeloid leukaemia, ATCC® CRL-1593.2™), C6 (rat glioma, ATCC® CCL-107™), and B16-F10 (murine melanoma, ATCC® CRL-6475™) cell lines were obtained from the American Type Culture Collection (ATCC) and European Collection of Authenticated Cell Cultures (ECACC).

HT-1080, MH-22A, A-549, NIH3T3, BHK-21, A-431, HeLa, and B16-F10 cells were cultured in Dulbecco's Modified Eagle's Medium (DMEM) with 2 mM glutamine and 10% foetal bovine serum (FBS). Hek293 cells were cultured in Eagle's Minimum Essential Medium with 10% foetal bovine serum. U937 cells were cultured in RPMI-1640 Medium with 2 mM glutamine and 10% foetal bovine serum. C6 cells were cultured in Ham's F12 Medium with 2 mM glutamine and 10% foetal bovine serum. Saos-2 cells were cultured in McCoy's 5a Medium with 2 mM glutamine and 10% foetal bovine serum. The cells were incubated at 37 °C with 5% CO_2, 95% air, and complete humidity. Once ~90% confluence was reached, they were detached using 0.05% trypsin/EDTA and counted. The cells were plated at optimally determined densities for the logarithmic phase of growth. These cells were then suspended at a concentration of 4000 cells/well for HT-1080, MH-22A, A-549, NIH3T3, BHK-21, A-431, HeLa, B16-F10, Hek293, and Saos-2 cells and a concentration of 9000 cells/well for U937, C6, Hek293, and GM08402 cells and added onto a 96-well plate. For background absorption, some wells remained cell free, i.e., as a blank control. Cells were incubated for 72 h at 37 °C and 5% CO_2 where they were exposed to a media containing serial dilutions of the investigated compounds (n = 3). Cell viability was measured using the 3-(4,5-dimethylthiazol-2-yl)-2,5-diphenyl-tetrazolium bromide (MTT) assay [71]. In brief, after incubating with compounds, the culture medium was removed and fresh medium with 0.2 mg/mL MTT was added to each well of the plate. After incubation (3 h, 37 °C, 5% CO_2), the medium with MTT solution was removed and replaced with 200 µL DMSO and 25 µL Sorenson's glycine buffer (glycine 0.1 M, NaCl 0.1 M, pH: 10.5 with 0.1 M NaOH). The plate was shaken for 15 min at room temperature and the optical density (OD) of the wells was determined using a multichannel spectrophotometer (Thermo Scientific Multiskan EX) at 540 nm. The quantity of the control cells was taken in calculations as 100%. The IC_{50} value was calculated using the program Graph Pad Prism® 3.0.

3.3. Evaluation of Basal Cytotoxicity

To assess basal cytotoxicity, the neutral red uptake (NRU) assay was conducted following a modified version of the standard protocol of Stokes' [71], as validated by the NICEATM-ECVAM study [72]. The NRU cytotoxicity assay determines the ability of viable cells to incorporate and bind a supravital dye, neutral red (NR).

NIH3T3 cells (9000 cells/well) were cultured in 96-well plates in DMEM medium supplemented with 5% FBS for 24 h. Subsequently, the cells were exposed to the test compounds over a range of seven concentrations (1000 µg/mL, 316 µg/mL, 100 µg/mL, 31 µg/mL, 10 µg/mL, 3.0 µg/mL, and 1.0 µg/mL) for 24 h. Untreated cells were used as a control. After the incubation period, the medium was removed from all plates and a solution of NR was added (250 µL, 0.05 mg/mL NR in DMEM 24 h pre-incubated at 37 °C and then filtered before use through a 0.22 μm syringe filter). The plates were incubated for 3 h, followed by three washes with phosphate-buffered saline (PBS). The cells were then treated with a mixture of ethanol (50%) and water (49%) supplemented with 1% acetic acid. The absorbance of NR was measured using a spectrophotometer multiplate reader (Infinite M1000; Tecan Austria GmbH, Salzburg, Austria) at 540 nm. The optical density (OD) was then calculated using the following formula:

$$\frac{OD_{\text{treated cells}} \times 100}{OD_{\text{control cells}}}.$$

The IC_{50} values were calculated using the program Graph Pad Prism® 3.0.

3.4. Estimation of LD_{50} from IC_{50} Values

In vitro test results were employed to estimate the starting dose for acute oral systemic toxicity tests in rodents (LD_{50}). The LD_{50} value [29,73] was calculated using the in vitro IC_{50} value and a following formula:

$$\log LD_{50}(\text{mM/kg}) = 0.439 \times \log IC_{50}(\text{mM}) + 0.621$$

The value was then converted to mg/kg and the compounds classified into four toxicity categories [74]. Category 1 (highly toxic): $LD_{50} \leq 5$ mg/kg; Category 2 (moderately toxic): $5 < LD_{50} \leq 50$ mg/kg; Category 3 (slightly toxic): $50 < LD_{50} \leq 300$ mg/kg; Category 4 (practically non-toxic): $300 < LD_{50} \leq 2000$ mg/kg. By employing an alternative in vitro method, it becomes possible to compare the potential toxicity of new compounds and select candidates for further study, significantly reducing the number of animal experiments.

3.5. Excitation/Emission Fluorescence Spectra Measurements

A concentration of 10 mM stock solution of compounds **6** or **7** was prepared by dissolving the compounds in DMSO. Working solutions were prepared by diluting the 10 mM stock solution of compounds with different solvents to reach the final concentration for the samples: DMSO (compound concentration 0.1 mM); for compound **6e**: in glycerol (concentration 0.01 mM), in water (concentration 0.5 mM), in 0.1 M HCl (concentration 0.5 mM), in 0.1 M NaOH (concentration 0.5 mM), or in 1% DMSO/99% water (concentration 5 mM). The fluorescence spectra were measured using a spectrophotometer multiplate reader (Tecan Infinite M1000; Tecan Austria GmbH, Salzburg, Austria). For excitation/emission fluorescence spectra measurements, the following measurement parameters were applied: excitation wavelength from 260 nm to 690 nm with a 10 nm sampling interval and emission wavelength from 300 nm to 840 nm with a 10 nm sampling interval. The fluorescence spectra of compound **6e** (concentration 5 mM) in EtOH–glycerol mixtures were measured with an Edinburgh Instruments FS5 Spectrofluorometer (Edinburgh Instruments Ltd., Kirkton Campus, UK). The data were analysed using Microsoft Excel software version 2308.

3.6. Assessment of the Self-Assembling Properties of Compounds 6 and 7 Using Dynamic Light Scattering (DLS) Measurements

The self-assembling properties of the tested compounds were estimated according to a published procedure with minor modifications [52,53]. Stock solutions of the compounds were prepared in EtOH at a concentration of 1 mM. Stock solutions of the compounds (0.3 mL) were injected into 2.7 mL of deionised water with maximum stirring (IKA Vortex 2 (IKA, Staufen, Germany)) to obtain a final concentration of 0.1 mM. The DLS measurements of the nanoparticles in the aqueous solution were conducted using a Zetasizer Nano ZSP (Malvern Panalytical Ltd., Malvern, UK) instrument with Malvern Instruments Ltd. Software 8.01.4906. The measurements were performed with the following settings—medium: water; refractive index: 1.33; viscosity: 0.8872 cP; temperature: 25 °C; dielectric constant: 78.5; nanoparticles: liposomes; refractive index of materials: 1.60; detection angle: 173° wavelength: 633 nm. The obtained data were analysed using the multimodal number distribution software provided with the instrument. To ensure reproducibility, the measurements were performed in triplicate.

3.7. Stability of Dye during Cell Washing Steps

BHK-21 cells were plated onto a 6-well plate and cultured in Glasgow's MEM (Cat. No. 21710025; Gibco, Life Technologies, Carlsbad, CA, USA) supplemented with 5% FBS, 10% tryptose phosphate broth solution (Cat. No. 18050039; Gibco, Life Technologies), 20 mM HEPES (Cat. No. 15630056; Gibco, Life Technologies), 2 mM L-glutamine (Cat. No. 25030024; Gibco, Life Technologies), 50 U/mL penicillin, and 50 mg/mL streptomycin (1% PEST; Cat. No. 15070063; Gibco, Life Technologies).

When cells reached 90% confluency, they were stained with 25 μM compound **6e** or 1 μM CellTracker™ Deep Red dye (Cat. No. C34565; Invitrogen) according to the manufacturer's instructions. Briefly, the 10 mM stock solution of SP compound **6e** was prepared in ethanol and stored at 4 °C in the dark. Then, the dye was dissolved in DMEM (Cat. No. 31966047; Gibco, Life Technologies) and the cells incubated with the dye solution in DMEM for 20 min at 37 °C. Then, the dye solution was replaced with complete medium. A total of 24 h after staining, the cells were collected for flow cytometry by trypsinisation using 0.05% trypsin solution (Cat. No. 15400054; Gibco, Life Technologies) and washed three times with PBS supplemented with 2–10% FBS. Stained cells were kept at 4 °C and analysed using a FACSAria BD Hardware flow cytometer with FACSDiva 6.1.3 software (BD Biosciences, San Jose, CA, USA). A 488 nm blue laser and 780/60 bandpass filter (PE-Cy7) were used to analyse the SP compound **6e** signal; a 633 nm red laser and 660/20 bandpass filter (APC) were used to analyse the Deep Red signal. The data were analysed using FlowJo 10.3 software (FlowJo LLC, Ashland, OR, USA) and presented as the mean of three independent experiments. The stain index was calculated using following equation:

$$\frac{Median_{Positive} - Median_{Negative}}{2 \times SD_{Negative}}$$

where $Median_{Positive}$ is the median intensity value of the positive population, $Median_{Negative}$ is the median intensity value of the negative population, and $SD_{Negative}$ is the standard deviation of the negative population intensity value.

3.8. Cell Passaging Assay

For the cell passaging, on day 0, BHK-21 cells were plated on a 6-well plate with a seeding density of 1.5 × 10^5 cells per well. On day 1, the cells were stained with 25 μM compound **6e** or 1 μM CellTracker™ Deep Red dye as described above. The cells were incubated in a humidified 5% CO_2 incubator at 37 °C and passaged on day 2 and day 4 using 0.05% trypsin solution. The fluorescent imaging was performed using a Leica DM-IL fluorescent microscope with N2.1 filter (excitation: 515–560 nm, emission: 580 nm). On day 6, cells were collected for flow cytometry analysis. The cells were kept at 4 °C and analysed as described above.

3.9. Isolation of Bone-Marrow-Derived Macrophages (BMDMs)

Mouse macrophages were derived from bone marrow cells following an established protocol employing the L929 cell-conditioned medium (CM) as a source of macrophage colony-stimulating factor (M-CSF) [65,66].

In summary, bone marrows were extracted from the femurs and tibiae of female BALB/c mice (8–12 weeks old). The bone marrow cells were flushed with RPMI-1640 medium supplemented with 10% FBS under sterile conditions and filtered through a 70 μm cell strainer (Cat. No. 800070; Bioswisstec, Schaffhausen, Switzerland). ACK lysis buffer (Cat. No. A10492-01; Gibco, Life Technologies) was employed to remove red blood cells via a 10 min incubation at room temperature. The remaining cells were then cultured in 10 cm non-tissue-culture-treated dishes (Cat. No. 0030702018; Eppendorf, Hamburg, Germany) in RPMI-1640 medium containing 30% L929 CM. The isolated bone marrow cells were cultured for 5 days, after which non-adherent cells were washed off with PBS. The adherent BMDMs were cultured for 24 h and then detached by incubation in PBS for 30 min at 4 °C followed by gentle flushing of the dish. The BMDMs were collected by centrifugation, frozen in Recovery™ Cell Culture Freezing Medium (Cat. No. 12648010; Gibco, New York, NY, USA), and stored at −150 °C for future experiments. Afer thawing, BMDMs were incubated for 4 days, detached, filtered through a 70 μM cell strainer, and used for experiments.

The experimental procedure was approved by the Latvian Animal Protection Ethical Committee of Food and Veterinary Service (Permit Nr. 93).

3.10. BMDMs' Viability Assay

To determine the cytotoxicity of the SP compound **6e**, the cells were stained with various concentrations of the dye (0, 0.1, 1, 5, 10, and 100 μM). Briefly, 2×10^4 cells were plated in a 96-well plate in 100 μL of the medium. After 24 h, the cells were washed once and a solution of compound **6e** in PBS was added to the cells. The cells were incubated in a humidified 5% CO_2 incubator at 37 °C for 15–20 min. Then, the dye solution was replaced with medium. The fluorescent imaging was performed using a Leica DM-IL fluorescent microscope with N2.1 filter (excitation: 515–560 nm, emission: 580 nm). After 24 h, the medium was replaced with 10 μL of CCK-8 reagent in 100 μL of fresh medium. The cells were incubated in a humidified 5% CO_2 incubator at 37 °C for 1–4 h. The absorbance at 450 nm was measured using a microplate spectrophotometer (BioTek Instruments, Winooski, VT, USA). Cell viability was calculated as a percentage of non-treated cell absorbance. IC_{50} values were calculated as the concentration at which 50% of the cells were viable.

3.11. Staining of Macrophages with SP Compound **6e**

For confocal microscopy, BMDMs were stained with 4 μM CellTracker™ CM-DiI dye (Cat. No. C7001; Invitrogen) or 5 μM compound **6e**. Briefly, the cells were washed and then a dye solution in PBS was added to them. The cells were incubated in a humidified 5% CO_2 incubator at 37 °C for 15–20 min. Then, the cells were stained with WGA Alexa Fluor 488 (Cat. No. W11261; Invitrogen) following the manufacturer's instructions. Briefly, the cells were incubated with 1 μg/mL WGA solution in PBS at 37 °C. After 10 min, the cells were washed several times, fixed with 3% PFA for 20 min, and stained with DAPI. The samples have been studied under confocal laser scanning microscopy using a Leica TCS SP8 (Leica Microsystems, Wetzlar, Germany, a 63.7 μm pinhole and a 400 Hz scan speed). The fluorescence was detected as follows: a DPSS (561 nm) laser was used for excitation of compound **6e** and CM-DiI, an argon (488 nm) laser was used for WGA, and a diode (405 nm) laser was used for DAPI. For fluorescence detection, HyD was used for compound **6e** and CM-DiI (578 nm–636 nm) and PMT detectors were used for WGA (493 nm–739 nm) and DAPI (417 nm–498 nm), pinhole at 95 μm. The images were analysed using Leica Application Suite X 3.1.5. software (Leica Microsystems, Germany).

3.12. Polarisation of Pre-Stained Macrophages to M1 Phenotype

For staining, the cells were seeded onto a 12-well untreated cell culture plate in complete BMDM cultivation medium (RPMI-1640 with 10% FBS and 30% L929-CM containing M-CSF, 2 mM L-glutamine, 1% penicillin/streptomycin) at a concentration of 4×10^5 cells in 1 mL and cultured for 2 days at 37 °C 5% CO_2. Then, the cell medium was replaced with 1 mL DMEM containing 0.5 μM, 15 μM, or 50 μM compound **6e**. The cells were incubated for 15 min at 37 °C and 5% CO_2, then the labelling solution was removed, the cells washed with 1.5 mL PBS three times, and 1 mL complete BMDM medium was added. The cells were further used for the functional analysis of pre-stained macrophage polarisation to the M1 phenotype.

For polarisation of macrophages to the pro-inflammatory M1 phenotype, the BMDM cells stained in a 12-well plate were used. Unstained BMDM cells were used as a positive control. The BMDM complete medium from stained and unstained cells was replaced with 1 mL BMDM medium (10% L929 CM containing M-CSF) supplemented with 50 ng/mL recombinant mouse IFNγ (Cat. No. BMS326; eBioscience, San Diego, CA, USA) and 100 ng/mL Pam3SCK4 (Cat. No. tlrl-pms; InvivoGen, Hong Kong, China), and the cells were incubated for 2 days at 37 °C and 5% CO_2. The BMDM cells cultured in a complete BMDM medium without the IFNγ and Pam3SCK4 activation ligands were incubated in parallel and used as a negative control (M0). After two days of activation, the nitric oxide levels (as a main signal of M1 polarisation) were measured in cell supernatants of activated (M1) and non-activated (M0) cells. The amount of nitric oxide was determined via a nitric oxide assay kit (Cat. No. EMSNO; Invitrogen). Briefly, 50 µL of cell culture media was collected from each well, clarified by centrifugation, and used for NO quantification. The nitrite standards provided by the kit were used for standard curve generation and NO quantification. The optical density was measured at 540 nm using a BioTek µQuant Microplate spectrophotometer (BioTek instruments, USA). The data are presented as the mean value of two independent experiments, and each measurement was performed in triplicate.

3.13. BMDM Co-Cultivation with Cancer Cells

For 3D cultivation, the BMDMs were detached from a 12-well plate, resuspended in DMEM at concentration 3×10^5 in 1 mL DMEM, and incubated in the presence of 5 µM compound **6e** for 15 min at 37 °C and 5% CO_2. Then, the cells were washed three times with PBS, resuspended in 100 µL of complete BMDM medium, and further cultivated in a 3D ultra-low attachment plate together with 4T1/eGFP cancer spheroids that had been prepared as previously described [37]. Briefly, 4T1/eGFP spheroids were generated from 4T1-Fluc-Neo/eGFP-Puro cells (4T1/eGFP; Imanis Life Sciences, Rochester, MN, USA) by seeding 3 000 cells into 96-well black round-bottom ultralow attachment plates (Cat. No. CLS4515, Corning, Life Sciences) according to the manufacturer's instructions. The next day, the stained macrophages (1.5×10^4 in 100 µL of BMDM medium) were added to the spheroids and the cells cultured for 1–2 days before confocal imaging. Confocal laser scanning microscopy (Leica Microsystems, Germany) was performed as described above: eGFP excitation laser = 488 nm and emission detector = PMT 2 (493–560 nm).

3.14. Statistical Analysis

Data statistical analysis was performed using GraphPad Prism 8.0.2. Data are presented as mean $\pm$ standard error of the mean (SEM). Statistical analysis was performed using a two-tailed *t*-test.

4. Concluding Remarks

In this study, we designed new styrylpyridinium derivatives with varying electron-donating substituents at the phenyl moiety, lengths of alkyl chain at the EWG *N*-alkylpyridinium fragment, and either vinyl or divinyl linkers between them and *bis*-styrylpyridinium derivatives with two styrylpyridinium moieties conjugated through the 1,3-trimethylene unit. A total of nine compounds were synthesised and subsequently evaluated for their spectral, self-assembling, and biological properties.

To reduce phototoxicity and background fluorescence from biomolecules (autofluorescence), a desired fluorescent molecule for bioimaging should be excitable and also emissive in the longer wavelength area, preferentially from the far-red to NIR region (650–950 nm). Currently, NIR-emitting cyanine dyes are widely employed on the market for bioimaging; however, these dyes have disadvantages such as photobleaching, toxicity, and the propensity to produce nonfluorescent aggregates. The new class of fluorescent compounds developed in this study possess relatively high cytotoxicity towards the cancerous cell lines HT-1080 and MH-22A, and at the same time their basal cytotoxicity to the NIH3T3 cell line is evaluated as being from toxic to harmful. The compounds also have intensive fluorescence, especially in lipophylic media. The introduction of a divinyl linker and a powerful electron-donating julolidine moiety into the molecules resulted in intensive fluorescence in the NIR region (maximum at ~760 nm) for compounds **6c**, **e**, and **7b**.

Compound **6e**, comprising a julolidine moiety as electron-donating part, a divinyl linker, and a *N*-hexadecylpyridinium group as the electron-withdrawing part, is proposed as a cell labelling reagent after evaluation using fluorescence microscopy and flow cytometry analysis. Compound **6e** exhibited the remarkable stability and a high stain index even after cell washing, with a stain index of (92 ± 19) before cell washing and (77 ± 15) after cell washing. Furthermore, it was observed that compound **6e** localises within the cell cytoplasm and does not stain the cell nucleus. Based on our findings, it can be inferred that compound **6e** is predominantly taken up by cells through passive transport mechanisms. Amphiphiles, such as SP compound **6e**, exhibit unique properties, allowing them to interact with and stain a diverse range of molecules (both hydrophilic and hydrophobic), making them versatile binding agents. The dye's ability to remain within the cell further confirms its stability and potential suitability for various experimental applications.

Additionally, staining primary BMDMs with compound **6e** was performed. The staining only slightly affected their viability and functional properties. Importantly, throughout the staining process the BMDMs remained functionally active, as demonstrated by the functional assay results showing no significant impact on gene expression, particularly at low concentrations. This finding underscores the reliability and biocompatibility of SP compound **6e** as a staining agent, making it a valuable tool for cellular studies with promising potential in cell biology research and various biomedical applications.

The development of new fluorescent probes may involve the design of theranostic formulations with multiple functions, including targeted drug delivery and molecular tracking properties.

Supplementary Materials: The following supporting information can be downloaded at: https://www.mdpi.com/article/10.3390/ph16091245/s1, Scheme S1. Synthesis of 4-methylpyridinium salts; Scheme S2. Synthesis of acrylaldehydes; Scheme S3. Synthesis of compounds **6** and **7**; Table S1. Substituents of compounds **6** and **7**; methods of synthesis; Table S2. Cytotoxicity (IC_{50}, μg/mL) on 9 cancer and 3 non-cancerous cell lines and calculated toxicity (LD_{50}) of styryl derivatives of **6**; Figure S1. ^{1}H NMR spectrum of 1-dodecyl-4-(3,4,5-trimethoxystyryl)pyridin-1-ium bromide (**6b**); Figure S2. ^{13}C NMR spectrum of 1-dodecyl-4-(3,4,5-trimethoxystyryl)pyridin-1-ium bromide (**6b**); Figure S3. HRMS data of 1-dodecyl-4-(3,4,5-trimethoxystyryl)pyridin-1-ium bromide (**6b**); Figure S4. ^{1}H NMR spectrum of 1-hexadecyl-4-(4-(4-methoxyphenyl)buta-1,3-dien-1-yl)pyridin-1-ium bromide (**6c**); Figure S5. ^{13}C NMR spectrum of 1-hexadecyl-4-(4-(4-methoxyphenyl)buta-1,3-dien-1-yl)pyridin-1-ium bromide (**6c**); Figure S6. HRMS data of 1-hexadecyl-4-(4-(4-methoxyphenyl)buta-1,3-dien-1-yl)pyridin-1-ium bromide (**6c**); Figure S7. ^{1}H NMR spectrum of 1-dodecyl-4-(2-(2,3,6,7-tetrahydro-1H,5H-pyrido[3,2,1-ij]quinolin-9-yl)vinyl)pyridin-1-ium bromide (**6d**); Figure S8. ^{13}C NMR spectrum of 1-dodecyl-4-(2-(2,3,6,7-tetrahydro-1H,5H-pyrido[3,2,1-ij]quinolin-9-yl)vinyl)pyridin-1-ium bromide (**6d**); Figure S9. HRMS data of 1-dodecyl-4-(2-(2,3,6,7-tetrahydro-1H,5H-pyrido[3,2,1-ij]quinolin-9-yl)vinyl)pyridin-1-ium bromide (**6d**); Figure S10. ^{1}H NMR spectrum of 1-hexadecyl-4-(4-(2,3,6,7-tetrahydro-1H,5H-pyrido[3,2,1-ij]quinolin-9-yl)buta-1,3-dien-1-yl)pyridin-1-ium bromide (**6e**); Figure S11. ^{13}C NMR spectrum of 1-hexadecyl-4-(4-(2,3,6,7-tetrahydro-1H,5H-pyrido[3,2,1-ij]quinolin-9-yl)buta-1,3-dien-1-yl)pyridin-1-ium bromide (**6e**); Figure S12. HRMS data of 1-hexadecyl-4-(4-(2,3,6,7-tetrahydro-1H,5H-pyrido[3,2,1-ij]quinolin-9-yl)buta-1,3-dien-1-yl)pyridin-1-ium bromide (**6e**); Figure S13. ^{1}H NMR spectrum of 1,1′-(propane-1,3-diyl)bis(4-(4-(2,3,6,7-tetrahydro-1H,5H-pyrido[3,2,1-ij]quinolin-9-yl)buta-1,3-dien-1-yl)pyridin-1-ium) dibromide (**7b**); Figure S14. ^{13}C NMR spectrum of 1,1′-(propane-1,3-diyl)bis(4-(4-(2,3,6,7-tetrahydro-1H,5H-pyrido[3,2,1-ij]quinolin-9-yl)buta-1,3-dien-1-yl)pyridin-1-ium) dibromide (**7b**); Figure S15. HRMS data of 1,1′-(propane-1,3-diyl)bis(4-(4-(2,3,6,7-tetrahydro-1H,5H-pyrido[3,2,1-ij]quinolin-9-yl)buta-1,3-dien-1-yl)pyridin-1-ium) dibromide (**7b**); Figure S16. Three-dimensional excitation emission fluorescent spectral scan of compound **6e**: (**A**) in DMSO (0.5 mM); (**B**) in glycerol (0.01 mM); (**C**) in water (0.5 mM); (**D**) in 0.1 M HCl (0.5 mM); (**E**) in 0.1 M NaOH (0.5 mM) and **DFT calculations**. References [8,38–40,75–85] are cited in the supplementary materials.

Author Contributions: Conceptualisation was conducted by A.Z., L.B., G.D. and A.P.; methodology and experimental works were conducted by R.P., K.K., M.K., L.K., M.R., M.P., Z.R., J.J. and I.D.; data curation and analysis were conducted by R.P., K.K., O.N., M.K., I.D., Z.R. and F.R.; writing—original draft preparation were conducted by I.D., K.P., A.Z. and G.D.; writing—review and editing were conducted by A.P., A.Z. and A.S.; visualisation, A.Z., K.K., M.K., R.P., J.J., K.P. and A.S.; supervision was conducted by A.Z.; project administration and funding acquisition were conducted by A.P. and A.Z. All authors have read and agreed to the published version of the manuscript.

Funding: This research was funded by the Latvian Council of Science project lzp-2021/1-0283 and PostDocLatvia Project No 1.1.1.2/VIAA/3/19/587 (K. Pajuste).

Institutional Review Board Statement: Not applicable.

Informed Consent Statement: Not applicable.

Data Availability Statement: Data is contained within the article.

Acknowledgments: We are indebted to Marina Petrova for the recording the NMR spectra, to Solveiga Grinberga for the mass spectral analysis, Kaspars Leduskrasts for recording of fluorescence spectra and determination of photoluminescence quantum yields, Artis Kinens for theoretical calculations of the excitation and emission mechanism, and Dace Skrastina for assistance in macrophage isolation.

Conflicts of Interest: The authors declare no conflict of interest.

References

1. Bhurta, D.; Bharate, S.B. Styryl Group, a Friend or Foe in Medicinal Chemistry. *ChemMedChem* **2022**, *17*, e202100706. [CrossRef]
2. Ulrik, C.S.; Lange, P. Targeting small airways in asthma: Improvement in clinical benefit? *Clin. Respir. J.* **2011**, *5*, 125–130. [CrossRef] [PubMed]
3. Putra, A.U.; Çakmaz, D.; Seferoğlu, N.; Barsella, A.; Seferoğlu, Z. Styryl-based new organic chromophores bearing free amino and azomethine groups: Synthesis, photophysical, NLO, and thermal properties. *Beilstein J. Org. Chem.* **2020**, *16*, 2282–2296. [CrossRef]
4. Zhang, L.X.; Li, C.X.; Kakar, M.U.; Khan, M.S.; Wu, P.F.; Amir, R.M.; Dai, D.F.; Naveed, M.; Li, Q.Y.; Saeed, M.; et al. Resveratrol (RV): A pharmacological review and call for further research. *Biomed. Pharmacother.* **2021**, *143*, 112164. [CrossRef]
5. Hewlings, S.J.; Kalman, D.S. Curcumin: A review of its effects on human health. *Foods* **2017**, *6*, 92. [CrossRef] [PubMed]
6. Dubur, G.Y.; Dobretsov, G.E.; Deme, A.K.; Dubure, R.R.; Lapshin, E.N.; Spirin, M.M. Fluorescent probes based on styrylpyridinium derivatives: Optical properties and membrane binding. *J. Biochem. Biophys. Methods* **1984**, *10*, 123–134. [CrossRef]
7. Xu, X.; Sun, Z.; Fan, K.; Jiang, Y.; Huang, R.; Wen, Y.; He, Q.; Ao, T. Conversion of 4-*N*,*N*-dimethylamino-4′-*N*′-methyl-stilbazolium tosylate (DAST) from a Simple Optical Material to a Versatile Optoelectronic Material. *Sci. Rep.* **2015**, *5*, 12269. [CrossRef] [PubMed]
8. Vaitkienė, S.; Kuliešienė, N.; Sakalauskaitė, S.; Bekere, L.; Krasnova, L.; Vigante, B.; Duburs, G.; Daugelavičius, R. Antifungal activity of styrylpyridinium compounds against Candida albicans. *Chem. Biol. Drug Des.* **2021**, *97*, 253–265. [CrossRef]
9. Vaitkienė, S.; Daugelavičius, R.; Sychrová, H.; Kodedová, M. Styrylpyridinium Derivatives as New Potent Antifungal Drugs and Fluorescence Probes. *Front. Microbiol.* **2020**, *11*, 2077. [CrossRef]
10. Panchenko, P.A.; Efremenko, A.V.; Feofanov, A.V.; Ustimova, M.A.; Fedorov, Y.V.; Fedorova, O.A. Ratiometric detection of mercury (Ii) ions in living cells using fluorescent probe based on bis(styryl) dye and azadithia-15-crown-5 ether receptor. *Sensors* **2021**, *21*, 470. [CrossRef]
11. Ma, B.; Tian, D.H.; Yan, S.; Li, X.C.; Dai, F.; Zhou, B. Developing a styrylpyridinium-based fluorescent probe with excellent sensitivity for visualizing basal H2S levels in mitochondria. *Sens. Actuators B Chem.* **2021**, *327*, 128937. [CrossRef]
12. Wangngae, S.; Ngivprom, U.; Khrootkaew, T.; Worakaensai, S.; Lai, R.Y.; Kamkaew, A. Cationic styryl dyes for DNA labelling and selectivity toward cancer cells and Gram-negative bacteria. *RSC Adv.* **2023**, *13*, 2115–2122. [CrossRef] [PubMed]
13. Abeywickrama, C.S.; Wijesinghe, K.J.; Stahelin, R.V.; Pang, Y. Bright red-emitting pyrene derivatives with a large Stokes shift for nucleus staining. *Chem. Commun.* **2017**, *53*, 5886–5889. [CrossRef] [PubMed]
14. Abeywickrama, C.S.; Pang, Y. Synthesis of a bis[2-(2′-hydroxyphenyl)benzoxazole]pyridinium derivative: The fluoride-induced large spectral shift for ratiometric response. *New J. Chem.* **2021**, *45*, 9102–9108. [CrossRef]
15. Peng, L.; Domanov, Y.; Ioffe, V.; Finashin, A.; Gorbenko, G. Interaction of a series of fluorescent probes with glucose oxidase. *J. Biol. Phys. Chem.* **2006**, *6*, 77–81.
16. Matsui, M.; Ooiwa, K.; Okada, A.; Kubota, Y.; Funabiki, K.; Sato, H. Solid-state fluorescence of pyridinium styryl dyes. *Dye. Pigment.* **2013**, *99*, 916–923. [CrossRef]
17. Padalkar, V.S.; Patil, V.S.; Sekar, N. Synthesis and photo-physical properties of fluorescent 1,3,5-triazine styryl derivatives. *Chem. Cent. J.* **2011**, *5*, 77. [CrossRef]

18. Cesaretti, A.; Foggi, P.; Fortuna, C.G.; Elisei, F.; Spalletti, A.; Carlotti, B. Uncovering Structure-Property Relationships in Push-Pull Chromophores: A Promising Route to Large Hyperpolarizability and Two-Photon Absorption. *J. Phys. Chem. C* **2020**, *124*, 15739–15748. [CrossRef]
19. Mencaroni, L.; Cesaretti, A.; Carlotti, B.; Alebardi, M.; Elisei, F.; Ratković, A.; Škorić, I.; Spalletti, A. Tuning the Photophysics of Two-Arm Bis[(dimethylamino)styryl]benzene Derivatives by Heterocyclic Substitution. *Molecules* **2022**, *27*, 8725. [CrossRef]
20. Rucins, M.; Kaukulis, M.; Plotniece, A.; Pajuste, K.; Pikun, N.; Sobolev, A. 1,1′-{[3,5-Bis(dodecyloxycarbonyl)-4-(naphthalen-2-yl)-1,4-dihydropyridine-2,6-diyl]bis(methylene)}bis{4-[(E)-2-(naphthalen-2-yl)vinyl]pyridin-1-ium}dibromide. *Molbank* **2022**, *2022*, M1396. [CrossRef]
21. Xia, Z.; Guo, X.; Zhu, Y.; Wang, Y.; Wang, J. A julolidine-fused anthracene derivative: Synthesis, photophysical properties, and oxidative dimerization. *RSC Adv.* **2018**, *8*, 13588–13591. [CrossRef] [PubMed]
22. Varejão, J.O.S.; Varejão, E.V.V.; Fernandes, S.A. Synthesis and Derivatization of Julolidine: A Powerful Heterocyclic Structure. *Eur. J. Org. Chem.* **2019**, *2019*, 4273–4310. [CrossRef]
23. Enoki, T.; Matsuo, K.; Ohshita, J.; Ooyama, Y. Synthesis and optical and electrochemical properties of julolidine-structured pyrido[3,4-b] indole dye. *Phys. Chem. Chem. Physic* **2017**, *19*, 3565–3574. [CrossRef] [PubMed]
24. Wyrzykiewicz, E.; Wendzonka, M.; Kedzia, B. Synthesis and antimicrobial activity of new (E)-4-[piperidino (4′-methylpiperidino-, morpholino-) N-alkoxy]stilbenes. *Eur. J. Med. Chem.* **2006**, *41*, 519–525. [CrossRef]
25. Wyrzykiewicz, E.; Prukała, W.; Kedzia, B. Synthesis and antimicrobial properties of N-substituted derivatives of (E)-alpha-(or gamma)-azastilbenols. *Farmaco* **1994**, *49*, 127–131.
26. Kluska, M.; Krajewska, E.; Jabłońska, J.; Prukała, W. New Applications and Analysis of (E)-Azastilbenes in Environmental Samples. *Crit. Rev. Anal. Chem.* **2019**, *49*, 395–402. [CrossRef]
27. Vaitkienė, S.; Bekere, L.; Duburs, G.; Daugelavičius, R. Interaction of styrylpyridinium compound with pathogenic candida albicans yeasts and human embryonic kidney hek-293 cells. *Microorganisms* **2021**, *9*, 48. [CrossRef]
28. Dumat, B.; Bordeau, G.; Aranda, A.I.; Mahuteau-Betzer, F.; El Harfouch, Y.; Metgé, G.; Charra, F.; Fiorini-Debuisschert, C.; Teulade-Fichou, M.P. Vinyl-triphenylamine dyes, a new family of switchable fluorescent probes for targeted two-photon cellular imaging: From DNA to protein labeling. *Org. Biomol. Chem.* **2012**, *10*, 6054–6061. [CrossRef]
29. Berdnikova, D.V.; Sosnin, N.I.; Fedorova, O.A.; Ihmels, H. Governing the DNA-binding mode of styryl dyes by the length of their alkyl substituents-from intercalation to major groove binding. *Org. Biomol. Chem.* **2018**, *16*, 545–554. [CrossRef]
30. Yeo, S.J.; Cuc, B.T.; Kim, S.A.; Kim, D.T.H.; Bao, D.T.; Tien, T.T.T.; Anh, N.T.V.; Choi, D.Y.; Chong, C.K.; Kim, H.S.; et al. Rapid detection of avian influenza A virus by immunochromatographic test using a novel fluorescent dye. *Biosens. Bioelectron.* **2017**, *94*, 677–685. [CrossRef]
31. Sehlangia, S.; Nayak, N.; Garg, N.; Pradeep, C.P. Substituent-Controlled Structural, Supramolecular, and Cytotoxic Properties of a Series of 2-Styryl-8-nitro and 2-Styryl-8-hydroxy Quinolines. *ACS Omega* **2022**, *7*, 24838–24850. [CrossRef] [PubMed]
32. Davis, T.J.; Soto-Ortega, D.D.; Kotarek, J.A.; Gonzalez-Velasquez, F.J.; Sivakumar, K.; Wu, L.; Wang, Q.; Moss, M.A. Comparative study of inhibition at multiple stages of amyloid-β self-assembly provides mechanistic insight. *Mol. Pharmacol.* **2009**, *76*, 405–413. [CrossRef]
33. Cueto-Díaz, E.J.; Ebiloma, G.U.; Alfayez, I.A.; Ungogo, M.A.; Lemgruber, L.; González-García, M.C.; Giron, M.D.; Salto, R.; Fueyo-González, F.J.; Shiba, T.; et al. Synthesis, biological, and photophysical studies of molecular rotor-based fluorescent inhibitors of the trypanosome alternative oxidase. *Eur. J. Med. Chem.* **2021**, *220*, 113470. [CrossRef]
34. Etrych, T.; Janoušková, O.; Chytil, P. Fluorescence imaging as a tool in preclinical evaluation of polymer-based nano-DDS systems intended for cancer treatment. *Pharmaceutics* **2019**, *11*, 471. [CrossRef]
35. Perrin, J.; Capitao, M.; Mougin-Degraef, M.; Guérard, F.; Faivre-Chauvet, A.; Rbah-Vidal, L.; Gaschet, J.; Guilloux, Y.; Kraeber-Bodéré, F.; Chérel, M.; et al. Cell Tracking in Cancer Immunotherapy. *Front. Med.* **2020**, *7*, 458234. [CrossRef]
36. Fukushima, H.; Matikonda, S.S.; Usama, S.M.; Furusawa, A.; Kato, T.; Štacková, L.; Klán, P.; Kobayashi, H.; Schnermann, M.J. Cyanine Phototruncation Enables Spatiotemporal Cell Labeling. *J. Am. Chem. Soc.* **2022**, *144*, 11075–11080. [CrossRef]
37. Trofimova, O.; Korotkaja, K.; Skrastina, D.; Jansons, J.; Spunde, K.; Isaguliants, M.; Zajakina, A.; Tripp, R.A. Alphavirus-Driven Interferon Gamma (IFNg) Expression Inhibits Tumor Growth in Orthotopic 4T1 Breast Cancer Model. *Vaccines* **2021**, *9*, 1247. [CrossRef] [PubMed]
38. Stöter, M.; Biersack, B.; Reimer, N.; Herling, M.; Stock, N.; Schobert, R.; Breu, J. Ordered heterostructures of two strictly alternating types of nanoreactors. *Chem. Mater.* **2014**, *26*, 5412–5419. [CrossRef]
39. Guo, L.; Li, C.; Shang, H.; Zhang, R.; Li, X.; Lu, Q.; Cheng, X.; Liu, Z.; Sun, J.Z.; Yu, X. A side-chain engineering strategy for constructing fluorescent dyes with direct and ultrafast self-delivery to living cells. *Chem. Sci.* **2020**, *11*, 661–670. [CrossRef]
40. Jedrzejewska, B.; Kabatc, J.; Ośmiałowski, B.; Paczkowski, J. Bischromophoric styrylpyridinium dyes. Spectroscopic properties of 1,3-bis-[4-(p-*N*,*N*-dialkylaminostyryl)pyridinyl]propane dibromides. *Spectrochim. Acta Part A Mol. Biomol. Spectrosc.* **2007**, *67*, 306–315. [CrossRef]
41. Skrzydlewski, P.; Twarużek, M.; Grajewski, J. Cytotoxicity of Mycotoxins and Their Combinations on Different Cell Lines: A Review. *Toxins* **2022**, *14*, 244. [CrossRef]
42. Kritis, A.A.; Stamoula, E.G.; Paniskaki, K.A.; Vavilis, T.D. Researching glutamate—Induced cytotoxicity in different cell lines: A comparative/collective analysis/study. *Front. Cell. Neurosci.* **2015**, *9*, 131867. [CrossRef]

43. Sharmin, S.; Rahaman, M.M.; Martorell, M.; Sastre-Serra, J.; Sharifi-Rad, J.; Butnariu, M.; Bagiu, I.C.; Bagiu, R.V.; Islam, M.T. Cytotoxicity of synthetic derivatives against breast cancer and multi-drug resistant breast cancer cell lines: A literature-based perspective study. *Cancer Cell Int.* **2021**, *21*, 612. [CrossRef] [PubMed]
44. Jabłońśtski, A. Efficiency of anti-stokes fluorescence in dyes. *Nature* **1933**, *131*, 839–840. [CrossRef]
45. Maillard, J.; Klehs, K.; Rumble, C.; Vauthey, E.; Heilemann, M.; Fürstenberg, A. Universal quenching of common fluorescent probes by water and alcohols. *Chem. Sci.* **2021**, *12*, 1352–1362. [CrossRef]
46. Dahal, D.; Ojha, K.R.; Pokhrel, S.; Paruchuri, S.; Konopka, M.; Liu, Q.; Pang, Y. NIR-emitting styryl dyes with large Stokes' shifts for imaging application: From cellular plasma membrane, mitochondria to zebrafish neuromast. *Dye. Pigment.* **2021**, *194*, 109629. [CrossRef] [PubMed]
47. Collot, M.; Boutant, E.; Fam, K.T.; Danglot, L.; Klymchenko, A.S. Molecular Tuning of Styryl Dyes Leads to Versatile and Efficient Plasma Membrane Probes for Cell and Tissue Imaging. *Bioconjug. Chem.* **2020**, *31*, 875–883. [CrossRef]
48. Abeywickrama, C.S.; Bertman, K.A.; Mcdonald, L.J.; Alexander, N.; Dahal, D.; Baumann, H.J.; Salmon, C.R.; Wesdemiotis, C.; Konopka, M.; Tessier, C.A.; et al. Synthesis of highly selective lysosomal markers by coupling 2-(2′-hydroxyphenyl)benzothiazole (HBT) with benzothiazolium cyanine (Cy): The impact of substituents on selectivity and optical properties. *J. Mater. Chem. B* **2019**, *7*, 7502–7514. [CrossRef]
49. Abeywickrama, C.S.; Wijesinghe, K.J.; Plescia, C.B.; Fisher, L.S.; Goodson, T.; Stahelin, R.V.; Pang, Y. A pyrene-based two-photon excitable fluorescent probe to visualize nuclei in live cells. *Photochem. Photobiol. Sci.* **2020**, *19*, 1152–1159. [CrossRef] [PubMed]
50. Abeywickrama, C.S.; Baumann, H.J.; Alexander, N.; Shriver, L.P.; Konopka, M.; Pang, Y. NIR-emitting benzothiazolium cyanines with an enhanced stokes shift for mitochondria imaging in live cells. *Org. Biomol. Chem.* **2018**, *16*, 3382–3388. [CrossRef]
51. Charcosset, C.; Juban, A.; Valour, J.P.; Urbaniak, S.; Fessi, H. Preparation of liposomes at large scale using the ethanol injection method: Effect of scale-up and injection devices. *Chem. Eng. Res. Des.* **2015**, *94*, 508–515. [CrossRef]
52. Krapivina, A.; Lacis, D.; Rucins, M.; Plotniece, M.; Pajuste, K.; Sobolev, A.; Plotniece, A. Synthesis and Characterization of Novel Amphiphilic N-Benzyl 1,4-Dihydropyridine Derivatives—Evaluation of Lipid Monolayer and Self-Assembling Properties. *Materials* **2023**, *16*, 4206. [CrossRef] [PubMed]
53. Ozolins, R.; Plotniece, M.; Pajuste, K.; Putralis, R.; Pikun, N.; Sobolev, A.; Plotniece, A.; Rucins, M. 1,1′-{[3,5-Bis((dodecyloxycarbonyl)-4-phenyl-1,4-dihydropyridine-2,6-diyl]bis(methylene)}bis[4-(anthracen-9-yl)pyridin-1-ium] Dibromide. *Molbank* **2022**, *2022*, M1438. [CrossRef]
54. Singh, H.; Sareen, D.; Sharma, S.; Kim, J.S.; Sharma, A. Molecular Theranostics for Cancer Therapy. In *Nanoengineering of Biomaterials*; John Wiley & Sons, Ltd.: Hoboken, NJ, USA, 2022; pp. 181–214.
55. He, H.; Zhang, X.; Du, L.; Ye, M.; Lu, Y.; Xue, J.; Wu, J.; Shuai, X. Molecular imaging nanoprobes for theranostic applications. *Adv. Drug Deliv. Rev.* **2022**, *186*, 114320. [CrossRef]
56. Zhang, J.; Ning, L.; Huang, J.; Zhang, C.; Pu, K. Activatable molecular agents for cancer theranostics. *Chem. Sci.* **2020**, *11*, 618–630. [CrossRef] [PubMed]
57. Liu, Z.; Song, F.; Shi, W.; Gurzadyan, G.; Yin, H.; Song, B.; Liang, R.; Peng, X. Nitroreductase-Activatable Theranostic Molecules with High PDT Efficiency under Mild Hypoxia Based on a TADF Fluorescein Derivative. *ACS Appl. Mater. Interfaces* **2019**, *11*, 15426–15435. [CrossRef]
58. Koo, S.; Bobba, K.N.; Cho, M.Y.; Park, H.S.; Won, M.; Velusamy, N.; Hong, K.S.; Bhuniya, S.; Kim, J.S. Molecular Theranostic Agent with Programmed Activation for Hypoxic Tumors. *ACS Appl. Bio Mater.* **2019**, *2*, 4648–4655. [CrossRef] [PubMed]
59. Petrichenko, O.; Rucins, M.; Vezane, A.; Timofejeva, I.; Sobolev, A.; Cekavicus, B.; Pajuste, K.; Plotniece, M.; Gosteva, M.; Kozlovska, T.; et al. Studies of the physicochemical and structural properties of self-assembling cationic pyridine derivatives as gene delivery agents. *Chem. Phys. Lipids* **2015**, *191*, 25–37. [CrossRef] [PubMed]
60. Vasilevska, J.; De Souza, G.A.; Stensland, M.; Skrastina, D.; Zhulenvovs, D.; Paplausks, R.; Kurena, B.; Kozlovska, T.; Zajakina, A. Comparative protein profiling of B16 mouse melanoma cells susceptible and non-susceptible to alphavirus infection: Effect of the tumor microenvironment. *Cancer Biol. Ther.* **2016**, *17*, 1035–1050. [CrossRef]
61. Zajakina, A.; Vasilevska, J.; Zhulenkovs, D.; Skrastina, D.; Spaks, A.; Plotniece, A.; Kozlovska, T. High efficiency of alphaviral gene transfer in combination with 5-fluorouracil in a mouse mammary tumor model. *BMC Cancer* **2014**, *14*, 460. [CrossRef]
62. Chen, S.; Saeed, A.F.U.H.; Liu, Q.; Jiang, Q.; Xu, H.; Xiao, G.G.; Rao, L.; Duo, Y. Macrophages in immunoregulation and therapeutics. *Signal Transduct. Target. Ther.* **2023**, *8*, 207. [CrossRef]
63. Spunde, K.; Korotkaja, K.; Zajakina, A. Recombinant Viral Vectors for Therapeutic Programming of Tumour Microenvironment: Advantages and Limitations. *Biomedicines* **2022**, *10*, 2142. [CrossRef]
64. Larochelle, A.; Bellavance, M.A.; Michaud, J.P.; Rivest, S. Bone marrow-derived macrophages and the CNS: An update on the use of experimental chimeric mouse models and bone marrow transplantation in neurological disorders. *Biochim. Biophys. Acta Mol. Basis Dis.* **2016**, *1862*, 310–322. [CrossRef]
65. Kurena, B.; Müller, E.; Christopoulos, P.F.; Johnsen, I.B.; Stankovic, B.; Øynebråten, I.; Corthay, A.; Zajakina, A. Generation and functional in vitro analysis of Semliki Forest virus vectors encoding TNF-α and IFN-γ. *Front. Immunol.* **2017**, *8*, 1667. [CrossRef]
66. Müller, E.; Speth, M.; Christopoulos, P.F.; Lunde, A.; Avdagic, A.; Øynebråten, I.; Corthay, A. Both type I and type II interferons can activate antitumor M1 macrophages when combined with TLR stimulation. *Front. Immunol.* **2018**, *9*, 2520. [CrossRef]
67. Stout, R.D.; Suttles, J. Functional plasticity of macrophages: Reversible adaptation to changing microenvironments. *J. Leukoc. Biol.* **2004**, *76*, 509–513. [CrossRef]

68. Murray, P.J.; Allen, J.E.; Biswas, S.K.; Fisher, E.A.; Gilroy, D.W.; Goerdt, S.; Gordon, S.; Hamilton, J.A.; Ivashkiv, L.B.; Lawrence, T.; et al. Macrophage Activation and Polarization: Nomenclature and Experimental Guidelines. *Immunity* **2014**, *41*, 14–20. [CrossRef] [PubMed]
69. Stout, R.D.; Jiang, C.; Matta, B.; Tietzel, I.; Watkins, S.K.; Suttles, J. Macrophages Sequentially Change Their Functional Phenotype in Response to Changes in Microenvironmental Influences. *J. Immunol.* **2005**, *175*, 342–349. [CrossRef] [PubMed]
70. Korotkaja, K.; Jansons, J.; Spunde, K.; Rudevica, Z.; Zajakina, A. Establishment and Characterization of Free-Floating 3D Macrophage Programming Model in the Presence of Cancer Cell Spheroids. *Int. J. Mol. Sci.* **2023**, *24*, 10763. [CrossRef] [PubMed]
71. Mosmann, T. Rapid colorimetric assay for cellular growth and survival: Application to proliferation and cytotoxicity assays. *J. Immunol. Methods* **1983**, *65*, 55–63. [CrossRef]
72. ICCVAM. *ICCVAM Test Method Evaluation Report (TMER): In Vitro Cytotoxicity Test Methods for Estimating Starting Doses For Acute Oral Systemic Toxicity Testing*; NIH Publication No: 07-4519; NIH: Bethesda, MD, USA, 2006.
73. National Institutes of Health (NIH). *Background Review Document: In Vitro Cytotoxicity Test Methods for Estimating Acute Oral Systemic Toxicity*; National Institutes of Health (NIH): Stapleton, NY, USA, 2006.
74. The European Parliament and the Council of the European Union. *Regulation (EC) No 1272/2008 of the European Parliament and of the Council of 16 December 2008 on Classification, Labelling and Packaging of Substances and Mixtures, Amending and Repealing Directives 67/548/EEC and 1999/45/EC, and Amending Regulation (EC)*; The European Parliament and the Council of the European Union: Brussels, Belgium, 2008; Volume L 269.
75. Tonga, M.; Lahti, P.M. Designing conjugation-extended viologens for high molar absorptivity with longer wavelength absorption. *Synth. Met.* **2019**, *254*, 75–84. [CrossRef]
76. Salajkova, S.; Benkova, M.; Marek, J.; Sleha, R.; Prchal, L.; Malinak, D.; Dolezal, R.; Sepčíc, K.; Gunde-Cimerman, N.; Kuca, K.; et al. Wide-Antimicrobial Spectrum of Picolinium Salts. *Molecules* **2020**, *25*, 2254. [CrossRef]
77. Baran, A.; Plotniece, A.; Sobolev, A.; Vigante, B.; Gosteva, M.; Olkhovik, V. Synthesis and photoluminescent properties of new cationic carbazole-containing luminophores. *Chem. Heterocycl. Compd.* **2012**, *48*, 287–295. [CrossRef]
78. Van Deun, R.; Nockemann, P.; Parac-Vogt, T.N.; Van Hecke, K.; Van Meervelt, L.; Görller-Walrand, C.; Binnemans, K. Near-infrared photoluminescence of lanthanide complexes containing the hemicyanine chromophore. *Polyhedron* **2007**, *26*, 5441–5447. [CrossRef]
79. Ibrahim, M.A.M.; Messali, M.; Moussa, Z.; Alzahrani, A.Y.; Alamry, S.N.; Hammouti, B. Corrosion inhibition of carbon steel by imidazolium and pyridinium cations ionic liquids in acidic environment. *Port. Electrochim. Acta* **2011**, *29*, 375–389. [CrossRef]
80. Pecchioli, T.; Muthyala, M.K.; Haag, R.; Christmann, M. Multivalent polyglycerol supported imidazolidin-4-one organocatalysts for enantioselective Friedel-Crafts alkylations. *Beilstein J. Org. Chem.* **2015**, *11*, 730–738. [CrossRef]
81. Battistuzzi, G.; Cacchi, S.; Fabrizi, G. An efficient palladium-catalyzed synthesis of cinnamaldehydes from acrolein diethyl acetal and aryl iodides and bromides. *Org. Lett.* **2003**, *5*, 777–780. [CrossRef]
82. Bachollet, S.P.J.T.; Addi, C.; Pietrancosta, N.; Mallet, J.M.; Dumat, B. Fluorogenic Protein Probes with Red and Near-Infrared Emission for Genetically Targeted Imaging. *Chem. A Eur. J.* **2020**, *26*, 14467–14473. [CrossRef] [PubMed]
83. Coe, B.J.; Fielden, J.; Foxon, S.P.; Harris, J.A.; Helliwell, M.; Brunschwig, B.S.; Asselberghs, I.; Clays, K.; Garín, J.; Orduna, J. Diquat derivatives: Highly active, two-dimensional nonlinear optical chromophores with potential redox switchability. *J. Am. Chem. Soc.* **2010**, *132*, 10498–10512. [CrossRef] [PubMed]
84. Coe, B.J.; Rusanova, D.; Joshi, V.D.; Sánchez, S.; Vávra, J.; Khobragade, D.; Severa, L.; Císařová, I.; Šaman, D.; Pohl, R.; et al. Helquat Dyes: Helicene-like Push-Pull Systems with Large Second-Order Nonlinear Optical Responses. *J. Org. Chem.* **2016**, *81*, 1912–1920. [CrossRef]
85. Frisch, M.J.; Trucks, G.W.; Schlegel, H.B.; Scuseria, G.E.; Robb, M.A.; Cheeseman, J.R.; Scalmani, G.; Barone, V.; Mennucci, B.; Petersson, G.A.; et al. *Gaussian 09*; Revision D.01; Gaussian, Inc.: Wallingford, CT, USA, 2013.

pharmaceuticals

MDPI

Article

Synthesis and Biological Evaluation of 6-*O*-Sucrose Monoester Glycolipids as Possible New Antifungal Agents

Michele Verboni, Maurizio Sisti, Raffaella Campana, Serena Benedetti, Francesco Palma, Lucia Potenza, Simone Lucarini * and Andrea Duranti

Department of Biomolecular Sciences, University of Urbino Carlo Bo, 61029 Urbino, PU, Italy
* Correspondence: simone.lucarini@uniurb.it; Tel.: +39-0722-303333

Abstract: A small library of 6-*O*-sucrose monoester surfactants has been synthesized and tested against various microorganisms. The synthetic procedure involved a modified Mitsunobu reaction, which showed improved results compared to those present in the literature (higher yields and larger scope). The antifungal activities of most of these glycolipids were satisfactory. In particular, sucrose palmitoleate (URB1537) showed good activity against Candida albicans ATCC 10231, *Fusarium* spp., and *Aspergillus fumigatus* IDRAH01 (MIC value: 16, 32, 64 μg/mL, respectively), and was further characterized through radical scavenging, anti-inflammatory, and biocompatibility tests. URB1537 has been shown to control the inflammatory response and to have a safe profile.

Keywords: glycolipids; sucrose monoesters; sugar-based surfactants; minimum inhibitory concentration (MIC); antifungal activity; anti-inflammatory activity

Citation: Verboni, M.; Sisti, M.; Campana, R.; Benedetti, S.; Palma, F.; Potenza, L.; Lucarini, S.; Duranti, A. Synthesis and Biological Evaluation of 6-*O*-Sucrose Monoester Glycolipids as Possible New Antifungal Agents. *Pharmaceuticals* **2023**, *16*, 136. https://doi.org/10.3390/ph16020136

Academic Editors: Amélia Pilar Rauter, Klaus Kopka, Thomas Efferth, Joachim Jose, Maria Emília De Sousa and Guangshun Wang

Received: 23 December 2022
Revised: 12 January 2023
Accepted: 13 January 2023
Published: 17 January 2023

1. Introduction

Fungal invasive infections represent a serious problem in human health and are associated with at least 1.5–2 million deaths worldwide [1–4]. In this scenario, opportunistic mycotic pathogens such as *Candida* spp., *A. fumigatus* and *C. neoformans* are the most important causes of invasive fungal infections, accounting for more than 90% of fungal deaths [4]. These infections are mainly common in immunocompromised patients subjected to anticancer therapy, long-term corticosteroid treatments or organ transplant as well as immunosuppressive infections. Moreover, fungi are responsible for infections on skin and mucosal surfaces with a greater incidence compared to invasive fungal ones. In addition, the development of antifungal drugs faces off with the fact that fungi are eukaryotes as their hosts and thus the potential targets may also be found in human cells with a substantial toxicity risk. Nowadays, the antifungals on the market are represented by various classes of drugs called polyenes (i.e., amphotericin B), azoles (i.e., fluconazole), allyl-/benzylamines (i.e., terbinafine), benzofurancyclohexenes (i.e., griseofulvin), echinocandins (i.e., caspofungin), antimetabolites (flucytosine), nikkomycines (nikkomycine Z), thiocarbamates (tolnaftate), and morpholines (amorolfine) [5,6]. Unfortunately, most of the antifungals and antibiotics actually commercially available, in particular those systemically and *per os* administered, show some limits, such as toxicity, the spectrum of activity, safety, pharmacokinetic properties, and drug-resistant strains [7]. Consequently, there is an urgent need to develop new antifungals as well as new antibiotics to fill this therapeutic gap.

Sucrose fatty acid esters (SFAE) are the most common glycolipids derived from polyalcohol sucrose (polar head) and naturally occurring fatty acid (hydrophobic tail). They are non-ionic surfactants and considered "green" because they are non-toxic and biodegradable. Moreover, they are non-irritant, odorless, and possess a safer biocompatibility profile in comparison with other amphiphiles such as polysorbates and sulfates [8]. Most of the time, SFAEs are synthesized from inexpensive and renewable agricultural products [8,9], and some of them are commercially available [8]. Moreover, they possess excellent emulsifying characteristics [10–12], antibacterial [8,10,13], antifungal [8] and insecticidal [8,14] activities, and permeability-enhancing properties [10]. For all these reasons, sucrose esters are very important commodity compounds in the agriculture, food, nutraceutical, cosmetic, dental, and pharmaceutical industries. In the 2010s, the world production of SFAE was above 6000 tons per year [8] and their market is projected to grow from USD 76 million in 2019 to USD 106 million by 2025 [15].

As part of our ongoing investigations on the synthesis, biological activities, and applications of sugar derivatives [16–24], a series of 6-*O*-sucrose monoesters were designed and synthesized by using an opportunely modified Mitsunobu-type methodology [25] starting from free sucrose and lipophilic acids. This coupling, compared to the others reported in the literature [8,10], such as esterification (acyl chlorides, anhydrides) and transesterification with activated esters (vinyl or enol esters) [26–29], and/or enzymatic preparation methods [30–32], showed higher conversions/yields, high regioselectivity and larger scope including saturated, mono- and polyunsaturated fatty acids or aromatic and alkyl aromatic acids.

The small library of sucrose esters was explored for antibacterial and antifungal activities and the profile of the most interesting glycolipid (URB1537) was deepened through the study of its antioxidant, anti-inflammatory, and biocompatibility properties.

2. Results and Discussion

2.1. Chemistry

A series of glycolipids characterized by the 6-*O*-sucrose hydrophilic head and a lipophilic tail of various sizes and nature has been designed and synthesized with the aim of exploring the influence of the hydrophobic portion on their biological properties. For this purpose, both linear saturated (caprylic (C8)), monounsaturated (undecylenic (C11:1), undecylinic (C11:1), palmitoleic (C16:1), oleic (C18:1), and nervonic (C24:1)) or polyunsaturated (linoleic (C18:2) and linolenic (C18:3)), and aromatic (benzoic and *p*-phenylbenzoic) or alkyl aromatic (phenyl-, *p*-biphenyl- and *p*-triphenylacetic) fatty acid derivatives substituents have been used. The most important physicochemical constants in the landscape of surfactants are hydrophilic–lipophilic balance (HLB), octanol–water portion coefficient (logP) and the polar surface area (TPSA). However, TPSA could not constitute a discriminant indicator when the polar head is maintained constant in all the designed surfactants, as in our case (calculated TPSA = 196 $Å^2$ for sucrose esters). The glycolipids were indeed designed with the aim of keeping the HLB and logP values in a non-strict range to try to understand if a relationship between the biological activity and physicochemical properties is present.

With the only exception of sucrose nervonate (HLB = 8.6, Table 1, Entry 8), all the derivatives have HLB values greater than 9.5 (9.7–13.4) classifying the molecules as hydrophilic surfactants and are consequently able to act as oil-in-water emulsifiers. On the other hand, the designed sucrose-based glycolipids are molecules showing a widespread calculated logP (clogP) ranging from a negative value −2.7 (sucrose benzoate, Entry 9) to a positive 2.9 (sucrose oleate, Entry 5). As above, the only exception of the series is the sucrose nervonate that showed a clogP out of the Lipinski's rules of five (5.9, Entry 8) due to its very lipophilic long tail.

The designed 6-*O*-sucrose-based monoesters could be synthesized by several reported procedures starting from unprotected sucrose [8,10]. We tried several of them by using classical chemical esterification and transesterification procedures and also using appropriate enzymatic methodologies but without satisfactory results in terms of regioselectivity and conversion/yield (data not shown). On the contrary, the Mitsunobu methodology reported by Molinier et al. [25] immediately proved to be suitable for our purposes. Indeed, not protected sucrose (**1**) and undecylenic acid (C11:1) (**2b**) under the reported Mitsunobu conditions gave the corresponding 6-*O*-sucrose ester **3b** in 34% yield (Table 1, Scheme 1 and Entry 1) and other known side products including the diester **4** [25]. With the aim of ameliorating the yield of **3b**, the reaction reported in Scheme 1 was studied as a model reaction. All the conditions were investigated in detail and the results are reported in Table 2.

Table 1. Calculated physicochemical properties of the sucrose-based ester surfactants.

Entry	Sucrose Ester	MW	HLB [a]	LogP [b]
1	Caprylate C8	468.5	12.7	−1.4
2	Undecylenate C11:1	508.6	11.7	−0.2
3	Undecylinate C11:1	506.6	11.8	−0.9
4	Palmitoleate C16:1	578.7	10.3	2.0
5	Oleate C18:1	606.7	9.8	2.9
6	Linoleate C18:2	604.7	9.9	2.6
7	Linolenate C18:3	602.7	9.9	2.4
8	Nervonate C24:1	690.9	8.6	5.9
9	Benzoate	446.4	13.4	−2.7
10	*p*-Phenyl benzoate	522.5	11.4	−0.3
11	Phenyl acetate	460.4	12.9	−2.1
12	*p*-Biphenyl acetate	536.5	11.1	−0.4
13	*p*-Triphenyl acetate	612.6	9.7	0.6

[a] Calculated hydrophilic–lipophilic balance by Griffin's method for non-ionic surfactants [33] [HLB = 20 × (MW hydrophilic portion/MW)]. [b] Calculated octanol–water portion coefficient clogP (by OSIRIS Property Explorer) [34].

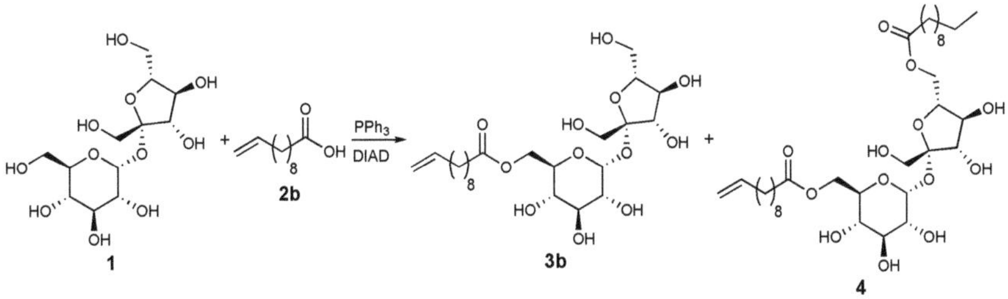

Scheme 1. Model reaction for methodology studies.

Table 2. Investigation of the conditions for Mitsunobu-type reaction between sucrose and undecylenic acid.

Entry [a]	Eq. 2b	Eq. DIAD:PPh_3	T (°C)	t (h)	Dry Solvent	Yield (%) 3b [b]	Yield (%) 4 [b]
1	2.5	2.7:2.7	20	24	DMF	34	38
2	1	2.5:2.5	20	24	DMF	19	trace
3	1.5	2.5:2.5	20	24	DMF	39	15
4	1.5	1.5:1.5	20	24	DMF	26	trace
5	1	1:1	20	24	DMF	trace	-
6	1.5	3:3	20	24	DMF	39	16
7	1.5	2.5:2.5	60	24	DMF	trace	41
8	1.5	2.5:2.5	20	6	DMF	12	2
9	1.5	2.5:2.5	20	12	DMF	18	9
10	1.5	2.5:2.5	20	24	THF	trace	-
11	1.5	2.5:2.5	20	24	dioxane	trace	-
12 [c]	1.5	2.5 [c]:2.5	20	24	DMF	35	19
13 [d]	1.5	2.5 [d]:2.5	20	24	DMF	19	trace

[a] Reaction conditions: reactions were carried out under nitrogen atmosphere with **1** (1.0 mmol, 1 eq.), and **2b** (x eq.), DIAD (y eq.), PPh_3 (y eq.) and solvent (10 mL) at T (°C) for certain time, t (h). [b] Yields are referred to isolated products. [c] Reaction was performed using DEAD. [d] Reaction was performed using DCAD. Eq. = equivalent.

As reported in Table 2 (Entry 1), the diester **4** represents the main product of the first attempt (38% yield vs. 34% of the desired product). A reduction of the acid could be decisive, with 1 equivalent of **2b**, almost no diester was obtained; however, the yield of 6-*O*-sucrose monoester was reduced to 19% (Entry 2). A 1.5 equivalent of the model acid gave the best result in term of yield of the desired product **3b** (39%, Entry 3) and lower production of 4 (15%). With the aim to reduce the 6-O-undecylenyl-3′,6′-anhydrosucrose side product [25], the equivalents of the Mitsunobu reagents DIAD and PPh_3 were also decreased (Entries 4 and 5). However, no good results were obtained. The yield of **3b** did not also increase with a higher amount of DIAD and PPh_3 (Entry 6). Increasing the temperature (Entry 7) or the reaction time to 48 h (data not shown) did not ameliorate the production of **3b** but enhanced the formation of the diester **4**. Moreover, the reduction of the reaction time worsened the yield of the desired monoester (Entries 8 and 9). Finally, other polar aprotic solvents, such as THF and dioxane, did not solubilize well sucrose and the reaction gave only a trace amount of the product (Entries 10 and 11). Other azodicarboxylates were also used under optimal reaction conditions as comparison with DIAD. By the use of DEAD, the reaction gave the desired product **3b** with comparable yield, albeit slightly lower, but with a higher production of diester **4** (Entry 12). Instead, the use of DCAD led to a clear lowering of the yield of 6-*O*-sucrose undecylenate (Entry 13).

The improved Mitsunobu-type procedure (Entry 3) was then used to synthesize all the designed 6-*O*-sucrose monoester surfactants starting from not protected sucrose and the corresponding commercially available acids (Scheme 2).

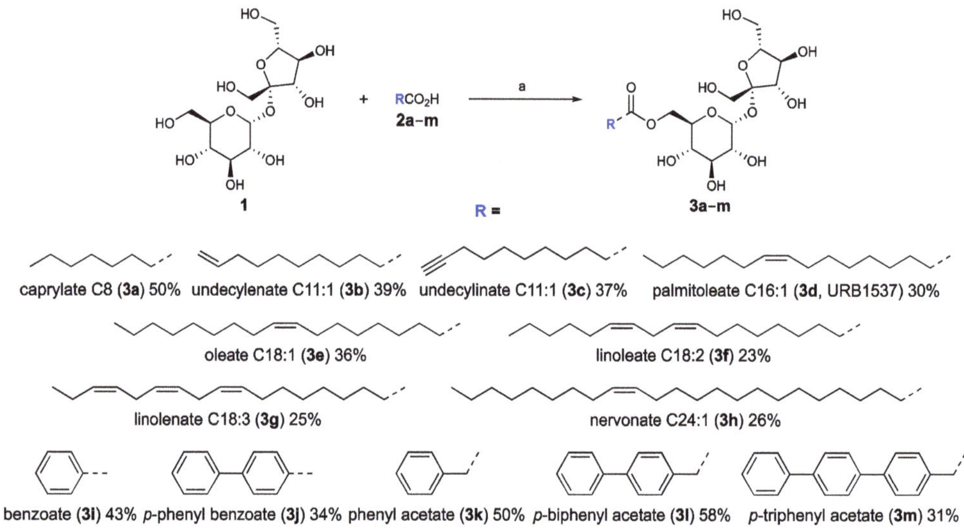

Scheme 2. Sucrose ester synthesis. Substrate scopes, reagents, conditions, and yields: (a) PPh_3, DIAD, dry DMF, rt, 24–30 h.

The ameliorated methodology worked well for all the acid substrates giving the corresponding products in low to moderate yields (Scheme 2). Although the yields for some substrates are not exciting, it should be noted that the procedure uses commercially available, inactivated, or unprotected reagents, avoiding the activation or protection and deprotection steps. Moreover, the reaction is regioselective for the six-position of sucrose and, conveniently, the desired products are easily purified from the side ones by a short flash column chromatography. Therefore, this improved Mitsunobu-type esterification has shown to be versatile and has widespread substrate scope, and could be used for the production of any kind of 6-*O*-sucrose monoesters in low to medium scale.

2.2. Antibacterial Activity

The antimicrobial activity of the different sucrose esters is summarized in Table 3. As shown, the activity of the tested compounds was prevalently observed against Gram-positive bacteria (*Enterococcus faecalis* ATCC 29212, *Staphylococcus aureus* ATCC 43387, *S.*

aureus ATCC 43300, *Listeria monocytogenes* ATCC 7644) but not against Gram-negative ones (*E. coli* O157:H7 ATCC 35150, *Klebsiella pneumoniae* ATCC 13883, *Pseudomonas aeruginosa* ATCC 9027, *Salmonella enteritidis* ATCC 13076). In detail, sucrose palmitoleate (**3d**) and sucrose oleate (**3e**) showed minimum inhibitory concentration (MIC) values of 1024 µg/mL against *E. faecalis* ATCC 29212, *S. aureus* ATCC 43387, *S. aureus* ATCC 43300 (MRSA) and *L. monocytogenes* ATCC 7644, while sucrose linoleate (**3f**) and sucrose linolenate (**3g**) reached MIC of 256 µg/mL and 512 µg/mL against *E. faecalis* ATCC 29212 and 1024 µg/mL against the other Gram-positive microorganisms. Sucrose nervonate (**3h**), sucrose benzoate (**3i**), and sucrose phenyl acetate (**3k**) were active only against *E. faecalis* ATCC 29212 (MIC 1024 µg/mL), while sucrose p-phenyl benzoate (**3j**) and sucrose undecylenate (**3b**) were active also against *L. monocytogenes* ATCC 7644 (in both cases MIC 1024 µg/mL). Sucrose caprylate (**3a**), sucrose undecylenate (**3c**), p-biphenyl acetate (**3l**), and p-triphenyl acetate (**3m**) were always inactive.

Table 3. Antibacterial activity of sucrose-based compounds (MIC values are reported in µg/mL).

Specie Target	3a	3b	3c	3d	3e	3f	3g	3h	3i	3j	3k	3l	3m
E. faecalis ATCC 29212	>1024	>1024	1024	1024	1024	256	512	1024	1024	1024	1024	>1024	>1024
E. coli O157:H7 ATCC 35150	>1024	>1024	>1024	>1024	>1024	>1024	>1024	>1024	>1024	>1024	>1024	>1024	>1024
S. aureus ATCC 29213	>1024	>1024	>1024	1024	1024	1024	>1024	>1024	>1024	>1024	>1024	>1024	>1024
S. aureus ATCC 43300	>1024	>1024	>1024	1024	1024	1024	>1024	>1024	>1024	>1024	>1024	>1024	>1024
K. pneumoniae ATCC 13883	>1024	>1024	>1024	>1024	>1024	>1024	>1024	>1024	>1024	>1024	>1024	>1024	>1024
L. monocytogenes ATCC 7644	>1024	>1024	1024	1024	1024	1024	1024	>1024	>1024	1024	>1024	>1024	>1024
S. enteritidis ATCC 13076	>1024	>1024	>1024	>1024	>1024	>1024	>1024	>1024	>1024	>1024	>1024	>1024	>1024
P. aeruginosa ATCC 9027	>1024	>1024	>1024	>1024	>1024	>1024	>1024	>1024	>1024	>1024	>1024	>1024	>1024

The mechanism of antibacterial action of sugar fatty acid esters is mainly based on the increase in permeability of the cell membrane and the consequent leakage of some cellular constituents, in particular proteins and sugars [35]. Polar compounds interact better with the bacterial cell wall, especially in Gram-positive bacteria, causing their damage. Effectively, our data showed that Gram-negative bacteria were more resistant to the examined sucrose esters compared to Gram-positive bacteria, probably due to their outer membrane, which limits the diffusion of sucrose fatty acid esters through the lipopolysaccharide covering. On the other hand, the literature data reports that the antibacterial activity of sugar fatty acid esters decreases rapidly as the chain length of fatty acid increases [36]. However, such a situation is not appreciable in this study because of the low activities of our tested compounds. Moreover, for the same reason, it is not possible to rationalize a correlation between the activity of the compounds and their physicochemical characteristics and lipophilic tails.

2.3. Antifungal Activity

The antifungal activity of the different sucrose esters is illustrated in Table 4. As a general trend, the tested compounds showed good antifungal activity and, interestingly, some of them (**3b**, **3d–f**) resulted active against all the selected fungi with MIC ranging from 16 to 1024 µg/mL. Among the molecules analyzed, the lowest MIC value of 16 µg/mL was observed for sucrose palmitoleate (C16:1) (**3d**), **3e** against *Candida albicans* ATCC 10231 and *Aspergillus fumigatus* IDRAH01, respectively, followed by **3f** (MIC 32 µg/mL against *A. fumigatus* IDRAH01). Further, **3b** resulted active against *Aspergillus niger* ATCC 9642 and *Fusarium* spp. (MIC 512 µg/mL) and **3g** showed MIC values of 128 µg/mL and 512 µg/mL for *Fusarium* spp. and *C. albicans* ATCC 10231, respectively. Similarly, **3i** and **3k** were active against *Fusarium* spp. and *C. albicans* ATCC 10231 with a MIC ranging from 256 to 1024 µg/mL. These two microorganisms showed to be sensitive also to the remaining sucrose esters, specifically **3b**, **3j**, **3l** (MIC 1024 µg/mL), confirming **3b** antimicrobial activity against *C. albicans*, reported in some previous research [37] As observed for bacteria, **3m** was inactive against all the examined fungi.

Table 4. Antifungal activity of sucrose-based compounds (MIC values are reported in μg/mL).

Specie Target	3a	3b	3c	3d	3e	3f	3g	3h	3i	3j	3k	3l	3m
A. fumigatus IDRAH01	1024	1024	>1024	64	16	32	1024	1024	>1024	>1024	1024	>1024	>1024
A. niger ATCC 9642	1024	512	>1024	512	1024	1024	>1024	>1024	>1024	>1024	>1024	>1024	>1024
Fusarium spp.	>1024	512	1024	32	128	128	128	1024	512	1024	256	1024	>1024
C. albicans ATCC 10231	1024	1024	1024	16	1024	1024	512	>1024	256	1024	1024	1024	>1024

The best series of these sucrose esters against the tested fungi were the mono and polyunsaturated fatty acid-containing tail compounds. In detail, **3d** showed a relevant activity against all four fungi strains (MIC from 16 to 512 μg/mL). By decreasing the chain length, **3b** is less active showing higher MIC in the range of 512–1024 μg/mL. The molecule obtained by substitution of the terminal double bond with a triple bond, that is **3c**, was detrimental to the activity (MIC 1024 or >1024 μg/mL). From a general point of view, an increase in chain length seems to produce sucrose ester surfactants less effective against the tested fungi strains. Indeed, **3e–h** showed higher MIC values than **3d** with only one exception. Indeed, **3e** and **3f** seem selective and very active against *A. fumigatus* (MIC 16 and 32 μg/mL, respectively). The only saturated fatty acid containing tail compound, that is **3a** is less effective than **3d** (MIC 1024 μg/mL or greater). Alkyl aromatic and aromatic sucrose esters (**3i–m**) are also less effective against the tested fungi compared to **3d**. They showed MIC of 1024 or >1024 μg/mL with the only exception of **3i** and **3k** which showed interesting activities against Fusarium spp. (MIC 512 and 256 μg/mL, respectively), and *C. albicans* (MIC 256 and 1024 μg/mL, respectively), strains.

2.4. Radical Scavenging Activity

The antioxidant activity of sucrose palmitoleate (**3d**, URB1537) was investigated by the 2,2-diphenyl-1-picrylhydrazyl hydrate (DPPH) radical scavenging assay. As indicated in Figure 1, URB1537 did not show scavenger effects against DPPH radicals within the range of concentrations tested (EC_{50} >> 512 μg/mL). This evidence agrees with previous findings showing that sugar esters possess lower radical scavenging properties than the reference antioxidant molecules [24,38,39], such as quercetin (calculated EC_{50} equal to 1.74 μg/mL). In particular, sugar esters show scavenger abilities at concentrations ranging from 1 to 10 mg/mL, higher than those herein tested for URB1537 (up to 512 μg/mL).

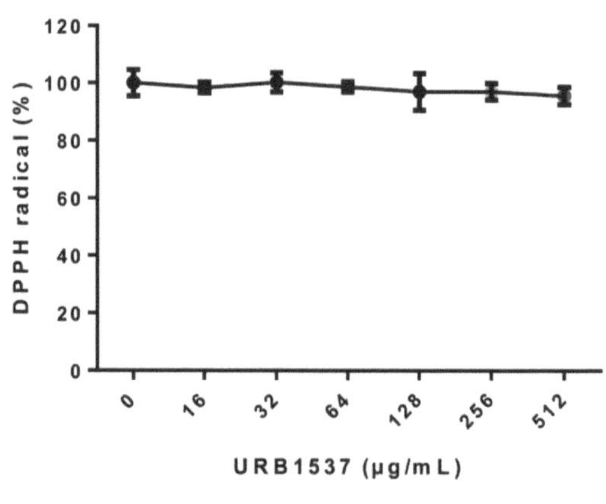

Figure 1. Evaluation of URB1537 scavenger activity against DPPH radicals. Data are expressed as the mean ± SD (n = 3).

2.5. Biocompatibility Assay

The evaluation of URB1537 cytotoxicity by 3-(4,5-dimethylthiazol-2-yl)-2,5-diphenyltet razolium bromide (MTT) assay revealed a significant reduction of cultured human keratinocyte (HaCaT) cell viability with 256 and 512 μg/mL concentrations (Figure 2). The calculated IC_{50} value was equal to 230 μg/mL. Based on cytotoxicity criteria, URB1537 could be classified as weakly cytotoxic [40]. This finding confirms that both carbon chain length and sucrose ester concentration may influence cell viability, as previously evidenced for lactose-based fatty acid monoesters [19]. However, the cytotoxicity of URB1537 in HaCaT is still higher than the MIC showed against most of the selected fungi strains and the compound is indeed safe at least up to 128 μg/mL.

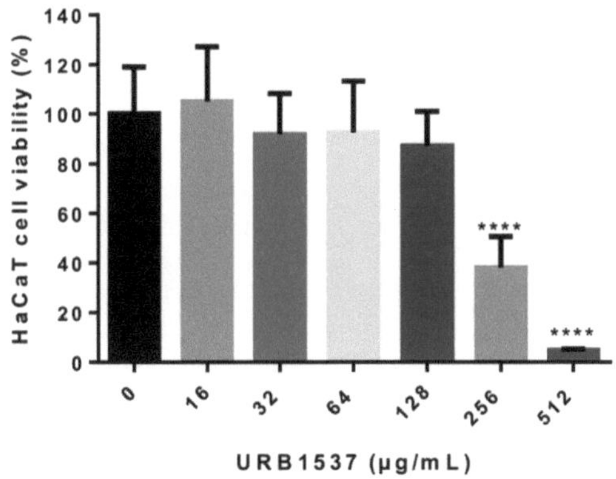

Figure 2. Cell viability evaluation after URB1537 administration to HaCaT cells for 24 h. Data are expressed as the mean ± SD (n = 3). **** $p < 0.0001$ vs. untreated control cells (Tukey's post hoc test).

2.6. Anti-Inflammatory Activity

As reported in Figure 3, the stimulation of RAW 264.7 cells (murine macrophages) by lipopolysaccharide (LPS) (CTR+) led to a strong extracellular release of the inflammatory response mediator nitric oxide (NO) as compared to untreated control cells (CTR-). When LPS-exposed cells were co-incubated with URB1537, a dose-dependent decrement of NO production was observed. The anti-inflammatory activity of URB1537 128 μg/mL was comparable to that of the reference drug dexamethasone 2 μg/mL, suggesting that this sugar ester might help to control the inflammatory response. NO release was not increased in non-stimulated RAW 264.7 cells after URB1537 administration, indicating that the compound did not lead to immune system activation (data not shown).

Overall, these data agree with previous observations demonstrating that sucrose fatty acid esters exhibit anti-inflammatory effects through different mechanisms, such as the inhibition of NF-κB activation involved in immunity and inflammation [41]. Similarly, we have recently observed that some lactose-based fatty acid esters may counteract NO production in LPS-stimulated macrophages [24].

The anti-inflammatory properties of URB1537 were also compared to those of its single constituents, i.e., sucrose (S) and palmitoleic acid (C16). As indicated in Figure 4, URB1537 and C16 presented a similar anti-inflammatory activity (p = ns URB1537 vs. C16), significantly reducing LPS-induced NO production. Sucrose alone did not show anti-inflammatory properties. This finding suggests that the anti-inflammatory capacity of URB1537 could be linked to the presence of the palmitoleate portion. Accordingly, it has been previously demonstrated that palmitoleic acid promotes anti-inflammatory effects in macrophages exposed to LPS by inhibiting the inflammasome pathway [42]. However, its real application could be limited by its oily nature, low solubility, and tendency to irritate (Safety Data Sheet). Indeed, URB1537 could be considered a safe potential prodrug of palmitoleic acid with better physicochemical and pharmacokinetic properties.

Noteworthily, no cytotoxic effects were observed after LPS, URB1537, and dexamethasone administration to RAW 264.7 cells compared to untreated control cells (Figure 5), confirming that URB1537 was safe at concentrations up to 128 μg/mL, as also evidenced on HaCaT cells.

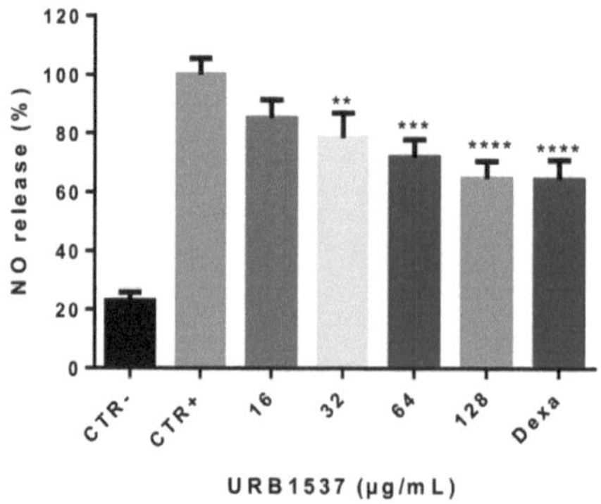

Figure 3. Extracellular NO release after RAW 264.7 stimulation by LPS for 24 h in the presence of URB1537 (16–128 μg/mL). CTR-: negative control (untreated cells); CTR+: positive control (LPS); Dexa: dexamethasone 2 μg/mL. Data are expressed as the mean ± SD ($n = 3$). ** $p < 0.01$, *** $p < 0.001$, **** $p < 0.0001$ vs. CTR+ (Tukey's post hoc test).

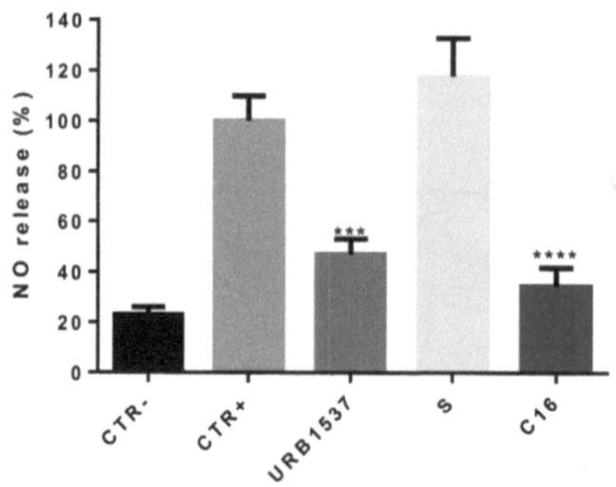

Figure 4. Extracellular NO release after RAW 264.7 stimulation by LPS for 24 h in the presence of URB1537, S, and C16 (128 μg/mL). CTR-: negative control (untreated cells); CTR+: positive control (LPS). Data are expressed as the mean ± SD ($n = 3$). *** $p < 0.001$, **** $p < 0.0001$ vs. CTR+ (Tukey's post hoc test).

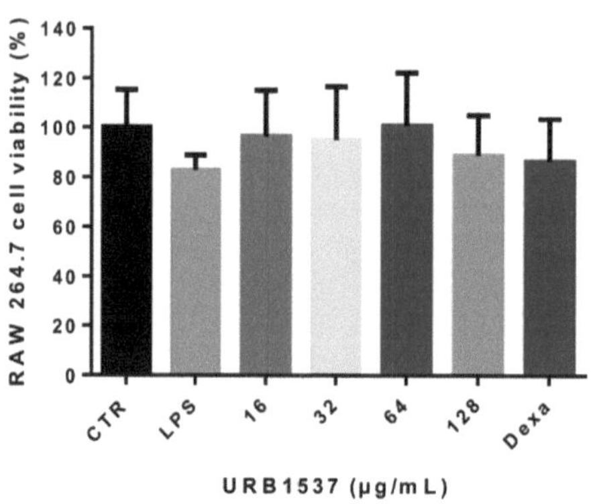

Figure 5. Cell viability evaluation after LPS (1 μg/mL), URB1537 (16–128 μg/mL), and dexamethasone (2 μg/mL) administration to RAW 264.7 cells for 24 h. Data are expressed as the mean ± SD ($n = 3$).

3. Materials and Methods

3.1. Chemicals

Caprylic and undecynoic acids were purchased from TCI (Zwijndrecht, Belgium). Undecylenic, palmitoleic, oleic, linoleic, linolenic, and nervonic acids, and sucrose were purchased from Fluorochem (Hadfield, UK). Diisopropyl azodicarboxylate (DIAD) and triphenylphosphine (PPh_3) were purchased from Alpha Aesar (Ward Hill, MA, USA). Benzoic, phenylacetic, phenylbenzoic, and biphenylacetic acids, dichloromethane (CH_2Cl_2), acetone [$CH_3C(O)CH_3$], methanol (CH_3OH), dimethyl sulfoxide (DMSO and DMSO-*d6*) were purchased from Sigma-Aldrich (Milan, Italy). The structures of compounds were unambiguously assessed by MS, ^{1}H NMR, ^{13}C NMR, and IR. ESI-MS spectra were recorded with a Waters Micromass ZQ spectrometer in a negative or positive mode using nebulizing nitrogen gas at 400 L/min and a temperature of 250 °C, cone flow 40 mL/min, capillary 3.5 kV and cone voltage 60 V; only molecular ions $[M-H]^-$ or $[M + NH_4]^+$ or $[M + Na]^+$ are given. HRMS spectra were performed by slow direct infusion (5 μL/min) of ≈ 0.1 μg/mL solution (acetonitrile/0.1% aqueous formic acid 1:1) of new compounds, using Orbitrap Exploris 240 mass spectrometer (Thermo Scientific, Waltham, MA, USA); only molecular ions $[M + Na]^+$ are given. ^{1}H NMR and ^{13}C NMR spectra were recorded on a Bruker AC 400 or 101, respectively, spectrometer and analyzed using the TopSpin 1.3 software package. Chemical shifts were measured by using the central peak of the solvent. Column chromatography purifications were performed under "flash" conditions using Merck 230–400 mesh silica gel. TLC was carried out on Merck silica gel 60 F254 plates, which were visualized by exposure to ultraviolet light and by exposure to an aqueous solution of ceric ammonium molybdate.

3.2. General Procedure for the Synthesis of Sucrose Ester Surfactants (3a–m, URB1480-1482, URB1534-1543)

Sucrose (**1**) (0.342 g, 1 mmol) was dissolved in dry DMF (7.9 mL) at 70 °C and stirred under N_2 atmosphere. The mixture was cooled to room temperature and then PPh_3 (0.656 g, 2.5 mmol), the appropriate carboxylic acid (**2a–m**) (1.5 mmol), and DMF (2.1 mL) were added. After complete dissolution, the mixture was cooled to 0 °C and DIAD (0.508 g, 0.493 mL, 2.5 mmol) was introduced. The mixture was stirred at room temperature until the total consumption of sucrose (24–30 h) and concentrated. Purification of the residue by column chromatography ($CH_2Cl_2/CH_3C(O)CH_3/MeOH/H_2O$ 78:10:10:1.5) gave **3a–m** as solids.

β-D-Fructofuranosyl 6-*O*-octanoyl-α-D-glucopyranoside, sucrose caprylate (**3a**, URB 1534) [43].

White solid. Yield: 50%. MS (ESI): 467 $[M-H]^-$, 486 $[M + NH_4]^+$, 491 $[M + Na]^+$. ^{1}H NMR (DMSO-d_6): δ = 0.86 (t, 3H, J = 7.0 Hz, CH_3) 1.22–1.30 [m, 8H, $(CH_2)_4$], 1.48–1.55 (m, 2H, CH_2CH_2COOR), 2.29–2.33 (m, 2H, CH_2COOR), 3.06 (ddd, 1H, $J_{H4\text{-}OH4}$ = 6.0 Hz, $J_{H4\text{-}H5}$ = 9.0 Hz, $J_{H4\text{-}H3}$ = 9.5 Hz, H4), 3.21 (ddd, 1H, $J_{H2\text{-}H1}$ = 3.5 Hz, $J_{H2\text{-}OH2}$ = 6.0 Hz, $J_{H2\text{-}H3}$ = 9.5 Hz, H2), 3.38–3.41 (m, 2H, H1′a, H1′b), 3.49 (ddd, 1H, $J_{H3\text{-}OH3}$ = 5.0 Hz, $J_{H3\text{-}H2} \approx J_{H3\text{-}H4}$ = 9.5 Hz, H3), 3.53–3.60 (m, 3H, H5′, H6′a, H6′b), 3.70–3.76 (m, 1H, H4′), 3.88 (dd, 1H, $J_{H3'\text{-}OH3'} \approx J_{H3'\text{-}H4'}$ = 8.0 Hz, H3′), 3.90 (ddd, 1H, $J_{H5\text{-}H6b}$ = 1.5 Hz, $J_{H5\text{-}H6a}$ = 6.0 Hz, $J_{H5\text{-}H4}$ = 9.0 Hz, H5), 4.02 (dd, 1H, $J_{H6a\text{-}H5}$ = 6.0 Hz, $J_{H6a\text{-}H6b}$ = 11.5 Hz, H6a), 4.23 (dd, 1H, $J_{H6b\text{-}H5}$ = 1.5 Hz, $J_{H6b\text{-}H6a}$ = 11.5 Hz, H6b), 4.38 (dd, 1H, $J_{OH6'\text{-}H6'a} \approx J_{OH6'\text{-}H6'b}$ = 6.0 Hz, OH6′), 4.55 (d, 1H, $J_{OH3'\text{-}H3'}$ = 8.0 Hz, OH3′), 4.80 (dd, 1H, $J_{OH1'\text{-}H1'a} \approx J_{OH1'\text{-}H1'b}$ = 6.5 Hz, OH1′), 4.88 (d, 1H, $J_{OH3\text{-}H3}$ = 5.0 Hz, OH3), 5.00 (d, 1H, $J_{OH4\text{-}H4}$ = 6.0 Hz, OH4), 5.12 (d, 1H, $J_{OH2\text{-}H2}$ = 6.0 Hz, OH2), 5.15 (d, 1H, $J_{OH4'\text{-}H4'}$ = 6.0 Hz, OH4′), 5.18 (d, 1H, $J_{H1\text{-}H2}$ = 3.5 Hz, H1) ppm. ^{13}C NMR (DMSO-d_6): δ = 14.4, 22.5, 24.9, 28.8, 28.9, 31.6, 33.8, 62.6, 63.1, 64.0, 70.4, 70.6, 72.0, 73.1, 75.0, 77.3, 83.2, 91.9, 104.3, 173.5 ppm.

β-D-Fructofuranosyl 6-*O*-undec-10-enoyl-α-D-glucopyranoside, sucrose undec-10-enoate (**3b**, URB1535).

White solid. Yield: 39%. MS (ESI): 507 $[M–H]^-$, 526 $[M + NH_4]^+$, 531 $[M + Na]^+$. ^1H NMR (DMSO-d_6): δ = 1.24–1.27 [m, 8H, $(CH_2)_4$], 1.33–1.35 (m, 2H, $CH_2CH_2CH{=}CH_2$), 1.49–1.53 (m, 2H, CH_2CH_2COOR), 1.98–2.04 (m, 2H, $CH_2CH{=}CH_2$), 2.28–2.32 (m, 2H, CH_2COOR), 3.06 (ddd, 1H, $J_{H4\text{-}OH4}$ = 5.0 Hz, $J_{H4\text{-}H3} \approx J_{H4\text{-}H5}$ = 9.5 Hz, H4), 3.20 (ddd, 1H, $J_{H2\text{-}H1}$ = 3.5 Hz, $J_{H2\text{-}OH2}$ = 6.0 Hz, $J_{H2\text{-}H3}$ = 9.5 Hz, H2), 3.37–3.40 (m, 2H, H1′a, H1′b), 3.48 (m, 1H, H3), 3.53–3.62 (m, 3H, H5′, H6′a, H6′b), 3.70–3.76 (m, 1H, H4′), 3.88 (dd, 1H, $J_{H3'\text{-}OH3'} \approx J_{H3'\text{-}H4'}$ = 8.0 Hz, H3′), 3.91 (ddd, 1H, $J_{H5\text{-}H6b}$ = 1.5 Hz, $J_{H5\text{-}H6a}$ = 6.0 Hz, $J_{H5\text{-}H4}$ = 9.5 Hz, H5), 4.01 (dd, 1H, $J_{H6a\text{-}H5}$ = 6.0 Hz, $J_{H6a\text{-}H6b}$ = 11.5 Hz, H6a), 4.23 (dd, 1H, $J_{H6b\text{-}H5}$ = 1.5 Hz, $J_{H6b\text{-}H6a}$ = 11.5 Hz, H6b), 4.39 (dd, 1H, $J_{OH6'\text{-}H6'a} \approx J_{OH6'\text{-}H6'b}$ = 5.0 Hz, OH6′), 4.56 (d, 1H, $J_{OH3'\text{-}H3'}$ = 8.0 Hz, OH3′), 4.81 (dd, 1H, $J_{OH1'\text{-}H1'a} \approx J_{OH1'\text{-}H1'b}$ = 6.0 Hz, OH1′), 4.89 (brs, 1H, OH3), 4.94 (dddd, 1H, $J_{gem} \approx J_1$ = 1.5 Hz, J_2 = 4.0 Hz, J_{cis} = 10.0 Hz, $HCH{=}CHCH_2$), 5.00 (dddd, 1H, $J_{gem} \approx J_1 \approx J_2$ = 1.5 Hz, J_{trans} = 17.0 Hz, $HCH{=}CHCH_2$), 5.01 (d, 1H, $J_{OH4\text{-}H4}$ = 5.0 Hz, OH4), 5.13 (d,1H, $J_{OH2\text{-}H2}$ = 6.0 Hz, OH2), 5.16 (d, 1H, $J_{OH4'\text{-}H4'}$ = 6.0 Hz, OH4′), 5.18 (d, 1H, $J_{H1\text{-}H2}$ = 3.5 Hz, H1), 5.80 (dddd, 1H, $J_1 \approx J_2$ = 7.0 Hz, J_{cis} = 10.0 Hz, J_{trans} = 17.0 Hz, $CH_2{=}CHCH_2$) ppm. ^{13}C NMR (DMSO-d_6): δ = 24.9, 28.7, 28.9, 29.1, 29.1, 29.2, 33.6, 33.8, 62.6, 63.1, 64.0, 70.5, 70.6, 72.0, 73.1, 75.0, 77.4, 83.2, 91.9, 104.4, 115.1, 139.3, 173.5 ppm. HRMS (ESI) m/z for $C_{23}H_{40}NaO_{12}$ $[M + Na]^+$: calcd 531.2412; found 531.2411.

β-D-Fructofuranosyl 6-*O*-undec-10-inoyl-α-D-glucopyranoside, sucrose undec-10-inoate (**3c**, URB1536).

White solid. Yield: 37%. MS (ESI): 505 $[M–H]^-$, 524 $[M + NH_4]^+$, 529 $[M + Na]^+$. ^1H NMR (DMSO-d_6): δ = 1.23–1.27 [m, 6H, $(CH_2)_3$], 1.30–1.37 (m, 2H, $CH_2CH_2CH_2CCH$), 1.40–1.47 (m, 2H, CH_2CH_2CCH), 1.49–1.55 (m, 2H, CH_2CH_2COOR), 2.14 (td, 2H, J_1 = 2.5 Hz, J_2 = 7.0 Hz, CH_2CCH), 2.28–2.33 (m, 2H, CH_2COOR), 2.72 (t, 1H, J = 2.5 Hz, CCH), 3.06 (ddd, 1H, $J_{H4\text{-}OH4}$ = 5.0 Hz, $J_{H4\text{-}H3} \approx J_{H4\text{-}H5}$ = 9.5 Hz, H4), 3.21 (ddd, 1H, $J_{H2\text{-}H1}$ = 3.5 Hz, $J_{H2\text{-}OH2}$ = 6.0 Hz, $J_{H2\text{-}H3}$ = 9.5 Hz, H2), 3.36–3.41 (m, 2H, H1′a, H1′b), 3.49 (ddd, 1H, $J_{H3\text{-}OH3}$ = 5.0 Hz, $J_{H3\text{-}H2} \approx J_{H3\text{-}H4}$ = 9.5 Hz, H3), 3.53–3.62 (m, 3H, H5′, H6′a, H6′b), 3.70–3.76 (m, 1H, H4′), 3.88 (dd, 1H, $J_{H3'\text{-}OH3'} \approx J_{H3'\text{-}H4'}$ = 8.0 Hz, H3′), 3.90 (ddd, 1H, $J_{H5\text{-}H6b}$ = 1.5 Hz, $J_{H5\text{-}H6a}$ = 6.0 Hz, $J_{H5\text{-}H4}$ = 9.5 Hz, H5), 4.02 (dd, 1H, $J_{H6a\text{-}H5}$ = 6.0 Hz, $J_{H6a\text{-}H6b}$ = 11.5 Hz, H6a), 4.23 (dd, 1H, $J_{H6b\text{-}H5}$ = 1.5 Hz, $J_{H6b\text{-}H6a}$ = 11.5 Hz, H6b), 4.38 (dd, 1H, $J_{OH6'\text{-}H6'a} \approx J_{OH6'\text{-}H6'b}$ = 5.0 Hz, OH6′), 4.56 (d, 1H, $J_{OH3'\text{-}H3'}$ = 8.0 Hz, OH3′), 4.80 (dd, 1H, $J_{OH1'\text{-}H1'a} \approx J_{OH1'\text{-}H1'b}$ = 6.0 Hz, OH1′), 4.88 (d, 1H, $J_{OH3\text{-}H3}$ = 5.0 Hz, OH3), 5.00 (d, 1H, $J_{OH4\text{-}H4}$ = 5.0 Hz, OH4), 5.13 (d, 1H, $J_{OH2\text{-}H2}$ = 6.0 Hz, OH2), 5.16 (d, 1H, $J_{OH4'\text{-}H4'}$ = 6.0 Hz, OH4′), 5.18 (d, 1H, $J_{H1\text{-}H2}$ = 3.5 Hz, H1) ppm. ^{13}C NMR (DMSO-d_6) δ 18.1, 24.9, 28.4, 28.5, 28.8, 28.9, 29.1, 33.8, 62.7, 63.0, 64.0, 70.5, 70.6, 71.5, 72.0, 73.1, 75.0, 77.4, 83.2, 85.0, 91.9, 104.4, 173.5 ppm. HRMS (ESI) m/z for $C_{23}H_{38}NaO_{12}$ $[M + Na]^+$: calcd 529.2256; found 529.2253.

β-D-Fructofuranosyl 6-*O*-hexadec-9-enoyl-α-D-glucopyranoside, sucrose palmitoleate (**3d**, URB1537) [25].

White solid. Yield: 30%. MS (ESI): 577 $[M–H]^-$, 596 $[M + NH_4]^+$, 601 $[M + Na]^+$. ^1H NMR (DMSO-d_6): δ = 0.86 (t, 3H, J = 7.0 Hz, CH_3), 1.26–1.30 [m, 16H, $(CH_2)_8$], 1.50–1.53 (m 2H, CH_2CH_2COOR), 1.97–2.00 (m, 4H, $CH_2CH{=}CHCH_2$), 2.28–2.32 (m, 2H, CH_2COOR), 3.06 (ddd, 1H, $J_{H4\text{-}OH4}$ = 5.0 Hz, $J_{H4\text{-}H3} \approx J_{H4\text{-}H5}$ = 9.5 Hz, H4), 3.19 (ddd, 1H, $J_{H2\text{-}H1}$ = 3.5 Hz, $J_{H2\text{-}OH2}$ = 4.0 Hz, $J_{H2\text{-}H3}$ = 9.5 Hz, H2), 3.38–3.39 (m, 2H, H1′a, H1′b), 3.49 (m, 1H, H3), 3.54–3.62 (m, 3H, H5′, H6′a, H6′b), 3.71–3.76 (m, 1H, H4′), 3.88 (dd, 1H, $J_{H3'\text{-}OH3'} \approx J_{H3'\text{-}H4'}$ = 8.0 Hz, H3′), 3.91 (ddd, 1H, $J_{H5\text{-}H6b}$ = 1.5 Hz, $J_{H5\text{-}H6a}$ = 6.0 Hz, $J_{H5\text{-}H4}$ = 9.5 Hz, H5), 4.03 (dd, 1H, $J_{H6a\text{-}H5}$ = 6.0 Hz, $J_{H6a\text{-}H6b}$ = 11.5 Hz, H6a), 4.23 (dd, 1H, $J_{H6b\text{-}H5}$ = 1.5 Hz, $J_{H6b\text{-}H6a}$ = 11.5 Hz, H6b), 4.38 (dd, 1H, $J_{OH6'\text{-}H6'a} \approx J_{OH6'\text{-}H6'b}$ = 5.0 Hz, OH6′), 4.55 (d, 1H, $J_{OH3'\text{-}H3'}$ = 8.0 Hz, OH3′), 4.80 (dd, 1H, $J_{OH1'\text{-}H1'a} \approx J_{OH1'\text{-}H1'b}$ = 6.0 Hz, OH1′), 4.88 (brs, 1H, OH3), 5.00 (d, 1H, $J_{OH4\text{-}H4}$ = 5.0 Hz, OH4), 5.12 (d, 1H, $J_{OH2\text{-}H2}$ = 4.0 Hz, OH2), 5.14 (d, 1H, $J_{OH4'\text{-}H4'}$ = 6.0 Hz, OH4′), 5.18 (d, 1H, $J_{H1\text{-}H2}$ = 3.5 Hz, H1), 5.31 (ddd, 1H, $J_1 \approx J_2$ = 6.0 Hz, J_3 = 11.0 Hz, CH=CH), 5.34 (ddd, 1H, $J_1 \approx J_2$ = 6.0 Hz, J_3 = 11.0 Hz, CH=CH) ppm.

^{13}C NMR (DMSO-d_6): δ = 14.4, 22.5, 24.8, 27.1 (2C), 28.7, 29.0, 29.1, 29.6, 31.6, 33.8, 62.6, 63.1, 64.0, 70.4, 70.6, 72.0, 73.1, 75.0, 77.4, 83.2, 91.9, 104.4, 130.1 (2C), 173.5 ppm.

β-D-Fructofuranosyl 6-*O*-octadec-9-enoyl-α-D-glucopyranoside, sucrose oleate (**3e**, URB1538) [44].

White solid. Yield: 36%. MS (ESI): 605 [M–H]$^-$, 624 [M + NH$_4$]$^+$, 629 [M + Na]$^+$. ^{1}H NMR (DMSO-d_6): δ = 0.86 (t, 3H, J = 6.5 Hz, CH$_3$) 1.22–1.33 [m, 20H, (CH$_2$)$_{10}$], 1.47–1.55 (m, 2H, C*H*$_2$CH$_2$COOR), 1.95–2.01 (m, 4H, C*H*$_2$CH=CHC*H*$_2$), 2.28–2.32 (m, 2H, C*H*$_2$COOR), 3.06 (ddd, 1H, $J_{H4\text{-}OH4}$ = 6.0 Hz, $J_{H4\text{-}H3} \approx J_{H4\text{-}H5}$ = 9.5 Hz, H4), 3.20 (ddd, 1H, $J_{H2\text{-}H1}$ = 3.5 Hz, $J_{H2\text{-}OH2}$ = 6.0 Hz, $J_{H2\text{-}H3}$ = 9.5 Hz, H2), 3.39–3.40 (m, 2H, H1′a, H1′b), 3.48 (ddd, 1H, $J_{H3\text{-}OH3}$ = 5.0 Hz, $J_{H3\text{-}H2} \approx J_{H3\text{-}H4}$ = 9.5 Hz, H3), 3.52–3.62 (m, 3H, H5′, H6′a, H6′b), 3.70–3.76 (m, 1H, H4′), 3.87 (dd, 1H, $J_{H3'\text{-}OH3'} \approx J_{H3'\text{-}H4'}$= 8.0 Hz, H3′), 3.91 (ddd, 1H, $J_{H5\text{-}H6b}$ = 1.5 Hz, $J_{H5\text{-}H6a}$ = 6.0 Hz, $J_{H5\text{-}H4}$ = 9.5 Hz, H5), 4.02 (dd, 1H, $J_{H6a\text{-}H5}$ = 6.0 Hz, $J_{H6a\text{-}H6b}$ = 11.5 Hz, H6a), 4.23 (dd, 1H, $J_{H6b\text{-}H5}$ = 1.5 Hz, $J_{H6b\text{-}H6a}$ = 11.5 Hz, H6b), 4.39 (dd, 1H, $J_{OH6'\text{-}H6'a} \approx J_{OH6'\text{-}H6'b}$ = 5.0 Hz, OH6′), 4.56 (d, 1H, $J_{OH3'\text{-}H3'}$ = 8.0 Hz, OH3′), 4.81 (dd, 1H, $J_{OH1'\text{-}H1'a} \approx J_{OH1'\text{-}H1'b}$ = 6.0 Hz, OH1′), 4.88 (d, 1H, $J_{OH3\text{-}H3}$ = 5.0 Hz, OH3), 5.00 (d, 1H, $J_{OH4\text{-}H4}$ = 6.0 Hz, OH4), 5.13 (d, 1H, $J_{OH2\text{-}H2}$ = 6.0 Hz, OH2), 5.16 (d, 1H, $J_{OH4'\text{-}H4'}$ = 6.0 Hz, OH4′), 5.18 (d, 1H, $J_{H1\text{-}H2}$ = 3.5 Hz, H1), 5.31 (ddd, 1H, $J_1 \approx J_2$ = 6.0 Hz, J_3 = 11.0 Hz, CH=CH), 5.35 (ddd, 1H, $J_1 \approx J_2$ = 6.0 Hz, J_3 = 11.0 Hz, CH=CH) ppm. ^{13}C NMR (DMSO-d_6): δ = 14.4, 22.5, 24.9, 27.0, 27.1, 28.95, 28.97, 29.03, 29.06, 29.12, 29.3, 29.6, 31.7, 33.7, 62.6, 63.0, 64.0, 70.4, 70.6, 72.0, 73.1, 75.0, 77.4, 83.2, 92.0, 104.4, 130.1(2C), 173.5 ppm.

β-D-Fructofuranosyl 6-*O*-octadec-9,12-enoyl-α-D-glucopyranoside, sucrose linoleate (**3f**, URB1539) [41].

White solid. Yield: 23%. MS (ESI): 603 [M–H]$^-$, 622 [M + NH$_4$]$^+$, 627 [M + Na]$^+$. ^{1}H NMR (DMSO-d_6): δ = 0.86 (t, 3H, J = 6.5 Hz, CH$_3$), 1.22–1.34 [m, 14H, (CH$_2$)$_7$], 1.49–1.54 (m 2H, C*H*$_2$CH$_2$COOR), 2.00–2.05 (m, 4H, C*H*$_2$CH=CHCH$_2$CH=CHC*H*$_2$), 2.28–2.32 (m, 2H, C*H*$_2$COOR), 2.74 (m, 2H, CH=CHC*H*$_2$CH=CH), 3.06 (ddd, 1H, $J_{H4\text{-}OH4}$ = 5.5 Hz, $J_{H4\text{-}H3} \approx J_{H4\text{-}H5}$ = 9.0 Hz, H4), 3.20 (ddd, 1H, $J_{H2\text{-}H1}$ = 3.5 Hz, $J_{H2\text{-}OH2}$ = 6.0 Hz, $J_{H2\text{-}H3}$ = 9.0 Hz, H2), 3.39–3.40 (m, 2H, H1′a, H1′b), 3.48 (m, 1H, H3), 3.53–3.61 (m, 3H, H5′, H6′a, H6′b), 3.71–3.77 (m, 1H, H4′), 3.87 (dd, 1H, $J_{H3'\text{-}OH3'} \approx J_{H3'\text{-}H4'}$ = 8.0 Hz, H3′), 3.90–3.93 (m, 1H, H5), 4.02 (dd, 1H, $J_{H6a\text{-}H5}$ = 6.0 Hz, $J_{H6a\text{-}H6b}$ = 11.5 Hz, H6a), 4.22 (dd, 1H, $J_{H6b\text{-}H6a}$ = 11.5 Hz, H6b), 4.38 (dd, 1H, $J_{OH6'\text{-}H6'a} \approx J_{OH6'\text{-}H6'b}$ = 5.0 Hz, OH6′), 4.56 (d, 1H, $J_{OH3'\text{-}H3'}$ = 8.0 Hz, OH3′), 4.81 (dd, 1H, $J_{OH1'\text{-}H1'a} \approx J_{OH1'\text{-}H1'b}$ = 6.0 Hz, OH1′), 4.89 (d, 1H, $J_{OH3\text{-}H3}$ = 4.5 Hz, OH3), 5.00 (d, 1H, $J_{OH4\text{-}H4}$ = 5.5 Hz, OH4), 5.12 (d,1H, $J_{OH2\text{-}H2}$ = 6.0 Hz, OH2), 5.14 (d, 1H, $J_{OH4'\text{-}H4'}$ = 6.5 Hz, OH4′), 5.18 (d, 1H, $J_{H1\text{-}H2}$ = 3.5 Hz, H1), 5.27–5.38 (m, 4H, C*H*=C*H*CH$_2$C*H*=C*H*) ppm. ^{13}C NMR (DMSO-d_6): δ = 14.4, 22.4, 24.9, 25.7, 27.0, 27.1, 28.95, 29.0, 29.1, 29.2, 29.5, 31.3, 33.8, 62.6, 63.0, 64.0, 70.4, 70.6, 72.0, 73.0, 75.0, 77.4, 83.2, 91.9, 104.4, 128.2(2C), 130.2(2C), 173.5 ppm.

β-D-Fructofuranosyl 6-*O*-octadec-9,12,15-enoyl-α-D-glucopyranoside, sucrose linolenate (**3g**, URB1540).

White solid. Yield: 25%. MS (ESI): 601 [M–H]$^-$, 620 [M + NH$_4$]$^+$, 625 [M + Na]$^+$. ^{1}H NMR (DMSO-d_6): δ = 0.93 (t, 3H, J = 7.5 Hz, CH$_3$) 1.19–1.26 [m, 8H, (CH$_2$)$_4$], 1.49–1.54 (m 2H, C*H*$_2$CH$_2$COOR), 2.02–2.08 (m, 4H, C*H*$_2$CH=CHCH$_2$CH=CHCH$_2$CH=CHC*H*$_2$), 2.28–2.32 (m, 2H, C*H*$_2$COOR), 2.73–2.79 (m, 4H, CH=CHC*H*$_2$CH=CHC*H*$_2$CH=CH), 3.06 (ddd, 1H, $J_{H4\text{-}OH4}$ = 6.0 Hz, $J_{H4\text{-}H3} \approx J_{H4\text{-}H5}$ = 9.5 Hz, H4), 3.21 (ddd, 1H, $J_{H2\text{-}H1}$ = 3.5 Hz, $J_{H2\text{-}OH2}$ = 6.0 Hz, $J_{H2\text{-}H3}$ = 9.5 Hz, H2), 3.38–3.39 (m, 2H, H1′a, H1′b), 3.49 (ddd, 1H, $J_{H3\text{-}OH3}$ = 5.0 Hz, $J_{H3\text{-}H2} \approx J_{H3\text{-}H4}$ = 9.5 Hz, H3), 3.54–3.61 (m, 3H, H5′, H6′a, H6′b), 3.70–3.76 (m, 1H, H4′), 3.86 (dd, 1H, $J_{H3'\text{-}OH3'} \approx J_{H3'\text{-}H4'}$ = 8.0 Hz, H3′), 3.90–3.93 (m, 1H, H5), 4.02 (dd, 1H, $J_{H6a\text{-}H5}$ = 6.0 Hz, $J_{H6a\text{-}H6b}$ = 11.5 Hz, H6a), 4.23 (dd, 1H, $J_{H6b\text{-}H5}$ = 1.5 Hz, $J_{H6b\text{-}H6a}$ = 11.5 Hz, H6b), 4.38 (dd, 1H, $J_{OH6'\text{-}H6'a} \approx J_{OH6'\text{-}H6'b}$ = 5.0 Hz, OH6′), 4.56 (d, 1H, $J_{OH3'\text{-}H3'}$ = 8.0 Hz, OH3′), 4.80 (dd, 1H, $J_{OH1'\text{-}H1'a} \approx J_{OH1'\text{-}H1'b}$ = 6.5 Hz, OH1′), 4.89 (d, 1H, $J_{OH3\text{-}H3}$ = 5.0 Hz, OH3), 5.00 (d, 1H, $J_{OH4\text{-}H4}$ = 6.00 Hz, OH4), 5.13 (d, 1H, $J_{OH2\text{-}H2}$ = 6.0 Hz, OH2), 5.15 (d, 1H, $J_{OH4'\text{-}H4'}$ = 6.0 Hz, OH4′), 5.18 (d, 1H, $J_{H1\text{-}H2}$ = 3.5 Hz, H1), 5.25–5.40 (m, 6H, C*H*=C*H*CH$_2$C*H*=C*H*CH$_2$C*H*=C*H*) ppm. ^{13}C NMR (DMSO-d_6): δ = 14.6, 20.5, 24.8, 25.6, 25.7, 27.1, 28.9, 29.0, 29.1, 29.5, 33.8, 62.6, 63.0, 64.0, 70.4, 70.6, 72.0, 73.1, 75.0, 77.4, 83.2, 92.0,

104.4, 127.4, 128.0, 128.3, 128.4, 130.4, 132.0, 173.5 ppm. HRMS (ESI) m/z for $C_{30}H_{50}NaO_{12}$ $[M + Na]^+$: calcd 625.3195; found 625.3191.

β-D-Fructofuranosyl 6-*O*-tetracos-15-enoyl-α-D-glucopyranoside, sucrose nervonate (**3h**, URB1541).

White solid. Yield: 26%. MS (ESI): 689 $[M–H]^-$ 708 $[M + NH_4]^+$, 713 $[M + Na]^+$. ^{1}H NMR (DMSO-d_6): δ = 0.86 (t, 3H, J = 7.0 Hz, CH_3), 1.24–1.29 [m, 32H, $(CH_2)_{16}$], 1.49–1.53 (m 2H, CH_2CH_2COOR), 1.96–2.00 (m, 4H, $CH_2CH{=}CHCH_2$), 2.30 (t, 2H, J = 7.5 Hz, CH_2COOR), 3.06 (ddd, 1H, $J_{H4\text{-}OH4}$ = 5.5 Hz, $J_{H4\text{-}H3} \approx J_{H4\text{-}H5}$ = 9.5 Hz, H4), 3.21 (ddd, 1H, $J_{H2\text{-}H1}$ = 3.5 Hz, $J_{H2\text{-}OH2}$ = 6.0 Hz, $J_{H2\text{-}H3}$ = 9.5 Hz, H2), 3.38–3.40 (m, 2H, H1′a, H1′b), 3.49 (dd, 1H, $J_{H3\text{-}H2} \approx J_{H3\text{-}H4}$ = 9.5 Hz, H3), 3.54–3.62 (m, 3H, H5′, H6′a, H6′b), 3.71–3.76 (m, 1H, H4′), 3.88 (dd, 1H, $J_{H3'\text{-}OH3'} \approx J_{H3'\text{-}H4'}$ = 8.0 Hz, H3′), 3.89–3.93 (m, 1H, H5), 4.02 (dd, 1H, $J_{H6a\text{-}H5}$ = 6.0 Hz, $J_{H6a\text{-}H6b}$ = 11.5 Hz, H6a), 4.23 (dd, 1H, $J_{H6b\text{-}H5}$ = 1.5 Hz, $J_{H6b\text{-}H6a}$ = 11.5 Hz, H6b), 4.37 (dd, 1H, $J_{OH6'\text{-}H6'a} \approx J_{OH6'\text{-}H6'b}$ = 5.5 Hz, OH6′), 4.56 (d, 1H, $J_{OH3'\text{-}H3'}$ = 8.0 Hz, OH3′), 4.80 (dd, 1H, $J_{OH1'\text{-}H1'a} \approx J_{OH1'\text{-}H1'b}$ = 6.5 Hz, OH1′), 4.89 (d, 1H, $J_{OH3\text{-}H3}$ = 3.5 Hz, OH3), 5.00 (d, 1H, $J_{OH4\text{-}H4}$ = 5.5 Hz, OH4), 5.13 (d, 1H, $J_{OH2\text{-}H2}$ = 6.0 Hz, OH2), 5.15 (d, 1H, $J_{OH4'\text{-}H4'}$ = 6.0 Hz, OH4′), 5.18 (d, 1H, $J_{H1\text{-}H2}$ = 3.5 Hz, H1), 5.31 (ddd, 1H, $J_1 \approx J_2$ = 6.0 Hz, J_3 = 11.0 Hz, CH=CH), 5.34 (ddd, 1H, $J_1 \approx J_2$ = 6.0 Hz, J_3 = 11.0 Hz CH=CH) ppm ^{13}C NMR (DMSO-d_6): δ = 14.3, 22.6, 24.9, 27.0, 29.0, 29.1, 29.29, 29.31, 29.4, 29.5, 31.75, 33.8, 62.7, 63.0, 64.0, 70.5, 70.6, 72.0, 73.1, 75.0, 77.4, 83.2, 92.0, 104.4, 130.1, 173.5 ppm. HRMS (ESI) m/z for $C_{36}H_{66}NaO_{12}$ $[M + Na]^+$: calcd 713.4447; found 713.4443.

β-D-Fructofuranosyl 6-*O*-benzoyl-α-D-glucopyranoside, sucrose benzoate (**3i**, URB1542) [45].

White solid. Yield: 43%. MS (ESI): 445 $[M–H]^-$, 428 $[M + NH_4]^+$, 433 $[M + Na]^+$. ^{1}H NMR (DMSO-d_6): δ = 3.26 (ddd, 1H, $J_{H4\text{-}OH4}$ = 6.0 Hz, $J_{H4\text{-}H3} \approx J_{H4\text{-}H5}$ = 9.5 Hz, H4), 3.28 (ddd, 1H, $J_{H2\text{-}H1}$ = 3.5 Hz, $J_{H2\text{-}OH2}$ = 6.0 Hz, $J_{H2\text{-}H3}$ = 9.5 Hz, H2), 3.41–3.43 (m, 2H, H1′a, H1′b), 3.50–3.61 (m, 4H, H3, H5′, H6′a, H6′b), 3.79 (ddd, 1H, $J_{H4'\text{-}OH4}$ = 6.00 Hz, $J_{H4'\text{-}H3'} \approx J_{H4'\text{-}H5'}$ = 8.00 Hz, H4′), 3.91 (dd, 1H, $J_{H3'\text{-}OH3'} \approx J_{H3'\text{-}H4'}$ = 8.0 Hz, H3′), 4.04–4.11 (m, 1H, H5), 4.34 (dd, 1H, $J_{H6a\text{-}H5}$ = 5.0 Hz, $J_{H6a\text{-}H6b}$ = 11.5 Hz, H6a), 4.37 (dd, 1H, $J_{OH6'\text{-}H6'a} \approx J_{OH6'\text{-}H6'b}$ = 5.5 Hz, OH6′), 4.45 (dd, 1H, $J_{H6b\text{-}H5}$ = 1.5 Hz, $J_{H6b\text{-}H6a}$ = 11.5 Hz, H6b), 4.65 (d, 1H, $J_{OH3'\text{-}H3'}$ = 8.0 Hz, OH3′), 4.83 (dd, 1H, $J_{OH1'\text{-}H1'a} \approx J_{OH1'\text{-}H1'b}$ = 6.0 Hz, OH1′), 4.92 (d, 1H, $J_{OH3\text{-}H3}$ = 5.0 Hz, OH3), 5.14 (d, 1H, $J_{OH4\text{-}H4}$ = 6.0 Hz, OH4), 5.15 (d, 1H, $J_{OH2\text{-}H2}$ = 6.0 Hz, OH2), 5.19 (d, 1H, $J_{OH4'\text{-}H4'}$ = 6.0 Hz, OH4′), 5.23 (d, 1H, $J_{H1\text{-}H2}$ = 3.5 Hz, H1), 7.51–7.57 (m, 2H, ArH), 7.67 [dddd, 1H, $J_1 \approx J_2$ = 1.0 Hz, $J_3 \approx J_4$ = 8.5 Hz, ArH(*p*)], 7.97–8.01 (m, 2H, ArH) ppm. ^{13}C NMR (DMSO-d_6): δ = 62.6, 63.0, 64.8, 70.5, 72.0, 73.1, 75.0, 77.4, 83.1, 92.2, 104.5, 129.2, 129.7, 130.1, 133.7, 173.5 ppm.

β-D-Fructofuranosyl 6-*O*-[2-(4-phenyl)benzoyl]-α-D-glucopyranoside, sucrose *p*-phenyl benzoate (**3j**, URB1481).

Pale yellow solid. Yield: 34%. MS (ESI): 521 $[M\text{-}H]^-$, 540 $[M + NH_4]^+$, 545 $[M + Na]^+$. ^{1}H NMR (DMSO-d_6): δ = 3.25–3.32 (m, 2H, H^4, H^2), 3.40–3.44 (m, 2H, $H^{1'a}$, $H^{1'b}$), 3.48 (ddd, 1H, $J_{H3\text{-}OH3}$ = 5.0 Hz, $J_{H3\text{-}H2} \approx J_{H3\text{-}H4}$ = 9.0 Hz, H^3), 3.52–3.62 (m, 3H, $H^{5'}$, $H^{6'a}$, $H^{6'b}$), 3.78–3.83 (m, 1H, $H^{4'}$), 3.92 (dd, 1H, $J_{H3'\text{-}OH3'} \approx J_{H3'\text{-}H4'}$ = 8.0 Hz, $H^{3'}$), 4.09 (ddd, 1H, $J_{H5\text{-}H6b}$ = 1.5 Hz, $J_{H5\text{-}H6a}$ = 5.0 Hz, $J_{H5\text{-}H4}$ = 9.0 Hz, H^5), 4.36 (dd, 1H, $J_{H6a\text{-}H5}$ = 5.0 Hz, $J_{H6a\text{-}H6b}$ = 12.0 Hz, H^{6a}), 4.41 (dd, 1H, $J_{OH6'\text{-}H6'a} \approx J_{OH6'\text{-}H6'b}$ = 6.0 Hz, $OH^{6'}$), 4.47 (dd, 1H, $J_{H6b\text{-}H5}$ = 1.5 Hz, $J_{H6b\text{-}H6a}$ = 12.0 Hz, H^{6b}), 4.68 (d, 1H, $J_{OH3'\text{-}H3'}$ = 8.0 Hz, $OH^{3'}$), 4.85 (dd, 1H, $J_{OH1'\text{-}H1'a} \approx J_{OH1'\text{-}H1'b}$ = 6.5 Hz, $OH^{1'}$), 4.96 (d, 1H, $J_{OH3\text{-}H3}$ = 5.0 Hz, OH^3), 5.17 (d, 1H, $J_{OH4\text{-}H4}$ = 5.0 Hz, OH^4), 5.18 (d,1H, $J_{OH2\text{-}H2}$ = 6.0 Hz, OH^2), 5.21 (d, 1H, $J_{OH4'\text{-}H4'}$ = 5.5 Hz, $OH^{4'}$), 5.24 (d, 1H, $J_{H1\text{-}H2}$ = 3.5 Hz, H^1), 7.42–7.46 (m, 1H, ArH), 7.50–7.54 (m, 2H, ArH), 7.74–7.76 (m, 2H, ArH), 7.82–7.85 (m, 2H, ArH), 8.05–8.08 (m, 2H, ArH) ppm. ^{13}C NMR (DMSO-d_6): δ = 62.5, 63.0, 64.8, 70.5, 70.6, 72.0, 73.1, 74.9, 77.3, 83.1, 92.2, 104.5, 127.4, 127.5, 128.9, 129.0, 129.6, 130.4, 139.4, 145.1, 166.1 ppm. HRMS (ESI) m/z for $C_{25}H_{30}NaO_{12}$ $[M + Na]^+$: calcd 545.1630; found 545.1636.

β-D-Fructofuranosyl 6-*O*-(2-phenylethanoyl)-α-D-glucopyranoside, sucrose phenyl acetate (**3k**, URB1480) [46].

White solid. Yield: 50%. MS (ESI): 459 $[M–H]^-$, 478 $[M + NH_4]^+$, 483 $[M + Na]^+$. ^1H NMR (DMSO-d_6): δ = 3.07 (ddd, 1H, $J_{H4\text{-}OH4}$ = 6.0 Hz, $J_{H4\text{-}H3} \approx J_{H4\text{-}H5}$ = 9.5 Hz, H^4), 3.21 (ddd, 1H, $J_{H2\text{-}H1}$ = 3.5 Hz, $J_{H2\text{-}OH2}$ = 6.0 Hz, $J_{H2\text{-}H3}$ = 9.5 Hz, H^2), 3.39–3.43 (m, 2H, $H^{1'a}$, $H^{1'b}$), 3.50 (ddd, 1H, $J_{H3\text{-}OH3}$ = 5.0 Hz, $J_{H3\text{-}H2} \approx J_{H3\text{-}H4}$ = 9.5 Hz, H^3), 3.58–3.64 (m, 3H, $H^{5'}$, $H^{6'a}$, $H^{6'b}$), 3.67 (d, J = 16.0 Hz, 1H, *H*CHAr), 3.72 (d, J = 16.0 Hz, 1H, HC*H*Ar), 3.76–3.82 (m, 1H, $H^{4'}$), 3.90 (dd, 1H, $J_{H3'\text{-}OH3'} \approx J_{H3'\text{-}H4'}$ = 8.0 Hz, $H^{3'}$), 3.95 (m, 1H, H^5), 4.05 (dd, 1H, $J_{H6a\text{-}H5}$ = 6.0 Hz, $J_{H6a\text{-}H6b}$ = 11.5 Hz, H^{6a}), 4.28 (dd, 1H, $J_{H6b\text{-}H5}$ = 1.5 Hz, $J_{H6b\text{-}H6a}$ = 11.5 Hz, H^{6b}), 4.43 (dd, 1H, $J_{OH6'\text{-}H6'a} \approx J_{OH6'\text{-}H6'b}$ = 5.5 Hz, $OH^{6'}$), 4.60 (d, 1H, $J_{OH3'\text{-}H3'}$ = 8.0 Hz, $OH^{3'}$), 4.83 (dd, 1H, $J_{OH1'\text{-}H1'a} \approx J_{OH1'\text{-}H1'b}$ = 6.5 Hz, $OH^{1'}$), 4.90 (d, 1H, $J_{OH3\text{-}H3}$ = 5.0 Hz, OH^3), 5.03 (d, 1H, $J_{OH4\text{-}H4}$ = 6.0 Hz, OH^4), 5.13 (d, 1H, $J_{OH2\text{-}H2}$ = 6.0 Hz, OH^2), 5.19 (d, 1H, $J_{H1\text{-}H2}$ = 3.5 Hz, H^1), 5.20 (d, 1H, $J_{OH4'\text{-}H4'}$ = 6.0 Hz, $OH^{4'}$), 7.24–7.35 (m, 5H, ArH) ppm. ^{13}C NMR (DMSO-d_6): δ = 21.2, 62.7, 63.1, 64.6, 70.6, 72.0, 73.1, 75.0, 77.4, 83.2, 92.0, 104.4, 127.2, 128.7, 129.9, 134.8, 171.7 ppm.

β-D-Fructofuranosyl 6-*O*-[2-(4-phenyl)phenylethanoyl]-α-D-glucopyranoside, sucrose *p*-biphenyl acetate (**3l**, URB1482).

White solid. Yield: 58%. MS (ESI): 527 $[M–H]^-$, 536 $[M + NH_4]^+$, 541 $[M + Na]^+$. ^1H NMR (DMSO-d_6): δ = 3.03 (ddd, 1H, $J_{H4\text{-}OH4}$ = 6.0 Hz, $J_{H4\text{-}H3} \approx J_{H4\text{-}H5}$ = 9.5 Hz, H^4), 3.13 (ddd, 1H, $J_{H2\text{-}H1}$ = 3.5 Hz, $J_{H2\text{-}OH2}$ = 6.0 Hz, $J_{H2\text{-}H3}$ = 9.5 Hz, H^2), 3.37–3.41 (m, 2H, $H^{1'a}$, $H^{1'b}$), 3.48 (ddd, 1H, $J_{H3\text{-}OH3}$ = 5.0 Hz, $J_{H3\text{-}H2} \approx J_{H3\text{-}H4}$ = 9.5 Hz, H^3), 3.56–3.63 (m, 3H, $H^{5'}$, $H^{6'a}$, $H^{6'b}$), 3.76–3.81 (m, 1H, $H^{4'}$), 3.88–3.96 (m, 2H, $H^{3'}$, H^5), 4.14 (dd, 1H, $J_{H6a\text{-}H5}$ = 5.0 Hz, $J_{H6a\text{-}H6b}$ = 11.5 Hz, H^{6a}), 4.33 (dd, 1H, $J_{H6b\text{-}H5}$ = 1.0 Hz, $J_{H6b\text{-}H6a}$ = 11.5 Hz, H^{6b}), 4.42 (dd, 1H, $J_{OH6'\text{-}H6'a} \approx J_{OH6'\text{-}H6'b}$ = 5.5 Hz, $OH^{6'}$), 4.62 (d, 1H, $J_{OH3'\text{-}H3'}$ = 8.0 Hz, $OH^{3'}$), 4.82 (dd, 1H, $J_{OH1'\text{-}H1'a} \approx J_{OH1'\text{-}H1'b}$ = 6.0 Hz, OH1′), 4.89 (d, 1H, $J_{OH3\text{-}H3}$ = 5.0 Hz, OH^3), 5.02 (d, 1H, $J_{OH4\text{-}H4}$ = 6.0 Hz, OH^4), 5.13 (d, 1H, $J_{OH2\text{-}H2}$ = 6.0 Hz, OH^2), 5.17 (d, 1H, $J_{H1\text{-}H2}$ = 3.5 Hz, H^1) 5.20 (d, 1H, $J_{OH4'\text{-}H4'}$ = 5.5 Hz, $OH^{4'}$), 7.24–7.28 (m, 2H, ArH), 7.29–7.37 (m, 7H, ArH) ppm. ^{13}C NMR (DMSO-d_6): δ = 56.1, 62.5, 63.0, 64.6, 70.3, 70.4, 71.9, 73.1, 75.0, 77.4, 83.2, 92.0, 104.5, 127.4, 127.5, 128.9, 130.0, 139.48, 139.52, 172.4 ppm. HRMS (ESI) m/z for $C_{26}H_{32}NaO_{12}$ $[M + Na]^+$: calcd 559.1786; found 559.1790.

β-D-Fructofuranosyl 6-*O*-[2-(4,4′-biphenyl)phenylethanoyl]-α-D-glucopyranoside, sucrose *p*-triphenylacetate (**3m**, URB1543).

White solid. Yield: 31%. MS (ESI): 611 $[M–H]^-$, 630 $[M + NH4]^+$, 635 $[M + Na]^+$. ^1H NMR (DMSO-d_6) δ: 3.08 (ddd, 1H, $J_{H4\text{-}OH4}$ = 5.0 Hz, $J_{H4\text{-}H3} \approx J_{H4\text{-}H5}$ = 9.5 Hz, H4), 3.23 (ddd, 1H, $J_{H2\text{-}H1}$ = 4.0 Hz, $J_{H2\text{-}OH2}$ = 6.0 Hz, $J_{H2\text{-}H3}$ = 9.5 Hz, H2), 3.38–3.43 (m, 2H, H1′a, H1′b), 3.46–3.54 (m, 1H, H3), 3.57–3.66 (m, 3H, H5′, H6′a, H6′b), 3.73 (d, J = 16.0 Hz, 1H, *H*CHAr), 3.77 (d, J = 16.0 Hz, 1H, HC*H*Ar), 3.79–3.82 (m, 1H, H4′), 3.90 (dd, 1H, $J_{H3'\text{-}OH3'} \approx J_{H3'\text{-}H4'}$ = 8.0 Hz, H3′), 3.97 (m, 1H, H5), 4.08 (dd, 1H, $J_{H6a\text{-}H5}$ = 6.0 Hz, $J_{H6a\text{-}H6b}$ = 11.5 Hz, H6a), 4.31 (dd, 1H, $J_{H6b\text{-}H5}$ = 1.5 Hz, $J_{H6b\text{-}H6a}$ = 11.5 Hz, H6b), 4.46 (dd, 1H, $J_{OH6'\text{-}H6'a} \approx J_{OH6'\text{-}H6'b}$ = 5.5 Hz, OH6′), 4.62 (d, 1H, $J_{OH3'\text{-}H3'}$ = 8.0 Hz, OH3′), 4.83 (dd, 1H, $J_{OH1'\text{-}H1'a} \approx J_{OH1'\text{-}H1'b}$ = 6.0 Hz, OH1′), 4.84 (brs, 1H, OH3), 5.06 (d, 1H, $J_{OH4\text{-}H4}$ = 5.0 Hz, OH4), 5.15 (d, 1H, $J_{OH2\text{-}H2}$ = 6.0 Hz, OH2), 5.21 (m, 2H, H1, OH4′), 7.35–7.42 (m, 3H, ArH), 7.45–7.51 (m, 2H, ArH), 7.66–7.78 (m, 8H, ArH) ppm. ^{13}C NMR (DMSO-d_6) δ: 62.7, 63.1, 64.7, 70.5, 70.6, 72.0, 73.1, 75.0, 77.3, 83.2, 91.9, 104.4, 126.9, 127.0, 127.5, 127.55, 127.65, 128.0, 129.4, 130.6, 134.2, 138.5, 139.3, 139.6, 140.1, 171.7 ppm. HRMS (ESI) m/z for $C_{32}H_{36}NaO_{12}$ $[M + Na]^+$: calcd 635.2099; found 635.2104.

3.3. Bacterial Strains and Culture Conditions

Eight reference human pathogens were used in this study: *E. faecalis* ATCC 29212, *E. coli* O157:H7 ATCC 35150, *S. aureus* ATCC 43387, *S. aureus* ATCC 43300 (MRSA), *L. monocytogenes* ATCC 7644, *K. pneumoniae* ATCC 13883, *P. aeruginosa* ATCC 9027, and *S. enteritidis* ATCC 13076. All the strains were maintained in TSA (tryptone soy agar) (VWR, Milan, Italy) at 37 °C, while the stock cultures were kept at −80 °C in nutrient broth with glycerol 15%.

3.4. Fungi and Culture Conditions

Three pathogenic filamentous fungi belonging to the strain collection of Pharmacology and Hygiene Section (Department of Biomolecular Sciences, University of Urbino Carlo Bo) *A. niger* ATCC 9642, *A. fumigatus* IDRAH01 and *Fusarium* spp., as well as the reference strain *C. albicans* ATCC 10231, were included. All the filamentous fungi were maintained on potato dextrose agar (PDA) (VWR) at 30 °C for 7 days, while *C. albicans* was grown at 37 °C for 24 h.

3.5. Minimum Inhibitory Concentration (MIC)

MIC of each compound was determined by standard micro-dilution method according to the National Committee for NCCLS (Clinical Laboratory Standards) document M100-S12 method (CLSI, 2016). First, stock solution of each compound (~12 mg) was prepared in 50:50 (*v*/*v*) of distilled sterile water and biological grade DMSO (final volume 1 mL). Several colonies of each bacterial strain were inoculated in 10 mL of sterile Mueller–Hinton broth II (MHB II) (VWR) and incubated at 37 °C for 24 h. At the end of the incubation period, each suspension was diluted in MHB II to obtain ca. 10^5 cfu/mL and 100 µL was added in wells of the 96-well plate together with the appropriate volumes of the test solutions (final concentration range: 4–1024 µg/mL). Two rows were used for positive (bacteria alone) and negative controls (MHB II alone), respectively. Preliminary assays with DMSO were carried out to exclude its possible bacteriostatic and/or bactericidal activity; in any case, the quantity of DMSO added in each well never exceeded 5% (*v*/*v*) of the final total volume. MIC was defined as the lowest concentration of compound able to inhibit bacterial growth after 24 h of incubation at 37 °C. All the experiments were performed in duplicate.

Regarding the filamentous fungi, spores were harvested from PDA plates by adding 2 mL of sterile 0.85% saline solution supplemented with 0.05% Tween 80; the surface was scraped with a sterile cotton swab and the suspension, transferred in a sterile tube, was left at room temperature for 5 min to allow the sedimentation of hyphal fragments. The upper homogeneous suspension was transferred into a new sterile tube, vortexed for 15 s and adjusted to an optical density (OD 530 nm) between 0.09 to 0.4 (about 10^6 spores/mL). For *C. albicans* the suspension was adjusted with the spectrophotometer to a turbidity of 0.12 (about 10^6 cfu/mL). Successively, 100 µL of each fungal suspension was diluted 1:50 in standard RPMI 1640 medium (Sigma-Aldrich, Milan, Italy) and inoculated into 96-well plates together with the appropriate volumes of each test solution as described above. Two rows were left for positive control growth and negative controls (medium only), respectively. Plates were incubated at 30 °C for 72 h; in the case of *C. albicans* plates were incubated at 37 °C for 24–48 h. MIC is defined as the lowest drug concentration that inhibits visible growth compared to the untreated control. In addition, the turbidity of the 96-wells plate was assessed by spectrophotometer (530 nm) (Multiskan EX, Thermo Scientific, Waltham, MA, USA).

3.6. DPPH Assays

The antioxidant capacity of URB1537 was evaluated in a cell-free system by the DPPH radical scavenging assay, as previously described (Verboni 2022). Concentrations up to 512 µg/mL were tested, and the EC_{50} value (i.e., the concentration necessary to achieve a 50% antioxidant effect) was calculated. Quercetin (up to 4.5 µg/mL) was used as a reference antioxidant molecule to check the procedure's correctness.

3.7. Cytotoxicity Assay

The cytotoxicity of URB1537 was investigated on HaCaT cells (immortalized human keratinocytes) from CLS-Cell Lines Service GmbH (Eppelheim, Germany) in vitro. Cells (5×10^3/well) were seeded in 96-well plates and incubated for 24 h with S-C16 (16–512 µg/mL). After incubation, cell viability was analyzed by the MTT assay. Color development was monitored at 570 nm in a multi-well plate reader (Thermo Scientific, Waltham, MA, USA), and data were expressed as cell viability (%) vs. untreated control cells. The IC_{50} value (i.e., the concentration required to reduce cell viability by 50%) was then calculated.

3.8. Griess Assay

The anti-inflammatory properties of URB1537 were evaluated in RAW 264.7 cells (murine macrophages) stimulated by lipopolysaccharide (LPS) (Sigma-Aldrich, Milan, Italy). Cells (3×10^4/well) were seeded in 96-well plates and treated for 24 h with LPS (1 μg/mL) in the presence of URB1537 (16–128 μg/mL). After that, NO release was determined in the culture medium by the Griess reagent (Sigma-Aldrich, Milan, Italy), as recently reported (Verboni 2022). The anti-inflammatory effects of URB1537 (128 μg/mL) in comparison to sucrose and palmitoleic acid were also assessed. Dexamethasone (2 μg/mL) was used as a reference anti-inflammatory drug. RAW 264.7 cell viability after LPS, URB1537, and dexamethasone administration were assessed by the MTT assay as described above.

3.9. Statistical Analysis

Comparisons between multiple means were performed via ANOVA followed by Tukey's post hoc test. Significance was set at $p < 0.05$. GraphPad Prism 6.0 (GraphPad Software, Inc., San Diego, CA, USA) was used for statistical analysis.

4. Conclusions

An effective variation of a Mitsunobu-type methodology reported in the literature has allowed obtaining a small library of 6-*O*-sucrose-based surfactants. The ameliorated procedure proved to be versatile and to have a large scope, starting from non-protected and commercially available substrates. The compounds were then screened for antimicrobic and antimycotic activities. Interestingly, some of the herein-reported sucrose monoesters showed interesting antifungal activities. In particular, sucrose palmitoleate (**3d**, URB1537) shows promising antifungal activity against *C. albicans*, *Fusarium* spp. and *A. fumigatus* with MIC values of 16, 32, and 64 μg/mL, respectively. For this reason, URB1537 was selected for further biological characterizations. The new surfactant is shown to be safe for human keratinocytes and murine macrophages up to 128 μg/mL. Although URB1537 did not show any antioxidant properties within the range of tested concentrations (up to 512 μg/mL), 6-*O*-sucrose palmitoleate presents an appreciable anti-inflammatory activity in LPS-activated macrophages, suggesting that the compound might help to control the inflammatory response. Thus, URB1537 could have therapeutic potential and could also be used in designing and developing innovative drugs such as those based on liposomes.

Author Contributions: Conceptualization, S.L. and A.D.; methodology, M.V., M.S., R.C., S.B., F.P. and L.P.; formal analysis, M.V.; investigation, M.V., M.S., R.C., S.B., F.P. and L.P.; resources, S.L. and A.D.; data curation, M.V., R.C., F.P. and L.P.; writing—original draft preparation, R.C., L.P., S.L. and A.D.; writing—review and editing, M.V., M.S., R.C., S.B., F.P., L.P., S.L. and A.D.; supervision, S.L. and A.D.; funding acquisition, S.L. and A.D. All authors have read and agreed to the published version of the manuscript.

Funding: This research was funded by the University of Urbino Carlo Bo.

Institutional Review Board Statement: Not applicable.

Informed Consent Statement: Not applicable.

Data Availability Statement: Data are contained within the article.

Acknowledgments: The authors would like to thank Michele Menotta and Federica Biancucci for HRMS technical support.

Conflicts of Interest: The authors declare no conflict of interest.

References

1. Almeida, F.; Rodrigues, M.L.; Coelho, C. The still underestimated problem of fungal diseases worldwide. *Front. Microbiol.* **2019**, *10*, 214. [CrossRef]
2. Liu, W.; Yuan, L.; Wang, S. Recent progress in the discovery of antifungal agents targeting the cell wall. *J. Med. Chem.* **2020**, *63*, 12429–12459. [CrossRef] [PubMed]
3. Heard, S.C.; Wu, G.; Winter, J.M. Antifungal natural products. *Curr. Opin. Biotechnol.* **2021**, *69*, 232–241. [CrossRef] [PubMed]
4. Dai, J.-K.; Dan, W.-J.; Wan, J.-B. Natural and synthetic β-carboline as a privileged antifungal scaffolds. *J. Med. Chem.* **2022**, *229*, 114057. [CrossRef]
5. Van Daele, R.; Spriet, I.; Wauters, J.; Maertens, J.; Mercier, T.; Van Hecke, S.; Brüggemann, R. Antifungal drugs: What brings the future? *Med. Mycol.* **2019**, *57*, S328–S343. [CrossRef]
6. Bouz, G.; Doležal, M. Advances in antifungal drug development: An up-to-date mini review. *Pharmaceuticals* **2021**, *14*, 1312. [CrossRef]
7. Aslam, B.; Wang, W.; Arshad, M.I.; Khurshid, M.; Muzammil, S.; Rasool, M.H.; Nisar, M.A.; Alvi, R.F.; Aslam, M.A.; Qamar, M.U.; et al. Antibiotic resistance: A rundown of a global crisis. *Infect. Drug Resist.* **2018**, *11*, 1645–1658. [CrossRef]
8. Teng, Y.; Stewart, S.G.; Hai, Y.W.; Li, X.; Banwell, M.G.; Lan, P. Sucrose fatty acid esters: Synthesis, emulsifying capacities, biological activities and structure-property profiles. *Crit. Rev. Food Sci. Nutr.* **2021**, *61*, 3297–3317. [CrossRef]
9. Farran, A.; Cai, C.; Sandoval, M.; Xu, Y.; Liu, J.; Hernaiz, M.J.; Linhardt, R.J. Green solvents in carbohydrate chemistry: From raw materials to fine chemicals. *Chem. Rev.* **2015**, *115*, 6811–6853. [CrossRef]
10. Polat, T.; Linhardt, R.J. Syntheses and applications of sucrose-based esters. *J. Surf. Deterg.* **2001**, *4*, 415–421. [CrossRef]
11. Garti, N.; Clement, V.; Fanun, M.; Leser, M.E. Some characteristics of sugar ester nonionic microemulsions in view of possible food applications. *J. Agric. Food Chem.* **2000**, *48*, 3945–3956. [CrossRef] [PubMed]
12. Megahed, M.G. Preparation of sucrose fatty acid esters as food emulsifiers and evaluation of their surface active and emulsification properties. *Grasas Aceites* **1999**, *50*, 280–282. [CrossRef]
13. Devulapalle, K.S.; Gomez de Segura, A.; Ferrer, M.; Alcalde, M.; Mooser, G.; Plou, F. Effect of carbohydrate fatty acid esters on *Streptococcus sobrinus* and glucosyltransferase activity. *Carbohydr. Res.* **2004**, *339*, 1029–1034. [CrossRef]
14. Chortyk, O.T.; Pomonis, J.G.; Johnson, A.W. Syntheses and characterizations of insecticidal sucrose esters. *J. Agric. Food Chem.* **1996**, *44*, 1551–1557. [CrossRef]
15. Markets and Markets. Sucrose Esters Market Worth $106 Million by 2025. Available online: https://www.marketsandmarkets.com/PressReleases/sucrose-esters.asp (accessed on 23 November 2022).
16. Verboni, M.; Lucarini, S.; Duranti, A. 6′-*O*-Lactose ester surfactants as an innovative opportunity in the pharmaceutical field: From synthetic methods to biological applications. *Pharmaceuticals* **2021**, *14*, 1306. [CrossRef] [PubMed]
17. Lucarini, S.; Fagioli, L.; Campana, R.; Cole, H.; Duranti, A.; Baffone, W.; Vllasaliu, D.; Casettari, L. Unsaturated fatty acids lactose esters: Cytotoxicity, permeability enhancement and antimicrobial activity. *Eur. J. Pharm. Biopharm.* **2016**, *107*, 88–96. [CrossRef]
18. Perinelli, D.R.; Lucarini, S.; Fagioli, L.; Campana, R.; Vllasaliu, D.; Duranti, A.; Casettari, L. Lactose oleate as new biocompatible surfactant for pharmaceutical applications. *Eur. J. Pharm. Biopharm.* **2018**, *124*, 55–62. [CrossRef]
19. Lucarini, S.; Fagioli, L.; Cavanagh, R.; Liang, W.; Perinelli, D.; Campana, M.; Stolnik, S.; Lam, J.; Casettari, L.; Duranti, A. Synthesis, structure–activity relationships and in vitro toxicity profile of lactose-based fatty acid monoesters as possible drug permeability enhancers. *Pharmaceutics* **2018**, *10*, 81. [CrossRef]
20. Campana, R.; Merli, A.; Verboni, M.; Biondo, F.; Favi, G.; Duranti, A.; Lucarini, S. Synthesis and evaluation of saccharide-based aliphatic and aromatic esters as antimicrobial and antibiofilm agents. *Pharmaceuticals* **2019**, *12*, 186. [CrossRef]
21. Lucarini, S.; Ciulla, M.G.; Mestichelli, P.; Duranti, A. Total Synthesis of Natural Disaccharide Sambubiose. *Pharmaceuticals* **2020**, *13*, 198. [CrossRef]
22. McCartney, F.; Perinelli, D.R.; Tiboni, M.; Cavanagh, R.; Lucarini, S.; Palmieri, G.P.; Casettari, L.; Brayden, D.J. Permeability-enhancing effects of three laurate-disaccharide monoesters across isolated rat intestinal mucosae. *Int. J. Pharm.* **2021**, *601*, 120593. [CrossRef]
23. Tiboni, M.; Elmowafy, E.; El-Derany, M.O.; Benedetti, S.; Campana, R.; Verboni, M.; Potenza, L.; Palma, F.; Citterio, B.; Sisti, M.; et al. A combination of sugar esters and chitosan to promote in vivo wound care. *Int. J. Pharm.* **2022**, *616*, 121508. [CrossRef]
24. Verboni, M.; Benedetti, S.; Campana, R.; Palma, F.; Potenza, L.; Sisti, M.; Duranti, A.; Lucarini, S. Synthesis and Biological Characterization of the New Glycolipid Lactose Undecylenate (URB1418). *Pharmaceuticals* **2022**, *15*, 456. [CrossRef] [PubMed]
25. Molinier, V.; Fitremann, J.; Bouchu, A.; Queneau, Y. Sucrose esterification under Mitsunobu conditions: Evidence for the formation of 6-*O*-acyl-3′,6′-anhydrosucrose besides mono and diesters of fatty acids. *Tetrahedron Asymmetry* **2004**, *15*, 1753–1762. [CrossRef]
26. Vlahov, I.R.; Vlahova, P.I.; Linhardt, R.J. Regioselective synthesis of sucrose monoesters as surfactants. *J. Carb. Chem.* **1997**, *16*, 1–10. [CrossRef]
27. Thévenet, S.; Wernicke, A.; Belniak, S.; Descotes, G.; Bouchu, A.; Queneau, Y. Esterification of unprotected sucrose with acid chlorides in aqueous medium: Kinetic reactivity versus acyl- or alkyloxycarbonyl-group migrations. *Carb. Res.* **1999**, *318*, 52–66. [CrossRef]
28. Cruces, M.A.; Plou, F.J.; Ferrer, M.; Bernabé, M.; Ballesteros, A. Improved synthesis of sucrose fatty acid monoesters. *J. Am. Oil Chem. Soc.* **2001**, *78*, 541–546. [CrossRef]

29. Jarosz, S.; Mach, M. Regio- and stereoselective transformations of sucrose at the terminal positions. *Eur. J. Org. Chem.* **2002**, *2002*, 769–780. [CrossRef]
30. Ferrer, M.; Cruces, M.A.; Bernabé, M.; Ballesteros, A.; Plou, F.J. Lipase-catalyzed regioselective acylation of sucrose in two solvent mixtures. *Biotechnol. Bioeng.* **1999**, *65*, 10–16. [CrossRef]
31. Shi, Y.-G.; Li, J.R.; Chu, Y.-H. Enzyme-catalyzed regioselective synthesis of sucrose-based esters. *J Chem. Tec. Biotechnol.* **2011**, *86*, 1457–1468. [CrossRef]
32. Possiel, C.; Bäuerle, A.; Seibel, J.J. A chemoenzymatic route to a class of sucrose esters. *Eur. J. Org. Chem.* **2017**, *2017*, 6335–6337. [CrossRef]
33. Griffin, W.C. Calculation of HLB values of non-ionic surfactants. *J. Soc. Cosmet. Chem.* **1954**, *5*, 249–256.
34. Organic Chemistry Portal. Available online: https://www.organic-chemistry.org/prog/ (accessed on 20 December 2022).
35. Zhao, L.; Zhang, H.; Hao, T.; Li, S. In vitro antibacterial activities and mechanism of sugar fatty acid esters against five food-related bacteria. *Food Chem.* **2015**, *187*, 370–377. [CrossRef] [PubMed]
36. Karlová, T.; Poláková, L.; Šmidrkal, J.; Filip, V. Antimicrobial effects of fatty acid fructose esters. *Czech J. Food Sci.* **2010**, *28*, 146–149. [CrossRef]
37. Petkova, N.; Vassilev, D.; Arabadzhieva, R.; Tumbarski, Y.; Vasileva, I.; Koleva, M.; Denev, P. "Green" synthesis of sucrose octaacetate and characterization of its physicochemical properties and antimicrobial activity. *Chem. Biochem. Eng. Q.* **2017**, *31*, 395–402. [CrossRef]
38. Merghni, A.; Dallel, I.; Noumi, E.; Kadmi, Y.; Hentati, H.; Tobji, S.; Ben Amor, A.; Mastouri, M. Antioxidant and antiproliferative potential of biosurfactants isolated from *Lactobacillus casei* and their anti-biofilm effect in oral *Staphylococcus aureus* strains. *Microb. Pathog.* **2017**, *104*, 84–89. [CrossRef]
39. Giri, S.S.; Ryu, E.C.; Sukumaran, V.; Park, S.C. Antioxidant, antibacterial, and anti-adhesive activities of biosurfactants isolated from *Bacillus* strains. *Microb. Pathog.* **2019**, *132*, 66–72. [CrossRef]
40. Geran, R.I.; Greenberg, N.H.; Macdonald, M.M.; Schumacher, A.M.; Abbott, B.J. Protocols for screening chemical agents and natural products against animal tumors and other biological systems (Third Edition). *Cancer Chemother. Rep.* **1972**, *3*, 59–61.
41. Li, W.; Sun, Y.N.; Yan, X.T.; Yang, S.Y.; Song, S.B.; Lee, Y.M.; Kim, Y.H. NF-κB Inhibitory activity of sucrose fatty acid esters and related constituents from *Astragalus membranaceus*. *J. Agric. Food Chem.* **2013**, *61*, 7081–7088. [CrossRef]
42. Souza, C.O.; Teixeira, A.A.S.; Biondo, L.A.; Silveira, L.S.; Calder, F.P.; Neto, J.C.R. Palmitoleic acid reduces the inflammation in LPS-stimulated macrophages by inhibition of NFκB, independently of PPARs. *Clin. Exp. Pharmacol. Physiol.* **2017**, *44*, 566–575. [CrossRef]
43. Abran, D.; Boucher, F.; Hamanaka, T.; Hiraki, K.; Kito, Y.; Koyama, K.; Leblanc, R.M.; Machida, H.; Munger, G.; Seidou, M.; et al. On some physicochemical properties of sucrose esters and the stability they confer to membrane proteins. *J. Colloid Interface Sci.* **1989**, *128*, 230–236. [CrossRef]
44. Youan, B.-B.C.; Hussain, A.; Nguyen, N.T. Evaluation of sucrose esters as alternative surfactants in microencapsulation of proteins by the solvent evaporation method. *AAPS PharmSci.* **2003**, *5*, 222–230. [CrossRef] [PubMed]
45. Li, J.; Jiang, Y.; Tu, P.-F. Tricornoses A−L, oligosaccharide multi-esters from the roots of *Polygala tricornis*. *J. Nat. Prod.* **2005**, *68*, 739–744. [CrossRef] [PubMed]
46. Furuya, T.; Ushiyama, M.; Asada, Y.; Yoshikawa, T.; Orihara, Y. Biotransformation of phenylacetic acid and 2-phenyl-propionic acid in suspension culture of *offea arabica*. *Phytochemistry* **1988**, *27*, 803–807. [CrossRef]

pharmaceuticals

Review

Photomodulation Approaches to Overcome Antimicrobial Resistance

Sofia N. Sarabando [1,2], Andreia Palmeira [1,3], Maria Emília Sousa [1,3,*], Maria Amparo F. Faustino [2] and Carlos J. P. Monteiro [2,*]

1 Laboratory of Organic and Pharmaceutical Chemistry, Chemical Sciences Department, Faculty of Pharmacy, University of Porto, 4050-313 Porto, Portugal; sofia.sarabando@ua.pt (S.N.S.); apalmeira@ff.up.pt (A.P.)
2 LAQV-Requimte and Department of Chemistry, University of Aveiro, 3010-193 Aveiro, Portugal; faustino@ua.pt
3 CIIMAR—Interdisciplinary Centre of Marine and Environmental Research, 4450-208 Porto, Portugal
* Correspondence: esousa@ff.up.pt (M.E.S.); cmonteiro@ua.pt (C.J.P.M.)

Abstract: Photopharmacology is an approach that aims to be an alternative to classical chemotherapy. Herein, the different classes of photoswitches and photocleavage compounds and their biological applications are described. Proteolysis targeting chimeras (PROTACs) containing azobenzene moieties (PHOTACs) and photocleavable protecting groups (photocaged PROTACs) are also mentioned. Furthermore, porphyrins are referenced as successful photoactive compounds in a clinical context, such as in the photodynamic therapy of tumours as well as preventing antimicrobial resistance, namely in bacteria. Porphyrins combining photoswitches and photocleavage systems are highlighted, taking advantage of both photopharmacology and photodynamic action. Finally, porphyrins with antibacterial activity are described, taking advantage of the synergistic effect of photodynamic treatment and antibiotic therapy to overcome bacterial resistance.

Keywords: photopharmacology; photoswitch; photocaged; porphyrin; antimicrobials

Citation: Sarabando, S.N.; Palmeira, A.; Sousa, M.E.; Faustino, M.A.F.; Monteiro, C.J.P. Photomodulation Approaches to Overcome Antimicrobial Resistance. *Pharmaceuticals* **2023**, *16*, 682. https://doi.org/10.3390/ph16050682

Academic Editor: Abdelwahab Omri

Received: 1 April 2023
Revised: 24 April 2023
Accepted: 30 April 2023
Published: 2 May 2023

1. Introduction

Photopharmacology is a new pharmacophore modality that is gaining attention as a promising alternative to classical chemotherapy drugs (Figure 1a) [1–3]. It is characterized by the combination of photochemistry and pharmacology, and it aims at solving issues related to poor drug selectivity, minimizing off-target effects and the emergence of drug resistance [4]. With the increased rate of antimicrobial resistance, there are fewer treatments available, so the World Health Organization (WHO) recommends developing new antimicrobial approaches to overcome this problem [5]. In this way, the use of light on the inactivation of microorganisms (MO) has been shown to be an outstanding tool, as it is non-evasive and capable of fast and controllable delivery to precise locations, paving the way for the use of photons for new medical treatment approaches, such as photopharmacology, an emerging field that utilizes light to control biological processes with high spatiotemporal precision, offering a promising strategy for overcoming antimicrobial resistance [2,6]. Its principle is the introduction of a photocaged or photoswitchable moiety into the molecular structure of the bioactive compound, where the light effect can be irreversible or reversible. Photocleavage molecules are an example of irreversible photoactivation by using photocleavable protecting groups (PPG) as "cages" [7]. PPG are small moieties that can be released upon irradiation, exposing the bioactive molecule in its active form [7]. On the other hand, photoswitches are molecules that can be interconverted upon the action of light, in a reversible and fast process, into different structural configurations (*cis* or *trans*), in which one of the configurations is the bioactive form [8]. Light offers the possibility to control both the pharmacokinetic (PK) and pharmacodynamic (PD) properties of molecules, switching the drug from a low to high affinity with the biological target (*cis* and *trans*, in case of photoswitches) and allowing molecular activation in case of PPG. Therefore, light can improve selectivity, being particularly suitable when the disease is localized, and consequently reduces adverse effects by adjusting its wavelength and intensity [4,9,10]. One of the major challenges in photopharmacology, using short-wavelength light (ultraviolet (UV), below 350 nm), is the limited penetration into tissue and off-target toxicity due to the fact that UV light is toxic to healthy cells (Figure 1b,c). At a first glance, utilizing UV light may not appear to be a favorable strategy. Nevertheless, the primary advantage of this approach is the mitigation of cytotoxicity once the photoswitches lose their activity after the therapeutic effect (Figure 1c). A high-precision photopharmacological approach utilizing long-wavelength light (near infrared, 650–900 nm) can effectively overcome the UV light problem. This approach has the added advantage of deeper tissue penetration without harming healthy cells, as it specifically targets only the desired treatment area (Figure 1d,e). Photoswitches, due to their reversible process, can also minimize and prevent side effects, such as generalized cytotoxicity and antimicrobial resistance (Figure 1e).

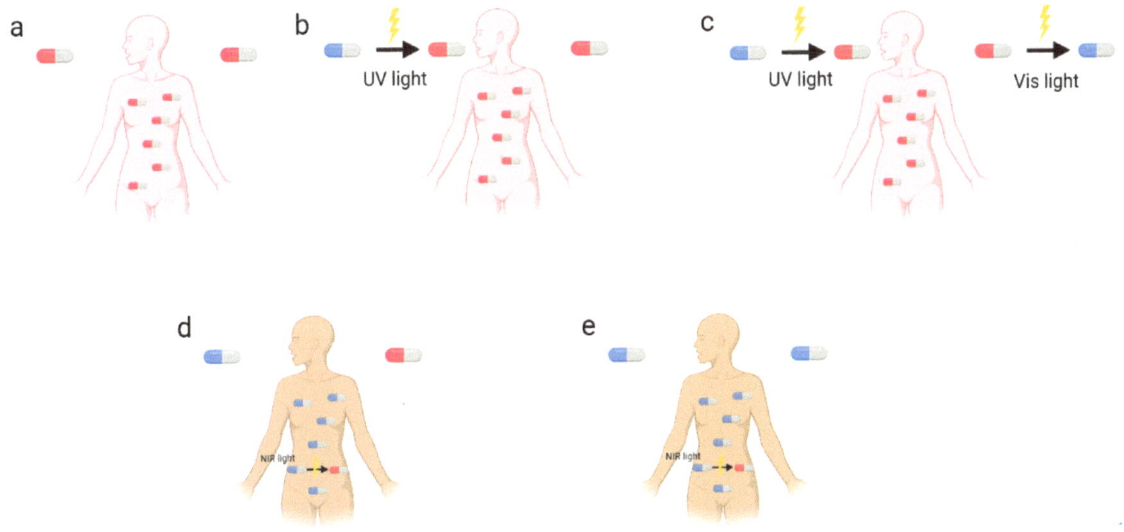

Figure 1. Comparison of the principle behind (**a**) classical chemotherapy, (**b**) photopharmacology using UV light—PPG, (**c**) photopharmacology using UV and visible light—photoswitches, (**d**) photopharmacology using IR light—PPG, and (**e**) high-precision photopharmacological chemotherapy using IR-light—photoswitches. Red capsule denotes cytotoxic effects and blue capsule is an inactive form and not cytotoxic.

Photodynamic therapy (PDT) is a spatiotemporal therapeutic modality approved by the Food and Drug Administration (USA-FDA) and European Medicines Agency (EMA) to treat skin diseases and other organs where light can reach [1,11–16]. It requires the administration of a photosensitizer (PS), which is a non-toxic drug that is accumulated in the target cells and activated by adequate visible light irradiation in presence of dioxygen (O_2) to produce reactive oxygen species (ROS) that cause cytotoxic effects at the therapeutic target. Briefly, the mechanism of action of PDT is based on PS promotion by light absorption from the ground state (^{1}PS) to an excited state (^{1}PS*). ^{1}PS* can return to the ground state (^{1}PS) directly through the emission of light (fluorescence) or heat (internal conversation), or indirectly by intersystem crossing, converting to the excited triplet state (^{3}PS*). In this state, ^{3}PS* can interact with O_2 by two different mechanisms. The type I mechanism involves electron transfer to lead to the production of free radicals, which interact with O_2 and produce ROS, such as hydrogen peroxide (H_2O_2), hydroxyl radicals ($HO^\bullet$), and superoxide radicals ($O_2^{\bullet-}$). The type II mechanism involves energy transfer between the

PS excited and the ground state dioxygen ($^3O_2$), producing singlet oxygen ($^1O_2$). As $^1O_2$ is an uncharged molecule, it can diffuse through plasma and biological membranes and will trigger a chain of successive oxidations of biological molecules (e.g., DNA, lipids, and proteins) [17–21]. It is believed that the production of ROS through the type II reaction is predominant in PDT. This photodynamic approach was also used with great success on the photoinactivation of microorganisms and demonstrates great advantages when compared with the conventional antimicrobial agents, namely (i) the non-induction of resistance mechanisms of the microbial cells and (ii) either resistant and sensitive strains are responsive to the photodynamic treatment and (iii) can be applied to inactivate a range of microbial entities (bacteria, fungi, viruses, protozoa, parasites [11,12,22–29], among others). The versatility of PDT is not limited to clinical applications [23,30]. This technique is valuable for clinical [31–33] and non-clinical applications, such as removing biofilms from medical devices and implants [34] and environmental water treatments [35–37].

Herein, different classes of photoswitches and photocleavage compounds and their biological applications will be described. Furthermore, porphyrins will be referred to as successful photoactive compounds for use in antimicrobial photodynamic treatment (aPDT), as well as promising approaches when combined with photoswitches or photocleavage to take advantage of both photopharmacology and PDT. Finally, examples combining aPDT and antibiotic therapy will be reported, highlighting this application to overcome resistance.

2. Different Classes of Photoswitches and Some Examples in Antibacterial Applications

Photoswitches are compounds that can be interconverted into two different structural configuration forms by using light. These compounds are of great importance in several fields, including molecular electronics, phytopharmacology, and catalysis [38]. In 1937, Hartley discovered that *trans* azobenzene could come from the *cis* isomer by UV-light and, conversely, by keeping the molecule in the dark, or irradiating it with blue light, the thermodynamically more stable *trans* isomer could be recovered [39]. The *trans* $\rightarrow$ *cis* photoisomerization of azobenzenes can occur by n $\rightarrow \pi^*$ excitation (lower energy) at 380–500 nm, or $\pi \rightarrow \pi^*$ excitation (higher energy) at 280–380 nm [40].

By coupling a photoswitch moiety with a biologically active compound, it is possible to turn ON (high-affinity form) and OFF (low-affinity form) this bioactive compound. In fact, azobenzene-based photoswitches are the most widely used due to their efficacy, reversibility, and repeatability. However, photoswitchable molecules show incomplete photoisomerization, insufficient light sensitivity for some applications, and difficulties in fine-tuning to produce a maximal difference between ON and OFF states of the compounds [41].

Ideal photoswitches are molecules that absorb visible light and have a long half-life time of the metastable isomers [42]. Although UV-light is typically used, studies have shown that azoarenes can be activated by visible light [42] and near infrared (NIR) light, which is required to penetrate tissues [39]. Diarylethenes [4], Schiff base-like [43], oxazolone-like [44], hydantoin-like [45], pyrrolinone-like [46], and stilbene [47] compounds are also types of photochromic compounds (Figure 2).

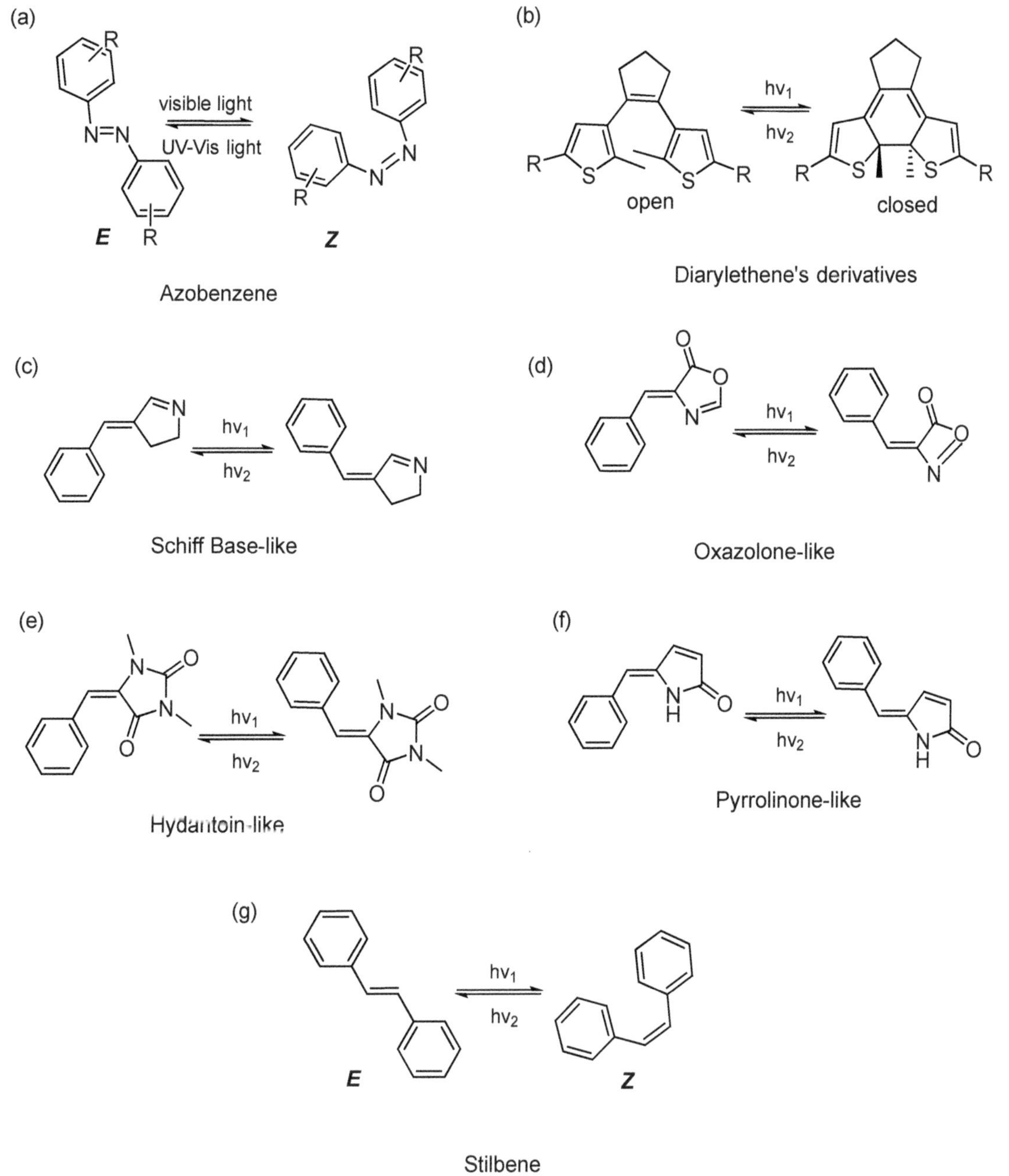

Figure 2. Structures of some photoswitches and their photoisomerization process. hν is the photon energy (h—Planck's constant; ν—frequency; $\nu_1 < \nu_2$).

2.1. Azobenzenes

Azobenzenes are a class of compounds that present an aromatic molecular structure, characterized by the presence of sensitivity to external stimuli -N=N- double bond, with the appropriate features for incorporation into most pharmacophores [48,49]. Feringa et al. [9]

investigated the possibility of obtaining pharmacological agents by conjugating photoswitches with ciprofloxacin (Figure 3) using azobenzenes to treat local diseases, such as a solid tumor or local inflammation. Structure activity relationship (SAR) studies indicated that the secondary amine in the piperazine ring of ciprofloxacin can be modified without loss of activity; therefore, this position was chosen for conjugation with the azobenzene moiety (Figure 3). This approach allowed for the preparation of an antibiotic with photoswitchable activity (azofloxacin). Before irradiation, azofloxacin occurs in 100% as the *trans*-isomer (compound **1**), but after irradiation with light at 365 nm, 61% of azofloxacin converts to the *cis*-isomer (compound **2**). The non-total conversion of *trans* azofloxacin to *cis* azofloxacin might be due to the aggregation between azobenzene and molecules in an aqueous environment, not allowing complete photoisomerization [9].

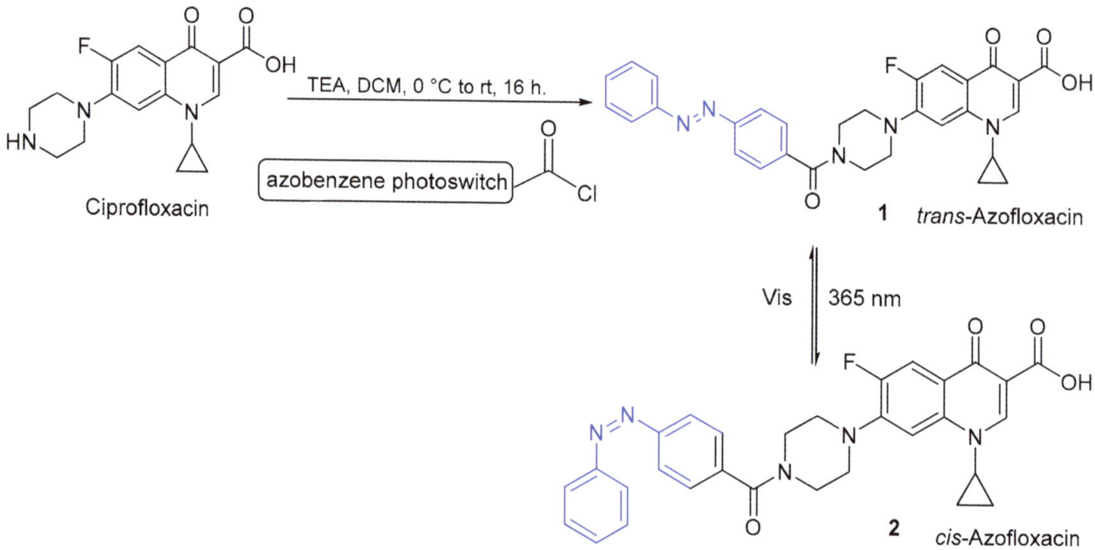

Figure 3. Molecular structure of ciprofloxacin and its photoswitchable analogue, azofloxacin.

The bacterial activity of azofloxacin was assessed on *Escherichia coli (E. coli)* CS1562 and *Micrococcus luteus* (*M. luteus*) ATCC 9341, before and after irradiation with light at 365 nm. Ciprofloxacin was chosen as the control. The azofloxacin analogue proved to be more efficient than ciprofloxacin in *M. luteus* (Table 1), specially before irradiation (*trans*-azofloxacin, compound **1**), which can be modulated by exposure to light, giving rise to *cis*-azofloxacin (compound **2**). While no significant changes were observed on the Gram-negative bacterial model *E. coli*, the antibacterial activity of *trans*-azofloxacin on the Gram-positive bacterial model *M. luteus* was found to be about 50 times greater than the native drug ciprofloxacin.

Table 1. Minimum inhibitory concentration (MIC) values of ciprofloxacin and their derivatives.

Compound	MIC (μM)			
	Before Irradiation Compound 1		**After Irradiation Compound 2**	
	E. coli	***M. luteus***	***E. coli***	***M. luteus***
Azofloxacin	0.50	0.250	0.50	0.50
Ciprofloxacin	0.0125	12.0	0.0125	12.0

The introduction of the azobenzene moiety into an antibacterial drug allows the light to spatiotemporally control the antibacterial activity and presents plenty of opportunities to explore new targets in photopharmacology. According to the authors hypothesis, although on *E. coli* the results are not promising, it is still an alternative to treatment, even when used in combination with chemotherapy. Clearly, further studies would be needed to confirm this hypothesis.

Wegener et al. [50] reported for the first time a red-shifted responsive azobenzene photoswitch containing the antibiotic trimethoprim as a core (Figure 4). Trimethoprim interferes with the biosynthesis of folate by inhibition of dihydrofolate reductase (DHFR), which catalyses the reduction of dihydrofolate to the active cofactor tetrahydrofolate [51,52]. Therefore, the biosynthesis of the amino acids glycine and methionine, as well as purines and thymidine triphosphate are impaired. Of note, trimethoprim has a higher selectivity to bacterial DHRF than mammalian DHFR and it is active against both Gram-positive and Gram-negative bacteria to treat urinary and respiratory tract infections. Nevertheless, with the emergence of bacterial resistance, it is important to develop new therapeutic entities, such as photoresponsive analogues, a strategy that can be undertaken by modifying the structure of trimethoprim without reducing bacterial DHFR affinity. Based on the aforementioned results, the authors prepared a red-shifted analogue (azobenzene substituted with fluorine or chlorine at ortho-positions), where the photoisomerization with visible light was possible, while maintaining the same antibacterial properties (Figure 4). These substituents allowed the effective use of lower-energy n $\rightarrow \pi^*$ excitation to trigger *trans* $\rightarrow$ *cis* photoisomerization. The photostationary state (PSS) of tetra-fluoro-substituted azobenzene was found to have an 89:11 ratio of *cis:trans* at 527 nm (visible light). In the case of tetra-*o*-chloroazobenzene, irradiation with 652 nm (near-infrared light) produced photoisomerization to a PSS of 87:13 proportion of *cis:trans*.

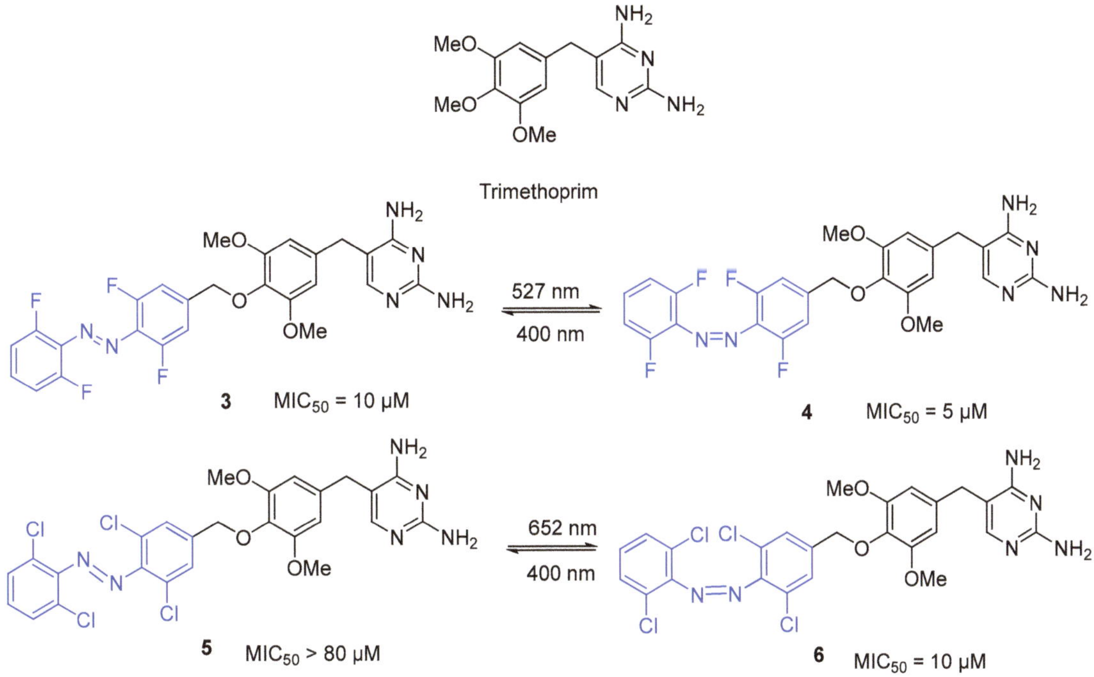

Figure 4. Red-shifted responsive trimethoprim derivatives and respective MIC_{50} values.

The effectiveness of antibacterial activity against *E. coli* was evaluated before and after irradiation. Even though compounds **5** and **6**, which contain trimethoprim tetra-*o*-chloroazobenzene substitution, exhibited less photoisomerization than compounds **3** and **4**, which contain tetra-*o*-fluoro-substituted azobenzene, they were still of interest. Compound **5** displayed no activity with MIC_{50} > 80 μM, nevertheless it was responsive to irradiation (652 nm), giving compound **6**, which induced bacteriostasis with MIC_{50} = 10 μM.

However, tetra-*o*-fluoro-substituted azobenzene is better at inducing bacteriostasis with an MIC_{50} = 5 μM. Furthermore, it was important to study the photoswitching reversibility to the *trans* isomer (inactive) found when blue light at 400 nm was used.

Amidohydrolase enzymes are virulence factors found in various pathogenic bacteria responsible for a great number of hospital-acquired infections and deaths. Weston and co-workers [53] reported inhibitors of these enzymes using azobenzenes as photoswitches (Figure 5). Amidohydrolase enzymes are homologous to the human histone deacetylases (HDACs); therefore, with the knowledge of the pharmacophore of HDAC inhibitors [54–56], the crystal structure of the enzyme active site of the HDAC family [57–59], and knowing the bacterial enzymes B/A-HDAH (histone deacetylase-like amidohydrolase from Bordetella/Alcaligenes [59]), APAHs (acetylpolyamine amidohydrolases from *Pseudomonas aeruginosa* (*P. aeruginosa*), PA1409 and PA0321 [58]), and PA-HDAHs (acetylpolyamine amidohydrolases from *P. aeruginosa*, PA3774), the authors designed and conjugated a pharmacophoric moiety with a zinc chelating group (hydroxamic acid), a phenyl capping unit, and an azobenzene linker.

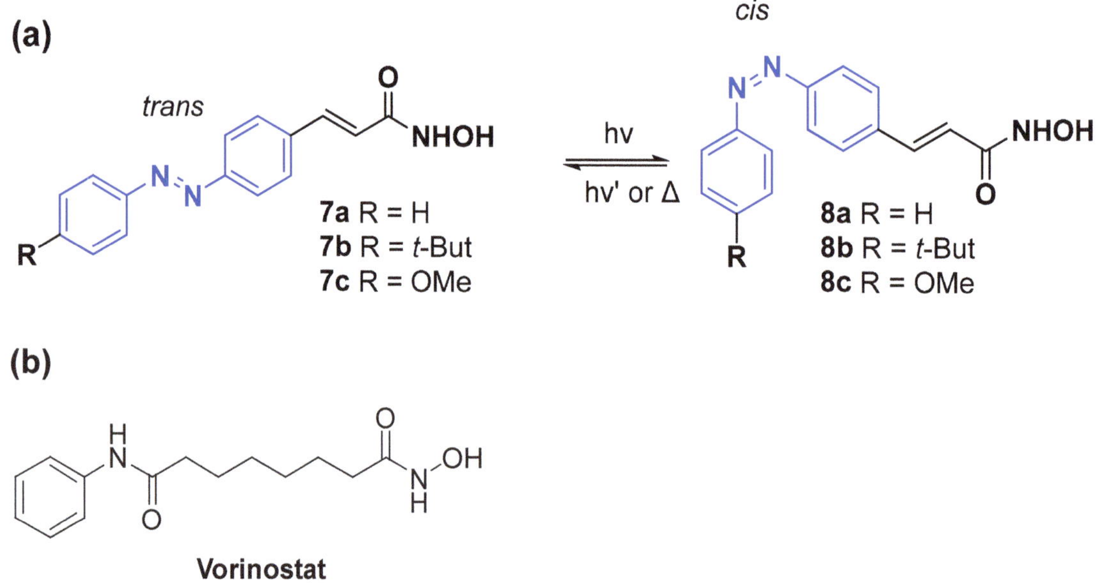

Figure 5. Figure **5.** (**a**) Structures of *trans* and *cis* azobenzene HDAH inhibitors and (**b**) structure of Vorinostat, inhibitor of histone deacetylation, 2 h half-life.

With the photoisomerization of the azobenzene compounds **8** activated by a UV-A lamp at 365 nm, the half-maximal inhibitory concentration values (IC_{50}) against the bacterial enzymes B/A-HDAH, PA-HDAH, and HPAHs were assessed using SAHA (suberoylanilide hydroxamic acid, also known as vorinostat, Figure 5b) as a positive control (Table 2). It should be noted that incomplete photoswitching is an issue, and about >10% of the former isomer was present upon irradiation.

Table 2. IC_{50} values for *trans* and *cis* azobenzenes against bacterial HDACs.

			IC_{50} (μM)		
Compound	**R**	**B/A-HDAH**	**PA-HDAH**	**APAH PA0321**	**APAH PA1409**
7a (*trans*)	H	0.042	0.0056	0.79	2.0
8a (*cis*)	H	0.062	0.0066	0.40	0.21
7b (*trans*)	*t*-But	0.37	0.072	2.2	>25
8b (*cis*)	*t*-But	0.38	0.094	1.8	3.5
7c (*trans*)	OMe	0.059	0.051	2.3	5.1
8c (*cis*)	OMe	0.075	0.090	1.4	0.65
Vorinostat		0.095	0.037	0.50	0.30

In general, compounds **7a** and **7c** showed better activity than the positive control against B/A-HDAH. However, treatment with UV-A light was not successful as the cis-isomer was more effective than the *trans*-isomer. Compound **8a** revealed a better efficacy than the positive control against the PA-HDAH and APAHs. Here, it was demonstrated that the efficacy of the cis-isomer obtained after UV-A light treatment is greater than the trans-isomer. Compound **7b** exhibits a greater steric effect due to the presence of the *tert*-butyl group, which leads to an unfavourable steric fit. As a consequence, its inhibitory activity is weaker than that of compounds **7a** and **7c**.

Initially, the authors proposed that the *cis*-isomers would induce more unfavourable steric interactions, reducing inhibitory activity, so the *trans*-isomer would be the most active. However, this was not observed, and the results showed that the most active isomer depends on the inhibitor substituent and the enzyme (the *cis*-isomers were found to be better in APAHs).

Gramicidin S is a cyclic peptide known for its antibacterial activity [60]; therefore, Yuan et al. [61] developed an approach based on an azobenzene moiety containing gramicidin S to modulate, reversibly, the antibacterial activity (Figure 6).

Peptide **9** was designed by replacing proline and D-phenylalanine amino acid residues with an azobenzene moiety, maintaining the amphiphilic side chain for biological activity. Peptide **10a** was obtained by removing the leucine residue in peptide **9**. Arginine (basic amino acid residue) and glutamic acid (acidic amino acid residue) were replaced in peptide **10a**, to produce peptides **10b** and **10c**, respectively, aiming to investigate possible differences in the antibacterial activity due to the positively or negatively charged amino acid residues. To study the amphiphilicity on the antibacterial activity, basic amino acid residues were incorporated into the peptide backbone by replacing arginine with leucine and valine residues (peptides **11a** and **11b**). When cis-isomers of peptides were irradiated with blue light (405 nm), *trans*-isomers are obtained. On the other hand, the *trans*-enriched PSS of **9–11** were irradiated by UV-A light of 352 nm for 2 h to convert to the respective *cis*-enriched PSS.

The MIC values of gramicidin S and its peptides were determined against *Staphylococcus aureus (S. aureus)* ATCC 49775 (Table 3).

Peptide **9** only has two intramolecular strong hydrogen bonds linking the upper and lower strand, showing a distorted β-strand secondary structure, proved in a molecular modelling study. Gramicidin S has four strong hydrogen bonds, while the *cis*-isomers of peptides **10** and **11** have three strong hydrogen bonds, causing a well-defined secondary structure. The *trans*-isomers of peptides **10** and **11** lack intramolecular hydrogen bonding and, therefore, show a weak secondary structure. The *cis*-isomers of peptides **10a** and **10b** proved to be more efficient than *trans*-isomers, with *cis*-enriched **10b** exhibiting the highest antibacterial potential in this study. Similar to peptide **10c**, peptides **11a** and **11b** were inactive, which was attributed to the replacement of hydrophobic residues (valine and leucine), which may disrupt the amphiphilic nature of peptides, decreasing penetration through the bacterial membrane. In this study, the use of light was not essential to activate the ON effect, since the *trans*-isomer was not more efficient than the *cis*-isomer. However, light would be a good strategy to generate an OFF effect and, therefore, reduce side effects.

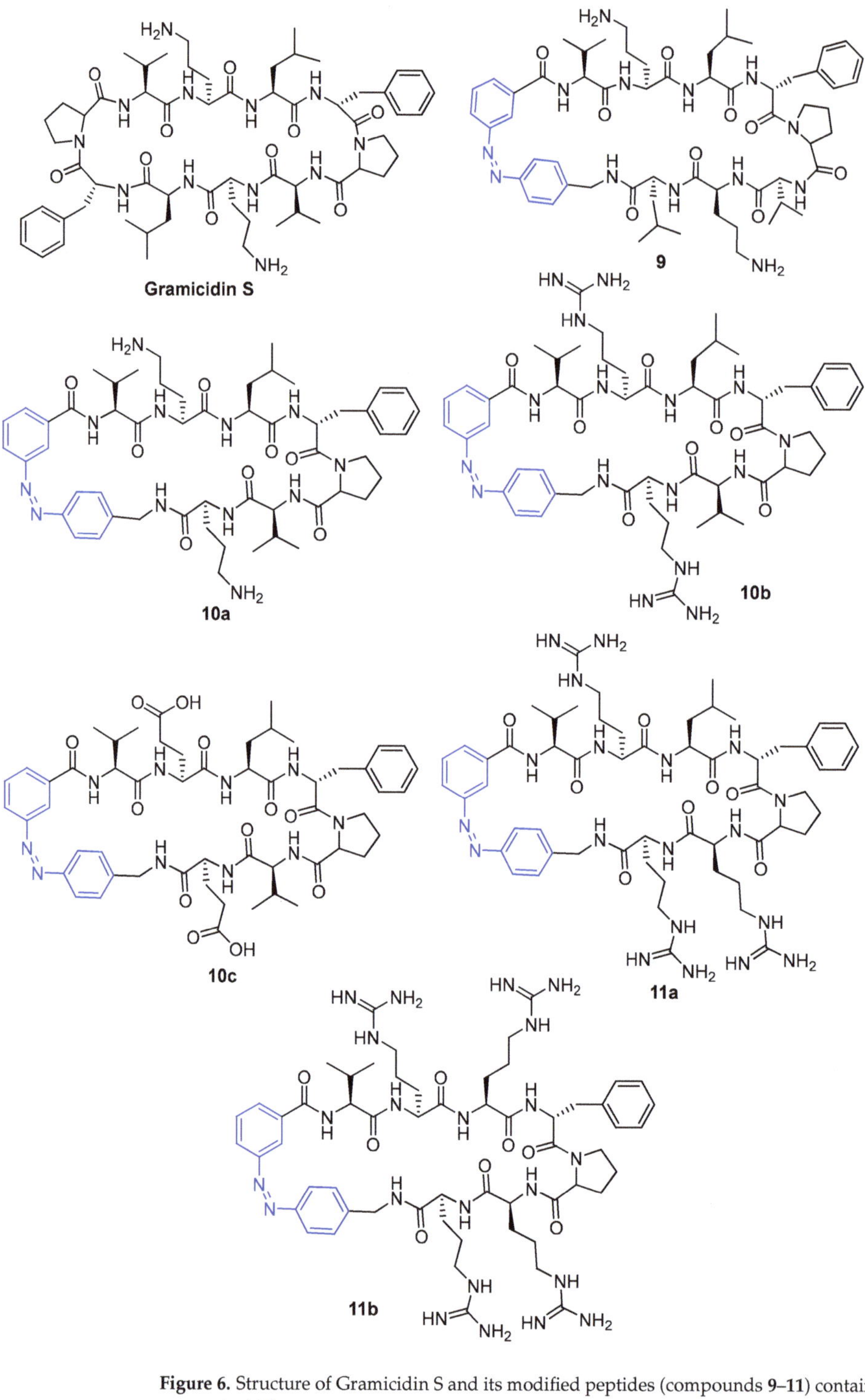

Figure 6. Structure of Gramicidin S and its modified peptides (compounds **9–11**) containing azobenzene units [61].

Table 3. MIC values of Gramicidin S and its mimetics peptides against *S. aureus* ATCC 49775.

	MIC (μg mL^{-1})	
Compound	*cis*	*trans*
9	64	64
10a	64	256
10b	32	128
10c	>256	>256
11a	256	>256
11b	>256	>256
Gramicidin S	2	

2.2. Diarylethenes

Similar to azobenzenes, diarylethenes have an aromatic molecular structure that allows perfect incorporation into most pharmacophores. Thus, diarylethenes have been studied and applied as photoswitch molecules that can be interconverted by modulation between open and closed forms [62,63]. Li et al. [64] developed an approach of molecular hybridization by introducing a diarylethene moiety into fluoroquinolone (norfloxacin, R^2 = ethyl and ciprofloxacin, R^2 = cyclopropyl) to obtain switchable antibacterial agents (Figure 7). As mentioned before, SAR studies showed that the secondary amine in the piperazine ring of fluoroquinolones can be modified without loss of antibacterial activity, so this position was chosen for conjugation with a diarylethene moiety.

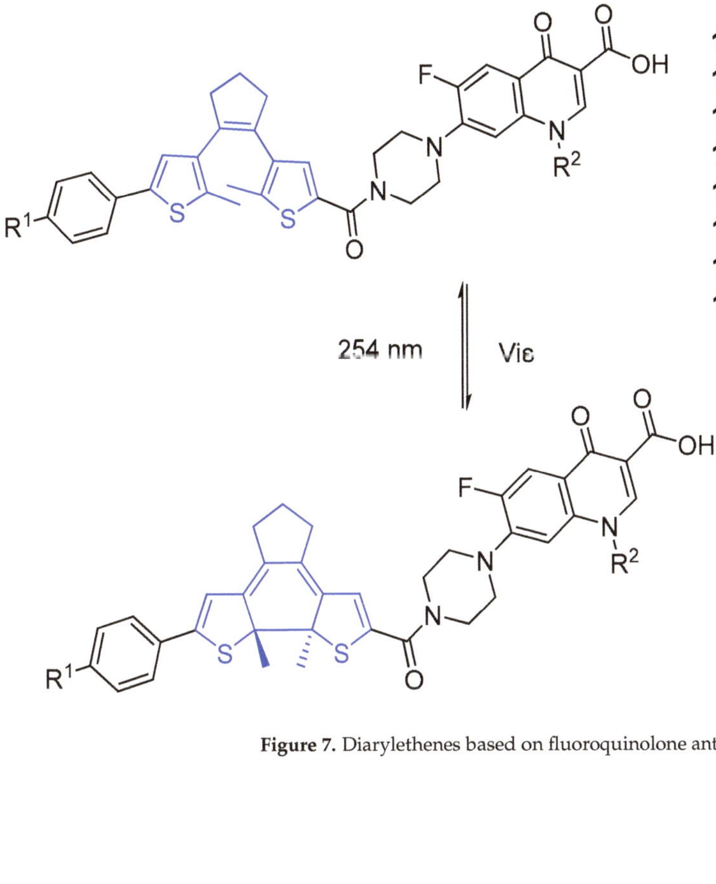

12a: R^1 = H; R^2 = ethyl
12b: R^1 = H; R^2 = cyclopropyl
13a: R^1 = CF_3; R^2 = ethyl
13b: R^1 = CF_3; R^2 = cyclopropyl
14a: R^1 = OCH_3; R^2 = ethyl
14b: R^1 = OCH_3; R^2 = cyclopropyl
15a: R^1 = NMe_2; R^2 = ethyl
15b: R^1 = NMe_2; R^2 = cyclopropyl

Figure 7. Diarylethenes based on fluoroquinolone antibiotics [64].

The antibacterial activity of the diarylethene derivative was assessed before and after irradiation with UV-C light (254 nm) by performing MIC values on *E. coli* and *S. aureus* (Table 4).

Table 4. MIC values of diarylethene-based switchable antibacterial agents before and after UV-C (254 nm) light irradiation.

Compound	MIC (μg mL^{-1})			
	Open-Isomer (before)		Closed-Isomer (after)	
	E. coli	*S. aureus*	*E. coli*	*S. aureus*
12a	32	8	16	8
12b	32	16	2	16
13a	16	32	16	32
13b	16	32	16	32
14a	32	16	8	16
14b	32	16	2	16
15a	32	16	16	16
15b	16	16	4	16
Norfloxacin	0.125	0.125	0.125	0.125
Ciprofloxacin	0.125	0.125	0.125	0.125

All the open-isomers showed a lower antibacterial activity on *E. coli* (MIC = 16–32 μg mL^{-1}) than closed-isomers. A similar activity against *S. aureus* (MIC = 8–32 μg mL^{-1}) is shown in both the open and closed forms. However, all of the compounds were less active than the native drugs, norfloxacin and ciprofloxacin (MIC = 0.125 μg mL^{-1}). The activity decrease can be related to the addition of the switchable moiety. Particularly, compounds **12b** and **14b** showed a great difference in the antibacterial activity against *E. coli* before and after irradiation, with the closed form being 16-fold more efficient than the open form. The ring-open form in diarylethenes shows a flexible conformation due to free rotation around the C–C bonds joining the thiophene to the central cyclopentene ring, consequently adopting various geometries. This flexible conformation was hypothesized not to allow the formation of stable complexes with DNA gyrase, decreasing the open-isomer antibacterial activity. Furthermore, the closed-form of compounds **12b** and **14b** appear to have a rigid conformation, which allows the formation of stable complexes with the DNA gyrase of *E. coli* and, consequently, derivatives were shown to increase the antibacterial activity up to 16-fold. Despite **12b** and **14b** being less effective than the parent drug ciprofloxacin, they showed the potential to be used as novel switchable antibacterial agents [64].

2.3. PHOTACs

PROTACs (**PRO**teolysis **TA**rgeting **C**himeras) are bifunctional small molecules that have been extensively studied as a novel method to treat diseases that result from the aberrant expression of specific disease-associated proteins [65]. PROTACs can degrade these proteins selectively by ubiquitination. This approach is promising for the treatment of cancer, neurodegenerative diseases, inflammatory diseases, and viral infections. By the combination of PROTACs with photopharmacology, a new approach arises—**PH**otoswitchable proteolysis **TA**rgeting **C**himeras (PHOTACs), which are bifunctional small molecules that target a protein of interest (POI) for ubiquitylation by an E3 ubiquitin ligase complex, promoting proteasome degradation (Figure 8). This approach allows the change in the reversibly endogenous protein levels, being helpful for potential treatments of diseases, such as cancer [66] and eventually Alzheimer's disease [67]. Several studies [68–70] have already designed and synthesized PHOTACs by incorporating photoswitches, such as azobenzenes, into PROTACs.

Pfaff et al. [70] were the first to report a novel approach based on PROTACs, including *o*-F_4-azobenzene linkers. AVR-771 is a PROTAC that degrades BRD2 and BRD4 (BET proteins), so it has (+)-JQ1 as a POI ligand and VHL (Von Hippel-Lindau) as a ligand of E3 ubiquitin ligase (Figure 9a). Based on AVR-771, a novel PHOTAC was designed (Figure 9b). The biological activity of PHOTACs was tested in Ramos cells and proved to be promising. While the *trans*-isomer was active in decreasing BRD2 levels, the *cis*-isomer did not induce protein degradation, hypothesized due to the shorter distance between BRD2 and VHL ligands because of the ''compact form'' of the *cis*-isomer (Figure 10). Despite the ARV-771 PROTAC being able to degrade both BRD2 and BRD4, degradation of BRD4 was not observed for photoPROTAC. This is hypothesized due to the reversed amide bond between (+)-JQ-1 and the *o*-F_4-azobenzene moiety.

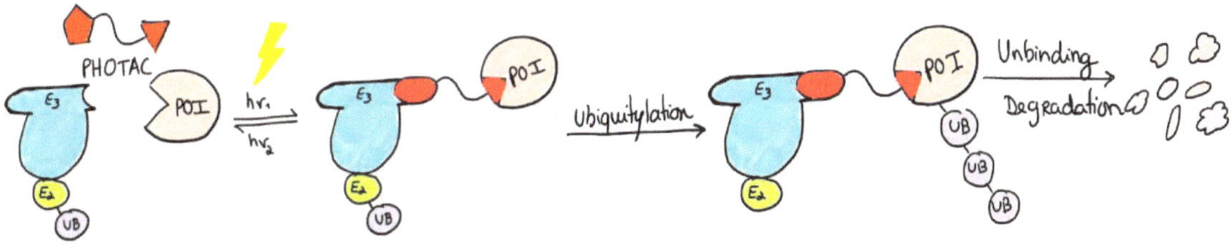

Figure 8. Schematic of the working model of a PHOTAC. The molecules toggle between an inactive form and an active form upon irradiation. Red pentagon does not fit in E3 ubiquitin ligase complex until irradiation occurs. There, the conformation changes, the ligand fits, and the ubiquitylation process occurs. Adapted with permission from [66]. Copyright, 2020, Elsevier.

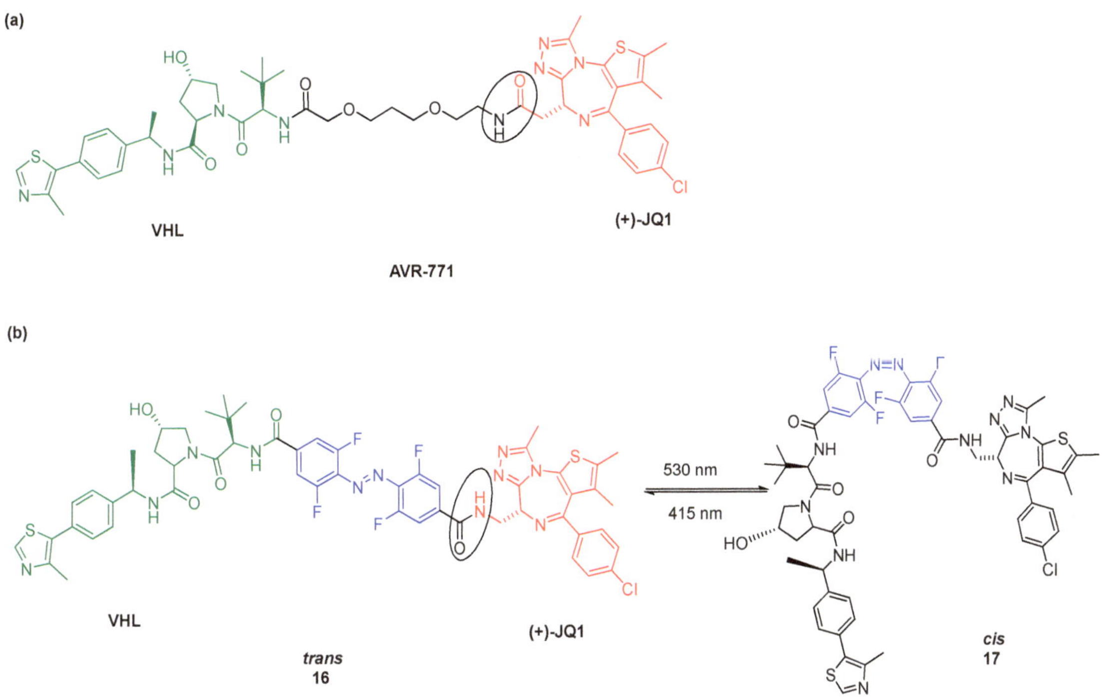

Figure 9. Figure **9.** (**a**) structure of PROTAC AVR-771 and (**b**) schematic representation of photoisomerization between *trans*-isomer (active, compound **16**) and *cis*-isomer (inactive, compound **17**) [70].

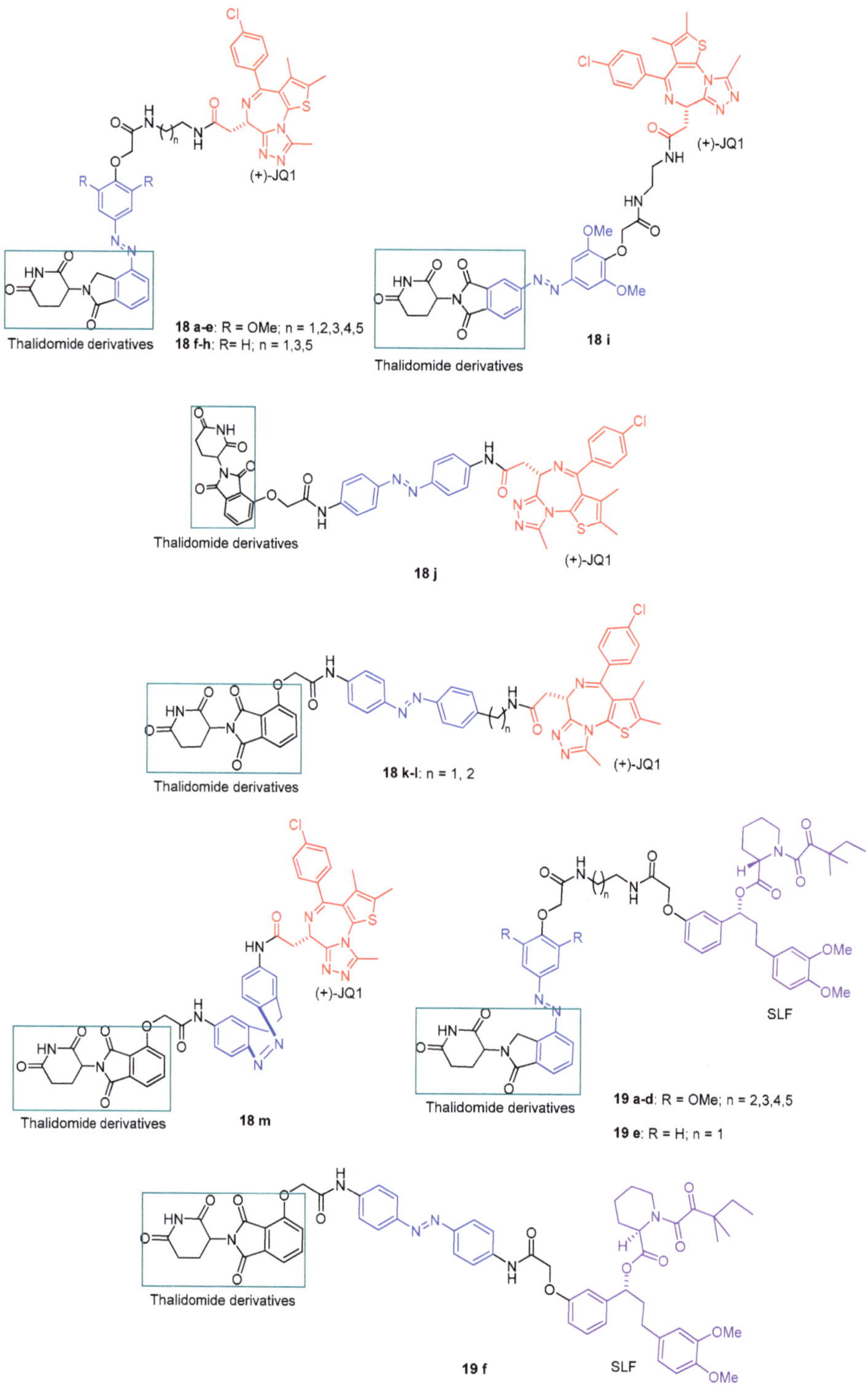

Figure 10. Structures of PHOTACs **18** and **19** series.

Reynders et al. [68] also reported the photopharmacological application to target protein degradation by incorporating azobenzenes into PROTACs, promoting light-dependent proteolysis. In the dark, the molecules do not have proteolytic activity, but upon irradiation with blue-violet light (380–440 nm), they were found to be activated. The azobenzene moiety was introduced on the phthalimide-conjugated ligand to bind cereblon E3 ubiquitin ligase complex to allow the affinity change. Based on the previously reported PROTAC dBET1 (Figure 10a) [71], a POI ligand, (+)-JQ1, was chosen for the design of the compound **18** series. (+)-JQ1 is an inhibitor of BET proteins BRD2-4. For the design of the compound **19** series, a synthetic ligand of FK506-binding protein (SLF) was chosen as the POI ligand based on PROTAC dFKBP-1 (Figure 10b), reported previously. The SLF ligand is a synthetic ligand for FKBP inhibitors. Figure 10 summarizes the small library of PHOTACs synthesized in this work.

The compound **18** series was evaluated on the viability of RS4;11 lymphoblast leukemic cells, in which compound **18c** (Figure 11c), with a 1,4diaminobuthyl spacer, proved to be the most effective. Therefore, BET protein (BRD2-4) levels were analysed in the presence of compound **18c**. A decrease in BRD4, BRD3, and BRD2 levels upon compound **18c** irradiation (390 nm) was observed, but not in the dark. A light dependent BRD4 degradation was also demonstrated in breast cancer cell lines (MBMDA231 and MBMDA468) using compound **18c**. Upon irradiation with UV-A light (390 nm), *trans*-**18c** was triggered to *cis*, obtaining a PSS of >90%. When the *cis*-isomer was irradiated with green light (500 nm), a PSS of >70% *trans* was obtained (Figure 11c). The compound **19** series was tested on FKBP12 levels; however, a slight activity in the dark was observed.

Jin et al. [69] designed PHOTACs (Figure 12a) to knockdown the BCR-ABL fusion protein to interact with E3 ubiquitin ligase cereblon (CRBN) using desatinib, a second-generation tyrosine kinase inhibitor, to target BCR-ABL. The compounds were assessed in a myelogenous leukaemia cell line K-562 and compound **20c** proved to be the most active in reducing BCR-ABL levels selectively by UV irradiation (Figure 12b). The *cis*-isomer did not induce protein degradation.

In general, contrasting with the examples of antibacterial compounds that incorporate azobenzenes (compounds **1–2** and **7–11**), light does not have an activating effect in some of them. However, in the case of PHOTACs, light triggers a protein destruction reaction. Nevertheless, it is still possible to reverse the process if it is desired to stop such a reaction. Although this is a promising strategy, it is still preliminary and has only been tested in cancer cells.

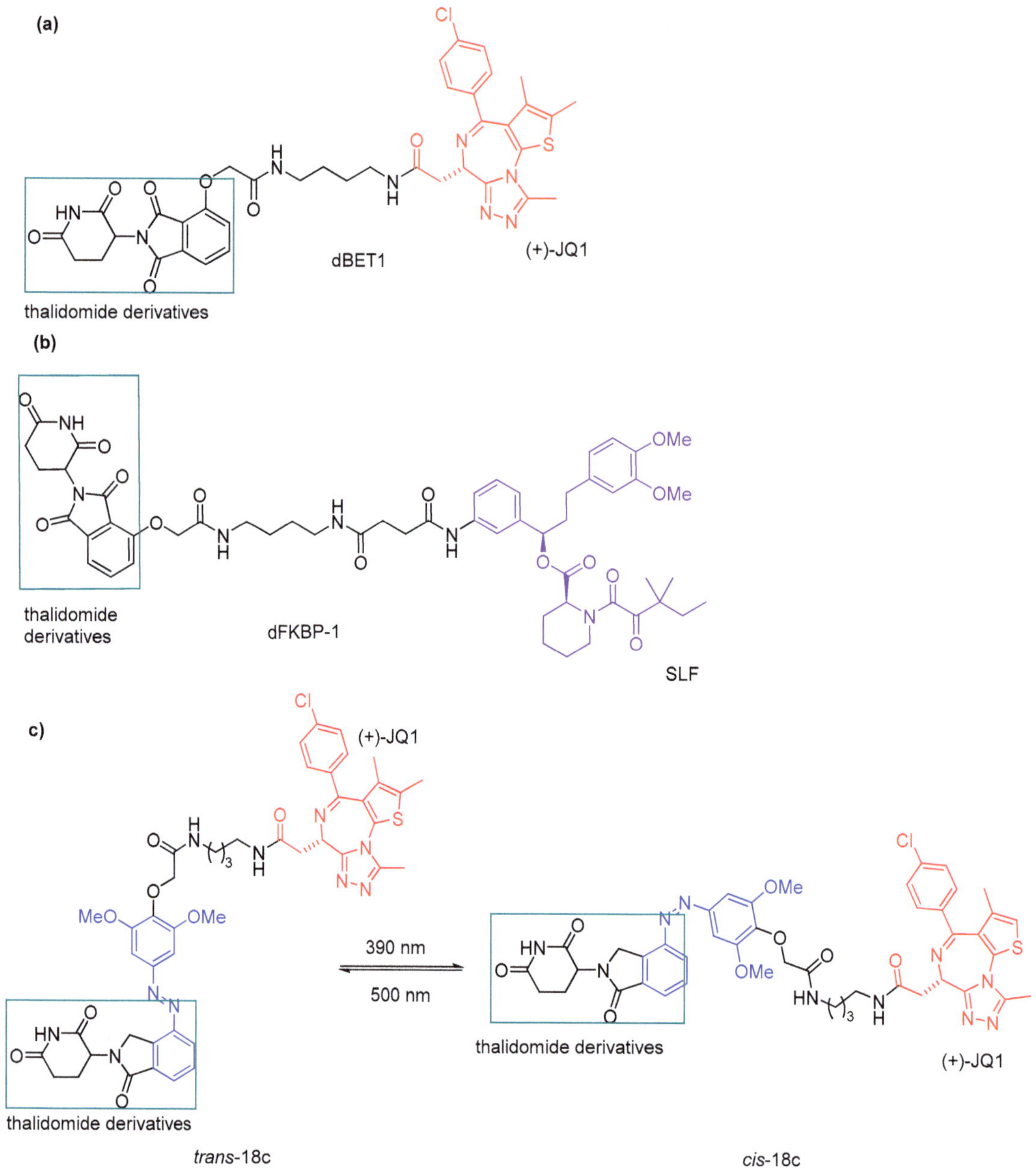

Figure 11. Structure of PROTACs (**a**) bBET1 and (**b**) dFKBP-1. (**c**) Photoswitching of compound **18c**.

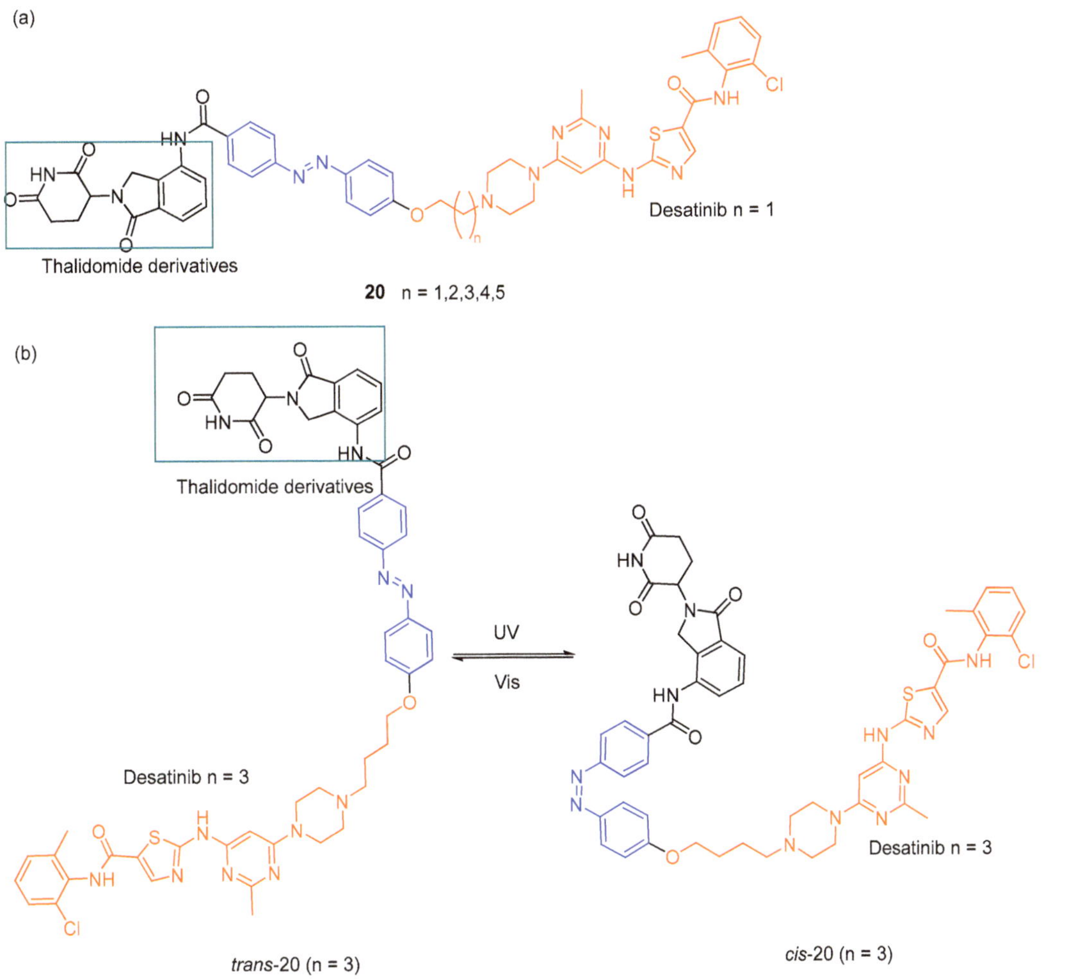

Figure 12. (**a**) Structures of PHOTACs synthesized and (**b**) schematic representation of PHOTAC **20c** (n = 3) and its isomerization [69].

3. Different Classes of Photocleavage and Some Examples in Antibacterial Applications

Photocleavage control is another category classified as light-controlled drug activity [72]. In photocleavage, the irradiation triggers the cleavage of a covalent bond between the drug and a light-responsive moiety that is responsible for modulating the drug's activity (Figure 13).

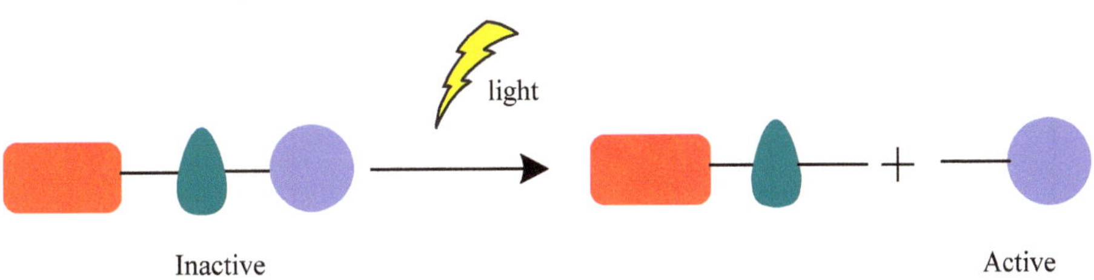

Figure 13. Representation of photocleavage control. Adapted from Friedman [72].

Photocleavage reactions need photoremovable protecting groups (PPG). PPG are small light-responsive units that can be covalently linked to the bioactive molecule. By light irradiation, PPG releases, irreversibly, the molecule in its active form [7]. Six classes of PPGs have already been investigated, including *o*-nitrobenzyl groups, coumarin-4-ylmethyl groups, arylmethyl groups, arylcarbonylmethyl groups, boron-dipyrromethene (BODIPY), and Ru complexes (Figure 14).

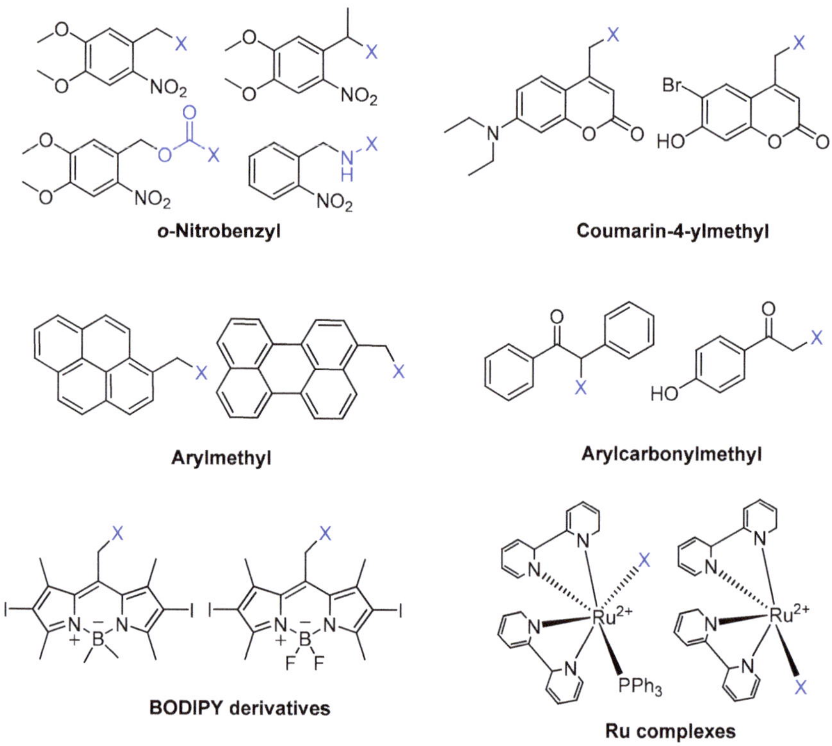

Figure 14. Different types of PPGs. X unit means leaving groups released by photocleavage [7].

As mentioned above, for light-controlled drug delivery, it is important to achieve photocleavage groups at long-wavelength light (between 650 nm and 900 nm), such as red light or near infra-red light (NIR), for clinical use [73]. Therefore, BODIPY [74] and Ru complexes [74] have attracted research for having long-wavelength light PPGs and their photolysis process occurring under visible or NIR light irradiation. However, BODIPY derivatives have shown cytotoxic effects due to $^1O_2$ generation, but there are studies in which after irradiation, this cytotoxicity has been applied to exert a synergistic effect combined with chemotherapy drugs [75].

The use of PPGs has the greatest interest in some biomedical applications, such as to treat of skin infections, cancer, neuronal and antibacterial diseases, and other cellular processes [76].

3.1. BODIPY

Kumari and co-workers [77] designed a BODIPY complexed with levofloxacin (BODIPY-Levo) to deliver fluoroquinolone upon visible blue light (470 nm) irradiation (Figure 15).

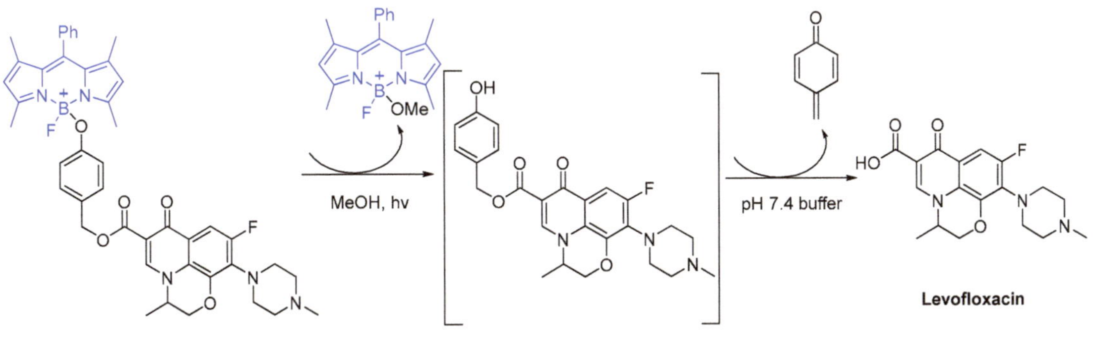

Figure 15. Mechanism proposed for generation of levofloxacin after photoirradiation with blue light (470 nm).

The antimicrobial activity was assessed against *E. coli* and *S. aureus* bacterial strains. Before irradiation, compound **21** did not contribute to the inhibitory effects towards *E. coli* and *S. aureus*. Upon light conditions (470 nm), after releasing levoflaxacin (5.0 µM), the optical density measurement at 600 nm (OD_{600}) decreased significantly in comparison to the compound **21** (p-value < 0.001), corroborating the rationale of the design [77].

3.2. *o-Nitrobenzyl Derivatives*

Shchelik et al. [78] investigated two "caged" antibiotics (vancomycin and cephalosporin, both of which inhibit bacterial cell wall biosynthesis), where the drug was planned to be photoreleased upon UV light exposure (365 nm, 5 min), using *o*-nitrobenzyl as a PPG group and polyethylene glycol (PEG) as a linker (Table 5). PEGylation was an approach to allow steric hindrance, which could suppress vancomycin's activity and prevent the binding to the amino acid residues of the peptide chain during cell wall synthesis. Cephalosporin PEGylation was expected to prevent binding to penicillin-binding protein or prevent transport issues.

The antibacterial activity was evaluated against *Bacillus subtilis* (*B. subtilis*) ATCC 6633, *S. aureus* ATCC 29,213, MRSA, *E. coli* ATCC 25922, and *P. aeruginosa* ATCC 27853, with MIC values calculated (Table 6).

Compound **22** containing PEG did not show significant activity. However, compound **24** (without PEG) revealed excellent activity against *B. subtilis* and *S. aureus*, as well as vancomycin (compound **23**). In the case of the cephalosporin series, contrary to expectations, PEG decreased the antibiotic activity (compound **25**). Compound **27** exhibited better activity than cephalosporin against all the strains except *P. aeruginosa*, but the activity was revealed to be more efficient in Gram-negative strains, which could be explained by the thiadiazole cephalosporin side chain and its zwitterionic properties that allow higher penetration through the outer membrane (present only on Gram-negative strains). The PEG linker decreased the antibacterial activity against all the tested strains, both in vancomycin and cephalosporin series.

Wong and co-workers [79] reported a delivery mechanism targeted to Gram-negative bacteria based on ciprofloxacin photoreleased by a cell wall-targeted dendrimer nanoconjugate (Figure 16a). The delivery system is a dendrimer conjugated with polymyxin B (PMB) or ethanolamine (EA) that acts as a carrier for ciprofloxacin (inhibitor of DNA gyrase), using the *o*-nitrobenzyl as PPG (ONB-cipro). When PMB or EA binds to lipopolysaccharide (LPS), 80% of ciprofloxacin is photoreleased from ONB-cipro by irradiating UV-A light (365 nm) for 30 min. The antibacterial activity of the conjugates (**29** and **30**, Figure 16b) was evaluated and MIC_{50} values were 2.0 µM (in *E. coli*) approximately, which may be due to an incomplete drug release. The MIC_{50} value of ciprofloxacin is 0.0125 µM. However, the molecules did not show phototoxicity in human cells and exhibited good selectivity for *E. coli*.

Table 5. Vancomycin (**22**,**23**, and **24**) and cephalosporin (**26**, **27**, and **28**) derivatives.

	Vancomycin Derivatives	Cephalosporin Derivatives
Target Compounds	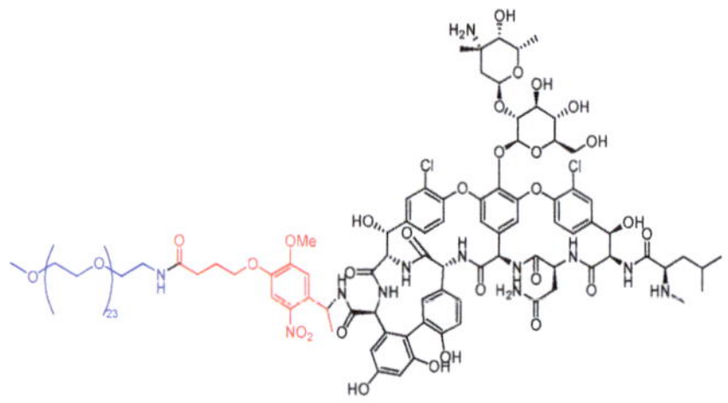 **22**	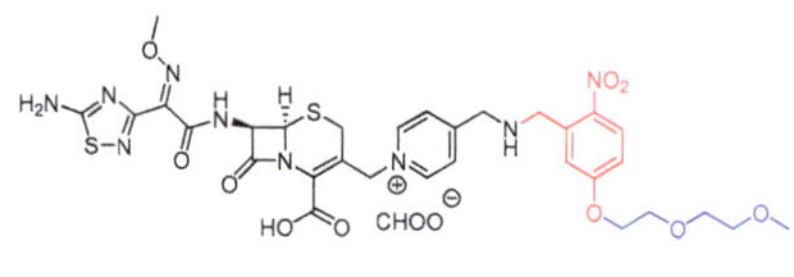 **25**
Release Compounds	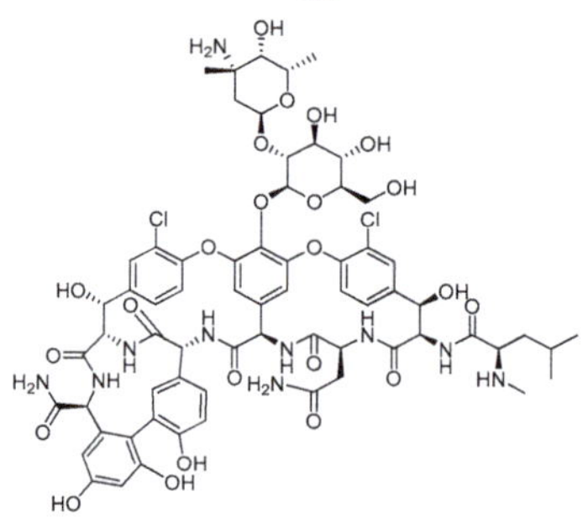 **23**	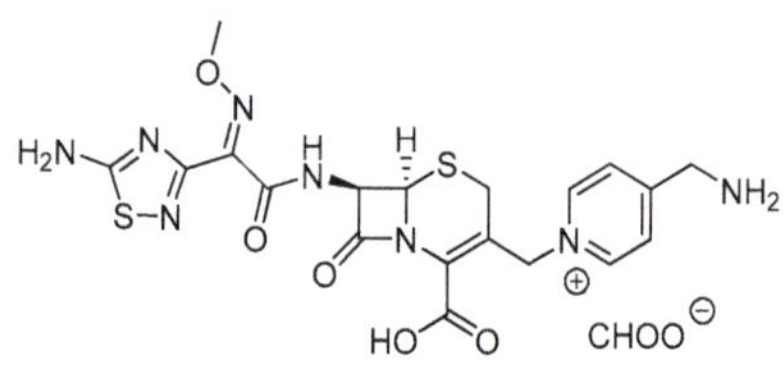 **26**
Control Compounds	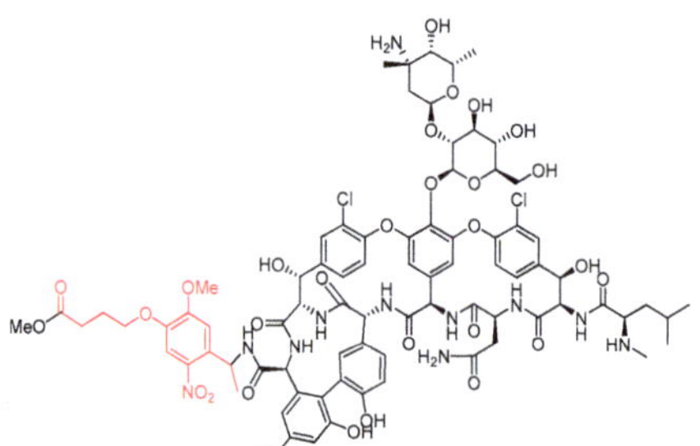 **24**	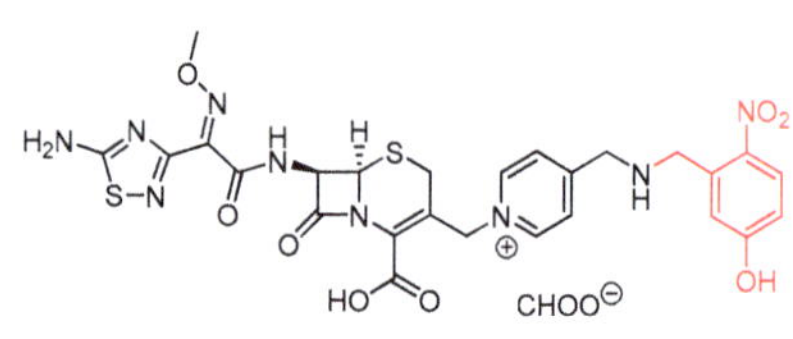 **27**

Table 6. MIC values (μg mL^{-1}) of vancomycin and cephalosporin derivatives.

Strains	MIC (μg mL^{-1})					
	Vancomycin Series			Cephalosporin Series		
	22	23	24	25	26	27
E. coli ATCC 25922	-	-	-	8	1–2	1
P. aeruginosa ATCC 27853	-	-	-	64	2–4	32
B. subtilis ATCC 6633	32	0.06–0.125	0.125	8	2–4	1
S. aureus ATCC 6633	>64	0.5–1	0.5	32	8	4
S. aureus ATCC 43300	>64	1–2	1	64	32	16

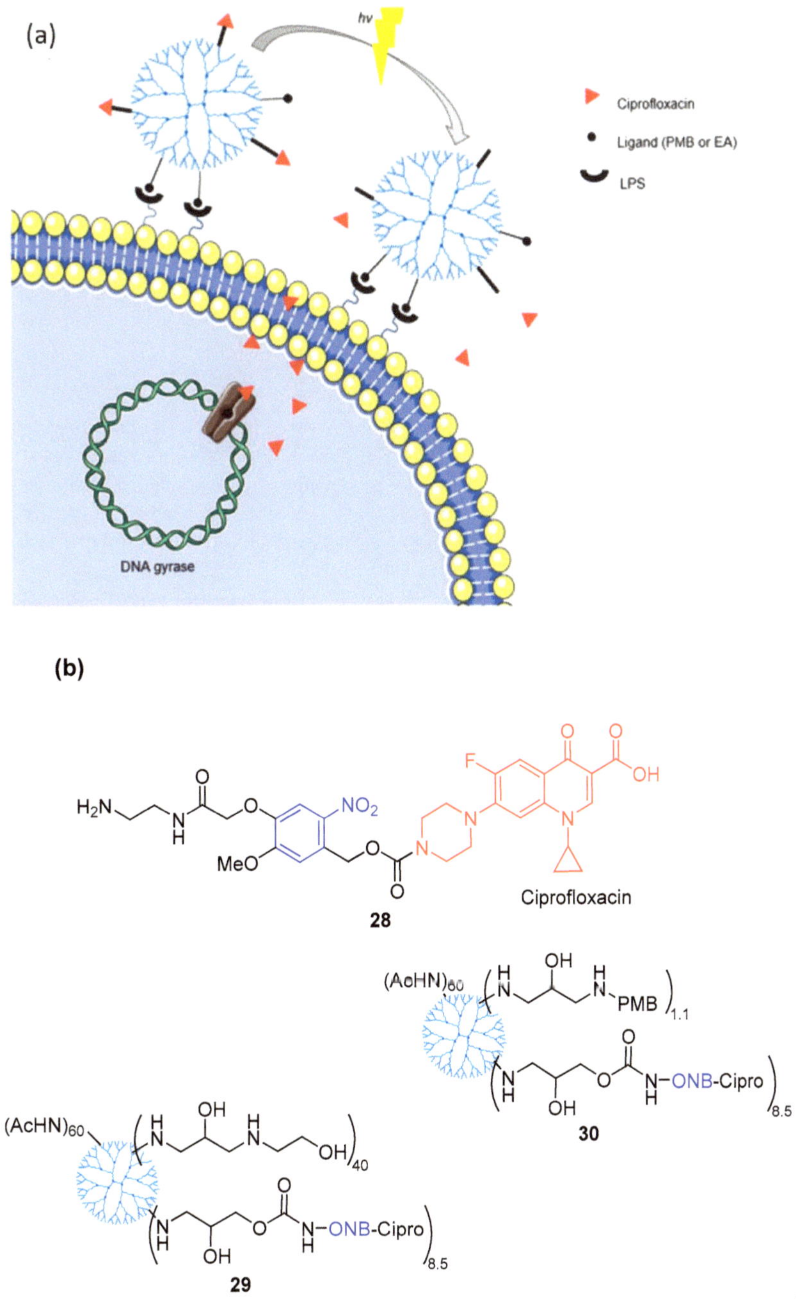

Figure 16. (**a**) Scheme for light-controlled targeted delivery of photocaged ciprofloxacin carried by lipopolysaccharide (LPS) adsorbed to the outer membrane of the Gram-negative bacterium (*E. coli*) cell wall. Internalized ciprofloxacin inhibits DNA gyrase. (**b**) Structures of ciprofloxacin, photocaged ciprofloxacin (**28**), dendrimers conjugated with ONB-ciprofloxacin and cell wall targeting ligand—EA (**29**), or PMB (**30**). Adapted with permission from [79]. Copyright, 1996, Royal Society of Chemistry.

3.3. Photocaged PROTACs

As in the case of photoswitches, photocaged PROTACs are an approach to photocleavage in which E3 ubiquitin ligase (a linker with a photocleavage moiety) and the POI, form a ternary complex to lead proteasomal degradation [80] (Figure 17). However, photocaged PROTACs are irreversible, unlike photoswitchable PROTACs, which are reversible. Several researchers reported photocaged PROTACs to induce protein degradation by light.

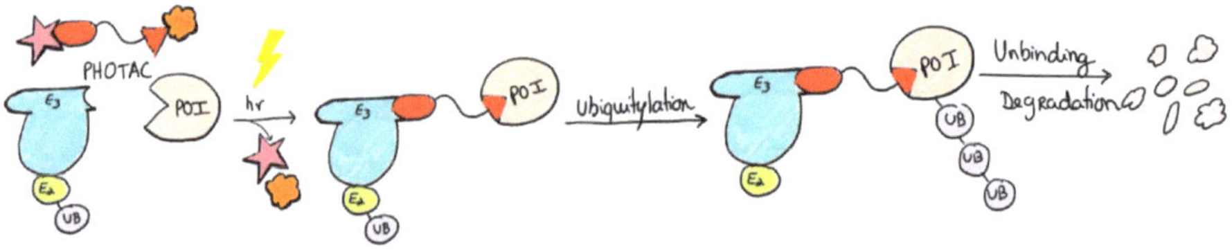

Figure 17. Schematic of the working model of a photocaged PROTAC. The caging moiety attached to either end of the PROTAC is removed upon irradiation. The red geometric figures are in the presence of the PPG groups (pink and orange figures) that prevent the structure from binding to the E3 ubiquitin ligase and POI. Upon irradiation, the PPG groups leave, allowing the structure to bind and, therefore, the ubiquitylation process occurs. Adapted with permission from [80]. Copyright, 2020, John Wiley and Sons.

Xue et al. [81], for the first time, designed two photocaged PROTACs based on dBET1 PROTAC [71], which used thalidomide as a ligand of E3 ubiquitin ligase cereblon (CRBN) and (+)-JQ1 as a ligand of BRD4. The 4,5-dimethoxy-2-nitrobenzyl (DMNB) introduced is a PPG that is cleaved upon irradiation (365 nm). The photocaged PROTAC **31** (Figure 18a) has the PPG group linked to the amide of (+)-JQ1, and the photocaged PROTAC **32** (Figure 18b) has the PPG group through the imide of thalidomide. Upon exposure to irradiation, photocaged PROTAC **31** generated approximately 50% of dBET1 production, whereas no dBET1 production was observed with photo-caged PROTAC **32**.

(a)

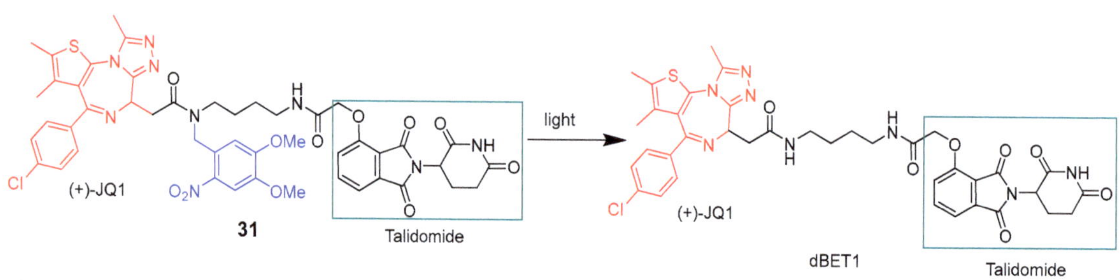

(b)

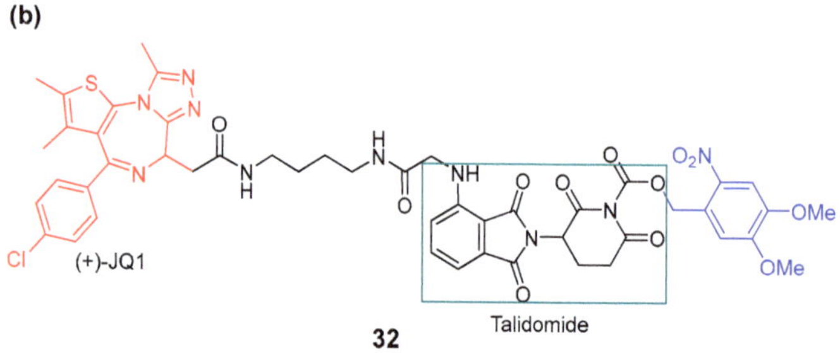

Figure 18. (**a**) Uncaging reaction of photocaged PROTAC **31** in dBET and (**b**) structure of photocaged PROTAC **32**.

Ramos cells were used to assess in vitro BRD4 degradation. BRD4 levels decreased when photocaged PROTAC **31** was irradiated (365 nm, 3 min), being almost completely degraded after 4 h. A zebrafish model was used to assess the in vivo activity of photocaged PROTAC **31,** which showed effectiveness.

Naro et al. [82] also designed other two different photocaged PROTACs using two different PPG groups and ligands recruiting E3 ubiquitin ligases (VHL and CRBN). Photocaged PROTAC **33** (Figure 19a) contains a VHL ligand, targets ERRα (estrogen related receptor α), and uses a diethylaminocoumarin as a PPG group. Photocaged PROTAC **34** (Figure 19b) has a CRBN ligand and targets BRD4 using a 6-nitropiperonyloxymethyl as a PPG group.

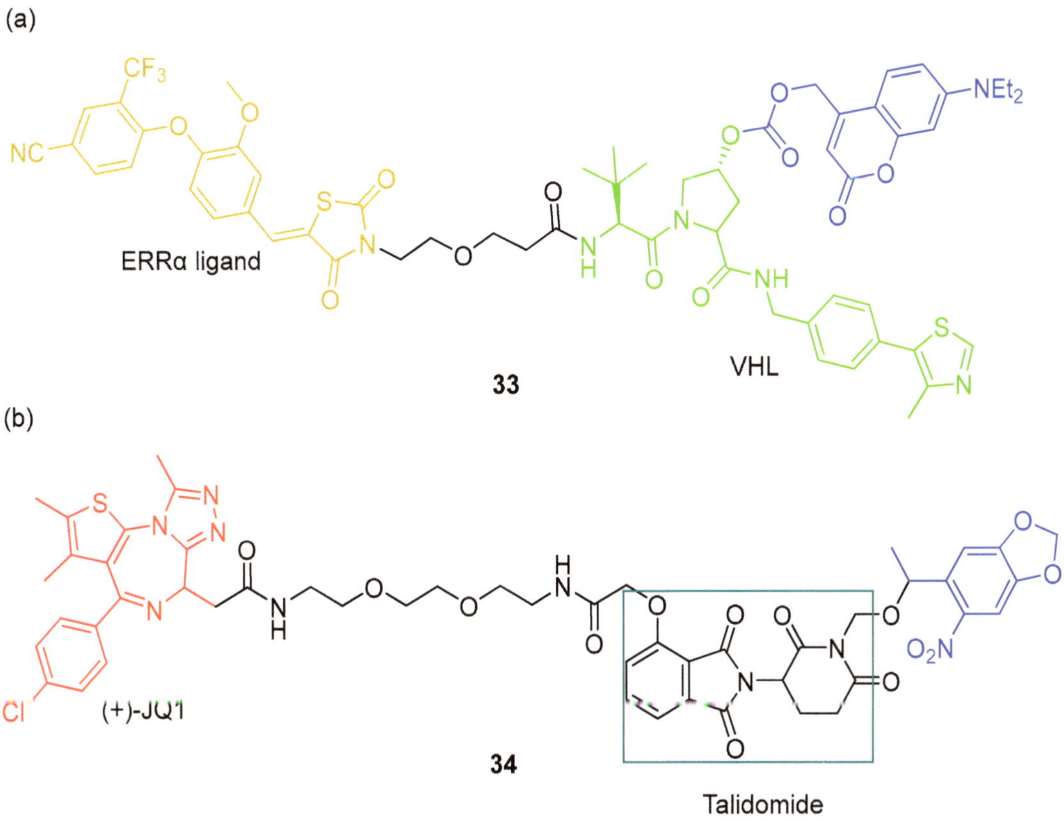

Figure 19. Chemical structures of (**a**) photocaged PROTAC **33** and (**b**) photocaged PROTAC **34**. Upon irradiation, PPG groups are released [82].

MCF-7 cells were used to test the ability of the PPG group of the photocaged PROTAC **33** to block ERRα degradation, and it was verified that the degradation of ERRα was blocked before irradiation. In the case of photocaged PROTAC **34,** HEK293T cells were used to test the degradation of BRD4. In the absence of light, no degradation was verified. Upon exposure to UV-A (365 nm) light, BRD4 was degraded, corroborating the expected results of the designed study.

Jing et al. [67] added a PPG group on pomalidomide based on dBET and dALK PROTACs—photocaged PROTAC **35** (Figure 20a) and photocaged PROTAC **36** (Figure 20b). Dimethoxy-2-nitrobenzyl was used as a PPG group that will prevent pomalidomide from binding to CRBN E3 ubiquitin ligase. In this way, protein degradation will not occur. This will only occur when there is UV-A irradiation, in which PPG leaves and thus pomalidomide can already recruit CRBN ligase for degradation to occur.

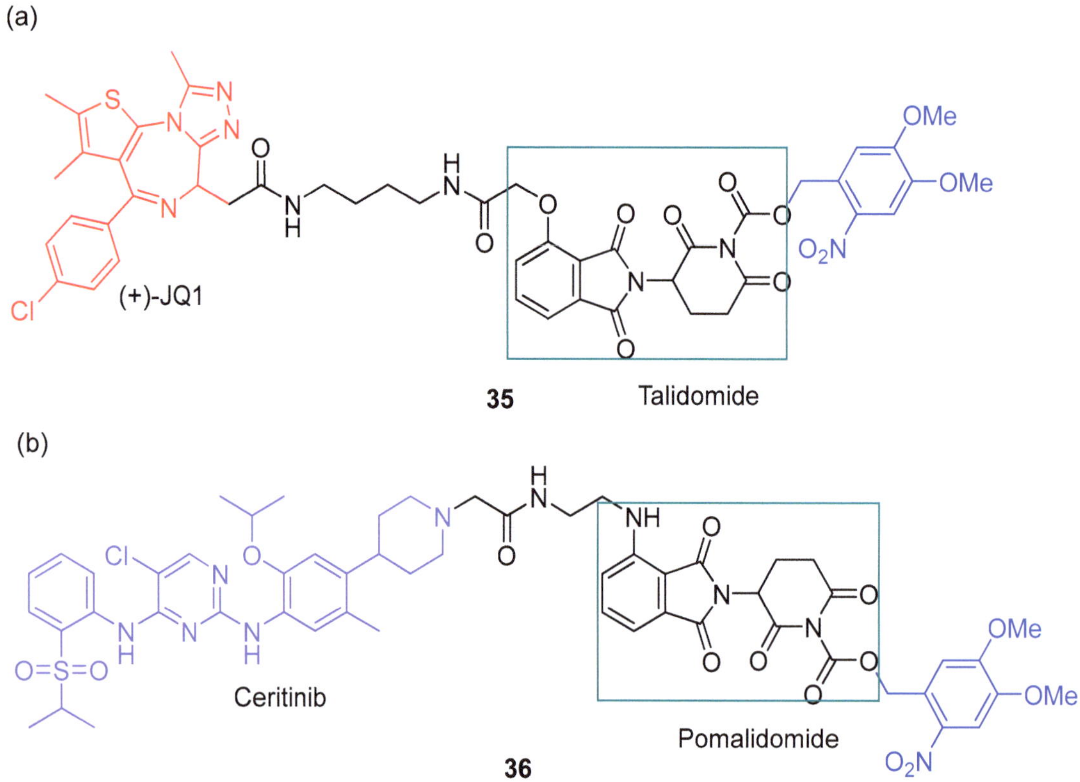

Figure 20. Chemical structures of photocaged PROTAC 35 (**a**) and photocaged PROTAC 36 (**b**).

HEK293FT cells were used to test the effect of photocaged PROTAC **35** to degrade BRD3-4 by UV-A irradiation. SU-DHL-1 cells were used to assess the effect of opto-dALK to degrade anaplastic lymphoma kinase (ALK, a fusion protein) by UV-A irradiation. Upon UV-A irradiation, photocaged PROTAC **35** is uncaged and leads to the induction of BRD3-4 degradation. On the other hand, when photocaged PROTAC **36** is irradiated with UV-A light, it is also uncaged and promotes the degradation of ALK.

Kounde and co-workers [83] designed a photocaged PROTAC (Figure 21a) using a DMNB group binding to VHL E3 ligase-recruiter ligand and (+)-JQ1 as an inhibitor of BET. This approach was based on PROTAC MZ1 [84] (Figure 21b). The DMNB moiety was used to block the recruitment of VHL E3 ubiquitin ligase without irradiation, but it was removed upon UV-A light (365 nm) irradiation.

The ability of photocaged PROTAC **37** to degrade BRD4 was tested in HeLa cells and its knockdown in real time was performed using live-cell fluorescence imaging (HEK293 cells). The results of this study proved to be good as BRD4 degradation was achieved.

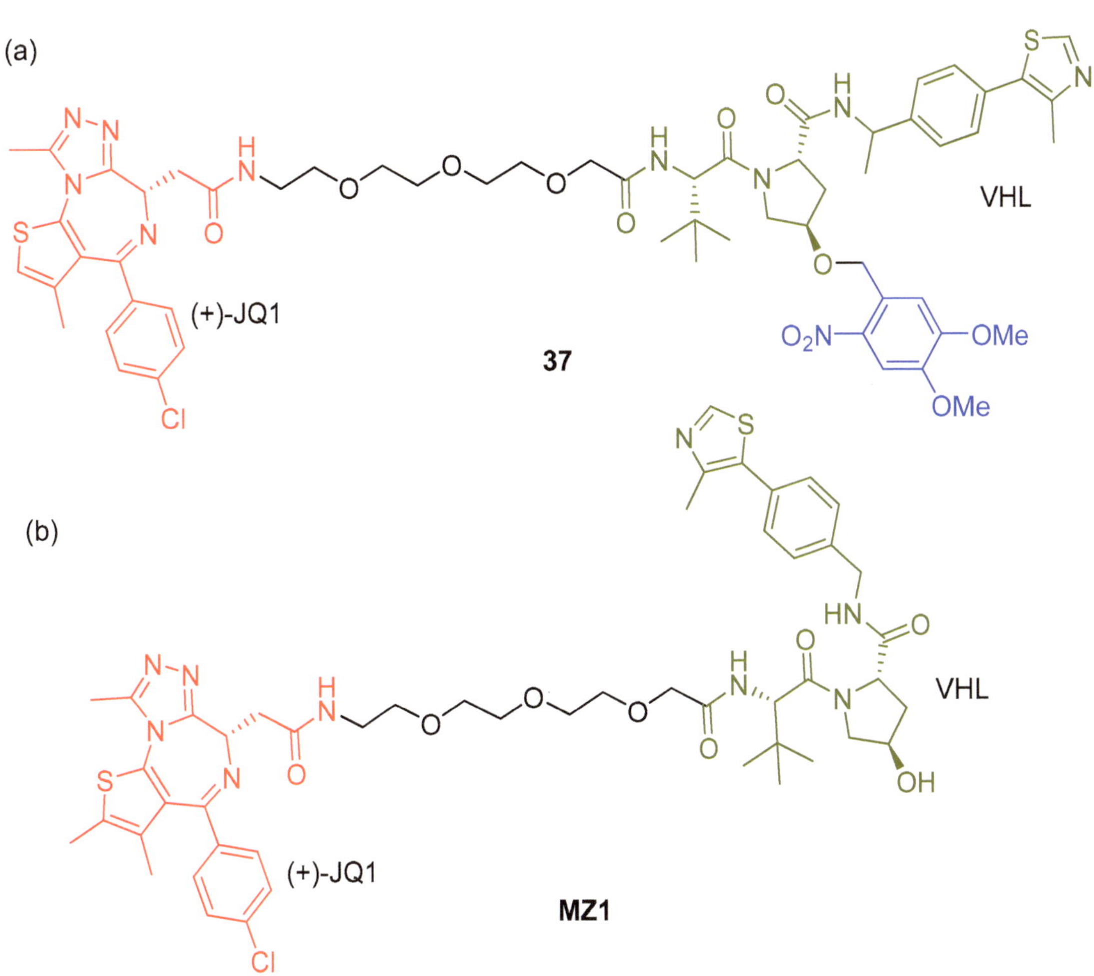

Figure 21. Chemical structures of photocaged PROTAC **37** (**a**) and PROTAC MZ1 (**b**).

4. Porphyrins Combining Photoswitch Systems and Some Application Examples

Usually, gadolinium complexes are applied to improve magnetic resonance imaging (MRI) structural contrast in tissues. Light can be used as a functional MRI contrast with high-spatiotemporal resolution. A light-responsive contrast agent for magnetic resonance imaging based on a Ni(II) porphyrin derivative and a switch moiety has previously been reported in the literature [85]. This agent can be applied non-invasively to switch magnetic resonance imaging contrast ON and OFF with a high-spatiotemporal resolution by light stimulus (Figure 22).

Porphyrin derivative **40** was found not to be soluble in water due to the first generation of small glycerol dendrimers, so the hydrophobic character of the Ni(II) porphyrin prevails. However, compound **42** (second generation glycerol dendrimers) prevents porphyrin aggregation and thus increases water solubility. Non-dendronized porphyrins (compounds **38** and **43**) form dimers and with increasing concentration become paramagnetic, decreasing the efficacy of contrast switching. Second generation dendrimers prevent this phenomenon.

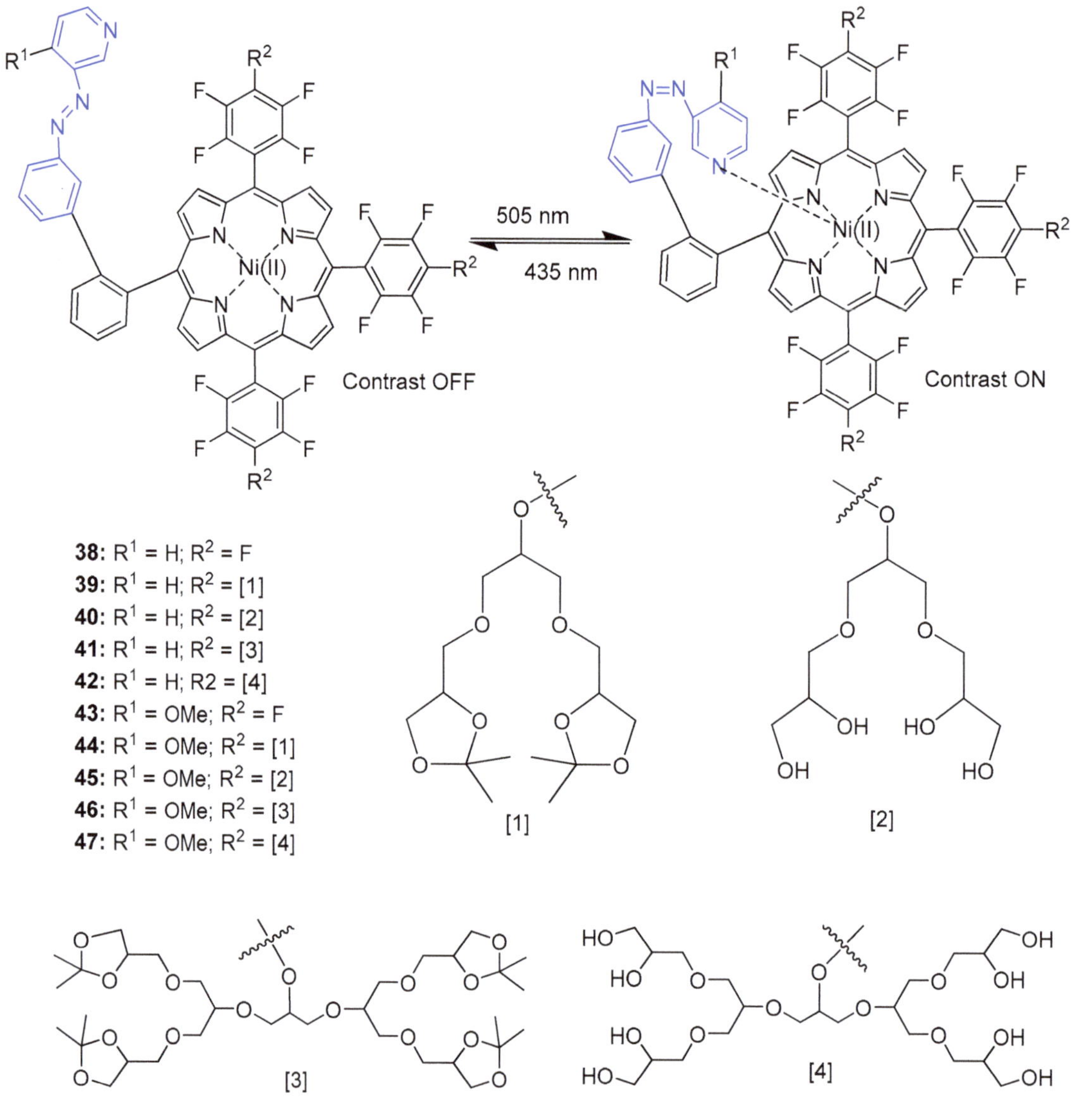

Figure 22. Structures of porphyrins irradiated with 505 nm (contrast ON) and reversible process (435 nm, contrast OFF) [85].

The *trans*-isomer was found not to coordinate intramolecularly, being diamagnetic and thus is MRI silent (contrast OFF), whereas since the *cis*-isomer coordinates intramolecularly, it is paramagnetic and MRI active (contrast ON). Substitutions on R^1 with pyridine (R^1 = H) furnished porphyrins with a PSS of 65% *cis*-isomer, while 4-methoxyl (R^1 = OMe) showed a PSS > 95% *cis*-isomer, allowing the retention of diamagnetic to paramagnetic switching. It was concluded that compound **47** provided the best structures in this study.

In the literature, photoswitchable micelles were applied for the control of singlet oxygen ($^1O_2$) generation in PDT, using a zinc(II) complex of 5,10,15,20-tetraphenylporphyrin (ZnTPP) as a photosensitizer and 1,2-bis(5-(4-carbonxyphenyl)-2-methylthien-3-yl)cyclopent-1-ene (BDTE) as a photoswitch (compound **48**, Figure 23). Compound **48** was encapsulated in micelles to cause tumour and bacterial cell death or prevent cell damage upon irradiation through $^1O_2$ generation [86]. The photoswitching was evaluated in a subcutaneous tissue model.

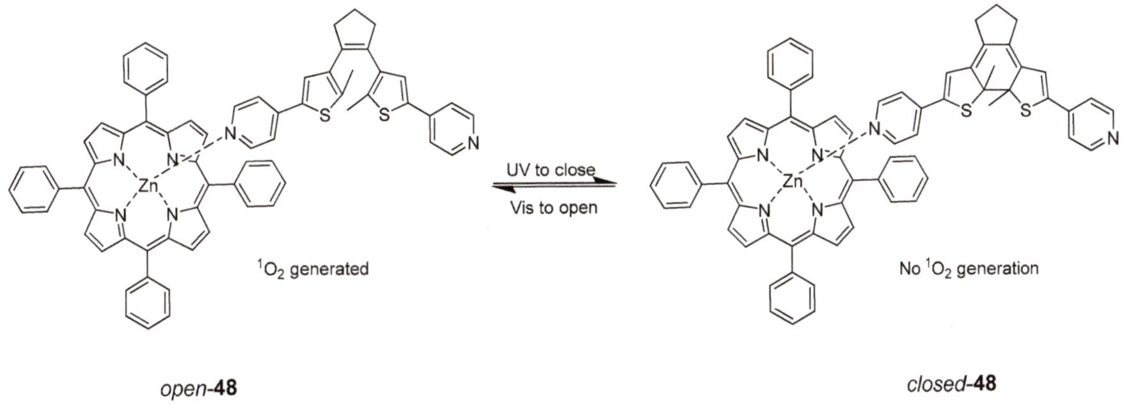

Figure 23. Singlet-oxygen control upon opening and closing photoswitch. The open form (open-**48**) generates $^1O_2$, while the closed is not active in producing $^1O_2$.

5. Porphyrins Combining Photocleavage Systems and Some Application Examples

Lin et al. [87] reported for the first time in 2008, an anticancer prodrug (compound **49**, Figure 24) based on an *o*-nitrobenzyl group coupled with a porphyrin. The porphyrin anticancer prodrug is composed of three parts, such as a porphyrin, a PPG (*o*-nitrobenzyl group), and a parent anticancer drug (Tegafur). First, the drug is released by irradiation with UV-A light (350 nm). After 12 min, it achieved a 50% conversion of compound **49** into tegafur. 5-Fluorouracil as an anticancer drug was used as a control, and it was released in the same way as Tegafur.

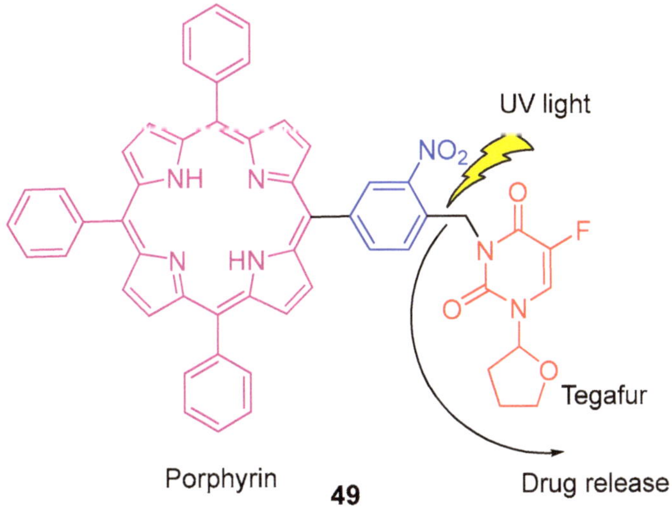

Figure 24. Schematic representation of compound **49** (anticancer prodrug composed by a porphyrin, a PPG, and a parent anticancer drug, tegafur) and its release upon UV-A irradiation [87].

The authors used MCF-7 human breast adenocarcinoma cells to evaluate the cytotoxicity of compound **49** in the absence and presence of irradiation. In the absence of UV-A light, tegafur induced 91% of cell death while the compound induced only 7% of cell death (less cytotoxic). When this entity is irradiated and, therefore, Tegafur is released, 69% of cell death is observed. Although there is less cell death, this approach allows for minimizing the side effects due to the tumour-affinity property of porphyrin and the light-controllable system.

Tessaro et al. [88] reported a novel photocontrolled nanoplatform comprising two photoresponsive components—a benzochlorin PS (Figure 25a) for PDT and a coumarin-photocaged chlorambucil (compound **50**, Figure 25b)—into core-shell micelles to amplify the anticancer activity against MCF-7 human breast adenocarcinoma cells. The authors demonstrated that with simultaneous irradiation with visible light in the micelles, benzochlorin produces $^1O_2$ as a PS and compound **50** acts as a prodrug by leaving of PPG group (Figure 25b).

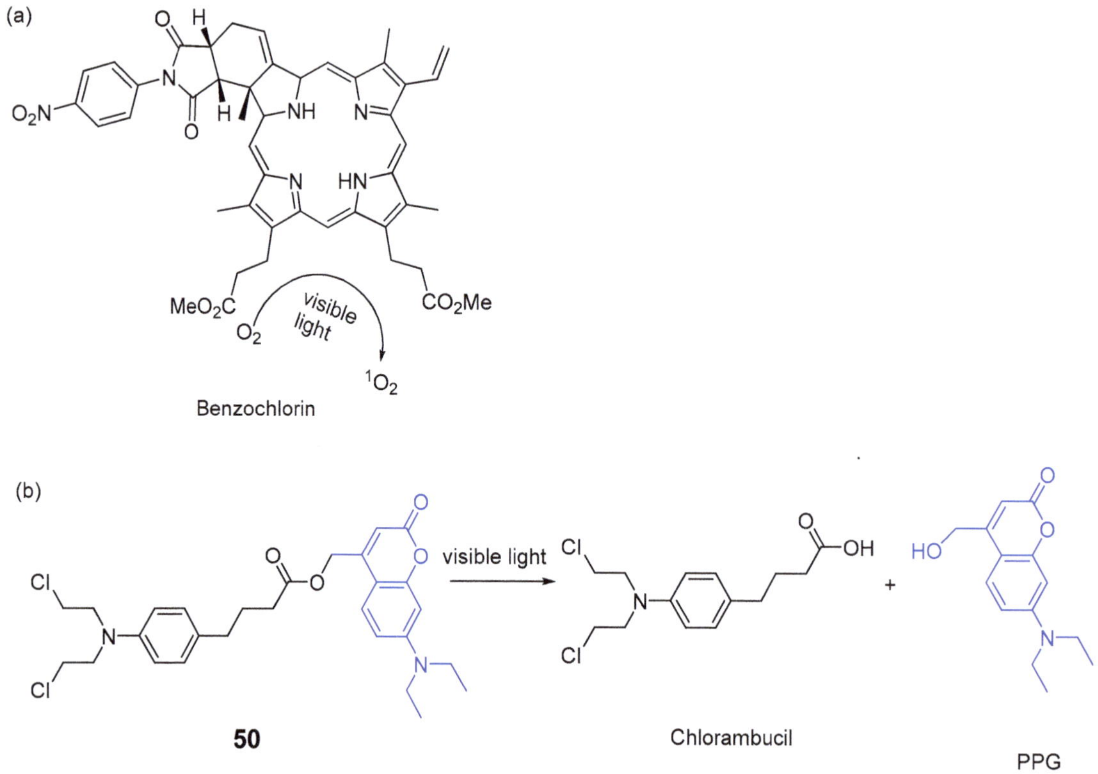

Figure 25. Benzochlorin used as a PS for PDT (**a**) and coumarin-photocaged (compound **50**) release chlorambucil (**b**) [88].

Results showed chlorambucil used alone induces 25% cell death, while micelles containing compound **50** and the sensitizer benzochlorin induce 75% cell death after 20 min of blue irradiation (400 nm). This promising multimodal therapy destroys the tumour in two ways—generating $^1O_2$ and releasing chlorambucil during the irradiation procedure, showing a synergistic effect.

6. Porphyrins with Antibacterial Activity

Antimicrobial photodynamic treatment (aPDT) requires, besides an adequate PS, a sufficient PS concentration and an adequate light dose to guarantee full microbial inactivation. However, achieving these conditions is not always easy. To overcome this problem, an ongoing approach that has been presented by the scientific community is the combination of PDT and chemotherapy, as such high concentrations of PS are not necessary because the antibiotic is also used in combination. Thus, the synergistic effect between aPDT and antibiotic therapy is a promising approach as it is possible to increase aPDT efficiency at concentrations below the MIC [89,90]. This section will be presented as an overview of using porphyrins as PS for aPDT and antibiotic therapy combination.

Xing et al. [91] conjugated vancomycin, a glycopeptide antibiotic, with a porphyrin derivative to generate a divalent dimeric system (Figure 26), which has a rigid structure and provides a steric hindrance necessary for interactions between vancomycin and the bacterial strain. The antibacterial activity of compound **51** was tested in *B. subtilis* (ATCC 33,677), *Enterococcus faecium (E. faecium)* ATCC 51559, and *Enterococcus faecalis (E. faecalis)* ATCC 51,299, vancomycin-resistant enterococci strains. The results showed a strong photoinactivation activity for the conjugate when compared to vancomycin and porphyrin alone.

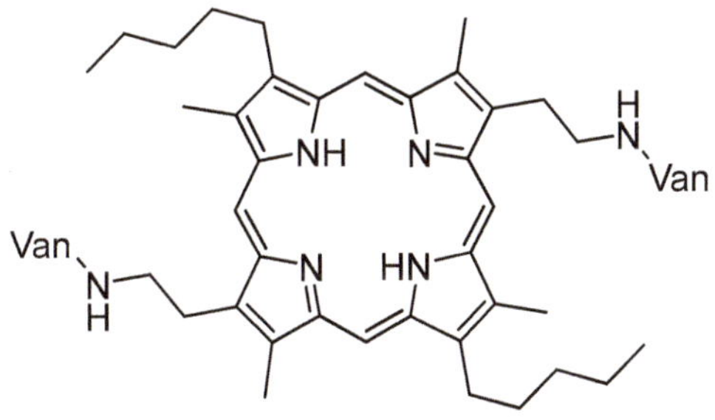

51

Figure 26. Chemical structure of the conjugated Van-porphyrin. Van = vancomycin [91].

Liu et al. [92] also developed an approach based on the bioconjugation of a protoporphyrin IX (PpIX) as a PS with lipopolysaccharide (LPS) and an antibacterial peptide (YI13WF) to photoinactivate Gram-negative bacterial strains (Figure 27).

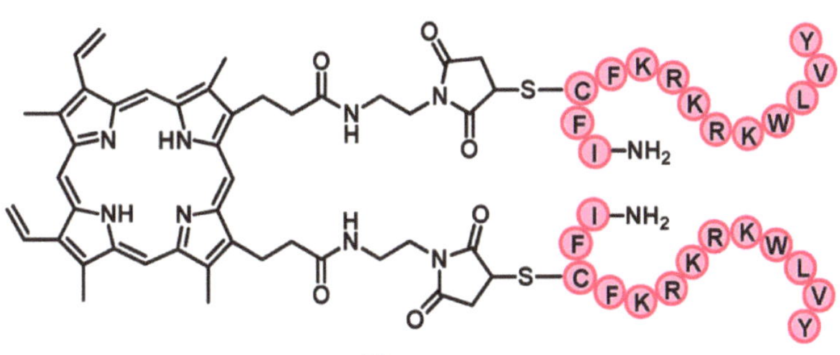

52

Figure 27. Dimeric PpIX−peptide conjugate. Reprinted with permission from [92]. Copyright, 2012, American Chemical Society.

The MIC values (Table 7) were calculated in *E. coli DH5a* (ATCC 53868), *S. enterica* (ATCC 14028), *E. coli* BL21, and *K. pneumoniae* (ATCC 700603). The antibacterial peptide sequence (YI13WF) binds strongly towards lipids at the outer membrane of Gram-negative bacteria, showing potent MIC values against the strains. However, the MIC values of the dimeric conjugate (compound **52**) demonstrated greater aPDT. The antibacterial activity of PpIX alone as a control was also evaluated. The results of the conjugates proved to be significantly better than these components alone.

Table 7. MIC values, expressed in μM, of the conjugate free peptide YI13WF and PpIX.

	MIC Values (μM)			
	E. coli **DH5a**	*S. enterica*	*E. coli* **BL21**	*K. pneumoniae*
PpIX−Peptide Conjugate	2.0	4.0	4.0	8.0
YI13WF	8	~24	32	32
PpIX	>64	>64	>64	>64

Dastgheyb and co-workers [93] tested a 5,10,15,20-tetrakis(4-aminophenyl)porphyrin (TAPP) combined with antibiotics used for the treatment of bacterial infections, assuming that combining TAPP with antibiotics would have a complementary effect (Figure 28). TAPP specifically targets bacterial membranes, while tobramycin and chloramphenicol inhibit protein synthesis, and ceftriaxone and vancomycin impair membrane integrity. Results indicated that TAPP with chloramphenicol or tobramycin had an additive effect on *S. aureus* and *E. coli* and a synergic effect on MRSA and *S. epidermidis*. On the other hand, vancomycin and ceftriaxone show only modest effects when combined with TAPP.

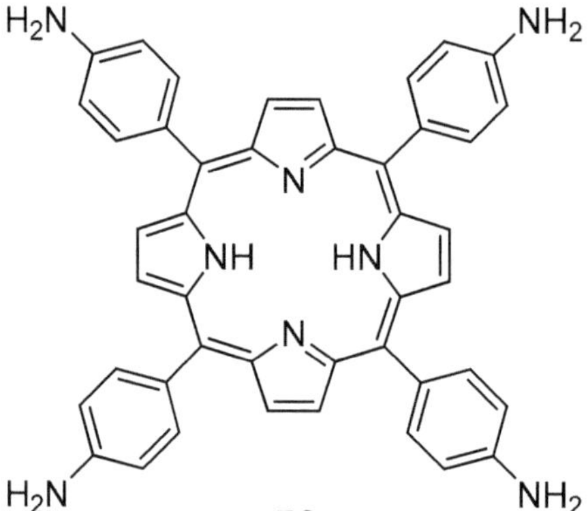

53

Figure 28. Chemical structure of 5,10,15,20-tetrakis(4-aminophenyl)porphyrin (TAPP).

Almeida et al. [94] aimed to allow the efficient treatment of aPDT on multidrug-resistant bacteria in hospital wastewaters. Therefore, a 5,10,15,20-tetrakis(1-methylpyridinium-4-yl)porphyrin tetra-iodide **54** (TMPyP, Figure 29) as the PS was used in combination with chloramphenicol. The antibacterial activity of the conjugate was performed in *E. coli* and it was proved that the combination effect of the conjugates was more efficient than TMPyP used alone. The synergic effect could be due to the aPDT cell membrane and the cell wall destabilization, allowing an easier entry of chloramphenicol.

Branco and co-workers [90] compared the effect of aPDT and aPDT combined with antibiotics to treat skin infected with *S. aureus* ATCC 6538. As well as in the previous work, TMPyP (compound **54**, Figure 29) was used as the PS. The antibacterial activity was performed in vitro and ex vivo (porcine skin), using compound **54** combined with and without antibiotics. In vitro, a bacterial reduction of ~8 log after 180 min of white light irradiation at an irradiance of 4.0 mW cm^{-2} was attained using 5.0 μM of **54** without antibiotic. When tested, the combined action of 0.5 and 1.0 μg mL^{-1} ampicillin with 5.0 μM of **54** increased the efficacy, observing a reduction of ~8 log after 60 min and 30 min of white light irradiation, respectively. When the ampicillin was tested alone, no colony-forming unit (CFU) reduction was observed, even at the highest tested concentration, 1.0 μg mL^{-1}. In ex vivo experiments, using 50 μM of **54** after 180 min of irradiation with white light at an irradiance of 150 W m^{-2}, a decrease of ~4 log (99.99% reduction) of bacterial abundance was observed. However, when treated the porcine skin simultaneously with TMPyP (50 μM) and with 5.0 μg mL^{-1} ampicillin, a bacterial reduction of ~5.6 log (>99.9995% reduction) after 180 min of white light irradiation was observed. Using the antibiotic alone at this concentration, only a CFU reduction of ~1 log was noted.

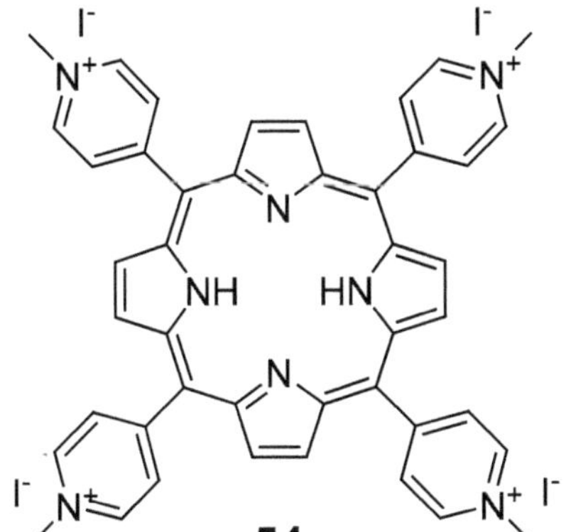

54

Figure 29. Chemical structure of the cationic porphyrin **54**, also known as TMPyP or Tetra-Py^{+}-Me.

Iluz et al. [95] searched for a synergistic effect between deuteroporphyrin **55** (Figure 30) and antibiotics usually used to treat infections with *S. aureus* ATCC 25,923 and MRSA. Although gentamicin, vancomycin, rifampin, and fusidic acid treatments were tested in combination with aPDT, only oxacillin (1 μg mL^{-1}) proved to have a synergistic effect in combination with deuteroporphyrin (4 μM), during 24 h, at a light dose of 15 J·cm^{-2}, showing a CFU reduction of ~5.5 log (99.9995% reduction). Cultures treated with monotherapy (oxacillin or 55 alone) showed no log decrease (0% reduction), even though there was an increase during 2h of treatment.

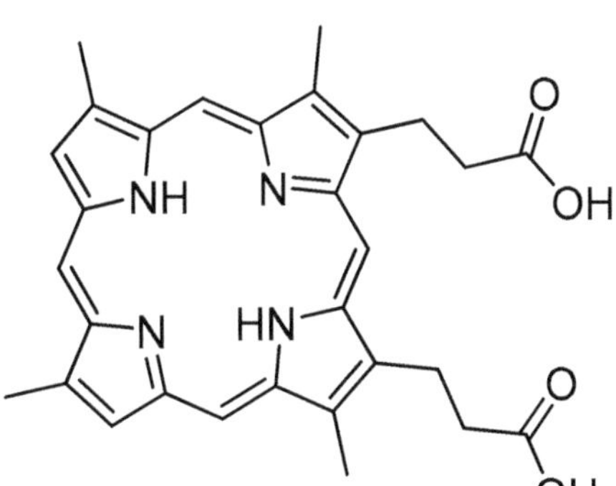

55

Figure 30. Chemical structure of the deuteroporphyrin **55** (DP).

7. Concluding Remarks and Future Challenges

The problem of antimicrobial and chemotherapeutic resistance is that is results frequently in therapeutic failure. With the emergence of new strains of resistant bacteria, the search for effective antibiotics has become a critical global health priority and has motivated several research groups around the world to find effective alternatives.

Photodynamic treatments, photoswitches, photocaged groups, and recent PHOTACs and photocaged PROTACs strategies have been explored as promising approaches to tackle the AMR issue. As it was demonstrated (compounds **3–6** and **12–15**), photoswitches have the advantage of being reversible compared to photocaged compounds. However, the design of these molecules is not always easy or straightforward because they were related to a series of azobenzene-based compounds where light would not have the activating effect but rather result in deactivation, which is very important for deactivation processes. Despite this, these deactivation approaches will be more relevant in the case of PHOTACs to stop the degradation protein process if necessary. However, many approaches reported using UV-A-C irradiation, which is harmful to healthy cells, or with visible blue light, which has limited tissue penetration. As far as we know, only one example was found in the literature that reports the use of NIR light in azobenzene-based compounds (compounds **5** and **6**). Therefore, it is necessary to improve strategies of incorporation of red-shifted chromophores to address the tissue penetration problem, even in PHOTACs (not just restricted to photoswitchable compounds). Apparently, there are not as many studies done on photocaged compounds as there are on photoswitchable compounds, which could be due to the fact that the process is not reversible. Photocaged PROTACs do not seem to be as promising because of they do not allow protein degradation to be stopped if necessary. Nevertheless, PHOTACs and photocaged PROTACs are reported for cancer treatment, but it should be important to prove their effectiveness for other diseases, such as antimicrobial infections, to combat resistant strains. Furthermore, porphyrins and their derivatives (e.g., chlorins) have been demonstrating a synergistic effect when used with antibiotic therapy. However, the most recent study on this topic was published several years ago, so it is critical to continue researching in this field, specially involving porphyrin derivatives with the antimicrobial activity in aPDT, since no resistance mechanism seems to be developed by the biological entities. It is important to note that, to date, examples of porphyrin derivatives combining photoswitchable or photocaged systems with antimicrobial activity are not described in the literature and there is still room to develop interesting photoswitchable or photocaged systems.

Photoswitches and photocleavage compounds have shown promising potential for use in various clinical applications. However, as with any new therapeutic agent or its combined action, there are potential safety concerns associated with their use that must be evaluated [96]. One of the main safety concerns is the potential risk for these compounds to elicit adverse effects or toxicity. To mitigate these potential safety risks in a clinical context, it is crucial to conduct comprehensive preclinical studies to evaluate toxicity and to find the optimal therapeutical conditions in order to carefully control and avoid toxicity. Additionally, the delivery method must be optimized to ensure that the desired target compounds are only delivered to the target tissues or organs. [97]. Overall, the successful translation of externally activated drug delivery systems from preclinical studies to human clinical trials requires a comprehensive understanding of the complex interactions between the delivery system and the target biological environment [98]. To address these challenges, it is important to employ animal models, assess drug dose and administration routes based on in vitro experiments, and develop effective methods for monitoring the long-term toxicity and pharmacokinetics of the delivery system in vivo [99]. While external triggers pose additional challenges, the imaging capabilities of many triggered delivery systems offer unique opportunities for better monitoring the fate of drugs and drug carriers in vivo [100].

Our aim for this review is to offer new perspectives on the use of photopharmacology, not only for bacterial inactivation to mitigate antibiotic resistance but also for cancer treatment, motivating the medical and scientific communities to devote greater attention and resources to these issues. This, in turn, will enhance the comprehension of the problem, enabling the identification of potential solutions that could lead to improved public health outcomes.

Author Contributions: Conceptualization, C.J.P.M., M.E.S., A.P. and M.A.F.F.; writing—original draft preparation, S.N.S., C.J.P.M., M.E.S., A.P. and M.A.F.F.; writing—review and editing, C.J.P.M., M.E.S., A.P. and M.A.F.F. All authors have read and agreed to the published version of the manuscript.

Funding: This work received support from PT national funds (FCT/MCTES, Fundação para a Ciência e a Tecnologia and Ministério da Ciência, Tecnologia e Ensino Superior) through the projects UIDB/50006/2020 and UIDP/50006/2020 (LAQV-REQUIMTE) and UIDB/04423/2020, UIDP/04423/2020 (Group of Marine Natural Products and Medicinal Chemistry CIIMAR).

Data Availability Statement: Data sharing not applicable.

Acknowledgments: The authors thank the University of Aveiro and FCT/MCT for the financial support provided to LAQV-REQUIMTE (UIDB/50006/2020 and UIDP/50006/2020), Group of Marine Natural Products and Medicinal Chemistry CIIMAR (UIDB/04423/2020, UIDP/04423/2020), through national funds (OE), and where applicable, co-financed by the FEDER Operational Thematic Program for Competitiveness and Internationalization-COMPETE 2020, within the PT2020 Partnership Agreement.

Conflicts of Interest: The authors declare no conflict of interest. The funders had no role in the design of the study; in the collection, analyses, or interpretation of data; in the writing of the manuscript; or in the decision to publish the results.

References

1. Reessing, F.; Szymanski, W. Beyond Photodynamic Therapy: Light-Activated Cancer Chemotherapy. *Curr. Med. Chem.* **2017**, *24*, 4905–4950. [CrossRef]
2. Jia, S.; Sletten, E.M. Spatiotemporal Control of Biology: Synthetic Photochemistry Toolbox with Far-Red and Near-Infrared Light. *ACS Chem. Biol.* **2022**, *17*, 3255–3269. [CrossRef] [PubMed]
3. Broichhagen, J.; Frank, J.A.; Trauner, D. A Roadmap to Success in Photopharmacology. *Acc. Chem. Res.* **2015**, *48*, 1947–1960. [CrossRef] [PubMed]
4. Velema, W.A.; Szymanski, W.; Feringa, B.L. Photopharmacology: Beyond Proof of Principle. *J. Am. Chem. Soc.* **2014**, *136*, 2178–2191. [CrossRef] [PubMed]
5. World Health Organization: Antimicrobial Resistance. Available online: https://www.who.int/news-room/fact-sheets/detail/antimicrobial-resistance (accessed on 21 March 2023).
6. Weinstain, R.; Slanina, T.; Kand, D.; Klan, P. Visible-to-NIR-Light Activated Release: From Small Molecules to Nanomaterials. *Chem. Rev.* **2020**, *120*, 13135–13272. [CrossRef]
7. Liu, J.Z.; Kang, W.R.; Wang, W.P. Photocleavage-based Photoresponsive Drug Deliver. *Photochem. Photobiol.* **2022**, *98*, 288–302. [CrossRef]
8. Hull, K.; Morstein, J.; Trauner, D. In Vivo Photopharmacology. *Chem. Rev.* **2018**, *118*, 10710–10747. [CrossRef]
9. Velema, W.A.; Hansen, M.J.; Lerch, M.M.; Driessen, A.J.M.; Szymanski, W.; Feringa, B.L. Ciprofloxacin-Photoswitch Conjugates: A Facile Strategy for Photopharmacology. *Bioconjugate Chem.* **2015**, *26*, 2592–2597. [CrossRef]
10. Mosinger, J.; Lang, K.; Kubát, P. Photoactivatable Nanostructured Surfaces for Biomedical Applications. In *Light-Responsive Nanostructured Systems for Applications in Nanomedicine*; Sortino, S., Ed.; Springer International Publishing: Cham, Switzerland, 2016; pp. 135–168. [CrossRef]
11. Mesquita, M.Q.; Dias, C.J.; Neves, M.G.P.M.S.; Almeida, A.; Faustino, M.A.F. The Role of Photoactive Materials Based on Tetrapyrrolic Macrocycles in Antimicrobial Photodynamic Therapy. In *Handbook of Porphyrin Science*; World Scientific: Singapore, 2022; Volume 46, pp. 201–277.
12. Q Mesquita, M.; J Dias, C.; P M S Neves, M.G.; Almeida, A.; F Faustino, M.A. Revisiting Current Photoactive Materials for Antimicrobial Photodynamic Therapy. *Molecules* **2018**, *23*, 2424. [CrossRef]
13. Henderson, B.W. *Photodynamic Therapy: Basic Principles and Clinical Applications*; CRC Press: Boca Raton, FL, USA, 1992.
14. De Silva, P.; Saad, M.A.; Thomsen, H.C.; Bano, S.; Ashraf, S.; Hasan, T. Photodynamic therapy, priming and optical imaging: Potential co-conspirators in treatment design and optimization—A Thomas Dougherty Award for Excellence in PDT paper. *J. Porphyr. Phthalocyanines* **2020**, *24*, 1320–1360. [CrossRef]
15. Santos, L.L.; Oliveira, J.; Monteiro, E.; Santos, J.; Sarmento, C. Treatment of Head and Neck Cancer with Photodynamic Therapy with Redaporfin: A Clinical Case Report. *Case Rep. Oncol.* **2018**, *11*, 769–776. [CrossRef] [PubMed]

16. Van Straten, D.; Mashayekhi, V.; de Bruijn, H.S.; Oliveira, S.; Robinson, D.J. Oncologic Photodynamic Therapy: Basic Principles, Current Clinical Status and Future Directions. *Cancers* **2017**, *9*, 19. [CrossRef]
17. Kwiatkowski, S.; Knap, B.; Przystupski, D.; Saczko, J.; Kędzierska, E.; Knap-Czop, K.; Kotlińska, J.; Michel, O.; Kotowski, K.; Kulbacka, J. Photodynamic therapy—Mechanisms, photosensitizers and combinations. *Biomed. Pharmacother.* **2018**, *106*, 1098–1107. [CrossRef]
18. Plaetzer, K.; Krammer, B.; Berlanda, J.; Berr, F.; Kiesslich, T. Photophysics and photochemistry of photodynamic therapy: Fundamental aspects. *Lasers Med. Sci.* **2009**, *24*, 259–268. [CrossRef]
19. Rkein, A.M.; Ozog, D.M. Photodynamic Therapy. *Dermatol. Clin.* **2014**, *32*, 415–425. [CrossRef] [PubMed]
20. Kim, M.; Jung, H.Y.; Park, H.J. Topical PDT in the Treatment of Benign Skin Diseases: Principles and New Applications. *Int. J. Mol. Sci.* **2015**, *16*, 23259–23278. [CrossRef] [PubMed]
21. Castano, A.P.; Demidova, T.N.; Hamblin, M.R. Mechanisms in photodynamic therapy: Part one—Photosensitizers, photochemistry and cellular localization. *Photodiagn. Photodyn. Ther.* **2004**, *1*, 279–293. [CrossRef]
22. Almeida, A.; Faustino, M.A.F.; Neves, M.G.P.M.S. Antimicrobial Photodynamic Therapy in the Control of COVID-19. *Antibiotics* **2020**, *9*, 320. [CrossRef]
23. Alves, E.; Faustino, M.A.F.; Neves, M.G.P.M.S.; Cunha, Â.; Nadais, H.; Almeida, A. Potential applications of porphyrins in photodynamic inactivation beyond the medical scope. *J. Photochem. Photobiol. C Photochem. Rev.* **2015**, *22*, 34–57. [CrossRef]
24. Youf, R.; Müller, M.; Balasini, A.; Thétiot, F.; Müller, M.; Hascoët, A.; Jonas, U.; Schönherr, H.; Lemercier, G.; Montier, T.; et al. Antimicrobial Photodynamic Therapy: Latest Developments with a Focus on Combinatory Strategies. *Pharmaceutics* **2021**, *13*, 1995. [CrossRef]
25. Pucelik, B.; Dąbrowski, J.M. Chapter Three—Photodynamic inactivation (PDI) as a promising alternative to current pharmaceuticals for the treatment of resistant microorganisms. In *Advances in Inorganic Chemistry*; van Eldik, R., Hubbard, C.D., Eds.; Academic Press: Cambridge, MA, USA, 2022; Volume 79, pp. 65–108.
26. Varzandeh, M.; Mohammadinejad, R.; Esmaeilzadeh-Salestani, K.; Dehshahri, A.; Zarrabi, A.; Aghaei-Afshar, A. Photodynamic therapy for leishmaniasis: Recent advances and future trends. *Photodiagn. Photodyn. Ther.* **2021**, *36*, 102609. [CrossRef] [PubMed]
27. Sadraeian, M.; da Cruz, E.F.; Boyle, R.W.; Bahou, C.; Chudasama, V.; Janini, L.M.R.; Diaz, R.S.; Guimarães, F.E.G. Photoinduced Photosensitizer–Antibody Conjugates Kill HIV Env-Expressing Cells, Also Inactivating HIV. *ACS Omega* **2021**, *6*, 16524–16534. [CrossRef] [PubMed]
28. Wu, X.; Hu, Y. Photodynamic Therapy for the Treatment of Fungal Infections. *Infect. Drug Resist.* **2022**, *15*, 3251–3266. [CrossRef]
29. Barbosa, A.F.S.; Santos, I.P.; Santos, G.M.P.; Bastos, T.M.; Rocha, V.P.C.; Meira, C.S.; Soares, M.B.P.; Pitta, I.R.; Pinheiro, A.L.B. Anti–Trypanosoma cruzi effect of the photodynamic antiparasitic chemotherapy using phenothiazine derivatives as photosensitizers. *Lasers Med. Sci.* **2020**, *35*, 79–85. [CrossRef] [PubMed]
30. Yoo, S.W.; Oh, G.; Ahn, J.C.; Chung, E. Non-Oncologic Applications of Nanomedicine-Based Phototherapy. *Biomedicines* **2021**, *9*, 113. [CrossRef]
31. Niculescu, A.-G.; Grumezescu, A.M. Photodynamic Therapy—An Up-to-Date Review. *Appl. Sci.* **2021**, *11*, 3626. [CrossRef]
32. Oyama, J.; Fernandes Herculano Ramos-Milaré, Á.C.; Lopes Lera-Nonose, D.S.S.; Nesi-Reis, V.; Galhardo Demarchi, I.; Alessi Aristides, S.M.; Juarez Vieira Teixeira, J.; Gomes Verzignassi Silveira, T.; Campana Lonardoni, M.V. Photodynamic therapy in wound healing in vivo: A systematic review. *Photodiagn. Photodyn. Ther.* **2020**, *30*, 101682. [CrossRef]
33. Diogo, P.; F. Faustino, M.A.; PMS Neves, M.G.; Palma, P.J.; P. Baptista, I.; Gonçalves, T.; Santos, J.M. An Insight into Advanced Approaches for Photosensitizer Optimization in Endodontics—A Critical Review. *J. Funct. Biomater.* **2019**, *10*, 44. [CrossRef]
34. Li, X.; Sun, L.; Zhang, P.; Wang, Y. Novel Approaches to Combat Medical Device-Associated BioFilms. *Coatings* **2021**, *11*, 294. [CrossRef]
35. Almeida, A. Photodynamic Therapy in the Inactivation of Microorganisms. *Antibiotics* **2020**, *9*, 138. [CrossRef]
36. Bartolomeu, M.; Neves, M.G.P.M.S.; Faustino, M.A.F.; Almeida, A. Wastewater chemical contaminants: Remediation by advanced oxidation processes. *Photochem. Photobiol. Sci.* **2018**, *17*, 1573–1598. [CrossRef] [PubMed]
37. Bartolomeu, M.; Oliveira, C.; Pereira, C.; Neves, M.G.P.M.S.; Faustino, M.A.F.; Almeida, A. Antimicrobial Photodynamic Approach in the Inactivation of Viruses in Wastewater: Influence of Alternative Adjuvants. *Antibiotics* **2021**, *10*, 767. [CrossRef] [PubMed]
38. Toro, P.M.; Jara, D.H.; Klahn, A.H.; Villaman, D.; Fuentealba, M.; Vega, A.; Pizarro, N. Spectroscopic Study of theE/ZPhotoisomerization of a New Cyrhetrenyl Acylhydrazone: A Potential Photoswitch and Photosensitizer(dagger). *Photochem. Photobiol.* **2021**, *97*, 61–70. [CrossRef] [PubMed]
39. Dong, M.X.; Babalhavaeji, A.; Samanta, S.; Beharry, A.A.; Woolley, G.A. Red-Shifting Azobenzene Photoswitches for in Vivo Use. *Acc. Chem. Res.* **2015**, *48*, 2662–2670. [CrossRef] [PubMed]
40. Ciminelli, C.; Granucci, G.; Persico, M. The photoisomerization mechanism of azobenzene: A semiclassical simulation of nonadiabatic dynamics. *Chem. A Eur. J.* **2004**, *10*, 2327–2341. [CrossRef] [PubMed]
41. Gutzeit, V.A.; Acosta-Ruiz, A.; Munguba, H.; Hafner, S.; Landra-Willm, A.; Mathes, B.; Mony, J.; Yarotski, D.; Borjesson, K.; Liston, C.; et al. A fine-tuned azobenzene for enhanced photopharmacology in vivo. *Cell Chem. Biol.* **2021**, *28*, 1648–1663.e16. [CrossRef]

42. Mukadum, F.; Nguyen, Q.; Adrion, D.M.; Appleby, G.; Chen, R.; Dang, H.; Chang, R.; Garnett, R.; Lopez, S.A. Efficient Discovery of Visible Light-Activated Azoarene Photoswitches with Long Half-Lives Using Active Search. *J. Chem. Inf. Model.* **2021**, *61*, 5524–5534. [CrossRef]
43. Blanco-Lomas, M.; Martinez-Lopez, D.; Campos, P.J.; Sampedro, D. Tuning of the properties of rhodopsin-based molecular switches. *Tetrahedron Lett.* **2014**, *55*, 3361–3364. [CrossRef]
44. Blanco-Lomas, M.; Funes-Ardoiz, I.; Campos, P.J.; Sampedro, D. Oxazolone-Based Photoswitches: Synthesis and Properties. *Eur. J. Org. Chem.* **2013**, *2013*, 6611–6618. [CrossRef]
45. Martinez-Lopez, D.; Yu, M.L.; Garcia-Iriepa, C.; Campos, P.J.; Frutos, L.M.; Golen, J.A.; Rasapalli, S.; Sampedro, D. Hydantoin-Based Molecular Photoswitches. *J. Org. Chem.* **2015**, *80*, 3929–3939. [CrossRef]
46. Carcia-Iriepa, C.; Ernst, H.A.; Liang, Y.; Unterreiner, A.N.; Frutos, L.M.; Sampedro, D. Study of Model Systems for Bilirubin and Bilin Chromophores: Determination and Modification of Thermal and Photochemical Properties. *J. Org. Chem.* **2016**, *81*, 6292–6302. [CrossRef]
47. Waldeck, D.H. Photoisomerization dynamics of stilbenes. *Chem. Rev.* **1991**, *91*, 415–436. [CrossRef]
48. Di Martino, M.; Sessa, L.; Di Matteo, M.; Panunzi, B.; Piotto, S.; Concilio, S. Azobenzene as Antimicrobial Molecules. *Molecules* **2022**, *27*, 5643. [CrossRef] [PubMed]
49. Jerca, F.A.; Jerca, V.V.; Hoogenboom, R. Advances and opportunities in the exciting world of azobenzenes. *Nat. Rev. Chem.* **2022**, *6*, 51–69. [CrossRef]
50. Wegener, M.; Hansen, M.J.; Driessen, A.J.M.; Szymanski, W.; Feringa, B. Photocontrol of Antibacterial Activity: Shifting from UV to Red Light Activation. *J. Am. Chem. Soc.* **2017**, *139*, 17979–17986. [CrossRef]
51. Burchall, J.J. Mechanism of Action of Trimethoprim-Sulfamethoxazole—II. *J. Infect. Dis.* **1973**, *128*, S437–S441. [CrossRef]
52. Hitchings, G.H. Mechanism of Action of Trimethoprim-Sulfamethoxazole—I. *J. Infect. Dis.* **1973**, *128*, S433–S436. [CrossRef] [PubMed]
53. Weston, C.E.; Kramer, A.; Colin, F.; Yildiz, O.; Baud, M.G.J.; Meyer-Almes, F.J.; Fuchter, M.J. Toward Photopharmacological Antimicrobial Chemotherapy Using Photoswitchable Amidohydrolase Inhibitors. *ACS Infect. Dis.* **2017**, *3*, 152–161. [CrossRef] [PubMed]
54. Jung, M.; Hoffmann, K.; Brosch, G.; Loidl, P. Analogues of trichostatin A and trapoxin B as histone deacetylase inhibitors. *Bioorg. Med. Chem. Lett.* **1997**, *7*, 1655–1658. [CrossRef]
55. Miller, T.A.; Witter, D.J.; Belvedere, S. Histone deacetylase inhibitors. *J. Med. Chem.* **2003**, *46*, 5097–5116. [CrossRef]
56. Mai, A.; Massa, S.; Rotili, D.; Cerbara, I.; Valente, S.; Pezzi, R.; Simeoni, S.; Ragno, R. Histone deacetylation in epigenetics: An attractive target for anticancer therapy. *Med. Res. Rev.* **2005**, *25*, 261–309. [CrossRef] [PubMed]
57. Vannini, A.; Volpari, C.; Filocamo, G.; Casavola, E.C.; Brunetti, M.; Renzoni, D.; Chakravarty, P.; Paolini, C.; De Francesco, R.; Gallinari, P.; et al. Crystal structure of a eukaryotic zinc-dependent histone deacetylase, human HDAC8, complexed with a hydroxamic acid inhibitor. *Proc. Natl. Acad. Sci. USA* **2004**, *101*, 15064–15069. [CrossRef] [PubMed]
58. Kramer, A.; Herzer, J.; Overhage, J.; Meyer-Almes, F.J. Substrate specificity and function of acetylpolyamine amidohydrolases from Pseudomonas aeruginosa. *BMC Biochem.* **2016**, *17*, 4. [CrossRef]
59. Hildmann, C.; Ninkovic, M.; Dietrich, R.; Wegener, D.; Riester, D.; Zimmermann, T.; Birch, O.M.; Bernegger, C.; Loidl, P.; Schwienhorst, A. A new amidohydrolase from Bordetella or Alcaligenes strain FB188 with similarities to histone deacetylases. *J. Bacteriol.* **2004**, *186*, 2328–2339. [CrossRef]
60. Yao, H.; Wynendaele, E.; Xu, X.L.; Kosgei, A.; De Spiegeleer, B. Circular dichroism in functional quality evaluation of medicines. *J. Pharm. Biomed. Anal.* **2018**, *147*, 50–64. [CrossRef] [PubMed]
61. Yeoh, Y.Q.; Yu, J.X.; Polyak, S.W.; Horsley, J.R.; Abell, A.D. Photopharmacological Control of Cyclic Antimicrobial Peptides. *Chembiochem* **2018**, *19*, 2591–2597. [CrossRef] [PubMed]
62. Irie, M.; Fulcaminato, T.; Matsuda, K.; Kobatake, S. Photochromism of Diarylethene Molecules and Crystals: Memories, Switches, and Actuators. *Chem. Rev.* **2014**, *114*, 12174–12277. [CrossRef] [PubMed]
63. Matsuda, K.; Irie, M. Diarylethene as a photoswitching unit. *J. Photochem. Photobiol. C Photochem. Rev.* **2004**, *5*, 169–182. [CrossRef]
64. Huang, Y.-Y.; Wintner, A.; Seed, P.C.; Brauns, T.; Gelfand, J.A.; Hamblin, M.R. Antimicrobial photodynamic therapy mediated by methylene blue and potassium iodide to treat urinary tract infection in a female rat model. *Sci. Rep.* **2018**, *8*, 7257. [CrossRef]
65. Zhao, L.; Zhao, J.; Zhong, K.H.; Tong, A.P.; Jia, D. Targeted protein degradation: Mechanisms, strategies and application. *Signal Transduct. Target. Ther.* **2022**, *7*, 113. [CrossRef]
66. Wang, Z.W.; Liu, Y.; Zhu, X.Q. PhotoPROTACs: A Novel Biotechnology for Cancer Treatment. *Trends Cell Biol.* **2020**, *30*, 749–751. [CrossRef] [PubMed]
67. Liu, J.; Chen, H.; Ma, L.N.; He, Z.X.; Wang, D.; Liu, Y.; Lin, Q.; Zhang, T.H.; Gray, N.; Kaniskan, H.U.; et al. Light-induced control of protein destruction by opto-PROTAC. *Sci. Adv.* **2020**, *6*, eaay5154. [CrossRef]
68. Reynders, M.; Matsuura, B.S.; Berouti, M.; Simoneschi, D.; Marzio, A.; Pagano, M.; Trauner, D. PHOTACs enable optical control of protein degradation. *Sci. Adv.* **2020**, *6*, eaay5064. [CrossRef] [PubMed]
69. Jin, Y.H.; Lu, M.C.; Wang, Y.; Shan, W.X.; Wang, X.Y.; You, Q.D.; Jiang, Z.Y. Azo-PROTAC: Novel Light-Controlled Small-Molecule Tool for Protein Knockdown. *J. Med. Chem.* **2020**, *63*, 4644–4654. [CrossRef] [PubMed]
70. Pfaff, P.; Samarasinghe, K.T.G.; Crews, C.; Carreira, E. Reversible Spatiotemporal Control of Induced Protein Degradation by Bistable PhotoPROTACs. *ACS Cent. Sci.* **2019**, *5*, 1682–1690. [CrossRef]

71. Winter, G.E.; Buckley, D.L.; Paulk, J.; Roberts, J.M.; Souza, A.; Dhe-Paganon, S.; Bradner, J.E. Phthalimide conjugation as a strategy for in vivo target protein degradation. *Science* **2015**, *348*, 1376–1381. [CrossRef]
72. Sharma, M.; Friedman, S.H. The Issue of Tissue: Approaches and Challenges to the Light Control of Drug Activity. *ChemPhotoChem* **2021**, *5*, 611–618. [CrossRef]
73. Yang, Y.M.; Mu, J.; Xing, B.G. Photoactivated drug delivery and bioimaging. *Wiley Interdiscip. Rev. Nanomed. Nanobiotechnol.* **2017**, *9*, e1408. [CrossRef] [PubMed]
74. Peterson, J.A.; Wijesooriya, C.; Gehrmann, E.J.; Mahoney, K.M.; Goswami, P.P.; Albright, T.R.; Syed, A.; Dutton, A.S.; Smith, E.A.; Winter, A.H. Family of BODIPY Photocages Cleaved by Single Photons of Visible/Near-Infrared Light. *J. Am. Chem. Soc.* **2018**, *140*, 7343–7346. [CrossRef]
75. Liu, M.; Meng, J.Q.; Bao, W.E.; Liu, S.Y.; Wei, W.; Ma, G.H.; Tian, Z.Y. Single-Chromophore-Based Therapeutic Agent Enables Green-Light-Triggered Chemotherapy and Simultaneous Photodynamic Therapy to Cancer Cells. *ACS Appl. Bio Mater.* **2019**, *2*, 3068–3076. [CrossRef]
76. Silva, J.M.; Silva, E.; Reis, R.L. Light-triggered release of photocaged therapeutics—Where are we now? *J. Control. Release* **2019**, *298*, 154–176. [CrossRef]
77. Kumari, P.; Kulkarni, A.; Sharma, A.K.; Chakrapani, H. Visible-Light Controlled Release of a Fluoroquinolone Antibiotic for Antimicrobial Photopharmacology. *ACS Omega* **2018**, *3*, 2155–2160. [CrossRef]
78. Shchelik, I.S.; Tomio, A.; Gademann, K. Design, Synthesis, and Biological Evaluation of Light-Activated Antibiotics. *ACS Infect. Dis.* **2021**, *7*, 681–692. [CrossRef]
79. Wong, P.T.; Tang, S.Z.; Mukherjee, J.; Tang, K.; Gam, K.; Isham, D.; Murat, C.; Sun, R.; Baker, J.R.; Choi, S.K. Light-controlled active release of photocaged ciprofloxacin for lipopolysaccharide-targeted drug delivery using dendrimer conjugates. *Chem. Commun.* **2016**, *52*, 10357–10360. [CrossRef] [PubMed]
80. Verma, S.; Manna, D. Controlling PROTACs with Light. *ChemMedChem* **2020**, *15*, 1258–1261. [CrossRef]
81. Xue, G.; Wang, K.; Zhou, D.L.; Zhong, H.B.; Pan, Z.Y. Light-Induced Protein Degradation with Photocaged PROTACs. *J. Am. Chem. Soc.* **2019**, *141*, 18370–18374. [CrossRef]
82. Naro, Y.; Darrah, K.; Deiters, A. Optical Control of Small Molecule-Induced Protein Degradation. *J. Am. Chem. Soc.* **2020**, *142*, 2193–2197. [CrossRef]
83. Kounde, C.S.; Shchepinova, M.M.; Saunders, C.N.; Muelbaier, M.; Rackham, M.D.; Harling, J.D.; Tate, E.W. A caged E3 ligase ligand for PROTAC-mediated protein degradation with light. *Chem. Commun.* **2020**, *56*, 5532–5535. [CrossRef]
84. Zengerle, M.; Chan, K.H.; Ciulli, A. Selective Small Molecule Induced Degradation of the BET Bromodomain Protein BRD4. *ACS Chem. Biol.* **2015**, *10*, 1770–1777. [CrossRef] [PubMed]
85. Dommaschk, M.; Grobner, J.; Wellm, V.; Hovener, J.B.; Riedel, C.; Herges, R. Dendronised Ni(II) porphyrins as photoswitchable contrast agents for MRI. *Phys. Chem. Chem. Phys.* **2019**, *21*, 24296–24299. [CrossRef] [PubMed]
86. Zhai, Y.; Busscher, H.J.; Liu, Y.; Zhang, Z.K.; van Kooten, T.G.; Su, L.Z.; Zhang, Y.M.; Liu, J.J.; Liu, J.F.; An, Y.L.; et al. Photoswitchable Micelles for the Control of Singlet-Oxygen Generation in Photodynamic Therapies. *Biomacromolecules* **2018**, *19*, 2023–2033. [CrossRef] [PubMed]
87. Lin, W.Y.; Peng, D.W.; Wang, B.; Long, L.; Guo, C.; Yuan, J. A model for light-triggered porphyrin anticancer prodrugs based on an o-nitrobenzyl photolabile group. *Eur. J. Org. Chem.* **2008**, *2008*, 793–796. [CrossRef]
88. Tessaro, A.L.; Fraix, A.; Failla, M.; Cardile, V.; Graziano, A.C.E.; Estevao, B.M.; Rescifina, A.; Sortino, S. Light-Controlled Simultaneous "On Demand" Release of Cytotoxic Combinations for Bimodal Killing of Cancer Cells. *Chem. A Eur. J.* **2018**, *24*, 7664–7670. [CrossRef]
89. Feng, Y.F.; Tonon, C.C.; Ashraf, S.; Hasan, T. Photodynamic and antibiotic therapy in combination against bacterial infections: Efficacy, determinants, mechanisms, and future perspectives. *Adv. Drug Deliv. Rev.* **2021**, *177*, 113941. [CrossRef]
90. Branco, T.M.; Valerio, N.C.; Jesus, V.I.R.; Dias, C.J.; Neves, M.; Faustino, M.A.F.; Almeida, A. Single and combined effects of photodynamic therapy and antibiotics to inactivate Staphylococcus aureus on skin. *Photodiagn. Photodyn. Ther.* **2018**, *21*, 285–293. [CrossRef]
91. Xing, B.G.; Jiang, T.T.; Bi, W.G.; Yang, Y.M.; Li, L.H.; Ma, M.L.; Chang, C.K.; Xu, B.; Yeow, E.K.L. Multifunctional divalent vancomycin: The fluorescent imaging and photodynamic antimicrobial properties for drug resistant bacteria. *Chem. Commun.* **2011**, *47*, 1601–1603. [CrossRef] [PubMed]
92. Liu, F.; Ni, A.S.Y.; Lim, Y.; Mohanram, H.; Bhattacharjya, S.; Xing, B.G. Lipopolysaccharide Neutralizing Peptide-Porphyrin Conjugates for Effective Photoinactivation and Intracellular Imaging of Gram-Negative Bacteria Strains. *Bioconjugate Chem.* **2012**, *23*, 1639–1647. [CrossRef]
93. Dastgheyb, S.S.; Eckmann, D.M.; Composto, R.J.; Hickok, N.J. Photo-activated porphyrin in combination with antibiotics: Therapies against Staphylococci. *J. Photochem. Photobiol. B Biol.* **2013**, *129*, 27–35. [CrossRef]
94. Almeida, J.; Tome, J.P.C.; Neves, M.; Tome, A.C.; Cavaleiro, J.A.S.; Cunha, A.; Costa, L.; Faustino, M.A.F.; Almeida, A. Photodynamic inactivation of multidrug-resistant bacteria in hospital wastewaters: Influence of residual antibiotics. *Photochem. Photobiol. Sci.* **2014**, *13*, 626–633. [CrossRef]
95. Iluz, N.; Maor, Y.; Keller, N.; Malik, Z. The synergistic effect of PDT and oxacillin on clinical isolates of Staphylococcus aureus. *Lasers Surg. Med.* **2018**, *50*, 535–551. [CrossRef]

96. Thangudu, S.; Kaur, N.; Korupalli, C.; Sharma, V.; Kalluru, P.; Vankayala, R. Recent advances in near infrared light responsive multi-functional nanostructures for phototheranostic applications. *Biomater. Sci.* **2021**, *9*, 5432–5443. [CrossRef]
97. Wang, X.; Xuan, Z.; Zhu, X.; Sun, H.; Li, J.; Xie, Z. Near-infrared photoresponsive drug delivery nanosystems for cancer photo-chemotherapy. *J. Nanobiotechnol.* **2020**, *18*, 108. [CrossRef]
98. Vargason, A.M.; Anselmo, A.C.; Mitragotri, S. The evolution of commercial drug delivery technologies. *Nat. Biomed. Eng.* **2021**, *5*, 951–967. [CrossRef] [PubMed]
99. Lu, C.; Di, L. In vitro and in vivo methods to assess pharmacokinetic drug– drug interactions in drug discovery and development. *Biopharm. Drug Dispos.* **2020**, *41*, 3–31. [CrossRef] [PubMed]
100. Said, S.S.; Campbell, S.; Hoare, T. Externally Addressable Smart Drug Delivery Vehicles: Current Technologies and Future Directions. *Chem. Mater.* **2019**, *31*, 4971–4989. [CrossRef]

pharmaceuticals

Article

Exploring Mannosylpurines as Copper Chelators and Cholinesterase Inhibitors with Potential for Alzheimer's Disease

Ignazio Schino [1,2,3], Mariangela Cantore [1,2,3], Modesto de Candia [3], Cosimo D. Altomare [3], Catarina Maria [1,4], João Barros [1], Vasco Cachatra [1], Patrícia Calado [1], Karina Shimizu [5], Adilson A. Freitas [5], Maria C. Oliveira [5], Maria J. Ferreira [5], José N. C. Lopes [5], Nicola A. Colabufo [3,*] and Amélia P. Rauter [1,*]

[1] Centro de Química Estrutural, Institute of Molecular Sciences, Faculdade de Ciências, Universidade de Lisboa, Ed C8, Piso 5, Campo Grande, 1749-016 Lisboa, Portugal
[2] Biofordrug Srl, Via Dante 95, 70019 Triggiano, Italy
[3] Department of Pharmacy-Pharmaceutical Sciences, University of Bari Aldo Moro, Via E. Orabona 4, 70125 Bari, Italy
[4] Departamento de Química, Faculdade de Ciências e Tecnologia, Universidade Nova de Lisboa, Largo da Torre, 2825-149 Caparica, Portugal
[5] Centro de Química Estrutural, Institute of Molecular Sciences Instituto Superior Técnico, Universidade de Lisboa, Av. Rovisco Pais, 1049-001 Lisboa, Portugal
* Correspondence: nicolaantonio.colabufo@uniba.it (N.A.C.); aprauter@fc.ul.pt (A.P.R.); Tel.: +351-964-408-824 (A.P.R.)

Abstract: Alzheimer's Disease (AD) is characterized by a progressive cholinergic neurotransmission imbalance, with a decrease of acetylcholinesterase (AChE) activity followed by a significant increase of butyrylcholinesterase (BChE) in the later AD stages. BChE activity is also crucial for the development of Aβ plaques, the main hallmarks of this pathology. Moreover, systemic copper dyshomeostasis alters neurotransmission leading to AD. In the search for structures targeting both events, a set of novel 6-benzamide purine nucleosides was synthesized, differing in glycone configuration and N^7/N^9 linkage to the purine. Their AChE/BChE inhibitory activity and metal ion chelating properties were evaluated. Selectivity for human BChE inhibition required N^9-linked 6-deoxy-α-D-mannosylpurine structure, while all three tested β-D-derivatives appeared as non-selective inhibitors. The N^9-linked L-nucleosides were cholinesterase inhibitors except the one embodying either the acetylated sugar or the *N*-benzyl-protected nucleobase. These findings highlight that sugar-enriched molecular entities can tune bioactivity and selectivity against cholinesterases. In addition, selective copper chelating properties over zinc, aluminum, and iron were found for the benzyl and acetyl-protected 6-deoxy-α-L-mannosyl N^9-linked purine nucleosides. Computational studies highlight molecular conformations and the chelating molecular site. The first dual target compounds were disclosed with the perspective of generating drug candidates by improving water solubility.

Keywords: cholinesterase inhibitors; copper chelators; purine nucleosides; Alzheimer's disease

Citation: Schino, I.; Cantore, M.; de Candia, M.; Altomare, C.D.; Maria, C.; Barros, J.; Cachatra, V.; Calado, P.; Shimizu, K.; Freitas, A.A.; et al. Exploring Mannosylpurines as Copper Chelators and Cholinesterase Inhibitors with Potential for Alzheimer's Disease. *Pharmaceuticals* **2023**, *16*, 54. https://doi.org/10.3390/ph16010054

Academic Editor: Jean-Pierre Bazureau

Received: 5 December 2022
Revised: 23 December 2022
Accepted: 26 December 2022
Published: 30 December 2022

1. Introduction

Alzheimer's Disease (AD) is a devastating disease with a multifactorial pathogenesis. The main histopathological hallmarks of AD are the generation and deposition of senile plaques containing extracellular aggregates of neurotoxic amyloid-β (Aβ) peptide, along with intracellular neurofibrillary tangles of hyperphosphorylated τ-protein and the impairment of the cholinergic neurotransmission, due to a decrease of the level of the neurotransmitter acetylcholine (ACh), leading to a degeneration of cholinergic neurons and neuronal loss [1,2]. In physiological conditions, ACh is predominantly hydrolyzed by acetylcholinesterase (AChE), while butyrylcholinesterase (BChE) has a marginal role.

During AD progression, AChE activity decreases to 55–67% of normal values in the hippocampus and temporal cortex, while the activity of BChE gradually increases in the late stage of AD up to 120% of normal levels and accumulates in senile plaques, thus suggesting an extremely critical role for BChE in ACh hydrolysis and disease progression [3,4]. Therefore, selective BChE inhibitors may also display modulatory activity on Aβ, as found in tacrine and rivastigmine, both lowering Aβ levels in cell cultures [5]. Moreover, dual inhibitors of both AChE and BChE may, indeed, have the potential to interfere with AD along the course of this pathology, as recently reviewed for the clinical applications and limitations in the use of cholinesterase inhibitors for the treatment of AD [6].

Transition metals play a crucial role in vital processes, in particular Cu^{2+}, the third bivalent ion after Fe^{2+} and Zn^{2+} in terms of abundance in the human body; all three ions are involved in the regulation of neuronal functions. Studies carried out to assess the relationship between serum copper, zinc, and iron levels with the risk of AD, via a meta-analysis of case-control studies, supported the hypothesis of an association of a high serum copper level to AD onset risk. Serum copper levels were found to be significantly higher in AD patients than in healthy controls, while serum zinc and iron levels were significantly lower in AD patients than in controls [7]. Furthermore, Al^{3+} ions resulting mainly from the intoxication of human exposure were found co-localized with Aβ in senile plaques, but future research is required to recognize its role in AD [8].

The role of aluminum (Al^{3+}) and zinc (Zn^{2+}) is opposite in the onset and progression of Alzheimer's disease. In fact, several biological assays in an animal model demonstrated that aluminum chloride is able to reduce cognitive capacity in rats [9]. The mechanism involved is yet to be ascertained, but there is evidence that this ion interacts directly with APP metabolism and/or with APOE, a known risk factor in AD [10,11]. Controversially, it has been demonstrated that zinc carboxylate, evaluated towards acetylcholinesterase (AChE) and butyrylcholinesterase (BChE), displayed high inhibitory activity having a strong potency to block cognitive decline in Alzheimer's disease [12].

Cu^{2+} plays an important role in various physiological processes as a catalytic cofactor for a variety of metal enzymes (superoxide dismutase, cytochrome oxidase, lypoxidase, and tyrosinase). However, it is toxic in biological systems when exceeding cellular requirements by negatively affecting physiological processes, and among them, neurotransmission. Copper dyshomeostasis and accumulation are known to be involved in AD onset and progression [7,13–15], and a most rapid accumulation of amyloid plaques in AD patients has been detected [13–15]. It has been reported that Cu^{2+} coordination with Aβ modulates Aβ aggregation pattern, potentiating Aβ neurotoxicity and stabilizing smaller soluble Aβ oligomers [16]. The trend for copper-chelating drugs was recently highlighted [17], and a review on metal complexes, including copper complexes with ligands incorporating amyloid-binding moieties for Aβ plaque binding, has been published [18]. Therefore, the development of new probes able to selectively chelate Cu^{2+} remains a challenge towards the discovery of new agents for early AD diagnosis or treatment.

Recently, the first series of purine nucleosides displaying a promising cholinesterase inhibition has been reported by our group [19].

The nucleoside 6-chloro-9-(2,3,4,6-tetra-*O*-benzyl-α-D-mannopyranosyl)purine derivative (compound **1**, Figure 1) showed a K*i* of ca 2.5 μM for both BChE and AChE, while the analogue 2-acetamidopurine nucleoside exhibiting glycone N^7 ligation (compound **2**, Figure 1), showed a selective inhibition (K*i* of 50 nM) of BChE.

Inspired by these first results, we designed a new series of purine nucleosides, aiming to insert a copper chelating molecular moiety for the development of multitarget agents against AD acting as cholinesterase inhibitors and/or as copper chelators. Starting from the general structure in Figure 1, the new compounds were synthesized and assayed as cholinesterase inhibitors aiming to evaluate the role on compound bioactivity of anomeric stereochemistry, sugar D, L configuration, sugar 6-deoxygenation, and glycosyl group regioligation to the purine base at positions N^7 or N^9. For preliminary information about protecting group requirements, the peracetylated analogue of the most potent perbenzylated nucleoside was synthesized. New compounds were also tested for their copper chelating efficiency. The results, here presented and discussed, highlight the potent activity of this purine nucleoside family of compounds and demonstrate how copper chelation and anticholinesterase activity/selectivity can be tuned by the regiochemistry of glycone ligation, glycone conformation, and also by the anomeric absolute configuration.

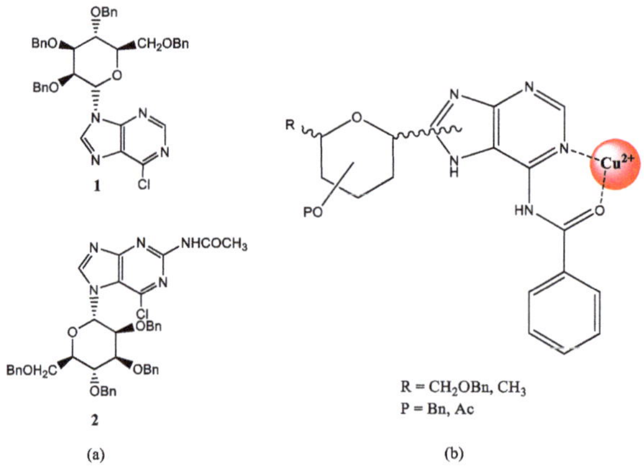

Figure 1. (**a**) Structure of the cholinesterase inhibitors 1 and 2 [19]; (**b**) General structure of the hypothesized purine nucleosides' copper complex.

2. Results

2.1. Chemistry

Synthesis of compounds **5-10, 12, 15** and **16** (Table 1, Schemes 1–3) was accomplished via TMSOTf catalyzed reaction of fully protected methyl glycosides with the purine base activated by reaction with BSA, a procedure adapted from a methodology previously described by us for other purine nucleosides [19]. The glycosyl donors were the perbenzylated methyl α-D-mannopyranoside (**3**), its 6-deoxy analogue (**4**), the peracetylated 6-deoxy-L-mannose (L-rhamnose) (**11**), and the perbenzylated methyl 6-deoxy-α-L-mannopyranoside (**14**). Reaction conditions optimization was required resulting from the very low yields obtained with the previously reported procedure. Several temperatures and reaction times were tried, under conventional and microwave-assisted syntheses, aiming to improve the overall nucleoside yield. Microwave irradiation at 150 W and 65 °C for 60 min gave the best results, and nucleoside isolated yields are given in Table 1. The four isomers differing in the anomeric configuration and linkage of the glycosyl group to the purine, either at the N^7 or the N^9 position, were formed by reacting the perbenzylated methyl α-D-mannopyranoside (**3**), but both the N^9 anomers were obtained in higher yield than the corresponding N^7 isomers. However, by reacting the perbenzylated methyl 6-deoxy-D-mannopyranoside donor (**4**), only the N^9 isomers were obtained in good yield, while the N^7 isomers were detected as traces ($\leq$1%) by ^{1}H NMR. Replacing the benzyl by the acetyl protecting group, the peracetylated L-rhamnose gave the N^9-linked α-nucleoside (**12**) as the single product, isolated in 63% yield (Table 1, Scheme 2). Compound **13** was then obtained in 15% isolated yield by reacting **12** with sodium hydroxide, tetrabutylammonium iodide, and benzyl bromide. Finally, with the perbenzylated methyl 6-deoxy-α-L-mannoside (14) as the starting material, only the N^9 isomers were obtained, as also observed for the reaction with compound **4** (Table 1, Scheme 3).

Table 1. Isolated yield and structural features of the new synthesized and tested purine nucleosides, accessed by microwave-assisted reaction of the methyl glycoside precursors with *N*-(9*H*-purin-6-yl)benzamide (150 W, 65 °C, 1 h).

Compound nr.	N^7/N^9 Purine Ligation	α,β Anomer	Glycone	Isolated Yield (%)
5	N^9	α	2,3,4,6-Tetra-*O*-benzyl-α-D-mannosyl	22
6	N^9	β	2,3,4,6-Tetra-*O*-benzyl-β-D-mannosyl	16
7	N^7	α	2,3,4,6-Tetra-*O*-benzyl-α-D-mannosyl	6
8	N^7	β	2,3,4,6-Tetra-*O*-benzyl-β-D-mannosyl	12
9	N^9	α	2,3,4-Tri-*O*-benzyl-6-deoxy-α-D-mannosyl	33
10	N^9	β	2,3,4-Tri-*O*-benzyl-6-deoxy-β-D-mannosyl	40
12	N^9	α	2,3,4-tri-*O*-acetyl-6-deoxy-α-L-mannosyl	63
13	N^9	α	2,3,4-tri-*O*-benzyl-6-deoxy-α-L-mannosyl [a]	15
15	N^9	α	2,3,4-tri-*O*-benzyl-6-deoxy-α-L-mannosyl	8
16	N^9	β	2,3,4-tri-*O*-benzyl-6-deoxy-β-L-mannosyl	36

[a] Linked to a purine embodying a *N*-benzylbenzamide group.

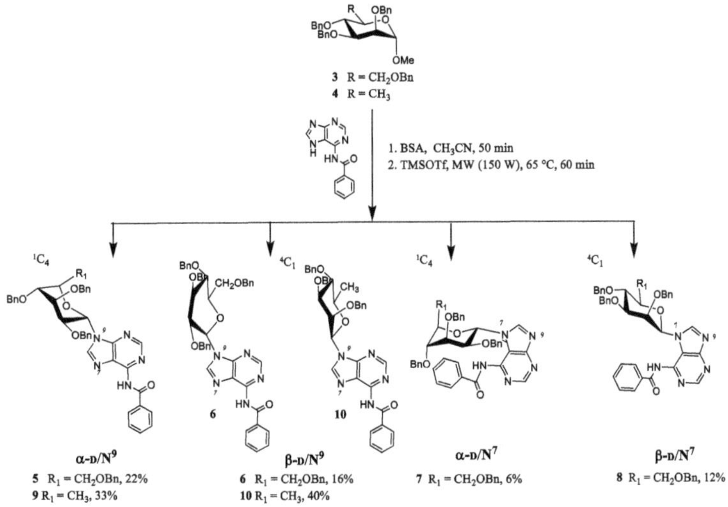

Scheme 1. Synthesis of nucleosides **5-10**, obtained by microwave-assisted reaction of glycosyl donors **3** or **4** with activated purine base catalyzed by TMSOTf, with conformations as deduced by NMR data and computational studies.

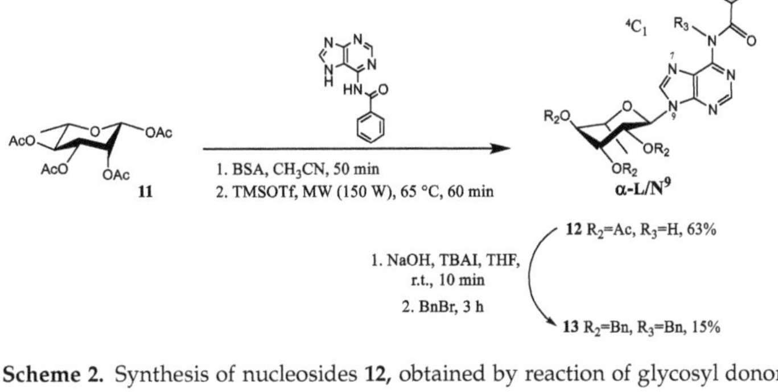

Scheme 2. Synthesis of nucleosides **12**, obtained by reaction of glycosyl donor **11** with activated purine base catalyzed by TMSOTf, and **13**, obtained by benzylation reaction of compound **12**, with conformations confirmed by NMR data and computational studies.

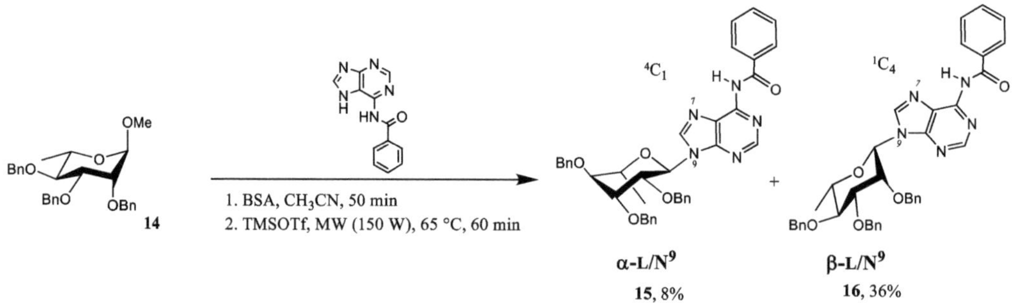

Scheme 3. Synthesis of nucleosides **15** and **16**, obtained by microwave-assisted reaction of glycosyl donor **14** and activated purine base catalyzed by TMSOTf, with conformations given by computational studies in agreement with NMR data.

Both the anomer configuration and the ligation at N^7 or N^9 were determined based on ^1H and ^{13}C NMR data and COSY, HMQC, and HMBC experiments. The chemical shift of purine carbon 5 is characteristic and permits to distinguish between the kinetic regioisomer N^7 and the N^9 thermodynamic one [19]. These values were between δ 118 and 112 ppm for compounds **7** and **8** (N^7 isomers), while ligation at N^9 resulted in chemical shifts of C-5 between ca δ 129 and 122 ppm (as in compounds **5**, **6**, **9**, **10**, **12**, **13**, and **15**, **16**).

The anomeric configuration was assigned from the coupling constant $^3J_{1',2'}$ and confirmed by COSY and NOESY experiments. The coupling constants $^3J_{1',2'}$ for compounds **5**, **7**, **12**, and **15** were 8.0 Hz, 9.3 Hz, 6.3 Hz, and 7.1 Hz, respectively, thus suggesting the presence of *trans*-diaxial like protons as in $^1C_4$ conformation for the D series and $^4C_1$ for the L series. The coupling constant $^3J_{1',2'}$ of 7.4 Hz for the β-anomer (**6**) was unexpected, indicating a conformation possibly bearing H-1′ and H-2′ approaching a parallel orientation, tentatively depicted in Scheme 1. ^1H NMR spectra of the β-anomers **8**, **10** and **16** show coupling constants $^3J_{1',2'}$ lower than 1 Hz, as expected.

By the NMR spectra analysis, the structures of ten nucleosides were proposed, six of them being unexpected. Usually, the most stable glycosyl conformation presents its voluminous protecting groups in equatorial positions, which does not happen in nucleosides **5**, **7**, **9**, **12**, and **15**. Moreover, compound **6** seems to present a non-usual β-conformation, and **12** (α-anomer) was obtained as a single product. These unexpected results led us to perform computational studies, which unambiguously confirmed the proposed structures deduced from NMR data. Figure 2 and Tables S1–S7 in Supplementary Materials show the comparison of experimental and calculated NMR data, and the formation of **12** as a single product (Table 2) was unambiguously confirmed. The tables comparing experimental and calculated NMR data can be found in the Supplementary Materials (Tables S1–S7), the latter in line with the experimental results. Interestingly, according to the acquired NMR data and the knowledge to date, all the studied nucleosides exhibited a preferred equatorial orientation of the purine base, which is in full agreement with the exoanomeric effect known for *N*-nucleosides [20].

Table 2. Energy difference between the α and the β anomers for compound **12**.

Anomer	Compounds 12 (α Anomer) and 12a (β Anomer)		
	E/kcal·mol^{-1}	ΔE (α-β)	Boltzmann Population
α	−1,132,494.89	−6.30	1.00
β	−1,132,488.59		0.00

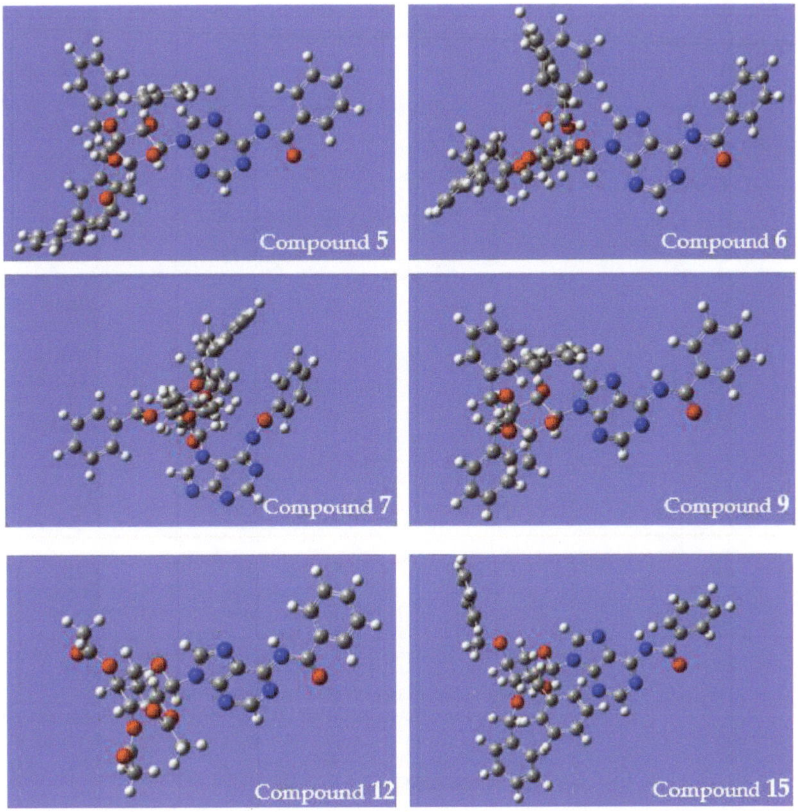

Figure 2. Representation of the final optimized geometries found for compounds **5**, **6**, **7**, **9**, **12**, and **15**, deduced in Gaussian16 with CAM-B3LYP/def2TZVP and IEFPCM for water.

2.2. Anticholinesterase Activity

The in vitro inhibitory activities of novel purine nucleosides **5-10**, **12**, **13** and **15**, and **16** were evaluated against both heterologous (electric eel eeAChE, horse eqBChE) and human (recombinant) cholinesterases isoforms, by applying the Ellmann colorimetric assay, with slight modifications [21,22]. The results are summarized in Table 3, thus reporting the half maximal inhibitory concentration (IC_{50}) values, or the percent of inhibition at the final tested concentration of 20 μM, with an exception for compounds **15** and **16**, which were tested at 5 μM due to their lower water solubility, but they were not able to produce more than 50% of inhibition for eeAChE and eqBChE. Compounds **6**, **8**, **9**, **10**, and **13** were the most interesting ones. Nucleosides **8** and **9** show eeAChE IC_{50} values of 2.6 and 4.7 μM, **6** and **13** show eqBChE IC_{50} values of 4.3 and 5.1 μM, respectively, whilst compound **10** proved moderate inhibition of BChE. Furthermore, compound **9** was an eeAChE selective inhibitor, while compounds **7** (IC_{50} = 10 μM) and **13** were selective for eqBChE. The most potent compound was the dual inhibitor compound **8**.

The reported activities highlight that sugar, the geometry of the glycosidic bond, and the regioisomerism of substitution pattern on the nucleoside affected the anticholinesterase activity. It has been, indeed, reported that depending on sugars stereochemistry, the acceptor/donor hydrogen bond properties of glycosylated compounds can significantly change and deeply affect binding properties to the target [23].

The relationship between glycone D-configuration, N-ligation, and selectivity is in agreement with our previous findings for related nucleosides with a 6-chloropurine in their structure, while those for the L-configuration are herein reported for the first time [19,24].

Table 3. Half maximal inhibitory concentration (IC_{50}) values of AChE and BChE inhibition by nucleosides **1–2, 5–10, 12–13, 15–16.**

Compound nr.	IC_{50} (μM) [a]			
	eeAChE [b]	hAChE [c]	eqBChE [d]	hBChE [c]
1	2.40 ± 0.30	-	2.80 ± 0.30	-
2	18.10 ± 4.80	-	0.05 ± 0.01	-
5	>20 [25 ± 5]	-	>20 [11 ± 7]	-
6	10.60 ± 1.10	17.00 ± 0.75	4.29 ± 1.70	4.58 ± 0.52
7	>20	-	10.00 ± 1.10	-
8	2.61 ± 1.23	4.81 ± 1.24	4.40 ± 1.30	4.55 ± 1.05
9	4.69 ± 0.51	>20 [25 ± 6]	>20 [28 ± 3]	15.4 ± 1.10
10	15.0 ± 2.1	17.8 ± 2.20	11.7 ± 1.2	8.70 ± 1.35
12	>20 [8 ± 5]	-	>20 [5 ± 3]	-
13	>20 [6 ± 3]	>20 [8 ± 5]	5.12 ± 0.25	>20 [25 ± 4]
15	>5 [31 ± 4]	>5 [28 ± 3]	>5 [24 ± 6]	>5 [18 ± 5]
16	>5 [18 ± 3]	>5 [16 ± 9]	>5 [26 ± 6]	>5 [30 ± 7]
Donepezil	0.021± 0.020	0.016 ± 0.002	2.75 ± 0.20	4.80 ± 0.05
Galantamine	0.56 ± 0.12	5.000 ± 0.071 [e]	12.0 ± 0.3	59.2 ± 1.7 [e]

[a] IC_{50} values determined by interpolation of the sigmoidal dose-response curves as obtained by regression with GraphPad Prism software (version 5.01) of at least seven different data points, or inhibition percentage at the higher tested concentration (20 μM) in parentheses; data are means ± SD of three independent measurements. Donepezil and Galantamine were used as positive controls. Reported data have been experimentally obtained, and values found accordingly with the literature. [b] Electric eel acetylcholinesterase. [c] Recombinant human isoform. [d] Horse serum butyrylcholinesterase. [e] Values obtained from the literature determined for brain AChE and serum BChE [25].

With the exception of the 2,3,4-tri-*O*-benzyl-6-deoxy-α-D-mannosyl purine derivative **9**, exhibiting a glycone $^1C_4$ conformation and reported as a selective AChE inhibitor, compounds bearing a 2,3,4,6-tetra-*O*-benzyl-β-D-mannosyl fragment (**6** and **8**) showed from moderate to good inhibition of heterologous AChE and BChE and resulted most active than the quite inactive corresponding α anomers. Moreover, regarding N^9 or N^7 ligation of sugar, the N^7-linked β-glycosyl regioisomer **8** proved a better activity than the N^9 regioisomers against both cholinesterases, thus resulting in compound **8** as a dual inhibitor against human cholinesterase isoforms as well.

Amongst the ten synthesized nucleosides, the peracetylated N^9-linked 6-deoxy-α-L-*manno* nucleoside **12** was not active towards any of the cholinesterases, in agreement with previous findings, which revealed that aromatic rings are preferred for the anti-cholinesterase activity of the studied purine nucleosides [24], and all the benzylated compounds, except the N^9-linked α-D-*manno* nucleoside **5**, were active at some extent. However, benzylation of amide nitrogen in N^9-linked *N*-benzyl 6-deoxy-α-L-*manno* nucleoside **13**, exhibiting a glycosyl $^4C_1$ conformation, deeply modified the activity profile, thus inducing a selective inhibition of the equine BChE, which disappeared in human BChE.

Substitution of D-mannosyl with L-mannosyl sugar moiety resulted in a dramatic reduction of aqueous solubility in compounds **15** and **16**, which were tested at a lower concentration (5 μM) without showing any significant inhibition.

In order to verify compound behavior against the human cholinesterases, the most potent inhibitors (**6, 8,** and **9**), dual inhibitor **10,** and nucleosides **15** and **16** were tested against hAChE and hBChE. The previously known dual inhibitors **6, 8,** and **10** seemed to keep this property, however, with a slight IC_{50} value variation. Compound **6** IC_{50} value for hAChE increased from 10.6 to 17.0 μM, while the value for hBChE did not undergo such a variation (from 4.3 to 4.6 μM), suggesting a higher affinity of this compound to

BChE. Nucleoside **8**, with IC_{50} values of 4.8 μM and 4.6 μM for hAChE and hBChE, respectively, seemed to keep the dual inhibition with no greater affinity for one of the enzymes. Furthermore, for compound **10**, a higher IC_{50} value for hAChE (IC_{50} = 17.8 μM) was found when compared to eeAChE. However, the IC_{50} value found for hBChE was lower than the one found for eqBChE. Nucleoside **9**, a selective eeAChE inhibitor, in the human enzymes inhibitory assays turned into a hBChE selective inhibitor, while compound **13** showed no activity against the human enzymes. For compounds **15** and **16**, the lower concentration which was tested was not able to produce 50% of enzyme inhibition. Finally, also in Table 3, the inhibition results for the latter compounds at 5 μM final concentration are disclosed. Nucleosides **15** seemed to have a slightly higher affinity to AChE, producing 31 and 28% inhibition of eeAChE and hAChE, respectively, while **16** seemed to present a higher inhibition towards BChE, with a higher inhibition value for the hBChE (30%). However, both seem to show potential as non-selective inhibitors. These results then suggest that by improving these compounds' water solubility, the inhibition rates could increase, and the nucleosides could become interesting candidates as cholinesterase inhibitors.

2.3. Chelating Activity

The chelating activity of all purine nucleoside probes was evaluated by the UV-Visible method in order to determine compounds' absorption spectra in the absence and in the presence of Cu^{2+} at concentrations ranging from 1 μM to 10 μM. Amongst nucleosides **5-10** and **12**, the latter displayed the most interesting profile (Figure 3). The resulting wavelength shift from λ = 280 to λ = 325 nm indicates a conformational rearrangement induced by interaction with Cu^{2+}. Furthermore, selectivity towards the biologically relevant ions Zn^{2+}, Fe^{2+}, and Al^{3+} was also investigated, thus demonstrating for most of them a negligible quenching.

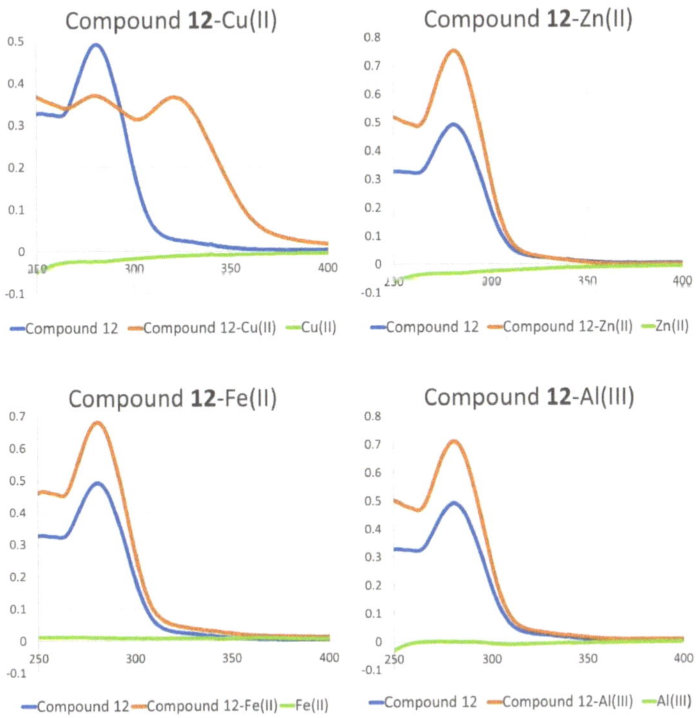

Figure 3. Absorbance spectra of compound **12** in the absence and in the presence of copper ion, iron ion, zinc ion, and aluminum ion (10 μM) and spectra of copper, iron, zinc aluminum ions, determined by UV-Visible spectroscopy with water/DMSO (90:10) as the solvent.

Hence, although **12**, the only acetylated nucleoside, did not show promising cholinesterase inhibition, it became the first nucleoside-based molecule with metal chelating properties and, therefore, became a starting point for the investigation of structural modifications that could lead to the desired dual-target compounds against AD.

In order to understand which structural alterations could be carried out to achieve our goal, computational studies were performed. Molecular electrostatic potential maps, in which the red and blue hues represent negatively and positively charged regions, respectively, and green tones depict neutral regions, were generated for compounds **5-7**, **9**, and **12**. In all cases, a significant electron density in the amide group of the purine base was found, and the benzyl units were found to be almost neutral. Moreover, compound **7**, exhibiting N^7 purine ligation, also exhibits high electron density between atoms N3 and N9, and compound **12** shows some negative density in the acetyl groups. Hereinafter, the generated maps for compounds **7** and **12** are disclosed (Figure 4), the remaining being found in the Supplementary Materials (Figures S48–S50).

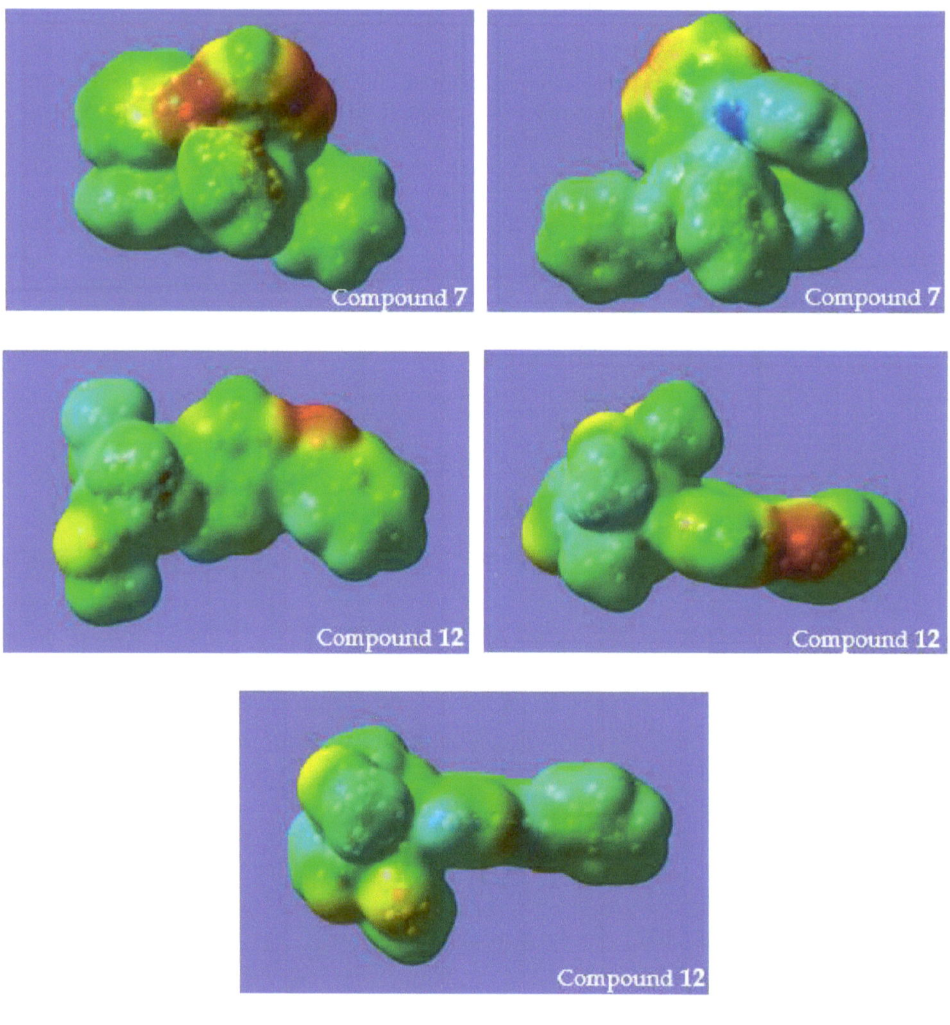

Figure 4. Different views of the molecular electrostatic potential surfaces for compounds **7** and **12**, deduced by Gaussian16 with CAM-B3LYP/def2TZVP with IEFPCM for water.

Based on the molecular electrostatic potential map generated for compound **12**, the geometry for the complex compound **12**-Cu^{2+} was optimized, starting with the copper ion placed near the amide group of the purine. The electronic transitions in water for the two possible conformers, "b" and "c," of compound **12** in the absence and in the presence of Cu^{2+} ions are presented in Figure 5. In comparison with the experimental results, the calculated transitions are displaced to shorter wavelengths; however, the bathochromic shift observed experimentally with the addition of Cu^{2+} is also noticed in the calculated results. The larger red shift in λ_{max} from 254 to 284 nm is observed for conformer "b." Also, the energy difference between both complexes listed in Table 4 shows that complex **12a**-Cu^{2+} is more stable. Finally, the optimized molecular geometry of complexes **12**-Cu^{2+} is shown in Figure 6, and the electrostatic potential map for the most stable complex, the **12a**-Cu^{2+} complex, is presented in Figure 7.

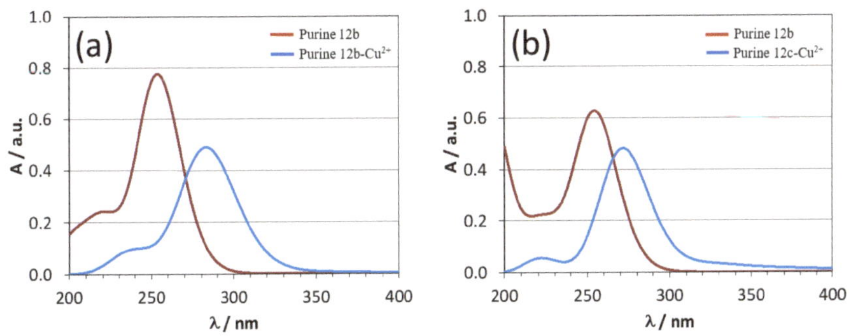

Figure 5. Electronic transitions evaluated for compound **12** in the absence and in the presence of Cu^{2+}; (**a**) being the conformer "b" and (**b**) being the conformer "c," determined in Gaussian16 with CAM/B3LYP/def2TZVP and IEFPCM for water.

Table 4. Energy difference between conformers "b" and "c" for complexes with Cu^{2+}.

Conformer-Cu^{2+}	Complex 12-Cu^{2+}		
	E/kcal·mol^{-1}	ΔE(12a–12b)/kcal·mol^{-1}	Boltzmann Population
12b-Cu^{2+}	−2,161,780.22	−0.99	0.84
12c-Cu^{2+}	2,161,779.23		0.16

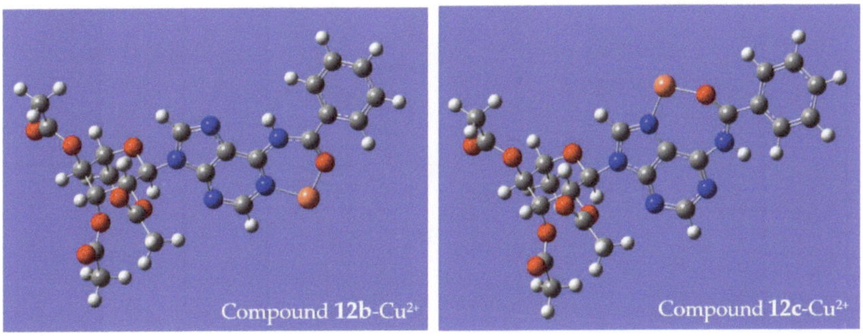

Figure 6. Final optimized geometry of the complexes **12**-Cu^{2+} determined in Gaussian16 with CAM/B3LYP/def2TZVP and IEFPCM for water.

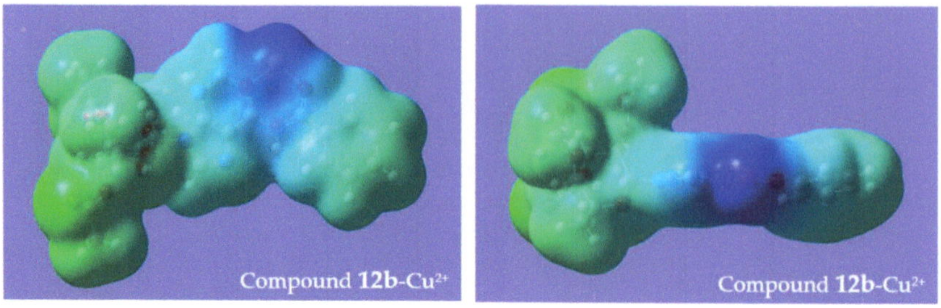

Figure 7. Different views of the molecular electrostatic potential surfaces for complex **12b**-Cu^{2+}determined in Gaussian16 with CAM/B3LYP/def2TZVP and IEFPCM for water.

Furthermore, having noticed that the glycosyl groups have a small contribution to the molecule electrostatic potential, the interaction with the metal ions of the purine in conformer "b" was studied after replacing the glycosyl unit of compound **12** with a hydrogen atom at N9. In Figure 8, the simulated absorption spectra based on electronic transitions calculated for 6-benzamidopurine in the absence and in the presence of the metal ions in water are presented. The spectra were simulated based on the assumption that all the solutions have the same concentration (10 μM) and that the metal ions are totally binded to the purine. As observed experimentally, the calculated absorption spectra indicate that the binding constant between the purine and Cu^{2+} is larger than the one with the other metal ions. Also, it is important to stress that the spin multiplicity of Cu^{2+} is doublet while all other metal ions' spin multiplicity is singlet.

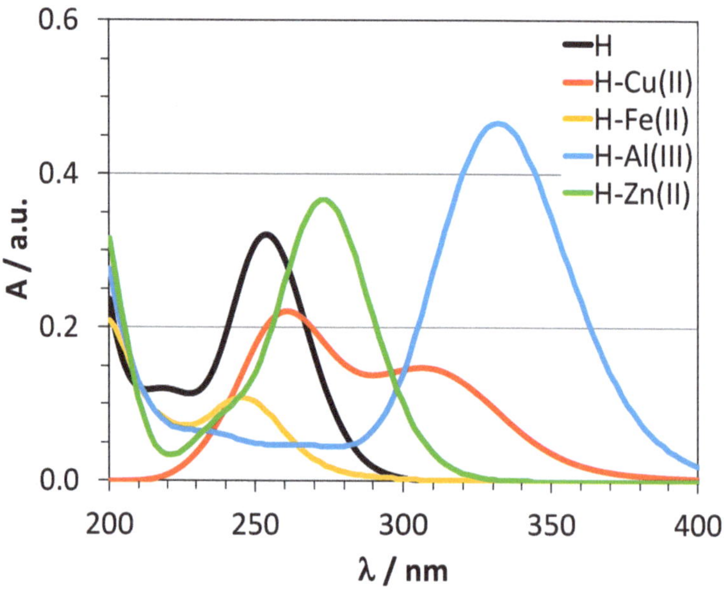

Figure 8. Electronic transitions of 6-benzamidopurine in the absence and in the presence of different metal ions, determined in Gaussian16 with CAM/B3LYP/def2TZVP and IEFPCM for water.

Having the computational studies performed, three new nucleosides, **13**, **15**, and **16**, were synthesized aiming at confirming the latter findings, **13** being a benzylated analogue of compound **12** with an additional *N*-benzyl group, **15** a benzylated analogue of **12** and **16** the β-anomer of **15**. Amongst these new compounds, **15** and **16** displayed the most interesting profiles since **13** did not show chelating activity towards any of the biologically relevant ions. Compound **15** was found to be a selective copper chelator, as it is shown in Figure 9. In the complex **15**-Cu^{2+} spectrum, there is a perceptible wavelength shift from $\lambda = 284$ to $\lambda = 334$ nm, indicating a conformational rearrangement due to the interaction with Cu(II). Through the analysis of the remaining spectra, it is possible to verify that there is no relevant interaction of this nucleoside with the other ions. Compound **16** was shown to be a non-selective chelator, as the resulting spectra showed conformational rearrangements induced by interaction with all the tested ions, as can be verified in Figure 10. Interactions with copper, zinc, iron and aluminum ions result in wavelength shifts from $\lambda = 290$ nm to $\lambda = 345$, 346, 339, and 351 nm, respectively.

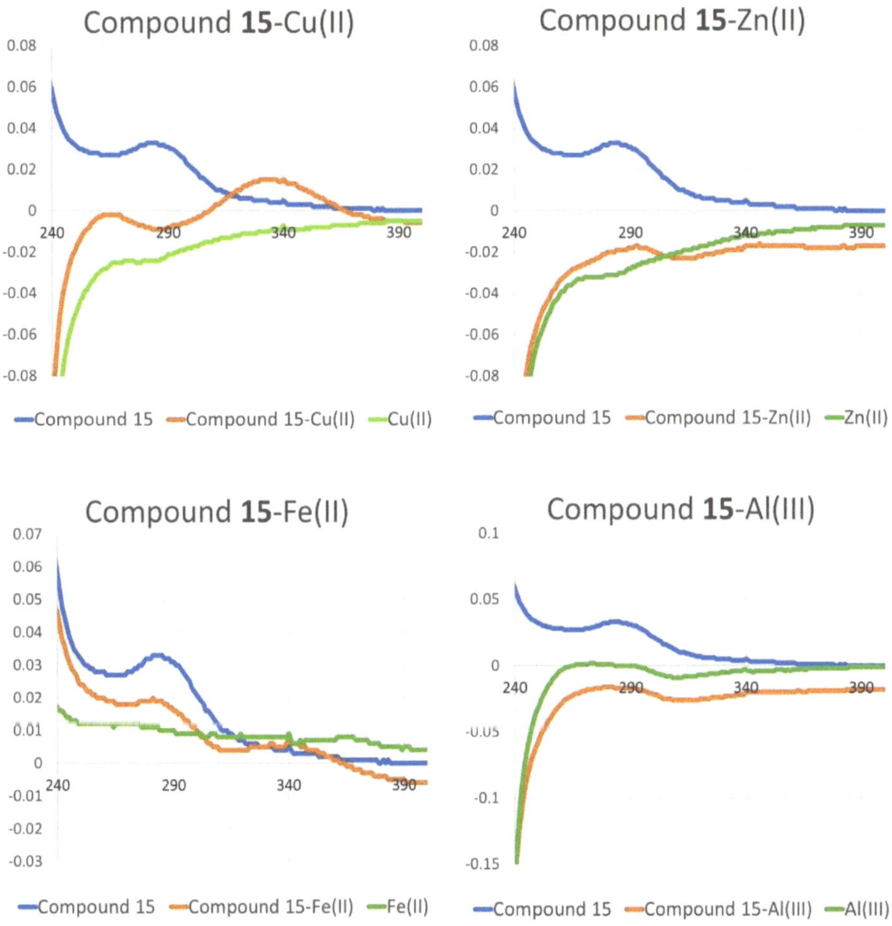

Figure 9. Absorbance spectra of compound **15** in the absence and in the presence of copper ion, zinc ion, iron ion, and aluminum ion (10 μM) and spectra of copper, zinc, iron, and aluminum ions, determined by UV-Visible spectroscopy with water/DMSO (90:10) as the solvent.

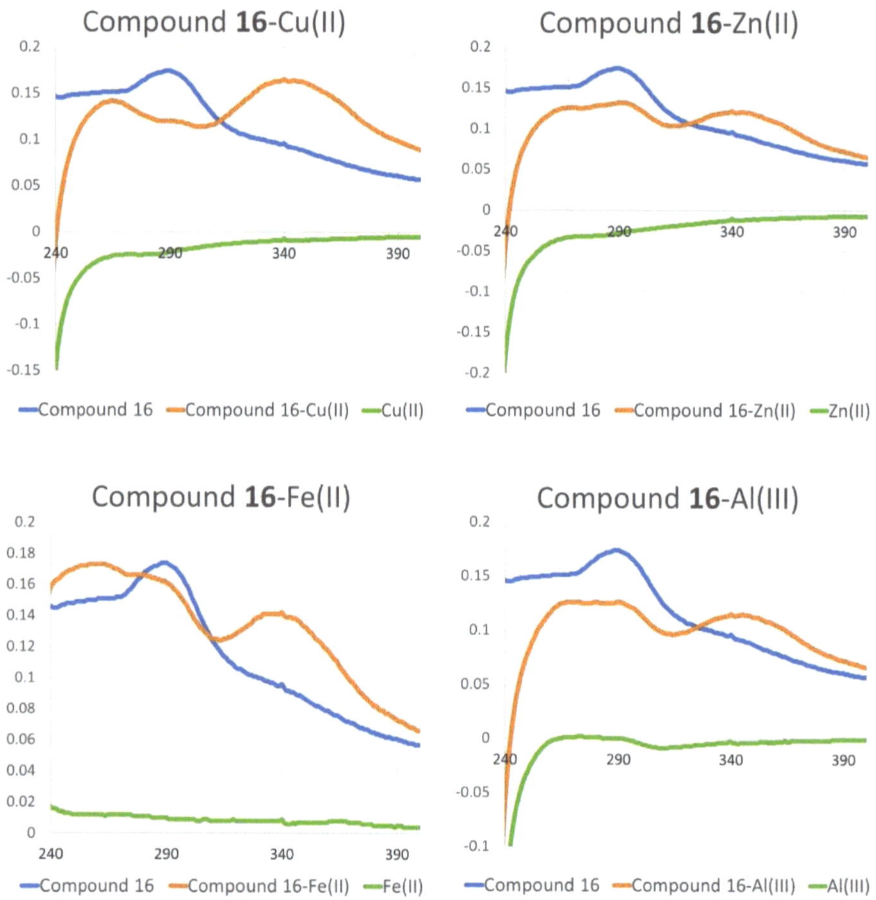

Figure 10. Absorbance spectra of compound **16** in the absence and in the presence of copper ion, zinc ion, iron ion, and aluminum ion (10 μM) and spectra of copper, zinc, iron, and aluminum ions, determined by UV-Visible spectroscopy with water/DMSO (90:10) as the solvent.

To conclude, the computational studies revealed that for metal chelation, the glycosyl units had a small contribution, being the amide group of the purine the most relevant. The results obtained with compounds **13** and **15** confirmed these findings. Nucleoside **15**, the benzylated analogue of **12**, also showed selective copper chelation over the remaining metals, proving that the glycosyl-protecting groups had no influence on this property. Moreover, compound **13**, being similar to **15** with an extra *N*-benzyl group in the purine moiety, showed no activity, showing that the metal chelation site should be found in the purine. Finally, the computational studies indicated that this purine base should have a higher affinity to Cu^{2+}; however, experimentally, compound **16** showed no selectivity over the four metal ions.

3. Discussion

A series of novel nucleosides were synthesized and tested, leading to new insights into the structural features required for BChE or AChE selective inhibition and Cu^{2+} binding. Starting from lead compounds 1 and 2, the replacement of 6-chloropurine by 6-benzamidopurine led to two novel and selective Cu^{2+} chelating agents 12 (with the peracetylated α-L-rhamnosyl group) and its analogue perbenzylated compound 15 (with the perbenzylated α-L-rhamnosyl group). Moreover, the latter modifications led to a novel metal chelator 16 ($^1C_4$ conformer), obtained with the same substrate as 15, with an affinity for Cu^{2+}, Zn^{2+}, Fe^{2+}, and Al^{3+}. Importantly, all synthesized nucleosides showed the equatorial orientation of the base, most probably resulting from the exoanomeric and steric effects. Amongst the new nucleosides tested, 12, the only acetylated compound, did not show cholinesterase inhibition, while all the benzylated compounds, except 5 (the N^9-linked α-D-mannosyl nucleoside), were active to some extent, highlighting the importance of aromatic rings for this property. With regard to eeAChE and eqBChE, the selective inhibitors exhibit a glycosyl $^1C_4$ conformation regardless of the ligation position to the purine base, while the dual inhibitors show a glycone $^4C_1$ conformation or derived thereof. Moreover, N^9 ligation is shown for AChE and N^7 ligation for BChE, in agreement with our previous results shown by the 6-chloropurine series [19]. However, for human cholinesterases, only compound 9 (N^9-linked 6-deoxy-α-D-*manno*) was selective for BChE inhibition, and no nucleoside showed AChE selective inhibition, while the known eeAChE and eqBChE dual inhibitors kept this property, with 6 (N^9-linked β-D-*manno*) and 10 (N^9-linked 6-deoxy-β-D-*manno*) having a higher affinity for BChE and 8 (N^7-linked β-D-*manno*) showing no greater affinity for any of the enzymes. Finally, compounds 15 and 16, with the glycone L configuration, N^9 purine ligation, and aromatic rings in their structure, showed metal chelation and some inhibition of the cholinesterases, both the animal and the human ones, at the tested concentration, becoming the first nucleoside-based molecules with potential to act as dual-target compounds against AD. This work illustrates the role of sugars by tuning bioactivity and selectivity, encouraging further investigation of this family of purine nucleosides towards glycone structure optimization to access anticholinesterase compounds and copper chelators as multitarget agents against AD.

4. Materials and Methods

Reagents were purchased from commercial suppliers (Sigma Aldrich, St. Louis, MI, USA) and used without further purification. The starting materials **3**, **4**, and **11** were available from Sigma Aldrich, Glentham Life Sciences (Corsham, UK), Toronto Research Chemicals (North York, ON, Canada). Compound **14** was obtained by benzylation of methyl α-D-rhamnoside purchased from Alfa Aesar. The physical and spectroscopic data of 14 was in full agreement with those previously reported [26]. Microwave-assisted synthesis was performed with a CEM Discover and Explorer SP. NMR spectra were recorded on a BRUKER Advance 400 spectrometer at 298 K or on a BRUKER Advance III 500 spectrometer at 298 K. Spectra were referenced internally to residual proton-solvent (^1H) or carbon-solvent (^{13}C) resonances, and reported relative to tetramethylsilane (0 ppm). Chemical shifts (δ) are given in ppm and coupling constants (J) in Hz. UV spectra were recorded on a SHIMADZU UV-1800 spectrometer with quartz cells. Optical rotations were determined on a Perkin-Elmer 341 polarimeter and on an Anton Paar MCP 100 polarimeter. HRMS analyses were performed on an Agilent 6530 Accurate Mass Q-TOF, only significant m/z peaks, with their percentage of relative intensity in parentheses, are reported. TLC was performed on silica gel (Merck 5554), and column chromatography for compound isolation was carried out with silica gel. Solvents were dried before use according to the conventional procedures.

4.1. *Synthesis*

N-glycosylation procedure: *N,O*-Bis(trismethylsilyl)acetamide (BSA) (3.0 equiv.) was added to the solution of *N*-(9*H*-purin-6-yl)benzamide (1.5 equiv.) in dry acetonitrile. The mixture was stirred at room temperature for 50 min. Then the fully protected methyl mannopyranoside (**3**, **4**, **11**, and **14**) (1 mmol) was dissolved in dry acetonitrile, and trimethylsilyl trifluoromethanesulfonate (TMSOTf) (8 equiv.) was added to the reaction mixture. The reaction was carried out under microwave irradiation (150 W, 65 °C, 60 min). The mixture was poured into dichloromethane, washed with a saturated solution of NaCl, and extracted with dichloromethane (4 × 25 mL). The combined organic phases were washed with brine, dried ($MgSO_4$), and concentrated. Compounds were isolated and purified by column chromatography (cyclohexane/EtOAc from 9:1 to 6:4).

Compound **12** benzylation procedure: Powdered sodium hydroxide (15.0 equiv.) and tetrabutylammonium iodide (TBAI) (0.75 equiv.) were added to a solution of **12** in THF (20 mL/mmol). The mixture was stirred for 10 min at room temperature. Then, benzyl bromide [12.0 equiv. (3.0 equiv./OH group)] was added, and the mixture was stirred for 3 h. The mixture was poured into dichloromethane, washed with water, and extracted with dichloromethane (3 × 100 mL). The combined organic phases were dried ($MgSO_4$), filtered, and concentrated. Compound **13** was isolated and purified by column chromatography (hexane/EtOAc from 10:0 to 6:1).

6-benzamido-9-(2,3,4,6-tetra-*O*-benzyl-α-D-mannopyranosyl)purine (**5**), 6-benzamido-9-(2,3,4,6-tetra-*O*-benzyl-β-D-mannopyranosyl)purine (**6**), 6-benzamido-7-(2,3,4,6-tetra-*O*-benzyl-α-D-mannopyranosyl)purine (**7**) and 6-benzamido-7-(2,3,4,6-tetra-*O*-benzyl-β-D-mannopyranosyl)purine (**8**).

Data for compound **5**, using compound **3** as starting material: Syrup. Yield = 22%. Rf = 0.85 (EtOAc/Hexane 4:1). $[\alpha]_D^{20} = -24°$ (*c* 2, MeOH). ^1H NMR (400 MHz, CD_3OD): δ 8.58 (s, 1H, H-2), 8.42 (s, 1H, H-8), 8.09 (d, 2H *ortho*, J = 7.0 Hz, PhCO), 7.65 (t, 1H *para*, J = 7.0 Hz, PhCO), 7.57 (t, 2H *meta*, J = 7.0 Hz, PhCO), 7.36–6.89 (m, 21H, Ph-Bn, NH), 6.13 (d, 1H, H-1′, $J_{1',2'}$ = 8.0 Hz), 4.78 (dd, 1H, H-2′, $J_{2',3'}$ = 2.8 Hz, superimposed to methanol contamination of the solvent), 4.71, 4.68, 4.66, 4.63 (AB system, 2H, Bn-CH_2, J = 11.7 Hz), 4.60-4.43 (m, 8H, H-5′, 3xAB system, 3x CH_2Ph, part A of AB system, CH_2Ph), 4.25, 4.22 (AB system, 1H, CH_2Ph, J = 12.0 Hz), 4.11 (t, 1H, H-3′, $J_{3',4'}$ = 2.6 Hz), 3.85 (dd, 1H, H-4′, $J_{4',5'}$ = 5.5 Hz), 3.76 (dd, 1H, H-6′a, $J_{6'a,5'}$ = 6.7 Hz, $J_{6'a,6'b}$ = 10.9 Hz), 3.65 (dd, 1H, H-6′b, $J_{6'b,5'}$ = 3.9 Hz). ^{13}C NMR (100 MHz, CD_3OD): δ 167.4 (CONH), 156.6 (C-4), 153.4 (C-2), 150.5 (C-6), 145.6 (C-8), 139.7(2) (Ph-Cq), 139.7 (Ph-Cq), 139.6 (Ph-Cq), 138.7 (Ph-Cq), 134.4 (Ph-Cq), 134.3 (Ph-Cq), 130.1, 129.9, 129.8, 129.7(3), 129.6(7), 129.6(0), 129.5(6), 129.4(3), 129.3(5), 129.3(1), 129.2(6), 129.2, 129.0, 127.9 (C-5), 81.9 (C-1′), 77.7 (CH_2Ph), 77.5 (CH_2Ph), 76.7 (C-4′), 75.8 (C-3′), 74.6 (CH_2Ph), 74.4 (C-2′), 74.1, 73.4, 72.7 (C-5′, 2x CH_2Ph), 70.2 (C-6′). HRMS (CI^+); calcd for $C_{46}H_{43}N_5O_6$ $(M+Na)^+$: 784.3106, found: 784.3162.

Data for compound **6** using compound **3** as starting material: Syrup. Yield = 16%. Rf = 0.70 (EtOAc/Hexane 3:2). $[\alpha]_D^{20} = -6°$ (*c* 2, MeOH). ^1H NMR (400 MHz, $CDCl_3$): δ 8.77 (s, 1H, H-2), 8.09 (s, 1H, H-8), 8.08 (d, 2H *ortho*, J = 6.0 Hz, PhCO), 7.63 (t, 1H *para*, J = 7.0 Hz, PhCO), 7.55 (t, 2H *meta*, J = 7.0 Hz, PhCO), 7.37–7.01 (m, 21H, Ph, NH), 6.08 (d, 1H, H-1′, $J_{1',2'}$ = 7.4 Hz), 4.97 (dd, 1H, H-2′, $J_{2',3'}$ = 2.9 Hz), 4.76, 4.73 (part A of AB system, 1H, CH_2Ph, J = 12.0 Hz), 4.68, 4.65 (part B of AB system, 1H, CH_2Ph, J = 12.0 Hz), 4.55–4.43 (m, 6H, H-5′, 2xAB system, CH_2Ph, part A of AB system, CH_2Ph), 4.30, 4.27 (part B of AB system, 1H, CH_2Ph, J = 12.0 Hz), 3.99 (dd, 1H, H-3′, $J_{3',4'}$ = 3.4 Hz), 3.91 (dd, 1H, H-4′, $J_{4',5'}$ = 6.0 Hz), 3.78 (dd, 1H, H-6′a, $J_{5',6'a}$ = 6.0 Hz, $J_{6'a,6'b}$ = 12.8 Hz), 3.71 (d, 1H, H-6′b, $J_{6'b,5'}$ = 3.8 Hz). ^{13}C NMR (100 MHz, $CDCl_3$): δ 164.6 (CONH), 152.6 (C-2), 151.6 (C-4), 149.6 (C-6), 142.8 (C-8), 137.9 (Ph-Cq), 137.9 (Ph-Cq), 137.7 (Ph-Cq), 137.5 (Ph-Cq), 136.8 (Ph-Cq), 136.8 (Ph-Cq), 128.9, 128.5, 128.4, 128.3, 128.0(0), 127.9(8), 127.9(5), 127.8 (Ph), 123.2 (C-5), 81.1 (C-1′), 75.7 (C-5′), 74.9, 74.8 (C-3′, C-4′), 73.2 (CH_2Ph, C-2′), 72.4 (CH_2Ph), 72.4 (CH_2Ph), 72.1 (CH_2Ph), 68.4 (C-6′). HRMS (CI^+); calcd for $C_{46}H_{43}N_5O_6$ $(M+H)^+$: 656.2867, found: 656.2856.

Data for compound **7** using compound **3** as starting material: Syrup. Yield = 6%. Rf = 0.40 (EtOAc/MeOH 4:1). $[\alpha]_D^{20} = +5°$ (*c* 3, MeOH). ^1H NMR (400 MHz, CD_3OD): δ 8.25 (s, 1H, H-2), 8.14 (s, 1H, H-8), 7.35-6.60 (m, 26H, Ph, NH), 5.84 (d, 1H, H-1′, $J_{1',2'}$ = 9.3 Hz), 4.66-4.40 (m, 9 H, H-3′, 4xAB system, CH_2Ph), 4.12-4.06 (m, 2H,H-4′, H-6′a), 3.91 (m, 1H, H-5′), 3.85 (d, 1H, H-2′), 3.75 (d, 1H, H-6′b, $J_{6'a,6'b}$ = 10.8 Hz); ^{13}C NMR (100 MHz, CD_3OD): δ 169.1 (CONH), 157.3 (C-4), 153.3, 153.2 (C-2, C-6),146.7 (C-8), 139.7, 139.4, 139.1, 137.8, 136.1, 134.5, 134.4, 130.7, 130.1, 130.0, 129.9, 129.8, 129.7, 129.6(6), 129.6(2), 129.5(5), 129.4, 129.3, 116.2 (C-5), 81.6 (C-1′), 80.3 (C-3′), 75.4 (CH_2Ph), 75.0 (CH_2Ph), 74.4 (C-2′), 74.3 (CH_2Ph), 73.8 (CH_2Ph), 73.7 (C-4′), 71.8 (C-5′), 67.9 (C-6′). HRMS (CI^+); calcd for $C_{46}H_{43}N_5O_6$ $(M+H)^+$: 762.3286, found: 762.3293.

Data for compound **8** using compound **3** as starting material: Syrup. Yield = 12%. Rf = 0.20 (EtOAc/MeOH 4:1). $[\alpha]_D^{20} = -5°$ (*c* 2.7, MeOH). ^1H NMR (400 MHz, CD_3OD): δ

8.31 (s, 1H, H-8), 8.16 (s, 1H, H-2), 7.43–7.06 (m, 24H, Ph, NH), 6.78 (d, 2H, Ph, J = 7.59 Hz), 6.08 (s, 1H, H-1′), 4.91–4.75 (m, 3H, AB system CH_2Ph-3′ + part A of AB system CH_2Ph-4′, superimposed by methanol contamination of the solvent), 4.65, 4.62 (part A of AB system, J = 11.0 Hz, 1H, CH_2Ph-2′), 4.64, 4.61 (part A of AB system, J = 11.0 Hz, 1H, CH_2Ph-6′), 4.58, 4.55 (part B of AB system, J = 10.0 Hz, 1H, CH_2Ph-4′), 4.46, 4.43 (part B of AB system, J = 11.0 Hz, 1H, CH_2Ph-6′), 4.31 (t, 1H, H-4′, $J_{4',3'}$ = 9.7 Hz, $J_{4',5'}$ = 9.6 Hz), 4.21 (br s, 1H, H-2′), 4.14, 4.12 (part B of AB system, J = 11.0 Hz, 1H, CH_2Ph-2′), 4.00 (dd, 1H, H-3′, $J_{3',2'}$ = 2.3 Hz), 3.84 (d, 1H, H-6′a, $J_{6'a,6'b}$ = 10.4 Hz), 3.78 (d, 1H, H-5′), 3.72 (dd, 1H, H-6′b). ^{13}C NMR (100 MHz, CD_3OD): δ 176.5 (CONH), 159.8 (C-4), 153.7 (C-6), 153.2 (C-2), 145.0 (C-8), 139.6(2) (Ph-Cq), 139.5(9) (Ph-Cq), 139.1 (Ph-Cq), 138.2 (Ph-Cq), 129.6, 129.4(0), 129.3(6), 129.3, 129.2, 129.0 (Ph), 113.6 (C-5), 86.9 (C-1′), 83.9 (C-3′), 79.1 (C-5′), 78.0 (C-2′), 76.3 (CH_2Ph), 76.2 (CH_2Ph), 74.3 (C-4′), 74.2 (CH_2Ph), 73.7 (CH_2Ph), 68.4 (C-6′). HRMS (CI^+); calcd for $C_{46}H_{43}N_5O_6$ $(M+H)^+$: 762.3286, found: 762.3276.

6-benzamido-9-(6-deoxy-2,3,4-tri-*O*-benzyl-α-D-mannopyranosyl)purine (**9**), 6-benzamido-9-(6-deoxy-2,3,4-tri-*O*-benzyl-β-D-mannopyranosyl)purine (**10**)

Data for compound **9** using **4** as starting material: Syrup. Yield = 33%. Rf = 0.8 (DCM/EtOAc 1:1). $[\alpha]_D^{20}$ = +18 (*c* 1, MeOH). ^1H NMR (400 MHz, CD_3OD): δ 8.58 (s, 1H, H-2), 8.42 (s, 1H, H-8), 8.08 (d, 2H, *ortho*, J = 7.3 Hz, PhCO), 7.64 (t, 1H, *para* J = 7.6 Hz, PhCO), 7.55 (t, 2H, *meta*, J = 7.6 Hz, PhCO), 7.40–7.27 (m, 11H, Ph, NH), 7.12–7.05 (m, 3H, Ph), 6.91 (d, 2H, Ph), 6.15 (d, 1H, H-1′, $J_{1',2'}$ = 8.1 Hz), 4.79 (dd, 1H, H-2′, $J_{2',3'}$ = 2.7 Hz), 4.75, 4.72, 4.70, 4.67 (AB system, 2H, OCH_2-3′, J = 12.1 Hz), 4.59, 4.56, 4.56, 4.53 (AB system, 2H, OCH_2-4′), 4.45, 4.42 (part A of AB system, 1H, OCH_2-2′, J = 12.3 Hz), 4.37 (t, 1H, H-5′, $J_{5',4'}$ = 5.6 Hz, $J_{5',6'}$ = 6.54 Hz), 4.25, 4.22 (part B of AB system, 1H, OCH_2-2′, J = 12.3 Hz), 4.10 (t, 1H, H-3′, $J_{3',4'}$ = $J_{3',2'}$ = 2.88 Hz), 3.62 (dd, 1H, H-4′), 1.34 (d, 3H, CH_3, J = 6.54 Hz). ^{13}C NMR (100 MHz, CD_3OD): δ 168.4 (CO), 153.7 (C-4), 153.4 (C-2), 153.4 (C-6), 145.5 (C-8), 139.8, 139.7, 138.8, 134.2, 130.1, 129.8(2), 129.8(1), 129.7, 129.6(2), 129.5(7), 129.5, 129.4, 129.2(9), 129.2(5) (Ph), 126.4 (C-5), 81.6 (C-1′), 81.4 (C-4′), 76.2 (C-3′), 74.9 (C-2′), 74.4 (C-5′), 74.1 (OCH_2-3′), 73.6 (OCH_2-4′), 72.7 (OCH_2-2′), 18.6 (CH_3). HRMS (CI+); calcd for $C_{39}H_{37}N_5O_5$ (M+H)+: 656.2867, found: 656.2856.

Data for compound **10** using **4** as starting material: Syrup. Yield = 40%. Rf = 0.7 (DCM/EtOAc 1:1). $[\alpha]_D^{20}$ = +4 (*c* 1, MeOH). ^1H NMR (400 MHz, CD_3OD): δ 8.54 (s, 1H, H-2), 8.32 (s, 1H, H-8), 8.08 (d, 2H, *ortho*, J = 7.4 Hz, PhCO), 7.64 (t, 1H, *para*, J = 7.5 Hz, PhCO), 7.55 (t, 2H, *meta*, J = 7.8 Hz, PhCO), 7.41 (d, 2H, J = 6.9 Hz, Ph), 7.34–7.25 (m, 9H, NH, Ph), 7.01–6.99 (m, 3H, Ph), 6.87–6.84 (m, 2H, Ph), 5.94 (s, 1H, H-1′), 4.95, 4.92 (part A of AB system, 1H, OCH_2-4′, J = 11.3 Hz), 4.86, 4.83 (part A of AB system, 1H, OCH_2-3′, J = 12 Hz), 4.79-4.64 (m, 3H, part B of AB system, OCH_2-3′, part B of AB system, OCH_2-4′, part A of AB system, OCH_2-2′), 4.30, 4.28 (part B of AB system, 1H, OCH_2-2′, J = 11.58 Hz), 4.23 (br s, 1H, H-2′), 3.97 (dd, 1H, H-3′, $J_{3',2'}$ = 1.4 Hz, $J_{3',4'}$ = 7.1 Hz), 3.74–3.64 (m, 2H, H-4′,H-5′), 1.45 (d, $J_{5',6'}$ = 5.8 Hz, 3H, CH_3). ^{13}C NMR (100 MHz, CD_3OD): δ 168.0 (CO), 152.9 (C-2), 151.8 (C-6), 150.9 (C-4), 144.1 (C-8), 139.8 (PhCq), 139.6 (PhCq), 138.1 (PhCq), 133.9 (PhCq), 129.8, 129.6(2), 129.5(6), 129.4(4), 129.3(5), 129.2, 129.1, 129.0, 128.8(9), 128.7(8) (Ph), 124.3 (C-5), 84.0 (C-3′), 83.2 (C-1′), 80.6 (C-4′), 76.4 (C-5′), 76.3 (OCH_2-2′), 75.5 (OCH_2-4′), 74.5 (C-2′), 73.8 (OCH_2-3′), 18.4 (C-6′). HRMS (CI^+); calcd for $C_{39}H_{37}N_5O_5$ $(M+H)^+$: 656.2867, found: 656.2870.

6-benzamido-9-(6-deoxy-2,3,4-tri-*O*-acetyl-α-L-mannopyranosyl)purine (**12**), 6-benzamido-*N*-benzyl-9-(6-deoxy-2,3,4-tri-*O*-benzyl-α-L-mannopyranosyl)purine(**13**)

Data for compound **12** using **11** as starting material: Syrup. Yield = 63%. Rf = 0.7 (DCM/EtOAc 1:1). $[\alpha]_D^{20}$ = −4 (*c* 1, MeOH). ^1H NMR (400 MHz, CD_3OD): δ 8.75 (s, 1H, H-2), 8.59 (s, 1H, H-8), 8.07 (d, 2H, *ortho*, J = 7.2 Hz, PhCO), 7.65 (t, 1H, *para*, J = 7.5 Hz, PhCO), 7.57–7.54 (m, 3H, *meta* PhCO, NH), 6.30 (d, 1H, H-1′, $J_{1',2'}$ = 6.3 Hz), 6.26 (dd, 1H, H-2′, $J_{2',3'}$ = 3.5 Hz), 5.65 (dd, 1H, H-3′, $J_{3',4'}$ = 5.6 Hz), 5.01 (t, 1H, H-4′, $J_{4',5'}$ = 5.6 Hz,), 4.28 (qd, 1H, H-5′, $J_{5',6'}$ = 5.2 Hz), 2.14 (s, 3H, CH_3), 2.12 (s, 3H, CH_3), 1.94 (s, 3H, CH_3), 1.41 (d, 3H, CH_3). ^{13}C NMR (100 MHz, CD_3OD): δ 171.4(5) (CO), 171.3 (CO), 171.0 (CO), 168.3 (COPh), 153.5 (C-4, C-2), 151.3 (C-6),144.7 (C-8), 134.9, 134.0 (Cq, Ph), 129.8 (Ph), 129.5 (Ph),

125.0 (C-5), 123.3 (Ph), 120.2 (Ph), 80.3 (C-1′), 73.2 (C-4′), 73.1 (C-5′), 70.3 (C-3′), 68.9 (C-2′), 20.7 (CH_3), 20.6 (CH_3), 20.4 (CH_3), 17.2 (CH_3). HRMS (CI^+); calcd for $C_{24}H_{25}N_5O_8$ $(M+H)^+$: 512.1776, found: 512.1775.

Data for compound **13** using **12** as starting material: Syrup. Yield = 15%. Rf = 0.8 (Hex/EtOAc 2:1). $[\alpha]_D^{20} = -90$ (*c* 0.1, MeOH). ^{1}H NMR (500 MHz, CD_3OD): δ8.53 (s, 1H, H-2), 8.28 (s, 1H, H-8), 7.48-7.01 (m, 25H, Ph), 6.08 (d, 1H, H-1′, $J_{1',2'}$= 8.26 Hz), 5.57 (s, 2H, NCH_2Ph,), 4.69, 4.67, 4.65, 4.63 (AB system, 2H, OCH_2-3′, J = 12.05 Hz), 4.64 (dd, 1H, H-2′, $J_{2',3'}$ = 2.81 Hz), 4.52 (s, 2H, OCH_2-4′), 4.23 (qd, 1H, H-5′, $J_{4',5'}$ = 5.23 Hz, $J_{5',6'}$ = 6.70 Hz), 4.16, 4.14, 4.07, 4.05 (AB system, 2H, OCH_2-2′, J = 11.98 Hz), 4.00 (br t, 1H, H-3′, $J_{3',4'}$=3.34 Hz), 3.59 (dd, 1H, H-4′), 1.33 (d, 1H, H-6′). ^{13}C NMR (100 MHz, CD_3OD): δ 174.1 (CO), 155.3 (C-6), 154.5 (C-4), 153.3 (C-2), 146.2 (C-8), 139.8, 139.7, 138.8, 138.7 (CqPh), 137.6, 132.4, 130.1, 129.8, 129.7(7), 129.7, 129.6, 129.5(9), 129.4, 129.3, 129.2, 128.9 (Ph), 128.7 (C-5), 80.9 (C-4′), 80.8 (C-1′), 76.1 (C-3′), 74.8 (C-2′), 74.6 (C-5′), 74.0 (OCH_2-3′), 73.5 (OCH_2-4′), 72.7 (OCH_2-2′), 52.7 (NCH_2Ph), 18.3 (C-6′). HRMS (CI^+); calcd for $C_{39}H_{38}N_5O_5$ $(M+H)_+$: 746.3354, found: 746.3337.

6-benzamido-9-(6-deoxy-2,3,4-tri-*O*-benzyl-α-L-mannopyranosyl)purine (**15**), 6-benzamido-9-(6-deoxy-2,3,4-tri-*O*-benzyl-β-L-mannopyranosyl)purine (**16**)

Data for compound **15** using **14** as starting material: Syrup. Yield = 8%. Rf = 0.9 (DCM/EtOAc 1:1). $[\alpha]_D^{20} = -75$ (*c* 0.2, MeOH). ^{1}H NMR (400 MHz, CD_3OD): δ 8.59 (s, 1H, H-2), 8.41 (s, 1H, H-8), 8.09 (d, 2H, *ortho*, J = 7.38 Hz, PhCO), 7.63 (t, 1H, *para*, J = 6.98 Hz, PhCO), 7.54 (t, 2H, *meta*, J = 7.86 Hz, PhCO), 7.41-7.24 (m, 11H, Ph, NH), 7.15-7.04 (3, 3H, Ph), 6.93 (dd, 2H, Ph, J = 1.46 Hz, J = 7.61 Hz), 6.17 (d, 1H, H-1′, $J_{1',2'}$ = 7.99 Hz), 4.82 (dd, 1H, H-2′, $J_{2',3'}$ = 2.49 Hz), 4.75, 4.72, 4.70, 4.67 (AB system, 2H, OCH_2-3′, J = 12.02 Hz), 4.60, 4.57, 4.56, 4.53 (AB system, 2H, OCH_2-4′, J = 11.91 Hz), 4.45, 4.42 (part A of AB system, 1H, OCH_2-2′, J = 12.00 Hz), 4.37 (qd, 1H, H-5′), 4.26, 4.23 (part B of AB system, 1H, OCH_2-2′, J = 12.00 Hz), 4.10 (t, 1H, H-3, $J_{3',4'}$ = 3.00 Hz), 3.64 (dd, 1H, H-4′, $J_{4',5'}$ = 5.11 Hz), 1.35 (d, 3H, $J_{6',5'}$ = 6.70 Hz). ^{13}C NMR (100 MHz, $CDCl_3$): δ 168.1 (CO), 153.4 (CqPh), 153.0 (C-2, C-4), 151.0 (C-6), 145.1 (C-8,Cq), 139.5, 139.4, 138.5 (CqPh), 138.5, 135.0, 133.9, 129.8, 129.5, 129.4, 129.3(5), 129.3, 129.2(5), 129.2, 129.1, 129.0, 128.9, 128.8(6), 128.7 (Ph), 125lç.1 (C-5), 81.3 (C-1′), 81.1 (C-4′), 76.0 (C-3′), 74.6 (C-2′), 74.1 (C-5′), 73.8 (OCH_2-3′), 73.3 (OCH_2-4′), 72.4 (OCH_2-2′), 18.3 (C-6′). HRMS (CI^+); calcd for $C_{39}H_{37}N_5NaO_5$ $(M+Na)^+$: 678.2687, found: 678.2695.

Data for compound **16** using **14** as starting material: Syrup. Yield = 36%. Rf = 0.7 (DCM/EtOAc 1:1). $[\alpha]_D^{20} = -29$ (*c* 1, MeOH). ^{1}H NMR (400 MHz, CD_3OD): δ 8.55 (s, 1H, H-2), 8.32 (s, 1H, H-8), 8.10 (d, 2H, *ortho* PhCO, J = 6.91 Hz), 7.63 (t, 1H, *para* PhCO, J = 7.30 Hz), 7.54 (t, 2H, *meta* PhCO, J = 7.50 Hz), 7.41 (d, 2H, Ph, J = 7.49 Hz), 7.35-7.23 (m, 9H, Ph, NH), 7.01 (dd, 3H, Ph, J = 2.00 Hz, J = 5.11 Hz), 6.88-6.83 (m, 2H, Ph), 5.92 (d, 1H, H-1′, $J_{1',2'}$ =0.72 Hz), 4.94, 4.91 (part A of AB system, 1H, OCH_2-4′, J = 11.05 Hz), 4.83, 4.80 (part A of AB system, 1H, OCH_2-3′, J = 12.05 Hz), 4.76, 4.73 (part B of AB system, 1H, OCH_2-3′, J = 12.05 Hz), 4.72, 4.69 (part B of AB system, 1H, OCH_2-4′), 4.69, 4.66 (part A of AB system, 1H, OCH_2-2′, J = 11.79 Hz), 4.29, 4.27 (part B of AB system, 1H, OCH_2-2′, J = 11.79 Hz), 4.23 (br s, 1H, H-2′), 3.95 (dd, 1H, H-3′, $J_{3',2'}$= 2.47 Hz, $J_{3',4'}$ = 9.21 Hz), 3.77-3.59 (m, 2H, H-4′, H-5′), 1.36 (d, 3H, H-6′, $J_{6',5'}$ = 5.87 Hz). ^{13}C NMR (100 MHz, CD_3OD): δ 167.9 (CO), 153.8 (C-2), 151.7 (C-4), 150.8 (C-6), 144.0 (C-8), 139.6, 139.5, 138.1, 135.0 (PhCq), 129.8, 129.6, 129.5, 129.4, 129.3, 129.2, 129.1, 128.9, 128.8(7), 128.8(5), 128.8 (Ph), 124.2 (C-5), 84.0 (C-3′), 83.2 (C-1′), 80.6 (C-4′), 76.3 (C-5′), 76.2 (OCH_2-4′), 75.5 (OCH_2-2′), 74.5 (C-2′), 73.8 (OCH_2-3′), 18.5 (C-6′). HRMS (CI^+); calcd for $C_{39}H_{37}N_5NaO_5$ $(M+Na)^+$: 678.2683, found: 678.2687.

4.2. Cholinesterase Inhibition Assay

The test compounds were assayed for their inhibitory activity toward *ee*AChE (electric eel AChE) and *eq*BChE (horse serum BChE), respectively, following Ellman's method with modifications [21,22]. The BChE activity was determined in a reaction mixture containing 20 µL of *eq*BChE solution (0.9 U/mL in 0.1 M pH 8.0 phosphate buffer, PB), 20 µL of 5,5-dithio-bis-(2-nitrobenzoic) acid solution (DTNB 3.3 mM in 0.1 M pH 7.0 PB, containing

0.1 mM $NaHCO_3$), 20 μL of a solution of test compound (five to seven concentrations, ranging 1×10^{-5} to 1×10^{-9} M (in 0.1 M pH 8.0 PB), and 120 μL of 0.1 M PB (pH 8.0). After 20 min of incubation at 25 °C, 20 μL of butyrylthiocholine iodide (BTC, 0.05 mM in 0.1 M pH 8.0 phosphate buffer, PB) was added as the substrate, and the hydrolysis rates of the substrate monitored at 412 nm for 5.0 min at 25 °C. For compounds **15** and **16**, this procedure was adapted to a final compound concentration of 5 μM. The concentration of compounds that produced 50% inhibition of BChE activity (IC_{50}) was calculated by nonlinear regression of response/concentration (log) curve by using Prisma Graph Pad software (vers. 5.01). AChE inhibitory activity was determined similarly by using a solution of AChE (0.9 U/mL in 0.1 M pH 8.0 phosphate buffer, PB) and acetylthiocholine iodide (ATC, 0.05 mM in 0.1 M pH 8.0 phosphate buffer, PB) as substrate. The inhibition data are reported as means of IC_{50}'s determined at least in three independent measurements.

4.3. *UV-Visible Chelating Studies*

UV-Visible spectroscopy offers the advantage of high sensitivity towards small changes that affect the electronic properties of ligand receptors [27], so absorption spectra of all reported compounds have been evaluated by recording UV-visible properties in subsequent conditions: 1.5 mL 1×10^{-4}M of the probe are added to 1.5 mL 2×10^{-5}M of each ion chloride salt's solution (Copper Chloride ($CuCl_2$), Zinc Chloride ($ZnCl_2$), Iron (II) Chloride ($FeCl_2$) and Aluminum Chloride ($AlCl_3$)). A mixture of distilled water/DMSO 95:5 has been selected as a solvent for compounds **5–10** and **12**. For compounds **14–16**, the selected solvent was a mixture of distilled water/DMSO 90:10 due to poor water solubility. Furthermore, a titration assay of each compound has been carried out by adding increasing concentration of selected ion (final ion concentration = 1 μM, 2 μM, 3 μM, 5 μM, and 10 μM).

4.4. *Computational Studies*

The initial geometries were optimized in Gaussian16 with CAM-B3LYP/6-31G(d) and integrated equation formalism polarized continuum model (IEFPCM) for water [28–30]. Final geometries were optimized with CAM-B3LYP/def2TZVP and IEFPCM for water. The vibrational frequencies were checked for true minima in all cases (no imaginary frequencies were found). To accelerate the search for the best configurations, the molecules were split into two parts: the sugar unit and the purine moiety. With respect to the purine moiety, the geometry optimization started from some initial purine geometries with small differences in the dihedral angle between the rings, and the optimized geometries always ended up in two conformers. The geometry optimization for the glycosyl groups followed the same procedure. The optimization was started from a series of different initial geometries for each sugar unit, and the most stable conformer was selected. For the final geometry optimizations, the best glycosyl conformer found was coupled with the two purine conformers, and a full optimization of the whole molecule was performed.

NMR parameters were evaluated from single-point calculations with CAM-B3LYP/def2TZVP and IEFPCM for water. The computed NMR shielding tensors were converted to chemical shifts with empirical scaling factors determined from a test set of 80 molecules at the same level of theory [31].

Supplementary Materials: The following supporting information can be downloaded at: https://www.mdpi.com/article/10.3390/ph16010054/s1.

Author Contributions: Paper conceptualization was made by A.P.R. and N.A.C. Methodology (synthetic approaches, metal chelation, and anticholinesterase activity were conceived by A.P.R., N.A.C. and C.D.A. Synthesis and structure characterization of the new nucleosides was carried out by I.S., M.C., C.M., J.B., P.C., V.C. and M.J.F. supervised by A.P.R., while chelation studies were carried out by I.S., C.M. and J.B supervised by N.A.C. and A.P.R. The anticholinesterase activity was evaluated by M.d.C. under the supervision of C.D.A. HRMS was conducted by M.C.O., while the computational studies were carried out by K.S. and A.A.F. and J.N.C.L. The original draft was written by I.S. and C.M., reviewed and edited by A.P.R. and N.A.C. All authors have read and agreed to the published version of the manuscript.

Funding: This research was funded by the EUROPEAN UNION project entitled "Diagnostic and Drug Discovery Initiative for Alzheimer's Disease" (D3i4AD), FP7-PEOPLE-2013-IAPP, grant agreement 612347. FUNDAçãO PARA A CIêNCIA E A TECNOLOGIA (FCT, PORTUGAL) also funded this research through the financial support of Centro de Química Estrutural (project UIDB/00100/2020), Institute of Molecular Sciences (project LA/P/0056/2020) and project PTDC/QUI-QFI/29527/2017.

Institutional Review Board Statement: Not applicable as studies did not involve humans nor animals.

Informed Consent Statement: Not applicable as studies did not involve humans.

Data Availability Statement: Data is contained within the article and supplementary material.

Conflicts of Interest: The authors declare no conflict of interest. The funders had no role in the design of the study; in the collection, analyses, or interpretation of data; in the writing of the manuscript, or in the decision to publish the results.

References

1. Polverino, A.; Grimaldi, M.; Sorrentino, P.; Jacini, F.; D'Ursi, A.M.; Sorrentino, G. Effects of Acetylcholine on β-Amyloid-Induced cPLA2, Activation in the TB Neuroectodermal Cell Line: Implications for the Pathogenesis of Alzheimer's Disease. *Cell Mol. Neurobiol.* **2018**, *38*, 817–826. [CrossRef] [PubMed]
2. Bohnen, N.I.; Grothe, M.J.; Ray, N.J.; Müller, M.L.T.M.; Teipel, S.J. Recent Advances in Cholinergic Imaging and Cognitive Decline-Revisiting the Cholinergic Hypothesis of Dementia. *Curr. Geriatr. Rep.* **2018**, *7*, 1–11. [CrossRef] [PubMed]
3. Li, Q.; Yang, H.; Chen, Y.; Sun, H. Recent progress in the identification of selective butyrylcholinesterase inhibitors for Alzheimer's disease. *Eur. J. Med. Chem.* **2017**, *132*, 294–309. [CrossRef]
4. Çokuğraş, A.N. Butyrylcholinesterase: Structure and Physiological Importance. *Turk. J. Biochem.* **2003**, *28*, 54–61.
5. Macdonald, I.R.; Rockwood, K.; Martin, E.; Darvesh, S. Cholinesterase Inhibition in Alzheimer's Disease: Is Specificity the Answer? *J. Alzheimer's Dis.* **2014**, *42*, 379–384. [CrossRef]
6. Sharma, K. Cholinesterase Inhibitors as Alzheimer's Therapeutics. *Mol. Med. Rep.* **2019**, *20*, 1479–1487. [CrossRef]
7. Li, D.-D.; Zhang, W.; Wang, Z.-Y.; Zhao, P. Serum Copper, Zinc, and Iron Levels in Patients with Alzheimer's Disease: A Meta-Analysis of Case-Control Studies. *Front. Aging Neurosci.* **2017**, *9*, 300. [CrossRef]
8. Exler, C.; Mold, M.J. Aluminium in human brain tissue: How much is too much? *J. Biol. Inorg. Chem.* **2019**, *24*, 1279–1282. [CrossRef]
9. Weng, M.H.; Chen, S.Y.; Li, Z.Y.; Yen, G.C. Camellia oil alleviates the progression of Alzheimer's disease in aluminum chloride-treated rats. *Free Radic. Biol. Med.* **2020**, *152*, 411–421. [CrossRef]
10. Mutter, J.; Naumann, J.; Schneider, R.; Walach, H. Mercury and Alzheimer's disease. *Fortschr. Neurol. Psych.* **2007**, *75*, 528–540. [CrossRef]
11. Armstrong, R.A. Risk factors for Alzheimer's disease. *Folia Neuropathol.* **2019**, *57*, 87–105. [CrossRef]
12. Zafar, R.; Zubair, M.; Ali, S.; Shahid, K.; Waseem, W.; Naureen, H.; Haider, A.; Jan, M.S.; Ullah, F.; Sirajuddin, M.; et al. Zinc metal carboxylates as potential anti-Alzheimer's candidate: In vitro anticholinesterase, antioxidant and molecular docking studies. *J. Biomol. Struct. Dyn.* **2020**, *13*, 1–11. [CrossRef]
13. Nam, E.; Nam, G.; Lim, M.H. Synaptic Copper, Amyloid-β, and Neurotransmitters in Alzheimer's Disease. *Biochemistry* **2020**, *59*, 15–17. [CrossRef]
14. Yu, H.-J.; Zhao, W.; Zhou, Y.; Chenge, G.-J.; Sun, M.; Wang, L.; Yu, L.; Liang, S.H.; Ran, C. Salen-based bifunctional chemosensor for copper (II) ions: Inhibition of copper-induced amyloid-β aggregation. *Anal. Chim. Acta* **2020**, *1097*, 144–152. [CrossRef]
15. Zhou, Y.; Sun, W.; Peng, J.; Yan, H.; Zhang, L.; Liu, X.; Zuo, Z. Design, synthesis and biological evaluation of novel copper-chelating acetylcholinesterase inhibitors with pyridine and *N*-benzylpiperidine fragments. *Bioorg. Chem.* **2019**, *93*, 103322. [CrossRef]
16. Rana, M.; Sharma, A.K. Cu and Zn interactions with Aβ peptides:consequence of coordination on aggregation and formation of neurotoxic soluble Aβ oligomers. *Metallomics* **2019**, *11*, 64–84. [CrossRef]
17. Lanza, V.; Milardi, D.; Di Natale, G.; Pappalardo, G. Repurposing of Copper (II)-chelating Drugs for the Treatment of Neurodegenerative Diseases. *Curr. Med. Chem.* **2018**, *25*, 525–539. [CrossRef]
18. Gomes, L.M.F.; Bataglioli, J.C.; Storr, T. Metal complexes that bind to the amyloid-b peptide of relevance to Alzheimer's disease. *Coord. Chem. Rev.* **2020**, *412*, 213255. [CrossRef]
19. Schwarz, S.; Csuk, R.; Rauter, A.P. Microwave-assisted synthesis of novel purine nucleosides as selective cholinesterase inhibitors. *Org. Biomol. Chem.* **2014**, *12*, 2446–2456. [CrossRef]
20. Milijkovic, M. *Electrostatic and Stereoelectronic Effects in Carbohydrate Chemistry*; Springer: Berlin/Heidelberg, Germany, 2014; pp. 181–196.
21. Purgatorio, R.; de Candia, M.; Catto, M.; Carrieri, A.; Pisani, L.; De Palma, A.; Toma, M.; Ivanova, O.A.; Voskressensky, L.G.; Altomare, C.D. Investigating 1,2,3,4,5,6-hexahydroazepino[4,3-*b*]indole as scaffold of butyrylcholinesterase-selective inhibitors with additional neuroprotective activities for Alzheimer's disease. *Eur. J. Med. Chem.* **2019**, *177*, 414–424. [CrossRef]

22. Purgatorio, R.; de Candia, M.; Catto, M.; Rullo, M.; Pisani, L.; Denora, N.; Carrieri, A.; Nevskaya, A.A.; Voskressensky, L.G.; Altomare, C.D. Evaluation of Water-Soluble Mannich Base Prodrugs of 2,3,4,5-Tetrahydroazepino[4,3-b]indol-1(6H)-one as Multitarget-Directed Agents for Alzheimer's Disease. *Chem. Med. Chem.* **2021**, *16*, 589–598. [CrossRef]
23. De Candia, M.; Fiorella, F.; Lopopolo, G.; Carotti, A.; Romano, M.R.; Lograno, M.D.; Martel, S.; Carrupt, P.-A.; Belviso, B.D.; Caliandro, R.; et al. Synthesis and Biological Evaluation of Direct Thrombin Inhibitors Bearing 4-(Piperidin-1-yl)pyridine at the P1 Position with Potent Anticoagulant Activity. *J. Med. Chem.* **2013**, *56*, 8696–8711. [CrossRef]
24. Marcelo, F.; Silva, F.V.M.; Goulart, M.; Justino, J.; Sinaÿ, P.; Bleriot, Y.; Rauter, A.P. Synthesis of novel purine nucleosides towards a selective inhibition of human butyrylcholinesterase. *Bioorg. Med. Chem.* **2009**, *17*, 5106–5116. [CrossRef]
25. Rakonczay, Z. Potencies and selectivities of inhibitors of acetylcholinesterase and its molecular forms in normal and Alzheimer's disease brain. *Acta. Biol. Hung.* **2003**, *54*, 183–189. [CrossRef]
26. Yin, Q.; Chen, Q.; Lu, L.; Han, B. Sugar-based micro/mesoporous hypercross-linked polymers with in situ embedded silver nanoparticles for catalytic reduction. *Beilstein J. Org. Chem.* **2017**, *13*, 1212–1221. [CrossRef]
27. Mathys, Z.K.; White, A.R. Copper and Alzheimer's Disease. *Adv. Neurobiol.* **2017**, *18*, 199–216.
28. Frisch, M.J.; Trucks, G.W.; Schlegel, H.B.; Scuseria, G.E.; Robb, M.A.; Cheeseman, J.R.; Scalmani, G.; Barone, V.; Petersson, G.A.; Nakatsuji, H.; et al. *Gaussian 16*; Revision, C.01; Gaussian, Inc.: Wallingford, CT, USA, 2016.
29. Yanai, T.; Tew, D.; Handy, N. A new hybrid exchange-correlation functional using the Coulomb-attenuating method (CAM-B3LYP). *Chem. Phys. Lett.* **2004**, *393*, 51–57. [CrossRef]
30. Scalmani, G.; Frisch, M.J. Continuous surface charge polarizable continuum models of solvation. I. General formalism. *J. Chem. Phys.* **2010**, *132*, 114110. [CrossRef]
31. Lodewyk, M.W.; Siebert, M.R.; Tantillo, D.J. Computational Prediction of 1H and 13C Chemical Shifts: A Useful Tool for Natural Product, Mechanistic, and Synthetic Organic Chemistry. *Chem. Rev.* **2012**, *112*, 1839–1862. [CrossRef]

pharmaceuticals

MDPI

Review

Future Treatment Strategies for Cancer Patients Combining Targeted Alpha Therapy with Pillars of Cancer Treatment: External Beam Radiation Therapy, Checkpoint Inhibition Immunotherapy, Cytostatic Chemotherapy, and Brachytherapy

Ruth Christine Winter [1,†], Mariam Amghar [1,†], Anja S. Wacker [2], Gábor Bakos [1], Harun Taş [1], Mareike Roscher [3], James M. Kelly [2] and Martina Benešová-Schäfer [1,*]

1 Research Group Molecular Biology of Systemic Radiotherapy/Translational Radiotheranostics, German Cancer Research Center (DKFZ), Im Neuenheimer Feld 280, 69120 Heidelberg, Germany; ruth.winter@dkfz-heidelberg.de (R.C.W.); mariam.amghar@dkfz-heidelberg.de (M.A.); gabor.bakos01@gmail.com (G.B.); harun.tas@dkfz-heidelberg.de (H.T.)

2 Department of Radiology, Molecular Imaging Innovations Institute (MI3), Weill Cornell Medicine, 413 East 69th Street, New York, NY 10021, USA; anw4010@med.cornell.edu (A.S.W.); jak2046@med.cornell.edu (J.M.K.)

3 Service Unit for Radiopharmaceuticals and Preclinical Studies, German Cancer Research Center (DKFZ), Im Neuenheimer Feld 280, 69120 Heidelberg, Germany; mareike.roscher@dkfz-heidelberg.de

* Correspondence: m.benesova@dkfz-heidelberg.de; Tel.: +49-6221-42-5355

† These authors contributed equally to this work.

Abstract: Cancer is one of the most complex and challenging human diseases, with rising incidences and cancer-related deaths despite improved diagnosis and personalized treatment options. Targeted alpha therapy (TαT) offers an exciting strategy emerging for cancer treatment which has proven effective even in patients with advanced metastatic disease that has become resistant to other treatments. Yet, in many cases, more sophisticated strategies are needed to stall disease progression and overcome resistance to TαT. The combination of two or more therapies which have historically been used as stand-alone treatments is an approach that has been pursued in recent years. This review aims to provide an overview on TαT and the four main pillars of therapeutic strategies in cancer management, namely external beam radiation therapy (EBRT), immunotherapy with checkpoint inhibitors (ICI), cytostatic chemotherapy (CCT), and brachytherapy (BT), and to discuss their potential use in combination with TαT. A brief description of each therapy is followed by a review of known biological aspects and state-of-the-art treatment practices. The emphasis, however, is given to the motivation for combination with TαT as well as the pre-clinical and clinical studies conducted to date.

Keywords: targeted radionuclide therapy; targeted alpha therapy; external beam radiation therapy; immunotherapy; checkpoint inhibitors; cytostatic chemotherapy; brachytherapy; combination therapy; alpha-emitters; pre-clinical and clinical studies

Citation: Winter, R.C.; Amghar, M.; Wacker, A.S.; Bakos, G.; Taş, H.; Roscher, M.; Kelly, J.M.; Benešová-Schäfer, M. Future Treatment Strategies for Cancer Patients Combining Targeted Alpha Therapy with Pillars of Cancer Treatment: External Beam Radiation Therapy, Checkpoint Inhibition Immunotherapy, Cytostatic Chemotherapy, and Brachytherapy. *Pharmaceuticals* **2024**, *17*, 1031. https://doi.org/10.3390/ph17081031

Academic Editor: Irina Velikyan

Received: 12 June 2024
Revised: 22 July 2024
Accepted: 25 July 2024
Published: 5 August 2024

1. Introduction

In 1937, three years after Irène and Frédéric Joliot-Curie discovered that new radioelements can be produced artificially from non-radioactive material [1], researchers at the Massachusetts Institute of Technology (MIT) injected 48 rabbits with iodine-128 (I-128, $\tau_{\frac{1}{2}}$ = 25 min) retrieved after irradiation of stable iodine-127 using a radium-beryllium source [2,3]. They followed the accumulation of this short-lived beta minus- (β^-, 93.1%) and positron- (β^+, 6.9%) emitter over a time course of 30 min. The subsequent publication contemplated the use of radioactive isotopes of iodine in hyperplastic and neoplastic thyroids as both a diagnostic and therapeutic tool [2,4]. Later, sodium [^{131}I]iodide (Na[^{131}I]I) became the first U.S. Food and Drug Administration (FDA)-approved radiopharmaceutical in 1951 and can still be found in the guidelines for treatment of differentiated thyroid cancer [5–7].

In the time that followed, numerous radiotherapeutic agents were developed, starting from the incorporation of radioactive iodine isotopes into smaller and larger (bio)molecules. Nevertheless, targeted radionuclide therapy (TRNT) was used as a last-line resort treatment for patients that had either not responded to or progressed after the state-of-the-art treatment. The remarkable successes of [^{177}Lu]Lu-DOTA-TATE (Lutathera®, FDA-approved in 2018, Novartis, Basel, Switzerland) in neuroendocrine tumors and [^{177}Lu]Lu-PSMA-617 (Pluvicto®, FDA-approved in 2022, Novartis) in metastatic castration-resistant prostate cancer (mCRPC) [8–11] promoted the integration of TRNT into standard-of-care (SoC) practices for an even greater variety of cancer types and patients. Many reviews provide excellent overviews on available radiotherapeutic compounds and delivery strategies, including small molecules, peptidomimetics, labeled antibodies (and antibody fragments), nanoparticles, and pre-targeting vectors used in radioimmunotherapy [12–14].

Notwithstanding the diversity of radiotherapeutic sizes, targets, and pharmacokinetics, the application of these compounds to treat cancer share some common features that define TRNT. By analogy to chemotherapy or biologically targeted therapy, radiotherapeutics are administered systemically and accumulate in the tissue of interest by exploiting properties unique to tumor cells. A resulting advantage is the opportunity to treat heavily metastasized disease that is inaccessible to external-beam radiotherapy [14] or any other local intervention, like surgery. In contrast to other systemically administered therapies, the applied amount of radiotherapeutic substance is miniscule (typically in the nmol to pmol range), limiting the toxicity of TRNT to mostly radiobiological effects [15]. Biological targets can be located on the cancer cells themselves as well as in the tumor microenvironment. They include, but are not restricted to, cell surface receptors, adhesion molecules, enzymes, transporters, and ion channels. Additionally, radiotherapeutics can capitalize on enhanced permeability effects and redox potentials [16,17]. Tuning the targeting of radiotherapeutics based on tumor characteristics puts TRNT in a position to effectively meet the growing demand for personalized treatment of cancer patients [18].

Direct effects of radiation have traditionally been quantified by the extent of DNA damage caused at the target site, revealing therein a multitude of different cell death mechanisms, senescence, and other coping strategies induced by radiation damage. More recently, radiation effects beyond DNA damage have drawn attention. Examples of these effects include sub-lethal damage to other cellular components, radiation-induced bystander effect (RIBE), and the role of the immune system [19]. The extent and nature of radiobiological effects differs between types of radiation. For a long time, TRNT was restricted to the use of β^--emitting radionuclides. Depending on their energies, the emitted electrons have a tissue penetration depth of up to a few millimeters. The long tissue range has been considered advantageous in tumors with a heterogenous target expression because of the substantial cross-fire effect, which describes the irradiation of tumor cells that do not directly interact with the radiotherapeutic. From a radiobiological standpoint, the linear energy transfer (LET) of approx. 0.1–2 keV/µm mainly results in the formation of reactive oxygen species (ROS) through the excitation of water (and oxygen) molecules. In a subsequent step, ROS can cause oxidative damage to cellular components, including the DNA. The direct induction of DNA single strand (SSBs) and double strand (DSBs) breaks is less common.

Targeted alpha therapy (TαT), which constitutes a significant subcategory of TRNT, has become a rapidly expanding field in more recent years. Patients currently receiving TαT have exhausted all other available treatment options and/or have become resistant to ^{177}Lu- or ^{90}Y-based TRNT [20–22]. The success of TαT in these patients is attributed to the physical properties of the α-particles, notably their increased LET of approx. 50–300 keV/µm and the decreased tissue range of 20–70 µm—which can equal a single up to a few cell diameters. Multiple reports of breaking the resistance to other types of radiation as well as hormone deprivation and chemotherapy have been published. Additionally, the shorter tissue penetration depth of α-particles allows the treatment of tumors embedded in radiosensitive tissue, and the large amounts of energy deposited are favorable even for small and micrometastases [23–27]. The primary impact on the cellular level is the direct induction of DSBs, which are often formed in complex clusters. This renders TαT highly efficient regardless of the cell's oxygenation status [28]. It is assumed that a single hit of an α-particle can cause sufficient DNA damage to kill a cell [29,30]. In comparison to TRNT with β^- emitters and according to the current consensus, TαT exhibits an up to 5-fold increased radiobiological effectiveness (RBE), which allows a reduction in the administered activities by a factor of 1000. An early, prominent example of successful TαT is a study from 2016, in which the first two patients with advanced mCRPC were treated with [^{225}Ac]Ac-PSMA-617 [25]. The first patient was not eligible for treatment with [^{177}Lu]Lu-PSMA-617 due to diffuse red marrow infiltration, whereas the second patient had progressed after two cycles of [^{177}Lu]Lu-PSMA-617. After four and three cycles of 100 kBq/kg [^{225}Ac]Ac-PSMA-617, respectively, both patients were in radiological as well as biochemical remission as determined by a prostate-specific antigen (PSA) value below the detectable level. In addition, the first studies providing information on long-term outcome of a median five cycles of TαT with 100–150 kBq/kg [^{225}Ac]Ac-DOTA-TATE in patients with metastatic gastroenteropancreatic neuroendocrine tumors were reported. [^{225}Ac]Ac-DOTA-TATE treatment prolonged the progression and overall survival in end-stage patients and appeared to be safe with transient, low-grade side effects [31].

Notwithstanding the exceptional successes not only of TRNT and herein especially TαT but also other single-line treatments (external beam radiation therapy (EBRT), immunotherapy with checkpoint inhibitors (ICI), cytostatic chemotherapy (CCT), and brachytherapy (BT)), cancer remains a highly heterogenous disease that becomes increasingly difficult to cure the more it has advanced. Some degree of resistance and disease progression are observed in every therapy currently available to patients with advanced disease. Efflux pumps, downregulation of target expression or mutation of the target to prevent binding of the therapeutic agent, increased DNA repair, and hypoxia are only a few examples of the biological effects observed. Additionally, parts of the primary tumor and/or secondary sites may be inaccessible due to lack of vascularization or impenetrability to therapeutic agents. Dormancy or reversible senescence can cause relapse at a later stage, and the side effects may render the repetition of a monotherapy impossible. In clinical settings, the administration of TαT has been associated with several early treatment-emergent side effects, including xerostomia, anemia, leukopenia, thrombocytopenia, and nephrotoxicity. To overcome the limitations of standalone therapies like TαT, the field of oncology is more frequently turning toward the combination of two or more treatment options.

Figure 1 presents an overview of therapies that are either state-of-the-art or otherwise routinely employed in the treatment and management of cancer, including EBRT, immunotherapy with ICI represented by PD-1/PD-L1 blockage, CCT involving taxane-based substances, and BT. Each of these therapies can potentially be combined with TαT to complement (synergistic effect) or even enhance (additive effect) the efficacy of α-particle-emitting radionuclides. This review will provide a brief introduction into each technique and opportunities for its combination with TαT. We point out and summarize promising combination strategies for patients and give an outlook on future perspective of further combinations. With this review, we aim to ignite ideas for promising treatment strategies which could be tested in pre-clinical settings.

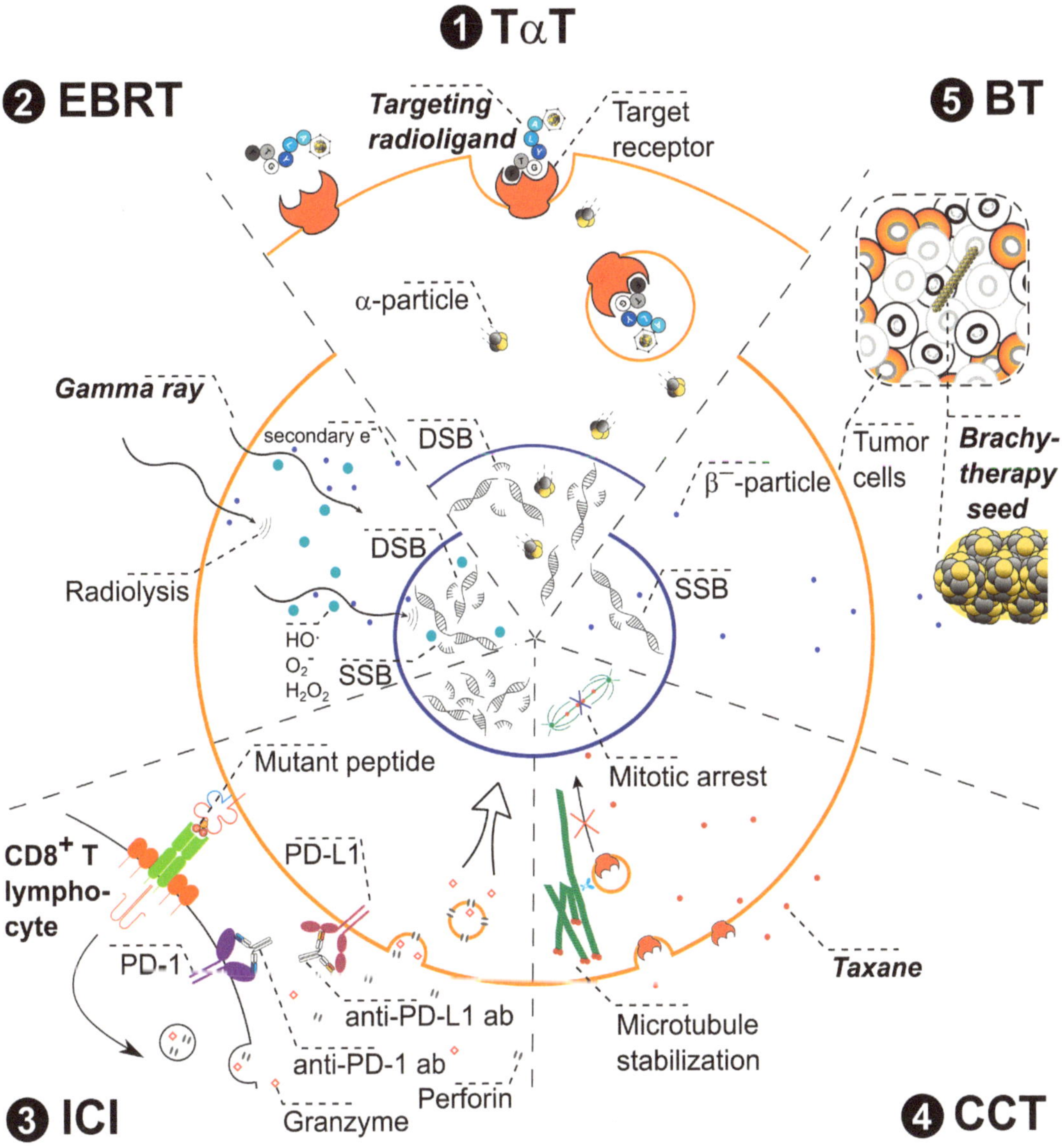

Figure 1. General overview of multiple local and systemic treatment strategies which could be beneficially combined with TαT. The upper middle section depicts TαT and its mechanism of action, while the remaining sections illustrate the four pillars of cancer therapies proposed to form a beneficial combination with TαT and their individual mechanism(s) of action. The orange circle represents a cell membrane and the blue circle a cell nucleus. Counterclockwise from top: **(1) Targeted alpha therapy (TαT)** employs ligands carrying α-emitting radionuclides which bind to cancer cells expressing a specific target. The high-LET radiation is responsible for a systemic cell killing via direct and indirect DNA damage. **(2) External beam radiation therapy (EBRT)** is based on the use of ionizing radiation from an external source. The beam facilitates cancer cell killing via direct and indirect DNA and protein damage, resulting from the generation of reactive oxygen species (ROS). **(3) Immune checkpoint inhibitor (ICI)**-based therapies facilitate cancer cell killing by inhibiting immune response escape mechanism, e.g. via the programmed cell death protein 1/programmed

death-ligand 1 (PD-1/PD-L1) system. **(4) Cytostatic chemotherapy (CCT)** relies on the use of cytostatic drugs, like taxanes, which induce mitotic arrest and interfere with other microtubule-related functions such as cellular trafficking. **(5) Brachytherapy (BT)** is based on the use of semi-solid therapeutic "seeds", bearing various radionuclides. The low- or high-LET radiation is responsible for a localized cell killing via direct and indirect DNA damage. Note: Due to space constraints, illustrated examples do not cover the full spectrum of possible vectors, mechanisms of action, and applications, but predominantly focus on the main approaches discussed in this review.

2. Comprehensive Overview of Targeted Alpha Therapy (TαT)

Key Facts:

- α-Emitters deposit high energies over short distances, classifying them as high-LET radiation.
- High-LET α-radiation induces complex DNA damage, resulting in direct ionization even in hypoxic tissue.
- Combining TαT with other treatments can address limitations such as target downregulation, tumor heterogeneity, as well as reduce side effects of either therapy alone.

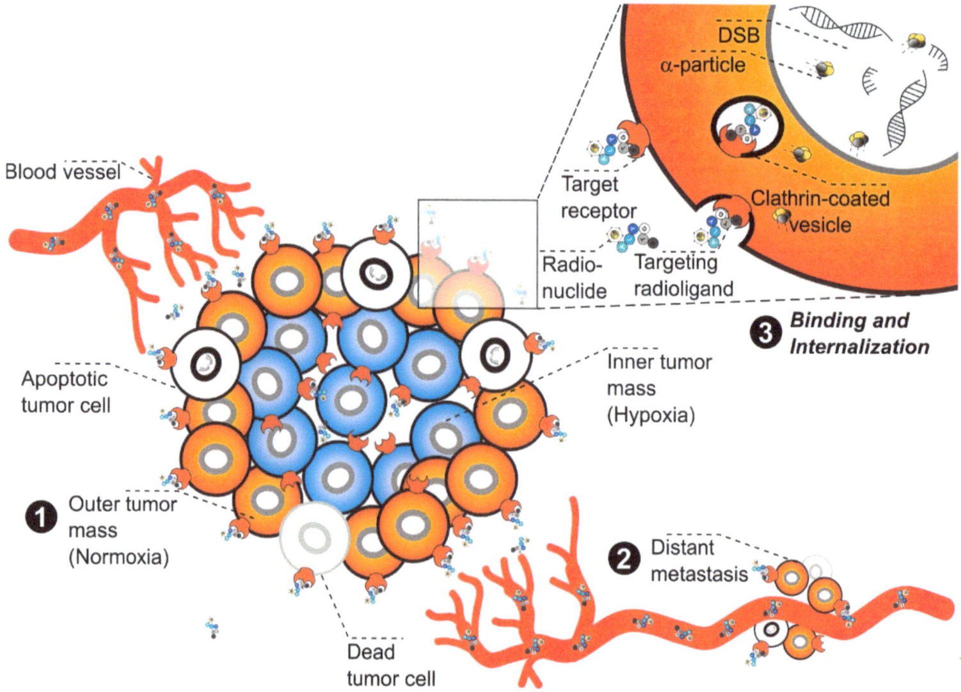

Figure 2. Graphical representation of the systemic TαT. The illustration depicts **tumor mass (1)**, **distant metastases (2)** along blood vessels, and a close-up on the **binding and internalization process** of targeting radioligands docking on target receptors **(3)**. Following intravenous injection, the radiolabeled ligand is carried to various cancer sites via the bloodstream. The targeting component enables selective binding to cancer cells through a specific target receptor. Following receptor binding, the ligand–receptor complex is internalized into the cancer cells. The high-LET radiation released upon decay damages the DNA of targeted cancer cells, mainly causing double-strand breaks (DSBs) that lead to apoptosis and other forms of cell death. Note: Due to space constraints, illustrated examples do not cover the full spectrum of possible vectors like antibodies, antibody fragments, nanoparticles, or non-conjugated α-emitters.

2.1. Physico-Chemical Properties and Therapeutic Implications of Alpha Particle-Emitting Radionuclides

With the exception of astatine-211 (At-211, $\tau_{\frac{1}{2}}$ = 7.21 h, 42.8% α-branching), which is a halogen and can be bound covalently to respective biovectors [32,33], all radionuclides used in TαT are radiometals. A typical pharmaceutical will therefore comprise a biovector addressing the target of interest and a chelating moiety to complex the radiometal. Most α-emitters with favorable properties for TαT have complex decay chains with several α- and β^-/β^+-emissions, leading to the designation of atomic nanogenerators or alpha in vivo nanogenerators [34,35]. Prominent examples include radium-223 (Ra-223, $\tau_{\frac{1}{2}}$ = 11.4 d, 4 α- and 2 β^--emissions), actinium-225 (Ac-225, $\tau_{\frac{1}{2}}$ = 9.9 d, 4 α- and 2 β^--emissions), and thorium-227 (Th-227, $\tau_{\frac{1}{2}}$ = 18.7 d, 5 α- and 2 β^--emissions). Other α-emitting radiometals commonly used in pre-clinical and clinical settings are lead-212/bismuth-212 (Pb-212/Bi-212, $\tau_{\frac{1}{2}}$ (Pb-212) = 10.64 h, β^--emission followed by α-decay with 35.9% probability), bismuth-213/polonium-213 (Bi-213/Po-213, $\tau_{\frac{1}{2}}$ (Bi-213) = 45.59 min, 2.1% α-branching and 97.9% β^--emission followed by α-decay), terbium-149 (Tb-149, $\tau_{\frac{1}{2}}$ = 4.12 h, 16.7% α-branching), and radium-224 (Ra-224, $\tau_{\frac{1}{2}}$ = 3.6 d, 4 α- and 2 β^--emissions) [36–38]. Emerging production and purification methods will likely increase the number of radionuclides available for TαT, as evidenced by recent studies with uranium-230 (U-230, $\tau_{\frac{1}{2}}$ = 20.2 d, 6 α- and 2 β^--emissions) [39,40]. Some of these radiometals form a sufficiently stable complex with the universal chelator 1,4,7,10-tetraazacyclododecane-1,4,7,10-tetraacetic acid (DOTA) for in vivo applications [38,41]. However, to achieve labeling at room temperature, which is paramount for heat-sensitive pharmaceuticals like antibodies used in radioimmunotherapy, tailored chelators have been developed for individual α-emitters. $[^{225}Ac]Ac^{3+}$, which is the largest trivalent cation in the periodic table, particularly benefits from macrocyclic chelators with increased ring size, including *N*,*N′*-bis[(6-carboxy-2-pyridil)methyl]-4,13-diaza-18-crown-6 (macropa) and its derivatives [42–44]. As another example, $[^{212}Pb]Pb^{2+}$ forms more stable complexes with 1,4,7,10-tetrakis(carbamoylmethyl)-1,4,7,10-tetraazacyclododecane (DOTAM/TCMC) than with DOTA [45,46]. For the formation of highly stable Th-227 conjugates, acyclic chelators like 3,2-hydroxypyridinone (3,2-HOPO) and derivatives are commonly used [47,48]. Chelation of Ra-223/224 has thus far proven challenging and with very few exceptions—macropa being the most prominent—no chelate complexes have the requisite stability for conjugation to a biovector and use in TαT. In vivo application of radium therefore remains largely limited to $[^{223}Ra]RaCl_2$, which mimics the binding behavior of calcium ions to areas of high bone turnover [49–52].

While in principle the molecular targets for TαT can be as versatile as those for TRNT, some additional considerations need to be made for radiometals with multiple α- and β^--emissions in the decay chain. The recoil energy of the first α-emission (>100 keV) is far greater than any chemical bond strength, causing release of the daughter radionuclide from the chelator [36,53–57]. Additionally, the daughter radionuclide may possess physical properties (i.e, charge, ionic radius) that are poorly matched with the chelating moiety. For example, the second daughter in the decay chain of Ac-225 is astatine-217 (At-217, $\tau_{\frac{1}{2}}$ = 32.63 ms), a halogen that cannot be complexed by the commonly used chelators. This is significant because redistribution of uncomplexed (grand)daughter radionuclides, which are still a source of highly toxic radiation, may otherwise contribute to long-term side effects for the patient due to unspecific accumulation in healthy organs and tissues. Internalization [58], local administration [59], or confinement of the radiopharmaceutical in a nanoparticle [60] are potentially advantageous strategies for confining ionizing radiation to the target site (Figure 2). The one exception to this paradigm is the bone-seeking agent Ra-223. Despite a half-life of 11.4 d and a decay chain comprising six consecutive α- and β^--decays, less than 2% of daughter radionuclides were released from the bone in mouse studies [61].

α-Emitting radionuclides decay by the emission of a helium nucleus (^4He)—consisting of 2 protons and 2 neutrons—and a +2 charge. The kinetic energy of these α-particles ranges from 5 to 9 MeV. By contrast, β^-- and β^+-particles reach a maximum kinetic energy of 2–3 MeV. Upon emission, α-particles travel at approx. 5–7% of the speed of light [62–66], only to rapidly lose their kinetic energy in a dense track of ionization events. Depending on the initial kinetic energy, the tissue penetration depth is typically between 20 and 70 μm, which is in the range of a few cell diameters. Taken together, α-emitters are thus classified as high-LET radiation. In the Karlsruhe Chart of Nuclides [67], the vast majority of radioisotopes that decay by α-emission are located among elements of greater atomic masses ($Z \geq 82$). In spite of the high energies, the low penetration depth in both organic and inorganic material presents a challenge for the direct detection of α-particles, and the necessary equipment is not readily commercially available. Hence, indirect detection of accompanying γ-emissions is often the method of choice [68]. Additionally, many radionuclides used in TαT are not stabilized by a single α-emission. The result is a complex decay chain and the transient formation of daughter nuclides, whose physical properties—including the physical half-life—differ greatly from the parent nuclide. Finally, in comparison to β^--emitters, for which the recoil energy rarely exceeds 25 eV, the recoil energy of an α-emission is non-negligible and, in many cases, greater than 100 keV [54]. This energy is deposited over a very short range (around 2000 Å), and the dose associated with the recoil energy therefore needs to be considered in dosimetry estimations.

2.2. Biological Mechanisms of Action and Cellular Response to TαT

Tumor response depends on the physical properties of the ionizing radiation and its radiobiological effects. First described for EBRT and brachytherapy, the four 'R's' describe the biological factors which influence the therapy response most: repair, repopulation, reoxygenation, and redistribution. The following paragraphs will explore the extent to which these factors, most prominently repair and reoxygenation, differ for α-particles compared to β^--emitters, thereby highlighting the unique potential of TαT for treating cancer.

Historically, the main biological effect of ionizing radiation has been considered to be damage to DNA [69]. In general, radiation can damage DNA in one of two ways: directly or indirectly. Low-LET radiation, including β^--emitters and EBRT, generates ROS through the ionization of water molecules [70]. As β^- energy is deposited over a range of a few millimeters, spanning tens to hundreds of cells, the number of ionization events per cell is relatively low. This is due to the small mass and charge of the electron emitted from the nucleus. Water is the most abundant molecule in a cell (averaging 70% by mass) and its ionization is therefore statistically favored. The generated ROS can then ionize other molecules in the cell, such as the DNA, which typically results in the formation of SSBs and less complex DSBs. DNA damage following exposure to low-LET radiation is enhanced in oxygenated tissue and suppressed in hypoxic areas of the tumor. Especially at low doses, TRNT with β^--emitters and EBRT are associated with a larger proportion of sub-lethal damage compared to TαT as SSBs are easier to repair than DSBs [71]. DNA damage inflicted by α-particles, on the other hand, is characterized by more complex and often clustered strand breaks. Because of the high energies deposited over a much shorter range, the ratio of ionization events per cell is greater, and the DNA can be directly ionized by α-particles, even in hypoxic tissue. Complex DSBs are challenging to repair, leading to a higher linearity in the relationship between dose and efficacy as the influence of sublethal damage decreases [72].

Even in the absence of radiation, the DNA is subject to damage from external and internal sources. In order to preserve integrity, the cell is equipped with a multitude of sophisticated repair mechanisms, each of which is tailored to repair a specific subset of DNA damage. In the context of DSBs caused by TRNT and TαT, the most important ones are homologous recombination repair (HR), non-homologous end joining (NHEJ), and alternative end joining (Alt-EJ, which includes several pathways) [73]. HR is the most accurate DNA repair pathway but is only active in the S- and G2-phases of the cell cycle as it requires an intact homologous strand of DNA [74]. NHEJ and Alt-EJ are potentially active throughout all phases of the cell cycle but are more prone to introduce errors during repair. DNA damage repair pathways are commonly impaired, especially in aggressive cancers [75].

DNA repair typically occurs in the time frame of a few minutes to several hours, depending on the pathway, the status of the cell cycle and the severity of the damage. There are multiple entities responsible for sensing DNA damage, some of which are used as biological markers for treatment response. The most widespread example is the formation of γH2AX foci, which can be detected by specific antibodies. As a response to DNA damage, the cell halts the progression of the cell cycle, causing accumulation of cells at the G2/M checkpoint. Cell cycle arrest is another factor that is often analyzed during preclinical evaluation. In normal cells, unrepaired damage results in cell death via p53- and apoptosis executioner caspases 3- and 7-mediated pathways. In cancer cells, this pathway is often defective, and the cells continue replicating without proper DNA repair. Chromosome breaks and additional chromosomal aberrations are the result, and these ultimately cause mitotic catastrophe and the formation of micronuclei. Other, less common cell death pathways which may be triggered in response to ionizing radiation include necrosis, necroptosis, autophagy-dependent death, pyroptosis, and ferroptosis. In addition, radiation-induced senescence has been observed, where it has been argued to contribute to a favorable outcome or to be an indicator of radiation resistance [76,77].

Several studies, many of them more recent, have argued that DNA cannot be the only target of ionizing radiation. Damage to the cellular membrane as well as organelles has been claimed to be an integral part of the biological response to radiation. It is also unlikely that all cells in a tumor mass die due to direct radiation exposure. On a macroscopic level, RIBE as well as activation of the immune system have been observed. While often confined to the immediate surrounding of the irradiated cells, even systemic effects can be elicited, including what is called the abscopal effect [78].

In summary, the physical differences of α-radiation in comparison to β^--, γ- and X-ray radiation shifts the importance of several biological factors in characterizing the response to TαT. The high-LET α-particles cause complex, often lethal DSBs independently of oxygenation status but typically traverse only 3–5 cells. Moreover, the complex decay chain and significant recoil energies typical of α-emitters add complexity to the planning of dosimetry calculations relevant to the treatment schedule.

2.3. *Current Advances and Innovations in TαT*

In May 2013, radium-223 dichloride ($^{223}RaCl_2$, Xofigo®, Bayer, Leverkusen, Germany) became the first FDA-approved radiotherapeutic for α-therapy in mCRPC patients with symptomatic bone metastases and no known visceral metastases [49,50]. Half a year later, the European Medicine Agency (EMA) granted its authorization for the use in Europe [79]. $^{223}Ra^{2+}$ mimics the physiological behavior of Ca^{2+} by binding hydroxyapatite and thus accumulates in areas of bone metastases associated with a high bone turnover [80]. Between 2014 and 2019, a global, prospective, observational study of Xofigo® in patients with mCRPC was carried out (REASSURE, NCT02141438) to evaluate long-term safety, clinical benefit, and treatment patterns. Counting from the first cycle of Xofigo® administration, median overall survival was 15.6 months. The final analysis will be reported in 2024 [81]. In addition to its use as monotherapy, the efficacy of Xofigo® in combination therapies is subject to ongoing clinical trials advancing the scope beyond prostate cancer (PCa) [49]. Combinations with chemotherapy, immunotherapy, and other types of radiation therapy have been reported and will be covered in greater detail in the individual sections of this review. As of 2024, no other TαT has been approved by the FDA or the EMA, but many are being explored in ongoing clinical trials [82].

In the early days of TαT, therapeutic antibodies conjugated to short-lived radioisotopes with a single α-emission (Pb-212, Bi-212, Bi-213, and At-211) were the most frequently used targeting vectors [80,83–86]. In recent years, longer-lived radioisotopes with more complex decay chains have superseded the shorter lived α-emitters. Their physical half-life aligns more efficiently with the biological half-life of the vector and the emission of multiple ionizing daughter nuclides increases the treatment efficacy whilst lowering the required doses. The value of Ac-225 for α-radioimmunotherapy was recognized as early as 1993 [87], resulting in a phase I clinical trial of a single cycle of [^{225}Ac]Ac-lintuzumab to treat 18 patients with relapsed or refractory acute myeloid leukemia [84,88,89]. Although treatment decreased tumor burden in a majority of patients, dose-limiting toxicities were observed in the highest dose groups [89,90]. In a subsequent phase I/II clinical trial of [^{225}Ac]Ac-lintuzumab in patients with untreated AML (NCT02575963), the prevalence of adverse effects required further dose reduction during the study [82,91]. This pioneering work spawned a number of subsequent studies of ^{225}Ac-labeled antibodies for TαT. For example, [^{225}Ac]Ac-J591, a monoclonal antibody targeting PSMA for treatment of mCRPC, has just completed a phase I dose escalation study (NCT03276572) in 32 patients. Although a biochemical response (PSA decrease $\geq$ 50%) was only observed in half of the patients, the maximum tolerated dose was not reached in this cohort [92]. Two additional monotherapy studies are also ongoing to assess the possibility of re-treatment for patients who responded to the first dose but then progressed (NCT04576871) and to explore the efficacy of dose fractionation over multiple treatment cycles (NCT04506567) [93]. Simultaneously, [^{225}Ac]Ac-J591 is under investigation for combination with [^{177}Lu]Lu-PSMA I&T (NCT04886986) and pembrolizumab, an androgen receptor signaling inhibitor (NCT04946370) [94], which will be covered in more detail in Section 4. Other examples include [^{225}Ac]Ac-FPI-1434, a humanized monoclonal antibody targeting the insulin-like growth factor type I receptor (NCT03746431) [95–98], and the ^{227}Th-labeled anti-HER2 antibody BAY 2701439, which just completed a first-in-human study (NCT04147819) [99]. Excellent and more thorough summaries of ongoing clinical trials are available in other reviews [33,82,83].

The FDA-approval of Lutathera® ([^{177}Lu]Lu-DOTATATE, Novartis) and Pluvicto® ([^{177}Lu]Lu-PSMA-617, Novartis) for TRNT has created a significant momentum for the potential translation of rapidly distributing vectors, e.g., small molecules and peptides, to TαT. Currently, the conjugation of Ac-225 to the aforementioned vectors is one of the most frequently used strategies. The net α-energy of 28 MeV exhibited by Ac-225 strongly exceeds the therapeutic capacity of Lu-177, which possesses a maximum β^--energy of 0.5 MeV. Furthermore, administered activities of the ^{225}Ac-labeled radiotherapeutic are significantly lower, with only one thousandth of the dose required for the corresponding ^{177}Lu-labeled compound (kBq/kg vs. MBq/kg). After the initial reports in 2016 of successful treatment of mCPRC patients with [^{225}Ac]Ac-PSMA-617 that no longer responded to other treatments [20,25], follow-up studies found this radiotherapeutic to be efficacious and safe as a last-line treatment in patients with mCRPC in a variety of settings [100,101]. These promising results led to an open label phase I multicycle dose escalation study seeking to recruit 60 patients with PCa, who will be subdivided in three groups according to prior treatment with chemotherapy, androgen receptor pathway inhibitors, and [^{177}Lu]Lu-PSMA-617 TRNT (AcTION, NCT04597411) [102]. In parallel, a phase II clinical trial is currently evaluating the efficacy of 100 kBq/kg [^{225}Ac]Ac-PSMA I&T (FPI-2265), another small molecule targeting PSMA, which will be administered to enrolled patients with mCRPC in four cycles every 8 weeks (TATCIST, NCT05219500).

Another example of small molecule Ac-225 TαT is the treatment of somatostatin receptor 2 (SSTR2)-expressing tumors. [^{225}Ac]Ac-DOTATATE (RYZ101), an SSTR2 agonist, is currently being evaluated in a phase Ib/III clinical trial in patients with SSTR2-positive gastro-enteropancreatic neuroendocrine tumors that have progressed after treatment with [^{177}Lu]Lu-labeled agonists and compared to SoC (ACTION-1, NCT05477576). Patients will receive up to four cycles every 8 weeks. The first part of the study is designed as a dose de-escalation study starting with 120 kBq/kg. In the first report released in May 2023, nine patients had been enrolled and no dose-limiting toxicities or serious treatment-related adverse effects were observed. Adverse effects requiring dose reduction occurred in two patients (decrease in grade 2 platelet count and grade 2 thrombocytopenia) [103]. The first dosimetry results of four patients were reported in June 2023 [104]. Aside from the comparison to SoC, [^{225}Ac]Ac-DOTATATE (RYZ101) is also being investigated as an addition to SoC consisting of carboplatin, etoposide, and atezolizumab (NCT05595460). The Pb-212 analogue, [^{212}Pb]Pb-DOTAMTATE (AlphaMedix), has recently completed a phase I study (NCT03466216) and is currently being evaluated in a multicenter phase II clinical trial (NCT05153772). Twenty patients were enrolled in the phase I study, none of whom had been pre-treated with TNRT. The trial was designed as a single ascending dose study, which turned into a multiple ascending dose study once response was observed. Ten patients received the highest dose (2.5 MBq/kg), with 80% of the patients presenting with a radiologic response. No serious treatment-emergent adverse effects were recorded, with only mild nausea, fatigue, and alopecia reported [105]. Moreover, four cycles of [^{212}Pb]Pb-DOTAMTATE therapy were effective even in patients with progression under prior treatment of Lu-177 or Y-90 TRNT [106].

In addition to the aforementioned examples, a number of small molecules or peptides conjugated to Pb-212, Bi-213, or Ac-225 are in active or recently completed clinical trials. Examples include a phase I dose escalation study of [^{212}Pb]Pb-DOTAM-GRPR1 in metastatic tumors positive for the gastrin releasing peptide receptor (NCT05283330), a phase I/IIa clinical trial using [^{212}Pb]Pb-VMT-α-NET in advanced SSTR2-positive neuroendocrine tumors (NCT05636618), and a first-in-human study of [^{212}Pb]VMT01 in patients with unresectable metastatic melanoma expressing the melanocortin sub-type 1 receptor (NCT05655312). A first-in-human phase I study in patients with advanced, metastatic, or recurrent solid tumors expressing the neurotensin receptor 1 is currently assessing the safety, tolerability, and dosimetry of [^{225}Ac]Ac-FPI-2059 (NCT05605522). [^{177}Lu]Lu- and [^{90}Y]Y-labeled small molecules targeting the fibroblast activation protein-α are well established in the clinical trial landscape [107]; however, their translation to TαT is still in the preclinical phase [108–111]. Substance P is the natural ligand of the neurokinin type 1 (NK-1) receptor and can be conjugated with a metal chelator and a suitable radioisotope [112]. Phase I and II studies have been undertaken with both ^{213}Bi- and ^{225}Ac-labeled Substance P analogues [113]. In 2019, 20 patients presenting with recurrent glioblastoma multiforme received between one and seven cycles of [^{213}Bi]Bi-DOTA-SP via local administration [114]. The median survival time from the first cycle was 7.5 months, and no severe adverse effect were reported. In 2021, Królicki et al. additionally reported the treatment of 21 glioblastoma patients with 10, 20, or 30 MBq of [^{225}Ac]Ac-DOTA-SP in one–six cycles. Overall survival from start of treatment was 9.0 months, and the only grade 3 adverse events (thrombopenia) occurred in patients in the 30 MBq group [59]. A recently published study comparing the efficacy of [^{213}Bi]Bi- with [^{225}Ac]Ac-DOTA-SP found no significant difference in survival between the groups. However, the median overall survival from primary diagnosis is increased to 35.0 month with both TαTs, compared to standard treatment (surgery, radiotherapy, and chemotherapy) with a median overall survival of up to 9–15 months [115].

The challenges presented by recoiling daughter nuclides, the complex decay chains of the longer-lived radioisotopes as well as the inefficient complexation of Ra-223 and Ra-224 by available chelators have initiated the advancement of targeted and non-targeted nanoparticles in TαT. In theory, the encapsulation of α-particle emitting radionuclides into nanoparticles followed by intracellular delivery due to endocytosis-like mechanisms would address many of the challenges currently encountered in TαT. However, the field of nanomedicine needs to overcome certain obstacles prior to successful translation to clinical practice. The biodistribution profile remains a major concern, as it is strongly dependent on the physical properties of applied nanoparticles, e.g., size, surface, shape, or material, and often leads to accumulation in the liver, spleen, and lungs. Other challenges include the lack of production infrastructure, regulatory issues, and toxicity considerations. The vast preclinical landscape of nanoparticles and its challenges regarding clinical translation have already been thoroughly discussed in excellent review articles published to date [116–118].

2.4. Rationale for Combining TαT with Other Cancer Therapies

TαT has proven highly successful as a last-line treatment. Tumor control or reduction of tumor burden have been accomplished even in patients whose disease had become resistant to chemotherapy and other types of radiotherapy. Furthermore, TαT demonstrated good anti-tumor effect as a first-line treatment in chemotherapy-naïve metastatic hormone-sensitive PCa [119]. This is especially relevant for patients of higher age and related comorbidities who do not tolerate chemotherapy treatment. Nevertheless, there are reports of patients who either responded poorly to TαT or who even appear to show an early onset of resistance against TαT [120]. Additional investigations in these patients found mutations in genes associated with DNA damage repair [121] leading to the hypothesis that combination therapy with other pillars of cancer treatment could present a strategy in overcoming such resistances.

As with every other cancer therapy in clinical use, TαT has its limitations. Being a subcategory of TRNT, some potential shortcomings of TαT are closely tied to the general nature of TRNT. These include the limitation that downregulation or mutations in the binding site of the target of interest will reduce the accumulation of the radiotherapeutic in the cancer tissue and thus the delivered dose. Lack of vascularization or tumor heterogeneity can lead to non-homogeneous irradiation of the cancer tissue with some parts receiving sublethal doses. Especially for TαT, the cost and lack of widespread availability of the α-emitters need to be considered [32,122,123], which limits the number of patients that can be treated. Additionally, the dose calculations are highly complex, especially for radionuclides with multiple α- and β^--emissions in the decay chain. Redistribution of (grand)daughter radionuclides not only are difficult to quantify but also contribute to potential side effects due to irradiation of off-target tissue and excreting organs. This is significant because acutely dose-limiting toxicities have been described for [^{225}Ac]Ac-PSMA-617 (salivary glands, irreversible xerostomia) [124], [^{225}Ac]Ac-DOTA-TOC (chronic renal toxicity) [125], and Xofigo® (substantial hematotoxicity) [126]. Moreover, long-term toxicities, which can occur up to a few years after the end of treatment, are still being assessed. Finally, although the incidence is currently lower than for other therapies, resistance to TαT and relapse shortly after treatment have been observed in some patients. The mechanism(s) of resistance, however, are not yet well understood and call for further elucidation.

The limitations of TαT can potentially be addressed by combination with other monotherapies. The choice of combination therapy will vary according to the problem at hand, but the variety of therapies available enhances the likelihood of an additive or even synergistic effect. As an example, the combination with EBRT or TRNT using β^--emitting radionuclides could lessen the effect of tumor heterogeneity on the therapy outcome. EBRT works independently of any target expression, either on the tumor cells themselves or in the tumor microenvironment. Heterogeneous target expression or its downregulation would therefore no longer confer an advantage to the cancer cells. As a second example, the use of TRNT with a β^--emitter in combination with TαT would increase the potential tumor cell kill via the cross-fire effect and more homogeneous tumor irradiation. Simultaneously, it may be possible to reduce the dose of each radiotherapeutic required, which would reduce any toxic side effects and, potentially, expand the availability of the TαT agent to a greater number of patients. On another perspective, EBRT cannot be used for treatment of oligometastasized patients; only the primary tumor and a few well-defined metastases can be irradiated. If combined with TαT, the smaller metastases would receive a simultaneous dose during EBRT treatment of larger tumor masses, circumventing the metastases' ability to grow out and cause a relapse. In analogy, this can be applied to TRNT as β^--emitters cannot deposit sufficient energy in small metastases to deliver a lethal dose, yet in combination with TαT, this could be possible. Numerous practical examples exist and will be discussed in greater detail in the subsequent sections.

The overarching theme of this review is to highlight the major impact that combining TαT with different therapeutic strategies can have on the treatment of cancer patients. Instead of starting with one therapy until it is no longer effective before switching to an alternative, the combination of several strategies might leave fewer loopholes for the cancer to escape treatment while simultaneously protecting healthy tissue. Additionally, the therapy can be better tailored to the patient's needs. For example, if a patient already has an impaired kidney function, TαT monotherapy might lead to renal failure. But if the dose could be substantially reduced by combination with EBRT, the kidneys may be spared while the effect on the tumor will be similar if not enhanced.

2.5. Pre-Clinical and Clinical Insights: TαT and TRNT

To date, only a handful of pre-clinical and clinical studies combining TRNT with a β^--emitter and TαT (or two therapeutic radionuclides in general) in so-called tandem therapies have been conducted, as outlined in Sections 2.3 and 2.4. In the following, the focus lies on the current status of combined radionuclide therapy (Table 1).

A phase I/II clinical trial combining [^{225}Ac]Ac-J591 and [^{177}Lu]Lu-PSMA I&T for men with progressive mCRPC is currently ongoing at Weill Cornell Medical College (NCT04886986). The dose escalation study (phase I) enrolling 18 patients was recently completed and the results reported at the yearly meeting of the American Society of Clinical Oncology (ASCO). Up to 35 kBq/kg [^{225}Ac]Ac-J591 in combination with 6.8 GBq [^{177}Lu]Lu-PSMA I&T were safely administered over two cycles with 8 weeks intermission. Increasing the dose of [^{225}Ac]Ac-J591 to 40 kBq/kg led to dose limiting toxicity (grade 2 or 3 thrombocytopenia) in two out of six patients after the first cycle. Seventeen of the eighteen patients had a PSA decline of any nature, and the decline was >50% in eleven patients. In addition to the therapeutic benefit of this combination approach, the study documented non-overlapping toxicities courtesy of the different pharmacokinetics of PSMA I&T (small molecule, ~1.5 kDa) and J591 (monoclonal antibody, ~150 kDa). [^{177}Lu]Lu-PSMA I&T transiently accumulates in organs with PSMA expression (e.g., tumor, salivary glands, kidneys, spleen) but clears in a few days. [^{225}Ac]Ac-J591, on the other hand, has a long circulation time, depositing a lower dose in the salivary glands and kidneys but a higher dose to the marrow. In the phase II, the effectiveness of the combined treatment will be assessed; study completion is expected in 2027 (NCT04886986). Taken together, both radiotherapeutics could contribute to the tumor dose while sparing other organs due to their different pharmacokinetic properties [127].

A second active clinical phase I/II trial is evaluating the safety of [^{177}Lu]Lu-PSMA I&T in combination with Xofigo® (AlphaBet, NCT05383079). The study is recruiting men with mCRPC which have progressed on second-generation androgen receptor agonists and were diagnosed with at least two untreated bone metastases. The hypothesis being tested is that [^{177}Lu]Lu-PSMA I&T will treat metastases in soft tissue while Xofigo® will simultaneously treat micrometastases in the bone, as these do not receive sufficient dose with [^{177}Lu]Lu-PSMA I&T alone. Patients are receiving 7.4 GBq [^{177}Lu]Lu-PSMA I&T for up to six cycles with 6 weeks intermission. A dose of 27.5 kBq/kg or 55 kBq/kg Xofigo® will be administered between day 1 and day 5 of every cycle [128]. A preclinical study using the intratibial LNCaP xenograft model previously demonstrated robust anti-tumor efficacy of the combination therapy [129].

Several clinical and preclinical studies have assessed the feasibility of combining [^{225}Ac]Ac-PSMA-617 and [^{177}Lu]Lu-PSMA-617. In a preceding dosimetry study, 35 MBq [^{177}Lu]Lu-PSMA-617 and 40 kBq [^{225}Ac]Ac-PSMA-617 were calculated to deliver an equivalent dose to a subcutaneous xenograft model with C4-2 cells. Treatment of tumor-bearing mice with either 40 kBq [^{225}Ac]Ac-PSMA-617 alone or a combination of 20 kBq [^{225}Ac]Ac-PSMA-617 and 17 MBq [^{177}Lu]Lu-PSMA-617 was found to significantly enhance overall survival of the mice, irrespective of the tumor size at the start of the treatment. By contrast, 35 MBq [^{177}Lu]Lu-PSMA-617 alone had a beneficial effect on survival when treatment was started 5 weeks after inoculation (macroscopic disease) but not after 3 weeks (microscopic disease) [130]. The combination of [^{177}Lu]Lu-PSMA-617 and [^{225}Ac]Ac-PSMA-617 was also explored in a pilot study involving 20 patients with mCRPC who had progressed after treatment with [^{177}Lu]Lu-PSMA-617 alone. All patients received one cycle of tandem therapy with activities ranging from 1.5 to 7.9 MBq [^{225}Ac]Ac-PSMA-617 and 5.0 to 11.6 GBq [^{177}Lu]Lu-PSMA-617. In patients with a good response, tandem therapy was followed up with [^{177}Lu]Lu-PSMA-617 monotherapy to further manage the disease. Only 2 patients were diagnosed with progressive disease while 16 patients showed a PSA response, and for 13 of these patients, the PSA decrease was >50%. The study demonstrates that a response to Ac-225/Lu-177 tandem therapy is still possible even if patients have acquired resistance to [^{177}Lu]Lu-PSMA monotherapy [131]. In a follow-up study, 15 TRNT-naïve mCRPC patients with visceral metastases, large tumor burden, and short PSA doubling times were subjected to two cycles of [^{177}Lu]Lu-PSMA-617, at least one of which was given in combination with [^{225}Ac]Ac-PSMA-617. Partial remission was observed in 8–10 patients, depending on the type of analysis (based on PSA decline or [^{68}Ga]Ga-PSMA-11 imaging). Two patients experienced grade 3 anemia and another two were diagnosed with grade 1/2 xerostomia [132].

In 2019, another study of combination therapy using 3–5 MBq [^{225}Ac]Ac-PSMA-617 and 3.5–7.5 GBq [^{177}Lu]Lu-PSMA-617 in 30 patients with end-stage PCa was reported. In this case, the objective was to reduce the side effects of Ac-225 monotherapy, especially with respect to dose limiting xerostomia. A total of 23 patients were monitored for 8 weeks after one cycle. Although two patients reported with G3 thrombocytopenia and G2 anemia/leukocytopenia, no severe xerostomia was observed. A PSA decline of >50% was found for 10 patients, and 3 patients experienced complete remission [133]. In a follow-up, salivary gland function was analyzed more closely in 18 patients who underwent the tandem therapy and were compared to patients who received [^{177}Lu]Lu-PSMA-617 or [^{177}Lu]Lu-PSMA I&T monotherapy. Using dryness of mouth assessment (CTCAE v.5.0), standardized questionnaire (sXI), salivary gland scintigraphy (SGS), and [^{68}Ga]Ga-PSMA-11 PET/CT as metrics, the investigators found that one cycle of Ac-225/Lu-177 tandem significantly impaired function while Lu-177 alone had only minor toxic effects. The tandem concept of co-administrating lower doses of [^{225}Ac]Ac-PSMA-617 in combination with [^{177}Lu]Lu-PSMA-617 can be a measure to decrease severe xerostomia [134,135].

Lastly, a pilot preclinical study has explored the combination of two α-emitters with different targeting properties. In this work, an EGFR-targeting monoclonal antibody (rituximab) was labeled with Pb-212 in a solution that also contained Ra-224. While the monoclonal antibody specifically targets (cancer) cells that overexpress EGFR, the additional bone-seeking properties of Ra-224 are able to simultaneously treat osseous lesions. The anti-tumor effect was confirmed in a metastatic breast cancer mouse model. The same combination also inhibited the growth of LNCaP spheroids [136,137].

2.6. Future Perspectives and Challenges

The benefits of targeted delivery of highly cytotoxic α-particles to metastasized lesions is already demonstrating efficacy in patients with advanced disease, even when resistance to other monotherapies has developed. Nevertheless, shortcomings are not to be neglected as TαT may be inefficient in targeting larger tumors or tumors with heterogeneous target expression. Furthermore, severe side-effects can greatly impact the quality of life, with the limited supply of α-emitting radionuclides crucially hindering availability to all eligible patients.

An amplified therapeutic use of TαT by combination with other treatments could effectively address these limitations and improve outcomes in cancer patients. This premise is already being evaluated by combination of TRNT and TαT, with early reports indicating that side effects can be reduced without compromising efficacy. A detailed summary of possible future combinations of TαT with other therapies is thoroughly discussed in the following sections.

Table 1. Overview of most recent and relevant publications on Targeted Alpha Therapy (TαT) as outlined in Section 2. To facilitate the orientation, the table groups references on TαT as a monotherapy first, followed by references on combination of TαT and other TRNT. Note: Only the most recent and relevant publications are included in the table to provide readers with a solid overview. More details can be found in Section 2.

Method	Author	Year	Study	Cancer Type	TRNT Agent	Main Findings	Ref.
TαT	Heynickx et al.	2021	Review	mCRPC	Various PSMA radioligands	Reviews salivary gland toxicity as a dose-limiting factor in PSMA-TRNT.	[124]
	Soldatos et al.	2019	Clinical	Bone metastases in PCa	Xofigo®	Toxicological profiling of Ra-223 for the treatment of bone metastases in PCa.	[126]
	Kratochwil et al.	2021	Clinical	SSTR tumors	[^{225}Ac]Ac-DOTA-TOC	Five-year follow-up study on hematological and renal toxicity in patients.	[125]
	Sathekge et al.	2023	Clinical	mCRPC	[^{225}Ac]Ac-PSMA-617	Treatment of chemotherapy-naïve patients: 20 patients (95%) any decline in PSA; 18 patients (86%) PSA decline of $\geq$50%	[119]
TαT & TRNT	Meyer et al.	2023	Pre-clinical	PCa (C4-2)	[^{225}Ac]Ac-PSMA-617, [^{177}Lu]Lu-PSMA-617	Combination therapy enhanced overall survival in mice models.	[130]
	Scholz et al.	2023	Pre-clinical	PCa (LNCaP bone metastasis)	Xofigo®, [^{177}Lu]Lu-PSMA-617	Demonstrated increased antitumor activity in combination therapy.	[129]
	Kulkarni et al.	2019	Clinical	mCRPC	[^{225}Ac]Ac-PSMA-617, [^{177}Lu]Lu-PSMA-617	Combination therapy reduced side effects of Ac-225 monotherapy.	[133]
	Khreish et al.	2019	Clinical	mCRPC	[^{225}Ac]Ac-PSMA-617, [^{177}Lu]Lu-PSMA-617	Demonstrated PSA response in patients progressing after Lu-177 monotherapy.	[131]
	Rosar et al.	2021	Clinical	mCRPC	[^{225}Ac]Ac-PSMA-617, [^{177}Lu]Lu-PSMA-617	Partial remission observed. Efficacy and safety established in patients with poor prognosis.	[132]
	Langbein et al.	2019, 2022	Clinical	mCRPC	[^{225}Ac]Ac-PSMA-617, [^{177}Lu]Lu-PSMA radioligands	Investigated salivary gland toxicity. Combination therapy significantly impaired function compared to Lu-177 alone.	[134,135]
	Kostos et al.	2022	Clinical	mCRPC	[^{177}Lu]Lu-PSMA I&T, Xofigo®	Evaluating safety and efficacy. Hypothesis: combined treatment will be more effective for treating metastases.	[128]
	Tagawa et al.	2023	Clinical	mCRPC	[^{225}Ac]Ac-J591, [^{177}Lu]Lu-PSMA I&T	Safe administration with observed PSA decline > 50% in 11 patients. Non-overlapping toxicities noted.	[92]

3. Integration of External Beam Radiation Therapy (EBRT) with Targeted Alpha Therapy (TαT)

Key Facts:

- Combination therapy of EBRT and TαT can increase the dose delivered to the tumor while protecting healthy tissue. Its effectiveness in hypoxic tissue can overcome resistance to EBRT in poorly oxygenated tumors as well as treat oligometastatic disease.
- Clinical studies combining Xofigo® (Ra-223) with SBRT in osteosarcoma and mCRPC demonstrate improved outcomes.

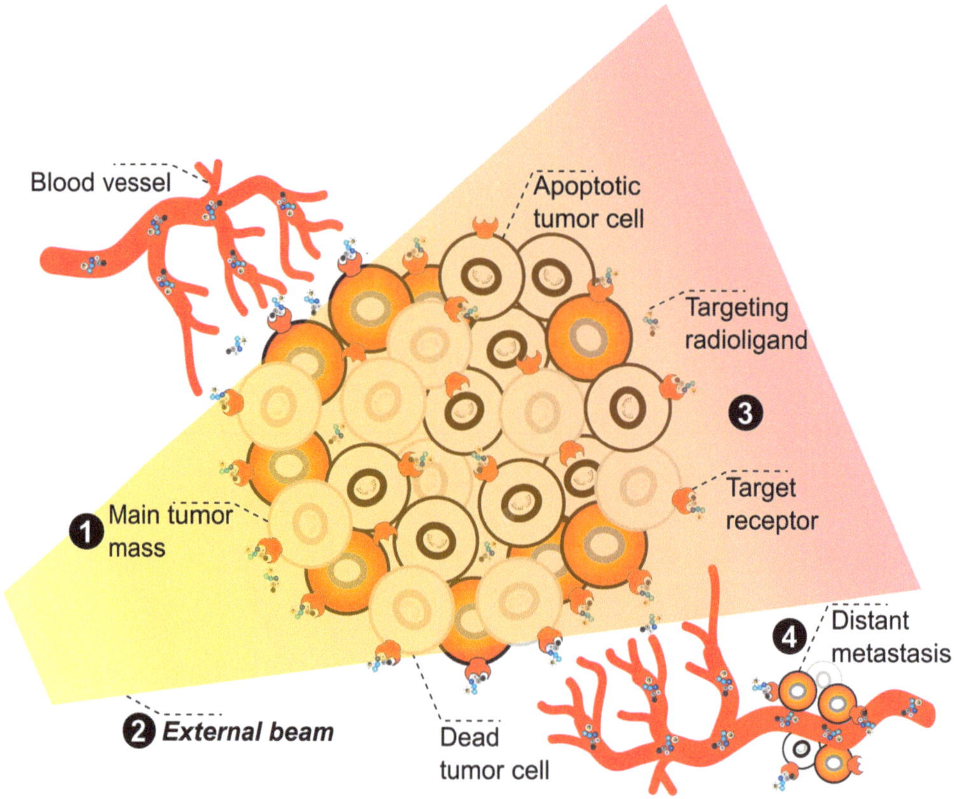

Figure 3. Graphical representation of local EBRT and systemic TαT combination therapy. The illustration depicts a **tumor mass (1)** irradiated by the **external beam** (photon irradiation) **(2)** in the form of a cropped triangle. **Targeting radioligands** docking on target receptors of the tumor cells **(3)** are addressing the main tumor mass as well as **distant metastases (4)**. Local EBRT indirectly induces DNA damage in the tumor mass. Along the radiation beam path, reactive oxygen species (ROS) are produced by ionization of water molecules. These ROS interact with the surrounding biomolecules and thereby generate DNA damage. Complex DNA damage leads to cell death. During its penetration into the tissue, the beam constantly loses energy by interaction with water (and other) molecules. As such, the penetration depth depends on the total beam energy and should be adapted to the tumor location. The beam also traverses any healthy tissue located between its entry point in the skin and the tumor as well as some tissue surrounding the tumor, which is subject to DNA damage and cell death. By contrast, the targeting compound of TαT is distributed via the blood stream and induces DNA damage primarily in those cells expressing the intended target. Moreover, systemic administration of the TαT agent will allow the targeting of distant metastases, micrometastases, and

circulating tumor cells (CTCs) untreatable by EBRT. To induce the highest damage possible and to overcome radioresistance related to hypoxic conditions in the inner tumor mass, parallel/combined treatment with EBRT and TαT of the primary or local recurrent tumor mass would be favorable. The pre-treatment with EBRT could render the inner tumor mass easier accessible for TαT and, by that, enhance its efficiency. By analogy, external irradiation of the primary or recurrent tumor mass could decrease the necessary dose for TαT together with the potential TαT-related side effects. The combination of EBRT with TαT could therefore increase the dose in the tumor volume while lowering the total dose delivered to the healthy tissue by reducing the overall dose of the EBRT. Note: Due to space constraints, illustrated examples do not cover single, well-defined metastases which could also be addressed by EBRT alone.

3.1. Comprehensive Overview of EBRT in Cancer Treatment

EBRT is an important pillar of cancer treatment [138] administered to approximately half of all patients in Europe diagnosed with cancer [139,140]. Additionally, EBRT is also one of the most cost-effective cancer treatments, making up only about 5% of the total cost of cancer treatments [141] in spite of its extensive use. Cancer patients have been treated with radiotherapy for over a century, with the first treatment [142] recorded shortly after the discovery of X-rays by Wilhelm Conrad Roentgen in 1896 [143]. In general, radiotherapy is available globally, although, due to the specialized infrastructure and the significant initial investments for the machines, radiotherapy treatment units are far more abundant (10.1 units per million population) in high-income regions like Europe and the USA as opposed to lower-income regions like Africa or Latin America (0.01–2.6 units per million population) [144,145]. About 30% of all radiotherapy centers are located in the USA and 19% in Europe [146].

In contrast to systematic therapies like TαT, EBRT represents a local treatment, only applied in the area of the primary or recurrent tumor or to treat a few, well-defined metastases. Its efficacy is based on the induction of highly complex and difficult-to-repair DNA damage, such as clustered DSB (cDSBs). The frequency of more complex damage including cDSBs increases if the beam contains heavy particles, due to their high LET. Additionally, particle therapy benefits from a more defined irradiated area due to the particularity of the Bragg Peak, which enables the dose deposition within a sharp maximum [147]. The most commonly used sources of external beam radiation for therapy are photons and electrons, although particle therapy with protons or heavier ions like carbon is an emerging strategy which will likely gain increased clinical relevance in the future [148,149]. The drawback is related to the need for more elaborate systems, which will most likely limit the possibilities for particle therapy to high-income countries only.

3.2. Mechanisms Underlying Radiobiological Effects of EBRT

As described in Section 2.2, low-LET radiation (e.g., photons, electrons) primarily interacts with water molecules present in the tissue, leading to the production of oxygen radicals such as superoxide anions or hydroxyl radicals. This process, known as water radiolysis, occurs whenever radiation interacts with biological matter. These radicals, called ROS, interact with biomolecules and DNA, causing damage to accumulate in cell organelles and the DNA itself. Thus, DNA damage is mainly induced indirectly [70]. In contrast to low-LET radiation, high-LET radiation (e.g., protons, carbon ions) directly damages DNA by striking the molecule like a "bullet ball". This causes complex DNA damage, consisting of cDSBs which are challenging to repair for the cell. Therefore, high-LET EBRT is insensitive to limiting factors, such as tumor hypoxia and cell cycle phase, which are responsible for radioresistance in many tumors. The mechanisms of DNA damage and cellular lethality are shared between high-LET EBRT and TαT. Protons and heavy ions release most of their energy at the end of the track, known as a Bragg peak. Such a specific dose distribution profile enables the majority of the dose to be delivered to the tumor while sparing the healthy surrounding tissue [150].

3.3. Current Advances and Innovations in EBRT

Photon therapy represents the standard of care in the field of external radiotherapy. Currently, the most advanced applications of photon therapy are 3D-conformal treatments, such as intensity-modulated radiotherapy (IMRT) [151] and image-guided radiation therapy (IGRT), enabling a tailored irradiation of tumors to, e.g., deliver a higher dose to hypoxic areas [152]. Stereotactic body radiation therapy (SBRT) has emerged as a novel treatment modality and is being evaluated in a growing number of clinical studies [153,154]. In SBRT, a biologically equivalent dose (compared to delivered doses by conventional EBRT) is delivered in a short time interval through a few high dose fractions. In one recent example, SBRT of PCa was performed in a patient with PCa using 35–36.25 Gy in five fractions. By comparison, a biologically equivalent dose of EBRT would require 40–45 fractions given a maximum dose of 2 Gy per fraction [155].

Proton therapy and carbon ion therapy are currently used to treat only specific cases due to their higher costs and limited availability. Indeed, the number of therapy centers in Europe offering carbon ion treatment is exceedingly small, namely four centers in Europe and nine in Asia as well as one center under construction in the USA (Mayo Clinic) [147,156]. Ultra-high-dose-rate external radiation therapy, named FLASH, is a promising, albeit experimental, approach to delivering EBRT. Pre-clinical studies have shown a reduction in normal tissue toxicity without compromising tumor control by delivering the conventional amount of dose in ultra-high dose rates in a short time interval (FLASH $\geq$ 40 Gy/s; EBRT < 0.1 Gy/s) [157]. Ongoing preclinical studies are investigating the FLASH techniques with electron, proton, and carbon ion beams [157], of which the molecular mechanisms are still unknown. Further advanced beam shaping techniques using mini-, micro-, or pencil beams also represent a highly promising treatment strategy in the future, which can be applied to photons as well as protons and heavy particles [158,159].

3.4. Rationale for Combining EBRT with TαT

Used in combination, TαT and EBRT could positively reinforce each other. EBRT itself is inadequate as a local treatment regime in progressive disease with multiple metastases. On the other hand, TαT may not be suitable for treating larger lesions and can cause side effects in any tissues in which the TαT agent accumulates, even if these tissues are remote from the tumor lesions. Ideally, EBRT could be used to treat the main tumor mass locally and TαT to treat the metastases systemically (Figure 3). Side effects of EBRT are mainly related to the damage induced in the healthy tissue surrounding the irradiated tumor site, as the conventional external photon beam needs to pass through this tissue to reach the tumor and a margin of error around the tumor site is needed to ensure the tumor receives the necessary dose. Combination therapy with TαT can enhance the total delivered dose in the tumor volume while sparing the surrounding healthy tissue. Significantly, the unique efficacy of TαT, even in hypoxic tissue, can overcome resistance to EBRT in poorly oxygenated tumors. First evidence of a synergistic combination of these therapies came from studies that demonstrated increased uptake of a TRNT agents ([^{131}I]I-cetuximab and [^{131}I]I-benzamide) after EBRT. The improved uptake was attributed to increased vascular permeability and penetrability of the tumors following fractionated EBRT [160]. By analogy, TαT treatment before and after EBRT could sensitize the tumor tissue to irradiation and enhance the effects of EBRT after therapy due to continuous delivery of radiation. Finally, the abscopal effect, stimulation of an immune response against tumor cells, is mainly associated with EBRT [78] but has been observed for TαT as well [161]. A combination of both treatments could thus enhance the likelihood of the abscopal effect and lead to greater efficacy, even in untreated tumor lesions or tumors receiving sub-lethal doses.

Further support for the potential additive effects of EBRT and TαT combination therapy can be extrapolated from preclinical studies investigating combination therapies with EBRT and β^--emitting TRNT agents (Table 2). These studies demonstrated that the simultaneous application of EBRT and TRNT in a fractionated scheme leads to effective tumor control [162]. A long delay between EBRT and TRNT showed a clear disadvantage.

Based on these data, Hartrampf et al. proposed that a "sandwich treatment", with TRNT ([^{177}Lu]Lu-DOTA-TOC and [^{177}Lu]Lu-DOTA-TATE) before and after EBRT, might be an even more effective treatment protocol [163]. Their hypothesis was based on high uptake of the TRNT agent in the pre-irradiation phase [164] coupled with EBRT-triggered enhancement of target expression that further increases TRNT agent uptake in the post-irradiation phase [165].

Cornelissen et al. combined the radionuclide indium-111 (In-111, EC, 100%, $\tau_{\frac{1}{2}}$ = 2.80 d) with the anti-γH2AX antibody coupled to the cell-penetrating peptide trans-activator of transcription (TAT) [166]. The ligand is targeting the DNA damage induced by EBRT treatment and marked by γH2AX. The combination of 60 MBq of [^{111}In]In-anti-γH2AX-TAT with 10 Gy EBRT (photons) enhanced DNA damage in breast cancer MDA-MB231/H2N xenograft tumors, as evidenced by a multiplication of γH2AX phosphorylation. Combination with α-emitters might be even more beneficial given similar treatment schedules and targeting vector combinations.

3.5. Pre-Clinical and Clinical Insights: EBRT and TαT Synergy

To date, and due to the relatively recent emergence of TαT as a treatment option, there is a paucity of preclinical and clinical evidence for the synergistic effect of EBRT and TαT combination therapy. Nevertheless, studies with TRNT agents support this new combination treatment approach, when the physical properties of α-particles are taken into account [162]. Furthermore, two clinical studies investigating the combination of α-emitting $^{223}RaCl_2$ (Xofigo®) with SBRT in osteosarcoma and mCRPC have served as a proof-of-concept for this approach. Xofigo® was administered at 55 kBq/kg and SBRT was delivered in five fractions (5–10 Gy per fraction) over sequential days, either concurrent with TαT and/or after TαT, depending on the individual patient. In total, a maximum of six Xofigo® infusions were delivered [167]. The patients who received combination therapy demonstrated a significantly improved outcome compared to those who received either monotherapy alone [167,168].

3.6. Future Perspectives and Challenges

Previous and current clinical studies with TRNT strongly indicate that the combination of EBRT and TαT is clinically feasible and beneficial for cancer patients. Despite offering a promising perspective, vital shortcomings remain and need to be overcome for widespread future clinical implementation. As such, protecting healthy tissue from radiation exposure persists to be a central challenge in this combinatorial approach. Hence, the dosimetry of TαT agents is a crucial variable in this regard. Utilizing a suitable diagnostic radiotracer for TαT purposes might enable superior personalized calculation of the optimal dose. However, this additional tool would increase the expenses of further prospective single photon emission computed tomography (SPECT) and positron emission tomography (PET) scans. A second challenge is the establishment of optimized dosing sequences, taking the heterogeneity of tumors within and between patients into account. FLASH radiotherapy or novel beam shaping techniques as microbeams are a potential asset in this regard, but further evaluation of these novel therapies is essential before future combinations can be considered. Last, but not least, the infrastructural requirements and costs of both therapies represent significant hurdles, which essentially might restrict the combination to a relatively small number of treatment centers throughout the world.

Table 2. Overview of most recent and relevant publications on External Beam Radiation Therapy (EBRT) as monotherapy and in combined treatment with TαT (refer to Section 3). To facilitate the orientation, the table groups references on EBRT as a monotherapy first, followed by references on combination of EBRT with TRNT and TαT. Note: Only the most recent and relevant publications are included in the table to provide readers with a solid overview. More details can be found in Section 3.

Method	Authors	Year	Study	Cancer Type	EBRT Modality	TRNT Agent	Main Findings	Ref.
EBRT	Bilski et al.	2021	Review	Neuroendocrine tumors	Photons	NA	Role of radiotherapy in treatment of neuroendocrine cancer of the lung.	[154]
	Lin et al.	2021	Review	Various	Photons (FLASH)	NA	Ultra-high dose rate FLASH radiotherapy reduces radiation-induced damage in healthy tissue without decreasing antitumor effectiveness.	[157]
	Kundapur et al.	2024	Review	Various	Photons (Minibeam)	NA	Potential of minibeams to induce immunomodulation augmenting a direct tumoricidal effect, leading to pronounced cell kill and subsequent immune surveillance.	[156]
EBRT & TRNT	Oddstig et al.	2011	Pre-clinical	Small Cell Lung Carcinoma	Photons (2–8 Gy)	$[^{177}Lu]$Lu-DOTA-TATE	Radiation induces up-regulation of somatostatin receptors in small cell lung cancer.	[165]
	Cornelissen et al.	2012	Pre-clinical	Breast cancer (MDA-MB-468)	Photons (10 Gy)	$[^{111}In]$In-DTPA-anti-γH2AX-TAT	Amplification of DNA damage by γH2AX-targeted radiopharmaceutical and exposure to external radiation.	[166]
	Melzig et al.	2018	Pre-clinical	Head and Neck tumors (A431)	Carbon ions/Photons	$[^{131}I]$I-cetuximab/$[^{131}I]$I-benzamide	Combination inhibited tumor growth effectively, attributed to reduced microvascular density and decreased proliferation index.	[160]
	Hartrampf et al.	2020	Clinical	Meningioma	Photons (IMRT)	$[^{177}Lu]$Lu-DOTATATE/-TOC	Combination of PRRT and fractionated EBRT resulted in disease stabilization in 7 of 10 patients with advanced symptomatic meningioma.	[163]
EBRT & TαT	Anderson et al.	2020	Clinical	Osteosarcoma	Photons (SBRT)	Xofigo	Combination therapy of Ra-223 and EBRT for metastatic osteosarcoma.	[167]
	Hasan et al.	2020	Clinical	PCa	SABR	Xofigo®	Randomized trial of Ra-223 and SABR versus SABR for oligometastatic PCa.	[168], NCT04037358

4. Integration of Immune Checkpoint Inhibitors (ICI) with Targeted Alpha Therapy (TαT)

Key Facts

- Combining ICI with TαT can enhance immune response and tumor inflammation by inducing immunogenic cell death, overcoming resistance in heterogenous or poorly accessible tumors.
- Studies show that combining ICI with TRNT, such as TαT, improves survival and immune response in various cancers. Ongoing clinical trials, like those combining [^{225}Ac]Ac-J591 with pembrolizumab, aim to validate these findings across different cancer types.

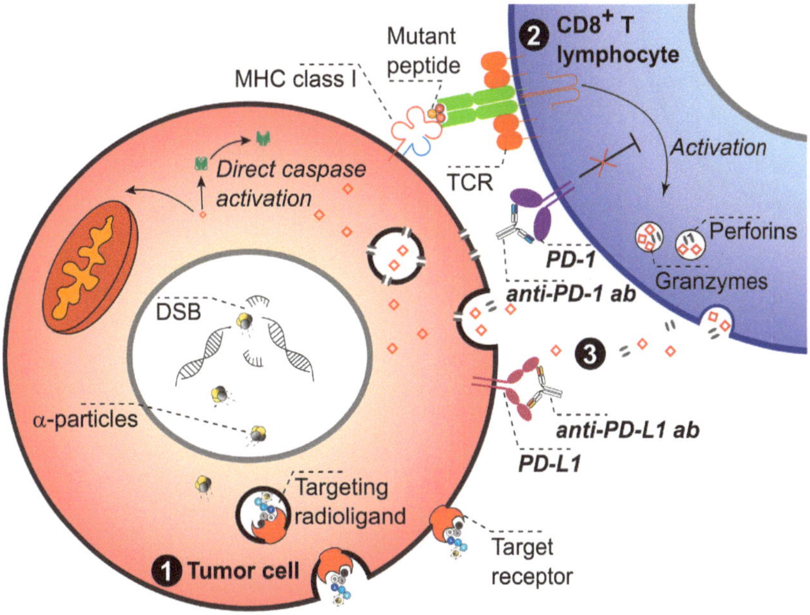

Figure 4. Graphical representation of systemic immunotherapy via ICI and systemic TαT combination therapy. The illustration depicts a single **tumor cell (red) (1)** and its interaction with a **CD8+ T lymphocyte (purple) (2)**. The immune checkpoint inhibitors **anti-PD-1 and anti-PD-L1 block the PD-1 and PD-L1 receptor–ligand complex (3)**. Such interactions facilitate cancer cell recognition and cancer cell killing by CD8+ T lymphocytes. The release of neo-antigens from tumor cells exposed to α-particle radiation enhances the recognition and killing of these cells by immune cells and activates dendritic cells, leading to improved tumor antigen presentation and immune response. At the same time, cancer cells can be irretrievably damaged by TαT, thereby killing cells which might not express PD-L1. Systemic immunotherapy with ICI activates the immune response for tumor cell killing. In parallel, systemic TαT directly kills tumor cells, even in tumors with a "cold" immune microenvironment in which immune response is suppressed. Furthermore, exposure to α-particles releases "neo-antigens" from tumor cells to stimulate both immune cell recognition and immune-mediated killing. Additionally, TαT promotes the production of many endogenous damage-associated molecular patterns (DAMPs). DAMPs activate dendritic cells, resulting in improved tumor antigen presentation, stimulation of cytotoxic T lymphocytes, and the release of cytokines and chemokines, all of which restore immune response. Therefore, the simultaneous immune checkpoint blockade and the enhancement of immune response by TαT could combine synergistically with the tumor killing effect of TαT. Note: Due to space constraints, illustrated examples do not cover the full spectrum of possible checkpoint inhibitors and focus exclusively on anti-PD-1 and anti-PD-L1 agents blocking the PD-1 and PD-L1 receptor–ligand complex.

4.1. Biological Basis of Immune Evasion and Restoration in ICI

The ability to elude host immunity is broadly accepted as one of the biological hallmarks of cancer. (Re)activation of the host immune system against tumors is a highly appealing strategy for treating cancer, and oncologists and immunologists have been working on new immunologic approaches for decades [169]. One reason this strategy is so appealing is that anti-tumor immune responses can, in theory, be highly specific. Blockade of T cell inhibitory molecules is one of the most promising methods for selectively enhancing immune response against cancer because it targets one of the pathways that tumor cells uniquely exploit to evade the host immune response [170]. As such, immunotherapy is perhaps the most tumor-specific treatment currently available [171].

In order to succeed, effector T cells must successfully infiltrate the tumor microenvironment (TME) and be activated upon detection of tumor antigens. Tumors have developed several strategies to disrupt this process and weaken anti-cancer immunity [172]. One strategy is to develop dense stroma that represents a physical barrier to tumor microenvironment (TME) infiltration. A second strategy is to overexpress proteins that modify T cell activation. CTLA-4, CD28, PD-1, and inducible co-stimulator (ICOS) and their ligands B7 (i.e., B7-1 or B7-2; CD80 or CD86), PD-L1, and ICOSL are members of the B7 family (within the immunoglobulin superfamily). These molecules modulate, as the so-called second signal (co-stimulation or co-inhibition), the intensity of the first signal delivered to T cells from the interaction of the T cell receptors (TCR) with the (tumor-) antigen presented in the major histocompatibility complex (MHC) [173]. The effect of the interaction is to inactivate the T cell. Disruption of the receptor–antigen binding by ICI ultimately prevents the inactivation of the T cell, enabling it to kill the tumor cell.

4.2. Comprehensive Overview of ICI in Cancer Treatment

Immune-checkpoint blockade is the most common strategy for restoring the host immune response for tumor cell killing. Drugs that exert their function based on this mechanism of action are known as immune checkpoint inhibitors (ICI). Anti-cytotoxic T lymphocyte antigen-4 (anti-CTLA-4), a monoclonal antibody specific for CTLA-4, the inhibitory receptor on T cells that binds the protein B7 on antigen-presenting cells, was the first ICI developed [174]. Anti-CTLA-4 immunotherapy is an FDA-approved treatment for advanced melanoma. Its mechanism of action may be to deplete T-regs, which constitutively express CTLA-4 [175]. Exhausted T cells also increase expression of CTLA-4 and are increasingly recognized as arising in cancer [176]. The programmed cell death protein 1/ligand 1 (PD-1/PD-L1) pathway is also implicated in the impaired T cell response to tumors. Antibody blockade of PD-1 or its ligand PD-L1 appears to be even more effective than anti-CTLA-4 in enhancing T cell tumor killing and preventing the progression of cancer in patients [177]. Anti-PD-1 and anti-PD-L1 antibodies also have fewer side effects than anti-CTLA-4 antibodies and are now authorized for the treatment of melanoma, lung carcinomas, renal carcinomas, bladder carcinomas, colon carcinomas, and Hodgkin lymphoma, amongst many other cancers [178]. However, these strategies do not uniformly benefit all patients. The efficacy of ICIs can be limited by factors such as the tumor microenvironment's immunosuppressive conditions, the genetic and epigenetic landscape of the tumor, and the diversity in T cell receptor repertoire among individuals [179].

4.3. Current Advances and Innovations in ICI

The first antibodies against CTLA-4 which were used for in vitro functional testing and in vivo modelling consisted of hamster anti-mouse CTLA-4 and mouse anti-human CTLA-4. For the initial clinical testing, however, human-sequence anti-human CTLA-4 was used to diminish any immunogenic effects against the antibodies themselves. These anti-human CTLA-4 antibodies were developed using transgenic mice comprising a germline configuration of human immunoglobulin genes that rearrange to direct a repertoire of B cells expressing human antibodies. To date, checkpoint blockade of CTLA-4 has relied on antibodies for blocking the interaction of CTLA-4 with its ligands [180]. This inhibition must not interfere with the CD28–B7 ligand interactions required for co-stimulation of primary T cell responses [181,182]. The most prominent example of a clinically approved anti-CTLA-4 antibody is ipilimumab (Yervoy®, Bristol-Myers Squibb, Princeton, New Jersey, USA), which received FDA-approval in 2011 for the treatment of metastatic melanoma [183]. Ipilimumab's approval was a milestone in cancer immunotherapy, demonstrating the potential of checkpoint inhibitors to significantly improve patient outcomes. This approach was awarded with the 2018 Nobel Prize in Physiology or Medicine for James Allison and Tasuku Honjo [184].

The first two PD-1-directed antibodies to enter the clinics, nivolumab (Opdivo®, Bristol Myers Squibb) and pembrolizumab (Keytruda®, Merck, Darmstadt, Germany), use the Fc domains of human immunoglobulin G4 (IgG4), comprising stabilizing hinge region mutations. Nivolumab was discovered in 2002 using human IgG transgenic mice, whereas pembrolizumab is the product of grafting the complementarity-determining regions of a mouse monoclonal antibody onto the homologous human antibody variable region. Nivolumab entered the clinics in 2006, and pembrolizumab as well as atezolizumab followed in 2011 [185]. Subsequently, cemiplimab (Libtayo®, Regeneron, Tarrytown, NY, USA), dostarlimab (Jemperli®, GlaxoSmithKline, Brentford, UK), toripalimab (Loqtorzi®, Coherus Bioscience, Redwood City, CA, USA), and tislelizumab (Tevimbra®, BeiGene, Cambridge, MA, USA) have been approved by the FDA as PD-1 binders for cancer treatment. Additionally, two antibodies targeting PD-L1 have also secured approval within the past 10 years. These drugs, durvalumab (Imfinzi®, AstraZeneca, Cambridge, UK) and avelumab (Bavencio®, EMD Serono, Rockland, MA, USA), also serve to disrupt the binding of PD-1 and PD-L1 for treatment of solid cancers [186].

4.4. Rationale for Combining ICI with TαT

Despite the number of approved ICIs and their use in a range of cancers, many tumors do not respond, or only partially respond, to this treatment. To overcome pre-existing or acquired escape mechanisms, ICI could be combined with TαT to render tumors even more T-cell inflamed [187]. Multiple studies have revealed that tumor cells become "immunogenic"—subject to both recognition by immune cells and immune-mediated killing—upon exposure to various radiation arts [185]. Cells undergoing apoptosis produce many damage-associated molecular patterns (DAMPs) which contribute to radiation-induced immunogenic cell death [188]. DAMPs activate dendritic cells, resulting in improved tumor antigen presentation, stimulation of cytotoxic T lymphocytes, and the release of cytokines and chemokines, all of which help the immune system respond to cancer cells [188,189]. TαT potentially induces significant cell damage and apoptotic cell death and might therefore be expected to enhance the effect of ICI immunotherapy by increasing T cell activation (Figure 4). Finally, because the mechanisms of tumor cell killing are not overlapping, ICI immunotherapy and TαT could combine to attack tumors that do not homogenously express the target of the TαT agent or are poorly accessible to T cells and immunotherapeutics.

4.5. Pre-Clinical and Clinical Insights: ICI and TαT Synergy

The journey of combining ICI with TRNT showcases a series of groundbreaking experimental and clinical endeavors that illuminate the immunostimulatory effects of β^--particles (Table 3). These initiatives have collectively demonstrated the enhanced efficacy of this combined approach, paving the way for novel cancer treatment strategies. The exploration into this combinatorial treatment began in 2018 with Jaeyeon et al., who reported the effective use of very late antigen-4 (VLA-4)-based TRNT with [^{177}Lu]Lu-LLP2A and ICI in the B16-F10 melanoma model. This study provided early evidence that combining TRNT with ICI could significantly improve survival rates in aggressive metastatic melanoma [190]. Building on this, in 2019 Chen et al. tested a novel therapeutic strategy combining anti-PD-L1 immunotherapy with peptide-based TRNT using [^{177}Lu]Lu-EB-RGD in a murine colon cancer model. This approach achieved an increase in PD-L1 expression in T cells and enhanced CD8+ T cell infiltration, thus improving local tumor control and overall survival [191]. Kim et al., in 2020, contributed further to the field with a phase I study on [^{177}Lu]Lu-DOTA-TATE (Lutathera®, Novartis) combined with nivolumab for neuroendocrine tumors of the lung, expanding the potential applications of TαT and ICI combinations in cancers not traditionally seen as highly immunogenic [192]. Another study by Rouanet et al., in 2020, showed that [^{131}I]ICF01012 induced immunogenic cell death and improved survival in immunocompetent mice while enhancing immune cell recruitment in the tumor microenvironment. The combination of [^{131}I]ICF01012 with ICI such as anti-CTLA-4, anti-PD-1, and anti-PD-L1 not only prolonged survival significantly (compared to TRNT alone) but also showed greater efficacy compared to ICIs alone without added toxicity [193].

Utilizing α-particles, Czernin et al. explored the synergistic effects of [^{225}Ac]Ac-PSMA-617 and PD-1 blockade in a mouse model of PCa, finding that these treatments significantly enhanced therapeutic outcomes. While both [^{225}Ac]Ac-PSMA-617 and anti-PD-1 antibody treatments individually prolonged time to progression (TTP) and survival, their combination further extended TTP to 47.5 days and increased survival to 51.5 days compared to either treatment alone (anti-PD1 alone 37 days, [^{225}Ac]Ac-PSMA-617 32 days). This suggests that combining PSMA-TαT with immunotherapy may be a promising approach for improving disease control in PCa which is per se also only poorly responding to ICI [194]. In 2021, the combination of TRNT using radiolabeled alkylphosphocholine (APC) analog [^{90}Y]Y-NM600 and ICI was shown to enhance the CD8+ immune response against T-cell non-Hodgkin lymphoma showed strong therapeutic effects, with 45 to 66% of mice exhibiting complete tumor response and developing tumor-specific T cell memory compared to none with either treatment alone [195]. Recent research has explored new treatment strategies for aggressive cancers by combining TRNT and immune checkpoint blockade (CP). One triple-negative breast cancer (TNBC) study revealed the combination of TCO-BSA labelled with Lu-177 and CP to extend survival and enhance the anti-tumor immune response in a murine model. These studies collectively highlight the promising potential of integrating TRNT with immunotherapy in treating resistant and aggressive cancer types [196]. Similarly, in a documented case report, the patient, having undergone multiple prior treatments, received a combined regimen of [^{177}Lu]Lu-PSMA-617 and immunotherapy. This approach effectively mitigated tumor progression in uterine leiomyosarcoma, thereby demonstrating the potential applicability of this strategy across a spectrum of PSMA-positive cancers [197].

Limited instances of preclinical or clinical studies have evaluated the combination of TαT and ICI immunotherapy to further confirm the combined approach to prolong overall survival compared to the respective monotherapies. The following clinical studies exemplify the ongoing exploration into the integration of TαT and immunotherapy for cancer treatment. The clinical trial NCT04946370 investigates the combination of [^{225}Ac]Ac-J591 with immune-modulating capabilities of pembrolizumab, further highlighting the innovative approach. To this date, no results were posted yet.

Multiple clinical trials are exploring the synergy between Ra-223 (Xofigo®) and various ICIs for advanced cancers. The NCT02814669 trial combined atezolizumab (targeting PD-L1) with Xofigo® for mCRPC. It reported an increase in adverse events, including severe reactions, and failed to demonstrate clinical benefits, raising questions about the effectiveness of this combination for mCRPC. The NCT03996473 trial investigated the use of pembrolizumab (targeting PD-1) with Xofigo® in stage IV NSCLC patients with bone metastases. The outcomes were mostly negative, with the majority of participants showing disease progression; only one achieved stable disease. Due to these disappointing results, the study was terminated early and did not move to phase II. Additional ongoing studies, such as NCT04071236 and NCT04109729, continue to test different ICIs like nivolumab and avelumab alongside Xofigo®. These trials remain active and are currently recruiting participants, aiming to uncover potentially improved treatment options for various types of cancer.

4.6. Future Perspectives and Challenges

The prevalence of immunologically "cold" tumors, as well as the exhaustion of T cells, upon persistent exposure to tumor antigens creates a need for a combination therapeutic that is efficacious in these tumors and (re)sensitizes tumor cells to the host immune system. TαT is ideally positioned to fulfill this role because it kills tumor cells independently of immune activity and also produces pro-inflammatory cytokines in the TME, generates new antigens, and promotes tumor infiltration and clonal expansion of CD8+ T cells. Preliminary preclinical studies support the hypothesis that TαT and ICI immunotherapy combine additively. Furthermore, the immune response can be further augmented by applying systemic immune-stimulating agents. This provides a rationale for further preclinical and clinical proof-of-concept studies to evaluate the therapeutic potential of TαT combinations with immunotherapies and immune-stimulating agents for specific and durable anti-tumor response.

Table 3. Overview of most recent and relevant publications on Immunotherapy with Checkpoint Inhibitors (ICI) as monotherapy and in combined treatment with TαT (refer to Section 4). To facilitate the orientation, the table groups references on ICI as a monotherapy first, followed by references on combination of ICI with TRNT. Note: Only the most recent and relevant publications are included in the table to provide readers with a solid overview. More details can be found in Section 4.

Method	Authors	Year	Study	Cancer Type	ICI Agent	TRNT Agent	Main Findings	Ref.
ICI	Bonaventura et al.	2019	Review	Various	Various	NA	Cold tumors: A therapeutic challenge for immunotherapy.	[187]
	Sun et al.	2020	Meta-analysis	Various	Anti-PD-1, Anti-PD-L1	NA	Clinical efficacy and safety of anti-PD-1/PD-L1 inhibitors for advanced or metastatic cancer.	[178]
	Arina et al.	2020	Review	Various	Various	NA	Radiotherapy and immunotherapy for cancer: From 'systemic' to 'multisite'.	[185]
	Yap et al.	2021	Review	Various	Various	NA	Development of immunotherapy combination strategies in cancer.	[171]
	Sharma et al.	2023	Review	Various	Various	NA	Immune checkpoint therapy—current perspectives and future directions.	
ICI & TRNT	Chen et al.	2019	Pre-clinical	Colon cancer (MC38)	Anti-PD-L1	$[^{177}Lu]$Lu-EB-RGD	Increased PD-L1 expression on T cells, enhanced CD8+ T cell infiltration, and improved local tumor control.	[191]
	Patel et al.	2021	Pre-clinical	Various (B78, 4T1, NXS2, Panc02, B16)	Anti-CTLA-4, Anti-PD-1, Anti-PD-L1	$[^{131}I]$I-CF01012	Enhanced CD8+ immune response.	[195]

5. Integration of Cytostatic Chemotherapy (CCT) with Targeted Alpha Therapy (TαT)

Key Facts:

- Combining CCT with TαT can enhance treatment effectiveness, increasing cytotoxicity due to high-LET radiation, potentially allowing for lower doses of each agent to reduce side effects.
- Studies show promise in enhancing the efficacy of taxanes through new formulations and combinations with other treatments. Early clinical trials indicate the potential of taxane and TαT combinations in PCa treatment, with ongoing studies aiming to confirm these findings.

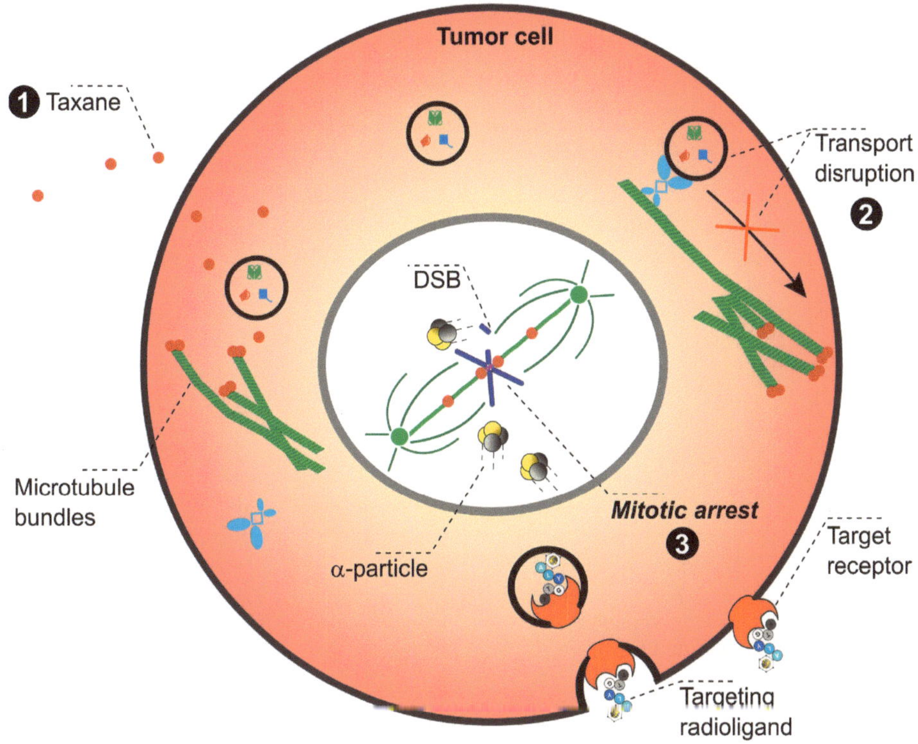

Figure 5. Graphical representation of systemic CCT and systemic TαT combination therapy. The illustration depicts single tumor cell under mitotic arrest and transport disruption. **Taxanes (1)** bind to microtubules and significantly increase their stability, leading to **transport disruption (2)**. This has a detrimental effect on the cell by inducing **mitotic arrest (3)**, DNA damage, chromosome segregation issues, and significant disturbance of cellular transport. The combination of TαT and taxane-based chemotherapy—docetaxel and cabazitaxel—is based on two different mechanisms which are potentially mutually reinforcing. TαT induces complex damage to the cellular DNA, which activates DNA repair pathways. Treatment with taxanes, however, disturbs the DNA repair machinery by induction of mitotic arrest and disruption of cellular trafficking. Therefore, the combination of the damage-inducing TαT with a DNA-repair-inhibiting CCT agent could lead to enhanced damage and ultimately cell death of tumor cells. Additionally, cell death induced by TαT is more specific to the tumor cells rather than rapidly-dividing cells—tumor or healthy alike—which are the primary targets of CCT. Furthermore, as both approaches represent systemic treatments, the combination might be highly promising for patients with a rather advanced disease stage and already developing treatment resistance. Note: Due to space constraints, illustrated examples do not cover the full spectrum of possible chemotherapeutic agents and focus exclusively on taxane-based drugs.

5.1. Comprehensive Overview of CCT in Cancer Treatment

Chemotherapy is an important and widely used treatment modality in many different cancer types. The use of chemical agents as potential anticancer drugs started in the 1940s with nitrogen mustard and quickly expanded over the following decades, both in terms of the number of chemotherapeutic agents and in terms of their frequency of use, to become one of the leading treatment modalities for cancer. As both the library of chemotherapeutic agents and their target cancer types are vast and the scope of this review is limited, we will mostly focus on PCa and on taxanes as therapeutic agents to illustrate the potential of combining various therapeutic approaches to potentially increase overall treatment efficiency [198–203].

In PCa, chemotherapy serves as a first and second line of treatment, usually alongside other treatment modalities [204]. Out of the various cytotoxic agents tested so far, only docetaxel (Taxotere®, Sanofi-Aventis, Paris, France; Docefrez®, Sun Pharma Global, Mumbai, India; Zytax®, Zydus, Ahmedabad, India), and cabazitaxel (Jevtana®, Sanofi-Aventis) showed clear overall survival benefits. Both docetaxel and cabazitaxel belong to the taxane family of microtubule-targeting agents (MTAs) [204–206]. These drugs contribute to tumor cell killing through multiple mechanisms summarized in Section 5.2.

5.2. Mechanistic Insights into Taxane-Induced Cytotoxicity

Taxanes stabilize microtubules against disassembly through binding β-tubulin to counteract the effect of hydrolysis of guanosine-5′-triphosphate (GTP) to guanosine-5′-diphosphate (GDP) and promote the assembly of tubulin dimers into microtubules [207]. This action leads to a disturbance in normal microtubule dynamics which—depending on the cell cycle phase—can lead to various catastrophic effects [208,209].

One of the most studied catastrophic effects is the induction of mitotic arrest. In mitotic cells, taxanes bind to, and stabilize, the mitotic spindle and prevent sister chromatid segregation. This in turn activates the spindle assembly checkpoint, which triggers a mitotic arrest [209]. Prolonged mitotic arrest is detrimental to cells, and it often leads to cell death. Another possible toxic effect of taxanes during mitosis is multipolar spindle formation. This happens at lower taxane concentrations, and while it does not trigger mitotic arrest, it can lead to serious chromosome mis-segregation and subsequent cell death [210]. Although the cell killing effect of taxanes was mainly linked to their mitotic effects in rapidly dividing cells—which describes both in vitro cell cultures and xenografts—the doubling time of human tumor cells is generally much longer than that of cell lines or xenografts [211]. As an example, human PCa cells have an average proliferative doubling time of 19–108 days, much slower than model cells cultured in vitro. Nevertheless, both docetaxel and cabazitaxel are effective against these cells [212]. This discrepancy led to the deeper exploration of taxane-related effects on interphase cells and turned up new, interphase-specific cytotoxic effects.

Recent studies showed that docetaxel and cabazitaxel induce microtubule bundling in interphase cells and cause major disruptions in cellular trafficking. As a result, essential proteins like DNA damage repair proteins or various receptors and transcription factors (e.g., androgen receptor) cannot be properly translocated to the nucleus [213–215]. Disrupted intracellular trafficking can severely disrupt normal signal transduction pathways and often leads to apoptotic cell death. Additionally, functional microtubule dynamics are necessary for DNA DSB repair. In interphase, this process happens via NHEJ, which requires a tightly regulated mobile DSB, for which a functional microtubule system is essential [216]. In the absence of proper microtubule dynamics, DSB accumulation or mis-regulated NHEJ can lead to severe genomic instability and cell death.

A lesser-known effect of docetaxel and cabazitaxel, which may also contribute to their efficacy in slower-growing prostate tumor cells, is the promotion of apoptosis in a microtubule-independent manner. The mechanism behind this is not fully understood; however, the observation that these drugs decrease the expression level of an anti-apoptotic protein Bcl-2 is likely relevant [217,218]. In the case of cabazitaxel, this effect is also paired with an increase in the expression of Bax, a pro-apoptotic member of the Bcl-2 family [219]. Together, these changes can lead to mitochondrial outer membrane permeabilization and to an increase in activated caspase-3 levels. Caspase-3 may be responsible for coordinating the apoptosis.

5.3. Current Advances and Innovations in CCT

Taxanes such as docetaxel remain a key component of SoC protocols in multiple cancer types [220]. However, they are systemic drugs that do not selectively target cancer cells and consequently have multiple side effects on dividing cells [204,221,222]. These side-effects are partially caused by non-specific cancer targeting, relatively low tumor accumulation, and the requirements for irritating surfactants to improve taxane solubility, which collectively limit the treatment doses, reduce efficacy at the administered dose, and can severely impact the patient's quality of life [220,221,223]. Therefore, much of the recent work with taxanes has focused on developing new formulations to improve tumor uptake and reduce access to off-target tissue as well as combining them with other chemotherapeutic drugs or even different treatment modalities. Among these efforts, nanoparticle formulations are playing a leading role [224,225]. For example, as part of a cross-over study, the plasma and intra-tumoral pharmacokinetics of CPC634—a nanoparticle entrapping docetaxel—was compared to classically formulated docetaxel [226]. The study showed a significant increase in intracellular docetaxel concentration (461% higher) and a significant decrease in plasma docetaxel levels (91% lower) compared to conventional docetaxel. The study also showed a marked decrease in serious grade 3 neutropenia in patients receiving CPC634 compared to conventional docetaxel; however, other grade 1–2 adverse effects such as skin toxicity and sensory neuropathy incidences were higher in the case of CPC634.

In principle, nanoparticles could facilitate selective tumor targeting by potentially increasing tumor uptake and by eliminating the use of some dose-limiting components of the current formulations [220,223]. Nano formulations are also appealing from a combination therapy perspective, as they could coordinate the selective targeting of tumor cells with both the chemotherapeutical drug and targeted radionuclide. This would potentially allow either or both agents to be administered at decreased dose levels, thereby decreasing the side effects and potentially increasing the patient's quality of life.

5.4. Rationale for Combining CCT with TαT

Although docetaxel and cabazitaxel can be highly effective against cells in interphase and mitosis, tumor cells can acquire resistance against these drugs, decreasing their effectiveness if used alone [214]. Therefore, combining various therapeutic approaches can significantly increase treatment effectiveness [198]. Based on their currently understood mechanisms of cytotoxicity, taxane-based chemotherapy and TαT are good candidates for combination therapy (Figure 5). Indeed, previous preclinical studies indicate that taxanes sensitize various tumors to low-LET radiation. Miyanaga et al. reported that low concentrations of paclitaxel and docetaxel (<1 nM) before irradiation (2–6 Gy X-rays) induced higher amount of DSBs and apoptosis. Zhang et al. observed enhanced cytotoxicity in clonogenic assays when cells were exposed to 20 nM of paclitaxel before radiation treatment (0–8 Gy X-rays) [227,228]. It is plausible that these CCT agents will have the same effect against high-LET radiation, further enhancing the cytotoxicity of TαT. Indeed, the effect may be even stronger using α-emitters because these agents induce DSBs at a higher rate than low-LET, rendering cells even more vulnerable to the impaired NHEJ repair induced by taxanes during interphase. Through improved efficacy, it might be possible to reduce the doses administered of each agent. To wit, the radiation sensitizing effect is evident even at a low taxane concentration [228], which would reduce the chemotherapeutic burden, while the cells would be sensitive to lower α-radiation doses, which would also reduce the radiation burden.

5.5. Pre-Clinical and Clinical Insights: CCT and TαT Synergy

So far, research on a combined taxane and TαT therapy in either pre-clinical or clinical settings is at its nascent stage. However, as stated in Section 5.3, there is a considerable effort to improve the efficiency and utility of taxanes in a clinical setting (Table 4).

BIND-014, a PSMA-targeted nanoparticle, has also been developed to improve CCT in mCRPC. In a phase II clinical trial, the safety and efficiency of BIND-014 was tested in 42 mCRPC patients [229]. Although the study did not compare BIND-014 directly to conventional docetaxel treatment, the authors saw some encouraging results. For example, the median radiographic progression-free survival (PFS) was 9.9 months, which was better than the pre-specified favorable outcome for the trial ($\geq$6 months). Additionally, almost a third of the patients showed a $\geq$50% decrease in PSA levels, while half of the patients with unfavorable circulating tumor cell counts saw a reduction to these values.

Yet another new formulation currently under investigation is albumin-bound paclitaxel (Nab-paclitaxel), a different member of the taxane family. In a recent phase III clinical trial, Nab-paclitaxel was compared to conventional docetaxel in non-small cell lung cancer (NSCLC) patients previously pre-treated with cytotoxic chemotherapy [230]. The study showed slightly increased overall survival (OS; 16.2 vs. 13.6 months) and increased PFS (4.2 vs. 3.4 months) for patients receiving Nab-paclitaxel vs. docetaxel. Interestingly, while conventional docetaxel treatment caused higher incidence of grade 3–4 neutropenia or febrile neutropenia than Nab-paclitaxel, Nab-paclitaxel treatment had a higher incidence of grade 3–4 peripheral sensory neuropathy.

In addition to these comparison studies, Nab-paclitaxel is also being used in various trials in combination with other treatment modalities. These include, but are not limited to, several phase III clinical trials focusing on triple-negative breast cancer (TNBC) or squamous non–small-cell lung cancer (sNSCLC) investigating the efficiency of combining Nab-paclitaxel with anti-PD-1 (pembrolizumab) and anti-PD-L1 (atezolizumab) antibodies [231–233]. In TNBC, the combination of Nab-paclitaxel with atezolizumab showed increased PFS and OS, albeit with grade 3–4 side effects occurring more frequently. These side effects seemed to be in line with the ones described in other studies [231]. Another study focusing on pembrolizumab also assessed the effect of combining the antibody with Nab-paclitaxel, paclitaxel, and gemcitabine as chemotherapeutic agents. Although the results were not broken down based on the chemotherapeutic agent used, the PFS and OS were higher in the combined therapy group than in the placebo-pembrolizumab group [232]. In the case of sNSCLC, the phase III trial focused on the combination of pembrolizumab and either carboplatin or paclitaxel, or carboplatin and Nab-paclitaxel. The study found a higher PFS and OS in the combination therapy groups vs. the placebo-chemotherapy group. Results were also comparable between the paclitaxel and Nab-paclitaxel groups.

To our knowledge, research on a combined taxane and TαT therapy has not been published. However, there are a handful of clinical studies of TRNT in combination with taxanes that support this approach. A recent case study highlighted the promise of taxane chemotherapy with targeted radiopharmaceutical therapeutics in a patient with PCa [234]. The patient received an initial course of [^{177}Lu]Lu-PSMA-617. After new bone lesions developed, a second sequence was initiated in combination with low doses of paclitaxel (Taxol®) to act as a radiation sensitizer. This second course led to remarkable and sustained tumor reduction with only transient, treatable side effects. A recently completed phase I trial (NCT00916123) in 15 men with PCa treated in combination with PSMA-targeting antibody [^{177}Lu]Lu-J591 and docetaxel concluded that TRNT could be safely conducted concurrently with standard taxane chemotherapy [235]. In parallel, a recently opened phase I/2 trial (LuCAB, NCT05340374) will assess the safety and efficacy of cabazitaxel in combination with [^{177}Lu]Lu-PSMA-617 in PCa patients [236].

5.6. Future Perspectives and Challenges

Multiple recent and ongoing clinical trials focus on testing new taxane formulations and combination therapies of taxanes with other treatment modalities. Many of these trials show that both of these approaches can offer increased survival benefits to patients. Given the rapid development of new taxane formulations, the safety of taxanes in combination with TRNT and the potentially additive interplay between TαT and taxanes, such a combination therapy, holds real potential to improve outcomes in cancer.

Table 4. Overview of most recent and relevant publications on Cytostatic Chemotherapy (CCT) as monotherapy and in combined treatment with TαT (refer to Section 5). To facilitate the orientation, the table groups references on CCT as a monotherapy first, followed by references on combination of CCT with TRNT. As mentioned in Section 5.5, there are currently no pre-clinical or clinical studies on combination of CCT and TαT. However, most TαT patients went through CCT before getting assigned to TαT. Note: Only the most recent and relevant publications are included in the table to provide readers with a solid overview. More details can be found in Section 5.

Method	Authors	Year	Study	Cancer Type	CCT Agent	TRNT Agent	Main Findings	Ref.
CCT	Xu et al.	2022	Pre-clinical	PCa	Cabazitaxel	NA	Cabazitaxel suppresses the proliferation and promotes apoptosis and radiosensitivity: suppression of PI3K/AKT pathway.	[218]
	Schmid et al.	2018	Clinical	Breast cancer (TNBC)	Nab-paclitaxel	NA	Atezolizumab and nab-paclitaxel combined prolonged progression-free survival in both the intention-to-treat population and PD-L1–positive subgroup.	[231]
	Paz-Ares et al.	2018	Clinical	Lung cancer (NSCLC)	Paclitaxel/Nab-paclitaxel	NA	Combination of pembrolizumab to chemotherapy with carboplatin plus paclitaxel/nab-paclitaxel lead to significantly longer overall survival and progression-free survival than chemotherapy alone.	[233]
	Yoneshima et al.	2021	Clinical	Lung cancer (NSCLC)	Nab-paclitaxel/Docetaxel	NA	Nab-paclitaxel was noninferior to docetaxel in terms of OS.	[230]
	Cortes et al.	2022	Clinical	Breast cancer (TNBC)	Paclitaxel/Gemcitabine–carboplatin	NA	Patients with high PD-L1 expression benefit from combination of pembrolizumab and chemotherapy: significantly longer overall survival than chemotherapy alone.	[232]
CCT & TRNT	Maharaj et al.	2021	Clinical	mCRPC	Docetaxel	[^{177}Lu]Lu-PSMA-617	Low-dose docetaxel as radiosensitizer with [^{177}Lu]Lu-PSMA-617 showed good response; no tumor resistance.	[234]
	Batra et al.	2020	Clinical	mCRPC	Docetaxel	[^{177}Lu]Lu-J591	Combination of [^{177}Lu]Lu-J591 (single fractionated cycle) with docetaxel was well tolerated.	[235]
	Kostos et al.	2023	Clinical	mCRPC	Cabazitaxel	[^{177}Lu]Lu-PSMA-617	Evaluation of cabazitaxel in combination with [^{177}Lu]Lu-PSMA-617 (LuCAB) to target micrometases.	[236] NCT05340374

6. Integration of Brachytherapy (BT) with Targeted Alpha Therapy (TαT)

Key Facts:

- Combining BT with TαT aims to enhance tumor cell killing by addressing regions with insufficient TαT dose, known as 'cold spots', leading to a uniform tumor control while sparing healthy tissues.
- New methods like diffusing α-emitters radiation therapy (DαRT) are being tested. Combining BT and TαT might enhance anti-tumor immune responses and improve therapeutic outcomes.

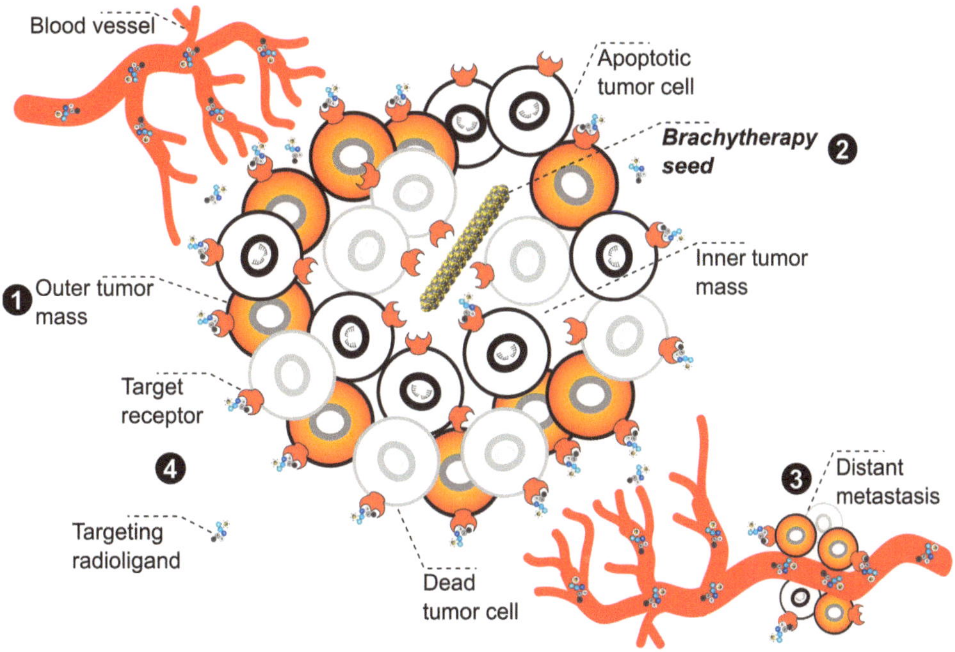

Figure 6. Graphical representation of the local BT and systemic TαT combination therapy. The illustration depicts **tumor mass (1)** with the **brachytherapy seed (2)** inserted in the middle of the tumor. The primary tumor and **distant metastases (3)** along blood vessels are treated by **targeting radioligand** docking on target receptors **(4)**. The inserted seed releases energy from radioactive decay along its path, efficiently eradicating the inner tumor mass and preventing relapse in tumors with 'cold spots' that lack target expression for TαT. Tumors expressing TαT targets enable the treatment of not only primary tumors but also metastatic lesions. Combination therapy with BT and TαT relies on the killing efficiency of various radionuclides. Specifically, the killing efficiency would be enhanced by attacking the tumor mass from two sites: cavital and systemic. The insertion of a radioactive seed inhibits the relapse in tumors that do not express specific targets for TαT but might have non-irradiated 'cold spots' in the tumor periphery that eventually lead to a relapse. In tumors that exhibit appropriate targets for TαT, this approach ensures comprehensive coverage of the tumor's 'cold spots'. When used in conjunction with BT, TαT can potentially lead to the complete eradication of both the primary tumor and any metastatic lesions.

6.1. Comprehensive Overview of BT in Cancer Treatment

Brachytherapy (BT) is a specific form of delivering radiation through sealed radioactive sources. This therapy is most commonly used to treat head and neck, breast, cervical, and prostate cancers. The radiation sources are implanted in the tumor via a catheter or, occasionally, intracavitary, intraluminal, or interstitial (needle) applicators. The efficacy of BT is based on the direct radiation of the tumor cells. In this respect, it is functionally analogous to TRNT and TαT [237].

6.2. Radiobiological Mechanisms Underlying BT

DNA is the primary cellular target of BT. If unrepaired, radiation-induced DNA damage leads to cell cycle redistribution, microenvironmental changes, and cell death. The therapeutic index of irradiation relies on the differential response between tumors and normal tissue, as tumor cells have a lower DNA repair capability compared to normal tissue cells. However, radiation-induced changes in organs at risk may lead to acute side effects and long-term functional sequelae. Therefore, the ideal technique should be able to deliver therapeutic doses to the tumor with doses as low as possible to the organs at risk [238].

6.3. Current Advances and Innovations in BT

Brachytherapy is an effective tool for obtaining high tumor doses while minimizing the exposure to organs at risk, as documented by dosimetry studies. The total dose can be delivered by one of three methods: (1) continuous low-dose-rate (LDR) irradiation, (2) pulse-dose-rate irradiation, whereby low-intensity pulses are repeated every hour for up to a few days, or (3) a few fractions that deliver high doses constantly (high-dose-rate [HDR] irradiation) [239]. Iridium-192 (Ir-192, $\tau_{1/2}$ = 73.8 d, 95.6% β^- decay), cobalt-60 (Co-60, $\tau_{1/2}$ = 5.27 years, 100% β^- decay), iodine-125 (I-125, $\tau_{1/2}$ = 59.4 d, 21 Auger electrons per decay), and palladium-103 (Pd-103, $\tau_{1/2}$ = 16.99 d, >24 Auger electrons per decay) are the radioisotopes that are commonly applied in modern brachytherapy [240]. The clinical use of brachytherapy has notably reduced over time. Among the reasons for its waning popularity are the lack of sophisticated training programs for physicians for placement of the implants and the invasive nature of the implantation, which requires a short procedure in the operating room to place the sources or the catheters and can cause pain when the catheter is removed. Additionally, any inaccurate positioning of the radiation sources may decrease the therapeutic index by exposing the tumor to underdosage or neighboring tissue to excessive dose [241].

6.4. Rationale for Combining BT with TαT

The conceptual basis for integrating BT and TαT lies in harnessing the cytotoxic capabilities of multiple radiation sources delivered through distinct methodologies. Although there are currently no clinical or pre-clinical studies directly investigating the combined effects of BT and TαT, parallels drawn from existing radiation combination strategies, such as those involving EBRT and TRNT, suggest potential efficacy. This analogy supports the hypothesis that combining BT and TαT could prove successful in enhancing therapeutic outcomes by exploiting synergistic effects of different radiation modalities to achieve more comprehensive tumor control.

Systemic administration of the TαT agent will cause it to kill cells expressing the desired target, but, through heterogenous target expression and vascularization or total lesion volume, not all regions of the tumor will receive a lethal dose. These 'cold spots' could be the genesis of relapse. When optimally combined, BT could be used to kill the tumor regions that receive insufficient doses from TαT therapy. By addressing these 'cold spots', this combination therapy could lead to more uniform tumor cell killing and lasting tumor control, all while sparing healthy surrounding tissues (Figure 6).

6.5. Pre-Clinical and Clinical Insights: BT and TαT Synergy

BT is commonly used alone or alongside conventional CCT, surgery, or EBRT in clinical settings. While BT has been explored in combination therapies for PCa, no studies have yet investigated its potential synergies with TαT (Table 5). Patients with low and intermediate-risk disease are eligible for LDR BT, which requires the permanent insertion of radioactive I-125 seeds (or, less frequently, Pd-103 seeds) into the prostate. The procedure offers a number of benefits over alternatives such as radical prostatectomy, including being less intrusive and associated with a lower risk of deficits in long-term quality of life. It can be performed as an outpatient procedure under spinal anesthesia, with a good recovery and a lower risk of dribbling [242]. The method of delivering prostate HDR BT over a few sessions involves introducing catheters inside the gland while using high-activity radioactive sources of Ir-192. In this setting, BT may be used in combination with EBRT to reduce the dose requirement. As evidence for this hypothesis, a randomized clinical trial in patients with localized or locally advanced PCa comparing the survival and toxicity for HDR BT in combination with EBRT and EBRT alone found that treatment with HDR BT before EBRT reduced the required dose of EBRT to achieve a comparable survival outcome at significantly lower side effects [243]. To further highlight the benefits of combined BT and EBRT, patients that underwent the combined therapy required a shorter duration of androgen deprivation therapy (ADT) treatment [244].

After guidelines for BT and radiation safety were issued in 2003, the clinical use of permanent BT using I-125 seeds to treat PCa has rapidly grown in Japan. Studies in Japanese patients confirmed that the clinical outcomes of PCa patients receiving combined BT and EBRT were significantly improved. Notably, patients who underwent combined BT and EBRT without adjuvant ADT had a higher probability of improved survival [245]. In the largest trial to evaluate BT and EBRT, the ASCENDE-RT (Androgen Suppression Combined with Elective Nodal and Dose Escalated Radiation Therapy) study, 398 patients with intermediate-risk and high-risk cancer were enrolled. That trial showed that the probability of biochemical failure at a median follow-up of 6.5 years was twice as high among patients who received an EBRT boost compared with those who received a BT boost [246]. Finally, Oshikane et al. recently reported that combined HDR BT and EBRT led to a significantly improved therapeutic outcome compared to EBRT alone in PCa patients receiving ADT [247].

A new iteration of BT, diffusing α-emitters radiation therapy (DαRT), is now being evaluated in clinical trials as a potential treatment for solid tumors [248]. DαRT utilizes interstitial seeds of Ra-224 at 10–1000 kBq activity levels which emit short-lived α-particles that disperse throughout the tumor via diffusion. The administration of DαRT treatment, which ostensibly combines BT and TαT in a single agent, elicits specific and systemic antitumor immune responses, ultimately activating the immune system against the tumor. The pro-immunogenic TME is characterized by an augmented infiltration of T lymphocytes and the secretion of granzyme B.

6.6. Future Perspectives and Challenges

BT, alone or as part of a combination therapy approach, has historically been an important tool for disease control, curative therapy, and palliative care in a small number of cancers. We have highlighted the combination of HDR BT and EBRT as a particularly effective therapeutic strategy in these cancers and suggest that TRNT, and in particular TαT, could potentially act in synergy with HDR BT or BT to specifically target the tumor and reduce side effects. The benefit of this strategy over existing combinations is the ability to target the tumor mass from its core (through BT) and from its periphery and metastatic sites (through TαT). This combination could lead to more complete tumor cell killing while reducing off-target effects, resulting in a new, safe, and highly potent approach for cancer treatment. The recent emergence of DαRT provides further insight into how BT and TαT could be effectively combined and represents an exciting avenue for study.

Table 5. Overview of most recent and relevant publications on Brachytherapy (BT) as monotherapy and in combined treatment with TαT (refer to Section 6). To facilitate the orientation, the table groups references on BT as a monotherapy first, followed by references on combination of BT with EBRT. As mentioned in Section 5.5, there are currently no pre-clinical or clinical studies on combination of BT and TαT. Note: Only the most recent and relevant publications are included in the table to provide readers with a solid overview. More details can be found in Section 5.

Method	Authors	Year	Study	Cancer Type	Main Findings	Ref.
BT	Annede et al.	2020	Review	Various	Foundations and new insights in radiobiology modelling for brachytherapy effects.	[238]
	Fonseca et al.	2020	Review	Various	Requirements and future directions for in vivo dosimetry in brachytherapy.	[239]
	Ito et al.	2018	Clinical	PCa	Analysis of survival outcomes for permanent I-125 seed implantation in PCa.	[245]
	Arazi et al.	2020	Modeling Study	Various Solid Tumors	Modelling of macroscopic α-particle dose for Ra-224 in DαRT.	[248]
	Tamihardja et al.	2022	Clinical	PCa	Comparison between Ir-192 and Co-60 sources in HDR brachytherapy for PCa.	[240]
	Morris et al.	2017	Clinical	PCa	Patients with LDR brachytherapy were twice as likely to be free of biochemical failure at a median follow-up in comparison to those receiving dose-escalated EBRT.	[246]
BT & EBRT	Mori et al.	2021	Clinical	PCa	High risk patients benefit from trimodal therapy with HDR brachytherapy, hypofractionated EBRT, and ADT.	[244]
	Oshikane et al.	2021	Clinical	PCa	HDR brachytherapy boost combined with EBRT has significantly higher biochemical-free survival rate than EBRT alone in high-risk PCa.	[247]

7. Outlook

This manuscript presents an overview of TαT to treat cancer and highlights a wide array of established therapeutic strategies that could potentially be combined with TαT for improved outcomes. While some of these combination approaches are undergoing preliminary (pre)clinical evaluation in a range of cancers, others represent untapped potential whose synergistic or additive therapeutic outcomes we hypothesize based on the mechanisms of action of the respective monotherapies. As much as combination therapy with TαT could lead to significantly improved treatment outcomes, it is equally important to acknowledge the biological, logistical, and economic obstacles to the translation of these approaches. Additionally, combined treatment requires the willingness of different medical disciplines to join forces for the greater benefit of the patient. However, the exchange and collaboration between different disciplines is often challenging. This is also related to the economics of the current health care system, which would require different disciplines to share the income for different combined treatments.

Furthermore, we would like to briefly discuss the current status of radiopharmaceutical production as well, with a particular focus on the prevailing shortage of α-emitting radionuclides, infrastructure considerations, and the optimal treatment sequencing required to incorporate TαT into existing treatment paradigms. It is our expectation that breakthroughs in production methodologies, purification strategies, and radioisotope distribution networks will soon be achieved, thereby facilitating the expansion of TαT to larger numbers of patients and medical centers. With concurrent investment in infrastructure, including increased capacity of treatment centers, education programs for patients and staff to be exposed to new sources of radiation, and radioactive waste management solutions, TαT combination therapy could be administered in a cost-effective and resource-efficient manner. An essential aspect that needs to be investigated is the optimal sequencing of therapies. It is crucial to determine whether a single dose or a fractionated approach would yield superior outcomes. The starting points for these studies should be based on the mechanism of action of TαT, its side effect profile, and the potential synergies it may have with other treatment modalities.

Finally, the exploration of triple combination therapies involving TαT opens up intriguing possibilities to further improve treatment outcomes and reduce side effects. For example, some cancers might benefit from EBRT + ICI + TαT to eradicate the primary tumor and metastases through tumor cell irradiation and immune system activation. By examining the synergistic effects and additive benefits of multiple treatment modalities, new treatment cocktails could be devised to truly cure some types of advanced cancers. The successful implementation of such complex strategies would require coordinated efforts from diverse scientific and medical specialties. Nevertheless, the rapid progress of TαT from bench to bedside should foster these collaborations and give us great optimism for future progress in the field.

8. Conclusions

In conclusion, this comprehensive review illuminates the foundational principles of five distinct therapeutic strategies as monotherapies and as potential combinations with TαT. The combination of EBRT, BT, CCT, or ICI with TαT all represent feasible treatment strategies that could ultimately be tailored based on each patient's disease and prior treatment history. Strategic combination of these complementary, but unique, therapeutic approaches on an individualized basis may well dramatically improve the prognosis for advanced-stage cancer patients through robust and durable disease control and improvements in the patients' quality of life. The collaborative efforts of clinicians, researchers, and practitioners in integrating these therapies demonstrate a commitment to advancing cancer care beyond conventional boundaries. This review sets the stage for future exploration and encourages continued research and clinical innovation aimed at expanding and defining the scope of combined therapeutic interventions for the benefit of cancer patients worldwide.

Author Contributions: Writing—original draft preparation, R.C.W., M.A., A.S.W. and G.B.; writing—review and editing, H.T., M.R., J.M.K. and M.B.-S. All authors have read and agreed to the published version of the manuscript.

Funding: R.C.W. was financially supported by a PhD grant (104515) from the Helmholtz International Graduate School (HIGS) for Cancer Research. M.A. was financially supported by a PhD grant (Ca 201) from the German Cancer Research Center (DKFZ) & Ministry of Science, Technology and Space (MOST) joint funding.

Acknowledgments: Co-funding of the research group Molecular Biology of Systemic Radiotherapy was made available by industrial collaboration with Bayer AG, Berlin, Germany.

Conflicts of Interest: Patent applications on various PSMA ligands have been filed by the German Cancer Research Center (DKFZ) in Heidelberg, Germany. M.B.-S. is listed as a co-inventor in those patents. In addition, patent applications covering ligands with applications to TαT have been filed by Cornell University and licensed to Ratio Therapeutics, Inc. J.M.K. is listed as a co-inventor in those patents and has financial interest in Ratio Therapeutics, Inc. No other potential conflicts of interest relevant to this article were reported.

References

1. Joliot, F.; Curie, I. Artificial Production of a New Kind of Radio-Element. *Nature* **1934**, *133*, 201–202. [CrossRef]
2. Becker, D.V.; Sawin, C.T. Radioiodine and thyroid disease: The beginning. *Semin. Nucl. Med.* **1996**, *26*, 155–164. [CrossRef]
3. Fahey, F.H.; Grant, F.D.; Thrall, J.H. Saul Hertz, MD, and the birth of radionuclide therapy. *EJNMMI Phys.* **2017**, *4*, 15. [CrossRef]
4. Hertz, S.; Roberts, A.; Evans, R.D. Radioactive Iodine as an Indicator in the Study of Thyroid Physiology. *Proc. Soc. Exp. Biol. Med.* **1938**, *38*, 510–513. [CrossRef]
5. Luster, M.; Clarke, S.E.; Dietlein, M.; Lassmann, M.; Lind, P.; Oyen, W.J.G.; Tennvall, J.; Bombardieri, E. Guidelines for radioiodine therapy of differentiated thyroid cancer. *Eur. J. Nucl. Med. Mol. Imaging* **2008**, *35*, 1941–1959. [CrossRef]
6. Haugen, B.R.; Alexander, E.K.; Bible, K.C.; Doherty, G.M.; Mandel, S.J.; Nikiforov, Y.E.; Pacini, F.; Randolph, G.W.; Sawka, A.M.; Schlumberger, M.; et al. American Thyroid Association Management Guidelines for Adult Patients with Thyroid Nodules and Differentiated Thyroid Cancer: The American Thyroid Association Guidelines Task Force on Thyroid Nodules and Differentiated Thyroid Cancer. *Thyroid®* **2015**, *26*, 1–133. [CrossRef]
7. Tuttle, R.M.; Ahuja, S.; Avram, A.M.; Bernet, V.J.; Bourguet, P.; Daniels, G.H.; Dillehay, G.; Draganescu, C.; Flux, G.; Führer, D.; et al. Controversies, Consensus, and Collaboration in the Use of 131I Therapy in Differentiated Thyroid Cancer: A Joint Statement from the American Thyroid Association, the European Association of Nuclear Medicine, the Society of Nuclear Medicine and Molecular Imaging, and the European Thyroid Association. *Thyroid* **2019**, *29*, 461–470. [CrossRef]
8. Hennrich, U.; Kopka, K. Lutathera®: The First FDA- and EMA-Approved Radiopharmaceutical for Peptide Receptor Radionuclide Therapy. *Pharmaceuticals* **2019**, *12*, 114. [CrossRef]
9. Fallah, J.; Agrawal, S.; Gittleman, H.; Fiero, M.H.; Subramaniam, S.; John, C.; Chen, W.; Ricks, T.K.; Niu, G.; Fotenos, A.; et al. FDA Approval Summary: Lutetium Lu 177 Vipivotide Tetraxetan for Patients with Metastatic Castration-Resistant Prostate Cancer. *Clin. Cancer Res.* **2023**, *29*, 1651–1657. [CrossRef]
10. Szponar, P.; Petrasz, P.; Brzeźniakiewicz-Janus, K.; Drewa, T.; Zorga, P.; Adamowicz, J. Precision strikes: PSMA-targeted radionuclide therapy in prostate cancer—A narrative review. *Front. Oncol.* **2023**, *13*, 1239118. [CrossRef]
11. Al-Toubah, T.; Strosberg, J.; Hallanger-Johnson, J.; El-Haddad, G. Targeted radionuclide therapy in endocrine-related cancers: Advances in the last decade. *Front. Endocrinol.* **2023**, *14*, 1187870. [CrossRef] [PubMed]
12. Sgouros, G.; Bodei, L.; McDevitt, M.R.; Nedrow, J.R. Radiopharmaceutical therapy in cancer: Clinical advances and challenges. *Nat. Rev. Drug Discov.* **2020**, *19*, 589–608. [CrossRef] [PubMed]
13. Ersahin, D.; Doddamane, I.; Cheng, D. Targeted Radionuclide Therapy. *Cancers* **2011**, *3*, 3838–3855. [CrossRef] [PubMed]
14. Lepareur, N.; Ramée, B.; Mougin-Degraef, M.; Bourgeois, M. Clinical Advances and Perspectives in Targeted Radionuclide Therapy. *Pharmaceutics* **2023**, *15*, 1733. [CrossRef] [PubMed]
15. Pouget, J.P.; Lozza, C.; Deshayes, E.; Boudousq, V.; Navarro-Teulon, I. Introduction to radiobiology of targeted radionuclide therapy. *Front. Med.* **2015**, *2*, 12. [CrossRef] [PubMed]
16. Floberg, J.M.; Wang, L.; Bandara, N.; Rashmi, R.; Mpoy, C.; Garbow, J.R.; Rogers, B.E.; Patti, G.J.; Schwarz, J.K. Alteration of Cellular Reduction Potential Will Change (64)Cu-ATSM Signal with or without Hypoxia. *J. Nucl. Med.* **2020**, *61*, 427–432. [CrossRef] [PubMed]
17. Wu, J. The Enhanced Permeability and Retention (EPR) Effect: The Significance of the Concept and Methods to Enhance Its Application. *J. Pers. Med.* **2021**, *11*, 771. [CrossRef] [PubMed]
18. Dhoundiyal, S.; Srivastava, S.; Kumar, S.; Singh, G.; Ashique, S.; Pal, R.; Mishra, N.; Taghizadeh-Hesary, F. Radiopharmaceuticals: Navigating the frontier of precision medicine and therapeutic innovation. *Eur. J. Med. Res.* **2024**, *29*, 26. [CrossRef] [PubMed]

19. Pouget, J.-P.; Santoro, L.; Piron, B.; Paillas, S.; Ladjohounlou, R.; Pichard, A.; Poty, S.; Deshayes, E.; Constanzo, J.; Bardiès, M. From the target cell theory to a more integrated view of radiobiology in Targeted radionuclide therapy: The Montpellier group's experience. *Nucl. Med. Biol.* **2022**, *104–105*, 53–64. [CrossRef]
20. Kratochwil, C.; Giesel, F.L.; Stefanova, M.; Benešová, M.; Bronzel, M.; Afshar-Oromieh, A.; Mier, W.; Eder, M.; Kopka, K.; Haberkorn, U. PSMA-Targeted Radionuclide Therapy of Metastatic Castration-Resistant Prostate Cancer with 177Lu-Labeled PSMA-617. *J. Nucl. Med.* **2016**, *57*, 1170–1176. [CrossRef]
21. Feuerecker, B.; Tauber, R.; Knorr, K.; Heck, M.; Beheshti, A.; Seidl, C.; Bruchertseifer, F.; Pickhard, A.; Gafita, A.; Kratochwil, C.; et al. Activity and Adverse Events of Actinium-225-PSMA-617 in Advanced Metastatic Castration-resistant Prostate Cancer After Failure of Lutetium-177-PSMA. *Eur. Urol.* **2021**, *79*, 343–350. [CrossRef]
22. Ilhan, H.; Gosewisch, A.; Böning, G.; Völter, F.; Zacherl, M.; Unterrainer, M.; Bartenstein, P.; Todica, A.; Gildehaus, F.J.; Ilhan, H.; et al. Response to 225Ac-PSMA-I&T after failure of long-term 177Lu-PSMA RLT in mCRPC. *Eur. J. Nucl. Med. Mol. Imaging* **2020**, *48*, 1262–1263. [CrossRef]
23. Friesen, C.; Glatting, G.; Koop, B.; Schwarz, K.; Morgenstern, A.; Apostolidis, C.; Debatin, K.-M.; Reske, S.N. Breaking Chemoresistance and Radioresistance with [213Bi]anti-CD45 Antibodies in Leukemia Cells. *Cancer Res.* **2007**, *67*, 1950–1958. [CrossRef]
24. Song, G.; Cheng, L.; Chao, Y.; Yang, K.; Liu, Z. Emerging Nanotechnology and Advanced Materials for Cancer Radiation Therapy. *Adv. Mater.* **2017**, *29*, 1700996. [CrossRef] [PubMed]
25. Kratochwil, C.; Bruchertseifer, F.; Giesel, F.L.; Weis, M.; Verburg, F.A.; Mottaghy, F.; Kopka, K.; Apostolidis, C.; Haberkorn, U.; Morgenstern, A. 225Ac-PSMA-617 for PSMA-Targeted α-Radiation Therapy of Metastatic Castration Resistant Prostate Cancer. *J. Nucl. Med.* **2016**, *57*, 1941–1944. [CrossRef]
26. Patel, C.M.; Wadas, T.J.; Shiozawa, Y.; Patel, C.M.; Wadas, T.J.; Shiozawa, Y. Progress in Targeted Alpha-Particle-Emitting Radiopharmaceuticals as Treatments for Prostate Cancer Patients with Bone Metastases. *Molecules* **2021**, *26*, 2162. [CrossRef] [PubMed]
27. Alan-Selcuk, N.; Beydagi, G.; Demirci, E.; Ocak, M.; Celik, S.; Oven, B.B.; Toklu, T.; Karaaslan, I.; Akcay, K.; Sonmez, O.; et al. Clinical Experience with [225Ac]Ac-PSMA Treatment in Patients with [177Lu]Lu-PSMA–Refractory Metastatic Castration-Resistant Prostate Cancer. *J. Nucl. Med.* **2023**, *64*, 1574–1580. [CrossRef]
28. Wulbrand, C.; Seidl, C.; Gaertner, F.C.; Bruchertseifer, F.; Morgenstern, A.; Essler, M.; Senekowitsch-Schmidtke, R. Alpha-Particle Emitting 213Bi-Anti-EGFR Immunoconjugates Eradicate Tumor Cells Independent of Oxygenation. *PLoS ONE* **2013**, *8*, e64730. [CrossRef] [PubMed]
29. Pouget, J.-P.; Constanzo, J. Revisiting the Radiobiology of Targeted Alpha Therapy. *Front. Med.* **2021**, *8*, 692436. [CrossRef]
30. Dekempeneer, Y.; Keyaerts, M.; Krasniqi, A.; Puttemans, J.; Muyldermans, S.; Lahoutte, T.; D'huyvetter, M.; Devoogdt, N. Targeted alpha therapy using short-lived alpha-particles and the promise of nanobodies as targeting vehicle. *Expert Opin Biol. Ther.* **2016**, *16*, 1035–1047. [CrossRef]
31. Bal, C.; Ballal, S.; Yadav, M. Long-term outcome of ^{225}Ac-DOTATATE Targeted Alpha Therapy in Patients with Metastatic Gastroenteropancreatic Neuroendocrine Tumors. *J. Nucl. Med.* **2021**, *62* (Suppl. S1), 19.
32. Lindegren, S.; Albertsson, P.; Bäck, T.; Jensen, H.; Palm, S.; Aneheim, E. Realizing Clinical Trials with Astatine-211: The Chemistry Infrastructure. *Cancer Biother. Radiopharm.* **2020**, *35*, 425–436. [CrossRef] [PubMed]
33. Albertsson, P.; Bäck, T.; Bergmark, K.; Hallqvist, A.; Johansson, M.; Aneheim, E.; Lindegren, S.; Timperanza, C.; Smerud, K.; Palm, S. Astatine-211 based radionuclide therapy: Current clinical trial landscape. *Front. Med.* **2023**, *9*, 1076210. [CrossRef] [PubMed]
34. McDevitt, M.R.; Ma, D.; Lai, L.T.; Simon, J.; Borchardt, P.; Frank, R.K.; Wu, K.; Pellegrini, V.; Curcio, M.J.; Miederer, M.; et al. Tumor Therapy with Targeted Atomic Nanogenerators. *Science* **2001**, *294*, 1537–1540. [CrossRef]
35. Roscher, M.; Bakos, G.; Benešová, M. Atomic Nanogenerators in Targeted Alpha Therapies: Curie's Legacy in Modern Cancer Management. *Pharmaceuticals* **2020**, *13*, 76. [CrossRef] [PubMed]
36. Poty, S.; Francesconi, L.C.; McDevitt, M.R.; Morris, M.J.; Lewis, J.S. α-Emitters for Radiotherapy: From Basic Radiochemistry to Clinical Studies—Part 1. *J. Nucl. Med.* **2018**, *59*, 878–884. [CrossRef]
37. Eychenne, R.; Chérel, M.; Haddad, F.; Guérard, F.; Gestin, J.-F. Overview of the Most Promising Radionuclides for Targeted Alpha Therapy: The "Hopeful Eight". *Pharmaceutics* **2021**, *13*, 906. [CrossRef] [PubMed]
38. Nelson, B.J.B.; Andersson, J.D.; Wuest, F. Targeted Alpha Therapy: Progress in Radionuclide Production, Radiochemistry, and Applications. *Pharmaceutics* **2020**, *13*, 49. [CrossRef]
39. Morgenstern, A.; Lebeda, O.; Stursa, J.; Bruchertseifer, F.; Capote, R.; McGinley, J.; Rasmussen, G.; Sin, M.; Zielinska, B.; Apostolidis, C. Production of 230U/226Th for Targeted Alpha Therapy via Proton Irradiation of 231Pa. *Anal. Chem.* **2008**, *80*, 8763–8770. [CrossRef]
40. Woods, J.J.; Unnerstall, R.; Hasson, A.; Abou, D.S.; Radchenko, V.; Thorek, D.L.J.; Wilson, J.J. Stable Chelation of the Uranyl Ion by Acyclic Hexadentate Ligands: Potential Applications for 230U Targeted α-Therapy. *Inorg. Chem.* **2022**, *61*, 3337–3350. [CrossRef]
41. Kostelnik, T.I.; Orvig, C. Radioactive Main Group and Rare Earth Metals for Imaging and Therapy. *Chem. Rev.* **2019**, *119*, 902–956. [CrossRef] [PubMed]
42. Thiele, N.A.; Brown, V.; Kelly, J.M.; Amor-Coarasa, A.; Jermilova, U.; MacMillan, S.N.; Nikolopoulou, A.; Ponnala, S.; Ramogida, C.F.; Robertson, A.K.H.; et al. An Eighteen-Membered Macrocyclic Ligand for Actinium-225 Targeted Alpha Therapy. *Angew. Chem. Int. Ed.* **2017**, *56*, 14712–14717. [CrossRef] [PubMed]

43. Reissig, F.; Zarschler, K.; Novy, Z.; Petrik, M.; Bendova, K.; Kurfurstova, D.; Bouchal, J.; Ludik, M.-C.; Brandt, F.; Kopka, K.; et al. Modulating the pharmacokinetic profile of Actinium-225-labeled macropa-derived radioconjugates by dual targeting of PSMA and albumin. *Theranostics* **2022**, *12*, 7203. [CrossRef] [PubMed]
44. King, A.P.; Gutsche, N.T.; Raju, N.; Fayn, S.; Baidoo, K.E.; Bell, M.M.; Olkowski, C.S.; Swenson, R.E.; Lin, F.I.; Sadowski, S.M.; et al. 225Ac-MACROPATATE: A Novel α-Particle Peptide Receptor Radionuclide Therapy for Neuroendocrine Tumors. *J. Nucl. Med.* **2023**, *64*, 549–554. [CrossRef] [PubMed]
45. Kokov, K.V.; Egorova, B.V.; German, M.N.; Klabukov, I.D.; Krasheninnikov, M.E.; Larkin-Kondrov, A.A.; Makoveeva, K.A.; Ovchinnikov, M.V.; Sidorova, M.V.; Chuvilin, D.Y. 212Pb: Production Approaches and Targeted Therapy Applications. *Pharmaceutics* **2022**, *14*, 189. [CrossRef] [PubMed]
46. Shannon, R.D. Revised effective ionic radii and systematic studies of interatomic distances in halides and chalcogenides. *Acta Cryst. A* **1976**, *32*, 751–767. [CrossRef]
47. Hagemann, U.B.; Wickstroem, K.; Hammer, S.; Bjerke, R.M.; Zitzmann-Kolbe, S.; Ryan, O.B.; Karlsson, J.; Scholz, A.; Hennekes, H.; Mumberg, D.; et al. Advances in Precision Oncology: Targeted Thorium-227 Conjugates As a New Modality in Targeted Alpha Therapy. *Cancer Biother. Radiopharm.* **2020**, *35*, 497–510. [CrossRef] [PubMed]
48. Ramdahl, T.; Bonge-Hansen, H.T.; Ryan, O.B.; Larsen, Å.; Herstad, G.; Sandberg, M.; Bjerke, R.M.; Grant, D.; Brevik, E.M.; Cuthbertson, A.S. An efficient chelator for complexation of thorium-227. *Bioorganic Med. Chem. Lett.* **2016**, *26*, 4318–4321. [CrossRef] [PubMed]
49. Franchi, S.; Asti, M.; Di Marco, V.; Tosato, M. The Curies' element: State of the art and perspectives on the use of radium in nuclear medicine. *EJNMMI Radiopharm. Chem.* **2023**, *8*, 38. [CrossRef]
50. Poeppel, T.D.; Handkiewicz-Junak, D.; Andreeff, M.; Becherer, A.; Bockisch, A.; Fricke, E.; Geworski, L.; Heinzel, A.; Krause, B.J.; Krause, T.; et al. EANM guideline for radionuclide therapy with radium-223 of metastatic castration-resistant prostate cancer. *Eur. J. Nucl. Med. Mol. Imaging* **2017**, *45*, 824–845. [CrossRef]
51. Abou, D.S.; Ulmert, D.; Doucet, M.; Hobbs, R.F.; Riddle, R.C.; Thorek, D.L.J. Whole-Body and Microenvironmental Localization of Radium-223 in Naïve and Mouse Models of Prostate Cancer Metastasis. *JNCI J. Natl. Cancer Inst.* **2016**, *108*, djv380. [CrossRef] [PubMed]
52. Henriksen, G.; Hoff, P.; Larsen, R.H. Evaluation of potential chelating agents for radium. *Appl. Radiat. Isot.* **2002**, *56*, 667–671. [CrossRef] [PubMed]
53. Kozempel, J.; Mokhodoeva, O.; Vlk, M. Progress in Targeted Alpha-Particle Therapy. What We Learned about Recoils Release from In Vivo Generators. *Molecules* **2018**, *23*, 581. [CrossRef] [PubMed]
54. Alekseev, I.; Miroslavov, A.; Stepanova, E. Post-effects of radioactive decay in ligands on biologically active transport platforms. *Radiat. Phys. Chem.* **2019**, *162*, 96–106. [CrossRef]
55. Trujillo-Nolasco, M.; Morales-Avila, E.; Cruz-Nova, P.; Katti, K.V.; Ocampo-García, B. Nanoradiopharmaceuticals Based on Alpha Emitters: Recent Developments for Medical Applications. *Pharmaceutics* **2021**, *13*, 1123. [CrossRef] [PubMed]
56. Kruijff, R.M.d.; Raavé, R.; Kip, A.; Molkenboer-Kuenen, J.; Morgenstern, A.; Bruchertseifer, F.; Heskamp, S.; Denkova, A.G. The in vivo fate of 225Ac daughter nuclides using polymersomes as a model carrier. *Sci. Rep.* **2019**, *9*, 11671. [CrossRef] [PubMed]
57. Singh Jaggi, J.; Kappel, B.J.; McDevitt, M.R.; Sgouros, G.; Flombaum, C.D.; Cabassa, C.; Scheinberg, D.A. Efforts to Control the Errant Products of a Targeted In vivo Generator. *Cancer Res.* **2005**, *65*, 4888–4895. [CrossRef] [PubMed]
58. Begum, N.J.; Glatting, G.; Wester, H.-J.; Eiber, M.; Beer, A.J.; Kletting, P. The effect of ligand amount, affinity and internalization on PSMA-targeted imaging and therapy: A simulation study using a PBPK model. *Sci. Rep.* **2019**, *9*, 20041. [CrossRef]
59. Królicki, L.; Bruchertseifer, F.; Kunikowska, J.; Koziara, H.; Pawlak, D.; Kuliński, R.; Rola, R.; Merlo, A.; Morgenstern, A. Dose escalation study of targeted alpha therapy with [(225)Ac]Ac-DOTA-substance P in recurrence glioblastoma—Safety and efficacy. *Eur. J. Nucl. Med. Mol. Imaging* **2021**, *48*, 3595–3605. [CrossRef]
60. Holzwarth, U.; Ojea Jimenez, I.; Calzolai, L. A random walk approach to estimate the confinement of α-particle emitters in nanoparticles for targeted radionuclide therapy. *EJNMMI Radiopharm. Chem.* **2018**, *3*, 9. [CrossRef]
61. Henriksen, G.; Fisher, D.R.; Roeske, J.C.; Bruland, Ø.S.; Larsen, R.H. Targeting of Osseous Sites with α-Emitting ^{223}Ra: Comparison with the β-Emitter ^{89}Sr in Mice. *J. Nucl. Med.* **2003**, *44*, 252–259. [PubMed]
62. Briggs, G.H.; Rutherford, E. A determination of the absolute velocity of the alpha-particles from radium C′. *Proc. R. Soc. Lond. Ser. A-Math. Phys. Sci.* **1936**, *157*, 183–194. [CrossRef]
63. Curie, I. Recherches sur les rayons α du polonium, oscillation de parcours, vitesse d'émission, pouvoir ionisant. *Ann. De Phys.* **1925**, *10*, 299–401. [CrossRef]
64. Rosenblum, S.; Dupouy, G. Mesure absolue des vitesses des principaux groupes de rayons Alpha. *J. Phys. Radium* **1933**, *4*, 262–268. [CrossRef]
65. Briggs, G.H.; Rutherford, E. Measurements of the relative velocities of the α-particles from radon, radium A, and radium C′. *Proc. R. Soc. Lond. Ser. A Contain. Pap. A Math. Phys. Character* **1934**, *143*, 604–617. [CrossRef]
66. Laurence, G.C.; Rutherford, E. Relative velocities of the alpha-particles emitted by certain radioactive elements. *Proc. R. Soc. Lond. Ser. A Contain. Pap. A Math. Phys. Character* **1929**, *122*, 543–551. [CrossRef]
67. Sóti, Z.; Magill, J.; Dreher, R. Karlsruhe Nuclide Chart—New 10th edition 2018. *EPJ Nucl. Sci. Technol.* **2019**, *5*, 11. [CrossRef]

68. Rasheed, R.; Usmani, S.; Naqvi, S.A.R.; Alkandari, F.; Marafi, F. Alpha Therapy with 225Actinium Labeled Prostate Specific Membrane Antigen: Reporting New Photopeak of 78 Kilo-electron Volts for Better Image Statistics. *Indian J. Nucl. Med.* **2019**, *34*, 76–77. [CrossRef] [PubMed]
69. Rodney Withers, H. Biological basis of radiation therapy for cancer. *Lancet* **1992**, *339*, 156–159. [CrossRef]
70. Desouky, O.; Ding, N.; Zhou, G. Targeted and non-targeted effects of ionizing radiation. *J. Radiat. Res. Appl. Sci.* **2015**, *8*, 247–254. [CrossRef]
71. Wilkinson, B.; Hill, M.A.; Parsons, J.L. The Cellular Response to Complex DNA Damage Induced by Ionising Radiation. *Int. J. Mol. Sci.* **2023**, *24*, 4920. [CrossRef]
72. Goodhead, D.T.; Weinfeld, M. Clustered DNA Damage and its Complexity: Tracking the History. *Radiat. Res.* **2024**. [CrossRef] [PubMed]
73. Huang, R.; Zhou, P.K. DNA damage repair: Historical perspectives, mechanistic pathways and clinical translation for targeted cancer therapy. *Signal Transduct. Target. Ther.* **2021**, *6*, 254. [CrossRef]
74. Nickoloff, J.A.; Sharma, N.; Allen, C.P.; Taylor, L.; Allen, S.J.; Jaiswal, A.S.; Hromas, R. Roles of homologous recombination in response to ionizing radiation-induced DNA damage. *Int. J. Radiat. Biol.* **2023**, *99*, 903–914. [CrossRef] [PubMed]
75. Filippi, L.; Palumbo, B.; Bagni, O.; Frantellizzi, V.; De Vincentis, G.; Schillaci, O. DNA Damage Repair Defects and Targeted Radionuclide Therapies for Prostate Cancer: Does Mutation Really Matter? A Systematic Review. *Life* **2023**, *13*, 55. [CrossRef] [PubMed]
76. Huang, R.X.; Zhou, P.K. DNA damage response signaling pathways and targets for radiotherapy sensitization in cancer. *Signal Transduct. Target. Ther.* **2020**, *5*, 60. [CrossRef] [PubMed]
77. Groelly, F.J.; Fawkes, M.; Dagg, R.A.; Blackford, A.N.; Tarsounas, M. Targeting DNA damage response pathways in cancer. *Nat. Rev. Cancer* **2023**, *23*, 78–94. [CrossRef]
78. Craig, D.J.; Nanavaty, N.S.; Devanaboyina, M.; Stanbery, L.; Hamouda, D.; Edelman, G.; Dworkin, L.; Nemunaitis, J.J. The abscopal effect of radiation therapy. *Future Oncol.* **2021**, *17*, 1683–1694. [CrossRef]
79. European Medicines Agency: Xofigo—Radium Ra223 Dichloride. Available online: https://www.ema.europa.eu/en/medicines/human/EPAR/xofigo (accessed on 11 March 2024).
80. Parker, C.; Lewington, V.; Shore, N.; Kratochwil, C.; Levy, M.; Lindén, O.; Noordzij, W.; Park, J.; Saad, F. Targeted Alpha Therapy, an Emerging Class of Cancer Agents: A Review. *JAMA Oncol.* **2018**, *4*, 1765–1772. [CrossRef]
81. Higano, C.S.; George, D.J.; Shore, N.D.; Sartor, O.; Miller, K.; Conti, P.S.; Sternberg, C.N.; Saad, F.; Sade, J.P.; Bellmunt, J.; et al. Clinical outcomes and treatment patterns in REASSURE: Planned interim analysis of a real-world observational study of radium-223 in metastatic castration-resistant prostate cancer. *eClinicalMedicine* **2023**, *60*, 101993. [CrossRef]
82. Jang, A.; Kendi, A.T.; Johnson, G.B.; Halfdanarson, T.R.; Sartor, O. Targeted Alpha-Particle Therapy: A Review of Current Trials. *Int. J. Mol. Sci.* **2023**, *24*, 11626. [CrossRef]
83. Pallares, R.M.; Abergel, R.J. Development of radiopharmaceuticals for targeted alpha therapy: Where do we stand? *Front. Med.* **2022**, *9*, 1020188. [CrossRef]
84. Sgouros, G.; Roeske, J.C.; McDevitt, M.R.; Palm, S.; Allen, B.J.; Fisher, D.R.; Brill, A.B.; Song, H.; Howell, R.W.; Akabani, G. MIRD Pamphlet No. 22 (Abridged): Radiobiology and Dosimetry of α-Particle Emitters for Targeted Radionuclide Therapy. *J. Nucl. Med.* **2010**, *51*, 311–328. [CrossRef]
85. Jurcic, J.G.; Larson, S.M.; Sgouros, G.; McDevitt, M.R.; Finn, R.D.; Divgi, C.R.; Ballangrud, Å.M.; Hamacher, K.A.; Ma, D.; Humm, J.L.; et al. Targeted α particle immunotherapy for myeloid leukemia. *Blood* **2002**, *100*, 1233–1239. [CrossRef] [PubMed]
86. Zalutsky, M.R.; Reardon, D.A.; Akabani, G.; Coleman, R.E.; Friedman, A.H.; Friedman, H.S.; McLendon, R.E.; Wong, T.Z.; Bigner, D.D. Clinical Experience with α-Particle–Emitting 211At: Treatment of Recurrent Brain Tumor Patients with 211At-Labeled Chimeric Antitenascin Monoclonal Antibody 81C6. *J. Nucl. Med.* **2008**, *49*, 30–38. [CrossRef]
87. Geerlings, M.W.; Kaspersen, F.M.; Apostolidis, C.; Der Hout, R.V. The feasibility of 225 Ac as a source of α-particles in radioimmunotherapy. *Nucl. Med. Commun.* **1993**, *14*, 121–125. [CrossRef] [PubMed]
88. Jurcic, J.G.; Rosenblat, T.L.; McDevitt, M.R.; Pandit-Taskar, N.; Carrasquillo, J.A.; Chanel, S.M.; Zikaras, K.; Frattini, M.G.; Maslak, P.G.; Cicic, D.; et al. Phase I Trial of the Targeted Alpha-Particle Nano-Generator Actinium-225 (225Ac)-Lintuzumab (Anti-CD33; HuM195) in Acute Myeloid Leukemia (AML). *Blood* **2011**, *118*, 768. [CrossRef]
89. Jurcic, J.G.; Rosenblat, T.L. Targeted alpha-particle immunotherapy for acute myeloid leukemia. In *American Society of Clinical Oncology Educational Book*; Annual Meeting; American Society of Clinical Oncology: Alexandria, VA, USA, 2014; pp. e126–e131.
90. Rosenblat, T.L.; McDevitt, M.R.; Carrasquillo, J.A.; Pandit-Taskar, N.; Frattini, M.G.; Maslak, P.G.; Park, J.H.; Douer, D.; Cicic, D.; Larson, S.M.; et al. Treatment of Patients with Acute Myeloid Leukemia with the Targeted Alpha-Particle Nanogenerator Actinium-225-Lintuzumab. *Clin. Cancer Res.* **2022**, *28*, 2030–2037. [CrossRef] [PubMed]
91. Atallah, E.; Berger, M.; Jurcic, J.; Roboz, G.; Tse, W.; Mawad, R.; Rizzieri, D.; Begna, K.; Orozco, J.; Craig, M.; et al. A Phase 2 Study of Actinium-225 (225Ac)-lintuzumab in Older Patients with Untreated Acute Myeloid Leukemia (AML). *J. Med. Imaging Radiat. Sci.* **2019**, *50*, S37. [CrossRef]
92. Tagawa, S.T.; Thomas, C.; Sartor, A.O.; Sun, M.; Stangl-Kremser, J.; Bissassar, M.; Vallabhajosula, S.; Castellanos, S.H.; Nauseef, J.T.; Sternberg, C.N.; et al. Prostate-Specific Membrane Antigen–Targeting Alpha Emitter via Antibody Delivery for Metastatic Castration-Resistant Prostate Cancer: A Phase I Dose-Escalation Study of 225Ac-J591. *J. Clin. Oncol.* **2023**, *42*, 842–851. [CrossRef]

93. Nauseef, J.T.; Sun, M.P.; Thomas, C.; Bissassar, M.; Patel, A.; Tan, A.; Fernandez, E.; Davidson, Z.; Chamberlain, T.; Earle, K.; et al. A phase I/II dose-escalation study of fractionated 225Ac-J591 for progressive metastatic castration-resistant prostate cancer (mCRPC) in patients with prior treatment with 177Lu-PSMA. *J. Clin. Oncol.* **2023**, *41* (Suppl. S6), TPS288. [CrossRef]
94. Sun, M.P.; Nauseef, J.T.; Palmer, J.; Thomas, J.E.; Stangl-Kremser, J.; Bissassar, M.; Castellanos, S.H.; Osborne, J.; Molina, A.M.; Sternberg, C.N.; et al. Phase I results of a phase I/II study of pembrolizumab and AR signaling inhibitor (ARSI) with 225Ac-J591. *J. Clin. Oncol.* **2023**, *41* (Suppl. S6), 181. [CrossRef]
95. Juergens, R.A.; Zukotynski, K.A.; Juneau, D.; Krnezich, L.; Simms, R.; Forbes, J.; Burak, E.S.; Valliant, J.; Krnezich, L.; Stafford, L.; et al. A Phase I Study of [225Ac]-FPI-1434 Radioimmunotherapy in Patients with IGF-1R Expressing Solid Tumors. *J. Clin. Oncol.* **2019**, *37*, TPS3152. [CrossRef]
96. Juneau, D.; Saad, F.; Berlin, A.; Metser, U.; Puzanov, I.; Lamonica, D.; Sparks, R.; Burak, E.; Simms, R.; Rhoden, J.; et al. Preliminary Dosimetry Results from a First-in-Human Phase I Study Evaluating the Efficacy and Safety of [^{225}Ac]-FPI-1434 in Patients with IGF-1R Expressing Solid Tumors. *J. Nucl. Med.* **2021**, *62* (Suppl. S1), 74.
97. Scott, A.; Saad, F.; Juneau, D.; Brown, M.; Kirkwood, I.; Nottage, M.; Rhoden, J.; Kazakin, J.; Schindler, J.; Baca, N.; et al. Impact of Pre-Administration of Anti-IGF-1R Antibody FPI 1175 on the Dosimetry, Tumor Uptake, and Pharmacokinetics of the IGF-1R Targeted Theranostic Imaging Agent [111In] FPI 1547 in Patients with Solid Tumors. *J. Nucl. Med.* **2022**, *63* (Suppl. S2), 2275.
98. Pandit-Taskar, N.; Wong, J.; Subbiah, V.; Chenard-Poirier, M.; Pryma, D.; Berlin, A.; Juergens, R.; Puzanov, I.; Jacene, H.; Schindler, J.; et al. Dose-escalation Study of [225Ac]-FPI-1434 (FPI-1434) in Patients (pts) with IGF-1R Expressing Advanced Solid Tumors: Preliminary Pharmacology and Dosimetry Results. *J. Nucl. Med.* **2023**, *64* (Suppl. S1), P630.
99. Karlsson, J. Abstract ND07: HER2-TTC: A Targeted thorium conjugate to treat HER2 expressing cancers with potent alpha radiation. *Cancer Res.* **2021**, *81* (Suppl. S13), ND07. [CrossRef]
100. Ma, J.; Zhang, C. Efficacy and Safety of 225Ac-PSMA-617-Targeted Alpha Therapy in Metastatic Castration-Resistant Prostate Cancer: A Systematic Review and Meta-Analysis. *Front. Oncol.* **2022**, *12*, 796657. [CrossRef] [PubMed]
101. Feuerecker, B.; Kratochwil, C.; Ahmadzadehfar, H.; Morgenstern, A.; Eiber, M.; Herrmann, K.; Pomykala, K.L. Clinical Translation of Targeted α-Therapy: An Evolution or a Revolution? *J. Nucl. Med.* **2023**, *64*, 685–692. [CrossRef]
102. Sathekge, M.; Crumbaker, M.; Joshua, A.; Bruchertseifer, F.; Kreisl, T.; Emineni, S.; Wehbe, J.; Korn, M.; Morgenstern, A.; Emmett, L. AcTION: A phase 1 study of [225Ac]Ac-PSMA-617 in men with PSMA-positive prostate cancer with or without prior [177Lu]Lu-PSMA-617 radioligand therapy. *Eur. J. Nucl. Med. Mol. Imaging* **2023**, *50*, S152–S153 OP-335. [CrossRef]
103. Morris, M.; Ulaner, G.A.; Halperin, D.M.; Mehr, S.H.; Li, D.; Soares, H.P.; Anthony, L.B.; Kotiah, S.D.; Jacene, H.; Pavel, M.E.; et al. ACTION-1 phase Ib/3 trial of RYZ101 in somatostatin receptor subtype 2–expressing (SSTR2+) gastroenteropancreatic neuroendocrine tumors (GEP-NET) progressing after 177Lu somatostatin analogue (SSA) therapy: Initial safety analysis. *J. Clin. Oncol.* **2023**, *41* (Suppl. S16), 4132. [CrossRef]
104. Sgouros, G.; Ulaner, G.; Delie, T.; Kotiah, S.; Ferreira, D.; Ma, K.; Rearden, J.; Moran, S.; Sneeden, E.; He, B.; et al. 225Ac-DOTATATE Dosimetry Results from Part 1 of the ACTION-1 Trial. *J. Nucl. Med.* **2023**, *64* (Suppl. S1), P129.
105. Delpassand, E.S.; Tworowska, I.; Esfandiari, R.; Torgue, J.; Hurt, J.; Shafie, A.; Núñez, R. Targeted α-Emitter Therapy with 212Pb-DOTAMTATE for the Treatment of Metastatic SSTR-Expressing Neuroendocrine Tumors: First-in-Humans Dose-Escalation Clinical Trial. *J. Nucl. Med.* **2022**, *63*, 1326–1333. [CrossRef] [PubMed]
106. Delpassand, E.; Esfandiari, R.; Tworowska, I.; Torgue, J.; Hurt, J.D.; Nunez, R. Targeted alpha-emitter therapy with 212Pb-DOTAMTATE in neuroendocrine tumor subjects who progressed following prior 177Lu/90Y-PRRT. *J. Clin. Oncol.* **2022**, *40* (Suppl. S16), 4128. [CrossRef]
107. Privé, B.M.; Boussihmad, M.A.; Timmermans, B.; van Gemert, W.A.; Peters, S.M.B.; Derks, Y.H.W.; van Lith, S.A.M.; Mehra, N.; Nagarajah, J.; Heskamp, S.; et al. Fibroblast activation protein-targeted radionuclide therapy: Background, opportunities, and challenges of first (pre)clinical studies. *Eur. J. Nucl. Med. Mol. Imaging* **2023**, *50*, 1906–1918. [CrossRef] [PubMed]
108. Ma, H.; Li, F.; Shen, G.; Pan, L.; Liu, W.; Liang, R.; Lan, T.; Yang, Y.; Yang, J.; Liao, J.; et al. In vitro and in vivo evaluation of 211At-labeled fibroblast activation protein inhibitor for glioma treatment. *Bioorganic Med. Chem.* **2022**, *55*, 116600. [CrossRef] [PubMed]
109. Aso, A.; Nabetani, H.; Matsuura, Y.; Kadonaga, Y.; Shirakami, Y.; Watabe, T.; Yoshiya, T.; Mochizuki, M.; Ooe, K.; Kawakami, A.; et al. Evaluation of Astatine-211-Labeled Fibroblast Activation Protein Inhibitor (FAPI): Comparison of Different Linkers with Polyethylene Glycol and Piperazine. *Int. J. Mol. Sci.* **2023**, *24*, 8701. [CrossRef] [PubMed]
110. Liu, Y.; Watabe, T.; Kaneda-Nakashima, K.; Shirakami, Y.; Naka, S.; Ooe, K.; Toyoshima, A.; Nagata, K.; Haberkorn, U.; Kratochwil, C.; et al. Fibroblast activation protein targeted therapy using [177Lu]FAPI-46 compared with [225Ac]FAPI-46 in a pancreatic cancer model. *Eur. J. Nucl. Med. Mol. Imaging* **2021**, *49*, 871–880. [CrossRef]
111. Watabe, T.; Liu, Y.; Kaneda-Nakashima, K.; Shirakami, Y.; Lindner, T.; Ooe, K.; Toyoshima, A.; Nagata, K.; Shimosegawa, E.; Haberkorn, U.; et al. Theranostics Targeting Fibroblast Activation Protein in the Tumor Stroma: ^{64}Cu- and ^{225}Ac-Labeled FAPI-04 in Pancreatic Cancer Xenograft Mouse Models. *J. Nucl. Med.* **2020**, *61*, 563–569. [CrossRef]
112. Kneifel, S.; Cordier, D.; Good, S.; Ionescu, M.; Ghaffari, A.; Hofer, S.; Kretschmar, M.; Tolnay, M.; Apostolidis, C.; Waser, B.; et al. Local Trageting of Malignant Gliomas by the Difussible Peptidic Vector DOTAGA-substance P. *Clin. Cancer Res.* **2006**, *12*, 3843–3850. [CrossRef]

113. Królicki, L.; Kunikowska, J.; Bruchertseifer, F.; Koziara, H.; Królicki, B.; Jakuciński, M.; Pawlak, D.; Rola, R.; Morgenstern, A.; Rosiak, E.; et al. 225 Ac- and 213 Bi-Substance P Analogues for Glioma Therapy. *Semin. Nucl. Med.* **2020**, *50*, 141–151. [CrossRef] [PubMed]
114. Królicki, L.; Bruchertseifer, F.; Kunikowska, J.; Koziara, H.; Królicki, B.; Jakuciński, M.; Pawlak, D.; Apostolidis, C.; Mirzadeh, S.; Rola, R.; et al. Safety and efficacy of targeted alpha therapy with 213Bi-DOTA-substance P in recurrent glioblastoma. *Eur. J. Nucl. Med. Mol. Imaging* **2018**, *46*, 614–622. [CrossRef] [PubMed]
115. Królicki, L.; Kunikowska, J.; Bruchertseifer, F.; Kulinski, R.; Pawlak, D.; Koziara, H.; Rola, R.; Morgenstern, A.; Merlo, A. Locoregional Treatment of Glioblastoma with Targeted Alpha Therapy [213Bi]Bi-DOTA-Substance P Versus [225Ac]Ac-DOTA-Substance P-Analysis of Influence Parameters. *Clin. Nucl. Med.* **2023**, *48*, 387–392. [CrossRef] [PubMed]
116. Trencsényi, G.; Csikos, C.; Képes, Z. Targeted Radium Alpha Therapy in the Era of Nanomedicine: In Vivo Results. *Int. J. Mol. Sci.* **2024**, *25*, 664. [CrossRef] [PubMed]
117. Kleynhans, J.; Sathekge, M.; Ebenhan, T. Obstacles and Recommendations for Clinical Translation of Nanoparticle System-Based Targeted Alpha-Particle Therapy. *Materials* **2021**, *14*, 4784. [CrossRef] [PubMed]
118. Davis, M.E.; Chen, Z.G.; Shin, D.M. Nanoparticle therapeutics: An emerging treatment modality for cancer. *Nat. Rev. Drug Discov.* **2008**, *7*, 771–782. [CrossRef] [PubMed]
119. Sathekge, M.; Bruchertseifer, F.; Vorster, M.; Lawal, I.O.; Mokoala, K.; Reed, J.; Maseremule, L.; Ndlovu, H.; Hlongwa, K.; Maes, A.; et al. 225Ac-PSMA-617 radioligand therapy of de novo metastatic hormone-sensitive prostate carcinoma (mHSPC): Preliminary clinical findings. *Eur. J. Nucl. Med. Mol. Imaging* **2023**, *50*, 2210–2218. [CrossRef] [PubMed]
120. Kratochwil, C.; Bruchertseifer, F.; Rathke, H.; Hohenfellner, M.; Giesel, F.L.; Haberkorn, U.; Morgenstern, A. Targeted α-Therapy of Metastatic Castration-Resistant Prostate Cancer with ^{225}Ac-PSMA-617: Swimmer-Plot Analysis Suggests Efficacy Regarding Duration of Tumor Control. *J. Nucl. Med.* **2018**, *59*, 795. [CrossRef] [PubMed]
121. Kratochwil, C.; Giesel, F.L.; Heussel, C.-P.; Kazdal, D.; Endris, V.; Nientiedt, C.; Bruchertseifer, F.; Kippenberger, M.; Rathke, H.; Leichsenring, J.; et al. Patients Resistant Against PSMA-Targeting α-Radiation Therapy Often Harbor Mutations in DNA Damage-Repair–Associated Genes. *J. Nucl. Med.* **2020**, *61*, 683–688. [CrossRef]
122. Zalutsky, M.R.; Pruszynski, M. Astatine-211: Production and availability. *Curr. Radiopharm.* **2011**, *4*, 177–185. [CrossRef]
123. Radchenko, V.; Morgenstern, A.; Jalilian, A.R.; Ramogida, C.F.; Cutler, C.; Duchemin, C.; Hoehr, C.; Haddad, F.; Bruchertseifer, F.; Gausemel, H.; et al. Production and Supply of α-Particle–Emitting Radionuclides for Targeted α-Therapy. *J. Nucl. Med.* **2021**, *62*, 1495–1503. [CrossRef] [PubMed]
124. Heynickx, N.; Herrmann, K.; Vermeulen, K.; Baatout, S.; Aerts, A. The salivary glands as a dose limiting organ of PSMA- targeted radionuclide therapy: A review of the lessons learnt so far. *Nucl. Med. Biol.* **2021**, *98–99*, 30–39. [CrossRef] [PubMed]
125. Kratochwil, C.; Apostolidis, L.; Rathke, H.; Apostolidis, C.; Bicu, F.; Bruchertseifer, F.; Choyke, P.L.; Haberkorn, U.; Giesel, F.L.; Morgenstern, A. Dosing (225)Ac-DOTATOC in patients with somatostatin-receptor-positive solid tumors: 5-year follow-up of hematological and renal toxicity. *Eur. J. Nucl. Med. Mol. Imaging* **2021**, *49*, 54–63. [CrossRef] [PubMed]
126. Soldatos, T.G.; Iakovou, I.; Sachpekidis, C. Retrospective Toxicological Profiling of Radium-223 Dichloride for the Treatment of Bone Metastases in Prostate Cancer Using Adverse Event Data. *Medicina* **2019**, *55*, 149. [CrossRef] [PubMed]
127. Tagawa, S.T.; Sun, M.P.; Nauseef, J.T.; Thomas, C.; Castellanos, S.H.; Thomas, J.E.; Davidson, Z.; Stangl-Kremser, J.; Bissassar, M.; Palmer, J.; et al. Phase I dose-escalation results of prostate-specific membrane antigen-targeted radionuclide therapy (PSMA-TRT) with alpha-radiolabeled antibody 225Ac-J591 and beta-radioligand 177Lu-PSMA I&T. *J. Clin. Oncol.* **2023**, *41* (Suppl. S16), 5018. [CrossRef]
128. Kostos, L.; Buteau, J.P.; Yeung, T.; Iulio, J.D.; Xie, J.; Cardin, A.; Chin, K.Y.; Emmerson, B.; Owen, K.L.; Parker, B.S.; et al. AlphaBet: Combination of Radium-223 and [177Lu]Lu-PSMA-I&T in men with metastatic castration-resistant prostate cancer (clinical trial protocol). *Front. Med.* **2022**, *9*, 1059122. [CrossRef]
129. Scholz, A.; Knuuttila, M.; Zdrojewska, J.; Suominen, M.I.; Alhoniemi, E.; Schatz, C.A.; Zitzmann-Kolbe, S.; Käkönen, S.-M.; Hagemann, U.B. Abstract 5043: Radium-223 demonstrates increased antitumor activity in combination with 177Lu-PSMA-617 in the intratibial LNCaP xenograft model of bone metastatic prostate cancer. *Cancer Res.* **2023**, *83* (Suppl. S7), 5043. [CrossRef]
130. Meyer, C.; Stuparu, A.; Lueckerath, K.; Calais, J.; Czernin, J.; Slavik, R.; Dahlbom, M. Tandem Isotope Therapy with ^{225}Ac- and ^{177}Lu-PSMA-617 in a Murine Model of Prostate Cancer. *J. Nucl. Med.* **2023**, *64*, 1772–1778. [CrossRef]
131. Khreish, F.; Ebert, N.; Ries, M.; Maus, S.; Rosar, F.; Bohnenberger, H.; Stemler, T.; Saar, M.; Bartholomä, M.; Ezziddin, S. 225Ac-PSMA-617/177Lu-PSMA-617 tandem therapy of metastatic castration-resistant prostate cancer: Pilot experience. *Eur. J. Nucl. Med. Mol. Imaging* **2019**, *47*, 721–728. [CrossRef]
132. Rosar, F.; Krause, J.; Bartholomä, M.; Maus, S.; Stemler, T.; Hierlmeier, I.; Linxweiler, J.; Ezziddin, S.; Khreish, F. Efficacy and Safety of [225Ac]Ac-PSMA-617 Augmented [177Lu]Lu-PSMA-617 Radioligand Therapy in Patients with Highly Advanced mCRPC with Poor Prognosis. *Pharmaceutics* **2021**, *13*, 722. [CrossRef]
133. Kulkarni, H.; Zhang, J.; Langbein, T.; Schuchardt, C.; Singh, A.; Mueller, D.; Baum, R. Radioligand therapy using combination of Ac-225 and Lu-177 labelled PSMA ligands for progressive end-stage metastatic prostate cancer: Effective trade-off between response and toxicity. *J. Nucl. Med.* **2019**, *60* (Suppl. S1), 464.
134. Langbein, T.; Kulkarni, H.R.; Schuchardt, C.; Mueller, D.; Volk, G.F.; Baum, R.P. Salivary Gland Toxicity of PSMA-Targeted Radioligand Therapy with 177Lu-PSMA and Combined 225Ac- and 177Lu-Labeled PSMA Ligands (TANDEM-PRLT) in Advanced Prostate Cancer: A Single-Center Systematic Investigation. *Diagnostics* **2022**, *12*, 1926. [CrossRef] [PubMed]

135. Langbein, T.; Kulkarni, H.; Singh, A.; Schuchardt, C.; Zhang, J.; Baum, R. Combined PRLT using Ac-225 and Lu-177 labeled PSMA-617 (TANDEM-PRLT) in end-stage metastatic prostate cancer: A concept to reduce salivary gland toxicity? *J. Nucl. Med.* **2019**, *60* (Suppl. S1), 77.
136. Juzeniene, A.; Stenberg, V.Y.; Bruland, Ø.S.; Revheim, M.-E.; Larsen, R.H. Dual targeting with 224Ra/212Pb-conjugates for targeted alpha therapy of disseminated cancers: A conceptual approach. *Front. Med.* **2023**, *9*, 1051825. [CrossRef] [PubMed]
137. Tornes, A.J.K.; Stenberg, V.Y.; Larsen, R.H.; Bruland, Ø.S.; Revheim, M.-E.; Juzeniene, A. Targeted alpha therapy with the 224Ra/212Pb-TCMC-TP-3 dual alpha solution in a multicellular tumor spheroid model of osteosarcoma. *Front. Med.* **2022**, *9*, 1058863. [CrossRef]
138. Delaney, G.; Jacob, S.; Featherstone, C.; Barton, M. The role of radiotherapy in cancer treatment: Estimating optimal utilization from a review of evidence-based clinical guidelines. *Cancer* **2005**, *104*, 1129–1137. [CrossRef] [PubMed]
139. Borras, J.M.; Lievens, Y.; Dunscombe, P.; Coffey, M.; Malicki, J.; Corral, J.; Gasparotto, C.; Defourny, N.; Barton, M.; Verhoeven, R.; et al. The optimal utilization proportion of external beam radiotherapy in European countries: An ESTRO-HERO analysis. *Radiother. Oncol.* **2015**, *116*, 38–44. [CrossRef] [PubMed]
140. Borras, J.M.; Lievens, Y.; Barton, M.; Corral, J.; Ferlay, J.; Bray, F.; Grau, C. How many new cancer patients in Europe will require radiotherapy by 2025? An ESTRO-HERO analysis. *Radiother. Oncol.* **2016**, *119*, 5–11. [CrossRef] [PubMed]
141. Norlund, A. Costs of Radiotherapy. *Acta Oncol.* **2003**, *42*, 411–415. [CrossRef]
142. Grubbé, E.H. Priority in the Therapeutic Use of X-rays. *Radiology* **1933**, *21*, 156–162. [CrossRef]
143. Röntgen, W.C. Ueber eine neue Art von Strahlen. *Ann. Der Phys.* **1898**, *300*, 12–17. [CrossRef]
144. Zubizarreta, E.; Van Dyk, J.; Lievens, Y. Analysis of Global Radiotherapy Needs and Costs by Geographic Region and Income Level. *Clin. Oncol.* **2017**, *29*, 84–92. [CrossRef] [PubMed]
145. Laskar, S.G.; Sinha, S.; Krishnatry, R.; Grau, C.; Mehta, M.; Agarwal, J.P. Access to Radiation Therapy: From Local to Global and Equality to Equity. *JCO Glob. Oncol.* **2022**, *8*, e2100358. [CrossRef] [PubMed]
146. DIRAC. The IAEA Directory of Radiotherapy Centres (DIRAC). Available online: https://dirac.iaea.org/ (accessed on 30 May 2024).
147. Grau, C.; Durante, M.; Georg, D.; Langendijk, J.A.; Weber, D.C. Particle therapy in Europe. *Mol. Oncol.* **2020**, *14*, 1492–1499. [CrossRef] [PubMed]
148. Baumann, M.; Krause, M.; Overgaard, J.; Debus, J.; Bentzen, S.M.; Daartz, J.; Richter, C.; Zips, D.; Bortfeld, T. Radiation oncology in the era of precision medicine. *Nat. Rev. Cancer* **2016**, *16*, 234–249. [CrossRef] [PubMed]
149. Laskar, S.G.; Kakoti, S. Modern Radiation Oncology: From IMRT to Particle Therapy—Present Status and the Days to Come. *Indian J. Med. Paediatr. Oncol.* **2022**, *43*, 47–51. [CrossRef]
150. Byun, H.K.; Han, M.C.; Yang, K.; Kim, J.S.; Yoo, G.S.; Koom, W.S.; Kim, Y.B. Physical and Biological Characteristics of Particle Therapy for Oncologists. *Cancer Res. Treat. Off. J. Korean Cancer Assoc.* **2021**, *53*, 611. [CrossRef] [PubMed]
151. Veldeman, L.; Madani, I.; Hulstaert, F.; De Meerleer, G.; Mareel, M.; De Neve, W. Evidence behind use of intensity-modulated radiotherapy: A systematic review of comparative clinical studies. *Lancet Oncol.* **2008**, *9*, 367–375. [CrossRef] [PubMed]
152. Verellen, D.; Ridder, M.D.; Linthout, N.; Tournel, K.; Soete, G.; Storme, G. Innovations in image-guided radiotherapy. *Nat. Rev. Cancer* **2007**, *7*, 949–960. [CrossRef]
153. Lo, S.S.; Fakiris, A.J.; Chang, E.L.; Mayr, N.A.; Wang, J.Z.; Papiez, L.; Teh, B.S.; McGarry, R.C.; Cardenes, H.R.; Timmerman, R.D. Stereotactic body radiation therapy: A novel treatment modality. *Nat. Rev. Clin. Oncol.* **2009**, *7*, 44–54. [CrossRef]
154. Bilski, M.; Mertowska, P.; Mertowski, S.; Sawicki, M.; Hymos, A.; Niedźwiedzka-Rystwej, P.; Grywalska, E. The Role of Conventionally Fractionated Radiotherapy and Stereotactic Radiotherapy in the Treatment of Carcinoid Tumors and Large-Cell Neuroendocrine Cancer of the Lung. *Cancers* **2021**, *14*, 177. [CrossRef] [PubMed]
155. Arcangeli, S.; Scorsetti, M.; Alongi, F. Will SBRT replace conventional radiotherapy in patients with low-intermediate risk prostate cancer? A review. *Crit. Rev. Oncol. Hematol.* **2012**, *84*, 101–108. [CrossRef] [PubMed]
156. Kron, T. Cui Bono, Proton Radiotherapy? *Clin. Oncol.* **2022**, *34*, 258–260. [CrossRef] [PubMed]
157. Lin, B.; Gao, F.; Yang, Y.; Wu, D.; Zhang, Y.; Feng, G.; Dai, T.; Du, X. FLASH Radiotherapy: History and Future. *Front. Oncol.* **2021**, *11*, 644400. [CrossRef] [PubMed]
158. Kundapur, V.; Torlakovic, E.; Auer, R.N. The Story Behind the First Mini-BEAM Photon Radiation Treatment: What is the Mini-Beam and Why is it Such an Advance? *Semin. Radiat. Oncol.* **2024**, *34*, 337–343. [CrossRef] [PubMed]
159. Hyer, D.E.; Bennett, L.C.; Geoghegan, T.J.; Bues, M.; Smith, B.R. Innovations and the use of collimators in the delivery of pencil beam scanning proton therapy. *Int. J. Part. Ther.* **2021**, *8*, 73–83. [CrossRef] [PubMed]
160. Melzig, C.; Golestaneh, A.F.; Mier, W.; Schwager, C.; Das, S.; Schlegel, J.; Lasitschka, F.; Kauczor, H.U.; Debus, J.; Haberkorn, U.; et al. Combined external beam radiotherapy with carbon ions and tumor targeting endoradiotherapy. *Oncotarget* **2018**, *9*, 29985. [CrossRef] [PubMed]
161. Gorin, J.-B.; Guilloux, Y.; Morgenstern, A.; Chérel, M.; Davodeau, F.; Gaschet, J. Using α radiation to boost cancer immunity? *OncoImmunology* **2014**, *3*, e954925. [CrossRef]
162. Dietrich, A.; Koi, L.; Zöphel, K.; Sihver, W.; Kotzerke, J.; Baumann, M.; Krause, M. Improving external beam radiotherapy by combination with internal irradiation. *Br. J. Radiol.* **2015**, *88*, 20150042. [CrossRef]
163. Hartrampf, P.E.; Hänscheid, H.; Kertels, O.; Schirbel, A.; Kreissl, M.C.; Flentje, M.; Sweeney, R.A.; Buck, A.K.; Polat, B.; Lapa, C. Long-term results of multimodal peptide receptor radionuclide therapy and fractionated external beam radiotherapy for treatment of advanced symptomatic meningioma. *Clin. Transl. Radiat. Oncol.* **2020**, *22*, 29–32. [CrossRef]

164. Ruan, S.; O'Donoghue, J.A.; Larson, S.M.; Finn, R.D.; Jungbluth, A.; Welt, S.; Humm, J.L. Optimizing the Sequence of Combination Therapy with Radiolabeled Antibodies and Fractionated External Beam. *J. Nucl. Med.* **2000**, *41*, 1905–1912. [PubMed]
165. Oddstig, J.; Bernhardt, P.; Nilsson, O.; Ahlman, H.; Forssell-Aronsson, E. Radiation Induces Up-Regulation of Somatostatin Receptors 1, 2, and 5 in Small Cell Lung Cancer In Vitro Also at Low Absorbed Doses. *Cancer Biother. Radiopharm.* **2011**, *26*, 759–765. [CrossRef] [PubMed]
166. Cornelissen, B.; Darbar, S.; Kersemans, V.; Allen, D.; Falzone, N.; Barbeau, J.; Smart, S.; Vallis, K.A. Amplification of DNA damage by a γH2AX-targeted radiopharmaceutical. *Nucl. Med. Biol.* **2012**, *39*, 1142–1151. [CrossRef] [PubMed]
167. Anderson, P.M.; Scott, J.; Parsai, S.; Zahler, S.; Worley, S.; Shrikanthan, S.; Subbiah, V.; Murphy, E. 223-Radium for metastatic osteosarcoma: Combination therapy with other agents and external beam radiotherapy. *ESMO Open* **2020**, *5*, e000635. [CrossRef] [PubMed]
168. Hasan, H.; Deek, M.P.; Phillips, R.; Hobbs, R.F.; Malek, R.; Radwan, N.; Kiess, A.P.; Dipasquale, S.; Huang, J.; Caldwell, T.; et al. A phase II randomized trial of RAdium-223 dichloride and SABR Versus SABR for oligomEtastatic prostate caNcerS (RAVENS). *BMC Cancer* **2020**, *20*, 492. [CrossRef]
169. Yu, J.W.; Bhattacharya, S.; Yanamandra, N.; Kilian, D.; Shi, H.; Yadavilli, S.; Katlinskaya, Y.; Kaczynski, H.; Conner, M.; Benson, W.; et al. Tumor-immune profiling of murine syngeneic tumor models as a framework to guide mechanistic studies and predict therapy response in distinct tumor microenvironments. *PLoS ONE* **2018**, *13*, e0206223. [CrossRef]
170. Bates, J.P.; Derakhshandeh, R.; Jones, L.; Webb, T.J. Mechanisms of immune evasion in breast cancer. *BMC Cancer* **2018**, *18*, 556. [CrossRef]
171. Yap, T.A.; Parkes, E.E.; Peng, W.; Moyers, J.T.; Curran, M.A.; Tawbi, H.A. Development of Immunotherapy Combination Strategies in Cancer. *Cancer Discov.* **2021**, *11*, 1368–1397. [CrossRef]
172. Khalil, D.N.; Smith, E.L.; Brentjens, R.J.; Wolchok, J.D. The future of cancer treatment: Immunomodulation, CARs and combination immunotherapy. *Nat. Rev. Clin. Oncol.* **2016**, *13*, 273–290. [CrossRef]
173. Lu, L.; Barbi, J.; Pan, F. The regulation of immune tolerance by FOXP3. *Nat. Rev. Immunol.* **2017**, *17*, 703–717. [CrossRef]
174. Sharma, P.; Goswami, S.; Raychaudhuri, D.; Siddiqui, B.A.; Singh, P.; Nagarajan, A.; Liu, J.; Subudhi, S.K.; Poon, C.; Gant, K.L.; et al. Immune checkpoint therapy—Current perspectives and future directions. *Cell* **2023**, *186*, 1652–1669. [CrossRef] [PubMed]
175. Boutros, C.; Tarhini, A.; Routier, E.; Lambotte, O.; Ladurie, F.L.; Carbonnel, F.; Izzeddine, H.; Marabelle, A.; Champiat, S.; Berdelou, A.; et al. Safety profiles of anti-CTLA-4 and anti-PD-1 antibodies alone and in combination. *Nat. Rev. Clin. Oncol.* **2016**, *13*, 473–486. [CrossRef] [PubMed]
176. Seidel, J.A.; Otsuka, A.; Kabashima, K. Anti-PD-1 and Anti-CTLA-4 Therapies in Cancer: Mechanisms of Action, Efficacy, and Limitations. *Front. Oncol.* **2018**, *8*, 86. [CrossRef] [PubMed]
177. Doroshow, D.B.; Bhalla, S.; Beasley, M.B.; Sholl, L.M.; Kerr, K.M.; Gnjatic, S.; Wistuba, I.I.; Rimm, D.L.; Tsao, M.S.; Hirsch, F.R. PD-L1 as a biomarker of response to immune-checkpoint inhibitors. *Nat. Rev. Clin. Oncol.* **2021**, *18*, 345–362. [CrossRef] [PubMed]
178. Sun, L.; Zhang, L.; Yu, J.; Zhang, Y.; Pang, X.; Ma, C.; Shen, M.; Ruan, S.; Wasan, H.S.; Qiu, S. Clinical efficacy and safety of anti-PD-1/PD-L1 inhibitors for the treatment of advanced or metastatic cancer: A systematic review and meta-analysis. *Sci. Rep.* **2020**, *10*, 2083. [CrossRef] [PubMed]
179. Lei, Q.; Wang, D.; Sun, K.; Wang, L.; Zhang, Y. Resistance Mechanisms of Anti-PD1/PDL1 Therapy in Solid Tumors. *Front. Cell Dev. Biol.* **2020**, *8*, 672. [CrossRef] [PubMed]
180. Hodi, F.S.; O'Day, S.J.; McDermott, D.F.; Weber, R.W.; Sosman, J.A.; Haanen, J.B.; Gonzalez, R.; Robert, C.; Schadendorf, D.; Hassel, J.C.; et al. Improved Survival with Ipilimumab in Patients with Metastatic Melanoma. *N. Engl. J. Med.* **2010**, *363*, 711–723. [CrossRef] [PubMed]
181. Letendre, P.; Monga, V.; Milhem, M.; Zakharia, Y. Ipilimumab: From preclinical development to future clinical perspectives in melanoma. *Future Oncol.* **2017**, *13*, 625–636. [CrossRef] [PubMed]
182. Leach, D.R.; Krummel, M.F.; Allison, J.P. Enhancement of antitumor immunity by CTLA-4 blockade. *Science* **1996**, *271*, 1734–1736. [CrossRef]
183. Graziani, G.; Tentori, L.; Navarra, P. Ipilimumab: A novel immunostimulatory monoclonal antibody for the treatment of cancer. *Pharmacol. Res.* **2012**, *65*, 9–22. [CrossRef]
184. Zang, X. 2018 Nobel Prize in medicine awarded to cancer immunotherapy: Immune checkpoint blockade—A personal account. *Genes Dis.* **2018**, *5*, 302–303. [CrossRef]
185. Arina, A.; Gutiontov, S.I.; Weichselbaum, R.R. Radiotherapy and Immunotherapy for Cancer: From "Systemic" to "Multisite". *Clin. Cancer Res.* **2020**, *26*, 2777–2782. [CrossRef]
186. Sharma, P.; Allison, J.P. The future of immune checkpoint therapy. *Science* **2015**, *348*, 56–61. [CrossRef]
187. Bonaventura, P.; Shekarian, T.; Alcazer, V.; Valladeau-Guilemond, J.; Valsesia-Wittmann, S.; Amigorena, S.; Caux, C.; Depil, S. Cold Tumors: A Therapeutic Challenge for Immunotherapy. *Front. Immunol.* **2019**, *10*, 168. [CrossRef] [PubMed]
188. Martin, J.D.; Cabral, H.; Stylianopoulos, T.; Jain, R.K. Improving cancer immunotherapy using nanomedicines: Progress, opportunities and challenges. *Nat. Rev. Clin. Oncol.* **2020**, *17*, 251–266. [CrossRef]
189. Hu, Z.I.; McArthur, H.L.; Ho, A.Y. The Abscopal Effect of Radiation Therapy: What Is It and How Can We Use It in Breast Cancer? *Curr. Breast Cancer Rep.* **2017**, *9*, 45–51. [CrossRef] [PubMed]

190. Choi, J.; Beaino, W.; Fecek, R.J.; Fabian, K.P.L.; Laymon, C.M.; Kurland, B.F.; Storkus, W.J.; Anderson, C.J. Combined VLA-4–Targeted Radionuclide Therapy and Immunotherapy in a Mouse Model of Melanoma. *J. Nucl. Med.* **2018**, *59*, 1843–1849. [CrossRef] [PubMed]
191. Chen, H.; Zhao, L.; Fu, K.; Lin, Q.; Wen, X.; Jacobson, O.; Sun, L.; Wu, H.; Zhang, X.; Guo, Z.; et al. Integrin αvβ3-targeted radionuclide therapy combined with immune checkpoint blockade immunotherapy synergistically enhances anti-tumor efficacy. *Theranostics* **2019**, *9*, 7948–7960. [CrossRef]
192. Kim, C.; Liu, S.V.; Subramaniam, D.S.; Torres, T.; Loda, M.; Esposito, G.; Giaccone, G. Phase I study of the (177)Lu-DOTA(0)-Tyr(3)-Octreotate (lutathera) in combination with nivolumab in patients with neuroendocrine tumors of the lung. *J. ImmunoTherapy Cancer* **2020**, *8*, e000980. [CrossRef]
193. Rouanet, J.; Benboubker, V.; Akil, H.; Hennino, A.; Auzeloux, P.; Besse, S.; Pereira, B.; Delorme, S.; Mansard, S.; D'Incan, M.; et al. Immune checkpoint inhibitors reverse tolerogenic mechanisms induced by melanoma targeted radionuclide therapy. *Cancer Immunol. Immunother.* **2020**, *69*, 2075–2088. [CrossRef]
194. Czernin, J.; Current, K.; Mona, C.E.; Nyiranshuti, L.; Hikmat, F.; Radu, C.G.; Lückerath, K. Immune-Checkpoint Blockade Enhances 225Ac-PSMA617 Efficacy in a Mouse Model of Prostate Cancer. *J. Nucl. Med.* **2021**, *62*, 228–231. [CrossRef] [PubMed]
195. Patel, R.B.; Hernandez, R.; Carlson, P.; Grudzinski, J.; Bates, A.M.; Jagodinsky, J.C.; Erbe, A.; Marsh, I.R.; Arthur, I.; Aluicio-Sarduy, E.; et al. Low-dose targeted radionuclide therapy renders immunologically cold tumors responsive to immune checkpoint blockade. *Sci. Transl. Med.* **2021**, *13*, eabb3631. [CrossRef] [PubMed]
196. Vito, A.; Rathmann, S.; Mercanti, N.; El-Sayes, N.; Mossman, K.; Valliant, J. Combined Radionuclide Therapy and Immunotherapy for Treatment of Triple Negative Breast Cancer. *Int. J. Mol. Sci.* **2021**, *22*, 4843. [CrossRef] [PubMed]
197. Digklia, A.; Boughdad, S.; Homicsko, K.; Dromain, C.; Trimech, M.; Dolcan, A.; Peters, S.; Prior, J.; Schaefer, N. First communication on the efficacy of combined 177Lutetium-PSMA with immunotherapy outside prostate cancer. *J. ImmunoTherapy Cancer* **2022**, *10*, e005383. [CrossRef] [PubMed]
198. Chabner, B.A.; Roberts, T.G. Chemotherapy and the war on cancer. *Nat. Rev. Cancer* **2005**, *5*, 65–72. [CrossRef] [PubMed]
199. Biller, L.H.; Schrag, D. Diagnosis and Treatment of Metastatic Colorectal Cancer. *JAMA* **2021**, *325*, 669–685. [CrossRef] [PubMed]
200. Shien, T.; Iwata, H. Adjuvant and neoadjuvant therapy for breast cancer. *Jpn. J. Clin. Oncol.* **2020**, *50*, 225–229. [CrossRef] [PubMed]
201. Fisusi, F.A.; Akala, E.O. Drug Combinations in Breast Cancer Therapy. *Pharm. Nanotechnol.* **2019**, *7*, 3–23. [CrossRef]
202. Baxevanos, P.; Mountzios, G. Novel chemotherapy regimens for advanced lung cancer: Have we reached a plateau? *Ann. Transl. Med.* **2018**, *6*, 139. [CrossRef]
203. Kumar, L.; Harish, P.; Malik, P.S.; Khurana, S. Chemotherapy and targeted therapy in the management of cervical cancer. *Curr. Probl. Cancer* **2018**, *42*, 120–128. [CrossRef]
204. Nader, R.; El Amm, J.; Aragon-Ching, J.B. Role of chemotherapy in prostate cancer. *Asian J. Androl.* **2018**, *20*, 221–229. [CrossRef] [PubMed]
205. Teo, M.Y.; Rathkopf, D.E.; Kantoff, P. Treatment of Advanced Prostate Cancer. *Annu. Rev. Med.* **2019**, *70*, 479–499. [CrossRef] [PubMed]
206. De Nunzio, C.; Presicce, F.; Giacinti, S.; Bassanelli, M.; Tubaro, A. Castration-resistance prostate cancer: What is in the pipeline? *Minerva Urol. Nephrol.* **2018**, *70*, 22–41. [CrossRef] [PubMed]
207. Prota, A.E.; Lucena-Agell, D.; Ma, Y.; Estevez-Gallego, J.; Li, S.; Bargsten, K.; Josa-Prado, F.; Altmann, K.H.; Gaillard, N.; Kamimura, S.; et al. Structural insight into the stabilization of microtubules by taxanes. *eLife* **2023**, *12*, e84791. [CrossRef] [PubMed]
208. Pienta, K.J. Preclinical mechanisms of action of docetaxel and docetaxel combinations in prostate cancer. *Semin. Oncol.* **2001**, *28*, 3–7. [CrossRef] [PubMed]
209. Weaver, B.A. How Taxol/paclitaxel kills cancer cells. *Mol. Biol. Cell* **2014**, *25*, 2677–2681. [CrossRef] [PubMed]
210. Scribano, C.M.; Wan, J.; Esbona, K.; Tucker, J.B.; Lasek, A.; Zhou, A.S.; Zasadil, L.M.; Molini, R.; Fitzgerald, J.; Lager, A.M.; et al. Chromosomal instability sensitizes patient breast tumors to multipolar divisions induced by paclitaxel. *Sci. Transl. Med.* **2021**, *13*, eabd4811. [CrossRef] [PubMed]
211. Komlodi-Pasztor, E.; Sackett, D.; Wilkerson, J.; Fojo, T. Mitosis is not a key target of microtubule agents in patient tumors. *Nat. Rev. Clin. Oncol.* **2011**, *8*, 244–250. [CrossRef] [PubMed]
212. Werahera, P.N.; Glode, L.M.; La Rosa, F.G.; Lucia, M.S.; Crawford, E.D.; Easterday, K.; Sullivan, H.T.; Sidhu, R.S.; Genova, E.; Hedlund, T. Proliferative Tumor Doubling Times of Prostatic Carcinoma. *Prostate Cancer* **2011**, *2011*, 301850. [CrossRef]
213. Poruchynsky, M.S.; Komlodi-Pasztor, E.; Trostel, S.; Wilkerson, J.; Regairaz, M.; Pommier, Y.; Zhang, X.; Kumar Maity, T.; Robey, R.; Burotto, M.; et al. Microtubule-targeting agents augment the toxicity of DNA-damaging agents by disrupting intracellular trafficking of DNA repair proteins. *Proc. Natl. Acad. Sci. USA* **2015**, *112*, 1571–1576. [CrossRef]
214. Gjyrezi, A.; Xie, F.; Voznesensky, O.; Khanna, P.; Calagua, C.; Bai, Y.; Kung, J.; Wu, J.; Corey, E.; Montgomery, B.; et al. Taxane resistance in prostate cancer is mediated by decreased drug-target engagement. *J. Clin. Investig.* **2020**, *130*, 3287–3298. [CrossRef] [PubMed]
215. Darshan, M.S.; Loftus, M.S.; Thadani-Mulero, M.; Levy, B.P.; Escuin, D.; Zhou, X.K.; Gjyrezi, A.; Chanel-Vos, C.; Shen, R.; Tagawa, S.T.; et al. Taxane-Induced Blockade to Nuclear Accumulation of the Androgen Receptor Predicts Clinical Responses in Metastatic Prostate Cancer. *Cancer Res.* **2011**, *71*, 6019–6029. [CrossRef] [PubMed]

216. Ma, S.; Rong, Z.; Liu, C.; Qin, X.; Zhang, X.; Chen, Q. DNA damage promotes microtubule dynamics through a DNA-PK-AKT axis for enhanced repair. *J. Cell Biol.* **2021**, *220*, e201911025. [CrossRef] [PubMed]
217. Ruiz de Porras, V.; Wang, X.C.; Palomero, L.; Marin-Aguilera, M.; Solé-Blanch, C.; Indacochea, A.; Jimenez, N.; Bystrup, S.; Bakht, M.; Conteduca, V.; et al. Taxane-induced Attenuation of the CXCR2/BCL-2 Axis Sensitizes Prostate Cancer to Platinum-based Treatment. *Eur. Urol.* **2021**, *79*, 722–733. [CrossRef] [PubMed]
218. Xu, Z.; Xu, L.; Ge, Y.; Sun, H.; Zhu, J.; Dou, Q.; Jia, R. Cabazitaxel suppresses the proliferation and promotes the apoptosis and radiosensitivity of castration-resistant prostate cancer cells by inhibiting PI3K/AKT pathway. *Am. J. Transl. Res.* **2022**, *14*, 166–181. [PubMed]
219. Kale, J.; Osterlund, E.J.; Andrews, D.W. BCL-2 family proteins: Changing partners in the dance towards death. *Cell Death Differ.* **2017**, *25*, 65–80. [CrossRef] [PubMed]
220. Fraguas-Sánchez, A.I.; Lozza, I.; Torres-Suárez, A.I. Actively Targeted Nanomedicines in Breast Cancer: From Pre-Clinal Investigation to Clinic. *Cancers* **2022**, *14*, 1198. [CrossRef] [PubMed]
221. Rajappa, S.; Joshi, A.; Doval, D.C.; Batra, U.; Rajendranath, R.; Deo, A.; Biswas, G.; Bajpai, P.; Tilak, T.V.S.; Kane, S.; et al. [Experts' Opinion] Novel formulations of docetaxel, paclitaxel and doxorubicin in the management of metastatic breast cancer. *Oncol. Lett.* **2018**, *16*, 3757–3769. [CrossRef] [PubMed]
222. De Leo, S.; Trevisan, M.; Fugazzola, L. Recent advances in the management of anaplastic thyroid cancer. *Thyroid Res.* **2020**, *13*, 17. [CrossRef]
223. Jurczyk, M.; Kasperczyk, J.; Wrześniok, D.; Beberok, A.; Jelonek, K. Nanoparticles Loaded with Docetaxel and Resveratrol as an Advanced Tool for Cancer Therapy. *Biomedicines* **2022**, *10*, 1187. [CrossRef]
224. Le, P.N.; Huynh, C.K.; Tran, N.Q. Advances in thermosensitive polymer-grafted platforms for biomedical applications. *Mater. Sci. Eng. C* **2018**, *92*, 1016–1030. [CrossRef] [PubMed]
225. Zhang, Y.; Sun, T.; Jiang, C. Biomacromolecules as carriers in drug delivery and tissue engineering. *Acta Pharm. Sin. B* **2018**, *8*, 34–50. [CrossRef] [PubMed]
226. Atrafi, F.; van Eerden, R.A.G.; van Hylckama Vlieg, M.A.M.; Oomen-de Hoop, E.; de Bruijn, P.; Lolkema, M.P.; Moelker, A.; Rijcken, C.J.; Hanssen, R.; Sparreboom, A.; et al. Intratumoral Comparison of Nanoparticle Entrapped Docetaxel (CPC634) with Conventional Docetaxel in Patients with Solid Tumors. *Clin. Cancer Res.* **2020**, *26*, 3537–3545. [CrossRef] [PubMed]
227. Zhang, A.L.; Russell, P.J.; Knittel, T.; Milross, C. Paclitaxel enhanced radiation sensitization for the suppression of human prostate cancer tumor growth via a p53 independent pathway. *Prostate* **2007**, *67*, 1630–1640. [CrossRef] [PubMed]
228. Miyanaga, S.; Ninomiya, I.; Tsukada, T.; Okamoto, K.; Harada, S.; Nakanuma, S.; Sakai, S.; Makino, I.; Kinoshita, J.; Hayashi, H.; et al. Concentration-dependent radiosensitizing effect of docetaxel in esophageal squamous cell carcinoma cells. *Int. J. Oncol.* **2016**, *48*, 517–524. [CrossRef] [PubMed]
229. Autio, K.A.; Dreicer, R.; Anderson, J.; Garcia, J.A.; Alva, A.; Hart, L.L.; Milowsky, M.I.; Posadas, E.M.; Ryan, C.J.; Graf, R.P.; et al. Safety and Efficacy of BIND-014, a Docetaxel Nanoparticle Targeting Prostate-Specific Membrane Antigen for Patients With Metastatic Castration-Resistant Prostate Cancer: A Phase 2 Clinical Trial. *JAMA Oncol.* **2018**, *4*, 1344–1351. [CrossRef]
230. Yoneshima, Y.; Morita, S.; Ando, M.; Nakamura, A.; Iwasawa, S.; Yoshioka, H.; Goto, Y.; Takeshita, M.; Harada, T.; Hirano, K.; et al. Phase 3 Trial Comparing Nanoparticle Albumin-Bound Paclitaxel with Docetaxel for Previously Treated Advanced NSCLC. *J. Thorac. Oncol.* **2021**, *16*, 1523–1532. [CrossRef]
231. Schmid, P.; Adams, S.; Rugo, H.S.; Schneeweiss, A.; Barrios, C.H.; Iwata, H.; Diéras, V.; Hegg, R.; Im, S.-A.; Wright, G.S.; et al. Atezolizumab and Nab-Paclitaxel in Advanced Triple-Negative Breast Cancer. *N. Engl. J. Med.* **2018**, *379*, 2108–2121. [CrossRef]
232. Cortes, J.; Rugo, H.S.; Cescon, D.W.; Im, S.-A.; Yusof, M.M.; Gallardo, C.; Lipatov, O.; Barrios, C.H.; Perez-Garcia, J.; Iwata, H.; et al. Pembrolizumab plus Chemotherapy in Advanced Triple-Negative Breast Cancer. *N. Engl. J. Med.* **2022**, *387*, 217–226. [CrossRef]
233. Paz-Ares, L.; Luft, A.; Vicente, D.; Tafreshi, A.; Gümüş, M.; Mazières, J.; Hermes, B.; Şenler, F.Ç.; Csőszi, T.; Fülöp, A.; et al. Pembrolizumab plus Chemotherapy for Squamous Non–Small-Cell Lung Cancer. *N. Engl. J. Med.* **2018**, *379*, 2040–2051. [CrossRef]
234. Maharaj, M.; Heslop, L.; Govender, T.; Korowlay, N.; Singh, A.; Choudhary, P.; Sathekge, M. The Outcome and Safety of Re-challenge Lutetium-177 PSMA ((177)Lu-PSMA) Therapy with Low-Dose Docetaxel as a Radiosensitizer-a Promising Combination in Metastatic Castrate-Resistant Prostate Cancer (mCRPC): A Case Report. *Nucl. Med. Mol. Imaging* **2021**, *55*, 136–140. [CrossRef]
235. Batra, J.S.; Niaz, M.J.; Whang, Y.E.; Sheikh, A.; Thomas, C.; Christos, P.; Vallabhajosula, S.; Jhanwar, Y.S.; Molina, A.M.; Nanus, D.M.; et al. Phase I trial of docetaxel plus lutetium-177-labeled anti-prostate-specific membrane antigen monoclonal antibody J591 ((177)Lu-J591) for metastatic castration-resistant prostate cancer. *Urol. Oncol.* **2020**, *38*, 848.e9–848.e16. [CrossRef]
236. Kostos, L.K.; Buteau, J.P.; Kong, G.; Yeung, T.; Iulio, J.D.; Fahey, M.T.; Fettke, H.; Furic, L.; Hofman, M.S.; Azad, A. LuCAB: A phase I/II trial evaluating cabazitaxel in combination with [177Lu]Lu-PSMA-617 in patients (pts) with metastatic castration-resistant prostate cancer (mCRPC). *J. Clin. Oncol.* **2023**, *41* (Suppl. S6), TPS278. [CrossRef]
237. Dale, R.G.; Jones, B. The clinical radiobiology of brachytherapy. *Br. J. Radiol.* **1998**, *71*, 465–483. [CrossRef] [PubMed]
238. Annede, P.; Cosset, J.-M.; Van Limbergen, E.; Deutsch, E.; Haie-Meder, C.; Chargari, C. Radiobiology: Foundation and New Insights in Modeling Brachytherapy Effects. *Semin. Radiat. Oncol.* **2020**, *30*, 4–15. [CrossRef]
239. Fonseca, G.P.; Johansen, J.G.; Smith, R.L.; Beaulieu, L.; Beddar, S.; Kertzscher, G.; Verhaegen, F.; Tanderup, K. In vivo dosimetry in brachytherapy: Requirements and future directions for research, development, and clinical practice. *Phys. Imaging Radiat. Oncol.* **2020**, *16*, 1–11. [CrossRef] [PubMed]

240. Tamihardja, J.; Weick, S.; Lutyj, P.; Zimmermann, M.; Bratengeier, K.; Flentje, M.; Polat, B. Comparing Iridium-192 with Cobalt-60 sources in high-dose-rate brachytherapy boost for localized prostate cancer. *Acta Oncol.* **2022**, *61*, 714–719. [CrossRef]
241. Bachand, J.; Schroeder, S.R.; Desai, N.B.; Folkert, M.R. Prostate brachytherapy procedural training: Incorporation of related procedures in resident training and competency assessment. *J. Contemp. Brachytherapy* **2019**, *11*, 601–606. [CrossRef] [PubMed]
242. Koukourakis, G.; Kelekis, N.; Armonis, V.; Kouloulias, V. Brachytherapy for Prostate Cancer: A Systematic Review. *Adv. Urol.* **2009**, *2009*, 327945. [CrossRef]
243. Palvai, S.; Harrison, M.; Thomas, S.S.; Hayden, K.; Green, J.; Anderson, O.; Romero, L.; Lodge, R.; Burns, P.; Ahmed, I. Timing of High-Dose Rate Brachytherapy with External Beam Radiotherapy in Intermediate and High-Risk Localized Prostate CAncer (THEPCA) Patients and Its Effects on Toxicity and Quality of Life: Protocol of a Randomized Feasibility Trial. *JMIR Res. Protoc.* **2015**, *4*, e4462. [CrossRef]
244. Mori, K.; Sasaki, H.; Tsutsumi, Y.; Sato, S.; Takiguchi, Y.; Saito, S.; Nishi, E.; Ishii, G.; Yamamoto, T.; Koike, Y.; et al. Trimodal therapy with high-dose-rate brachytherapy and hypofractionated external beam radiation combined with long-term androgen deprivation for unfavorable-risk prostate cancer. *Strahlenther. Und Onkol.* **2021**, *197*, 976–985. [CrossRef] [PubMed]
245. Ito, K.; Saito, S.; Yorozu, A.; Kojima, S.; Kikuchi, T.; Higashide, S.; Aoki, M.; Koga, H.; Satoh, T.; Ohashi, T.; et al. Nationwide Japanese Prostate Cancer Outcome Study of Permanent Iodine-125 Seed Implantation (J-POPS): First analysis on survival. *Int. J. Clin. Oncol.* **2018**, *23*, 1148–1159. [CrossRef] [PubMed]
246. Morris, W.J.; Tyldesley, S.; Rodda, S.; Halperin, R.; Pai, H.; McKenzie, M.; Duncan, G.; Morton, G.; Hamm, J.; Murray, N. Androgen Suppression Combined with Elective Nodal and Dose Escalated Radiation Therapy (the ASCENDE-RT Trial): An Analysis of Survival Endpoints for a Randomized Trial Comparing a Low-Dose-Rate Brachytherapy Boost to a Dose-Escalated External Beam Boost for High- and Intermediate-risk Prostate Cancer. *Int. J. Radiat. Biol.* **2017**, *98*, 275–285. [CrossRef]
247. Oshikane, T.; Kaidu, M.; Abe, E.; Ohta, A.; Saito, H.; Nakano, T.; Honda, M.; Tanabe, S.; Utsunomiya, S.; Sasamoto, R.; et al. A comparative study of high-dose-rate brachytherapy boost combined with external beam radiation therapy versus external beam radiation therapy alone for high-risk prostate cancer. *J. Radiat. Res.* **2021**, *62*, 525–532. [CrossRef] [PubMed]
248. Arazi, L. Diffusing alpha-emitters radiation therapy: Approximate modeling of the macroscopic alpha particle dose of a point source. *Phys. Med. Biol.* **2020**, *65*, 015015. [CrossRef] [PubMed]

pharmaceuticals

Article

Assessing Nasal Epithelial Dynamics: Impact of the Natural Nasal Cycle on Intranasal Spray Deposition

Amr Seifelnasr [1], Xiuhua Si [2] and Jinxiang Xi [1,*]

[1] Department of Biomedical Engineering, University of Massachusetts, Lowell, MA 01854, USA; amr_seifelnasr@student.uml.edu
[2] Department of Mechanical Engineering, California Baptist University, Riverside, CA 92504, USA; asi@calbaptist.edu
* Correspondence: jinxiang_xi@uml.edu; Tel.: +1-978-934-3259

Abstract: This study investigated the intricate dynamics of intranasal spray deposition within nasal models, considering variations in head orientation and stages of the nasal cycle. Employing controlled delivery conditions, we compared the deposition patterns of saline nasal sprays in models representing congestion (N1), normal (N0), and decongestion (P1, P2) during one nasal cycle. The results highlighted the impact of the nasal cycle on spray distribution, with congestion leading to confined deposition and decongestion allowing for broader dispersion of spray droplets and increased sedimentation towards the posterior turbinate. In particular, the progressive nasal dilation from N1 to P2 decreased the spray deposition in the middle turbinate. The head angle, in conjunction with the nasal cycle, significantly influenced the nasal spray deposition distribution, affecting targeted drug delivery within the nasal cavity. Despite controlled parameters, a notable variance in deposition was observed, emphasizing the complex interplay of gravity, flow shear, nasal cycle, and nasal morphology. The magnitude of variance increased as the head tilt angle increased backward from upright to 22.5° to 45° due to increasing gravity and liquid film destabilization, especially under decongestion conditions (P1, P2). This study's findings underscore the importance of considering both natural physiological variations and head orientation in optimizing intranasal drug delivery.

Keywords: intranasal spray deposition; nasal cycle; turbinate; deposition patterns; targeted drug delivery; nasal morphology; nasal congestion; decongestion; nasal resistance

Citation: Seifelnasr, A.; Si, X.; Xi, J. Assessing Nasal Epithelial Dynamics: Impact of the Natural Nasal Cycle on Intranasal Spray Deposition. *Pharmaceuticals* **2024**, *17*, 73. https://doi.org/10.3390/ph17010073

Academic Editor: Nitesh K. Kunda

Received: 15 December 2023
Revised: 31 December 2023
Accepted: 5 January 2024
Published: 6 January 2024

1. Introduction

The nasal cycle is a phenomenon in which the majority of airflow initially passes through one side and then the other side of the nose, depending on the side experiencing the most resistance to airflow. In approximately 80% of adults, this cyclic variation in nasal resistance has a frequency ranging from 50 min to 4 h [1]. The nasal cycle is driven by an imbalance in blood flow, causing swelling of erectile cells in the nasal septum and inferior turbinate of one passage compared to the other. This uneven enlargement of tissue obstructs the airflow more in one passage, which ventilates different amounts of inhaled chemicals to the olfactory mucosa, facilitating the localization of the odor source [2]. Vasodilation on one side occurs simultaneously with vasoconstriction on the opposite side, causing the predominant airflow to shift between nasal passages while the overall resistance remains constant. During periods of increased respiratory activity, such as exercise or deep breathing, nasal resistance decreases, likely due to vessels becoming decongested [1].

The purpose of the nasal cycle is a subject of discussion. Some research suggests that the alternating nasal airflows function as a means for conditioning air, eliminating trapped pollutants, facilitating mucociliary clearance, and localizing the odor origin [3,4]. Conversely, other theories propose that the nasal cycle plays a role in safeguarding against respiratory infections, or it serves as an indicator of these physiological states [5–7]. The exact mechanisms of how the circadian rhythm governs the nasal cycle remain unclear. However, it is believed that the autonomic nervous system plays a vital part in this process. The occurrence of nasal cycle can vary with age, physical activity, and specific medication [8–12]. Furthermore, factors that cause the widening of nasal blood vessels, thickening of the nasal mucosa, and alterations in nasal secretion levels, like allergic, nonallergic, or infectious forms of rhinitis, also affect the nasal cycle's occurrence [13,14].

Intranasal drug administration is firmly established for treating nasal diseases and is gaining recognition as an effective alternative to oral and injectable methods. The nasal mucosa has increasingly been identified as a feasible pathway for systemic drug delivery [15]. The structure and function of the nasal cavity offer distinct benefits for reaching targets in both the respiratory and olfactory regions, facilitating targeted or systemic drug delivery. The efficacy of intranasal drugs relies on a combination of factors related to the drug, delivery device, and the patient. These encompass characteristics of the drug formulation, design of the delivery device, administration method, target of deposition, and health conditions of the nose [16,17]. The variability in nasal anatomy among individuals is especially significant in influencing intranasal drug delivery, as it has a significant impact on the drug distribution in the nose and, thus, the effectiveness of the drug.

Nasal anatomy and morphology impact the effectiveness of intranasal drug delivery and can be significantly affected by nasal diseases, nasal obstruction, and the periodic turbinate congestion/decongestion from the natural nasal cycle. Nasal obstruction, which limits airflow, can be caused by various pathologies, including the common cold, rhinosinusitis, nasal polyps, nonallergic rhinitis, and allergic rhinitis [18]. These factors pose challenges in achieving consistent and optimal drug absorption through the nasal mucosa, particularly when targeting specific regions in the nose for therapeutic effects. For instance, the effectiveness of nasal irrigation can be influenced by the nasal cycle, which alters nasal geometry and resistance [19].

Many intranasally administered medications, such as nasal sprays, rely on inspiratory flow to enhance their effectiveness in reaching target regions within the nasal cavity. Inspiratory flow helps facilitate the movement of the medication through the nasal passages and promotes optimal drug delivery to the desired areas for absorption or a therapeutic effect. Nasal obstruction or congestion can thus increase nasal airflow resistance, potentially hindering the optimal intranasal delivery to target regions. Measuring nasal airflow directly in living organisms is difficult, leading to the frequent use of computational fluid dynamics (CFD) and experimental techniques like particle image velocimetry (PIV) to study respiratory airflows [20–24]. The findings of several studies illustrate the impact of anatomical structures within the nasal cavity on nasal resistance [25–29]. Bickford et al. indicated that decongestants can notably decrease the resistance of the nasal airway [30]. Xi et al. investigated the nasal dilation effects on olfactory deposition and found that nasal dilation could double the delivery efficiency to the olfactory region [31]. However, there is limited quantification of the influence of the nasal cycle on nasal airflows and the dosimetry of intranasal drug delivery. Eccles et al. reported that the nasal cycle could induce a nearly fourfold alteration in the resistance of a unilateral nasal passage, suggesting a significant impact on the behaviors and fates of aerosolized drugs delivered intranasally [32].

According to a study by Xiao et al., nasal decongestants lead to a notable expansion in the passage posterior to the nasal valve by reducing the volume of the anterior turbinate [20]. Findings from their study suggest that, apart from reducing nasal cavity resistance by approximately 50%, decongestion also reduced the intersubject variability in nasal resistance across the test cohort. One noteworthy point is that the majority of patient-specific experimental or CFD studies utilize nasal cavity geometries acquired through medical imaging, capturing the airway at a single moment in time [33–35]. The objective of this study was to employ in vitro techniques to observe and investigate the impact of congestion/decongestion resulting from the natural nasal cycle on the penetration and distribution of administered nasal sprays within the nasal cavity. Specific aims included:

(1) Prepare a set of multi-sectioned transparent 3D-printed nasal cast models based on a patient-specific nasal geometry, incorporating varying degrees of turbinate thickness to reflect different levels of congestion/decongestion resulting from the nasal cycle.
(2) Visualize the distribution of deposited intranasally administered sprays within the various sections of each nasal model under the influence of inhalation flow and various head positions.
(3) Quantify the deposition in the various model sections: front nose, turbinate regions (anterior, middle, posterior), and nasopharynx, and evaluate the outcomes of the experimental scenarios.

The outcomes of this study will enhance our knowledge of how congestion and decongestion, which are part of the natural nasal cycle, affect the dosimetry of nasal sprays and the delivery of accurate doses to specific intranasal targets.

2. Results

2.1. Nasal Cycle Models and 3D-Printed Casts

Four nasal models were developed to represent nasal passage variations at different stages of the nasal cycle (Figure 1a). Here, N1 represents the passage shrinkage (i.e., negative) due to congestion and N0 represents the normal, while P1 and P2 represent the passage dilation (i.e., positive) due to increasing concha decongestions (Figure 1b). Figure 1c shows the morphological dimensions of the four models (N1, N0, P1, and P2). Because the congestion/decongestion occurred only on the conchae, the front nose and nasopharynx remained the same throughout the nasal cycle. To illustrate the three-dimensional structural change of the left inferior concha, the right side of the nose and the nasal septum were removed (see Figure 1d), revealing the gradual transition in the left inferior concha from a congested state (N1) to a normal condition (N0), and finally to decongested states (P1 and P2).

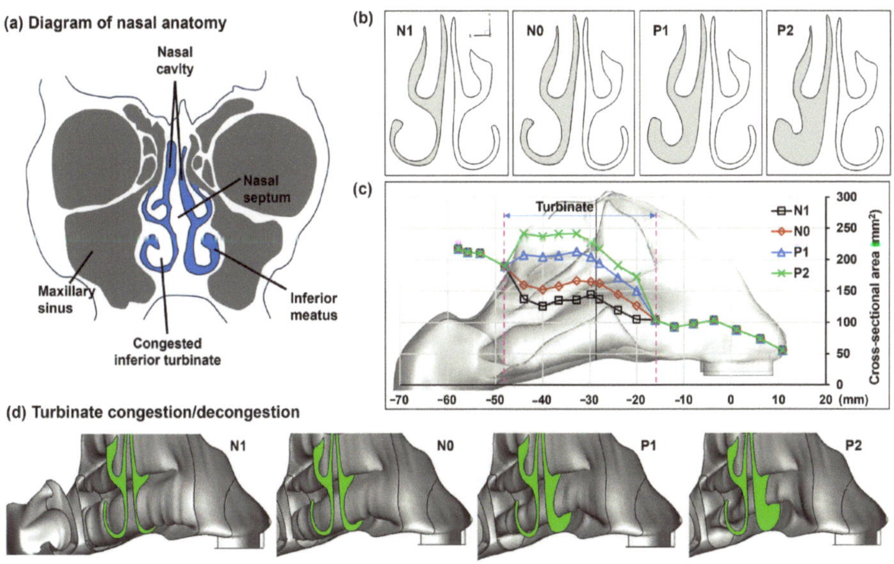

Figure 1. Variation in the left inferior meatus due to concha (turbinate) congestion/decongestion (N1–P2) within one nasal cycle: (**a**) diagram of nasal anatomy, (**b**) visualization of a selected cross section among N1–P2, (**c**) quantification of the cross-sectional area variation among N1–P2 in the turbinate region, and (**d**) turbinate morphological variation among N1–P2. N1: the left inferior meatus shrinks due to concha congestion; N0: normal; P1 and P2: the left inferior meatus dilates due to increasing concha decongestion.

To quantify spray deposition distributions, each nasal model was divided into five sections, namely, the front nose, anterior turbinate (T1), middle turbinate (T2), posterior turbinate (T3), and nasopharynx (NP), as illustrated in Figure 2a. Changes in the morphology of the four nasal models were reflected in sections T1, T2, and T3 (Figure 2b). Lap joints were generated at the connecting ends to facilitate robust assembly and secure sealing. The assembled nasal casts (N1, N0, P1, and P2) were subsequently used in spray deposition tests with three head angles, i.e., upright (0°), 22.5° back tilt, and 45° back tilt (Figure 2c).

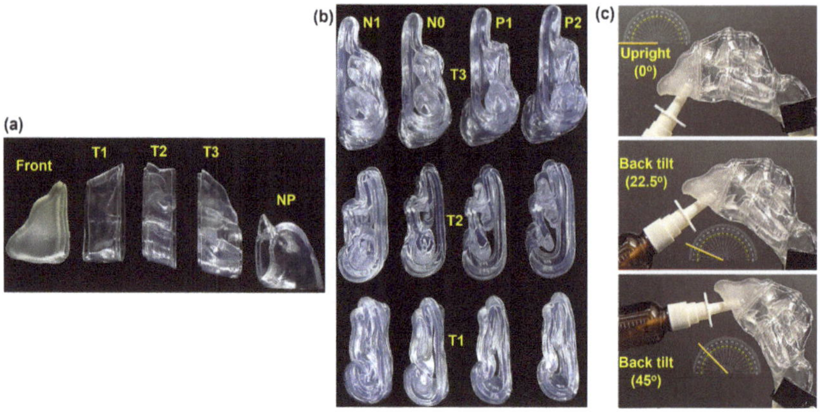

Figure 2. Nasal casts and experimental setup: (**a**) Nasal cast sections (front nose, T1, T2, T3, and nasopharynx [NP]) for deposition quantification. (**b**) Sectional views of the turbinate region sections of the four nasal casts (N1, N0, P1, and P2) illustrating varying inferior turbinate thickness. (**c**) Experimental setup depicting three head positions: upright, 22.5°, and 45° back tilts.

2.2. *Spray Deposition vs. Nasal Cycle at a 22.5° Backward Head Tilt*

2.2.1. Time-Sequence Visualization of Deposition in Model N1

Figure 3 depicts the time-sequence visualization of the liquid droplet and film deposition of intranasally administered saline nasal spray within the N1 nasal model, representing the most congested stage in a nasal cycle after one- and two-dose applications at a head orientation of 22.5° tilted backward.

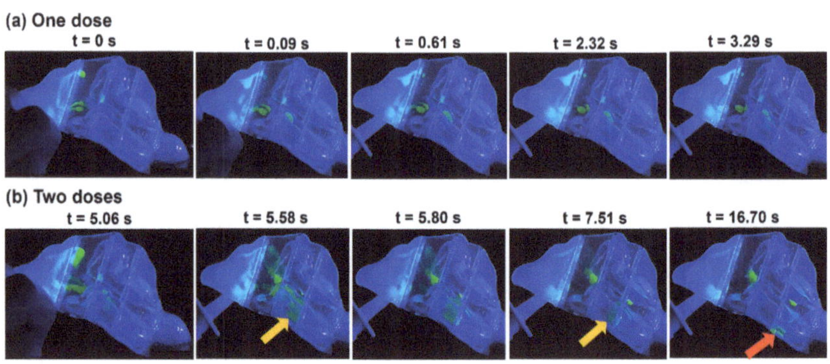

Figure 3. Time series visualization of droplet distribution, liquid film development, and translocation in the N1 model with a 22.5° backward tilt following: (**a**) a single spray pump dose and (**b**) two doses of spray applications.

Upon application of the initial dose at t = 0 s, a portion of the merged spray droplets settled in the superior front nose, and an almost equal volume was dragged by flow shear and gravity through the mid-anterior section of the turbinate region, depositing along the septum wall as well as the outer surface of the inferior turbinate anteriorly within sections T1 and T2 (at t = 0.09 s). Between 0.09 s and 3.29 s, the deposited film in section T1 translocated downwards and posteriorly, settling within the same section. In T2, the film spread wider downwards and, at t = 3.29 s, settled mostly in the narrow clearance between the inferior turbinate and septum wall.

After the second application at t = 5.06 s (Figure 3b), more doses were deposited in the roof of the front nose, adding to the liquid film that was deposited there after the first spray application. Furthermore, additional liquid film passed into T1 and merged with previously deposited liquid film in that section. Accumulated liquid film in the superior front nose was dragged by flow shear through the anterior section of the roof of the respiratory region, spreading across the septum wall and dragged further downwards and diagonally through T1 by gravity, passing the anterior and middle turbinate, and ultimately merging between t = 5.06 s and 5.58 s with film deposited inferiorly in T1 (Figure 3b). The film continued to spread and translocate diagonally under the combined effects of flow shear and gravity, only to extend within the very narrow cavity in T2 between the inferior turbinate and the septum (indicated by the orange arrow at t = 5.58 s). A large section of that liquid film detached at t = 5.80 s (Figure 3b), by which time inhalation flow had ceased, and translocated posteriorly through T2, maintaining its degree of spread across the very narrow passageway until reaching the very end of T2 (white arrow at t = 7.51 s, Figure 3b) and continued mobilizing into T3. Upon entering the T3 section of the turbinate region, the sectional area of the passage widened, causing the film to split into separate parts along the lining of the inferior turbinate and septum. With the lack of inspiratory airflow by the time the film entered T3, flow shear effects vanished. The vertical component of gravity was strong enough to overcome the liquid-surface adhesive force, detaching and splitting the liquid film and ultimately leading to its settlement on the nasal cavity floor in T3 (red arrow).

2.2.2. Comparison of Spray Deposition Patterns among N1, N0, P1, and P2

The spray deposition distribution after one- and two-dose applications within the four models (N1, N0, P1, and P2), representing various degrees of congestion/decongestion throughout one nasal cycle, can be seen in Figure 4.

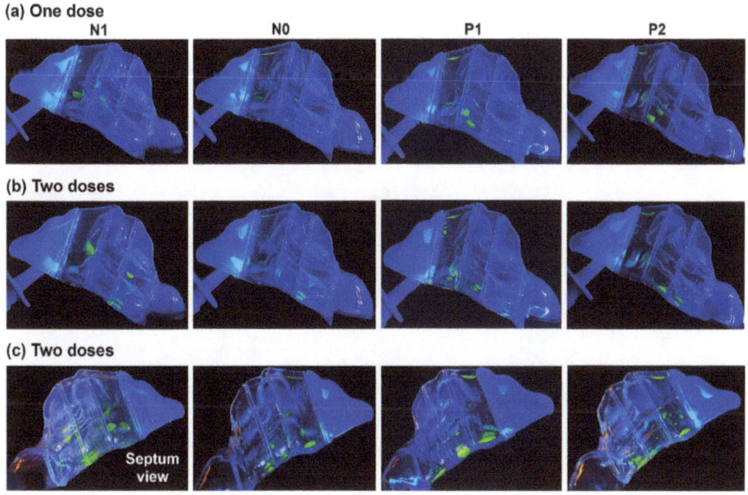

Figure 4. Liquid drop and film deposition within the various models with a 22.5° backward head tilt: (**a**) after one dose of spray application, (**b**) after two doses, and (**c**) after two doses captured from the septum side of each model.

The first two rows depict the results of one- and two-dose applications, respectively, viewed from the lateral left side. The third row shows the captures from the nasal septum side. At the backward head tilt of 22.5°, in all models, a single-dose application led to a scatter of spray droplets and liquid film throughout the middle as well as the inferior regions of the T1 and T2 turbinate region sections, with some deposition in the front nose and the roof of anterior T1. P1 and P2 exhibited more deposition near or on the floor of T2 compared to models N1 and N0, where the deposition in that section was in the middle region between the outer surface of the inferior turbinate and septum. This is attributed to the wider inferior passages in P1 and P2, allowing for the wider spread of the spray droplets and sedimentation further towards the floor of T2. In contrast, the narrower passageway in the T2 section of N0, and even more so in N1, resulted in more liquid film settling in the tight space between the lining of the inferior turbinate and septum.

Upon administering a second dose, in addition to increased deposition in the front nose and slight additions to the superior region of T1, larger liquid film formations developed. Flow shear and gravity effects dragged more liquid posteriorly within the nasal cavity, resulting in deposition in the rearmost T3 section. It was observed that the narrower inferior passageways in N1 and N0 induced greater flow shear effects compared to P1 and P2, owing to higher flow velocities in those passages than in P1 and P2. When viewed from the septum side, a larger deposition distribution on or near the floor of the nasal cavity was evident in all sections of the turbinate region (T1, T2, and T3) of models P1 and P2 compared to N1 and N0. This is primarily attributed to the wider inferior passages of P1 and P2, as explained above.

2.3. *Spray Deposition vs. Nasal Cycle at Upright and 45° Backward Head Tilt*

Figure 5 showcases the final deposition distribution after the application of two nasal spray doses in all four models (N1, N0, P1, and P2) at both an upright head orientation and a backward head tilt of 45°.

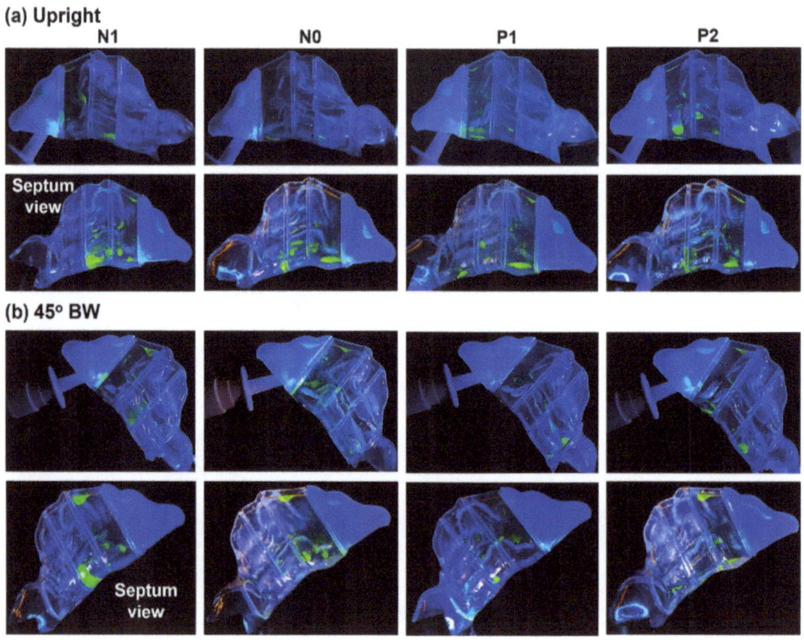

Figure 5. Liquid drop and film deposition within the various models after two doses of spray application with an (**a**) upright (0°) head angle and (**b**) 45° backward head tilt. The upper and lower panels in (**a**,**b**) are views from the lateral and septum side, respectively.

In an upright head position, liquid deposition in the turbinate region was mostly confined to the T1 and T2 sections, as well as at the interface between T2 and T3. Deposition in these regions resulted from the combined effects of inertial impaction, gravitational sedimentation, wall film formation, and flow shear drag. However, the effects of gravity, coupled with the convoluted morphology of the nasal cavity within the T1 and T2 sections in all the models, facilitated the settling of liquid droplets and film in the middle and inferior turbinate regions, with the majority of deposition close to or at the cavity floor. The narrower clearance between the outer surface of the inferior turbinate and the nasal septum in models N1 and N0 resulted in a wider scattering of spray droplets and film, both within and across this confined space. The dispersion extended from the middle zone of sections T1 and T2 downwards towards the nasal cavity floor. This is clearly depicted in the nasal septum view in the second row of Figure 5.

At a 45° back tilt head orientation, the steeper angle accentuated the gravitational component in the direction of the flow, which assisted flow shear in dragging administered spray droplets and film formations deeper into the entire nasal cavity, mostly through the middle meatus. This resulted in deposition distributions dispersed throughout all three turbinate region sections (T1, T2, T3, and T4) in all four models (N1, N0, P1, and P2). Moreover, compared to the upright head position, there was a larger film formation along the superior cavity of section T1, covering the entire length of this section. It is also worth mentioning that the seemingly large deposition in section T2 in the N1 model, as seen from its septum side, is actually due to the spread of liquid film formation within and across the very narrow passage clearance between the congested inferior turbinate's outer lining and that of the nasal septum wall.

2.4. *Quantification of Regional Deposition vs. Nasal Cycle and Head Angle*

Figure 6 shows the deposition variation vs. nasal cycle (N1, N0, P1, and P2) in different regions of the nose (front nose, T1, T2, and T3) with an upright (0°) head position. Insignificant differences in deposition among N1–P2 are observed in the front nose (Figure 6a), which is expected because the nasal cycle affects only the turbinate region, not the front nose.

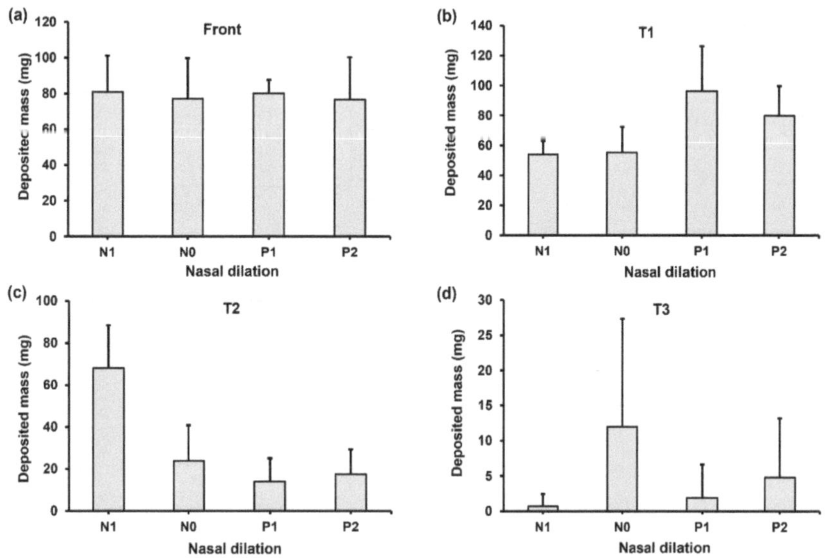

Figure 6. Deposition variation vs. nasal cycle (N1, N0, P1, and P2) in different regions of the nose with an upright head position: (**a**) front nose, (**b**) T1, (**c**) T2, and (**d**) T3.

Considering the deposition in the turbinate, the deposition in the middle turbinate region section (T2) gradually decreases from N1 to P2 (Figure 6c), while the trend appears to be irregular in both T1 (anterior turbinate section, Figure 6b) and T3 (posterior turbinate section, Figure 6d). This irregularity is more pronounced in T3, with much higher deposition in model N0 than in the other three models (Figure 6d). Large variances in deposition are also observed in T3, indicating a high level of liquid film instability in this region. It is noted that the spray deposition depends on both the initial deposition and subsequent translocation of the deposited liquid films, which are sensitive to gravity and local anatomical topologies. The gradual turbinate decongestion from N1 to P2 has at least three effects on the spray deposition: (1) the turbinate shrinks, reducing the surface area for droplet deposition and liquid film attachment; (2) the inferior meatus widens, allowing the spray plume to penetrate deeper; and (3) the turbinate topology changes, affecting the stability of the deposited liquid film. Thus, the gradual decrease in T2 deposition may result from the decreased turbinate area. In T1 deposition, the gradual increase from N1 to P1 may result from the widening meatus, while the slight decrease from P1 to P2 can be attributed to the shrinking turbinate. It is also noted that interactions exist among the local deposition mass, local turbinate topology, and associated gravity, leading to a nonlinear deposition distribution in T1, T2, and T3. In this test case, the abnormally higher T3 deposition in model N0 is a direct result of its drastic deposition decreases in T2 (Figure 6d vs. Figure 6c).

Figures 7 and 8 show the deposition variation vs. nasal cycle (N1, N0, P1, and P2) with a head position back tilt of 22.5° and 45°, respectively. Together with Figure 6 with an upright head position (i.e., 0°), it is observed that the head angle exerts a significant effect on the spray deposition distribution and noticeably regulates the influence of the nasal cycle. In Figure 7 with a 22.5° back tilt head angle, the front nose dose peaks in model P1 (Figure 7a). Within a nasal cycle with a decongesting turbinate (N1–P2), the regional dose persistently increases in T1 while it decreases in T2 (Figure 7b,c). Again, large dose variances are noted in T3, reaffirming the high level of instability of the liquid film in upstream regions.

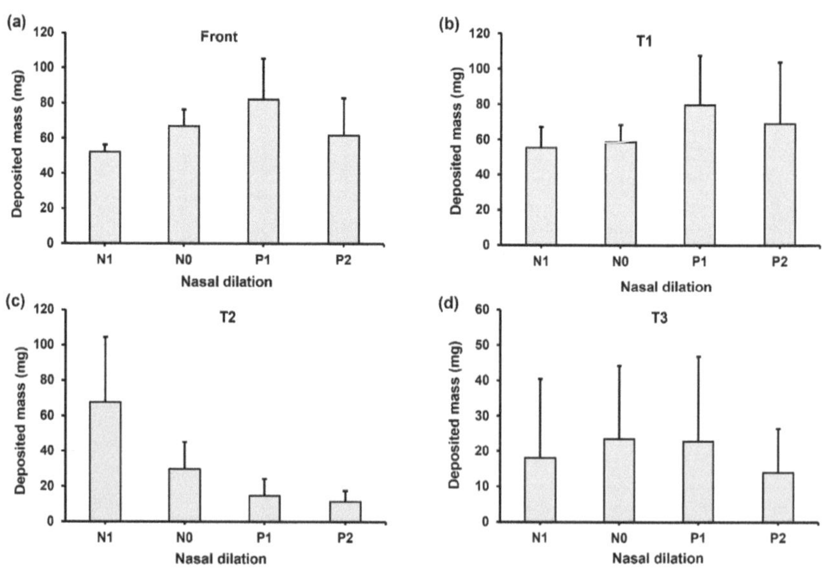

Figure 7. Deposition variation vs. nasal cycle (N1, N0, P1, and P2) in different regions of the nose with a 22.5° back tilt head position: (**a**) front nose, (**b**) T1, (**c**) T2, and (**d**) T3.

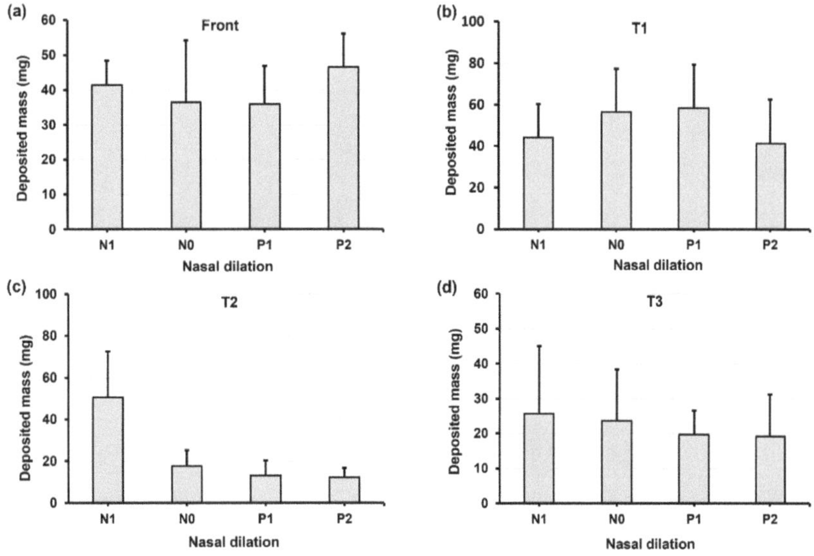

Figure 8. Deposition variation vs. nasal cycle (N1, N0, P1, and P2) in different regions of the nose with a 45° back tilt head position: (**a**) front nose, (**b**) T1, (**c**) T2, and (**d**) T3.

When increasing the head angle to 45° (Figure 8), we observe different patterns in deposition distribution among N1–P2 compared to those at 0° and 22.5°, i.e., with a relatively flat pattern in the front nose (Figure 8a), a modal pattern in T1 (Figure 8b), and a constant decrease pattern in T2 (Figure 8c) and T3 (Figure 8d). These patterns can be attributed to the predominant gravitational effect at 45°, which drives the deposited liquid film downstream of the nasal passage and, at the same time, leaves a thinner liquid film attaching onto the turbinate wall. This competing effect explains the dose increase from N1 to P1 (gravity-driven film translocation) and decrease from P1 to P2 (decreasing turbinate area). In T2 and T3 (Figure 8c,d), the constant decrease from N1 to P2 comes from the progressively shrinking turbinate. It is also observed that certain doses exit the nasopharynx in models P1 and P2 due to the pronounced gravity-driven translation at 45°.

2.5. Factorial Analyses of Effects of Nasal Cycle and Head Angle

2.5.1. Total Deposition

Figure 9 displays both the violin and mean plots illustrating the total deposition concerning head angle and nasal dilation due to the nasal cycle. The statistical analysis of total deposition, considering both factors (head angle and nasal cycle), is prominently featured in Figure 9. The violin plot further highlights the variances in total deposited mass concerning head angle (Figure 9a) and nasal cycle (Figure 9b), revealing substantial deposition variations in both factors. Significant deposition variances are evident in both factors, with the head angle exerting a more pronounced impact relative to the nasal cycle effect. This is evidenced by a noticeable decrease in deposition from the upright nose position (0°) to a 22.5° back tilt, followed by a drastic decrease from 22.5° to a 45° back tilt. Moreover, the magnitude of variance escalates as the back tilt angle increases, causing a diverse range of movements of the liquid film within the nasal cavity.

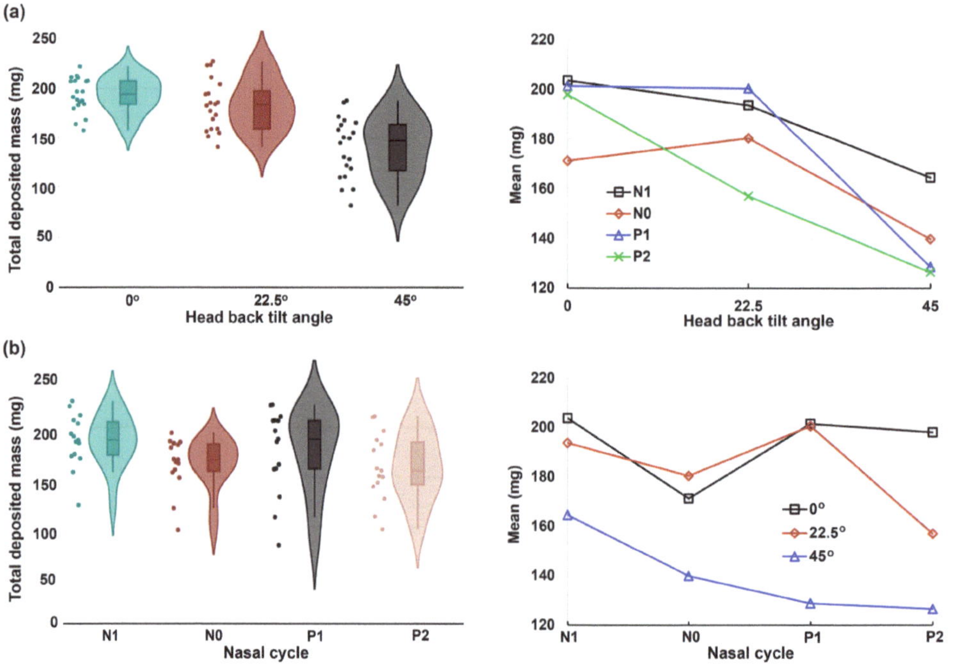

Figure 9. The violin and mean plots of the total deposited mass in four nasal models (N1–P2) and three head orientations: (**a**) head angle effect and (**b**) nasal dilation effect.

The mean plots in the right panels of Figure 9a,b provide insight into the two-factor ANOVA concerning head angle and nasal cycle effects along the x-axis, respectively. The corresponding p-values are listed in Table 1. As expected, a highly significant difference (p-value < 0.001) emerges in total deposition across the three head angles (0°, 22.5°, and 45°). Additionally, a very significant difference (p-value = 0.003) surfaces among the four models, reflecting nasal cycle-induced turbinate congestion/decongestion (N1, N0, P1, and P2). Notably, the interaction between the head angle and nasal cycle effect significantly influences total deposition, as indicated by a p-value of 0.031.

Table 1. The p-values from two-factor analyses of variance vs. the head angle and nasal dilation.

Region	Total	Front	Turbinate	NP	T1	T2	T3
Head angle	<0.001	<0.001	0.034	0.001	0.001	0.244	0.001
Nasal dilation	0.003	0.223	0.001	0.061	0.003	<0.001	0.78
Interaction	0.031	0.074	0.364	0.075	0.164	0.794	0.941

2.5.2. Regional Deposition

Figure 10 displays the mean and violin plots for regional deposition concerning the head angle on the x-axis, while Figure 11 represents the nasal dilation on the x-axis.

The regions considered include the front nose, turbinate, and nasopharynx (NP), with the turbinate further divided into T1, T2, and T3 (Figures 10a and 11a). The corresponding p-values for individual regions are listed in Table 1. The analysis reveals that the head angle significantly influences deposition in the front nose, T1, T3, and NP, with highly significant p-values (p-value $\leq$ 0.001). This influence is visually illustrated in the violin plots for the front nose and NP in Figure 10b. However, the head angle's effect on turbinate deposition is significant (p-value = 0.034) but insignificant on T2 (middle turbinate), which is consistent with spray reaching the middle turbinate in all test cases post-dose applications.

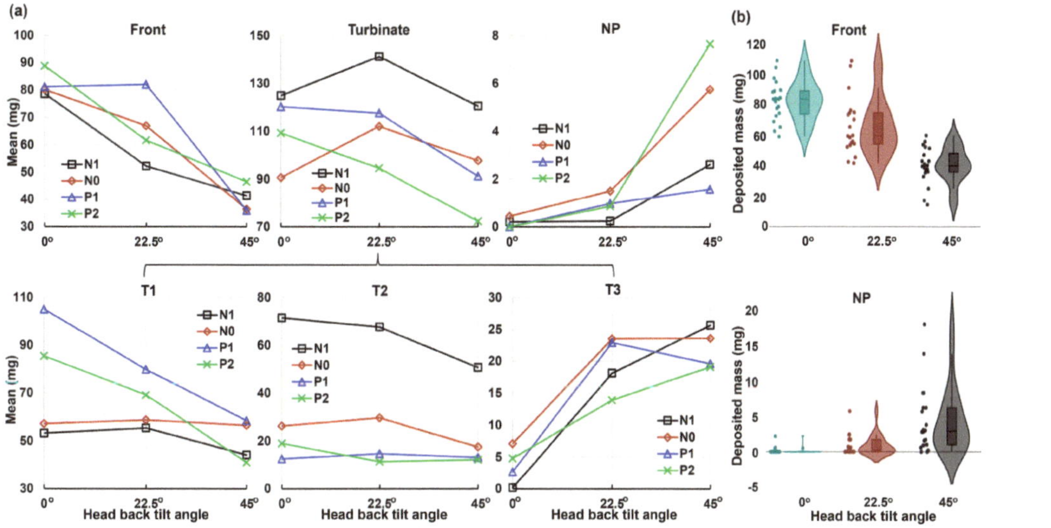

Figure 10. Factorial analyses vs. head angle for regional deposition in four nasal models (N1–P2): (**a**) mean plot and (**b**) violin plot.

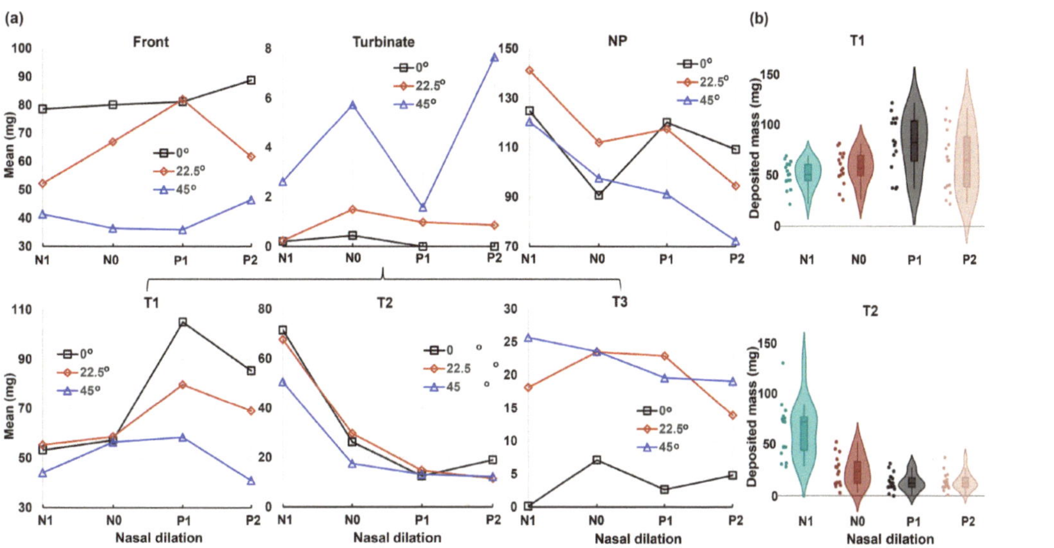

Figure 11. Factorial analyses vs. nasal cycle for regional deposition with three head orientations: (**a**) mean plot vs. nasal dilation and (**b**) violin plot.

Conversely, the nasal cycle has a highly significant effect (p-value ≤ 0.001) on deposition in both the turbinate and the T2 section, as illustrated in the lower panel of Figure 11b. The nasal cycle effect is also very significant (p-value = 0.003) for spray deposition in T1 (anterior turbinate), as shown in Table 1 and the upper panel of Figure 11b. However, the nasal cycle effect is not significant for spray deposition in the front nose (p-value = 0.223), T3 (posterior turbinate, p-value = 0.78), and NP (p-value = 0.061). These results align with the expectation that the congestion/decongestion induced by the nasal cycle primarily affects the turbinate region, leaving the front and back of the nose largely unaffected by passage morphology. Considering the mutual effect between the head angle and nasal cycle, insignificant interactions are observed in all individual regions considered, with p-values exceeding 0.05 (Table 1). This contrasts with the significant interaction between the two factors for the entire region (Figure 9). The observed difference is speculated to be due to the overall slight decrease in regional deposition with increasing head angles (i.e., increasing gravitational effect), cumulatively resulting in a significant difference in total deposition.

3. Discussion

This study investigated the effects of the nasal cycle, in conjunction with varying head angles, on the deposition patterns of intranasally administered nasal sprays. The results highlight several key findings that contribute to our understanding of the complex dynamics involved.

3.1. Effects of Nasal Cycle on Spray Deposition Distribution

The nasal cycle, characterized by cyclic changes in nasal airflow resistance, plays a crucial role in shaping the deposition patterns of intranasal sprays. This study delves into the intricate relationship between the nasal cycle and nasal spray deposition, utilizing models representing stages of congestion and decongestion (N1, N0, P1, and P2). Consistent parameters such as spray formulation, pump nozzle angle, insertion depth, and inspiratory flow rate were maintained. Despite this, significant variations in deposition were observed across nasal cycle stages. During congestion (N1), the increased resistance in one passage led to a more confined deposition space in the narrow clearance between the inferior turbinate and septum. Decongestion (P1, P2) widened the passage, allowing for broader dispersion of spray droplets and increased sedimentation towards the floor of turbinate sections T1, T2, and T3 (congested N1 model in the first column vs. decongested P2 in the second column, Figures 4 and 5).

3.2. Effects of Head Angle and Interactions with Nasal Cycle

Head orientation, combined with the nasal cycle, played a crucial role in spray distribution. In an upright position, deposition was confined to specific turbinate sections, namely, T1 and T2 within the congested N1 and normal N0 model, due to inertial impaction and gravitational sedimentation (first and second columns, Figure 5a). Flow shear effects were small in drawing fluid deeper into the narrow nasal passages. However, a 22.5° back tilt increased gravitational effects, aiding flow shear in dragging droplets deeper into the nasal cavity. A 45° tilt accentuated these effects and facilitated deeper penetration of spray droplets, predominantly through the middle meatus, dispersing spray throughout all the turbinate sections, T1, T2, and T3 (Figure 5b). The impact of head angle on deposition was substantial, and its interaction with the nasal cycle significantly influenced the overall spray distribution. Both factors contributed to the complexity of the deposition pattern observed in this study.

3.3. Deposition Variance Related to Nasal Cycle and Head Angle

Despite efforts to maintain controlled conditions, a notable variance in deposition was observed, as illustrated by the charts of deposition mass distributions in the four nasal models (N1–P2) with the various head positions (upright, 22.5°, and 45° back tilt head positions, Figures 6–8). Figure 9, depicting the violin and mean plots of the total deposited mass throughout the various stages of congestion and decongestion, shows that the magnitude of variance increased as the head tilt angle increased backward from upright to 22.5° to 45°. This change occurred due to the increasing influence of gravity, leading to the destabilization of the liquid film and inducing a variety of movements in the liquid layer. This variability persisted across both nasal cycle stages and different head angles.

Gravitational sedimentation, inertial impaction, flow shear, and the complex nasal cavity morphology contributed to this observed variability. The nonlinear nature of the variance suggests a multi-faceted interplay of factors. The nasal cycle, head angle, and convoluted nasal cavity geometry collectively contribute to the intricate and variable nature of the spray deposition within the nasal cavity.

3.4. Implications for Intranasal Spray Administration

The intricate interplay between the natural nasal cycle and the distribution patterns of intranasal sprays has significant implications for targeted drug delivery within the nasal cavity. The nasal cycle, marked by cyclic variations in nasal airflow resistance, exerts a substantial influence on intranasal spray distribution. Despite maintaining consistent parameters, the models representing congestion and decongestion stages (N1, N0, P1, and P2) display notable variations in deposition. During congestion (N1), elevated nasal resistance confines spray deposition, while decongestion (P1, P2) widens the nasal passage, leading to broader dispersion of spray droplets, especially towards the posterior turbinate section (T3) and nasopharynx. These regions are the target for mucosal immunization against respiratory infectious diseases, as well as the initial site of viral infections such as COVID-19 [36–41].

Furthermore, head orientation, in conjunction with the nasal cycle, proves to be a critical factor in spray distribution. In an upright position, deposition is confined to specific turbinate sections (T1 and T2), whereas back tilts (22.5° and 45°) increase gravitational effects, facilitating deeper droplet penetration. A 45° tilt notably enhances dispersion throughout turbinate sections T1, T2, and T3. The interaction of the head angle with the nasal cycle significantly influences overall spray distribution, impacting the targeted delivery to specific nasal regions.

3.5. Limitations

This study is subject to several limitations that should be considered in the interpretation of results. Firstly, the focus on morphological variation was confined to the inferior concha, specifically affecting the left inferior meatus. This choice simplifies the investigation, neglecting the cyclic changes in congestion and decongestion typically involving all turbinate regions (inferior, middle, and superior) within one nasal passage. Secondly, the study's scope is limited by the relatively small variety of nasal models used. Although variations in the left inferior turbinate were explored across different phases of the nasal cycle, a more extensive range of patient-specific nasal models reflecting diverse nasal passage morphologies would provide a more accurate assessment of intersubject variability.

Additionally, the connection interfaces between different sections of the model utilized tongue-and-groove connections that were not watertight. In some instances, this design allowed the deposited saline liquid to seep through tight spaces at the connections via capillary action, leading to minor inaccuracies in deposition quantifications. Although these discrepancies were deemed insignificant, they should be acknowledged. Furthermore, this study utilized nasal cavity models constructed with SLA resin through 3D printing. This material choice differs from the in vivo compliant airway [42]. A life-condition nasal cavity comprises squamous epithelial tissue in the anterior (front) region and nasal mucosa in the turbinate region. A comprehensive evaluation of the liquid-surface properties of the nasal mucosa compared to the resin surface is warranted to understand the degree of similarity and potential implications for this study's findings.

4. Materials and Methods

4.1. Study Design

To assess the impact of mucosal congestion and decongestion during the natural nasal cycle on intranasal spray deposition, four multi-piece anatomically accurate nasal models were developed, each with distinct turbinate profiles (N1, N0, P1, and P2; Figure 1b,c). To accurately assess the effects of turbinate congestion/decongestion resulting from the nasal cycle, it was necessary to incorporate factors facilitating the penetration of administered nasal sprays deep into the nasal passage. This approach allowed for a more comprehensive evaluation of the impact of nasal passage clearance along the entire length of the nasal cavity. Key influencing factors included the nasal spray delivery device, liquid formulation, administration angle, device nozzle insertion depth, number of spray applications (doses), head orientation, and inhalation flow rate. The nasal delivery device used was a refillable multi-dose nasal spray pump (Hengni) loaded with a saline solution comprising 0.65% *w/v* sodium chloride, benzalkonium chloride, monosodium phosphate, phenylcarbinol, and disodium phosphate. The rationale for using a saline solution in this study stems from its status as one of the most used formulations for nasal sprays, employed for purposes such as nasal irrigation, moisturizing, and clearing nasal passages.

The nozzle of the nasal spray pump was fixed at an angle of 30° from the nostril's normal, pointing toward the internal nasal valve in all the experimental runs, and inserted at a fixed depth of approximately 10 mm. Two applications were administered for each test run, with a single application delivering an average dose of 95 mg of saline solution. Three different head orientations were considered to assess the influence of gravity on the spray deposition distribution in the various regions of the nasal cavity at various positions of the nasal cavity under the effects of the nasal morphology due to turbinate congestion and decongestion. Among the many factors affecting intranasal deposition, gravitational sedimentation can contribute to the deposition of larger droplets. The translocation paths of liquid film, developed from the merging of multiple droplets within the nasal cavity, are also susceptible to gravitational effects, which vary based on head position. Thus, the experimental runs utilized the following head orientations: upright, 22.5°, and 45° back tilts from the horizontal (Figure 2). These orientations aimed to evaluate the influence of gravity on spray distribution within various nasal cavity regions, corresponding to different head positions affected by nasal morphology changes due to mucosal congestion and decongestion. To generate inhalation flow, each model was connected to a vacuum via a flexible tube (Robinair 3 CFM, Warren, MI, USA). The flow rate was set at a constant 16 L/min, corresponding to deep inhalation, only during nasal spray applications for all test cases and was regulated using a flow meter (Omega, FL-510, Stamford, CT, USA).

4.2. Nasal Cast Models

The nasal model N0, representing a normal condition, was previously developed from MRI scans of a 53-year-old male with no reported respiratory diseases [43,44]. The remaining three models (N1, P1, and P2) were modified from model N0, with changes made to the left inferior turbinate to reflect various degrees of congestion and decongestion. Thus, the four nasal models had a gradually widening nasal passage from a congested state (N1) to a normal condition (N0) and finally to decongested states (P1 and P2) throughout a nasal cycle, as illustrated in Figure 1. The geometrical modification was implemented using SolidWorks (Dassault Systems, Waltham, MA, USA).

The nasal casts were printed using a Formlabs 3D printer (Form 3B+, Somerville, MA, USA). The printing material was a clear stereolithography (SLA) rigid resin (Formlabs Clear Resin, FLGPCL04), creating transparent nasal casts (Figure 2). In contrast, the front nose section was printed using elastic SLA resin (Formlabs Elastic 50A Resin, FLELCL01) using the same 3D printer. This choice was made to simulate the flexible nature of the real front nose and allow for dilation, enabling the nasal spray pump's nozzle to be inserted deeper into the nostril to achieve the desired insertion depth. A major benefit of employing transparent casts is the capability to visually monitor the real-time movement of liquid sprays administered intranasally and the resulting film formation within the nasal cast.

The rationale behind creating the nasal casts with only a single passage and an incorporated nasal septum was to facilitate clear observation and visualization of the spray deposition and liquid film translocation through the passage. This included views not only from the cheek side but also from the septum side, enabling visualization from the opposite side of the turbinate. This aspect would have been obscured if a complete nasal model had been used.

4.3. Experimental Protocol

In conducting experiments for each head orientation, the sections (Front, T1, T2, T3, and NP) of each nasal model (N1, N0, P1, and P2) were assembled using clear adhesive tape wrapped around the outside surfaces, securely and firmly fastening the parts together. The assembled models were then positioned to correspond to one of the three head orientations outlined in the study design, namely, upright, 22.5° back tilt, and 45° back tilt. After being filled with saline solution and properly primed, the nasal spray pump was fixed into place to achieve the desired administration angle and insertion depth. For each experimental run, the vacuum, connected to the model's nasopharynx (NP) via a flexible tube, was activated for a short duration, simulating a period of deep inhalation. During this phase, two spray doses were administered successively, with a brief pause of a few seconds between applications.

A fluorescent green dye (GLO Effex, Murrieta, CA, USA) was used to visualize the dynamics of the liquid spray and subsequent liquid film translocation. A camera facing the side of the nasal model captured real-time intranasal spray dynamics, as well as the final deposition distribution on the nasal walls. An LED light with wavelengths ranging from 385 nm to 395 nm was used to illuminate the fluorescent dye. The deposited mass of liquid spray in various sections (Front, T1, T2, T3, and NP) was quantified for each model by measuring the mass of each section before and after each experimental run. This was accomplished with a Bonvoisin electronic scale with a precision of 0.1 mg (San Jose, CA, USA). A minimum of five repetitions were conducted for each experimental scenario to account for variability.

4.4. Statistical Analysis

The statistical analysis was performed using Minitab software 21.4 (State College, PA, USA). To assess result variability, a one-way analysis of variance (ANOVA) was conducted. The doses delivered in each test case were expressed as a range and presented as mean ± standard deviation. Deposition data statistics were illustrated using violin plots, which are akin to box plots and display the data's probability density at various values, smoothed by a kernel density estimator [45]. ANOVA mean effect analysis was employed to evaluate the comparative influence of the nasal cycle and head angle. Statistical significance was determined when a p-value was less than 0.05, while a p-value ≤ 0.01 indicated a very significant difference and a p-value < 0.001 denoted a highly significant difference.

5. Conclusions

In conclusion, this comprehensive investigation into intranasal spray deposition sheds light on critical factors influencing targeted drug delivery within the nasal cavity. This study underscores the intricate relationship between the natural nasal cycle, head orientation, and spray distribution patterns. Notably, congestion and decongestion stages induce significant variations in deposition, emphasizing the relevance of considering physiological conditions for optimal drug delivery. The impact of head angle, particularly in back tilts, is substantial, influencing gravitational effects and spray dispersion. The observed variance in deposition, even under controlled conditions, reflects the complex interplay of gravity, nasal cycle, flow shear, and cavity morphology. These findings provide valuable insights for enhancing targeted drug delivery strategies and emphasize the necessity of accounting for individual variability in nasal physiology.

Author Contributions: Conceptualization, A.S., X.S. and J.X.; methodology, A.S., X.S. and J.X.; validation, A.S., X.S. and J.X.; formal analysis, A.S., X.S. and X.J; investigation, A.S. and J.X.; resources, X.S. and J.X.; data curation, J.X.; writing—original draft preparation, A.S. and J.X.; writing—review and editing, X.S. All authors have read and agreed to the published version of the manuscript.

Funding: This research received no external funding.

Institutional Review Board Statement: Not applicable.

Informed Consent Statement: Not applicable.

Data Availability Statement: Data is contained within the article.

Conflicts of Interest: The authors declare no conflicts of interest.

References

1. Ulusoy, S.; Bayar Muluk, N.; Scadding, G.K.; Passali, G.S.; Dilber, M.; Gevaert, P.; Passali, D.; Resuli, A.S.; Van Gerven, L.; Kalogjera, L.; et al. The intranasal trigeminal system: Roles in rhinitis (allergic and non-allergic). *Eur. Rev. Med. Pharmacol. Sci.* **2022**, *26*, 25–37. [PubMed]
2. Kahana-Zweig, R.; Geva-Sagiv, M.; Weissbrod, A.; Secundo, L.; Soroker, N.; Sobel, N. Measuring and characterizing the human nasal cycle. *PLoS ONE* **2016**, *11*, e0162918. [CrossRef] [PubMed]
3. Soane, R.J.; Carney, A.S.; Jones, N.S.; Frier, M.; Perkins, A.C.; Davis, S.S.; Illum, L. The effect of the nasal cycle on mucociliary clearance. *Clin. Otolaryngol. Allied Sci.* **2001**, *26*, 9–15. [CrossRef] [PubMed]
4. White, D.E.; Bartley, J.; Nates, R.J. Model demonstrates functional purpose of the nasal cycle. *Biomed. Eng. Online* **2015**, *14*, 38. [CrossRef] [PubMed]
5. Eccles, R. A role for the nasal cycle in respiratory defence. *Eur. Respir. J.* **1996**, *9*, 371–376. [CrossRef]
6. Eccles, R.; Reilly, M.; Eccles, K.S. Changes in the amplitude of the nasal cycle associated with symptoms of acute upper respiratory tract infection. *Acta Otolaryngol.* **1996**, *116*, 77–81. [CrossRef]
7. Hanif, J.; Jawad, S.S.; Eccles, R. The nasal cycle in health and disease. *Clin. Otolaryngol. Allied Sci.* **2000**, *25*, 461–467. [CrossRef]
8. Braun, T.; Rich, M.; Berghaus, A.; Kramer, M.F. Effects of oxymetazoline nasal spray on the nasal cycle assessed by long-term rhinoflowmetry. *Rhinology* **2012**, *50*, 370–375. [CrossRef]
9. Van Cauwenberge, P.B.; Deleye, L. Nasal cycle in children. *Arch. Otolaryngol.* **1984**, *110*, 108–110. [CrossRef]
10. Williams, M.R.; Eccles, R. The nasal cycle and age. *Acta Otolaryngol.* **2015**, *135*, 831–834. [CrossRef]
11. Lindemann, J.; Weindel, C.; Hoffmann, T.K.; Sommer, F.; Scheithauer, M.O.; Stupp, F.; Reins, E.F. Re-evaluating the nasal cycle by long-term rhinoflowmetry: Most individuals demonstrate a 'mixed' nasal cycle. *Rhinology* **2021**, *59*, 538–544. [CrossRef] [PubMed]
12. Baraniuk, J.N.; Kim, D. Nasonasal reflexes, the nasal cycle, and sneeze. *Curr. Allergy Asthma Rep.* **2007**, *7*, 105–111. [CrossRef] [PubMed]
13. Letzel, J.; Darbinjan, A.; Hummel, T. The nasal cycle before and after nasal septoplasty. *Eur. Arch. Otorhinolaryngol.* **2022**, *279*, 4961–4968. [CrossRef] [PubMed]
14. Pendolino, A.L.; Scarpa, B.; Ottaviano, G. Relationship between nasal cycle, nasal symptoms and nasal cytology. *Am. J. Rhinol. Allergy* **2019**, *33*, 644–649. [CrossRef] [PubMed]
15. Cingi, C.; Bayar Muluk, N.; Mitsias, D.I.; Papadopoulos, N.G.; Klimek, L.; Laulajainen-Hongisto, A.; Hytönen, M.; Toppila-Salmi, S.K.; Scadding, G.K. The nose as a route for therapy: Part 1. pharmacotherapy. *Front. Allergy* **2021**, *2*, 638136. [CrossRef] [PubMed]
16. Blaiss, M.S.; Benninger, M.S.; Fromer, L.; Gross, G.; Mabry, R.; Mahr, T.; Marple, B.; Stoloff, S. Expanding choices in intranasal steroid therapy: Summary of a roundtable meeting. *Allergy Asthma Proc.* **2006**, *27*, 254–264. [CrossRef] [PubMed]
17. Merkus, P.; Ebbens, F.A.; Muller, B.; Fokkens, W.J. Influence of anatomy and head position on intranasal drug deposition. *Eur. Arch. Otorhinolaryngol.* **2006**, *263*, 827–832. [CrossRef]
18. Naclerio, R.M.; Bachert, C.; Baraniuk, J.N. Pathophysiology of nasal congestion. *Int. J. Gen. Med.* **2010**, *3*, 47–57. [CrossRef]
19. Salati, H.; Bartley, J.; White, D.E. Nasal saline irrigation—A review of current anatomical, clinical and computational modelling approaches. *Respir. Physiol. Neurobiol.* **2020**, *273*, 103320. [CrossRef]
20. Xiao, Q.; Bates, A.J.; Cetto, R.; Doorly, D.J. The effect of decongestion on nasal airway patency and airflow. *Sci. Rep.* **2021**, *11*, 14410. [CrossRef]
21. Basu, S.; Holbrook, L.T.; Kudlaty, K.; Fasanmade, O.; Wu, J.; Burke, A.; Langworthy, B.W.; Farzal, Z.; Mamdani, M.; Bennett, W.D.; et al. Numerical evaluation of spray position for improved nasal drug delivery. *Sci. Rep.* **2020**, *10*, 10568. [CrossRef] [PubMed]
22. Sun, Q.; Dong, J.; Zhang, Y.; Tian, L.; Tu, J. Numerical study of the effect of nasopharynx airway obstruction on the transport and deposition of nanoparticles in nasal airways. *Exp. Comput. Multiph. Flow* **2022**, *4*, 399–408. [CrossRef]
23. Shang, Y.; Dong, J.; He, F.; Inthavong, K.; Tian, L.; Tu, J. Detailed comparative analysis of environmental microparticle deposition characteristics between human and monkey nasal cavities using a surface mapping technique. *Sci. Total Environ.* **2022**, *853*, 158770. [CrossRef] [PubMed]
24. Li, C.; Maza, G.; Farag, A.A.; Krebs, J.P.; Deshpande, B.; Otto, B.A.; Zhao, K. Asymptomatic vs. symptomatic septal perforations: A computational fluid dynamics examination. *Int. Forum Allergy Rhinol.* **2019**, *9*, 883–890. [CrossRef] [PubMed]
25. Frank-Ito, D.O.; Kimbell, J.S.; Borojeni, A.A.T.; Garcia, G.J.M.; Rhee, J.S. A hierarchical stepwise approach to evaluate nasal patency after virtual surgery for nasal airway obstruction. *Clin. Biomech.* **2019**, *61*, 172–180. [CrossRef] [PubMed]
26. Moghaddam, M.G.; Garcia, G.J.M.; Frank-Ito, D.O.; Kimbell, J.S.; Rhee, J.S. Virtual septoplasty: A method to predict surgical outcomes for patients with nasal airway obstruction. *Int. J. Comput. Assist. Radiol. Surg.* **2020**, *15*, 725–735. [CrossRef] [PubMed]

27. Rhee, J.S.; Cannon, D.E.; Frank, D.O.; Kimbell, J.S. Role of virtual surgery in preoperative planning: Assessing the individual components of functional nasal airway surgery. *Arch. Facial Plast. Surg.* **2012**, *14*, 354–359. [CrossRef] [PubMed]
28. Vanhille, D.L.; Garcia, G.J.M.; Asan, O.; Borojeni, A.A.T.; Frank-Ito, D.O.; Kimbell, J.S.; Pawar, S.S.; Rhee, J.S. Virtual surgery for the nasal airway: A preliminary report on decision support and technology acceptance. *JAMA Facial Plast. Surg.* **2018**, *20*, 63–69. [CrossRef]
29. Zhang, Y.; Zhou, X.; Lou, M.; Gong, M.; Zhang, J.; Ma, R.; Zhang, L.; Huang, F.; Sun, B.; Zhu, K.; et al. Computational fluid dynamics (CFD) investigation of aerodynamic characters inside nasal cavity towards surgical treatments for secondary atrophic rhinitis. *Math. Probl. Eng.* **2019**, *2019*, 6240320. [CrossRef]
30. Bickford, L.; Shakib, S.; Taverner, D. The nasal airways response in normal subjects to oxymetazoline spray: Randomized double-blind placebo-controlled trial. *Br. J. Clin. Pharmacol.* **1999**, *48*, 53–56. [CrossRef]
31. Xi, J.; Wang, Z.; Si, X.A.; Zhou, Y. Nasal dilation effects on olfactory deposition in unilateral and bi-directional deliveries: In Vitro tests and numerical modeling. *Eur. J. Pharm. Sci.* **2018**, *118*, 113–123. [CrossRef] [PubMed]
32. Flanagan, P.; Eccles, R. Spontaneous changes of unilateral nasal airflow in man. A re-examination of the 'nasal cycle'. *Acta Otolaryngol.* **1997**, *117*, 590–595. [CrossRef]
33. Cherobin, G.B.; Voegels, R.L.; Gebrim, E.; Garcia, G.J.M. Sensitivity of nasal airflow variables computed via computational fluid dynamics to the computed tomography segmentation threshold. *PLoS ONE* **2018**, *13*, e0207178. [CrossRef] [PubMed]
34. Kim, S.K.; Chung, S.K. Investigation on the respiratory airflow in human airway by PIV. *J. Vis.* **2009**, *12*, 259–266. [CrossRef]
35. Wang, Y.D.; Lee, H.P.; Gordon, B.R. Impacts of fluid dynamics simulation in study of nasal airflow physiology and pathophysiology in realistic human three-dimensional nose models. *Clin. Exp. Otorhinolaryngol.* **2012**, *5*, 181–187. [CrossRef] [PubMed]
36. Xi, J. Development and challenges of nasal spray vaccines for short-term COVID-19 protection. *Curr. Pharm. Biotechnol.* **2022**, *23*, 1671–1677. [CrossRef] [PubMed]
37. Alu, A.; Chen, L.; Lei, H.; Wei, Y.; Tian, X.; Wei, X. Intranasal COVID-19 vaccines: From bench to bed. *EBioMedicine* **2022**, *76*, 103841. [CrossRef] [PubMed]
38. Xi, J.; Lei, L.R.; Zouzas, W.; April Si, X. Nasally inhaled therapeutics and vaccination for COVID-19: Developments and challenges. *MedComm* **2021**, *2*, 569–586. [CrossRef]
39. Wu, S.; Huang, J.; Zhang, Z.; Wu, J.; Zhang, J.; Hu, H.; Zhu, T.; Zhang, J.; Luo, L.; Fan, P.; et al. Safety, tolerability, and immunogenicity of an aerosolised adenovirus type-5 vector-based COVID-19 vaccine (Ad5-nCoV) in adults: Preliminary report of an open-label and randomised phase 1 clinical trial. *Lancet Infect. Dis.* **2021**, *21*, 1654–1664. [CrossRef]
40. Nian, X.; Zhang, J.; Huang, S.; Duan, K.; Li, X.; Yang, X. Development of nasal vaccines and the associated challenges. *Pharmaceutics* **2022**, *14*, 1983. [CrossRef]
41. Xi, J.; Si, X.A. A next-generation vaccine for broader and long-lasting COVID-19 protection. *MedComm* **2022**, *3*, e138. [CrossRef] [PubMed]
42. Wang, J.; Xi, J.; Han, P.; Wongwiset, N.; Pontius, J.; Dong, H. Computational analysis of a flapping uvula on aerodynamics and pharyngeal wall collapsibility in sleep apnea. *J. Biomech.* **2019**, *94*, 88–98. [CrossRef] [PubMed]
43. Seifelnasr, A.; Si, X.A.; Xi, J. Visualization and estimation of nasal spray delivery to olfactory mucosa in an image-based transparent nasal model. *Pharmaceutics* **2023**, *15*, 1657. [CrossRef]
44. Jamalabadi, M.Y.; Xi, J. Olfactory drug aerosol delivery with acoustic radiation. *Biomedicines* **2022**, *10*, 1347.
45. Hintze, J.L.; Nelson, R.D. Violin plots: A box plot-density trace synergism. *Am. Stat.* **1998**, *52*, 181–184.

pharmaceuticals

MDPI

Article

A Novel Flow Cytometry-Based Assay for the Identification of HCN4 CNBD Ligands

Magdalena N. Wojciechowski, Sebastian Schreiber and Joachim Jose *

University of Münster, Institute of Pharmaceutical and Medicinal Chemistry, Pharmacampus, 48149 Münster, Germany

* Correspondence: joachim.jose@uni-muenster.de

Abstract: Hyperpolarization-activated and cyclic nucleotide-gated (HCN) channels are promising therapeutic targets because of their association with the genesis of several diseases. The identification of selective compounds that alter cAMP-induced ion channel modulation by binding to the cyclic nucleotide-binding domain (CNBD) will facilitate HCN channel-specific drug development. In this study, a fast and protein purification-free ligand-binding approach with a surface-displayed HCN4 C-Linker-CNBD on *E. coli* is presented. 8-Fluo-cAMP ligand binding was monitored by single-cell analysis via flow cytometry, and a K_d-value of 173 ± 46 nM was determined. The K_d value was confirmed by ligand depletion analysis and equilibrium state measurements. Applying increasing concentrations of cAMP led to a concentration-dependent decrease in fluorescence intensity, indicating a displacement of 8-Fluo-cAMP. A K_i-value of 8.5 ± 2 μM was determined. The linear relationship of IC_{50} values obtained for cAMP as a function of ligand concentration confirmed the competitive binding mode: IC_{50}: 13 ± 2 μM/16 ± 3 μM/23 ± 1 μM/27 ± 1 μM for 50 nM/150 nM/250 nM/500 nM 8-Fluo-cAMP. A similar competitive mode of binding was confirmed for 7-CH-cAMP, and an IC_{50} value of 230 ± 41 nM and a K_i of 159 ± 29 nM were determined. Two established drugs were tested in the assay. Ivabradine, an approved HCN channel pore blocker and gabapentin, is known to bind to HCN4 channels in preference to other isoforms with an unknown mode of action. As expected, ivabradine had no impact on ligand binding. In addition, gabapentin had no influence on 8-Fluo-cAMP's binding to HCN4-CNBD. This is the first indication that gabapentin is not interacting with this part of the HCN4 channel. The ligand-binding assay as described can be used to determine binding constants for ligands such as cAMP and derivatives. It could also be applied for the identification of new ligands binding to the HCN4-CNBD.

Keywords: autodisplay; HCN channels; CNBD; ligand binding

Citation: Wojciechowski, M.N.; Schreiber, S.; Jose, J. A Novel Flow Cytometry-Based Assay for the Identification of HCN4 CNBD Ligands. *Pharmaceuticals* **2023**, *16*, 710. https://doi.org/10.3390/ph16050710

Academic Editor: Liqiang Chen

Received: 29 March 2023
Revised: 25 April 2023
Accepted: 3 May 2023
Published: 7 May 2023

1. Introduction

Hyperpolarization-activated and cyclic nucleotide-gated (HCN) channels belong to the superfamily of voltage-gated pore loop channels such as CNG channels and EAG-like K^+ channels [1]. Upon hyperpolarization, a mixed Na^+/K^+ inward pacemaker ion current I_f (funny) is generated in the heart region and an I_h (hyperpolarization-activated) current is generated in the brain region [2]. HCN channels are encoded by four genes (HCN1-4) [3,4] and homo- or heterotetrametric proteins were formed within the membrane [5]. Crystal structures were obtained for HCN1 [6,7] and HCN4 [8] and it was revealed that each monomer consists of an N-terminal intracellular HCN-domain, followed by six α-helical transmembrane segments (S1–6). S1–4 constitute the voltage–sensor–domain (VSD) and S5–6 form the pore domain that includes the selectivity filter. The C-terminus is located intracellularly and comprises the cyclic nucleotide-binding domain (CNBD), which is connected to S6 by the C-Linker. Endogenous cyclic nucleotides, such as cAMP or cGMP, induce a voltage shift to more depolarized potentials upon binding to the CNBD and alter channel opening and closing kinetics. The extent of cyclic nucleotide modulation differs between the HCN subtypes [9–11]. Expression patterns of the isoforms are varying in the brain, heart, retina and peripheral nervous system [12–14]. HCN channels are involved in the regulation of cardiac and neuronal rhythmicity, generation of the resting membrane potential, dendritic integration or synaptic transmission [15]. Therefore, it appears evident that HCN channelopathies contribute to the pathogenesis of several diseases, including bradycardia [16], arrhythmia [17], epilepsy [18,19] or neuropathic pain disorder [20,21]. Although HCN channels were discovered more than 40 years ago [22–24] and their involvement in different diseases has been described in many studies [25,26], only a single drug has been approved until now, addressing HCN channels as targets. Ivabradine, a non-selective HCN channel pore blocker, was approved for the treatment of stable angina pectoris and heart failure disease [27]. Identification of new drugs addressing HCN channel isoforms selectively could be a promising option for novel specific therapies. At least two ways are possible by which a compound can target a HCN channel: either by blocking the pore like ivabradine or by interacting with the CNBD, and hence altering the cAMP-induced channel modulation. So far, drug discovery approaches have solely identified compounds targeting the HCN channel pore. By the remaining 7,8-dimethoxy-benzazepinone moiety and modification of the propyl chain of the non-selective pore blocker Zatebradine, subtype-selective HCN channel pore blockers as MEL57A, MEL55A and EC18 were developed [28–30]. Nakashima et al. (2021) identified novel HCN4 channel pore blockers, T-478, a methoxybenzenesulfonamide derivate, T-788, a tetrahydrooxazolopyrazinone derivate, and T-524, a thetrahydroisoquinoline derivate. However, these compounds were tested exclusively on HCN4 blocking, and selectivity tests with other HCN isoforms are still lacking [31]. Several other drugs were described to modulate HCN channel activity in addition to primary targets. This includes general and local anesthetics, anticonvulsive drugs and plant-derived compounds [32–34]. The application of such drugs for the treatment of HCN channel-dependent diseases is limited due to non-selectivity and hence risks severe side effects. To date, no compounds targeting the CNBD of HCN channels selectively have been described [32]. The first comprehensive study on a range of cyclic nucleotide derivatives binding to the CNBD of HCN1, 2 and 4 provided structural information about moieties required for ligand binding [35]. Studies with HCN C-Linker-CNBD applied surface plasmon resonance (SPR) [36], isothermal titration calorimetry (ITC) [9] or fluorescence anisotropy (FA) [35] for measuring the binding to the corresponding C-Linker-CNBD. For these purposes, protein purification was necessary. Successful purification of recombinant proteins can be impaired by several parameters including low yield, solubility and protein-folding issues or proteolytic degradation [37,38]. A strategy to circumvent these problems is to tag a second protein, e.g. the maltose-binding protein (MBP), to the target protein [38,39]. In this study, a new autodisplay-based [40,41] ligand-binding assay for the HCN4 C-Linker-CNBD was developed, which avoids the need for protein purification and provides protein stability. For this purpose, the HCN4 C-Linker-CNBD (Figure 1B) was displayed on the surface of *E. coli*. Fluorescent cAMP derivative 8-Fluo-cAMP (Figure 1A) was used to determine ligand binding to the CNBD by single-cell analysis via flow cytometry [42,43]. The assay conditions were rectified with regard to equilibrium state conditions, ligand depletion and non-specific ligand binding.

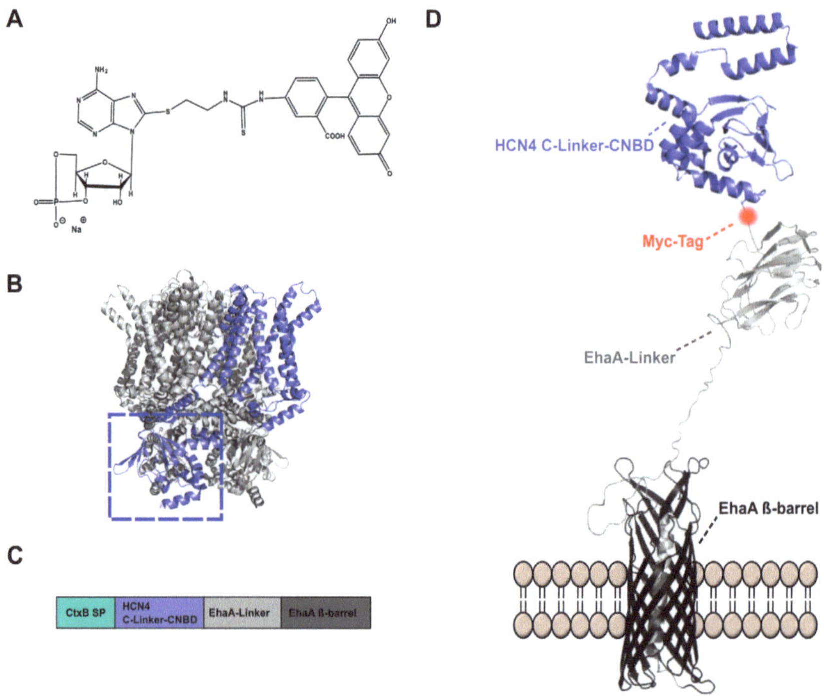

Figure 1. Graphical representation of surface-displayed HCN4 C-Linker-CNBD. (**A**) Structure of 8-Fluo-cAMP. (**B**) HCN4 tetrameric crystal structure (PDB:6GYN), one monomer (blue) consists of an N-terminal domain, transmembrane domains and an intracellular C-Linker-CNBD domain (framed). (**C**) Schematic representation of the domains required for autodisplay: CtxB signal peptide (SP), passenger domain consisting of HCN4 C-Linker-CNBD and the translocator domain consisting of EhaA-Linker and EhaA ß-barrel. (**D**) HCN4 C-Linker-CNBD (blue; PDB: 3OTF) presented on the surface of *E. coli* by the EhaA translocator domain consisting of an EhaA ß-barrel (black) and the EhaA-Linker (silver) with a Myc-Tag (red), provided by D. Gercke [44].

2. Results

2.1. Autodisplay of the HCN4 C-Linker-CNBD

Autodisplay is a technique that employs the natural type V autotransporter secretion pathway in Gram-negative bacteria to present proteins of interest on the bacterial surface [40]. The gene construct for surface display of HCN4 C-Linker-CNBD was based on the maximized autotransporter-mediated expression (MATE) system as described previously [45]. The artificial gene encoded a sequence for an N-terminal CtxB signal peptide, which is cleaved off after the translocation across the inner membrane, the passenger domain human HCN4 C-Linker-CNBD and the autotransporter translocator domain, consisting of an EhaA-Linker (including a Myc-tag) and the EhaA ß-barrel [45] (Figure 1C,D). Successful HCN4 C-Linker-CNBD surface display was confirmed via immunolabeling and flow cytometry (Figure S1). *E. coli* BL21 cells expressing the fusion protein were incubated with the primary anti-Myc-tag antibody and a secondary antibody conjugated to a DyLight^{TM} 633 fluorophore. *E. coli* BL21 cells without plasmid served as a control. Flow cytometer analysis indicated an increased mean fluorescence intensity (mFI) for HCN4 C-Linker-CNBD displaying cells when compared with control cells (Figure S1A). This was, on the one hand, to prove the surface display of the recombinant protein and on the other hand, it indicated the non-permeability of the fluorophore through the bacterial membrane. The incubation of such cells with Proteinase K leads to the digestion of surface-displayed proteins because the enzyme is not able to cross the bacterial outer membrane as has been shown by many examples before [46]. After the Proteinase K treatment, cells expressing the HCN4 C-Linker-CNBD fusion protein showed a decreased mFI compared with the same cells untreated (Figure S1B). This confirmed the surface display of the HCN4 C-Linker-CNBD. As a further control, cells displaying the target protein were treated with the secondary antibody only to exclude non-specific binding events. Flow cytometry analysis showed no increase in mFI for cells treated with the secondary antibody alone when compared to cells that were treated with the primary and secondary antibodies. This result confirmed the specific binding of the primary antibody (Figure S1C).

2.2. Binding of 8-Fluo-cAMP to Surface-Displayed HCN4 C-Linker-CNBD

The intrinsic fluorescence of 8-Fluo-cAMP enabled the direct measurement of binding without additional labelling steps. Bacterial cells displaying the HCN4 C-Linker-CNBD were incubated with 8-Fluo-cAMP in concentrations ranging from 1 nM to 4 µM. Whole-cell fluorescence (mFI) monitored by flow cytometry indicated a concentration-dependent increase in linear relationship to the 8-Fluo-cAMP concentration (Figure 2A). Saturation appeared to be attained at a concentration of around 500 nM 8-Fluo-cAMP. Plotting the ligand fraction bound, derived from the mFI values against the ligand concentration, resulted in an ordinary binding curve (Figure 2B). The binding curve obtained was used to determine the dissociation constant (K_d) of 8-Fluo-cAMP with the surface-displayed HCN4 C-Linker-CNBD and turned out to be 173 ± 46 nM. This was in good agreement with those binding affinities that have been reported before, as determined with fluorescence polarization (FP) ranging from 111 nM [47] to 189 nM [35], when bound to purified HCN4 MBP-C-Linker-CNBD fusion protein, and ranging from 167 nM [47] to 280 nM [9] when determined with purified HCN4 C-Linker-CNBD.

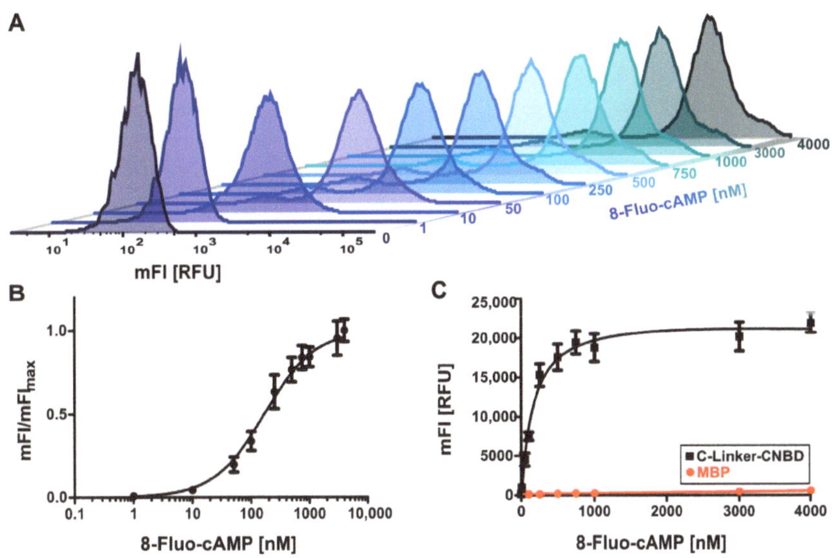

Figure 2. Flow cytometry-based ligand-binding assay with surface-displayed HCN4 C-Linker-CNBD and 8-Fluo-cAMP. (**A**) Flow cytometry histogram obtained for cells presenting the HCN4 C-Linker-CNBD after incubation with increasing concentrations of 8-Fluo-cAMP ranging from 1 nM–4 µM. (**B**) Ligand-binding curve of 8-Fluo-cAMP binding to cells presenting HCN4 C-Linker-CNBD. Mean fluorescence intensity (mFI) of each ligand concentration was normalized to the mFI of the highest ligand concentration used (mFI_{max}) and plotted against the ligand concentration. (**C**) Comparison of 8-Fluo-cAMP ligand binding to cells displaying HCN4 C-Linker-CNBD (black) or maltose-binding protein (MBP) (red). Absolute mFI obtained for each sample is plotted against the ligand concentration used.

2.3. Evaluation of Assay Conditions

To evaluate whether the conditions chosen were adequate, three scenarios were taken into account: unspecific ligand binding to the surface-displayed protein, non-equilibrium state conditions and a potential effect of ligand depletion. As a control for unspecific protein binding, surface-displayed MBP was used because 8-Fluo-cAMP is not supposed to have any affinity to MBP. When cells displaying MBP were treated with the same concentrations of 8-Fluo-cAMP as cells displaying HCN4 C-Linker-CNBD, only a marginal increase in mFI was observed (Figure 2C). In contrast, cells displaying the HCN4 C-Linker-CNBD showed a concentration-dependent increase in mFI with the saturation of binding attained at around 500 nM, as expected. This clearly indicated a specific binding of 8-Fluo-cAMP to the CNBD and a negligible unspecific protein binding.

To elucidate whether the ligand-binding assay was performed under equilibrium state conditions, the incubation time of 8-Fluo-cAMP with HCN4 C-Linker-CNBD displaying cells was doubled from 30 min to 60 min. It was to be expected that in the case of non-equilibrium state conditions, the ligand-binding curve as obtained should be shifted to lower values after 60 min incubation time [48]. The binding curves obtained with the same concentrations of 8-Fluo-cAMP after 30 min and after 60 min appeared to be identical (Figure S2A). This indicated that an incubation time of 30 min is sufficient to obtain the equilibrium state of 8-Fluo-cAMP binding to CNBD.

Generally, in ligand-binding studies, it is assumed that the total amount of ligand as applied is large enough to avoid the fact that the free ligand concentration is considerably altered upon ligand binding, i.e., the amount of bound ligand is, by order of magnitude, lower than the amount of ligand as added. Ligand depletion with a supposed impact on the result of the binding assay occurs when the free ligand concentration is substantially decreased upon ligand binding. To avoid ligand depletion conditions, the ratio of the total amount of ligand binding sites to the total amount of the ligand added to the system should not exceed 10% [48,49]. To investigate whether the amount of ligand as applied was sufficient to avoid ligand depletion, the reaction volume was increased from 100 μL to 200 μL, leaving the 8-Fluo-cAMP concentrations identical. The total number of cells presenting the target protein remained constant and the total amount of ligand added to the samples would need to be increased to obtain the same molar concentration in the 200 μL samples [48]. In the case of ligand depletion, the ligand-binding curve obtained for a 100 μL sample volume would be rightward shifted when compared to the binding curve obtained for a 200 μL sample volume. Flow cytometry analysis and the comparison of the binding curves obtained for samples with the same concentrations of 8-Fluo-cAMP in 100 μL and 200 μL reaction volumes showed no difference. This indicated that under the assay conditions as applied, ligand depletion appeared to not be of influence (Figure S2B).

To elucidate the affinity range of the assay, the amount of surface-displayed C-Linker-CNBD protein was determined by densitometry as described before by Tian et al., 2022 [50]. It was calculated to be 5.1×10^4 molecules per cell, which was in the same order of magnitude as reported before for other surface-displayed proteins [40,50,51]. Since a minimum number of 10^5 cells is required for a flow cytometry sample [48], a minimum number of approximately 10^9–10^{10} receptors can be calculated per sample. The entire number of ligands as applied needs to exceed the maximum amount of bound ligand (equal to 10^9–10^{10} receptors sites) by a factor of 10, as described above, the lower limit of K_d values as determinable with this assay appeared to be in the low nM range. The upper limit of the assay can be derived from the amount of unspecific bound 8-Fluo-cAMP to control cells. Unspecific binding to the cells depends on the ligand as such and has to be determined prior to K_d estimation. In the present study, no unspecific binding of 8-Fluo-cAMP to surface-displayed MBP (control) was observed up to 4 μM. Higher concentrations have not been applied. Hence, the affinity range of the assay is supposed to be in between the low nM and the medium or high μM range.

The robustness of the assay was estimated by comparing the values as obtained for 8-Fluo-cAMP in measurements performed as biological replicates. Here, it needs to be taken into account that vivid bacterial cells were applied, which indeed could exhibit an intra-assay and an inter-assay variance, and hence, the absolute values for the mFI of 8-Fluo-cAMP bound to the CNBD can differ on different days. However, when comparing mFI values normalized by mFI_{max} of the same series of measurements, this resulted in almost identical mFI_{norm} values, binding curves and calculated binding affinities (data not shown). This indicated a high robustness of the assay as described here.

2.4. Binding of cAMP to Surface-Displayed HCN4 C-Linker-CNBD

To measure the binding of cAMP to surface-displayed HCN4 C-Linker-CNBD in a quantitative manner, corresponding cells were incubated with a fixed concentration of 8-Fluo-cAMP (50 nM) and increasing concentrations of cAMP reaching from 100 nM to 1 mM. This resulted in decreasing mFI values with increasing cAMP concentration as measured by flow cytometry (not shown), indicating that both compounds, 8-Fluo-cAMP and cAMP, addressed the same binding pocket in the surface-displayed HCN4 C-Linker-CNBD. It indicated as well that the flow cytometer-based quantification of 8-Fluo-cAMP binding could be a method to determine the binding of compounds addressing the CNBD of HCN4 in a quantitative manner. To support this hypothesis, the mean fluorescence intensity for each cAMP concentration (mFI_{cAMP}) was normalized by the mFI of 50 nM 8-Fluo-cAMP ($mFI_{8\text{-Fluo-cAMP}}$) and plotted against the cAMP concentration. This resulted in a curve indicating the displacement of 8-Fluo-cAMP by cAMP (Figure 3A). Half of the maximum displacement was obtained at a concentration of 13 ± 2 µM, representing the IC_{50} value of cAMP for 8-Fluo-cAMP binding to surface-displayed CNBD (Figure 3A).

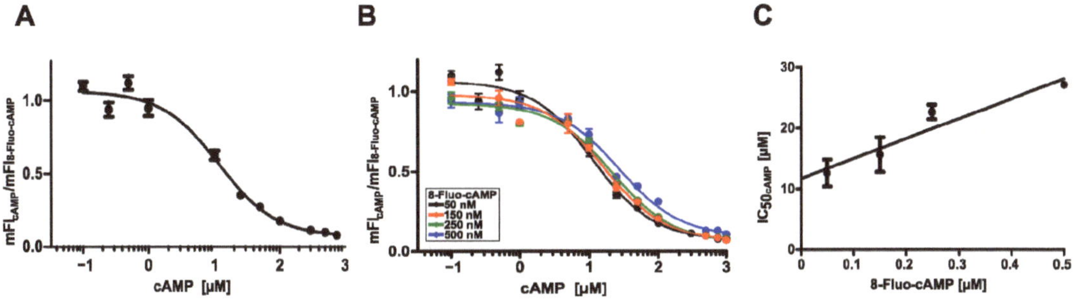

Figure 3. Competitive binding assay with 8-Fluo-cAMP and cAMP. (**A**) Displacement-binding curve obtained for cells displaying the HCN4 C-Linker-CNBD after incubation with 50 nM 8-Fluo-cAMP (ligand) and increasing concentrations of cAMP (competitor) ranging from 100 nM to 1 mM. (**B**) Ligand displacement-binding curves obtained at different ligand concentrations: 50 nM (black), 150 nM (red), 250 nM (green) or 500 nM (blue) 8-Fluo-cAMP. (**C**) Plot of the calculated IC_{50} values for cAMP against ligand concentration (R^2 = 0.86).

To evaluate the suitability of the assay in terms of ligand displacement studies, a second compound, 7-CH-cAMP (Figure S3A), was subjected to the same experiments. The concentration of 8-Fluo-cAMP was fixed at 50 nM as before, and different concentrations of 7-CH-cAMP ranging from 1 nM to 25 µM were applied. Again, the mFI value as obtained for each 7-CH-cAMP concentration ($mFI_{7\text{-CH-cAMP}}$) was normalized by that of 8-Fluo-cAMP ($mFI_{8\text{-Fluo-cAMP}}$) and plotted against the 7-CH-cAMP concentration. This resulted again in a displacement-binding curve (Figure S3B). An IC_{50} value for 7-CH-cAMP was determined to be 230 ± 41 nM and an inhibitory constant (K_i) was calculated as 159 ± 29 nM. This K_i-value determined for 7-CH-cAMP is about fivefold higher than the previously described K_d-value of 30 nM for 7-CH-cAMP, which was determined by ITC measurements with HCN4 MBP-C-Linker-CNBD fusion protein [35]. The difference could be due to the different assays applied but it may as well be the results of two different proteins used in the different assays.

2.5. Further Analysis of the Competitive Binding between 8-Fluo-cAMP and cAMP

To further confirm the competitive binding mechanism between cAMP and 8-Fluo-cAMP, cells displaying the CNBD were incubated with four different but fixed 8-Fluo-cAMP concentrations (50 nM, 150 nM, 250 nM, 500 nM) in the presence of increasing cAMP concentrations ranging from 100 nM to 1 mM, followed by flow cytometry and data analysis as described above. For each fixed concentration of 8-Fluo-cAMP, a typical displacement curve was obtained (Figure 3B). From the four different binding curves, an IC_{50} value for cAMP could be determined and turned out to be 13 $\pm$ 2 µM/16 $\pm$ 3 µM/23 $\pm$ 1 µM/27 $\pm$ 1 µM at 50 nM/150 nM/250 nM/500 nM 8-Fluo-cAMP, respectively. In the case of a competitive binding mode, increasing IC_{50} values of cAMP at higher 8-Fluo-cAMP concentrations were expected because higher competitor concentrations are necessary to reach a half-maximal displacement. Subsequently, the IC_{50} values obtained for cAMP were plotted against the concentration of 8-Fluo-cAMP. This resulted in a linear relationship, confirming the competitive binding mode between the two compounds (Figure 3C). Furthermore, the K_i-value for cAMP could be graphically determined as the *y*-intercept is supposed to be equal to the K_i of cAMP [52]. A K_i value of 11.7 µM as determined here graphically was in good agreement with a calculated K_i value of 8.5 $\pm$ 2 µM obtained by the competition experiments. Both values are in a similar range as the K_d values determined before for cAMP in other studies, ranging from 0.8 µM (determined by ITC) [9] and 1.5 µM (SPR) [36] to 1–9 µM in Saturation Transfer Difference (STD)-NMR experiments [53]. In these studies, the direct binding affinity (K_d) of cAMP to purified HCN4 C-Linker-CNBD was measured. In competitive binding studies, the binding affinity of the competitor is determined indirectly by fluorescently labeled ligand displacement. The calculated inhibitory constant K_i represents the dissociation constant of the inhibitor [54]. The obtained K_i values can be compared to the described K_d values because both represent the binding affinity of the compound. However, the way to determine binding affinities is different in both scenarios.

2.6. Investigating Ivabradine and Gabapentin in the C-Linker-CNBD Binding Assay

The effect of ivabradine and gabapentin on 8-Fluo-cAMP binding was investigated to find out whether these two approved drugs could have an influence on the affinity of 8-Fluo-cAMP to CNBD. The binding assay was performed with 50 nM 8-Fluo-cAMP and 100 µM either ivabradine or gabapentin. *E. coli* BL21 cells without plasmid were treated identically and served as a control. In a similar experiment, incubation with 50 nM 8-Fluo-cAMP together with 100 µM cAMP served as an additional control. As expected, the incubation with cAMP resulted in a decreased mFI of cells displaying the CNBD in comparison with the same cells incubated with 8-Fluo-cAMP alone (Figure 4A). Ivabradine, an approved HCN isoform unselective pore channel blocker, was chosen for the analysis, because it is known to block the HCN channels by interacting from the intracellular site with a cavity formed below the pore, and hence, no impact on 8-Fluo-cAMP binding should occur [55]. As expected, there was no influence of 100 µM ivabradine on the binding of 50 nM 8-Fluo-cAMP to the CNBD (Figure 4B). Gabapentin is a drug with a wide range of indications including epilepsy, neuropathic pain disorder and off-label use for the treatment of bipolar disorders or anxieties [56]. Tae et al. showed the selective modulation of HCN4 voltage dependence by gabapentin; however, the mode of action remained unveiled [57]. Therefore, in this study, the effect of gabapentin on 8-Fluo-cAMP binding to the CNBD of HCN4 was tested. The flow cytometer analysis showed a slight increase in the population of cells displaying the HCN4 C-Linker-CNBD binding 8-Fluo-cAMP and a similar decrease in the population of cells not binding 8-Fluo-cAMP in the same experiment (Figure 4C). This could be due to an experimental variability of 8-Fluo-cAMP binding to the CNBD. In this experiment, 100 µM gabapentin had no impact on the binding of 50 nM 8-Fluo-cAMP to the CNBD. This is the first indication that gabapentin is not interacting with the CNBD of HCN4 channels; however, further investigations are required to identify its mode of action with HCN4 channels.

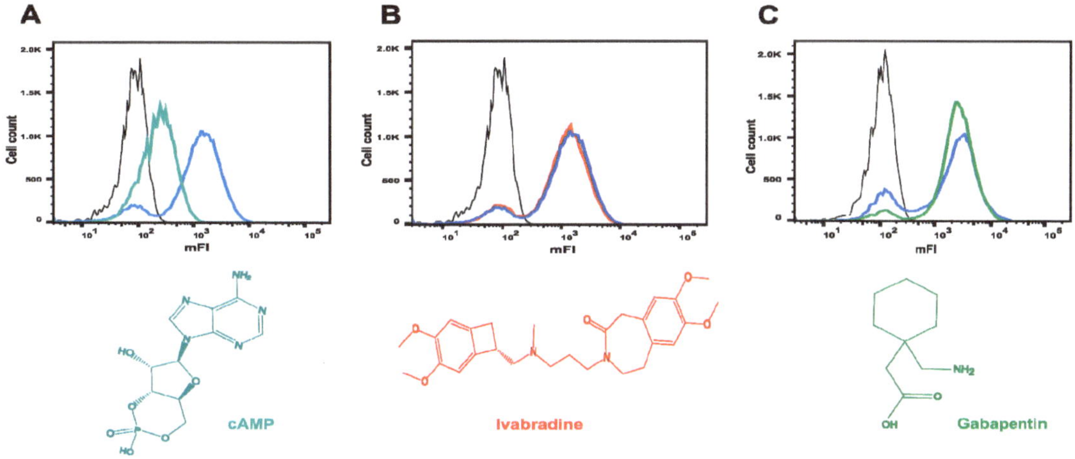

Figure 4. Flow cytometry histograms of the competitive binding assay with 50 nM 8-Fluo-cAMP and 100 µM competitor. *E. coli* BL21 cells without plasmid served as negative control (black). mFI of cells presenting the HCN4 C-Linker-CNBD after 8-Fluo-cAMP treatment are represented in blue. mFI measured after competitor addition are shown for (**A**) cAMP (aqua), (**B**) ivabradine (red) and (**C**) gabapentin (green).

2.7. Screening for Inhibitors

In total, 76 compounds from an in-house library were analyzed on 8-Fluo-cAMP displacement. The assay conditions were identical to those described for ivabradine and gabapentin. The selection from the in-house library contained compounds with different chemical scaffolds including benzimidazole derivatives [58] (Table S1), indenoindole and indenoindoledione derivatives [59–62] (Table S2), phenoxazine and phenothiazine derivatives [63,64] (Table S3), acridine derivatives [65] (Table S4), naphthofuran and naphthothiophene derivatives [66–68] (Table S5). Unfortunately, none of these compounds had an effect of more than 10% on the mFI values obtained with 8-Fluo-cAMP and hence, no novel HCN4 C-linker CNBD ligand could be identified.

3. Discussion

A flow cytometry-based assay with surface-displayed HCN4 C-Linker-CNBD was established as a method for the investigation of ligand binding to the CNBD. Compared to other methods described before, there is neither a need for protein purification nor the need for enhancing the stability and solubility of the C-linker-CNBD by an additional protein domain, such as MBP [35,39]. Protein stability appeared to be no issue when fused to the autotransporter translocator domains. As no protein purification is required, it appears that the costs of the assay as described are lower than those of other assay applied for the same purpose. In addition, it seems quite convenient because only bacterial cells, simple to cultivate, were incubated with a commercially available fluorescent ligand (8-Fluo-cAMP) and the compound which shall be analyzed on HCN4 C-linker-CNBD binding, with a fluorescence readout at the end. The binding constants determined for 8-Fluo-cAMP, 7-CH-cAMP and cAMP are in agreement with those determined before by other means [9,35,47,53].

In the current setup of the assay, around 35 compounds can be analyzed per day when samples are manually prepared. The throughput could be increased by performing the assay in, e.g., 96-well plates instead of reaction tubes as applied here and through the use of robotics. For flow cytometry, 50,000 events per sample were measured in approximately 30–45 s, which means that the throughput of the measurement was around 80–120 samples per hour.

The assay as described here could be used to screen compound libraries on new binders to the CNBD of HCN4, which then could be analyzed in further experiments, e.g., patch clamp on channel modulation. Furthermore, drugs already known to modulate HCN channels with an unclear mode of action can be specified as having a potential effect on the CNBD.

A clear limitation of the assay is that it can exclusively differentiate between binding and non-binding compounds. As only the intracellular ion channel part is expressed on the cell surface, it is not possible to distinguish between compounds inhibiting or modulating the cAMP-induced ion channel effects. Therefore, compounds identified by the assay must be tested in whole-cell experiments to investigate their impact on ion channel voltage-dependent properties. However, it appears that the assay as described can be helpful as a prescreening method before more expensive and time-consuming eukaryotic cell experiments are conducted. It has been described before that proteins expressed as monomers using autodisplay can form dimeric [69,70] or tetrameric [50,71] structures at the cell surface. This is supposed to be due to the mobility of the anchoring β-barrel domain within the outer membrane after transport [41] and the affinity of the protein subunits to each other as displayed. The C-Linker-CNBD of HCN4 is supposed to form tetramers. However, in our expression studies, we could not detect any hint of tetramerization. Therefore, for the time being, we assume the C-linker CNBD of HCN4 to be monomeric at the cell surface.

In a further embodiment, surface displays of the CNBDs of the other HCN channel subtypes 1, 2 and 3 could be performed and applied for compound selectivity testing. Moreover, mutational analysis of the binding pocket in the CNBD could be performed by standard methods to investigate the importance of different amino acids on ligand binding, which could serve as a basis for the design of subtype-specific compounds. These features of the assay described here could contribute to drug discovery approaches targeting the HCN CNBD and reveal new insights into the function of HCN channels. Overall, its robustness, the comparably low costs, a considerable throughput and the simple implementation of the assay make it a convenient screening method for HCN4 C-linker CNBD ligands.

4. Materials and Methods

4.1. Chemicals and Materials

7-Deazaadenosine-3,5′-cyclic monophosphate (7-CH-cAMP) sodium salt and 8-(2-[Fluoresc einyl]aminoethylthio)adenosine-3′,5′-cyclic monophosphate (8-[Fluo]-cAMP) sodium salt were obtained from Biolog (Bremen, Germany). 3-[(3-Cholamidopropyl)-dimethylammonio]-1-propansulfonate hydrate (CHAPS), Adenosine-3′,5′-cyclic monophosphate (cAMP), Phenylmethylsulfonylfluorid (PMSF) and NaCl were obtained from Sigma-Aldrich (Schnelldrof, Germany). Kanamycin sulfate, L-arabinose, Proteinase K and standard media ingredients were purchased from Carl Roth (Karlsruhe, Germany). Dipotassiumhydrogenphosphate (K_2HPO_4) was sourced from Merck (Darmstadt, Germany); Disodiumhydrogenphosphate (Na_2HPO_4) was obtained from VWR Chemicals (Darmstadt, Germany) and KCl from Applichem (Darmstadt, Germany). Myc Tag monoclonal mouse IgG1 antibody and the Goat anti-Mouse IgG (H + L) Secondary Antibody and DyLightTM 633 were obtained from Thermo Fisher Scientific (Braunschweig, Germany). The DNA string encoding the Maltose-binding-protein (MBP)-hHCN4-C-Linker-CNBD optimized for *E. coli* was synthesized by Invitrogen Thermo Fisher Scientific/Life Technologies GmbH (Darmstadt, Germany). Gabapentin was purchased from TCI (Zwijndrecht, Belgium). Ivabradine was provided by Prof. Seebohm (Institute for Genetics of Heart Diseases, IfGH), Department of Cardiovascular Medicine, University Hospital Münster, 48149 Münster, Germany).

4.2. Bacterial Strains and Plasmid Construction

Outer membrane protease (OmpT)-deficient *E. coli* BL21 [72,73] strain was used for the surface expression of HCN4 C-Linker-CNBD. To prepare the passenger domain, the synthesized DNA string encoding the MBP-HCN4-C-Linker-CNBD served as a template for PCR amplification of the HCN4 C-Linker-CNBD only. The corresponding primers used (5′-AAA ACT CGA GGA TAG CAG CCG TCG TCA GTA TC-3′ and 5′-TTT TGG TAC CAT GCA GCA GAA TAC-3′) were flanked with terminal XhoI and KpnI restriction sites. The amplified PCR product was inserted into the pDG01 [44] expression vector via restriction and ligation. The resulting plasmid pMJ03 carried the autotransporter fusion gene for the surface display of HCN4 C-Linker-CNBD containing a N-terminal His-Tag and C-terminal Myc-Tag. Using the primer 5′-GAG AAT CTT TAT TTT CAG GGC CTG ACC AAC AAT GGC ACG CTG ATG-3′ and 5′-GCC CTG AAA ATA AAG ATT CTC CAG ATC CTC TTC TGA GAT GAG TTT TTG TTC-3′, an additional Tobacco Etch Virus (TEV) cleavage site was introduced at the C-terminal site of the passenger protein, creating the plasmid pMJ13. As this led to an enhanced ligand-binding signal (data not shown), the work was continued with this plasmid. As a negative control, the plasmid pMJ22 was constructed encoding MBP for surface display. The MBP encoding sequence was amplified from the template DNA-string using the primer 5′-AAA ACT CGA GAA AAT CGA AGA GGG-3′ and 5′-TTT TGG TAC CGC TGC TGC TAT TGG TCT G-3′. The amplified DNA was inserted into the pMJ13 backbone via restriction and ligation using the XhoI/KpnI sites, hence replacing the HCN4-C-Linker-CNBD by the MBP encoding sequence.

4.3. Culture Conditions and Sample Preparation

Bacteria were cultivated in lysogeny broth (LB) medium (5 g/L yeast extract, 10 g/L peptone, 10 g/L NaCl) supplemented with 50 µg/mL kanamycin. An overnight culture was grown, followed by 1:100 inoculation into fresh LB medium. Cells were incubated at 37 °C (200 rpm) until an OD578 of 0.5–0.6 was reached. Gene expression was induced by the addition of 0.2% final concentration of L-arabinose for 2 h at 23 °C (200 rpm). Cells were harvested by centrifugation (3850× *g*, 4 °C, 5 min) and stored in 1 mL PBS (2.7 mM KCl, 137 mM NaCl, 2 mM KH_2PO_4, 10 mM Na_2HPO_4, pH 7.4) overnight at 4 °C.

4.4. Proteinase K Digestion

After culturing, cells were harvested by centrifugation (3850× *g*, 4 °C, 5 min) and suspended in 1 mL PBS. A total of 12.5 µL of Proteinase K (5 mg/mL) was added to the sample and it was incubated for 1 h at 37 °C (200 rpm). After the addition of 5 mM PMSF, cells were harvested and stored in 1 mL PBS overnight at 4 °C.

4.5. Immunolabeling

Bacterial cells were suspended and washed three times by sedimentation and suspension in 5 mL ice-cold sterile PBS (3850× *g*, 4 °C, 5 min). The sediment was suspended in 5 mL PBS and an OD_{578} of 0.35 was adjusted for each sample followed by centrifugation (13,000× *g*, 4 °C, 1 min). Cells were suspended in 100 µL PBS followed by the addition of 1 µL Myc-Tag monoclonal mouse IgG1 antibody. After incubation for 1 h at RT and 600 rpm, cells were harvested and washed three times by sedimentation and suspension with 500 µL PBS (13,000× *g*, 4 °C, 1 min). Cells were suspended in 100 µL PBS and 2 µL Goat anti-Mouse IgG (H + L) Secondary Antibody, DyLight™ 633 were added to the samples and incubated for 1 h at RT and 600 rpm. After the incubation, the cells were harvested by centrifugation and washed three times as described above. Cells were suspended in 200 µL PBS and analyzed by flow cytometry.

4.6. Ligand-Binding Assays

For ligand-binding studies, cells were adjusted to an OD578 of 0.35 followed by centrifugation (12,000× *g*, 4 °C, 2 min). The sediment was suspended in ice-cold sterile PBS containing 0.1% CHAPS. Corresponding volumes of 8-Fluo-cAMP were added to the samples to create a concentration series reaching from 1 nM to 4 µM. The final sample volume was 100 µL. Cells were incubated for 30 min at 30 °C with vigorous shaking (600 rpm). After harvesting by centrifugation, 100 µL ice-cold sterile PBS was added, followed by flow cytometry analysis. For the competitive binding assay, cells were preincubated with 50nM

8-Fluo-cAMP. After 10 min, the competitor was added and incubated to a final incubation time of 30 min at 30 °C (600 rpm). This was followed by the steps described previously. The concentration for cAMP ranged from 100 nM to 1 mM and for 7-CH-cAMP—from 1 nM to 25 µM. Ivabradine and gabapentin were tested at 100 µM.

4.7. Flow Cytometry Analysis

A total of 50,000 cells per sample were analyzed with a FACS Aria III flow cytometer (BD Biosciences, Franklin Lakes, NJ, USA). For immunolabeling, an excitation wavelength of 633 nm and an emission wavelength filter of 660/20 nm were used. Ligand-binding analysis was performed using an excitation wavelength of 488 nm and an emission wavelength filter of 530/30 nm.

4.8. Data Analysis

For data analysis, the GraphPad Prism software version 5.02 was used. For ligand-binding studies, the mean fluorescence intensity (mFI) of each sample was normalized to the mean maximum mFI (mFI_{max}) obtained for 4 µM 8-Fluo-cAMP as the highest concentration used. The resulting value corresponding to the ligand fraction bound was plotted against the ligand concentration. The K_d-value for 8-Fluo-cAMP was determined using Equation (1) following Hunter and Cochran (2016) [48].

$$y = \frac{x}{(K_d + x)} \tag{1}$$

For competitive binding studies, the mFI values resulting for 50 nM 8-Fluo-cAMP with competitor added ($mFI_{competitor}$) were normalized to mFI of 50 nM 8-Fluo-cAMP only ($mFI_{8\text{-}Fluo\text{-}cAMP}$). IC_{50} values were determined using the GraphPad Prism equation Binding-Competitive-One Site-Fit logIC_{50} (2).

$$y = y_{min} + (y_{max} - y_{min}) / \left(1 + 10^{x - logIC_{50}}\right) \tag{2}$$

The Binding-Competitive-One Site-Fit Ki Equation (3) was used to determine the competitor Ki values.

$$logIC_{50} = \log\left(10^{log\mathrm{Ki} \times (1 + \frac{c(ligandnM)}{K_d(ligandnM)})}\right) \tag{3}$$

The histogram plots were illustrated using the software FlowJo 10 (LLC, Ashland, OP, USA). Each histogram plot was recorded in three technical replicates for a single biological replicate. For each concentration, a total of three biological triplicates were analyzed accordingly.

Supplementary Materials: The following supporting information can be downloaded at: https://www.mdpi.com/article/10.3390/ph16050710/s1, Figure S1: Proof of HCN4 C-Linker-CNBD surface display via immunolabeling and flow cytometry, Figure S2: Analysis of equilibrium conditions and ligand depletion, Figure S3: Structure of 7-CH-cAMP, competitive binding assay with 8-Fluo-cAMP and 7-CH-cAMP, Table S1: benzimidazole derivatives; Table S2: indenoindole and indenoindoledione derivatives; Table S3: phenoxazine and phenothiazine derivatives; Table S4: acridine derivatives; Table S5: naphthofuran and naphthothiophene derivatives.

Author Contributions: Conceptualization, J.J.; Formal analysis, M.N.W.; Funding acquisition, J.J.; Investigation, M.N.W. and S.S.; Project administration, J.J.; Software, S.S.; Supervision, J.J.; Visualization, M.N.W.; Writing—original draft, M.N.W.; Writing—review and editing, J.J. All authors have read and agreed to the published version of the manuscript.

Funding: This work was supported by the Research Training Group "Chemical biology of ion channels (Chembion)" (M.N.W., J.J.) funded by the German Research Foundation (DFG), Project number: 404595355, GRK: 2515.

Institutional Review Board Statement: Not applicable.

Informed Consent Statement: Not applicable.

Data Availability Statement: Data is contained within the article and supplementary material.

Acknowledgments: The assistance of S. Kohaus for the flow cytometer experiments is gratefully acknowledged, as well as the support of H. Prinz by providing compounds and D. Gercke for providing the EhaA-Linker-EhaA-ß-barrel pdb file. We would like to thank S. Krylov for helpful discussions.

Conflicts of Interest: The authors declare no conflict of interest.

References

1. Yu, F.; Yarov-Yarovoy, V.; Gutman, G.A.; Catterall, W.A. Overview of Molecular Relationships in the Voltage-Gated Ion Channel Superfamily. *Pharmacol. Rev.* **2005**, *57*, 387–395. [CrossRef]
2. Wahl-Schott, C.; Biel, M. HCN channels: Structure, cellular regulation and physiological function. *Cell. Mol. Life Sci.* **2009**, *66*, 470–494. [CrossRef]
3. Santoro, B.; Liu, D.T.; Yao, H.; Bartsch, D.; Kandel, E.R.; Siegelbaum, S.; Tibbs, G.R. Identification of a Gene Encoding a Hyperpolarization-Activated Pacemaker Channel of Brain. *Cell* **1998**, *93*, 717–729. [CrossRef] [PubMed]
4. Ludwig, A.; Zong, X.; Jeglitsch, M.; Hofmann, F.; Biel, M. A family of hyperpolarization-activated mammalian cation channels. *Nature* **1998**, *393*, 587–591. [CrossRef] [PubMed]
5. Chen, S.; Wang, J.; Siegelbaum, S.A. Properties of Hyperpolarization-Activated Pacemaker Current Defined by Coassembly of Hcn1 and Hcn2 Subunits and Basal Modulation by Cyclic Nucleotide. *J. Gen. Physiol.* **2001**, *117*, 491–504. [CrossRef]
6. Lee, C.-H.; MacKinnon, R. Structures of the Human HCN1 Hyperpolarization-Activated Channel. *Cell* **2017**, *168*, 111–120.E11. [CrossRef] [PubMed]
7. Lee, C.-H.; MacKinnon, R. Voltage Sensor Movements during Hyperpolarization in the HCN Channel. *Cell* **2019**, *179*, 1582–1589.E7. [CrossRef]
8. Saponaro, A.; Bauer, D.; Giese, M.H.; Swuec, P.; Porro, A.; Gasparri, F.; Sharifzadeh, A.S.; Chaves-Sanjuan, A.; Alberio, L.; Parisi, G.; et al. Gating movements and ion permeation in HCN4 pacemaker channels. *Mol. Cell* **2021**, *81*, 2929–2943.E6. [CrossRef]
9. Xu, X.; Vysotskaya, Z.V.; Liu, Q.; Zhou, L. Structural Basis for the cAMP-dependent Gating in the Human HCN4 Channel. *J. Biol. Chem.* **2010**, *285*, 37082–37091. [CrossRef]
10. Wainger, B.J.; DeGennaro, M.; Santoro, B.; Siegelbaum, S.A.; Tibbs, G.R. Molecular mechanism of cAMP modulation of HCN pacemaker channels. *Nature* **2001**, *411*, 805–810. [CrossRef]
11. Almanza, A.; Luis, E.; Mercado, F.; Vega, R.; Soto, E. Molecular identity, ontogeny, and cAMP modulation of the hyperpolarization-activated current in vestibular ganglion neurons. *J. Neurophysiol.* **2012**, *108*, 2264–2275. [CrossRef] [PubMed]
12. Moosmang, S.; Stieber, J.; Zong, X.; Biel, M.; Hofmann, F.; Ludwig, A. Cellular expression and functional characterization of four hyperpolarization-activated pacemaker channels in cardiac and neuronal tissues. *Eur. J. Biochem.* **2001**, *268*, 1646–1652. [CrossRef] [PubMed]
13. Müller, F.; Scholten, A.; Ivanova, E.; Haverkamp, S.; Kremmer, E.; Kaupp, U.B. HCN channels are expressed differentially in retinal bipolar cells and concentrated at synaptic terminals. *Eur. J. Neurosci.* **2003**, *17*, 2084–2096. [CrossRef]
14. Tellez, J.O.; Dobrzynski, H.; Greener, I.D.; Graham, G.M.; Laing, E.; Honjo, H.; Hubbard, S.J.; Boyett, M.R.; Billeter, R. Differential Expression of Ion Channel Transcripts in Atrial Muscle and Sinoatrial Node in Rabbit. *Circ. Res.* **2006**, *99*, 1384–1393. [CrossRef] [PubMed]
15. Biel, M.; Wahl-Schott, C.; Michalakis, S.; Zong, X. Hyperpolarization-Activated Cation Channels: From Genes to Function. *Physiol. Rev.* **2009**, *89*, 847–885. [CrossRef]
16. Milanesi, R.; Baruscotti, M.; Gnecchi-Ruscone, T.; DiFrancesco, D. Familial Sinus Bradycardia Associated with a Mutation in the Cardiac Pacemaker Channel. *N. Engl. J. Med.* **2006**, *354*, 151–157. [CrossRef] [PubMed]
17. Ueda, K.; Nakamura, K.; Hayashi, T.; Inagaki, N.; Takahashi, M.; Arimura, T.; Morita, H.; Higashiuesato, Y.; Hirano, Y.; Yasunami, M.; et al. Functional Characterization of a Trafficking-Defective HCN4 Mutation, D553N, Associated with Cardiac Arrhythmia. *J. Biol. Chem.* **2004**, *279*, 27194–27198. [CrossRef]
18. Marini, C.; Porro, A.; Rastetter, A.; Dalle, C.; Rivolta, I.; Bauer, D.; Oegema, R.; Nava, C.; Parrini, E.; Mei, D.; et al. HCN1 mutation spectrum: From neonatal epileptic encephalopathy to benign generalized epilepsy and beyond. *Brain* **2018**, *141*, 3160–3178. [CrossRef]
19. Nava, C.; Dalle, C.; Rastetter, A.; Striano, P.; de Kovel, C.G.; Nabbout, R.; Cances, C.; Ville, D.; Brilstra, E.H.; Gobbi, G.; et al. De novo mutations in HCN1 cause early infantile epileptic encephalopathy. *Nat. Genet.* **2014**, *46*, 640–645. [CrossRef]
20. Luo, L.; Chang, L.; Brown, S.; Ao, H.; Lee, D.; Higuera, E.; Dubin, A.; Chaplan, S. Role of peripheral hyperpolarization-activated cyclic nucleotide-modulated channel pacemaker channels in acute and chronic pain models in the rat. *Neuroscience* **2007**, *144*, 1477–1485. [CrossRef]
21. Emery, E.C.; Young, G.T.; Berrocoso, E.M.; Chen, L.; McNaughton, P.A. HCN2 Ion Channels Play a Central Role in Inflammatory and Neuropathic Pain. *Science* **2011**, *333*, 1462–1466. [CrossRef] [PubMed]

22. Brown, H.F.; Difrancesco, D.; Noble, S.J. How does adrenaline accelerate the heart? *Nature* **1979**, *280*, 235–236. [CrossRef] [PubMed]
23. Brown, H.F.; Giles, W.; Noble, S.J. Membrane currents underlying activity in frog sinus venosus. *J. Physiol.* **1977**, *271*, 783–816. [CrossRef] [PubMed]
24. DiFrancesco, D. A new interpretation of the pace-maker current in calf Purkinje fibres. *J. Physiol.* **1981**, *314*, 359–376. [CrossRef] [PubMed]
25. Kessi, M.; Peng, J.; Duan, H.; He, H.; Chen, B.; Xiong, J.; Wang, Y.; Yang, L.; Wang, G.; Kiprotich, K.; et al. The Contribution of HCN Channelopathies in Different Epileptic Syndromes, Mechanisms, Modulators, and Potential Treatment Targets: A Systematic Review. *Front. Mol. Neurosci.* **2022**, *15*, 807202. [CrossRef]
26. Rivolta, I.; Binda, A.; Masi, A.; DiFrancesco, J.C. Cardiac and neuronal HCN channelopathies. *Pflug. Arch.* **2020**, *472*, 931–951. [CrossRef] [PubMed]
27. Oliphant, C.S.; Owens, R.E.; Bolorunduro, O.B.; Jha, S.K. Ivabradine: A Review of Labeled and Off-Label Uses. *Am. J. Cardiovasc. Drugs* **2016**, *16*, 337–347. [CrossRef]
28. Balducci, V.; Credi, C.; Sacconi, L.; Romanelli, M.N.; Sartiani, L.; Cerbai, E. The HCN channel as a pharmacological target: Why, where, and how to block it. *Prog. Biophys. Mol. Biol.* **2021**, *166*, 173–181. [CrossRef]
29. Romanelli, M.N.; Del Lungo, M.; Guandalini, L.; Zobeiri, M.; Gyökeres, A.; Árpádffy-Lovas, T.; Koncz, I.; Sartiani, L.; Bartolucci, G.; Dei, S.; et al. EC18 as a Tool To Understand the Role of HCN4 Channels in Mediating Hyperpolarization-Activated Current in Tissues. *ACS Med. Chem. Lett.* **2019**, *10*, 584–589. [CrossRef]
30. Del Lungo, M.; Melchiorre, M.; Guandalini, L.; Sartiani, L.; Mugelli, A.; Koncz, I.; Szel, T.; Varro, A.; Romanelli, M.N.; Cerbai, E. Novel blockers of hyperpolarization-activated current with isoform selectivity in recombinant cells and native tissue. *Br. J. Pharmacol.* **2012**, *166*, 602–616. [CrossRef]
31. Nakashima, K.; Nakao, K.; Matsui, H. Discovery of Novel HCN4 Blockers with Unique Blocking Kinetics and Binding Properties. *SLAS Discov.* **2021**, *26*, 896–908. [CrossRef] [PubMed]
32. Romanelli, M.N.; Sartiani, L.; Masi, A.; Mannaioni, G.; Manetti, D.; Mugelli, A.; Cerbai, E. HCN Channels Modulators: The Need for Selectivity. *Curr. Top. Med. Chem.* **2016**, *16*, 1764–1791. [CrossRef] [PubMed]
33. Tanguay, J.; Callahan, K.M.; D'avanzo, N. Characterization of drug binding within the HCN1 channel pore. *Sci. Rep.* **2019**, *9*, 465. [CrossRef] [PubMed]
34. Postea, O.; Biel, M. Exploring HCN channels as novel drug targets. *Nat. Rev. Drug Discov.* **2011**, *10*, 903–914. [CrossRef]
35. Möller, S.; Alfieri, A.; Bertinetti, D.; Aquila, M.; Schwede, F.; Lolicato, M.; Rehmann, H.; Moroni, A.; Herberg, F.W. Cyclic Nucleotide Mapping of Hyperpolarization-Activated Cyclic Nucleotide-Gated (HCN) Channels. *ACS Chem. Biol.* **2014**, *9*, 1128–1137. [CrossRef] [PubMed]
36. Hayoz, S.; Tiwari, P.B.; Piszczek, G.; Uren, A.; Brelidze, T.I. Investigating cyclic nucleotide and cyclic dinucleotide binding to HCN channels by surface plasmon resonance. *PLoS ONE* **2017**, *12*, e0185359. [CrossRef]
37. Goldenberg, D.P. Genetic Studies of Protein Stability and Mechanisms of Folding. *Annu. Rev. Biophys. Biophys. Chem.* **1988**, *17*, 481–507. [CrossRef]
38. Butt, T.R.; Edavettal, S.C.; Hall, J.P.; Mattern, M.R. SUMO fusion technology for difficult-to-express proteins. *Protein Expr. Purif.* **2005**, *43*, 1–9. [CrossRef]
39. Routzahn, K.M.; Waugh, D.S. Differential effects of supplementary affinity tags on the solubility of MBP fusion proteins. *J. Struct. Funct. Genom.* **2002**, *2*, 83–92. [CrossRef]
40. Jose, J.; Maas, R.M.; Teese, M.G. Autodisplay of enzymes—Molecular basis and perspectives. *J. Biotechnol.* **2012**, *161*, 92–103. [CrossRef]
41. Jose, J.; Meyer, T.F. The Autodisplay Story, from Discovery to Biotechnical and Biomedical Applications. *Microbiol. Mol. Biol. Rev.* **2007**, *71*, 600–619. [CrossRef] [PubMed]
42. McKinnon, K.M. Flow Cytometry: An Overview. *Curr. Protoc. Immunol.* **2018**, *120*, 5.1.1–5.1.11. [CrossRef] [PubMed]
43. Nickelsen, A.; Jose, J. Label-free flow cytometry-based enzyme inhibitor identification. *Anal. Chim. Acta* **2021**, *1179*, 338826. [CrossRef]
44. Gercke, D.; Furtmann, C.; Tozakidis, I.E.P.; Jose, J. Highly Crystalline Post-Consumer PET Waste Hydrolysis by Surface Displayed PETase Using a Bacterial Whole-Cell Biocatalyst. *Chemcatchem* **2021**, *13*, 3479–3489. [CrossRef]
45. Sichwart, S.; Tozakidis, I.E.P.; Teese, M.G.; Jose, J. Maximized Autotransporter Mediated Expression (MATE) for Surface Display and Secretion of Recombinant Proteins in *Escherichia coli*. *Food Technol. Biotechnol.* **2015**, *53*, 251–260. [CrossRef] [PubMed]
46. Schumacher, S.D.; Hannemann, F.; Teese, M.G.; Bernhardt, R.; Jose, J. Autodisplay of functional CYP106A2 in *Escherichia coli*. *J. Biotechnol.* **2012**, *161*, 104–112. [CrossRef]
47. Lolicato, M.; Nardini, M.; Gazzarrini, S.; Möller, S.; Bertinetti, D.; Herberg, F.W.; Bolognesi, M.; Martin, H.; Fasolini, M.; Bertrand, J.A.; et al. Tetramerization Dynamics of C-Terminal Domain Underlies Isoform-Specific cAMP Gating in Hyperpolarization-Activated Cyclic Nucleotide-Gated Channels. *J. Biol. Chem.* **2011**, *286*, 44811–44820. [CrossRef]
48. Hunter, S.; Cochran, J. Cell-Binding Assays for Determining the Affinity of Protein–Protein Interactions: Technologies and Considerations. *Methods Enzymol.* **2016**, *580*, 21–44. [CrossRef] [PubMed]
49. Hulme, E.C.; Trevethick, M.A. Ligand binding assays at equilibrium: Validation and interpretation. *Br. J. Pharmacol.* **2010**, *161*, 1219–1237. [CrossRef]

50. Tian, H.; Furtmann, C.; Lenz, F.; Srinivasamurthy, V.; Bornscheuer, U.T.; Jose, J. Enzyme cascade converting cyclohexanol into epsilon-caprolactone coupled with NADPH recycling using surface displayed alcohol dehydrogenase and cyclohexanone monooxygenase on *E. coli*. *Microb. Biotechnol.* **2022**, *15*, 2235–2249. [CrossRef]
51. Jose, J.; Chung, J.-W.; Jeon, B.-J.; Maas, R.M.; Nam, C.-H.; Pyun, J.-C. Escherichia coli with autodisplayed Z-domain of protein A for signal amplification of SPR biosensor. *Biosens. Bioelectron.* **2009**, *24*, 1324–1329. [CrossRef]
52. Newton, P.; Harrison, P.; Clulow, S. A novel method for determination of the affinity of protein: Protein interactions in homogeneous assays. *J. Biomol. Screen* **2008**, *13*, 674–682. [CrossRef] [PubMed]
53. Akimoto, M.; Zhang, Z.; Boulton, S.; Selvaratnam, R.; VanSchouwen, B.; Gloyd, M.; Accili, E.A.; Lange, O.F.; Melacini, G. A Mechanism for the Auto-Inhibition of Hyperpolarization-Activated Cyclic Nucleotide-Gated (HCN) Channel Opening and Its Relief by cAMP. *J. Biol. Chem.* **2014**, *289*, 22205–22220. [CrossRef] [PubMed]
54. Lazareno, S. Quantification of receptor interactions using binding methods. *J. Recept. Signal Transduct. Res.* **2001**, *21*, 139–165. [CrossRef]
55. Bucchi, A.; Baruscotti, M.; Nardini, M.; Barbuti, A.; Micheloni, S.; Bolognesi, M.; DiFrancesco, D. Identification of the Molecular Site of Ivabradine Binding to HCN4 Channels. *PLoS ONE* **2013**, *8*, e53132. [CrossRef] [PubMed]
56. Hong, J.S.W.; Atkinson, L.Z.; Al-Juffali, N.; Awad, A.; Geddes, J.R.; Tunbridge, E.M.; Harrison, P.J.; Cipriani, A. Gabapentin and pregabalin in bipolar disorder, anxiety states, and insomnia: Systematic review, meta-analysis, and rationale. *Mol. Psychiatry* **2022**, *27*, 1339–1349. [CrossRef]
57. Tae, H.-S.; Smith, K.M.; Phillips, A.M.; Boyle, K.A.; Li, M.; Forster, I.C.; Hatch, R.J.; Richardson, R.; Hughes, D.I.; Graham, B.A.; et al. Gabapentin Modulates HCN4 Channel Voltage-Dependence. *Front. Pharmacol.* **2017**, *8*, 554. [CrossRef] [PubMed]
58. Algul, O.; Kaessler, A.; Apcin, Y.; Yilmaz, A.; Jose, J. Comparative Studies on Conventional and Microwave Synthesis of Some Benzimidazole, Benzothiazole and Indole Derivatives and Testing on Inhibition of Hyaluronidase. *Molecules* **2008**, *13*, 736–748. [CrossRef]
59. Lindenblatt, D.; Nickelsen, A.; Applegate, V.M.; Hochscherf, J.; Witulski, B.; Bouaziz, Z.; Marminon, C.; Bretner, M.; Le Borgne, M.; Jose, J.; et al. Diacritic Binding of an Indenoindole Inhibitor by CK2α Paralogs Explored by a Reliable Path to Atomic Resolution CK2α′ Structures. *ACS Omega* **2019**, *4*, 5471–5478. [CrossRef]
60. Hundsdörfer, C.; Hemmerling, H.-J.; Götz, C.; Totzke, F.; Bednarski, P.; Le Borgne, M.; Jose, J. Indeno[1,2-b]indole derivatives as a novel class of potent human protein kinase CK2 inhibitors. *Bioorg. Med. Chem.* **2012**, *20*, 2282–2289. [CrossRef]
61. Di Pietro, A.; Gozzi, G.J.; Bouaziz, Z.; Winter, E.; Daflon-Yunes, N.; Honorat, M.; Guragossian, N.; Marminon, C.; Valdameri, G.; Bollacke, A.; et al. Phenolic indeno[1,2-b]indoles as ABCG2-selective potent and non-toxic inhibitors stimulating basal ATPase activity. *Drug Des. Dev. Ther.* **2015**, *9*, 3481–3495. [CrossRef] [PubMed]
62. Hundsdörfer, C.; Hemmerling, H.-J.; Hamberger, J.; Le Borgne, M.; Bednarski, P.; Götz, C.; Totzke, F.; Jose, J. Novel indeno[1,2-b]indoloquinones as inhibitors of the human protein kinase CK2 with antiproliferative activity towards a broad panel of cancer cell lines. *Biochem. Biophys. Res. Commun.* **2012**, *424*, 71–75. [CrossRef]
63. Prinz, H.; Chamasmani, B.; Vogel, K.; Böhm, K.J.; Aicher, B.; Gerlach, M.; Günther, E.G.; Amon, P.; Ivanov, I.; Müller, K. *N*-Benzoylated Phenoxazines and Phenothiazines: Synthesis, Antiproliferative Activity, and Inhibition of Tubulin Polymerization. *J. Med. Chem.* **2011**, *54*, 4247–4263. [CrossRef]
64. Prinz, H.; Ridder, A.-K.; Vogel, K.; Böhm, K.J.; Ivanov, I.; Ghasemi, J.B.; Aghaee, E.; Müller, K. *N*-Heterocyclic (4-Phenylpiperazin-1-yl)methanones Derived from Phenoxazine and Phenothiazine as Highly Potent Inhibitors of Tubulin Polymerization. *J. Med. Chem.* **2017**, *60*, 749–766. [CrossRef]
65. Waltemate, J.; Ivanov, I.; Ghasemi, J.B.; Aghaee, E.; Daniliuc, C.G.; Müller, K.; Prinz, H. 10-(4-Phenylpiperazine-1-carbonyl)acridin-9(10H)-ones and related compounds: Synthesis, antiproliferative activity and inhibition of tubulin polymerization. *Bioorg. Med. Chem. Lett.* **2021**, *32*, 127687. [CrossRef] [PubMed]
66. Bannwitz, S.; Krane, D.; Vortherms, S.; Kalin, T.; Lindenschmidt, C.; Golpayegani, N.Z.; Tentrop, J.; Prinz, H.; Müller, K. Synthesis and Structure–Activity Relationships of Lapacho Analogues. 2. Modification of the Basic Naphtho[2,3-*b*]furan-4,9-dione, Redox Activation, and Suppression of Human Keratinocyte Hyperproliferation by 8-Hydroxynaphtho[2,3-*b*]thiophene-4,9-diones. *J. Med. Chem.* **2014**, *57*, 6226–6239. [CrossRef]
67. Basoglu, A.; Dirkmann, S.; Golpayegani, N.Z.; Vortherms, S.; Tentrop, J.; Nowottnik, D.; Prinz, H.; Fröhlich, R.; Müller, K. Oxadiazole-substituted naphtho[2,3-b]thiophene-4,9-diones as potent inhibitors of keratinocyte hyperproliferation. Structure−activity relationships of the tricyclic quinone skeleton and the oxadiazole substituent. *Eur. J. Med. Chem.* **2017**, *134*, 119–132. [CrossRef]
68. Reichstein, A.; Vortherms, S.; Bannwitz, S.; Tentrop, J.; Prinz, H.; Müller, K. Synthesis and Structure–Activity Relationships of Lapacho Analogues. 1. Suppression of Human Keratinocyte Hyperproliferation by 2-Substituted Naphtho[2,3-*b*]furan-4,9-diones, Activation by Enzymatic One- and Two-Electron Reduction, and Intracellular Generation of Superoxide. *J. Med. Chem.* **2012**, *55*, 7273–7284. [CrossRef]
69. Jose, J.; Bernhardt, R.; Hannemann, F. Cellular surface display of dimeric Adx and whole cell P450-mediated steroid synthesis on *E. coli*. *J. Biotechnol.* **2002**, *95*, 257–268. [CrossRef]
70. Bopp, B.; Ciglia, E.; Ouald-Chaib, A.; Groth, G.; Gohlke, H.; Jose, J. Design and biological testing of peptidic dimerization inhibitors of human Hsp90 that target the C-terminal domain. *Biochim. Biophys. Acta* **2016**, *1860*, 1043–1055. [CrossRef] [PubMed]

71. Gratz, A.; Bollacke, A.; Stephan, S.; Nienberg, C.; Le Borgne, M.; Götz, C.; Jose, J. Functional display of heterotetrameric human protein kinase CK2 on *Escherichia coli*: A novel tool for drug discovery. *Microb. Cell Fact.* **2015**, *14*, 74. [CrossRef] [PubMed]
72. Studier, F.W.; Moffatt, B.A. Use of bacteriophage T7 RNA polymerase to direct selective high-level expression of cloned genes. *J. Mol. Biol.* **1986**, *189*, 113–130. [CrossRef] [PubMed]
73. Grodberg, J.; Dunn, J.J. ompT encodes the *Escherichia coli* outer membrane protease that cleaves T7 RNA polymerase during purification. *J. Bacteriol.* **1988**, *170*, 1245–1253. [CrossRef] [PubMed]

pharmaceuticals

Review

Identification and Development of BRD9 Chemical Probes

Ester Colarusso [1], Maria Giovanna Chini [2], Giuseppe Bifulco [1], Gianluigi Lauro [1] and Assunta Giordano [3,*]

[1] Department of Pharmacy, University of Salerno, Via Giovanni Paolo II 132, 84084 Fisciano, Salerno, Italy; ecolarusso@unisa.it (E.C.); bifulco@unisa.it (G.B.); glauro@unisa.it (G.L.)
[2] Department of Biosciences and Territory, University of Molise, Contrada Fonte Lappone, 86090 Pesche, Isernia, Italy; mariagiovanna.chini@unimol.it
[3] Institute of Biomolecular Chemistry (ICB), Consiglio Nazionale Delle Ricerche (CNR), Via Campi Flegrei 34, 80078 Pozzuoli, Napoli, Italy
* Correspondence: assunta.giordano@cnr.it

Abstract: The development of BRD9 inhibitors involves the design and synthesis of molecules that can specifically bind the BRD9 protein, interfering with the function of the chromatin-remodeling complex ncBAF, with the main advantage of modulating gene expression and controlling cellular processes. Here, we summarize the work conducted over the past 10 years to find new BRD9 binders, with an emphasis on their structure–activity relationships, efficacies, and selectivities in preliminary studies. BRD9 is expressed in a variety of cancer forms, hence, its inhibition holds particular significance in cancer research. However, it is crucial to note that the expanding research in the field, particularly in the development of new degraders, may uncover new therapeutic potentials.

Keywords: BRD9; inhibitors; small molecules; PROTACs; leukemia; drug discovery

1. Introduction

Bromodomains (BRDs) represent evolutionarily preserved protein modules with the capability to interpret modifications introduced by histone acetyltransferases (HATs). Specifically, they exhibit selective recognition and binding of ε-N-acetyl-lysine residues on histone tails, functioning as readers of the genetic code. Consequently, bromodomains play a crucial role in the regulation of gene expression.

The term "bromodomains" originates from the Brahma gene of Drosophila, in which the bromodomain sequence was initially discovered [1]. The human genome encompasses 61 bromodomains within 46 distinct proteins exhibiting catalytic functions. These proteins are implicated in various pathological processes and are ubiquitously present in most tissues [2]. Seventeen bromodomains are categorized into eight primary families (Figure 1A). Among these, the BET proteins (bromo- and extra-terminal domains) stand out as the most extensively studied. This group includes BRD2, BRD3, BRD4, and BRDT (bromodomain testis-specific protein). These proteins are encoded by paralogous genes, possibly originating from the duplication of an ancestral gene [3].

The BRD7 and BRD9 proteins, which belong to group IV of the family, are strictly related. Both of them are associated with two ATP-dependent chromatin-remodeling complexes, which constitute components of the switch/sucrose nonfermentable (SWI/SNF) complex, crucial for regulating gene expression by modification of the chromatin structure. BRD9 (bromodomain-containing protein 9) is primarily linked to the ncBAF (noncanonical BRG1/BRM-associated factor), consisting of a central ATPase (either BRG/SMARCA4 or BRM/SMARCA2) and several BRG-/BRM-associated factors (BAF subunits), including BRD9 [4]. On the other hand, the PBAF (polybromo-associated BAF) chromatin-remodeling complex, also characterized by its ATPase subunit, incorporates BRD7 instead of BRD9 [4].

Citation: Colarusso, E.; Chini, M.G.; Bifulco, G.; Lauro, G.; Giordano, A. Identification and Development of BRD9 Chemical Probes. *Pharmaceuticals* **2024**, *17*, 392. https://doi.org/10.3390/ph17030392

Academic Editors: Thomas Efferth, Klaus Kopka, Guangshun Wang, Maria Emília De Sousa, Joachim Jose and Amélia Pilar Rauter

Received: 16 February 2024
Revised: 12 March 2024
Accepted: 15 March 2024
Published: 19 March 2024

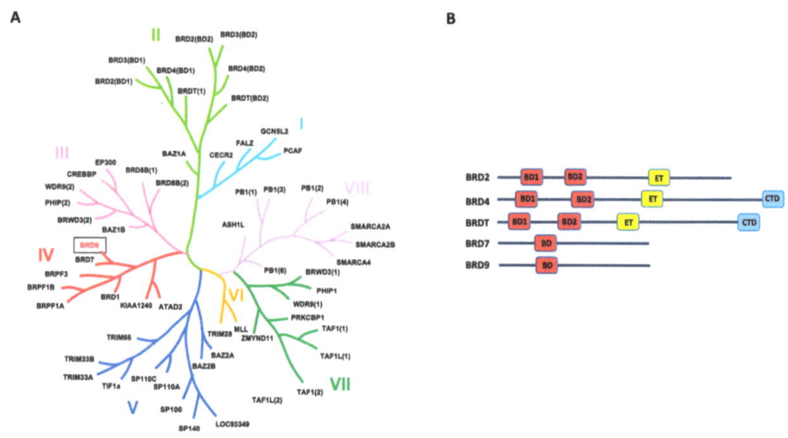

Figure 1. (**A**) Human bromodomain phylogenetic tree. The Roman numerals (I to VIII) represent the different families. (**B**) Modular structure of most representative BET proteins and bromodomain-containing proteins of family IV.

These complexes can exhibit distinct or even opposing functional roles. The ncBAF complexes present on chromatin are associated with enhancers, regulating various crucial biological processes, from self-renewal and pluripotency in embryonic stem cells to neural differentiation. Instead, the PBAF complex plays an essential role in maintaining genomic integrity during mitosis [5].

Similarly, BRD9 and BRD7, sharing a certain sequence homology, may have different functionalities or even antagonistic roles. BRD9 demonstrates recognition and binding capabilities for acetylated or butylated lysine residues, leading to post-translational modifications implicated in various diseases, especially cancer. In addition, numerous studies highlight the role of BRD7 as a tumor suppressor gene, implicating it in the p53 and BRCA1 pathways, particularly in the development of breast cancer [4].

BET proteins share a common architecture, characterized by the presence of two bromodomains (BD1 and BD2) and an extra-terminal (ET) domain (Figure 1B). In addition, BRD4 and BRDT contain a C-terminal domain (CTD). Differently, BRD7 and BRD9 proteins contain a unique bromodomain (BD) (Figure 1B). In terms of structural features, all domains share a conserved secondary structure known as a "BRD fold", characterized by four α-helices (αZ, αA, αB, and αC) and two loops (ZA and BC). The ZA loop connects αZ and αA helices, while the BC loop connects the αB and αC helices [6]. The acetyl-lysine-binding site (Kac region) comprises a conserved asparagine crucial for acetyl-lysine recognition in the ZA and BC loops, a hydrophobic region referred to as the "shelf", and the initial residue of the alpha helix αC, known as the gatekeeper. In the BET family, the gatekeeper typically consists of an isoleucine-valine, effectively limiting the size of the deep pocket. In contrast, for BRD7 and BRD9, the gatekeeper is a tyrosine residue [6].

Bromodomains are categorized on the basis of the amino acid sequence at the binding site, a crucial factor for ligand interaction. This classification becomes pivotal for evaluating the selectivity of inhibitors [7].

Additionally, each bromodomain has a specific histone peptide as substrate. The investigation into substrate specificity of BRDs has been investigated with different biophysical techniques, such as isothermal titration calorimetry (ITC), fluorescent polarization (FP) spectroscopy, surface plasmon resonance (SPR), and NMR. Subfamily IV of BRDs contains two AAA-domain-containing proteins (ATAD2A and KIAA1240), the bromodomain-containing proteins BRD7 and BRD9, and three bromodomain and PHD finger-containing proteins (BRD1, BRPF1, and BRPF3). The binding of these proteins to specific lysine residues of histone proteins is summarized in Table 1 [7–12].

Table 1. Specific substrates for the subfamily IV of BRDs.

Bromodomain	Substrate	Reference
ATAD2	H3K9ac, H3K14ac, H4K5ac, H4K12ac	[8]
BRD1	H3K14ac, H4K5acK8ac, H4K5prK8pr	[6–8]
BRD9	H4K5acK8ac, H4K5prK8pr, H4K5buK8bu	[9]
BRD7	H3K9ac H3K14ac H4K8ac H4K12ac H4K16ac	[10]
BRPF1	H2AK5ac H3K14ac H4K8ac H4K5ac H4K12ac	[11]
BRPF3	H3K14ac	[12]
KIAA1240	Not determined	

The inhibitors discovered for each bromodomain were evaluated using the same biophysical assays utilized for the specific substrates. Among these, the most commonly employed techniques are certainly AlphaScreen and DSF, which are utilized for screening large libraries of compounds due to their high throughput, high sensitivity, and rapid detection time. In particular, AlphaScreen technology stands out compared to others, as it not only assesses the binding with the protein but also the displacement of the natural ligand from the BRD binding site. However, each biophysical assay is often used in conjunction with orthogonal techniques to avoid false positive results and to validate the binding.

2. Bromodomain-Containing Protein 9 (BRD9): An Epigenetic Target Implicated in the Progression of Cancer

This review will specifically focus on bromodomain-containing protein 9 (BRD9). As already mentioned, BRD9 serves as a subunit within the ncBAF complex, and it is classified as a member of bromodomain family IV (Figure 1A) [13]. This protein functions as an epigenetic reader; BRD9 specifically recognizes acetylated lysines of histones and other proteins and recruits the ncBAF complex, thereby controlling gene transcription (Figure 2).

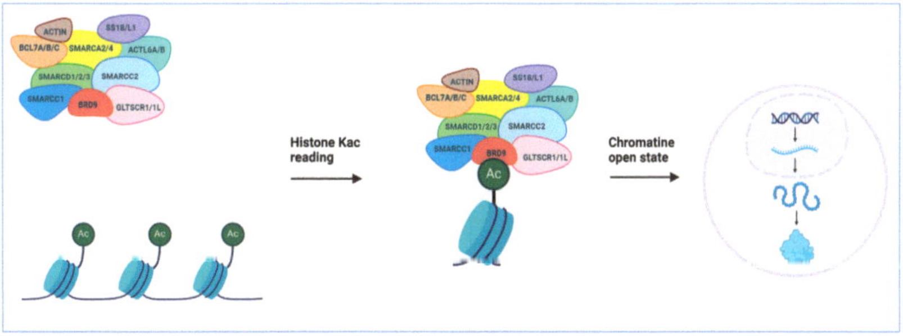

Figure 2. BRD9 in the SWI/SNF remodeling complex orchestrates gene regulation by recognizing acetylated histones, contributing to chromatin accessibility, thus enabling transcription.

Numerous research studies have consistently disclosed BRD9's crucial oncogenic role in various cancer types, influencing tumor proliferation and differentiation [4]. Furthermore, BRD9 is intricately involved in multiple signal transduction pathways, chromosomal activities, and nuclear organization [14].

In the Cancer Genome Atlas cohort, a recent study thoroughly examined the genomic changes in the genes encoding histone acetylation modulator proteins (HAMPs). The expression analysis revealed that the majority of HAMPs are widely expressed in all forms of cancer. Among the 63 possible therapeutic target HAMPs found by integrated genome analysis, BRD9 has the highest overall recurrent score, which suggests that this protein is a very attractive target for cancer therapy [15]. Notably, the BRD9-encoding gene, situated on chromosome 5p, exhibits overexpression in cervical cancer [16] and nonsmall cell lung cancer [17]. Moreover, BRD9 has been frequently identified with mutations in squamous cell lung cancer [18], prostatic adenocarcinoma [19], endometrial cancer [20], and hepatocellular carcinoma [21]. Significantly, various pieces of evidence have demonstrated that BRD9 knockdown effectively restrains the proliferation of human acute myeloid leukemia (AML), synovial sarcoma [22], and malignant rhabdoid tumor (MRT) [23]. Despite the acknowledged therapeutic potential of BRD9, the precise biological mechanisms in which it is involved remain elusive.

Therefore, since BRD9 is overexpressed in several types of cancer and involved in the dysregulation of gene expression, inhibiting BRD9 aims to interfere with the processes that promote cancer development and progression. BRD9 inhibitors strive for specificity, meaning they target the BRD9 protein, minimizing the impact on other bromodomain proteins that could have an opposite effect, such as BRD7, which has tumor-suppressive functions. This selectivity can be advantageous in reducing side effects and improving the overall safety profile of the therapeutic intervention. However, it is important to note that, while these potential advantages are promising, the development of BRD9 inhibitors is still an active area of research, and clinical trials are necessary to determine their safety and efficacy in human patients.

3. BRD9 Binders

BRD9 has been a target of recent exploration. Only in the last decade has the scientific community uncovered its involvement in cancer diseases and commenced investigations into ways to suppress it. LP99 was the first small molecule used as a BRD9 inhibitor in 2015 [24]. Following this, because of the structural similarity among various bromodomains, numerous nonselective inhibitors were initially developed [25–31]. I-BRD9, discovered in 2016, represents the first potent and selective inhibitor of this bromodomain, highlighting the distinct activity between BRD9 and BRD7 [32]. All BRD9 inhibitors share the common characteristic of being nontoxic to healthy cells. However, even though many of these inhibitors (both selective and nonselective) prove to be highly potent against the target of interest, sometimes their cellular activity is weak. For this reason, starting from these small molecules, PROTACs (protein-targeted cell activity enhancers) have been developed to enhance cellular activity.

3.1. Nonselective Binders

3.1.1. Quinolone Analogues

LP99 [24] development originated from a carefully chosen lead compound, 1-methylquinolone (**1**, Figure 3), demonstrated to function as an orthosteric ligand targeting the bromodomain of ATAD2, which shares similarity with the bromodomain of BRD9. This selection was deliberate, as the N-methyl amide portion serves as a mimetic of acetyl-lysine, forming hydrogen bonds with a water molecule and a conserved asparagine. A range of quinolones with diverse N-heterocycles were strategically incorporated at position C7, aiming to capitalize on the hydrophobic cavity between the ZA and BC loops for the selective inhibition of the protein. Among the various evaluated substitutes, valerolactam appears to be the most promising. Then, in order to form interactions with the hydrophobic surface and thereby enhance the compound's affinity, a methyl group was introduced at position 4 of the 1-methylquinolone (compound **2**). Further modifications were made to the valerolactam group to bolster selectivity: a 4-chlorophenyl group, interacting with another hydrophobic cavity, was introduced at C6, and a carbamate group creating an additional hydrogen bond between the NH of the group and the Gly43 carbonyl group, was added at C5 (compound **3**). Then, the carbamate-protecting group was eliminated, and the resultant amine underwent derivatization through reactions with a variety of acyl chlorides, chloroformates, isocyanates, and sulfonyl chlorides, yielding diverse amides, carbamates, ureas, and sulfonamides for subsequent testing. The three best compounds, **4–6**, all with different groups, were generated, and, finally, compounds were resolved into their enantiomers. Notably, compound **5** (Figure 3), also referred to as LP-99, seemed to be the most active of the series, with a K_D value of 99 nM against BRD9. Cocrystallization studies revealed that the enantiomer targeting BRD9 is the 2*R*,3*S* configuration [24].

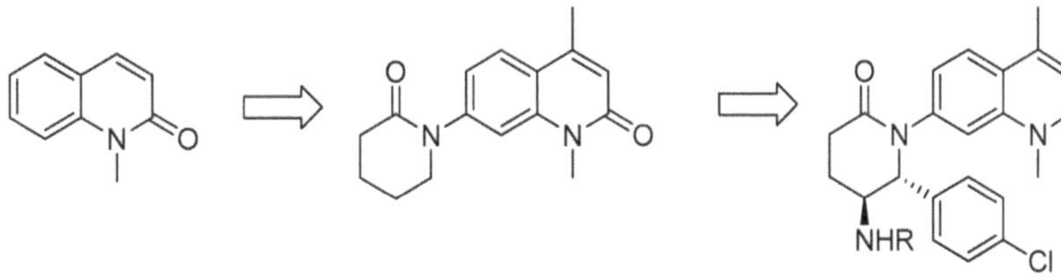

1 2 3-6

COMPOUND	R
3	Boc
4	Bz
5(LP99)	SO_2iBu
6	CONHPh

Figure 3. Structure of quinolone binders (compounds **3–6**) and their hit compounds (compounds **1** and **2**).

The compound was extensively profiled for BRD selectivity using differential scanning fluorimetry (DSF), revealing remarkable selectivity with minimal stabilization (<1 °C) of all expressible BRDs (48 out of 61 in the human genome), excluding BRD7/9. Finally, cytotoxicity tests in human osteosarcoma U2OS cells for 24 and 72 h indicated the nontoxic nature of the inhibitor at concentrations below 33 μM.

3.1.2. Indolizine Analogues

In 2015, Chen et al. [33] reported the development of GSK2801 (compound **7**, Figure 4), a potent and cell-active acetyl-lysine competitive inhibitor of BAZ2A and BAZ2B bromodomains. However, during the evaluation of selectivity by ITC, two off-target bromodomains, BRD9 and TAF1(L), were identified (K_D = 1.1 and 3.2 μM, respectively). Because of the observed off-target activity with BRD9, they designed a closely related control compound with a highly similar structure (compound **8**, GSK8573), which demonstrated no activity on BAZ2A/B and all other bromodomains, except BRD9 (K_D = 1.04 μM).

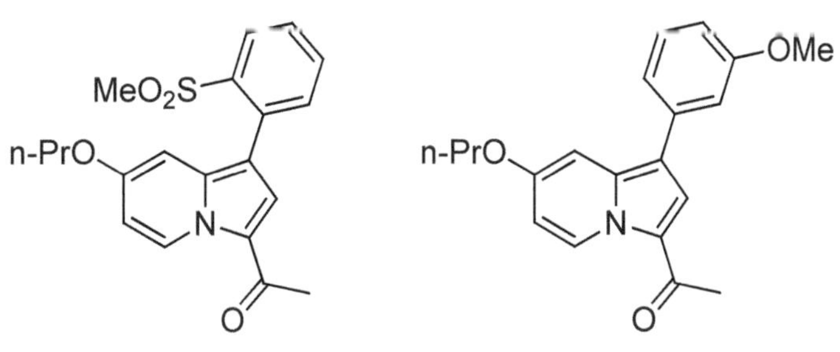

7 GSK2801 **8** GSK8573

Figure 4. Structure of GSK2801 (compound **7**) and GSK8573 (compound **8**).

Later, Hay et al. [25] moved from the indolizine scaffold of GSK2801 to another series of indolizine derivatives to improve the selectivity toward BRD9 (Figure 5). They found that the introduction of a pyridine ring in position C1 (R_2, Figure 5) led to an increase in potency and selectivity. Replacing the propoxy ether at position C7 (compounds **10** and **12**) with a morpholine substituent (compounds **9** and **11**) promoted affinity for BRD9, while modifications at R_3 generally determined a decrease in selectivity. Other analogues were then synthesized, introducing different heterocycles in R_2 position. Compound **13** with an imidazopyridine in R_2 position exhibited good affinity for BRD9 ($\Delta Tm = 4.5 \pm 0.19$ °C), while the C-7 piperazine analogue compound **14** showed poor selectivity against BRD4(BD1) but strong effectiveness against BRD9 ($\Delta Tm = 5.7 \pm 0.071$ °C). Compound **13** was shown to be very active against BRD9 ($K_D = 68 \pm 4.9$ nM) and significantly less potent against BRD7 ($K_D = 368$ nM), with low affinity for BRD proteins of subfamilies I–III and V–VIII. Compound **13** was also evaluated in cellular studies on U2OS cells and, in particular, it exhibited significant inhibitory potency in a ferric-reducing antioxidant power (FRAP) assay [25] involving BRD9/chromatin (Table 2).

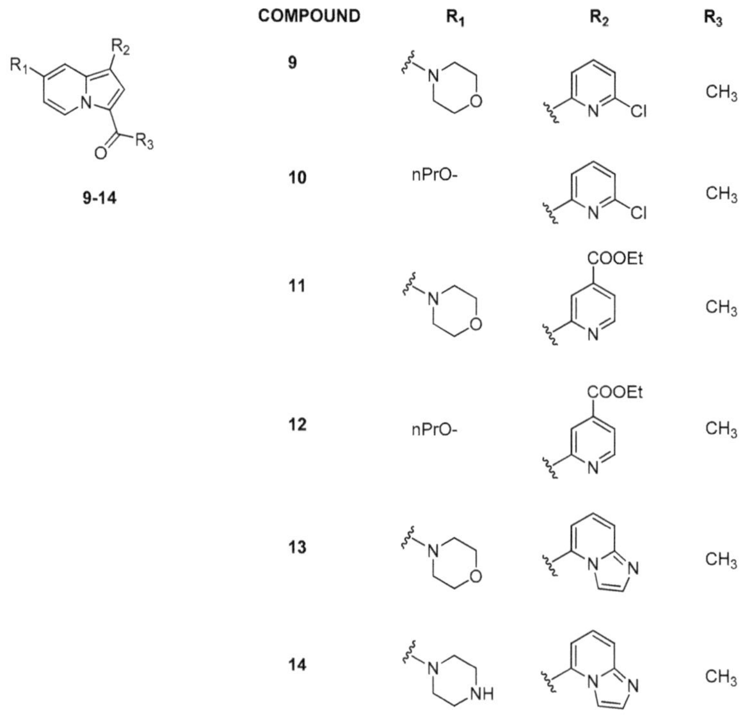

Figure 5. Structure of representative indolizine analogues (compounds **9–14**).

Table 2. FRAP assay results [(a)] and cytotoxicity results [(b)].

Compound	Cell Line	Conc (μM)	FRAP Recovery Time	Cytotoxicity
5 (LP99) [(a)]	U2OS	0.8	0.8	-
5 (LP99) [(b)]	U2OS	≤33	-	no
13 [(a)]	U2OS	1	0.8	-
19 [(a)]	U2OS	1	~5	-
20 [(a)]	U2OS	1	~5	-
21 [(a)]	U2OS	1	~3	-
22 (Bromosporine) [(a)]	U2OS	1	1	-
31 [(b)]	HEK293	≤33	-	no
75 (BI-7273) [(a)]	U2OS	1	~0.70	-
79 (BI-9564) [(a)]	U2OS	0.1	~0.80	-

3.1.3. From [1,2,4]triazolo[4,3-a]phthalazine Analogues to Bromosporine and Bromotriazine

Fedor et al. [26] developed the first example of binders of bromodomains outside the BET subfamily based on the structures and analysis of the binding mode of (+)-JQ1 and I-BET762 (compounds **15** and **16**, Figure 6), which are chemical probes of the BET family. They proposed that by preserving the 3-methyl-[1,2,4]-triazole motif while modifying the fused ring and pendant substituents, it might be possible to discover new compounds that maintained potency toward bromodomains while altering selectivity for non-BET proteins. Initially, several commercially available compounds containing triazole were assessed against 17 bromodomain-containing proteins of the bromodomain family using DSF. From this screening, it emerged that compound **17**, which is only a modest inhibitor, exhibits a certain selectivity for BRD9. On the basis of this result, a series of analogues were generated, including compounds **18–21**. When tested using DSF, these compounds also showed some activity against BRD9. Therefore, these data were confirmed through an AlphaScreen assay against four bromodomains. Compound **18** acted as a nonspecific inhibitor, displaying submicromolar IC_{50} values against BRD4, BRD9, CECR2, and CREBBP (pIC_{50} between 6.2 ± 0.62 and 6.8 ± 0.17). In contrast, compounds **19** and **20** exhibited at least 100-fold potency for BRD4, BRD9, and CREBBP compared to CECR2. Instead, compound **21**, showed a slight preference for CECR2 over BRD4 compared to BRD9 and CREBBP.

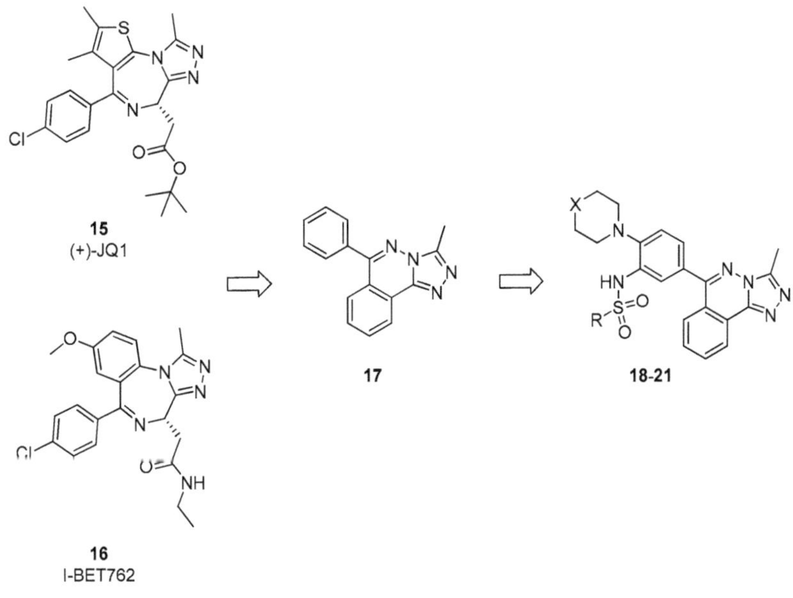

COMPOUND	R	X
18	CH_3	O
19	2-Cl-Ph	O
20	4-Cl-Ph	O
21	4-CH_3O-Ph	N-CH_3

Figure 6. Structure of [1,2,4]triazolo[4,3-*a*]phthalazine analogues (compounds **15–21**).

Moreover, these compounds demonstrated activity in cells, displaying potent cellular inhibition in a CREBBP and chromatin association FRAP model.

A year later, the same group analyzed the common binding modes of histone-derived peptides across different structural classes of BRDs. To create potent nonselective inhibitors, they opted for a similar triazolopyridazine di-cyclic core scaffold [34] based on previously discussed research on the tricyclic chemotype [26].

They reasoned that expanding the scaffold toward the identified binding groove and the BC loop might assist in avoiding subfamily-specific traits. Compounds derived from a specialized library of dicyclic chemotypes, modified at two positions (Figure 7), exhibited widespread activity in a thermal stability assay against diverse human bromodomain targets. Then, through multiple optimization cycles, they identified a potent inhibitor (compound **22**, Figure 7) with nanomolar potency against 13 BRDs and low micromolar activity against 12 additional BRDs. In particular, the best activity was highlighted with CECR2, TAF1(2), BRD4(BD2), BRDT(2), BRD9, and BRD4(BD1) bromodomains (K_D = 8.0 ± 1.0, 16.6 ± 2.7, 39.7 ± 2.2, 40.2 ± 2.8, 41.7 ± 3.8, and 41.8 ± 2.8 nM, respectively). They named this versatile inhibitor "bromosporine" (BSP), inspired by the nonselective kinase inhibitor staurosporine [35]. Moreover, they examined the impact of bromosporine on transcription in leukemic cell lines, a cancer type extensively investigated with BET inhibitors [36]. The ultimate findings revealed that BSP demonstrated effects on cell proliferation and clonogenic growth similar to those observed with (+)-JQ1 (Table 3).

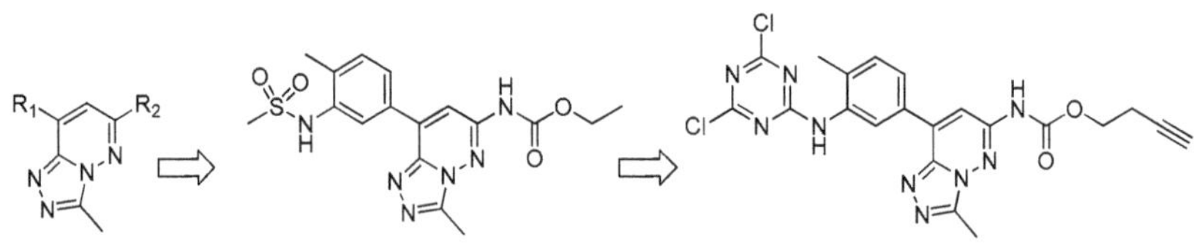

22
Starting Scaffold Bromosporine

23
Bromotriazine

Figure 7. Structure of bromosporine (compound **22**) and bromotriazine (compound **23**).

Table 3. Antiproliferative biological activity for the most promising compounds.

Compound	Cell Line	IC_{50} (μM)	EC_{50} (μM)	GI_{50} (μM)
22 (Bromosporine)	MV4;11	0.5793	n.d	n.d
	KASUMI-1	0.2067	n.d	n.d
	OCI-AML3	0.3990	n.d	n.d
44	CRF-CEM	50 ± 5	n.d	n.d
	K-562	90 ± 5	n.d	n.d
	HL-60	>100	n.d	n.d
	Kasumi1	97 ± 4	n.d	n.d
	THP-1	95 ± 6	n.d	n.d
	HaCaT	>100	n.d	n.d
51	CCRF-CEM	35 ± 4	n.d	n.d
	K-562	65 ± 4	n.d	n.d
	HL-60	81 ± 5.5	n.d	n.d
	Kasumi1	72 ± 5	n.d	n.d
	THP-1	60 ± 5	n.d	n.d
	HaCaT	>100	n.d	n.d
61 (I-BRD9)	NB4	n.d	n.d	n.d
	MV4-11	n.d	n.d	n.d
	SU-DHL4	n.d	n.d	n.d
75 (BI-7273)	EOL-1	n.d	1.4	n.d
79 (BI-9564)	EOL-1	n.d	0.8	n.d
84	EOL-1	1.76 ± 0.05	n.d	n.d
	A549	6.12 ± 0.18	n.d	n.d
88	SNB-75	n.d	n.d	0.182
	UO-31	n.d	n.d	0.479
	CAKI-1	n.d	n.d	0.708
	DU-145	n.d	n.d	13.2
	SW-620	n.d	n.d	12.0
	NCI-H460	n.d	n.d	10.2

Table 3. *Cont.*

Compound	Cell Line	IC_{50} (μM)	EC_{50} (μM)	GI_{50} (μM)
96	Jurkat	110 ± 9	n.d	n.d
	MCF-7	>500	n.d	n.d
	A375	450 ± 16	n.d	n.d
	Caco2	300 ± 14	n.d	n.d
	HaCaT	>500	n.d	n.d
97	Jurkat	145 ± 11	n.d	n.d
	MCF-7	>500	n.d	n.d
	A375	270 ± 18	n.d	n.d
	Caco2	300 ± 12	n.d	n.d
	HaCaT	>500	n.d	n.d

After, the goal of D'ascienzo et al. [27] was to develop a new covalent probe for bromodomains by replacing the sulfonamide scaffold of bromosporine with a reactive warhead and introducing a clickable alkyne handle. After discovering a dichlorotriazine reactive moiety capable of selectively labeling lysine residues in HeLa cells, this moiety was selected for the probe [37]. Bromosporine was redesigned to include the chosen warhead and clickable handle. The resulting bromotriazine (compound **23**, Figure 7) was tested for its ability to covalently modify BRDs, and mass spectrometry confirmed the formation of covalent adducts (especially with CECR2, TAF1(1), TAF1(2), BRD7, BRD9, BRD2(BD1), and BRD3(BD1)). Surprisingly, BRD9, lacking a lysine near the KAc binding site, was modified up to 77%. Attempts to map the exact interaction site through LC-MS/MS peptide mapping were inconclusive, prompting a crystallization experiment for further investigation.

3.1.4. Purine Scaffolds

In 2015, Picaud et al. [28] conducted initial molecular docking experiments on compounds **24**, **25**, and **26** (Figure 8) to assess the interaction of purine fragments with human BRDs. These experiments utilized the crystal structure of the BRD4 complex, and to confirm the computational predictions, a thermal shift assay was used to assess the binding to BRD4 and to five additional BRDs, to cover a broad spectrum of the human BRD phylogenetic tree.

Among these results, the most interesting finding was the interaction between compound **25** and BRD9, especially considering the limited presence of known inhibitors during that period. To delve deeper into the structure–activity relationships of 6-phenyl substituted 9*H*-purines, various compounds with distinct functional patterns were synthesized. Notably, compound **27** demonstrated significant binding to CREBBP when tested by isothermal titration calorimetry (Δt = 1.5 °C) and BRD9 (Δt = 2.9 °C, K_D = 641 ± 33 nM).

Subsequently, after a first set of inactive compounds, halide analogues at the meta position, while maintaining the ortho-methoxy functionality (compounds **28**–**30**), revealed improved binding with BRD9 (K_D = 351 ± 18, 297 ± 10, and 397 ± 19 nM). However, compounds **29** and **30** were found to be weak binders of BRD4 (K_D = 2.04 ± 0.12 and 4.7 ± 0.20 μM). Subsequent substitutions focused on the primary amine, and the addition of a methyl at position 8 did not yield satisfactory results. These results suggest that, in contrast to the previously described ligands, there are two functional groups capable of forming hydrogen bonds with Asn100, as follows: the primary amine and the nitrogen at position 3. Moreover, the functional groups mimicking KAc are unconventional. On the basis of the SAR studies and the evaluation of the crystallographic structure of BRD9, it was hypothesized that increasing the size of the phenyl fragment at position 7 would enhance activity on BRD9. The methoxyphenyl of compound **30** was cyclized to obtain a 2,3-dihydrobenzofuran (compound **31**, Figure 8). Compound **31** confirmed affinity for BRD9 at 278 ± 15 nM, selectively inhibiting chromatin interaction without affecting the histone binding of BRD4, for which compound **30** showed weak affinity (1.37 ± 0.03 μM). Finally, when these compounds were tested in a HEK293 cellular system, they did not exhibit cytotoxicity at concentrations up to 33 μM. Nevertheless, within the same concentration range in bioluminescence proximity assays, they were unable to displace full-length human BRD4 despite the relatively modest in vitro affinity difference between these two proteins.

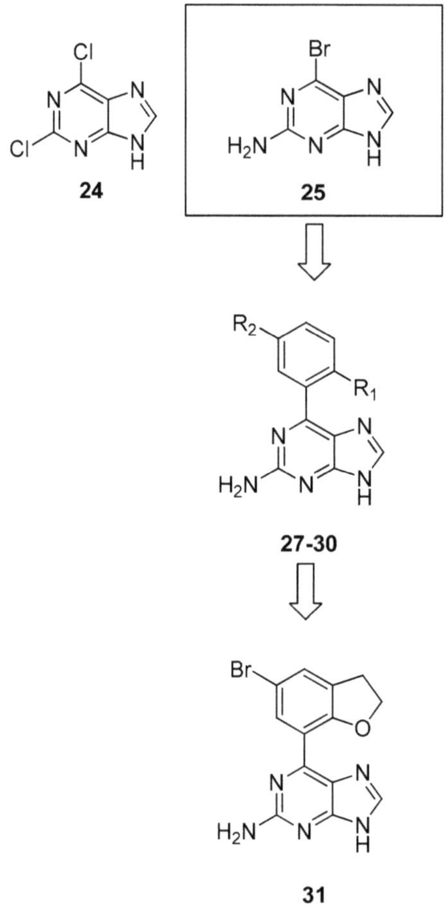

COMPOUND	R_1	R_2
27	OCH_3	H
28	OCH_3	F
29	OCH_3	Cl
30	OCH_3	Br

Figure 8. Structure of purine analogues (compounds **24**–**31**).

3.1.5. Alkyl-pyridazin-3(2H)-one Analogs

Clegg et al. [29] started from a previous study [38] on the development of inhibitors targeting the BET family, wherein unpublished data revealed a particular affinity of pyridazinone derivatives for BRD9 and PCAF bromodomains. In light of the high similarity between BRD9 and BRD7, the researchers hypothesized that inhibitors for BRD9 and BRD7 could be developed from the same pyridazinone template.

Compound **32** (Figure 9), reported as being representative of the series, demonstrated activity against both BRD9 and BRD4(BD1) due to a chloro group in the pyridazinone core acting as a KAc methyl mimetic. The substitution of Cl with CH_3 in compound **33** showed over 100-fold selectivity for BRD9 over BRD4(BD1). Then, to address issues of lipophilicity, the researchers replaced the benzyl ring with polar saturated heterocycles, resulting first in compound **34**. Moreover, the researchers decided to explore the impact of hydrophobic pockets induced by longer lipophilic KAc methyl mimetics.

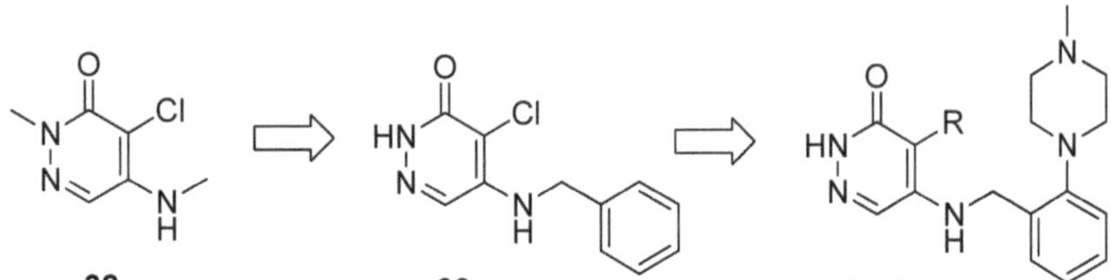

32 **33** **34-39**

COMPOUND	R
34	CH_3
35	$CH_2CH{=}CHCH_3$
36	CH_2CH_3
37	$CH_2CH_2CH_3$
38	$CH_2CH_2CH_2CH_3$
39	$CH_2CH_2CH_2CH_2CH_3$

Figure 9. Structure of alkyl-pyridazin-3(2*H*)-one analogues (compounds **32–39**).

Compound **35**, bearing an E-crotyl KAc methyl mimetic, showed a ≥40-fold reduction in BRPF1 potency while maintaining BRD9 activity. Exploring other unsaturated groups led to diminished BRD9 potency, underscoring the significance of the KAc methyl mimetic for effectiveness against both BRD9 and BRD4(BD1). Additionally, saturated KAc methyl mimetic derivatives (**36–39**) were synthesized and evaluated. Compounds **36** and **37**, bearing an ethyl and a propyl, respectively, were observed to significantly diminish BRPF1 potency while displaying a slight reduction in BRD9 activity. Switching from a three-carbon chain to a four-carbon chain with n-butyl 38 increased BRD9 potency (pKi = 7.2), although the compound remained inactive at BRD4(BD1). Extending the chain with n-pentyl in compound **39** resulted in a reduction in BRD9 potency, accompanied by an unexpected increase in BRD4(BD1) activity. Compound **35** was selected for the evaluation of the selectivity within the broader bromodomain family using the DiscoverX BROMOscan panel. Remarkably, it exhibited high selectivity against the closely related and highly homologous BRD7 (pKi = 6.3), showcasing excellent specificity against the BET family (>×500) and non-BET bromodomains (≥×280). Finally, assessments of cellular permeability demonstrated good permeability with slight intracellular accumulation, underscoring the compound's potential for effective cellular penetration.

3.1.6. Triazoloquinoxaline Analogues

In 2022, the building of BRD9 structure-based three-dimensional pharmacophore models provided an effective computational implement for the rapid and accurate identification of new potential chemotypes presenting the pharmacophoric features [30]. In particular, the application of the pharmacophore models during an in silico screening applied on an online "Anticancer" and "Anti-Inflammatory" ChemDiv library resulted in the successful identification of compound **40** (IC_{50} = 4.20 ± 1.92 μM) (Figure 10). Starting from the experimental data obtained for compound **40**, the 1-ethyl-[1,2,4]triazolo[4,3-a]quinoxaline moiety (Figure 10) was used for the in silico generation of a combinatorial library of compounds, with the modifications at the C4 position. After in silico investigations and the synthesis of the selected compounds, AlphaScreen technology highlighted the binding with BRD9 for compounds **41**–**43** (Figure 10) (IC_{50} = 10.06 ± 2.60, 4.79 ± 0.49, and 7.48 ± 2.20 μM, respectively). Moreover, in terms of selectivity, compound **42** proved to be the best of the series, as it is not able to bind any of the other investigated bromodomains (BRD2(BD1), BRD2(BD2), BRD3(BD1), BRD3(BD2), BRD4(BD1), BRD4(BD2), BRDT(BD1), BAZ2B, and CREBBP). Compound **43** showed a weak binding with BRD4(BD2) and BRDT(BD1), while compound **40** with BRD2(BD2), BRD3(BD1), BRD4(BD1), and BRD4(BD2). Instead, compound **41** was the less selective of all, being able to bind the bromodomains BRD3(BD1), BRD4(BD1), BRD4(BD2), and BRDT(BD1) with a residual binding percentage of around 50%.

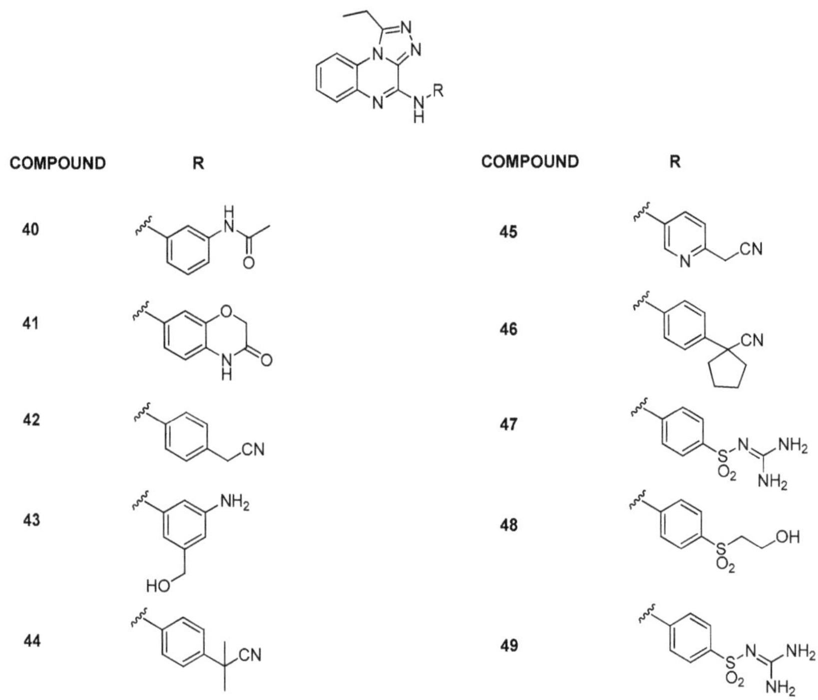

Figure 10. Structures of the first and the second series of triazoloquinoxaline analogues (compounds **40–49**).

In a follow-up study [31], a detailed exploration of the triazoloquinoxaline core was conducted by designing and synthesizing new derivatives, maintaining the central core while probing chemical variations around the C-4 position, situated in the ZA loop of the BRD9 binding site. Compounds with amine functionality demonstrated effective binding to BRD9, affirming the importance of the scaffold and NH group (compounds **44–49**, Figure 10, IC_{50} between 3.93 ± 0.54 µM and 8.31 ± 0.85 µM). Conversely, the removal or substitution of the linker resulted in a lack of binding. Then, the BromoMELT of compound **44** was performed on all 61 bromodomains as an orthogonal assay to BRD9 and to highlight the binding with other bromodomains. This compound exhibited significant binding affinity exclusively toward BRD9 and two other bromodomains belonging to subfamily 4, BRD7 and BRPF1. In light of this data, to enhance selectivity for BRD9, two new derivatives (compounds **50–51**, Figure 11) were designed with propyl and butyl groups at the C-1 position. Despite increased substituent volume, both compounds maintained considerable binding to BRD9. Compound **51**, with an IC_{50} of 6.73 ± 1.55 µM, showed comparable binding to the parent compound **44** (IC_{50} = 3.93 ± 0.54 µM), indicating tolerance for bulkier groups in the hydrophobic anchor region of BRD9.

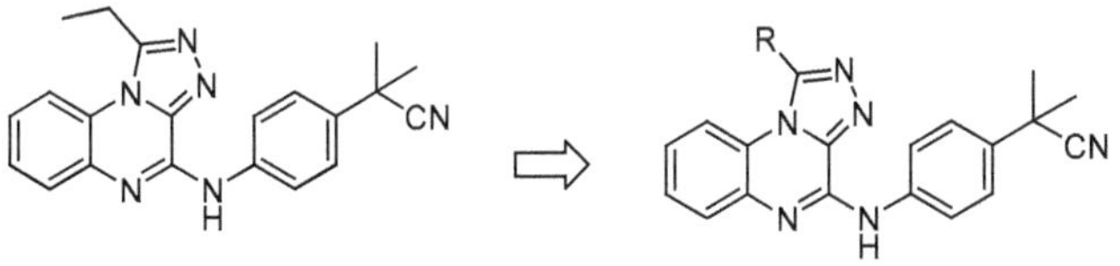

44 **50-51**

COMPOUND	R
50	$CH_2CH_2CH_3$
51	$CH_2CH_2CH_2CH_3$

Figure 11. Structures of compounds **50–51**.

AlphaScreen experiments confirmed the initial BromoMELT findings, revealing that compounds **44** and **51** bind BRPF1 within the low micromolar range (IC_{50} = 4.50 ± 0.31 μM and 1.48 ± 0.24 μM, respectively). Interestingly, they displayed lower affinity toward BRD7, particularly in the case of **51** (IC_{50} = 18.38 ± 2.06 μM and 95.76 ± 2.74 μM for **44** and **51**, respectively). These findings emphasized the importance of introducing a butyl group at the C-1 position, contributing significantly to the enhanced selectivity toward BRD9 over BRD7.

Then, the bioactivities of compounds **44** and **51** were validated using human leukemia cellular models, including THP-1, Kasumi-1, HL-60, K-562, and CCRF-CEM cells, showing concentration-dependent antiproliferative effects. Moreover, the best IC_{50} value (35 ± 4 μM) was detected for compound **51** on CCRF-CEM cells. These findings suggest prospective anticancer activity warranting further investigation, while no cytotoxicity was observed in healthy human cultures, indicating potential selectivity against high replicative cells.

3.2. Selective Binders

3.2.1. Thienopyridone Amidines and Their Amide Analogues

Starting from a cross-screening approach involving GSK internal compounds, Theodoulou et al. [32] identified compound **52** (Figure 12), with a thienopyridone scaffold, as a robust binder of BRD9 (as determined by the TR-FRET assay), boasting pIC_{50} values of 6.7 ± 0.12 and 4.7 ± 0.12 against BRD9 and BRD4(BD1), respectively. The potency and selectivity of thienopyridone **52** for the BRD9 bromodomain over BRD4(BD1) were elucidated through X-ray crystal structures for BRD9 and BRD4(BD1), respectively. The thienopyridone ring interactions were found to be similar in both bromodomains. Nevertheless, the detailed interactions of this selective compound exhibit distinctions. Notably, the substitution of BRD9 Ala54/Tyr106 for BRD4 Leu94/Ile46 introduces a differently shaped pocket, causing a tilt of the thienopyridone ring by approximately 10° away from Leu94 in BRD4. This alters the torsion angle of the thiophene 2-position carbonyl bond and influences the position of the pendant piperidinyl sulfonamide. In BRD4, the sulfonamide fails to hydrogen bond to the backbone of Lys141, unlike in BRD9.

To verify whether the amide carbonyl group was responsible for the selectivity over BRD4 of compound **52**, Theodoulou, N.H. et al. conducted investigations into amide modifications. The analysis of the data related to compound **52** revealed the significance of the methyl sulfonamide moiety for BRD9 potency and selectivity over BRD4(BD1). Substituting piperidine in **53** or dimethyl amide in **54** resulted in reduced BRD9 potency, with minimal impact on BRD4 activity. Surprisingly, secondary methyl amide **55**, although not selective over BRD4, restored the BRD9 potency observed in **52**, indicating positive interactions with both bromodomains. Cyclic sulfone amide **56** exhibited a 10-fold improvement in BRD9 activity compared to **55**, although selectivity over BRD4 was limited. X-ray crystallography of **56** in both BRD9 and BRD4 bromodomains provided insights into the observed data, highlighting the importance of amide modifications in influencing interactions and compound efficacy. Unexpectedly, compound **56** exhibited a significantly reduced selectivity window for BRD9 over BRD4 compared to **52**. This is due to the fact that the thiophene and amide of **56** are coplanar in both cases.

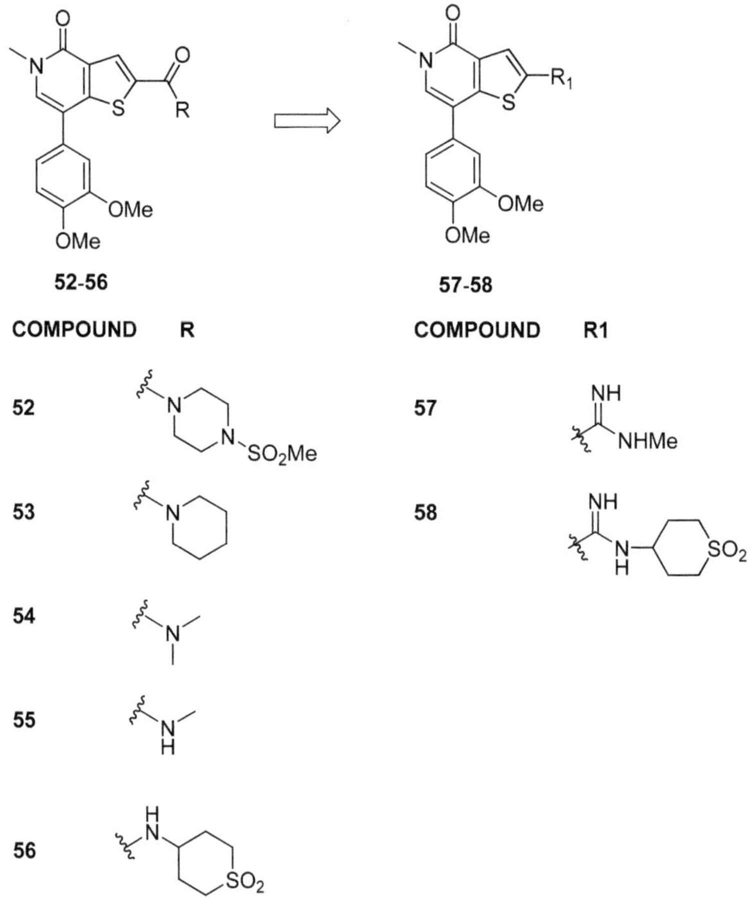

Figure 12. Structures of thienopyridone analogues (compounds **52–58**).

Considering the basic nature of the amidine moiety, the hypothesis was that its charged state would be more favorable in the less hydrophobic environment adjacent to Ala54 of BRD9 than beside Leu94 of BRD4(BD1). Consequently, amidine analogues of secondary amides **55** and **56** were designed and synthesized. Encouragingly, amidines **57** and 58 maintained the BRD9 activity of their direct amide counterparts (compounds **55** and **56**, respectively) while exhibiting improved selectivity over BRD4(BD1). The transformation of methyl amide **55** to amidine **57** resulted in a significant increase in selectivity from 2- to 16-fold. Furthermore, amidine **58** demonstrated a 50-fold selectivity over BRD4, surpassing the 4-fold selectivity observed for its amide analogue **56**.

To further improve the selectivity, modifications were explored at the 7-position of the thienopyridone core, resulting in 7-aryl substituted compounds. Additional modifications with electron-donating and electron-withdrawing substituents on the aryl ring were investigated, with nitrile-substituted compound **59** achieving a 60-fold selectivity, and trifluoromethyl compound **60** attaining a remarkable 160-fold selectivity window (Figure 13). The crystal structures of **60** in BRD9 and BRD4(BD1) explained the observed selectivity, showcasing the significance of electrostatic differences and confirming the role of the amidine. The N-ethyl analogue **61** maintaining $\geq$100-fold selectivity over BRD4(BD1), while the inclusion of the bulkier isopropyl substituent, compound **62**, led to a decrease in the activities of BRD9 and BRD4. In broader profiling, compound **61** (also known as I-BRD9) demonstrated nanomolar affinity at BRD9 (>700-fold selectivity over the BET family) and 200-fold selectivity over the BRD7 bromodomain; these findings highlight the first selective cellular chemical probe for BRD9. After, Kasumi-1 cells were subjected to treatment with I-BET and I-BRD9. The results indicate that the majority of genes exhibited selective regulation by I-BRD9. Further validation through qPCR identified four genes (CLEC1, DUSP6, FES, and SAMSN1) significantly downregulated by I-BRD9, without a similar effect by I-BET [32].

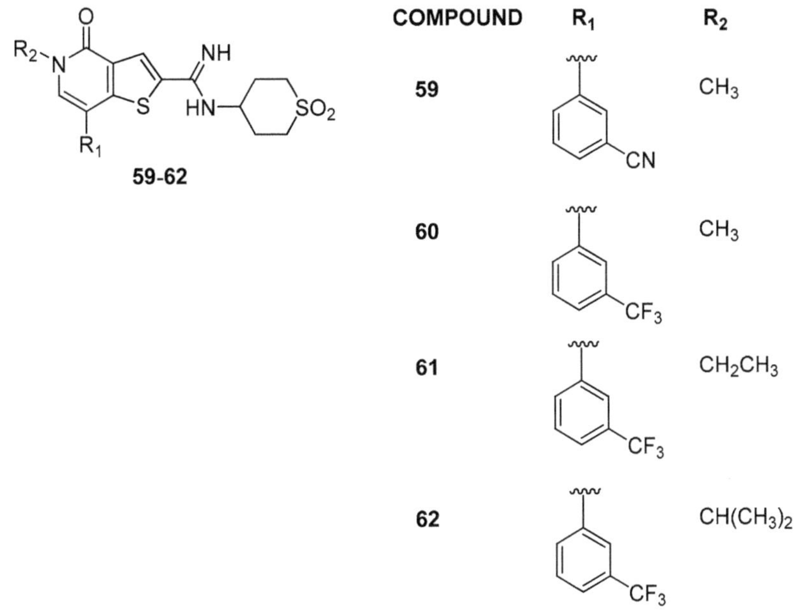

Figure 13. Structures of compounds **59–62**.

3.2.2. Isoquinolinone or Pyridinone Analogues

Thanks to a combined screening involving DSF, surface plasmon resonance (SPR), microscale thermophoresis (MST), and heteronuclear single-quantum coherence nuclear magnetic resonance (^{15}N HSQC NMR) assays, Martin et al. [39] identified the methylpyridopyrimidinone and dimethylpyridinone scaffolds as promising starting points for the development of new BRD9 inhibitors. The chosen primary compound **62** (Figure 14) engaged in T-stacking with Phe44, leading to the substitution of the amide group with methylene dimethylamine (compound **63**), resulting in an eight-fold increase in potency. The incorporation of additional electron-donating groups on the phenyl ring (compounds **64–68**) further heightened potency by strengthening T-stacking with Phe44. These modifications aimed to optimize the edge-to-face interaction for enhanced compound efficacy. Examining alterations in the ZA linker region, researchers noted enhanced selectivity toward BRD9 in comparison to its BRD7 counterpart. Potency enhancement was achieved by addressing the backbone carbonyl of His42 with hydroxyl or amine moieties (compounds **69** and **70**). The azetidine substituent proved optimal for interacting with the His42 carbonyl without disrupting T-stacking with Phe44. Compound **70** exhibited low nanomolar activity (IC_{50} = 9 nM) and displayed an induced fit with Phe47, contributing to improved selectivity against BRD4(BD1).

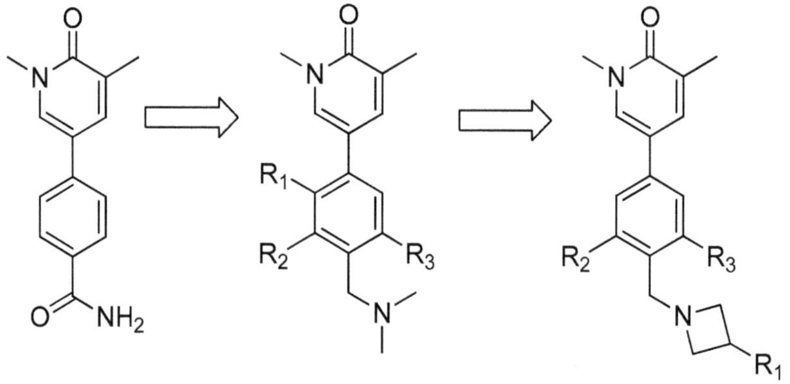

62 **63-68** **69-70**

COMPOUND	R_1	R_2	R_3
63	H	H	H
64	H	H	OMe
65	H	OMe	OMe
66	H	OMe	Me
67	H	OMe	Et
68	OMe	H	OMe
69	OH	OMe	OMe
70	NH_2	OMe	OMe

Figure 14. Structures of the first series of pyridinone analogues (compounds **63–70**).

Further modifications in the anchor region, introducing substituents at the 4 or 6 position on the pyridine-2-one core, led to enhanced selectivity against the BET family (compounds **71** and **72**, Figure 15). Employing a comparable strategy, the introduction of an aromatic ring into the pyridinone scaffold enhanced selectivity against the BET family, augmenting the π-stacking interaction with Tyr106 in the BRD9 anchor region for compounds **73–78**. The most effective inhibitor, 2-methyl-2,7-naphthyridin-1-one compound **75** (BI-7273), demonstrated a 3-fold increase in affinity for BRD9 and an impressive 50-fold increase in selectivity against BRD4(BD1) compared to compound **65**. Compound **75** additionally established a positive interaction with the carbonyl of Asn100 in BRD9, facilitated by the nitrogen atom at position 7 on the naphthyridinone ring. This interaction involved the acidification of the CH bond at position 8, enabling engagement with the carbonyl side chain of Asn100. Finally, compound **79** was obtained, which exhibited slightly lower potency against BRD9 compared to **75**, but demonstrated enhanced selectivity, being 45-fold more potent for BRD9 compared to BRD7.

In conclusion, the compounds **73**, **75**, and **79**, also named BI-7271, BI-7273, and BI-9564 (Figure 16), were revealed through the expansion of the dimethylpyridinone structure to design ring-fused compounds. These compounds exhibited robust potency and selectivity for BRD9 and displayed antitumor activity in an AML xenograft model.

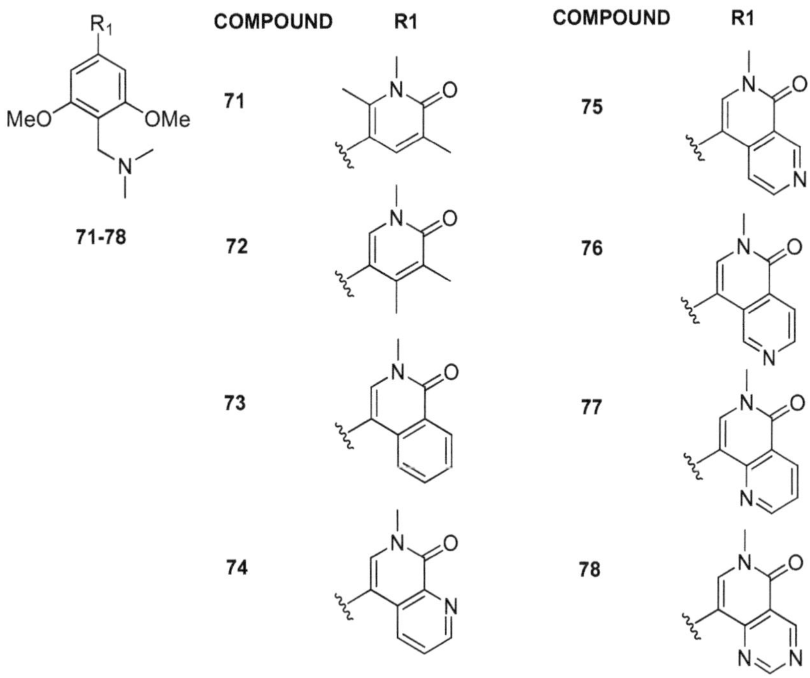

Figure 15. Structures of the second series of pyridinone (compounds **71** and **72**) and isoquinoline analogues (compounds **73–78**).

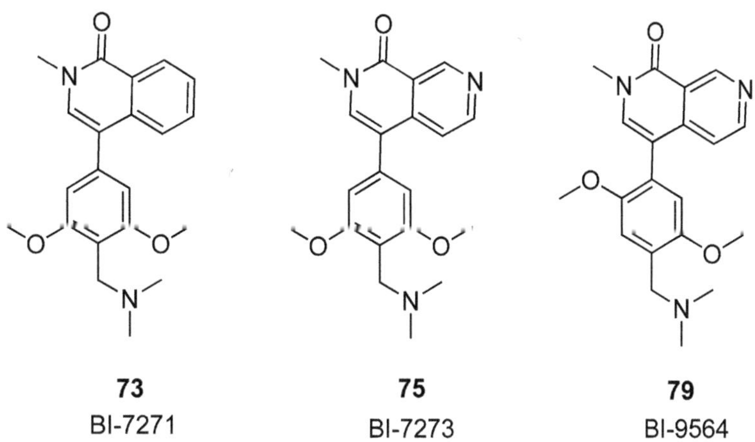

Figure 16. Structures of the most active compounds among the isoquinolinone and pyridinone analogues: BI-7271 (compound **73**), BI-7273 (compound **75**), and BI-9564 (compound **79**).

3.2.3. Imidazo[1,5-a]pyrazin-8(7H)-one Derivatives

Starting from the structure of compound **80** (Figure 17), initially developed as a BRD4 inhibitor (IC_{50} = 30 nM) [40], Zheng et al. [41] conducted structural modifications to enhance activity against BRD9 (IC_{50} = 10 μM). The authors designed a new series of derivatives based on the imidazo[1,5-a]pyrazin-8(7H)-one chemical scaffold by substituting the heterocyclic bicyclic ring with various groups previously reported for BRD9 inhibitors.

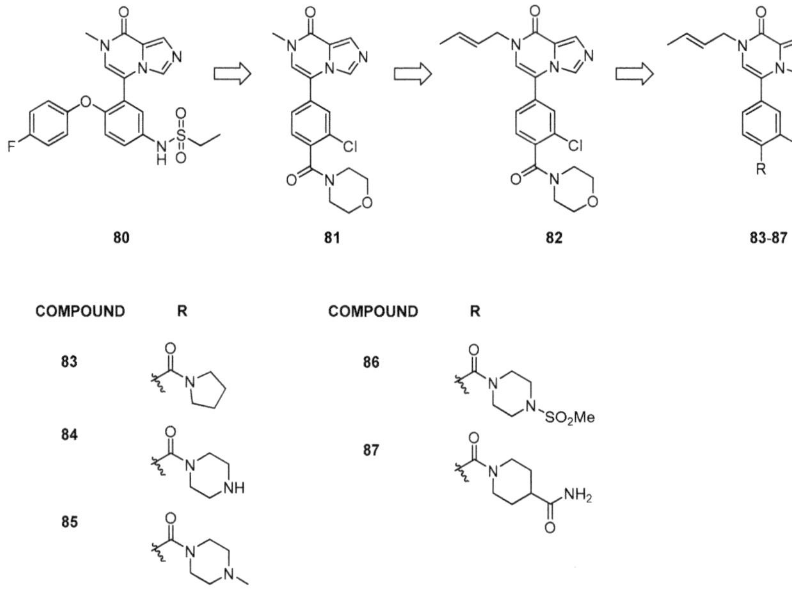

Figure 17. Structures of imidazo[1,5-a]pyrazin-8(7*H*)-one derivatives (compounds **80–87**).

These compounds exhibited moderate inhibitory activity against the target protein (IC_{50} ranging from 1.502 ± 0.421 to 7.410 ± 0.913 μM). Compound **81**, the most promising of the series, served as the lead compound, and different alkyl groups on the nitrogen of the imidazo[1,5-a]pyrazin-8(7H)-one moiety were assessed. Among these, compound **82** demonstrated the highest BRD9 inhibitory activity, achieving a 91% inhibition rate at 1 μM and an IC_{50} of 465 ± 13 nM. Its activity is attributed to the presence of double bonds on the alkyl group on the nitrogen and chlorine instead of fluorine in the meta position of the aromatic ring compared to the other compounds. Additionally, it was determined that chlorination in the meta potion was preferable to that in ortho.

Furthermore, modifications to the para position on the benzene ring led to a noticeable increase in the IC_{50} values (compounds **83–87**). Compounds featuring piperazine or its derivatives as substitutions exhibited enhanced activity. Notably, compounds **84** and **85** demonstrated significant potency in inhibiting BRD9, with IC_{50} values of 35 ± 7 and 103 ± 16 nM, respectively. All these exemplary compounds exhibited solely BRD9 inhibitory activity without demonstrating any BRD4 inhibitory effects.

Moreover, compound **84** displayed potent inhibition of cell proliferation in the A549 and EOL-1 cell lines (IC_{50} = 6.12 ± 0.18 μM and 1.76 ± 0.05 μM, respectively).

3.2.4. Pyrrole Analogs

A set of 4-acyl pyrroles [42] was initially designed for BRD4(BD1), and specific modifications were introduced to create active small molecules without solubility issues. Binding assays among bromodomain proteins revealed their affinity for the BET family proteins and BRD7, BRD9, BRPF1B, and CBP, as additional targets [42].

Notably, compounds containing a sulfonamide moiety display K_D values on BRD9 ranging between 67 and 530 nM. Compound **88** emerged as the most effective when evaluated on BRD9, prompting the design of close derivatives (compounds **89** and **90**, Figure 18) to enhance specificity toward BRD9 [43].

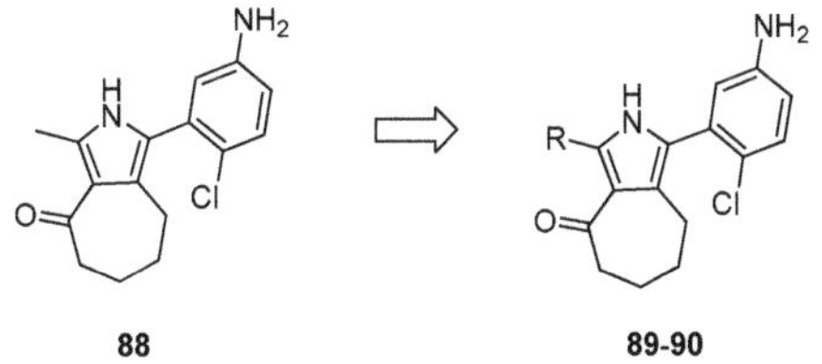

COMPOUND	R
89	$CH_2CH_2CH_3$
90	$CH_2CH_2CH_2CH_3$

Figure 18. Structures of pyrrole analogs (compounds **88–90**).

Compounds **89** and **90**, compared to the lead compound **88**, showed a 150-fold increased selectivity toward BRD9 compared to BRD7 (for compound **88**, BRD9 K_D = 150 nM, BRD7 = 1100 nM, BRD4 = 6900 nM, and BRPF1b = 7000 nM; for compound **90**, BRD9 Kd = 800 nM, BRD7 > 1100 nM, BRD4 > 6900 nM, and BRPF1b > 7000 nM). The binder has a broad spectrum of activity, showing inhibition of growth (GI_{50}) in nanomolar values in different cancer cell lines, such as SNB-75, UO-31, CAKI-1, DU-145, SW-620, and NCI-H460.

3.2.5. 6-Methylquinazolin-4(3H)-one Analogs

In 2022, through a multidisciplinary scientific approach integrating computational techniques, synthesis, and biological evaluation, new and selective inhibitors targeting BRD9, characterized by a 6-methylquinazolin-4(3*H*)-one chemical core, were discovered [44]. Specifically, utilizing a study by Combiglide [45], the 6-methylquinazolin-4(3*H*)-one was in silico functionalized with approximately 3000 commercially available benzaldehydes at position 2 and 570 boronic acids at position 8. This process yielded a large virtual library, which was further refined through in silico processing before undergoing molecular docking experiments. Drawing upon the recently developed 3D structure-based pharmacophore models specific to BRD9 [30], 16 compounds were identified for the synthesis step. Notably, compounds **91–95** (Figure 19) emerged as particularly potent binders to BRD9, exhibiting efficacy at low micromolar concentrations (IC_{50} between 2.5 ± 0.4 and 10.9 ± 0.3 μM). Additionally, given that quinazolin-4(3*H*)-one derivatives had already been documented as binders for the BET family member BRD4 [46], compounds **91–95** were assessed for their interaction with BRD4(BD1). Intriguingly, these compounds did not demonstrate significant binding to BRD4(BD1), establishing them as novel BRD9 binders with promising selective behavior.

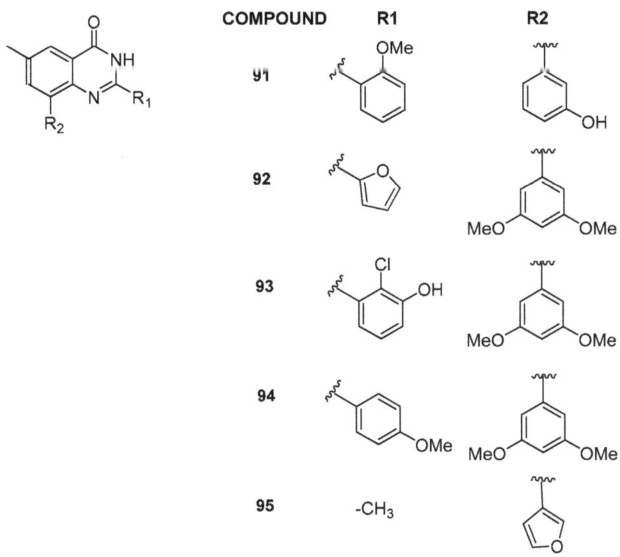

Figure 19. Structures of 6-methylquinazolin-4(3*H*)-one analogues (compounds **91–95**).

3.2.6. 2,4,5-Trisubstituted-2,4-dihydro-3H-1,2,4-triazol-3-one Analogs

Recently [47], employing computational methodologies, selective BRD9 inhibitors with a 2,4,5-trisubstituted-2,4-dihydro-3*H*-1,2,4-triazol-3-one chemical core were identified (Figure 20). Initially, six differently functionalized cores were virtually examined to assess their potential to bind the target. Subsequently, these six scaffolds were combined with 316 aryl isocyanates commercially available to introduce chemical variability at position 4, using the same computational workflow as previously discussed. Among the twenty-one molecules selected and synthesized, two (**96** and **97**) exhibited IC_{50} values in the low micromolar range (0.14 ± 0.03 and 0.35 ± 0.18 μM, respectively). To rationalize the biophysical data and design additional derivatives, the 3D structure-based pharmacophore model was applied, and seven of these new derivatives (**98–104**) effectively bound BRD9. Encouraged by these results, the selectivity of all the active compounds was evaluated across a panel of nine BRDs covering most of the related human phylogenetic tree (BRD2(BD1), BRD2(BD2), BRD3(BD1), BRD3(BD2), BRD4(BD1), BRD4(BD2), BRDT(BD1), BAZ2B, and CREBBP). In addition, for compound **97**, the selectivity was further assessed on eight additional bromodomains (including ATAD2, CECR2, SMARCA4, TAF1(BD2), BRPF3, BRD1, BPTF(BRD), and BRD7). This compound exhibited no binding with the most considered bromodomains (residual protein activity percentage between 92.6 ± 3.1 and 100%). However, with BRD7 and TAF(1), this percentage dropped to 75.8 and 77.7, respectively. The IC_{50} evaluation revealed that compound **97** is approximately 60 times more potent on BRD9 (0.14 ± 0.03 μM) compared to BRD7 (71.4 ± 1.1 μM) and similar to TAF(1). Moreover, all the active compounds were evaluated on a selected panel of human cells, composed of both healthy (keratinocytes and enterocytes) and cancer (leukemia, breast, melanoma, and colorectal) cell lines. At the highest concentrations tested, no cytotoxic responses were observed in healthy cells, with IC_{50} values exceeding 500 μM, indicating an absence of detectable biological effects. Preliminary data on human cancer models revealed a notable selectivity in action on leukemic cells. Notably, compounds **96** and **97** demonstrated IC_{50} values below 150 μM on Jurkat cells, indicating a modest yet noteworthy antiproliferative activity that warrants further investigation.

3.3. *Summary of Cellular Assays*

In the first works, the cellular assays were performed through the fluorescence recoveries after photobleaching technique (FRAP) by disruption of the interaction of full-length green fluorescent protein or full-length GFP-tagged BRD9 with acetylated chromatin. With this technique, the displacement of proteins from chromatin was observed, as evidenced by rapid recovery after bleaching. In the case of BRD9, cells were treated with the histone deacetylase inhibitor SAHA (suberoylanilide hydroxamic acid) to elevate acetylation levels and enhance binding. When examining a representative selection of bromodomains in the presence of inhibitors, a significant reduction in fluorescence recovery times was noted, indicative of the inhibition of chromatin-BRD interaction (Table 2).

For compounds **5** and **31** the cytotoxicity assay was also reported. Compounds were tested in different cell lines but in both cases they do not appear cytotoxic at a concentration lower or equal to 33 μM (Table 2).

In the most recent studies, the antiproliferative biological effects were assessed. This involved determining the IC_{50} (the concentration of a drug or inhibitor needed to inhibit a biological process or response by 50%) or EC_{50} (the concentration effective in producing 50% of the maximal response and is a convenient way of comparing drug potencies), or GI_{50} (the dose that inhibits the growth of cells by 50%) values (Table 3) on different cancer cell lines, in which BRD9 is more prominently expressed.

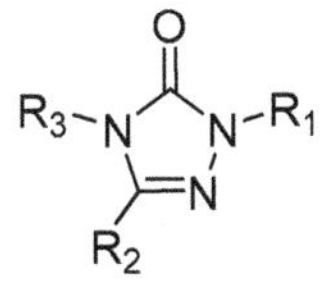

COMPOUND	R1	R2	R3
96	CH_3	CH_3	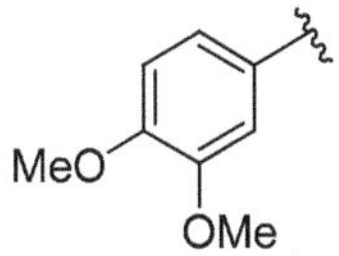
97	CH_3	CH_2CH_3	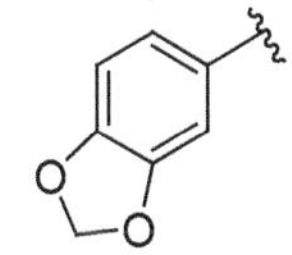
98	CH_3	CH_2CH_3	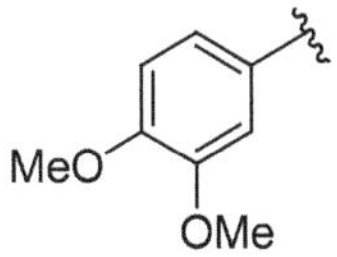
99	CH_2CH_3	CH_3	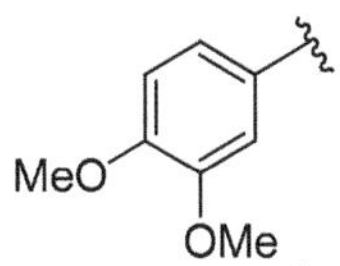
100	CH_3	CH_2CH_3	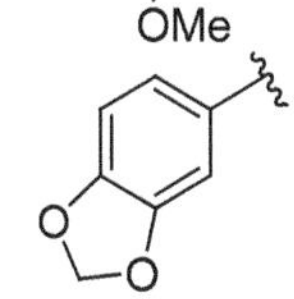
101	CH_2CH_3	CH_3	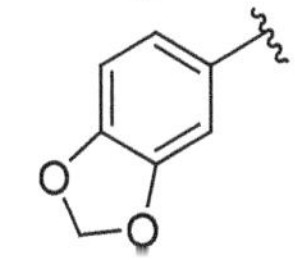
102	CH_3	CH_2CH_3	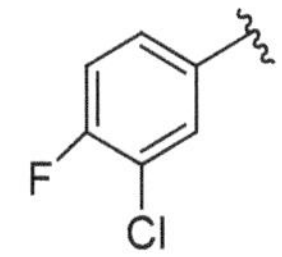
103	CH_2CH_3	CH_3	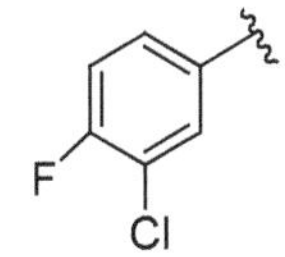
104	CH_2CH_3	CH_2CH_3	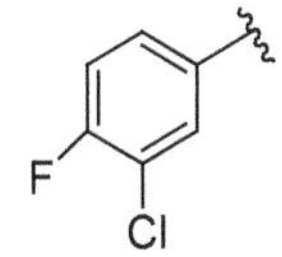

Figure 20. Structures of the 2,4,5-trisubstituted-2,4-dihydro-3*H*-1,2,4-triazol-3-one analogues (compounds **96–104**).

Nevertheless, numerous inhibitors have only been characterized through binding assays on BRD9 or assessments of selectivity against other bromodomains; cellular experiments have not consistently been conducted. Moreover, even though BRD9 operates within a complex, compounds exhibiting significant activity on isolated BRD9 protein might not translate into favorable outcomes in cell assays. To boost cellular activity and thus enhance biological impact, the most promising binders are presently serving as a starting point for chemical degraders (e.g., PROTACs) development.

4. BRD9 PROteolysis TArgeting Chimeras (PROTACs)

The conventional drug discovery paradigm has long relied on small molecules binding to specific sites on target proteins to modulate their activity [48]. However, many proteins lack such binding sites or catalytic activity, rendering modulation challenging [49]. Additionally, some proteins operate within large complexes that can compensate for reduced activity due to inhibition. Notably, PROteolysis TArgeting Chimeras (PROTACs) represents a revolutionary approach, particularly in cancer treatment, within the realm of small molecule therapeutics. PROTAC technology offers a novel strategy for drug discovery, capable of targeting proteins previously deemed undruggable by conventional methods [50–52]. More specifically, PROTACs are bifunctional degraders that hijack the intracellular ubiquitin-proteasome system (UPS) to induce the degradation and elimination of a target protein. An E3 ubiquitin ligase ligand is linked to a protein-of-interest ligand (also known as a "warhead") by a linker of varying lengths and chemical composition [53]. BRD9 inhibitors often demonstrate poor cellular activity [47] despite strong activity on the protein, attributable to BRD9's involvement in the SWI/SNF complex. Consequently, developing PROTACs from these known inhibitors may prove crucial in enhancing their biological efficacy on the target of interest, offering promising avenues for therapeutic intervention.

The first BRD9 heterobifunctional degrader [54] was developed using as a starting point the thienopyridone analogue of I-BRD9 (compound **61**, Figure 21), which displayed high binding affinity (IC_{50} = 7.9 nM) and the possibility of anchoring the E3 ligand by an ether linker attached to the methoxy substituent [28]. A first series of ligands was synthesized, exploring both binding to CRBN and VHL E3 ligase. Products were tested in human AML cell line for 4 h, and BRD9 protein levels were detected by immunoblot. Among the CBRN ligand-based compounds, the one presenting a linker with intermediate length resulted as the most promising (**106**), leading to the most efficient protein degradation; on the contrary, VHL-ligand conjugates (**108** and **109**) failed to induce BRD9 degradation.

To further expand the investigation, in the same work [54], other degraders based on the structure of the binder BI-7273 (compound **75**) and pomalidomide as CBRN ligand were prepared. The impact of nature and length of the linker was also evaluated to improve biochemical and cellular selectivity. Compound **110** (Figure 22) with PEG-linked pomalidomide conjugate (dBRD9) was able to efficiently degrade BRD9 in the MOLM-13 cell line, with reduced off-target binding activity toward other bromodomain containing proteins. The antiproliferative activity of dBRD9 was tested in human AML lines (EOL-1, MOLM-13, MV4), showing large improvement with respect to classical binders I-BRD9 and BI-7273.

In 2022, on the basis of the thienopyridone scaffold of I-BRD9 (compound 61), a novel binder, named EA-89, was developed, (compound **111**, Figure 23), [55] in which the propyl group linked to pyridone nitrogen accommodates in a hydrophobic pocket acetyl lysine pocket, and the 3-trifluoromethylphenyl ring is substituted with a 3-methylindolic group, that serves as the site for anchoring the linker. EA-89 was then used as a warhead to obtain a cereblon-targeting degrader QA-68 (compound **112**, Figure 23). Both the binder and the PROTAC were tested on several cellular lines of hematological malignancies, showing a strong selectivity for AML as well as B cell acute lymphoblastic leukemia (B-ALL) and multiple myeloma (MM). Interestingly, no impact on the viability of cell lines of the T lymphocytic lineage (acute monocytic leukemia and erythroleukemia) was observed, revealing a BRD9 dependency in acute leukemia and MM [55].

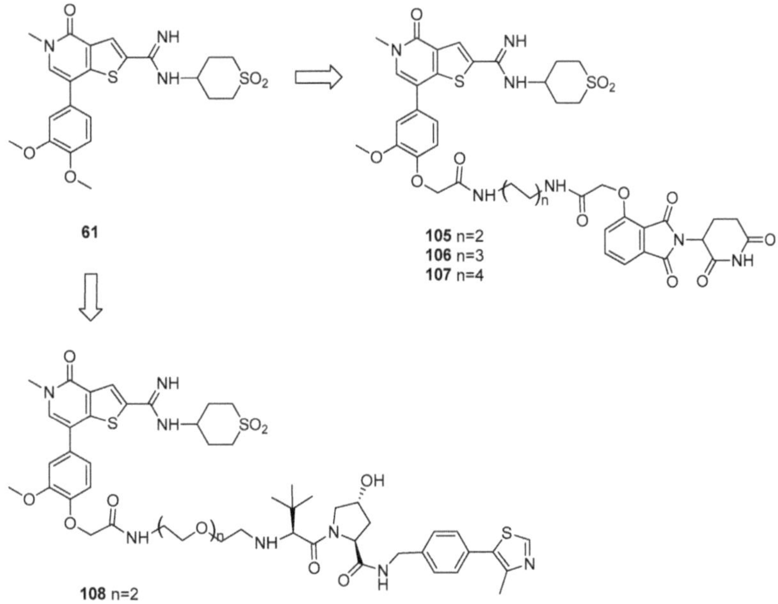

61

105 n=2
106 n=3
107 n=4

108 n=2
109 n=3

Figure 21. Structures of compounds **105–109**.

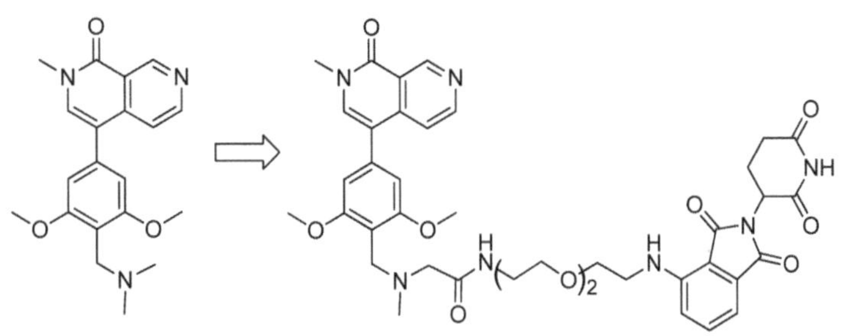

75
BI-7272

110
d-BRD9

Figure 22. Structure of compound **110**.

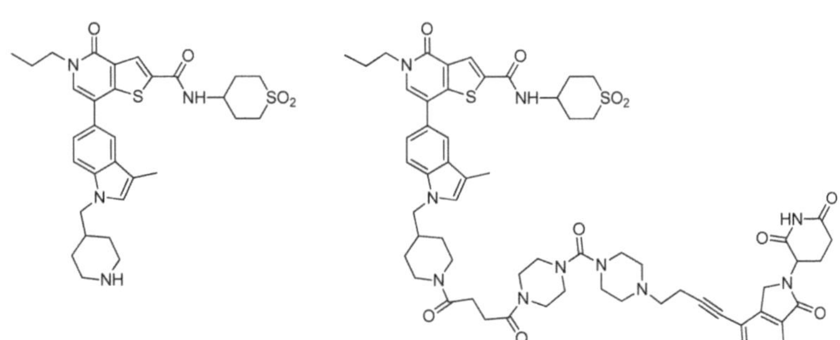

111
EA-89

112
QA-68

Figure 23. Structures of EA-89 (compound **111**) and QA-68 (compound **112**).

As highlighted in the study by Remillard et al. [54], CRBN-based PROTACs can exhibit off-target degradation of other proteins. From this perspective, Ciulli's group developed the first VHL ligand-based dual BRD7/BRD9 degrader [56]. A first set of compounds was synthesized using PEG linkers to connect a BRD binder in which a piperazine group replaced the dimethylamine group of compound BI-7273. As E3 ligase recruiting moiety, VH032 was selected, together with pomalidomide and DCAF 15. Differently from CRBN-based PROTACs, VHL-based degraders **114** and **115** (Figure 24) showed weak activity against both BRD7 and BRD9.

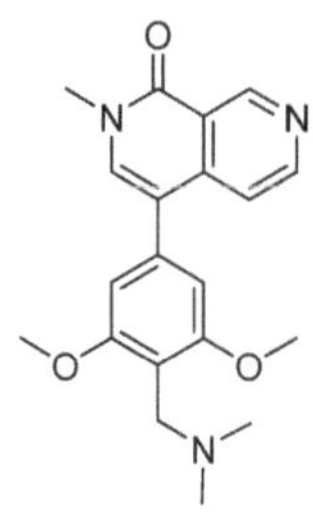

75
BI-7272

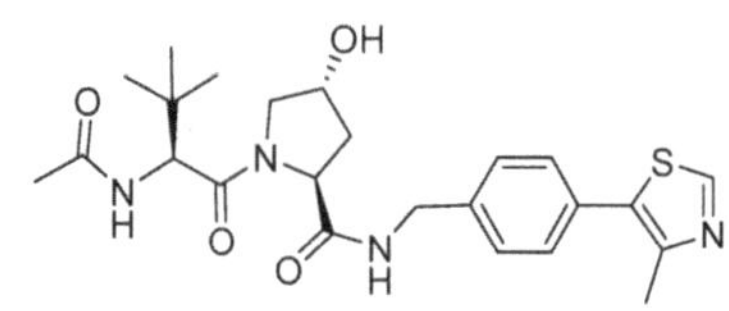

113
VH032

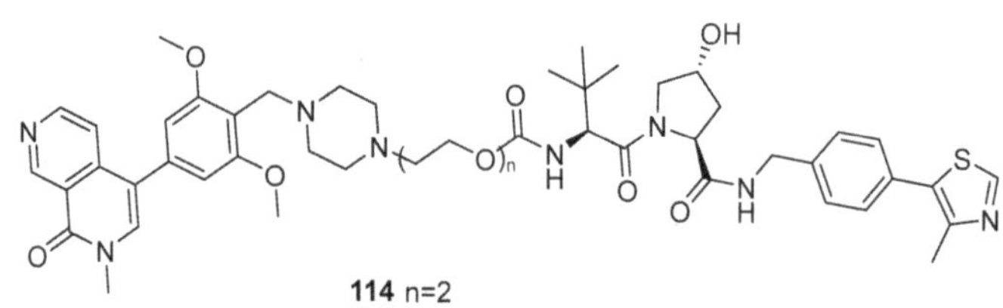

114 n=2
115 n=4

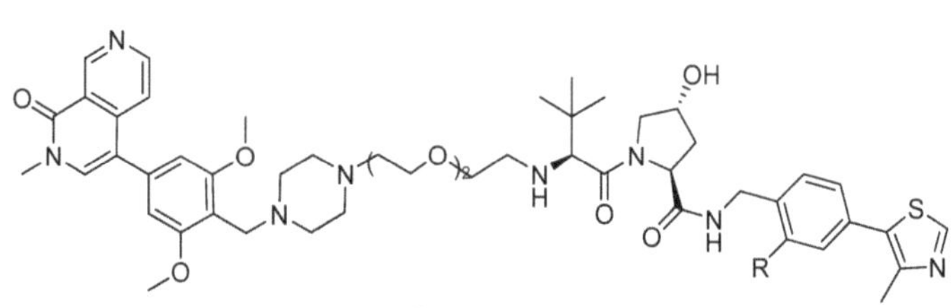

116 R= 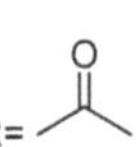

117 R= 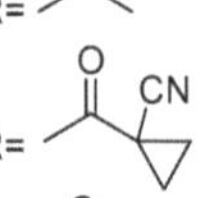

118 R= 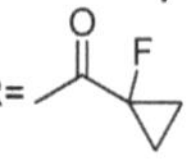

Figure 24. Structures of compounds **113**–**118**.

To enhance the formation of ternary complexes and the degrading activities of **114** and **115**, a second generation of PROTACs was created by altering the linker's length and composition, as well as the kind of E3 ligase binder used. Positive results were obtained with shorter and more lipophilic linkers. To enhance binding affinity, small modifications to the structure of the VHL ligand were introduced, using alternative warheads (compounds **116**–**118**, Figure 24). The screening for BRD9 and BRD7 degradation after treatment for 4 and

16 h showed compound **119** (Figure 25) as the most active compound, with a half-degrading concentration of 560 nM against BRD9. To further enhance the potency of compound **119**, a new series of compounds was designed, keeping the same warhead of 118 and changing the linker and the BRD7/BRD9 binder. After RI-1 cells were treated with 1 μM of the novel compounds for 2 and 8 h, BRD9 and BRD7 levels were analyzed using Western blot, and 120 was found to be the most potent degrader. Analysis at different concentrations gave a half-maximal degradation concentration (DC_{50}) of 1.76 nM and 4.5 nM against BRD9 and BRD7.

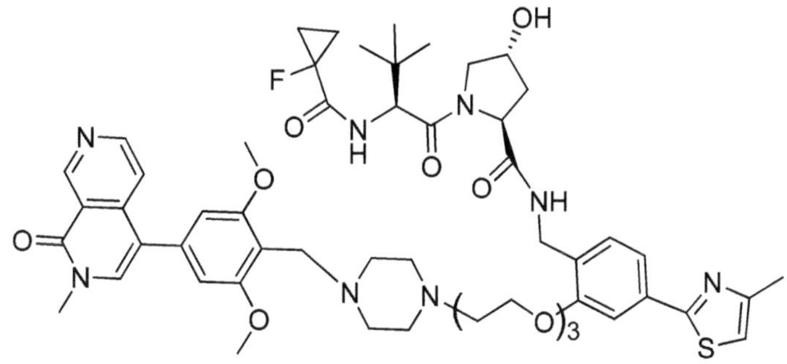

119

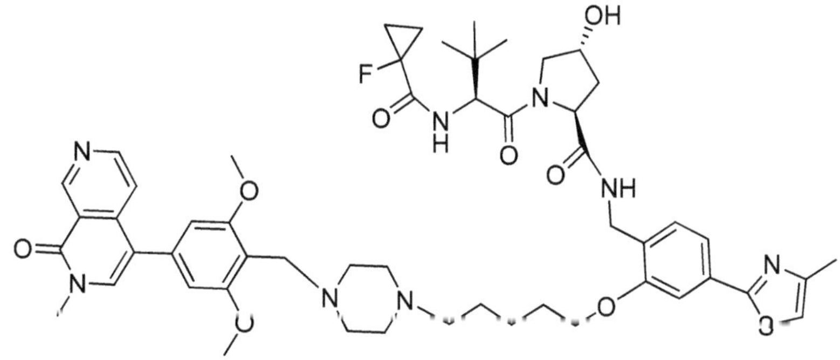

120

Figure 25. Structures of compounds **119** and **120**.

Recently, with the aim of obtaining an orally bioavailable BRD9-PROTAC, Zhang et al. [57] reported the synthesis of a new series of BRD9 degraders based on BI-7271 binder (compound **73**) and the cereblon ligand thalidomide. The authors started from the consideration that more rigid linkers increase oral activity, as shown by literature data. They designed the first series of dual binders, with linkers differing in number of rings and polarity, and attached to the C3 of thalidomide (compounds **121–126**, Figure 26A). Except for **125** and **126** which displayed poor BRD9 degradation profiles, compounds **121–124** gave excellent results in terms of protein degradation activity and cell proliferation inhibition, with IC_{50} values ranging from 3.32 ± 1.91 to 0.06 ± 0.05 nM against MV4-11 cells. The parameters calculated with ADMETlab to predict oral activity indicated compound **123** as the most promising. However, analysis of blood samples after a single oral administration of 20 mg/kg of 123 to ICR mice showed poor oral absorption.

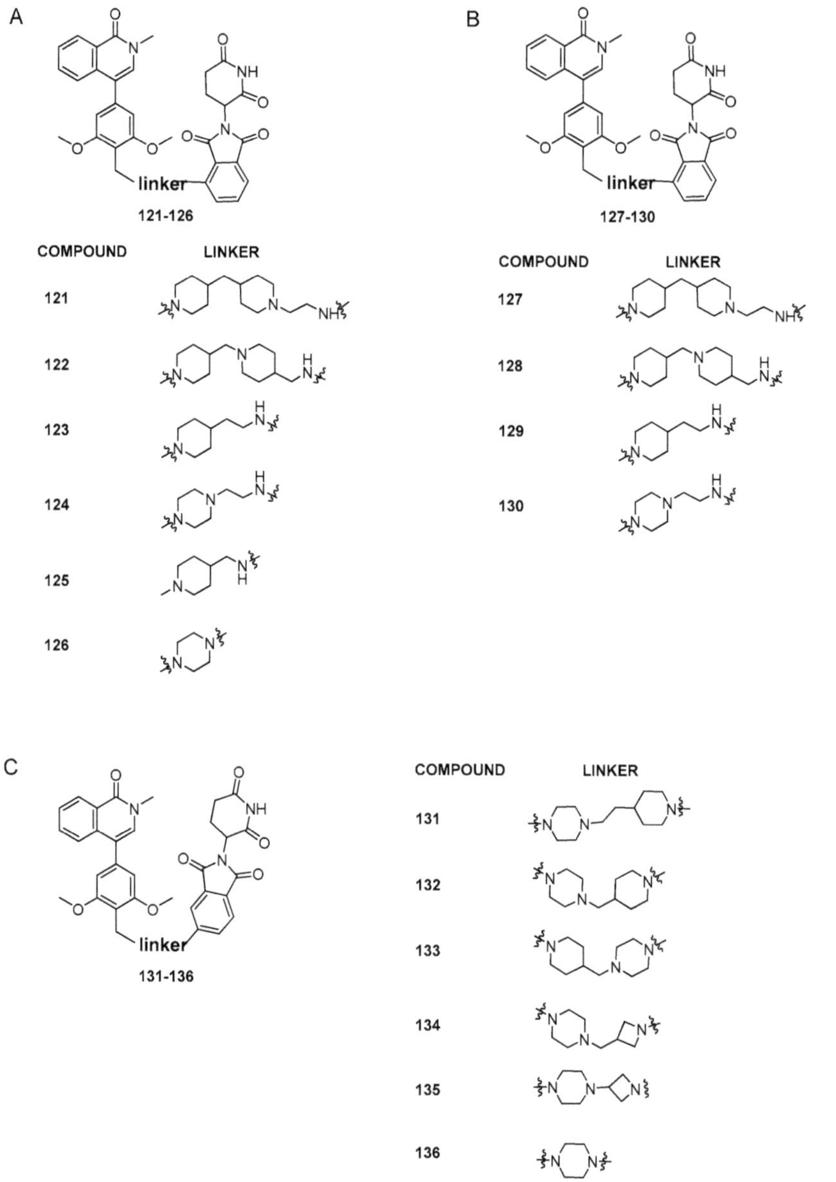

Figure 26. (**A**) Structures of compounds **121–126**; (**B**) structures of compounds **127–130**; (**C**) structures of compounds **131–136**.

In order to decrease the size of the BRD9 warhead, while retaining the E3 ligand and the best linkers from the first series (Figure 26A), a new series of degraders was created (compounds **127–130**, Figure 26B). In comparison to the related compounds **121–126**, the degradation activity experiments on **127–130** showed a decrease in potency, indicating the significance of the benzene ring structure in the BRD9 binding moiety for preserving pharmacological efficacy. The last series of compounds (Figure 26C) was then designed, keeping the BRD9 binder of the first series and connecting the linkers to the C-4 position of thalidomide. Among all the synthesized compounds (**131–136**), **136** demonstrated the most potent proliferation suppression and degradation activity with an IC_{50} of 3.69 ± 3.58 nM in

MV4-11 cells and a degradation rate of 93% at 10 nM and 99% at 100 nM. Interestingly, in MV4-11 cells, BRD4 and BRD7 proteins did not degrade at a dose of 100 nM, suggesting that C6 is selective for BRD9 protein. Lastly, **136** oral activity was studied in ICR mice, which showed good oral absorption characteristics with a Cmax value of 3436.95 ng/mL.

Recently, DCAF1 (also known as Vpr-binding protein (VprBP)) [58] was identified as a new interesting E3 ligase receptor for targeted protein degradation. Schröder et al. took advantage of a DCAF1 binder **137** previously developed by their group (Figure 27), [59] to design a new BRD9 PROTAC targeting this ligand [60]. The piperazine group attached to the quinazoline core was used to anchor an aliphatic carbon linker. As BRD9 binder, the structure of BI-9564 [39] was used, affording DBr-1 (compound **138**, Figure 27).

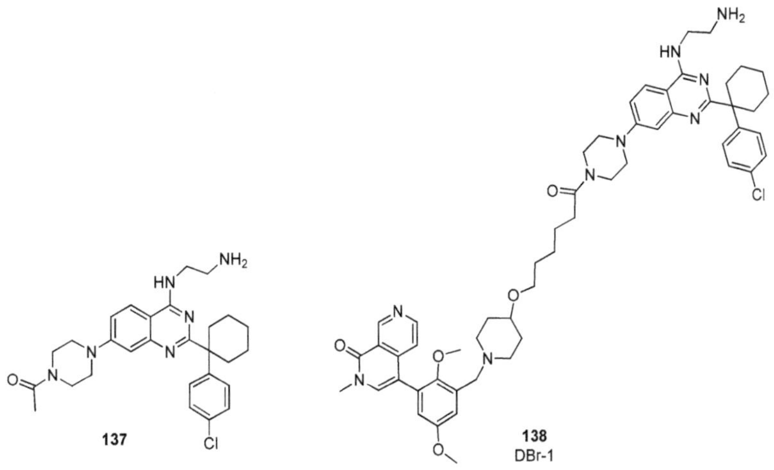

Figure 27. Structures of compounds **137–138**.

DBr-1 showed a slightly lower degrading potency with respect to VZ185 and dBRD9, with a DC_{50} value of 90 nM, and a selectivity for BRD9 over BRD7 similar to dBRD9. These results prove that E3 ligase receptor DCAF1 can be a useful target alternative to CRBN and VHL for the preparation of new PROTACs, with the aim of reducing the possibility of ligase-specific resistance mechanisms.

5. Conclusions

BRD9 inhibition can lead to the modulation of gene expression, potentially creating a more controlled and regulated cellular environment. This modulation proves to be beneficial in influencing specific pathways related to disease or cellular processes. Successful BRD9 inhibition could open up new avenues for therapeutic interventions, providing alternative treatment options for conditions associated with dysregulated chromatin remodeling.

BRD9 binders prove to be promising as potential therapeutics for a variety of malignant cancers, above all AML and MRT. Preclinical studies have demonstrated the efficacy of BRD9 inhibition in suppressing tumor growth, enhancing the sensitivity of cancer cells to chemotherapy, and modulating immune responses in inflammatory conditions.

However, some BRD9 inhibitors may lack the desired selectivity, affecting other bromodomain-containing proteins and causing off-target effects. This lack of specificity can limit the therapeutic potential of such inhibitors. Furthermore, despite the role of BRD9 in the related remodeling complex, compounds demonstrating robust activity on isolated BRD9 protein may not necessarily yield favorable results in cell assays. To enhance cellular activity and, consequently, the biological effect, the most promising binders are currently being utilized as a starting point for the development of PROTACs.

Recently, two BRD9 degraders entered clinical trials. FHD-609, developed by Foghorn Therapeutics, Inc., was in a phase 1 trial (NCT04965753) in patients with advanced synovial sarcoma and SMARCB1-deleted tumors. Unfortunately, FDA has placed a partial clinical hold on the phase 1 trial after adverse effects. On the other end, C4 Therapeutics, Inc also discontinued phase 1 clinical trial of CFT8634 (NCT05355753), a potent and selective oral heterobifunctional degrader of BRD9 for the treatment of synovial sarcoma and SMARCB1-deleted tumors, since BRD9 degradation did not produce sufficiently effective results in patients.

In conclusion, despite the prospect of BRD9 binders and degraders as therapeutic agents, several challenges remain to be addressed. Continued research efforts are needed to fully understand the therapeutic implications of targeting BRD9 and to realize the clinical potential of these compounds in treating various diseases.

Author Contributions: Conceptualization, A.G. and E.C.; methodology, E.C., M.G.C., G.B., G.L. and A.G.; writing—original draft preparation, E.C. and A.G.; writing—review and editing, E.C., M.G.C., G.B., G.L. and A.G.; supervision, A.G.; funding acquisition, G.L. All authors have read and agreed to the published version of the manuscript.

Funding: This research was funded by AIRC (MFAG 2017)—grant number ID: 20160—P.I. Lauro Gianluigi.

Institutional Review Board Statement: Not applicable.

Informed Consent Statement: Not applicable.

Data Availability Statement: Not applicable.

Conflicts of Interest: The authors declare no conflicts of interest. The funders had no role in the design of the study; in the collection, analyses, or interpretation of data; in the writing of the manuscript; or in the decision to publish the results.

References

1. Tamkun, J.W.; Deuring, R.; Scott, M.P.; Kissinger, M.; Pattatucci, A.M.; Kaufman, T.C.; Kennison, J.A. Brahma: A regulator of Drosophila homeotic genes structurally related to the yeast transcriptional activator SNF2SWI2. *Cell* **1992**, *68*, 561–572. [CrossRef] [PubMed]
2. Filippakopoulos, P.; Picaud, S.; Mangos, M.; Keates, T.; Lambert, J.-P.; Barsyte-Lovejoy, D.; Felletar, I.; Volkmer, R.; Müller, S.; Pawson, T.; et al. Histone recognition and large-scale structural analysis of the human bromodomain family. *Cell* **2012**, *149*, 214–231. [CrossRef] [PubMed]
3. Taniguchi, Y. The bromodomain and extra-terminal domain (BET) family: Functional anatomy of BET paralogous proteins. *Int. J. Mol. Sci.* **2016**, *17*, 1849–1873. [CrossRef]
4. Zhu, X.; Liao, Y.; Tang, L. Targeting BRD9 for Cancer Treatment: A New Strategy. *Onco Targets Ther.* **2020**, *13*, 13191–13200. [CrossRef] [PubMed]
5. Hodges, C.; Kirkland, J.G.; Crabtree, G.R. The Many Roles of BAF (mSWI/SNF) and PBAF Complexes in Cancer. *Cold Spring Harb. Perspect. Med.* **2016**, *6*, 1–24. [CrossRef] [PubMed]
6. Flynn, E.M.; Huang, O.W.; Poy, F.; Oppikofer, M.; Bellon, S.F.; Tang, Y.; Cochran, A.G. A subset of human bromodomains recognizes butyryllysine and crotonyllysine histone peptide modifications. *Structure* **2015**, *23*, 1801–1814. [CrossRef]
7. Filippakopoulos, P.; Knapp, S. Targeting bromodomains: Epigenetic readers of lysine acetylation. *Nat. Rev. Drug Discov.* **2014**, *13*, 337–356. [CrossRef]
8. Nayak, A.; Dutta, M.; Roychowdhury, A. Emerging oncogene ATAD2: Signaling cascades and therapeutic initiatives. *Life Sci.* **2021**, *276*, 119322. [CrossRef]
9. Mishima, Y.; Miyagi, S.; Saraya, A.; Negishi, M.; Endoh, M.; Endo, T.A.; Toyoda, T.; Shinga, J.; Katsumoto, T.; Chiba, T.; et al. The Hbo1-Brd1/Brpf2 complex is responsible for global acetylation of H3K14 and required for fetal liver erythropoiesis. *Blood* **2011**, *118*, 2443–2453. [CrossRef]
10. Sun, H.; Liu, J.; Zhang, J.; Shen, W.; Huang, H.; Xu, C.; Dai, H.; Wu, J.; Shi, Y. Solution structure of BRD7 bromodomain and its interaction with acetylated peptides from histone H3 and H4. *Biochem. Biophys. Res. Commun.* **2007**, *358*, 435–441. [CrossRef]
11. Poplawski, A.; Hu, K.; Lee, W.; Natesan, S.; Peng, D.; Carlson, S.; Shi, X.; Balaz, S.; Markley, J.L.; Glass, K.C. Molecular insights into the recognition of N-terminal histone modifications by the BRPF1 bromodomain. *J. Mol. Biol.* **2014**, *426*, 1661–1676. [CrossRef] [PubMed]
12. Feng, Y.; Vlassis, A.; Roques, C.; Lalonde, M.E.; González-Aguilera, C.; Lambert, J.P.; Lee, S.B.; Zhao, X.; Alabert, C.; Johansen, J.V.; et al. BRPF3-HBO1 regulates replication origin activation and histone H3K14 acetylation. *EMBO J.* **2016**, *35*, 176–192. [CrossRef] [PubMed]

13. Dreier, M.R.; De la Serna, I.L. SWI/SNF chromatin remodeling enzymes in melanoma. *Epigenomes* **2022**, *6*, 10–33. [CrossRef] [PubMed]
14. Euskirchen, G.M.; Auerbach, R.K.; Davidov, E.; Gianoulis, T.A.; Zhong, G.; Rozowsky, J.; Bhardwaj, N.; Gerstein, M.B.; Snyder, M. Diverse roles and interactions of the SWI/SNF chromatin remodeling complex revealed using global approaches. *PLoS Genet.* **2011**, *7*, e1002008. [CrossRef] [PubMed]
15. Hu, Z.; Zhou, J.; Jiang, J.; Yuan, J.; Zhang, Y.; Wei, X.; Loo, N.; Wang, Y.; Pan, Y.; Zhang, T.; et al. Genomic characterization of genes encoding histone acetylation modulator proteins identifies therapeutic targets for cancer treatment. *Nat. Commun.* **2019**, *10*, 733–750. [CrossRef] [PubMed]
16. Scotto, L.; Narayan, G.; Nandula, S.V.; Subramaniyam, S.; Kaufmann, A.M.; Wright, J.D.; Pothuri, B.; Mansukhani, M.; Schneider, A.; Arias-Pulido, H.; et al. Integrative genomics analysis of chromosome 5p gain in cervical cancer reveals target over-expressed genes, including Drosha. *Mol. Cancer* **2008**, *7*, 58. [CrossRef] [PubMed]
17. Kang, J.U.; Koo, S.H.; Kwon, K.C.; Park, J.W.; Kim, J.M. Gain at chromosomal region 5p15. 33, containing TERT, is the most frequent genetic event in early stages of non-small cell lung cancer. *Cancer Genet. Cytogen.* **2008**, *182*, 1–11. [CrossRef]
18. Network, C.G.A.R. Comprehensive genomic characterization of squamous cell lung cancers. *Nature* **2012**, *489*, 519–525.
19. Barbieri, C.E.; Baca, S.C.; Lawrence, M.S.; Demichelis, F.; Blattner, M.; Theurillat, J.-P.; White, T.A.; Stojanov, P.; Van Allen, E.; Stransky, N.; et al. Exome sequencing identifies recurrent SPOP, FOXA1 and MED12 mutations in prostate cancer. *Nat. Genet.* **2012**, *44*, 685–689. [CrossRef]
20. Kandoth, C.; Schultz, N.; Cherniack, A.D.; Akbani, R.; Liu, Y.; Shen, H.; Robertson, A.G.; Pashtan, I.; Shen, R.; Benz, C.C.; et al. Integrated genomic characterization of endometrial carcinoma. *Nature* **2013**, *497*, 67–73.
21. Cleary, S.P.; Jeck, W.R.; Zhao, X.; Chen, K.; Selitsky, S.R.; Savich, G.L.; Tan, T.X.; Wu, M.C.; Getz, G.; Lawrence, M.S.; et al. Identification of driver genes in hepatocellular carcinoma by exome sequencing. *Hepatology* **2013**, *58*, 1693–1702. [CrossRef]
22. Brien, G.L.; Remillard, D.; Shi, J.; Hemming, M.L.; Chabon, J.; Wynne, K.; Dillon, E.T.; Cagney, G.; Van Mierlo, G.; Baltissen, M.P.; et al. Targeted degradation of BRD9 reverses oncogenic gene expression in synovial sarcoma. *ELife* **2018**, *7*, e41305. [CrossRef]
23. Krämer, K.F.; Moreno, N.; Frühwald, M.C.; Kerl, K. BRD9 inhibition, alone or in combination with cytostatic compounds as a therapeutic approach in rhabdoid tumors. *Int. J. Mol. Sci.* **2017**, *18*, 1537–1549. [CrossRef]
24. Clark, P.G.; Vieira, L.C.; Tallant, C.; Fedorov, O.; Singleton, D.C.; Rogers, C.M.; Monteiro, O.P.; Bennett, J.M.; Baronio, R.; Müller, S.; et al. LP99: Discovery and Synthesis of the First Selective BRD7/9 Bromodomain Inhibitor. *Angew. Chem. Int. Ed.* **2015**, *54*, 6217–6221. [CrossRef]
25. Hay, D.A.; Rogers, C.M.; Fedorov, O.; Tallant, C.; Martin, S.; Monteiro, O.P.; Müller, S.; Knapp, S.; Schofield, C.J.; Brennan, P.E. Design and synthesis of potent and selective inhibitors of BRD7 and BRD9 bromodomains. *MedChemComm* **2015**, *6*, 1381–1386. [CrossRef]
26. Fedorov, O.; Lingard, H.; Wells, C.; Monteiro, O.P.; Picaud, S.; Keates, T.; Yapp, C.; Philpott, M.; Martin, S.J.; Felletar, I. [1, 2, 4] triazolo [4, 3-a] phthalazines: Inhibitors of diverse bromodomains. *J. Med. Chem.* **2014**, *57*, 462–476. [CrossRef]
27. D'Ascenzio, M.; Pugh, K.M.; Konietzny, R.; Berridge, G.; Tallant, C.; Hashem, S.; Monteiro, O.; Thomas, J.R.; Schirle, M.; Knapp, S. An Activity-Based Probe Targeting Non-Catalytic, Highly Conserved Amino Acid Residues within Bromodomains. *Angew. Chem. Int. Ed.* **2019**, *131*, 1019–1024. [CrossRef]
28. Picaud, S.; Strocchia, M.; Terracciano, S.; Lauro, G.; Mendez, J.; Daniels, D.L.; Riccio, R.; Bifulco, G.; Bruno, I.; Filippakopoulos, P. 9 H-purine scaffold reveals induced-fit pocket plasticity of the BRD9 bromodomain. *J. Med. Chem.* **2015**, *58*, 2718–2736. [CrossRef]
29. Clegg, M.A.; Bamborough, P.; Chung, C.-W.; Craggs, P.D.; Gordon, L.; Grandi, P.; Leveridge, M.; Lindon, M.; Liwicki, G.M.; Michon, A.-M. Application of atypical acetyl-lysine methyl mimetics in the development of selective inhibitors of the bromodomain-containing protein 7 (BRD7)/bromodomain-containing protein 9 (BRD9) bromodomains. *J. Med. Chem.* **2020**, *63*, 5816–5840. [CrossRef]
30. Pierri, M.; Gazzillo, E.; Chini, M.G.; Ferraro, M.G.; Piccolo, M.; Maione, F.; Irace, C.; Bifulco, G.; Bruno, I.; Terracciano, S.; et al. Introducing structure-based three-dimensional pharmacophore models for accelerating the discovery of selective BRD9 binders. *Bioorg. Chem.* **2022**, *118*, 105480. [CrossRef] [PubMed]
31. Gazzillo, E.; Pierri, M.; Colarusso, E.; Chini, M.G.; Ferraro, M.G.; Piccolo, M.; Irace, C.; Bruno, I.; Bifulco, G.; Terracciano, S.; et al. Exploring the chemical space of functionalized [1, 2, 4] triazolo [4, 3-a] quinoxaline-based compounds targeting the bromodomain of BRD9. *Bioorg. Chem.* **2023**, *139*, 106677. [CrossRef] [PubMed]
32. Theodoulou, N.H.; Bamborough, P.; Bannister, A.J.; Becher, I.; Bit, R.A.; Che, K.H.; Chung, C.W.; Dittmann, A.; Drewes, G.; Drewry, D.H.; et al. Discovery of I-BRD9, a Selective Cell Active Chemical Probe for Bromodomain Containing Protein 9 Inhibition. *J. Med. Chem.* **2016**, *59*, 1425–1439. [CrossRef] [PubMed]
33. Chen, P.; Chaikuad, A.; Bamborough, P.; Bantscheff, M.; Bountra, C.; Chung, C.-W.; Fedorov, O.; Grandi, P.; Jung, D.; Lesniak, R.; et al. Discovery and characterization of GSK2801, a selective chemical probe for the bromodomains BAZ2A and BAZ2B. *J. Med. Chem.* **2016**, *59*, 1410–1424. [CrossRef]
34. Picaud, S.; Leonards, K.; Lambert, J.P.; Dovey, O.; Wells, C.; Fedorov, O.; Monteiro, O.; Fujisawa, T.; Wang, C.Y.; Lingard, H.; et al. Promiscuous targeting of bromodomains by bromosporine identifies BET proteins as master regulators of primary transcription response in leukemia. *Sci. Adv.* **2016**, *2*, e1600760. [CrossRef]

35. Nakano, H.; Ōmura, S. Chemical biology of natural indolocarbazole products: 30 years since the discovery of staurosporine. *J. Antibiot.* **2009**, *62*, 17–26. [CrossRef]
36. Zuber, J.; Shi, J.; Wang, E.; Rappaport, A.R.; Herrmann, H.; Sison, E.A.; Magoon, D.; Qi, J.; Blatt, K.; Wunderlich, M.; et al. RNAi screen identifies Brd4 as a therapeutic target in acute myeloid leukaemia. *Nature* **2011**, *478*, 524–528. [CrossRef]
37. Shannon, D.A.; Banerjee, R.; Webster, E.R.; Bak, D.W.; Wang, C.; Weerapana, E. Investigating the proteome reactivity and selectivity of aryl halides. *JACS* **2014**, *136*, 3330–3333. [CrossRef]
38. Humphreys, P.G.; Bamborough, P.; Chung, C.-W.; Craggs, P.D.; Gordon, L.; Grandi, P.; Hayhow, T.G.; Hussain, J.; Jones, K.L.; Lindon, M.; et al. Discovery of a potent, cell penetrant, and selective p300/CBP-associated factor (PCAF)/general control nonderepressible 5 (GCN5) bromodomain chemical probe. *J. Med. Chem.* **2017**, *60*, 695–709. [CrossRef]
39. Martin, L.J.; Koegl, M.; Bader, G.; Cockcroft, X.-L.; Fedorov, O.; Fiegen, D.; Gerstberger, T.; Hofmann, M.H.; Hohmann, A.F.; Kessler, D.; et al. Structure-based design of an in vivo active selective BRD9 inhibitor. *J. Med. Chem.* **2016**, *59*, 4462–4475. [CrossRef]
40. Zhao, L.; Yang, Y.; Guo, Y.; Yang, L.; Zhang, J.; Zhou, J.; Zhang, H. Design, synthesis and biological evaluation of 7-methylimidazo [1, 5-a] pyrazin-8 (7*H*)-one derivatives as BRD4 inhibitors. *Bioorg. Med. Chem.* **2017**, *25*, 2482–2490. [CrossRef]
41. Zheng, P.; Zhang, J.; Ma, H.; Yuan, X.; Chen, P.; Zhou, J.; Zhang, H. Design, synthesis and biological evaluation of imidazo [1, 5-a] pyrazin-8 (7*H*)-one derivatives as BRD9 inhibitors. *Bioorg. Med. Chem.* **2019**, *27*, 1391–1404. [CrossRef]
42. Hügle, M.; Regenass, P.; Warstat, R.; Hau, M.; Schmidtkunz, K.; Lucas, X.; Wohlwend, D.; Einsle, O.; Jung, M.; Breit, B.; et al. 4-Acyl pyrroles as dual BET-BRD7/9 bromodomain inhibitors address BETi insensitive human cancer cell lines. *J. Med. Chem.* **2020**, *63*, 15603–15620. [CrossRef]
43. Warstat, R.; Pervaiz, M.; Regenass, P.; Amann, M.; Schmidtkunz, K.; Einsle, O.; Jung, M.; Breit, B.; Hügle, M.; Günther, S. A novel pan-selective bromodomain inhibitor for epigenetic drug design. *Eur. J. Med. Chem.* **2023**, *249*, 115139. [CrossRef]
44. Colarusso, E.; Gazzillo, E.; Boccia, E.; Giordano, A.; Chini, M.G.; Bifulco, G.; Lauro, G. 6-Methylquinazolin-4 (3H)-one Based Compounds as BRD9 Epigenetic Reader Binders: A Rational Combination of in silico Studies and Chemical Synthesis. *EurJoc* **2022**, *2022*, e202200868. [CrossRef]
45. *Schrödinger Release 2022-1: Combiglide*; Schrödinger LCC: New York, NY, USA, 2021.
46. Bhat, M.; Belagali, S.; Mamatha, S.; Sagar, B.; Sekhar, E.V. Importance of quinazoline and quinazolinone derivatives in medicinal chemistry. *Stud. Nat. Prod. Chem.* **2021**, *71*, 185–219.
47. Colarusso, E.; Ceccacci, S.; Monti, M.C.; Gazzillo, E.; Giordano, A.; Chini, M.G.; Ferraro, M.G.; Piccolo, M.; Ruggiero, D.; Irace, C.; et al. Identification of 2, 4, 5-trisubstituted-2, 4-dihydro-3*H*-1, 2, 4-triazol-3-one-based small molecules as selective BRD9 binders. *Eur. J. Med. Chem.* **2023**, *247*, 115018. [CrossRef] [PubMed]
48. Santos, R.; Ursu, O.; Gaulton, A.; Bento, A.P.; Donadi, R.S.; Bologa, C.G.; Karlsson, A.; Al-Lazikani, B.; Hersey, A.; Oprea, T.I.; et al. Comprehensive map of molecular drug targets. *Nat. Rev. Drug Discov.* **2017**, *16*, 19–34. [CrossRef] [PubMed]
49. Ciulli, A.; Trainor, N. A beginner's guide to PROTACs and targeted protein degradation. *Biochemistry* **2021**, *43*, 74–79. [CrossRef]
50. Zhou, X.; Dong, R.; Zhang, J.-Y.; Zheng, X.; Sun, L.-P. PROTAC: A promising technology for cancer treatment. *Eur. J. Med. Chem.* **2020**, *203*, 112539. [CrossRef] [PubMed]
51. Lai, A.C.; Crews, C.M. Induced protein degradation: An emerging drug discovery paradigm. *Nat. Rev. Drug Discov.* **2017**, *16*, 101–114. [CrossRef] [PubMed]
52. Collins, I.; Wang, H.; Caldwell, J.J.; Chopra, R. Chemical approaches to targeted protein degradation through modulation of the ubiquitin–proteasome pathway. *Biochem. J.* **2017**, *474*, 1127–1147. [CrossRef]
53. Salerno, A.; Seghetti, F.; Caciolla, J.; Uliassi, E.; Testi, E.; Guardigni, M.; Roberti, M.; Milelli, A.; Bolognesi, M.L. Enriching proteolysis targeting chimeras with a second modality: When two are better than one. *J. Med. Chem.* **2022**, *65*, 9507–9530. [CrossRef] [PubMed]
54. Remillard, D.; Buckley, D.L.; Paulk, J.; Brien, G.L.; Sonnett, M.; Seo, H.S.; Dastjerdi, S.; Wühr, M.; Dhe-Paganon, S.; Armstrong, S.A.; et al. Degradation of the BAF complex factor BRD9 by heterobifunctional ligands. *Angew. Chem. Int. Ed.* **2017**, *56*, 5738–5743. [CrossRef] [PubMed]
55. Weisberg, E.; Chowdhury, B.; Meng, C.; Case, A.E.; Ni, W.; Garg, S.; Sattler, M.; Azab, A.K.; Sun, J.; Muz, B.; et al. BRD9 degraders as chemosensitizers in acute leukemia and multiple myeloma. *Blood Cancer J.* **2022**, *12*, 110–120. [CrossRef] [PubMed]
56. Zoppi, V.; Hughes, S.J.; Maniaci, C.; Testa, A.; Gmaschitz, T.; Wieshofer, C.; Koegl, M.; Riching, K.M.; Daniels, D.L.; Spallarossa, A.; et al. Iterative design and optimization of initially inactive proteolysis targeting chimeras (PROTACs) identify VZ185 as a potent, fast, and selective von Hippel–Lindau (VHL) based dual degrader probe of BRD9 and BRD7. *J. Med. Chem.* **2018**, *62*, 699–726. [CrossRef] [PubMed]
57. Zhang, J.; Duan, H.; Gui, R.; Wu, M.; Shen, L.; Jin, Y.; Pang, A.; Yu, X.; Zeng, S.; Zhang, B.; et al. Structure-based identification of new orally bioavailable BRD9-PROTACs for treating acute myelocytic leukemia. *Eur. J. Med. Chem.* **2023**, *262*, 115872. [CrossRef]
58. Tao, Y.; Remillard, D.; Vinogradova, E.V.; Yokoyama, M.; Banchenko, S.; Schwefel, D.; Melillo, B.; Schreiber, S.L.; Zhang, X.; Cravatt, B.F. Targeted protein degradation by electrophilic PROTACs that stereoselectively and site-specifically engage DCAF1. *J. Am. Chem. Soc.* **2022**, *144*, 18688–18699. [CrossRef]

59. Vulpetti, A.; Holzer, P.; Schmiedeberg, N.; Imbach-Weese, P.; Pissot-Soldermann, C.; Hollingworth, G.J.; Radimerski, T.; Thoma, C.R.; Stachyra, T.M.; Wojtynek, M.; et al. Discovery of New Binders for DCAF1, an Emerging Ligase Target in the Targeted Protein Degradation Field. *ACS Med. Chem. Lett.* **2023**, *14*, 949–954. [CrossRef]
60. Schröder, M.; Renatus, M.; Liang, X.; Meili, F.; Zoller, T.; Ferrand, S.; Gauter, F.; Li, X.; Sigoillot, F.; Gleim, S.; et al. DCAF1-based PROTACs with activity against clinically validated targets overcoming intrinsic-and acquired-degrader resistance. *Nat. Commun.* **2024**, *15*, 275–294. [CrossRef]

MDPI AG
Grosspeteranlage 5
4052 Basel
Switzerland
Tel.: +41 61 683 77 34

Pharmaceuticals Editorial Office
E-mail: pharmaceuticals@mdpi.com
www.mdpi.com/journal/pharmaceuticals

www.ingramcontent.com/pod-product-compliance
Lightning Source LLC
Chambersburg PA
CBHW041424210926
51850CB00013B/3347